W9-BRQ-730

A First Course in Abstract Algebra

Seventh Edition

John B. Fraleigh
University of Rhode Island

Historical Notes by Victor Katz
University of District of Columbia

Boston San Francisco New York
London Toronto Sydney Tokyo Singapore Madrid
Mexico City Munich Paris Cape Town Hong Kong Montreal

Publisher: Greg Tobin
Acquisitions Editor: William Hoffman
Associate Editor: RoseAnne Johnson
Production Supervisor: Cynthia Cody
Marketing Coordinator: Weslie Lewis
Senior Manufacturing Buyer: Evelyn Beaton
Cover Designer: Andrea Toth
Compositor: TechBooks

Library of Congress Cataloging-in-Publication Data

Fraleigh, John B.
A first course in abstract algebra / John B. Fraleigh ; historial notes by Victor Katz.—7th ed.
p. cm.
Includes index.
ISBN 0-201-76390-7
1. Algebra, Abstract. I. Title.
QA162 .F7 2002
512′.02—dc21 2002019357

7 8 9 10 HP 09 08 07

Contents

III

Homomorphisms and Factor Groups 125

IV

Rings and Fields 167

V

Ideals and Factor Rings 237

VI

Extension Fields 265

VII

Advanced Group Theory 307

---

† Not required for the remainder of the text.

‡ This section is a prerequisite for Sections 17 and 36 only.

Instructor's Preface

This is an introduction to abstract algebra. It is anticipated that the students have studied calculus and probably linear algebra. However, these are primarily *mathematical maturity* prerequisites; subject matter from calculus and linear algebra appears mostly in illustrative examples and exercises.

As in previous editions of the text, my aim remains to teach students as much about groups, rings, and fields as I can in a first course. For many students, abstract algebra is their first extended exposure to an axiomatic treatment of mathematics. Recognizing this, I have included extensive explanations concerning what we are trying to accomplish, how we are trying to do it, and why we choose these methods. Mastery of this text constitutes a firm foundation for more specialized work in algebra, and also provides valuable experience for any further axiomatic study of mathematics.

Changes from the Sixth Edition

The amount of preliminary material had increased from one lesson in the first edition to four lessons in the sixth edition. My personal preference is to spend less time before getting to algebra; therefore, I spend little time on preliminaries. Much of it is review for many students, and spending four lessons on it may result in their not allowing sufficient time in their schedules to handle the course when new material arises. Accordingly, in this edition, I have reverted to just one preliminary lesson on sets and relations, leaving other topics to be reviewed when needed. A summary of matrices now appears in the Appendix.

The first two editions consisted of short, consecutively numbered sections, many of which could be covered in a single lesson. I have reverted to that design to avoid the cumbersome and intimidating triple numbering of definitions, theorems examples, etc. In response to suggestions by reviewers, the order of presentation has been changed so that the basic material on groups, rings, and fields that would normally be covered in a one-semester course appears first, before the more-advanced group theory. Section 1 is a new introduction, attempting to provide some feeling for the nature of the study.

In response to several requests, I have included the material on homology groups in topology that appeared in the first two editions. Computation of homology groups strengthens students' understanding of factor groups. The material is easily accessible; after Sections 0 through 15, one need only read about free abelian groups, in Section 38 through Theorem 38.5, as preparation. To make room for the homology groups, I have omitted the discussion of automata, binary linear codes, and additional algebraic structures that appeared in the sixth edition.

I have also included a few exercises asking students to give a one- or two-sentence synopsis of a proof in the text. Before the first such exercise, I give an example to show what I expect.

Some Features Retained

I continue to break down most exercise sets into parts consisting of computations, concepts, and theory. Answers to odd-numbered exercises not requesting a proof again appear at the back of the text. However, in response to suggestions, I am supplying the answers to parts a), c), e), g), and i) only of my 10-part true–false exercises.

The excellent historical notes by Victor Katz are, of course, retained. Also, a manual containing complete solutions for all the exercises, including solutions asking for proofs, is available for the instructor from the publisher.

A dependence chart with section numbers appears in the front matter as an aid in making a syllabus.

Acknowledgments

I am very grateful to those who have reviewed the text or who have sent me suggestions and corrections. I am especially indebted to George M. Bergman, who used the sixth edition and made note of typographical and other errors, which he sent to me along with a great many other valuable suggestions for improvement. I really appreciate this voluntary review, which must have involved a large expenditure of time on his part.

I also wish to express my appreciation to William Hoffman, Julie LaChance, and Cindy Cody of Addison-Wesley for their help with this project. Finally, I was most fortunate to have John Probst and the staff at TechBooks handling the production of the text from my manuscript. They produced the most error-free pages I have experienced, and courteously helped me with a technical problem I had while preparing the solutions manual.

Suggestions for New Instructors of Algebra

Those who have taught algebra several times have discovered the difficulties and developed their own solutions. The comments I make here are not for them.

This course is an abrupt change from the typical undergraduate calculus for the students. A graduate-style lecture presentation, writing out definitions and proofs on the board for most of the class time, will not work with most students. I have found it best to spend at least the first half of each class period answering questions on homework, trying to get a volunteer to give a proof requested in an exercise, and generally checking to see if they seem to understand the material assigned for that class. Typically, I spent only about the last 20 minutes of my 50-minute time talking about new ideas for the next class, and giving at least one proof. From a practical point of view, it is a waste of time to try to write on the board all the definitions and proofs. They are in the text.

I suggest that at least half of the assigned exercises consist of the computational ones. Students are used to doing computations in calculus. Although there are many exercises asking for proofs that we would love to assign, I recommend that you assign at most two or three such exercises, and try to get someone to explain how each proof is performed in the next class. I do think students should be asked to do at least one proof in each assignment.

Students face a barrage of definitions and theorems, something they have never encountered before. They are not used to mastering this type of material. Grades on tests that seem reasonable to us, requesting a few definitions and proofs, are apt to be low and depressing for most students. My recommendation for handling this problem appears in my article, *Happy Abstract Algebra Classes,* in the November 2001 issue of the *MAA FOCUS.*

At URI, we have only a single semester undergraduate course in abstract algebra. Our semesters are quite short, consisting of about 42 50-minute classes. When I taught the course, I gave three 50-minute tests in class, leaving about 38 classes for which the student was given an assignment. I always covered the material in Sections 0–11, 13–15, 18–23, 26, 27, and 29–32, which is a total of 27 sections. Of course, I spent more than one class on several of the sections, but I usually had time to cover about two more; sometimes I included Sections 16 and 17. (There is no point in doing Section 16 unless you do Section 17, or will be doing Section 36 later.) I often covered Section 25, and sometimes Section 12 (see the Dependence Chart). The job is to keep students from becoming discouraged in the first few weeks of the course.

Student's Preface

This course may well require a different approach than those you used in previous mathematics courses. You may have become accustomed to working a homework problem by turning back in the text to find a similar problem, and then just changing some numbers. That may work with a few problems in this text, but it will not work for most of them. This is a subject in which understanding becomes all important, and where problems should not be tackled without first studying the text.

Let me make some suggestions on studying the text. Notice that the text bristles with definitions, theorems, corollaries, and examples. The definitions are crucial. We must agree on terminology to make any progress. Sometimes a definition is followed by an example that illustrates the concept. Examples are probably the most important aids in studying the text. *Pay attention to the examples.* I suggest you skip the proofs of the theorems on your first reading of a section, unless you are really "gung-ho" on proofs. You should read the statement of the theorem and try to understand just what it means. Often, a theorem is followed by an example that illustrates it, a great aid in really understanding what the theorem says.

In summary, on your first reading of a section, I suggest you concentrate on what information the section gives, and on gaining a real understanding of it. If you do not understand what the statement of a theorem means, it will probably be meaningless for you to read the proof.

Proofs are very basic to mathematics. After you feel you understand the information given in a section, you should read and try to understand at least some of the proofs. Proofs of corollaries are usually the easiest ones, for they often follow very directly from the theorem. Quite a lot of the exercises under the "Theory" heading ask for a proof. Try not to be discouraged at the outset. It takes a bit of practice and experience. Proofs in algebra can be more difficult than proofs in geometry and calculus, for there are usually no suggestive pictures that you can draw. Often, a proof falls out easily if you happen to look at just the right expression. Of course, it is hopeless to devise a proof if you do not really understand what it is that you are trying to prove. For example, if an exercise asks you to show that given thing is a member of a certain set, you must *know* the defining criterion to be a member of that set, and then show that your given thing satisfies that criterion.

There are several aids for your study at the back of the text. Of course, you will discover the answers to odd-numbered problems not requesting a proof. If you run into a notation such as $\mathbb{Z}_n$ that you do not understand, look in the list of notations that appears after the bibliography. If you run into terminology like *inner automorphism* that you do not understand, look in the Index for the first page where the term occurs.

In summary, although an understanding of the subject is important in every mathematics course, it is really crucial to your performance in this course. May you find it a rewarding experience.

Narragansett, RI *J.B.F.*

Dependence Chart

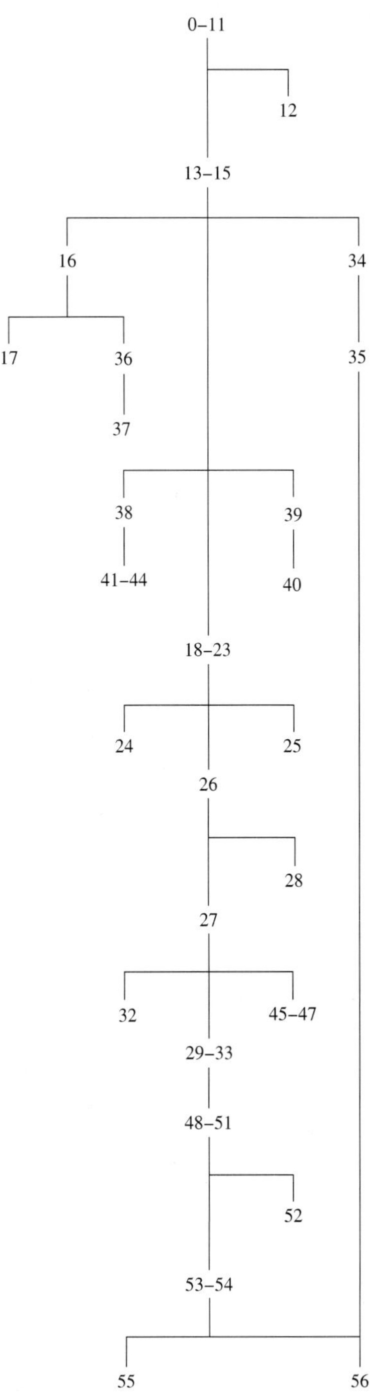

SECTION 0 Sets and Relations

On Definitions, and the Notion of a Set

Many students do not realize the great importance of definitions to mathematics. This importance stems from the need for mathematicians to communicate with each other. If two people are trying to communicate about some subject, they must have the same understanding of its technical terms. However, there is an important structural weakness.

> It is impossible to define every concept.

Suppose, for example, we define the term *set* as "A **set** is a well-defined collection of objects." One naturally asks what is meant by a *collection.* We could define it as "A collection is an aggregate of things." What, then, is an *aggregate?* Now our language is finite, so after some time we will run out of new words to use and have to repeat some words already examined. The definition is then circular and obviously worthless. Mathematicians realize that there must be some undefined or primitive concept with which to start. At the moment, they have agreed that *set* shall be such a primitive concept. We shall not define *set,* but shall just hope that when such expressions as "the set of all real numbers" or "the set of all members of the United States Senate" are used, people's various ideas of what is meant are sufficiently similar to make communication feasible.

We summarize briefly some of the things we shall simply assume about sets.

1. A set S is made up of **elements,** and if a is one of these elements, we shall denote this fact by $a \in S$.
2. There is exactly one set with no elements. It is the **empty set** and is denoted by $\varnothing$.
3. We may describe a set either by giving a characterizing property of the elements, such as "the set of all members of the United States Senate," or by listing the elements. The standard way to describe a set by listing elements is to enclose the designations of the elements, separated by commas, in braces, for example, $\{1, 2, 15\}$. If a set is described by a characterizing property $P(x)$ of its elements x, the brace notation $\{x \mid P(x)\}$ is also often used, and is read "the set of all x such that the statement $P(x)$ about x is true." Thus

$$\begin{aligned}\{2, 4, 6, 8\} &= \{x \mid x \text{ is an even whole positive number} \leq 8\}\\ &= \{2x \mid x = 1, 2, 3, 4\}.\end{aligned}$$

   The notation $\{x \mid P(x)\}$ is often called "set-builder notation."

4. A set is **well defined,** meaning that if S is a set and a is some object, then either a is definitely in S, denoted by $a \in S$, or a is definitely not in S, denoted by $a \notin S$. Thus, we should never say, "Consider the set S of some positive numbers," for it is not definite whether $2 \in S$ or $2 \notin S$. On the other hand, we

> can consider the set T of all prime positive integers. Every positive integer is definitely either prime or not prime. Thus $5 \in T$ and $14 \notin T$. It may be hard to actually determine whether an object is in a set. For example, as this book goes to press it is probably unknown whether $2^{(2^{65})} + 1$ is in T. However, $2^{(2^{65})} + 1$ is certainly either prime or not prime.

It is not feasible for this text to push the definition of everything we use all the way back to the concept of a set. For example, we will never define the number π in terms of a set.

> Every definition is an *if and only if* type of statement.

With this understanding, definitions are often stated with the *only if* suppressed, but it is always to be understood as part of the definition. Thus we may define an isosceles triangle as follows: "A triangle is **isosceles** if it has two sides of equal length," when we really mean that a triangle is isosceles *if and only if* it has two sides of equal length.

In our text, we have to define many terms. We use specifically labeled and numbered definitions for the main algebraic concepts with which we are concerned. To avoid an overwhelming quantity of such labels and numberings, we define many terms within the body of the text and exercises using boldface type.

> **Boldface Convention**
>
> A term printed **in boldface** in a sentence is being defined by that sentence.

Do not feel that you have to memorize a definition word for word. The important thing is to *understand* the concept, so that you can define precisely the same concept in your own words. Thus the definition "An **isosceles** triangle is one having two equal sides" is perfectly correct. Of course, we had to delay stating our boldface convention until we had finished using boldface in the preceding discussion of sets, because we do not define a set!

In this section, we do define some familiar concepts as sets, both for illustration and for review of the concepts. First we give a few definitions and some notation.

0.1 Definition A set B is a **subset of a set** A, denoted by $B \subseteq A$ or $A \supseteq B$, if every element of B is in A. The notations $B \subset A$ or $A \supset B$ will be used for $B \subseteq A$ but $B \neq A$. ■

Note that according to this definition, for any set A, A itself and $\varnothing$ are both subsets of A.

0.2 Definition If A is any set, then A is the **improper subset of** A. Any other subset of A is a **proper subset of** A. ■

0.3 Example Let $S = \{1, 2, 3\}$. This set S has a total of eight subsets, namely $\emptyset$, $\{1\}$, $\{2\}$, $\{3\}$, $\{1, 2\}$, $\{1, 3\}$, $\{2, 3\}$, and $\{1, 2, 3\}$. ▲

0.4 Definition Let A and B be sets. The set $A \times B = \{(a, b) \mid a \in A \text{ and } b \in B\}$ is the **Cartesian product** of A and B. ■

0.5 Example If $A = \{1, 2, 3\}$ and $B = \{3, 4\}$, then we have

$$A \times B = \{(1, 3), (1, 4), (2, 3), (2, 4), (3, 3), (3, 4)\}.$$ ▲

Throughout this text, much work will be done involving familiar sets of numbers. Let us take care of notation for these sets once and for all.

$\mathbb{Z}$ is the set of all integers (that is, whole numbers: positive, negative, and zero).

$\mathbb{Q}$ is the set of all rational numbers (that is, numbers that can be expressed as quotients m/n of integers, where $n \neq 0$).

$\mathbb{R}$ is the set of all real numbers.

$\mathbb{Z}^+$, $\mathbb{Q}^+$, and $\mathbb{R}^+$ are the sets of positive members of $\mathbb{Z}$, $\mathbb{Q}$, and $\mathbb{R}$, respectively.

$\mathbb{C}$ is the set of all complex numbers.

$\mathbb{Z}^*$, $\mathbb{Q}^*$, $\mathbb{R}^*$, and $\mathbb{C}^*$ are the sets of nonzero members of $\mathbb{Z}$, $\mathbb{Q}$, $\mathbb{R}$, and $\mathbb{C}$, respectively.

0.6 Example The set $\mathbb{R} \times \mathbb{R}$ is the familiar Euclidean plane that we use in first-semester calculus to draw graphs of functions. ▲

Relations Between Sets

We introduce the notion of an element a of set A being *related* to an element b of set B, which we might denote by $a \mathscr{R} b$. The notation $a \mathscr{R} b$ exhibits the elements a and b in left-to-right order, just as the notation (a, b) for an element in $A \times B$. This leads us to the following definition of a relation $\mathscr{R}$ as a *set*.

0.7 Definition A **relation** between sets A and B is a subset $\mathscr{R}$ of $A \times B$. We read $(a, b) \in \mathscr{R}$ as "a is related to b" and write $a \mathscr{R} b$. ■

0.8 Example **(Equality Relation)** There is one familiar relation between a set and itself that we consider every set S mentioned in this text to possess: namely, the equality relation $=$ defined on a set S by

$$= \text{ is the subset } \{(x, x) \mid x \in S\} \text{ of } S \times S.$$

Thus for any $x \in S$, we have $x = x$, but if x and y are different elements of S, then $(x, y) \notin\, =$ and we write $x \neq y$. ▲

We will refer to any relation between a set S and itself, as in the preceding example, as a **relation on** S.

0.9 Example The graph of the function f where $f(x) = x^3$ for all $x \in \mathbb{R}$, is the subset $\{(x, x^3) \mid x \in \mathbb{R}\}$ of $\mathbb{R} \times \mathbb{R}$. Thus it is a relation on $\mathbb{R}$. The function is completely determined by its graph. ▲

The preceding example suggests that rather than define a "function" $y = f(x)$ to be a "rule" that assigns to each $x \in \mathbb{R}$ exactly one $y \in \mathbb{R}$, we can easily describe it as a certain type of subset of $\mathbb{R} \times \mathbb{R}$, that is, as a type of relation. We free ourselves from $\mathbb{R}$ and deal with any sets X and Y.

0.10 Definition A **function** ϕ mapping X into Y is a relation between X and Y with the property that each $x \in X$ appears as the first member of exactly one ordered pair (x, y) in ϕ. Such a function is also called a **map** or **mapping** of X into Y. We write $\phi : X \to Y$ and express $(x, y) \in \phi$ by $\phi(x) = y$. The **domain** of ϕ is the set X and the set Y is the **codomain** of ϕ. The **range** of ϕ is $\phi[X] = \{\phi(x) \mid x \in X\}$. ■

0.11 Example We can view the addition of real numbers as a function $+ : (\mathbb{R} \times \mathbb{R}) \to \mathbb{R}$, that is, as a mapping of $\mathbb{R} \times \mathbb{R}$ into $\mathbb{R}$. For example, the action of $+$ on $(2, 3) \in \mathbb{R} \times \mathbb{R}$ is given in function notation by $+((2, 3)) = 5$. In set notation we write $((2, 3), 5) \in +$. Of course our familiar notation is $2 + 3 = 5$. ▲

Cardinality

The number of elements in a set X is the **cardinality** of X and is often denoted by $|X|$. For example, we have $|\{2, 5, 7\}| = 3$. It will be important for us to know whether two sets have the same cardinality. If both sets are finite there is no problem; we can simply count the elements in each set. But do $\mathbb{Z}$, $\mathbb{Q}$, and $\mathbb{R}$ have the same cardinality? To convince ourselves that two sets X and Y have the same cardinality, we try to exhibit a pairing of each x in X with only one y in Y in such a way that each element of Y is also used only once in this pairing. For the sets $X = \{2, 5, 7\}$ and $Y = \{?, !, \#\}$, the pairing

$$2 \leftrightarrow ?, \quad 5 \leftrightarrow \#, \quad 7 \leftrightarrow !$$

shows they have the same cardinality. Notice that we could also exhibit this pairing as $\{(2, ?), (5, \#), (7, !)\}$ which, as a subset of $X \times Y$, is a *relation* between X and Y. The pairing

1	2	3	4	5	6	7	8	9	10	$\cdots$
$\updownarrow$	$\updownarrow$	$\updownarrow$	$\updownarrow$	$\updownarrow$	$\updownarrow$	$\updownarrow$	$\updownarrow$	$\updownarrow$	$\updownarrow$	
0	-1	1	-2	2	-3	3	-4	4	-5	$\cdots$

shows that the sets $\mathbb{Z}$ and $\mathbb{Z}^+$ have the same cardinality. Such a pairing, showing that sets X and Y have the same cardinality, is a special type of relation $\leftrightarrow$ between X and Y called a **one-to-one correspondence.** Since each element x of X appears precisely once in this relation, we can regard this one-to-one correspondence as a *function* with domain X. The range of the function is Y because each y in Y also appears in some pairing $x \leftrightarrow y$. We formalize this discussion in a definition.

0.12 Definition *A function $\phi : X \to Y$ is **one to one** if $\phi(x_1) = \phi(x_2)$ only when $x_1 = x_2$ (see Exercise 37). The function ϕ is **onto** Y if the range of ϕ is Y. ■

---

* We should mention another terminology, used by the disciples of N. Bourbaki, in case you encounter it elsewhere. In Bourbaki's terminology, a one-to-one map is an **injection,** an onto map is a **surjection,** and a map that is both one to one and onto is a **bijection.**

If a subset of $X \times Y$ is a *one-to-one* function ϕ mapping X *onto* Y, then each $x \in X$ appears as the first member of exactly one ordered pair in ϕ and also each $y \in Y$ appears as the second member of exactly one ordered pair in ϕ. Thus if we interchange the first and second members of all ordered pairs (x, y) in ϕ to obtain a set of ordered pairs (y, x), we get a subset of $Y \times X$, which gives a one-to-one function mapping Y onto X. This function is called the **inverse function** of ϕ, and is denoted by ϕ^{-1}. Summarizing, if ϕ maps X one to one onto Y and $\phi(x) = y$, then ϕ^{-1} maps Y one to one onto X, and $\phi^{-1}(y) = x$.

0.13 Definition Two sets X and Y have the **same cardinality** if there exists a one-to-one function mapping X onto Y, that is, if there exists a one-to-one correspondence between X and Y. ■

0.14 Example The function $f : \mathbb{R} \to \mathbb{R}$ where $f(x) = x^2$ is not one to one because $f(2) = f(-2) = 4$ but $2 \neq -2$. Also, it is not onto $\mathbb{R}$ because the range is the proper subset of all nonnegative numbers in $\mathbb{R}$. However, $g : \mathbb{R} \to \mathbb{R}$ defined by $g(x) = x^3$ is both one to one and onto $\mathbb{R}$. ▲

We showed that $\mathbb{Z}$ and $\mathbb{Z}^+$ have the same cardinality. We denote this cardinal number by $\aleph_0$, so that $|\mathbb{Z}| = |\mathbb{Z}^+| = \aleph_0$. It is fascinating that a proper subset of an infinite set may have the same number of elements as the whole set; an **infinite set** can be defined as a set having this property.

We naturally wonder whether all infinite sets have the same cardinality as the set $\mathbb{Z}$. A set has cardinality $\aleph_0$ if and only if *all* of its elements could be listed in an infinite row, so that we could "number them" using $\mathbb{Z}^+$. Figure 0.15 indicates that this is possible for the set $\mathbb{Q}$. The square array of fractions extends infinitely to the right and infinitely downward, and contains all members of $\mathbb{Q}$. We have shown a string winding its way through this array. Imagine the fractions to be glued to this string. Taking the beginning of the string and pulling to the left in the direction of the arrow, the string straightens out and all elements of $\mathbb{Q}$ appear on it in an infinite row as $0, \frac{1}{2}, -\frac{1}{2}, 1, -1, \frac{3}{2}, \cdots$. Thus $|\mathbb{Q}| = \aleph_0$ also.

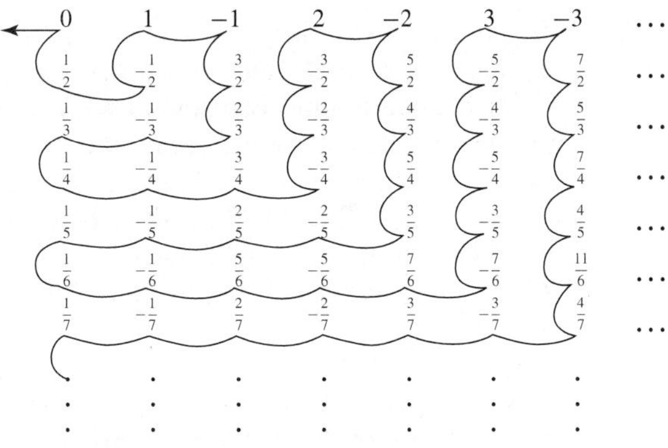

0.15 Figure

If the set $S = \{x \in \mathbb{R} \mid 0 < x < 1\}$ has cardinality $\aleph_0$, all its elements could be listed as unending decimals in a column extending infinitely downward, perhaps as

$$\begin{array}{c} 0.3659663426\cdots \\ 0.7103958453\cdots \\ 0.0358493553\cdots \\ 0.9968452214\cdots \\ \vdots \end{array}$$

We now argue that any such array must omit some number in S. Surely S contains a number r having as its nth digit after the decimal point a number different from 0, from 9, and from the nth digit of the nth number in this list. For example, r might start $.5637\cdots$. The 5 rather than 3 after the decimal point shows r cannot be the first number in S listed in the array shown. The 6 rather than 1 in the second digit shows r cannot be the second number listed, and so on. Because we could make this argument with *any list,* we see that S has too many elements to be paired with those in $\mathbb{Z}^+$. Exercise 15 indicates that $\mathbb{R}$ has the same number of elements as S. We just denote the cardinality of $\mathbb{R}$ by $|\mathbb{R}|$. Exercise 19 indicates that there are infinitely many different cardinal numbers even greater than $|\mathbb{R}|$.

Partitions and Equivalence Relations

Sets are **disjoint** if no two of them have any element in common. Later we will have occasion to break up a set having an algebraic structure (e.g., a notion of addition) into disjoint subsets that become elements in a related algebraic structure. We conclude this section with a study of such breakups, or *partitions* of sets.

0.16 Definition A **partition** of a set S is a collection of nonempty subsets of S such that every element of S is in exactly one of the subsets. The subsets are the **cells** of the partition. ■

When discussing a partition of a set S, we denote by $\bar{x}$ the cell containing the element x of S.

0.17 Example Splitting $\mathbb{Z}^+$ into the subset of even positive integers (those divisible by 2) and the subset of odd positive integers (those leaving a remainder of 1 when divided by 2), we obtain a partition of $\mathbb{Z}^+$ into two cells. For example, we can write

$$\overline{14} = \{2, 4, 6, 8, 10, 12, 14, 16, 18, \cdots\}.$$

We could also partition $\mathbb{Z}^+$ into three cells, one consisting of the positive integers divisible by 3, another containing all positive integers leaving a remainder of 1 when divided by 3, and the last containing positive integers leaving a remainder of 2 when divided by 3.

Generalizing, for each positive integer n, we can partition $\mathbb{Z}^+$ into n cells according to whether the remainder is $0, 1, 2, \cdots, n-1$ when a positive integer is divided by n. These cells are the **residue classes modulo** n in $\mathbb{Z}^+$. Exercise 35 asks us to display these partitions for the cases $n = 2, 3,$ and 5. ▲

Each partition of a set S yields a relation $\mathscr{R}$ on S in a natural way: namely, for $x, y \in S$, let $x \,\mathscr{R}\, y$ if and only if x and y are in the same cell of the partition. In set notation, we would write $x \,\mathscr{R}\, y$ as $(x, y) \in \mathscr{R}$ (see Definition 0.7). A bit of thought shows that this relation $\mathscr{R}$ on S satisfies the three properties of an *equivalence relation* in the following definition.

0.18 Definition An **equivalence relation** $\mathscr{R}$ on a set S is one that satisfies these three properties for all $x, y, z \in S$.

1. (Reflexive) $x \,\mathscr{R}\, x$.
2. (Symmetric) If $x \,\mathscr{R}\, y$, then $y \,\mathscr{R}\, x$.
3. (Transitive) If $x \,\mathscr{R}\, y$ and $y \,\mathscr{R}\, z$ then $x \,\mathscr{R}\, z$. ■

To illustrate why the relation $\mathscr{R}$ corresponding to a partition of S satisfies the symmetric condition in the definition, we need only observe that if y is in the same cell as x (that is, if $x \,\mathscr{R}\, y$), then x is in the same cell as y (that is, $y \,\mathscr{R}\, x$). We leave the similar observations to verify the reflexive and transitive properties to Exercise 28.

0.19 Example For any nonempty set S, the equality relation $=$ defined by the subset $\{(x, x) \mid x \in S\}$ of $S \times S$ is an equivalence relation. ▲

0.20 Example **(Congruence Modulo n)** Let $n \in \mathbb{Z}^+$. The equivalence relation on $\mathbb{Z}^+$ corresponding to the partition of $\mathbb{Z}^+$ into residue classes modulo n, discussed in Example 0.17, is **congruence modulo n**. It is sometimes denoted by $\equiv_n$. Rather than write $a \equiv_n b$, we usually write $a \equiv b \pmod{n}$, read, "a is congruent to b modulo n." For example, we have $15 \equiv 27 \pmod{4}$ because both 15 and 27 have remainder 3 when divided by 4. ▲

0.21 Example Let a relation $\mathscr{R}$ on the set $\mathbb{Z}$ be defined by $n \,\mathscr{R}\, m$ if and only if $nm \geq 0$, and let us determine whether $\mathscr{R}$ is an equivalence relation.

Reflexive $a \,\mathscr{R}\, a$, because $a^2 \geq 0$ for all $a \in \mathbb{Z}$.

Symmetric If $a \,\mathscr{R}\, b$, then $ab \geq 0$, so $ba \geq 0$ and $b \,\mathscr{R}\, a$.

Transitive If $a \,\mathscr{R}\, b$ and $b \,\mathscr{R}\, c$, then $ab \geq 0$ and $bc \geq 0$. Thus $ab^2c = acb^2 \geq 0$. If we knew $b^2 > 0$, we could deduce $ac \geq 0$ whence $a \,\mathscr{R}\, c$. We have to examine the case $b = 0$ separately. A moment of thought shows that $-3 \,\mathscr{R}\, 0$ and $0 \,\mathscr{R}\, 5$, but we do *not* have $-3 \,\mathscr{R}\, 5$. Thus the relation $\mathscr{R}$ is not transitive, and hence is not an equivalence relation. ▲

We observed above that a partition yields a natural equivalence relation. We now show that an equivalence relation on a set yields a natural partition of the set. The theorem that follows states both results for reference.

0.22 Theorem **(Equivalence Relations and Partitions)** Let S be a nonempty set and let $\sim$ be an equivalence relation on S. Then $\sim$ yields a partition of S, where

$$\bar{a} = \{x \in S \mid x \sim a\}.$$

Also, each partition of S gives rise to an equivalence relation $\sim$ on S where $a \sim b$ if and only if a and b are in the same cell of the partition.

Proof We must show that the different cells $\bar{a} = \{x \in S \mid x \sim a\}$ for $a \in S$ do give a partition of S, so that every element of S is in some cell and so that if $a \in \bar{b}$, then $\bar{a} = \bar{b}$. Let $a \in S$. Then $a \in \bar{a}$ by the reflexive condition (1), so a is in *at least one* cell.

Suppose now that a were in a cell $\bar{b}$ also. We need to show that $\bar{a} = \bar{b}$ as sets; this will show that a cannot be in more than one cell. There is a standard way to show that two sets are the same:

Show that each set is a subset of the other.

We show that $\bar{a} \subseteq \bar{b}$. Let $x \in \bar{a}$. Then $x \sim a$. But $a \in \bar{b}$, so $a \sim b$. Then, by the transitive condition (3), $x \sim b$, so $x \in \bar{b}$. Thus $\bar{a} \subseteq \bar{b}$. Now we show that $\bar{b} \subseteq \bar{a}$. Let $y \in \bar{b}$. Then $y \sim b$. But $a \in \bar{b}$, so $a \sim b$ and, by symmetry (2), $b \sim a$. Then by transitivity (3), $y \sim a$, so $y \in \bar{a}$. Hence $\bar{b} \subseteq \bar{a}$ also, so $\bar{b} = \bar{a}$ and our proof is complete. ◆

Each cell in the partition arising from an equivalence relation is an **equivalence class**.

EXERCISES 0

In Exercises 1 through 4, describe the set by listing its elements.

1. $\{x \in \mathbb{R} \mid x^2 = 3\}$
2. $\{m \in \mathbb{Z} \mid m^2 = 3\}$
3. $\{m \in \mathbb{Z} \mid mn = 60 \text{ for some } n \in \mathbb{Z}\}$
4. $\{m \in \mathbb{Z} \mid m^2 - m < 115\}$

In Exercises 5 through 10, decide whether the object described is indeed a set (is well defined). Give an alternate description of each set.

5. $\{n \in \mathbb{Z}^+ \mid n \text{ is a large number}\}$
6. $\{n \in \mathbb{Z} \mid n^2 < 0\}$
7. $\{n \in \mathbb{Z} \mid 39 < n^3 < 57\}$
8. $\{x \in \mathbb{Q} \mid x \text{ is almost an integer}\}$
9. $\{x \in \mathbb{Q} \mid x \text{ may be written with denominator greater than } 100\}$
10. $\{x \in \mathbb{Q} \mid x \text{ may be written with positive denominator less than } 4\}$
11. List the elements in $\{a, b, c\} \times \{1, 2, c\}$.
12. Let $A = \{1, 2, 3\}$ and $B = \{2, 4, 6\}$. For each relation between A and B given as a subset of $A \times B$, decide whether it is a function mapping A into B. If it is a function, decide whether it is one to one and whether it is onto B.
    - **a.** $\{(1, 4), (2, 4), (3, 6)\}$
    - **b.** $\{(1, 4), (2, 6), (3, 4)\}$
    - **c.** $\{(1, 6), (1, 2), (1, 4)\}$
    - **d.** $\{(2, 2), (1, 6), (3, 4)\}$
    - **e.** $\{(1, 6), (2, 6), (3, 6)\}$
    - **f.** $\{(1, 2), (2, 6), (2, 4)\}$
13. Illustrate geometrically that two line segments AB and CD of different length have the same number of points by indicating in Fig. 0.23 what point y of CD might be paired with point x of AB.

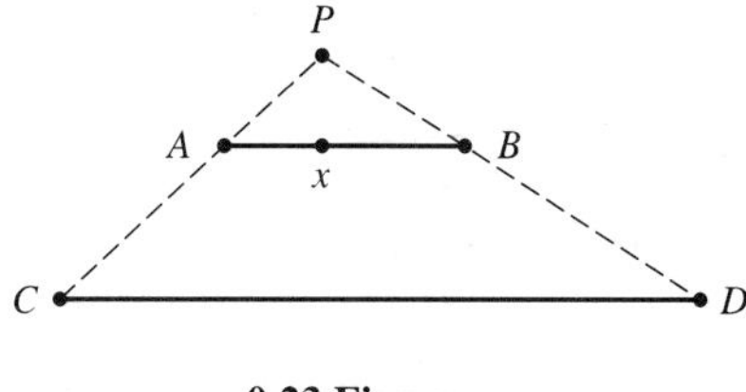

0.23 Figure

14. Recall that for $a, b \in \mathbb{R}$ and $a < b$, the **closed interval** $[a, b]$ in $\mathbb{R}$ is defined by $[a, b] = \{x \in \mathbb{R} \mid a \leq x \leq b\}$. Show that the given intervals have the same cardinality by giving a formula for a one-to-one function f mapping the first interval onto the second.

a. $[0, 1]$ and $[0, 2]$ **b.** $[1, 3]$ and $[5, 25]$ **c.** $[a, b]$ and $[c, d]$

15. Show that $S = \{x \in \mathbb{R} \mid 0 < x < 1\}$ has the same cardinality as $\mathbb{R}$. [*Hint:* Find an elementary function of calculus that maps an interval one to one onto $\mathbb{R}$, and then translate and scale appropriately to make the domain the set S.]

For any set A, we denote by $\mathscr{P}(A)$ the collection of all subsets of A. For example, if $A = \{a, b, c, d\}$, then $\{a, b, d\} \in \mathscr{P}(A)$. The set $\mathscr{P}(A)$ is the **power set** of A. Exercises 16 through 19 deal with the notion of the power set of a set A.

16. List the elements of the power set of the given set and give the cardinality of the power set.

a. $\varnothing$ **b.** $\{a\}$ **c.** $\{a, b\}$ **d.** $\{a, b, c\}$

17. Let A be a finite set, and let $|A| = s$. Based on the preceding exercise, make a conjecture about the value of $|\mathscr{P}(A)|$. Then try to prove your conjecture.

18. For any set A, finite or infinite, let B^A be the set of all functions mapping A into the set $B = \{0, 1\}$. Show that the cardinality of B^A is the same as the cardinality of the set $\mathscr{P}(A)$. [*Hint:* Each element of B^A determines a subset of A in a natural way.]

19. Show that the power set of a set A, finite or infinite, has too many elements to be able to be put in a one-to-one correspondence with A. Explain why this intuitively means that there are an infinite number of infinite cardinal numbers. [*Hint:* Imagine a one-to-one function ϕ mapping A into $\mathscr{P}(A)$ to be given. Show that ϕ cannot be onto $\mathscr{P}(A)$ by considering, for each $x \in A$, whether $x \in \phi(x)$ and using this idea to define a subset S of A that is not in the range of ϕ.] Is *the set of everything* a logically acceptable concept? Why or why not?

20. Let $A = \{1, 2\}$ and let $B = \{3, 4, 5\}$.

a. Illustrate, using A and B, why we consider that $2 + 3 = 5$. Use similar reasoning with sets of your own choice to decide what you would consider to be the value of

i. $3 + \aleph_0$, ii. $\aleph_0 + \aleph_0$.

b. Illustrate why we consider that $2 \cdot 3 = 6$ by plotting the points of $A \times B$ in the plane $\mathbb{R} \times \mathbb{R}$. Use similar reasoning with a figure in the text to decide what you would consider to be the value of $\aleph_0 \cdot \aleph_0$.

21. How many numbers in the interval $0 \leq x \leq 1$ can be expressed in the form .##, where each # is a digit $0, 1, 2, 3, \cdots, 9$? How many are there of the form .#####? Following this idea, and Exercise 15, decide what you would consider to be the value of $10^{\aleph_0}$. How about $12^{\aleph_0}$ and $2^{\aleph_0}$?

22. Continuing the idea in the preceding exercise and using Exercises 18 and 19, use exponential notation to fill in the three blanks to give a list of five cardinal numbers, each of which is greater than the preceding one.

$$\aleph_0, |\mathbb{R}|, \_\_, \_\_, \_\_.$$

In Exercises 23 through 27, find the number of different partitions of a set having the given number of elements.

23. 1 element **24.** 2 elements **25.** 3 elements

26. 4 elements **27.** 5 elements

28. Consider a partition of a set S. The paragraph following Definition 0.18 explained why the relation

$$x \mathscr{R} y \text{ if and only if } x \text{ and } y \text{ are in the same cell}$$

satisfies the symmetric condition for an equivalence relation. Write similar explanations of why the reflexive and transitive properties are also satisifed.

In Exercises 29 through 34, determine whether the given relation is an equivalence relation on the set. Describe the partition arising from each equivalence relation.

29. $n \mathscr{R} m$ in $\mathbb{Z}$ if $nm > 0$ **30.** $x \mathscr{R} y$ in $\mathbb{R}$ if $x \geq y$

31. $x \mathscr{R} y$ in $\mathbb{R}$ if $|x| = |y|$ **32.** $x \mathscr{R} y$ in $\mathbb{R}$ if $|x - y| \leq 3$

33. $n \mathscr{R} m$ in $\mathbb{Z}^+$ if n and m have the same number of digits in the usual base ten notation

34. $n \mathscr{R} m$ in $\mathbb{Z}^+$ if n and m have the same final digit in the usual base ten notation

35. Using set notation of the form $\{\#, \#, \#, \cdots\}$ for an infinite set, write the residue classes modulo n in $\mathbb{Z}^+$ discussed in Example 0.17 for the indicated value of n.

a. $n = 2$ **b.** $n = 3$ **c.** $n = 5$

36. Let $n \in \mathbb{Z}^+$ and let $\sim$ be defined on $\mathbb{Z}$ by $r \sim s$ if and only if $r - s$ is divisible by n, that is, if and only if $r - s = nq$ for some $q \in \mathbb{Z}$.

a. Show that $\sim$ is an equivalence relation on $\mathbb{Z}$. (It is called "congruence modulo n" just as it was for $\mathbb{Z}^+$. See part b.)

b. Show that, when restricted to the subset $\mathbb{Z}^+$ of $\mathbb{Z}$, this $\sim$ is the equivalence relation, *congruence modulo n*, of Example 0.20.

c. The cells of this partition of $\mathbb{Z}$ are *residue classes modulo n* in $\mathbb{Z}$. Repeat Exercise 35 for the residue classes modulo in $\mathbb{Z}$ rather than in $\mathbb{Z}^+$ using the notation $\{\cdots, \#, \#, \#, \cdots\}$ for these infinite sets.

37. Students often misunderstand the concept of a one-to-one function (mapping). I think I know the reason. You see, a mapping $\phi : A \rightarrow B$ has a *direction* associated with it, from A to B. It seems reasonable to expect a one-to-one mapping simply to be a mapping that carries one point of A into one point of B, in the direction indicated by the arrow. But of course, *every* mapping of A into B does this, and Definition 0.12 did not say that at all. With this unfortunate situation in mind, make as good a pedagogical case as you can for calling the functions described in Definition 0.12 *two-to-two functions* instead. (Unfortunately, it is almost impossible to get widely used terminology changed.)

PART I

Groups and Subgroups

SECTION 1 Introduction and Examples

In this section, we attempt to give you a little idea of the nature of abstract algebra. We are all familiar with addition and multiplication of real numbers. Both addition and multiplication combine two numbers to obtain one number. For example, addition combines 2 and 3 to obtain 5. We consider addition and multiplication to be *binary operations*. In this text, we abstract this notion, and examine sets in which we have one or more binary operations. We think of a binary operation on a set as giving an algebra on the set, and we are interested in the *structural properties* of that algebra. To illustrate what we mean by a structural property with our familiar set $\mathbb{R}$ of real numbers, note that the equation $x + x = a$ has a solution x in $\mathbb{R}$ for each $a \in \mathbb{R}$, namely, $x = a/2$. However, the corresponding multiplicative equation $x \cdot x = a$ does not have a solution in $\mathbb{R}$ if $a < 0$. Thus, $\mathbb{R}$ with addition has a different algebraic structure than $\mathbb{R}$ with multiplication.

Sometimes two different sets with what we naturally regard as very different binary operations turn out to have the same algebraic structure. For example, we will see in Section 3 that the set $\mathbb{R}$ with addition has the same algebraic structure as the set $\mathbb{R}^+$ of positive real numbers with multiplication!

This section is designed to get you thinking about such things informally. We will make everything precise in Sections 2 and 3. We now turn to some examples. Multiplication of complex numbers of magnitude 1 provides us with several examples that will be useful and illuminating in our work. We start with a review of complex numbers and their multiplication.

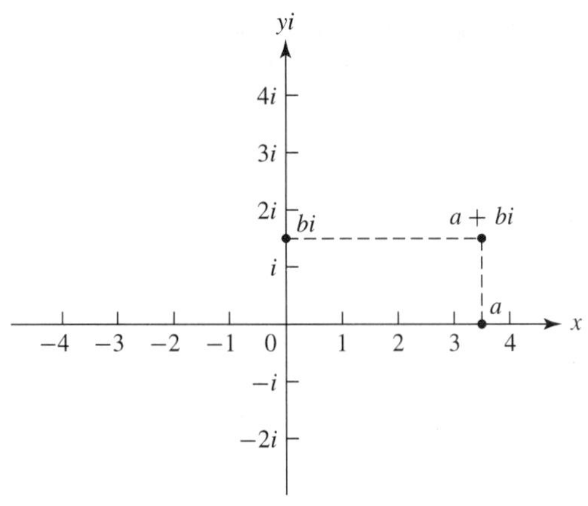

1.1 Figure

Complex Numbers

A real number can be visualized geometrically as a point on a line that we often regard as an x-axis. A complex number can be regarded as a point in the Euclidean plane, as shown in Fig. 1.1. Note that we label the vertical axis as the yi-axis rather than just the y-axis, and label the point one unit above the origin with i rather than 1. The point with Cartesian coordinates (a, b) is labeled $a + bi$ in Fig. 1.1. The set $\mathbb{C}$ of **complex numbers** is defined by

$$\mathbb{C} = \{a + bi \mid a, b \in \mathbb{R}\}.$$

We consider $\mathbb{R}$ to be a subset of the complex numbers by identifying a real number r with the complex number $r + 0i$. For example, we write $3 + 0i$ as 3 and $-\pi + 0i$ as $-\pi$ and $0 + 0i$ as 0. Similarly, we write $0 + 1i$ as i and $0 + si$ as si.

Complex numbers were developed after the development of real numbers. The complex number i was *invented* to provide a solution to the quadratic equation $x^2 = -1$, so we require that

$$i^2 = -1. \tag{1}$$

Unfortunately, i has been called an **imaginary number,** and this terminology has led generations of students to view the complex numbers with more skepticism than the real numbers. Actually, *all* numbers, such as $1, 3, \pi, -\sqrt{3}$, and i are inventions of our minds. There is no physical entity that *is* the number 1. If there were, it would surely be in a place of honor in some great scientific museum, and past it would file a steady stream of mathematicians, gazing at 1 in wonder and awe. A basic goal of this text is to show how we can invent solutions of polynomial equations when the coefficients of the polynomial may not even be real numbers!

Multiplication of Complex Numbers

The product $(a + bi)(c + di)$ is defined in the way it must be if we are to enjoy the familiar properties of real arithmetic and require that $i^2 = -1$, in accord with Eq. (1).

Namely, we see that we want to have

$$\begin{aligned}(a+bi)(c+di) &= ac + adi + bci + bdi^2 \\ &= ac + adi + bci + bd(-1) \\ &= (ac - bd) + (ad + bc)i.\end{aligned}$$

Consequently, we define multiplication of $z_1 = a + bi$ and $z_2 = c + di$ as

$$z_1 z_2 = (a+bi)(c+di) = (ac - bd) + (ad + bc)i, \qquad \textbf{(2)}$$

which is of the form $r + si$ with $r = ac - bd$ and $s = ad + bc$. It is routine to check that the usual properties $z_1z_2 = z_2z_1$, $z_1(z_2z_3) = (z_1z_2)z_3$ and $z_1(z_2 + z_3) = z_1z_2 + z_1z_3$ all hold for all $z_1, z_2, z_3 \in \mathbb{C}$.

1.2 Example Compute $(2 - 5i)(8 + 3i)$.

Solution We don't memorize Eq. (2), but rather we compute the product as we did to motivate that equation. We have

$$(2 - 5i)(8 + 3i) = 16 + 6i - 40i + 15 = 31 - 34i. \qquad \blacktriangle$$

To establish the geometric meaning of complex multiplication, we first define the **absolute value** $|a + bi|$ of $a + bi$ by

$$|a + bi| = \sqrt{a^2 + b^2}. \qquad \textbf{(3)}$$

This absolute value is a nonnegative real number and is the distance from $a + bi$ to the origin in Fig. 1.1. We can now describe a complex number z in the polar-coordinate form

$$z = |z|(\cos\theta + i\sin\theta), \qquad \textbf{(4)}$$

where θ is the angle measured counterclockwise from the x-axis to the vector from 0 to z, as shown in Fig. 1.3. A famous formula due to Leonard Euler states that

$$e^{i\theta} = \cos\theta + i\sin\theta.$$

Euler's Formula

We ask you to derive Euler's formula formally from the power series expansions for e^θ, $\cos\theta$ and $\sin\theta$ in Exercise 41. Using this formula, we can express z in Eq. (4) as

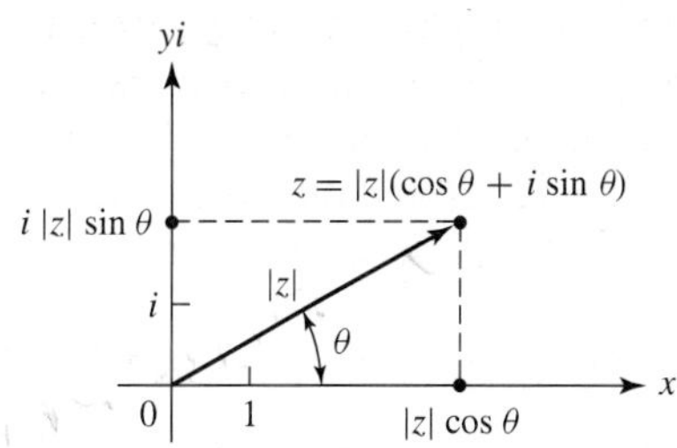

1.3 Figure

$z = |z|e^{i\theta}$. Let us set

$$z_1 = |z_1|e^{i\theta_1} \qquad \text{and} \qquad z_2 = |z_2|e^{i\theta_2}$$

and compute their product in this form, assuming that the usual laws of exponentiation hold with complex number exponents. We obtain

$$\begin{aligned} z_1 z_2 &= |z_1|e^{i\theta_1}|z_2|e^{i\theta_2} = |z_1||z_2|e^{i(\theta_1+\theta_2)} \\ &= |z_1||z_2|[\cos(\theta_1+\theta_2) + i\sin(\theta_1+\theta_2)]. \end{aligned} \tag{5}$$

Note that Eq. 5 concludes in the polar form of Eq. 4 where $|z_1 z_2| = |z_1||z_2|$ and the polar angle θ for $z_1 z_2$ is the sum $\theta = \theta_1 + \theta_2$. Thus, geometrically, we multiply complex numbers by multiplying their absolute values and adding their polar angles, as shown in Fig. 1.4. Exercise 39 indicates how this can be derived via trigonometric identities without recourse to Euler's formula and assumptions about complex exponentiation.

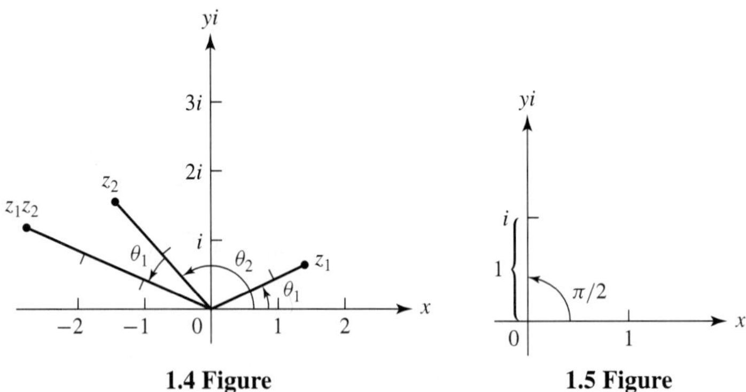

1.4 Figure **1.5 Figure**

Note that i has polar angle $\pi/2$ and absolute value 1, as shown in Fig. 1.5. Thus i^2 has polar angle $2(\pi/2) = \pi$ and $|1 \cdot 1| = 1$, so that $i^2 = -1$.

1.6 Example Find all solutions in $\mathbb{C}$ of the equation $z^2 = i$.

Solution Writing the equation $z^2 = i$ in polar form and using Eq. (5), we obtain

$$|z|^2(\cos 2\theta + i\sin 2\theta) = 1(0 + i).$$

Thus $|z|^2 = 1$, so $|z| = 1$. The angle θ for z must satisfy $\cos 2\theta = 0$ and $\sin 2\theta = 1$. Consequently, $2\theta = (\pi/2) + n(2\pi)$, so $\theta = (\pi/4) + n\pi$ for an integer n. The values of n yielding values θ where $0 \leq \theta < 2\pi$ are 0 and 1, yielding $\theta = \pi/4$ or $\theta = 5\pi/4$. Our solutions are

$$z_1 = 1\left(\cos\frac{\pi}{4} + i\sin\frac{\pi}{4}\right) \qquad \text{and} \qquad z_2 = 1\left(\cos\frac{5\pi}{4} + i\sin\frac{5\pi}{4}\right)$$

or

$$z_1 = \frac{1}{\sqrt{2}}(1+i) \qquad \text{and} \qquad z_2 = \frac{-1}{\sqrt{2}}(1+i).$$

▲

1.7 Example Find all solutions of $z^4 = -16$.

Solution As in Example 1.6 we write the equation in polar form, obtaining

$$|z|^4(\cos 4\theta + i\sin 4\theta) = 16(-1 + 0i).$$

Consequently, $|z|^4 = 16$, so $|z| = 2$ while $\cos 4\theta = -1$ and $\sin 4\theta = 0$. We find that $4\theta = \pi + n(2\pi)$, so $\theta = (\pi/4) + n(\pi/2)$ for integers n. The different values of θ obtained where $0 \leq \theta < 2\pi$ are $\pi/4, 3\pi/4, 5\pi/4$, and $7\pi/4$. Thus one solution of $z^4 = -16$ is

$$2\left(\cos\frac{\pi}{4} + i\sin\frac{\pi}{4}\right) = 2\left(\frac{1}{\sqrt{2}} + \frac{1}{\sqrt{2}}i\right) = \sqrt{2}(1+i).$$

In a similar way, we find three more solutions,

$$\sqrt{2}(-1+i), \qquad \sqrt{2}(-1-i), \qquad \text{and} \qquad \sqrt{2}(1-i).$$ ▲

The last two examples illustrate that we can find solutions of an equation $z^n = a + bi$ by writing the equation in polar form. There will always be n solutions, provided that $a + bi \neq 0$. Exercises 16 through 21 ask you to solve equations of this type.

We will not use addition or division of complex numbers, but we probably should mention that addition is given by

$$(a + bi) + (c + di) = (a + c) + (b + d)i. \tag{6}$$

and division of $a + bi$ by nonzero $c + di$ can be performed using the device

$$\begin{aligned}\frac{a+bi}{c+di} &= \frac{a+bi}{c+di}\cdot\frac{c-di}{c-di} = \frac{(ac+bd)+(bc-ad)i}{c^2+d^2}\\ &= \frac{ac+bd}{c^2+d^2} + \frac{bc-ad}{c^2+d^2}i.\end{aligned} \tag{7}$$

Algebra on Circles

Let $U = \{z \in \mathbb{C} \mid |z| = 1\}$, so that U is the circle in the Euclidean plane with center at the origin and radius 1, as shown in Fig. 1.8. The relation $|z_1z_2| = |z_1||z_2|$ shows that the product of two numbers in U is again a number in U; we say that U is *closed* under multiplication. Thus, we can view multiplication in U as providing algebra on the circle in Fig. 1.8.

As illustrated in Fig. 1.8, we associate with each $z = \cos\theta + i\sin\theta$ in U a real number $\theta \in \mathbb{R}$ that lies in the half-open interval where $0 \leq \theta < 2\pi$. This half-open interval is usually denoted by $[0, 2\pi)$, but we prefer to denote it by $\mathbb{R}_{2\pi}$ for reasons that will be apparent later. Recall that the angle associated with the product z_1z_2 of two complex numbers is the sum $\theta_1 + \theta_2$ of the associated angles. Of course if $\theta_1 + \theta_2 \geq 2\pi$

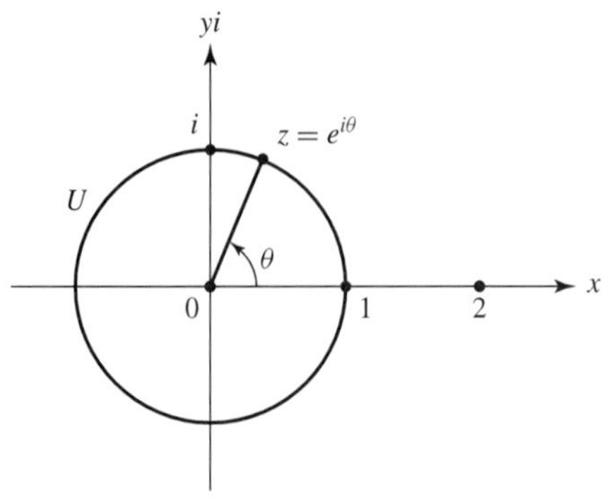

1.8 Figure

then the angle in $\mathbb{R}_{2\pi}$ associated with $z_1 z_2$ is $\theta_1 + \theta_2 - 2\pi$. This gives us an **addition modulo 2π** on $\mathbb{R}_{2\pi}$. We denote this addition here by $+_{2\pi}$.

1.9 Example In $\mathbb{R}_{2\pi}$, we have $\frac{3\pi}{2} +_{2\pi} \frac{5\pi}{4} = \frac{11\pi}{4} - 2\pi = \frac{3\pi}{4}$. ▲

There was nothing special about the number 2π that enabled us to define addition on the half-open interval $\mathbb{R}_{2\pi}$. We can use any half-open interval $\mathbb{R}_c = \{x \in \mathbb{R} \mid 0 \leq x < c\}$.

1.10 Example In $\mathbb{R}_{23}$, we have $16 +_{23} 19 = 35 - 23 = 12$. In $\mathbb{R}_{8.5}$, we have $6 +_{8.5} 8 = 14 - 8.5 = 5.5$. ▲

Now complex number multiplication on the circle U where $|z| = 1$ and addition modulo 2π on $\mathbb{R}_{2\pi}$ have the same *algebraic properties*. We have the natural one-to-one correspondence $z \leftrightarrow \theta$ between $z \in U$ and $\theta \in \mathbb{R}_{2\pi}$ indicated in Fig. 1.8. Moreover, we deliberately defined $+_{2\pi}$ so that

$$\text{if } \quad z_1 \leftrightarrow \theta_1 \quad \text{and} \quad z_2 \leftrightarrow \theta_2, \quad \text{then} \quad z_1 \cdot z_2 \leftrightarrow (\theta_1 +_{2\pi} \theta_2). \qquad \textbf{(8)}$$

isomorphism

The relation (8) shows that if we rename each $z \in U$ by its corresponding angle θ shown in Fig. 1.8, then the product of two elements in U is renamed by the sum of the angles for those two elements. Thus U with complex number multiplication and $\mathbb{R}_{2\pi}$ with addition modulo 2π must have the same algebraic properties. They differ only in the names of the elements and the names of the operations. Such a one-to-one correspondence satisfying the relation (8) is called an *isomorphism*. Names of elements and names of binary operations are not important in abstract algebra; we are interested in algebraic properties. We illustrate what we mean by saying that the algebraic properties of U and of $\mathbb{R}_{2\pi}$ are the same.

1.11 Example In U there is exactly one element e such that $e \cdot z = z$ for all $z \in U$, namely, $e = 1$. The element 0 in $\mathbb{R}_{2\pi}$ that corresponds to $1 \in U$ is the only element e in $\mathbb{R}_{2\pi}$ such that $e +_{2\pi} x = x$ for all $x \in \mathbb{R}_{2\pi}$. ▲

1.12 Example The equation $z \cdot z \cdot z \cdot z = 1$ in U has exactly four solutions, namely, $1, i, -1$, and $-i$. Now $1 \in U$ and $0 \in \mathbb{R}_{2\pi}$ correspond, and the equation $x +_{2\pi} x +_{2\pi} x +_{2\pi} x = 0$ in $\mathbb{R}_{2\pi}$ has exactly four solutions, namely, $0, \pi/2, \pi$, and $3\pi/2$, which, of course, correspond to $1, i, -1$, and $-i$, respectively. ▲

Because our circle U has radius 1, it has circumference 2π and the radian measure of an angle θ is equal to the length of the arc the angle subtends. If we pick up our half-open interval $\mathbb{R}_{2\pi}$, put the 0 in the interval down on the 1 on the x-axis and wind it around the circle U counterclockwise, it will reach all the way back to 1. Moreover, each number in the interval will fall on the point of the circle having that number as the value of the central angle θ shown in Fig. 1.8. This shows that we could also think of addition on $\mathbb{R}_{2\pi}$ as being computed by adding lengths of subtended arcs counterclockwise, starting at $z = 1$, and subtracting 2π if the sum of the lengths is 2π or greater.

If we think of addition on a circle in terms of adding lengths of arcs from a starting point P on the circle and proceeding counterclockwise, we can use a circle of radius 2, which has circumference 4π, just as well as a circle of radius 1. We can take our half-open interval $\mathbb{R}_{4\pi}$ and wrap it around counterclockwise, starting at P; it will just cover the whole circle. Addition of arcs lengths gives us a notion of algebra for points on this circle of radius 2, which is surely isomorphic to $\mathbb{R}_{4\pi}$ with addition $+_{4\pi}$. However, if we take as the circle $|z| = 2$ in Fig. 1.8, multiplication of complex numbers does not give us an algebra on this circle. The relation $|z_1 z_2| = |z_1||z_2|$ shows that the product of two such complex numbers has absolute value 4 rather than 2. Thus complex number multiplication is *not closed* on this circle.

The preceding paragraphs indicate that a little geometry can sometimes be of help in abstract algebra. We can use geometry to convince ourselves that $\mathbb{R}_{2\pi}$ and $\mathbb{R}_{4\pi}$ are isomorphic. Simply stretch out the interval $\mathbb{R}_{2\pi}$ uniformly to cover the interval $\mathbb{R}_{4\pi}$, or, if you prefer, use a magnifier of power 2. Thus we set up the one-to-one correspondence $a \leftrightarrow 2a$ between $a \in \mathbb{R}_{2\pi}$ and $2a \in \mathbb{R}_{4\pi}$. The relation (8) for isomorphism becomes

$$\text{if} \quad a \leftrightarrow 2a \quad \text{and} \quad b \leftrightarrow 2b \quad \text{then} \quad (a +_{2\pi} b) \leftrightarrow (2a +_{4\pi} 2b). \qquad \textbf{(9)}$$

isomorphism

This is obvious if $a + b \leq 2\pi$. If $a + b = 2\pi + c$, then $2a + 2b = 4\pi + 2c$, and the final pairing in the displayed relation becomes $c \leftrightarrow 2c$, which is true.

1.13 Example $x +_{4\pi} x +_{4\pi} x +_{4\pi} x = 0$ in $\mathbb{R}_{4\pi}$ has exactly four solutions, namely, $0, \pi, 2\pi$, and 3π, which are two times the solutions found for the analogous equation in $\mathbb{R}_{2\pi}$ in Example 1.12. ▲

There is nothing special about the numbers 2π and 4π in the previous argument. Surely, $\mathbb{R}_c$ with $+_c$ is isomorphic to $\mathbb{R}_d$ with $+_d$ for all $c, d \in \mathbb{R}^+$. We need only pair $x \in \mathbb{R}_c$ with $(d/c)x \in \mathbb{R}_d$.

Roots of Unity

The elements of the set $U_n = \{z \in \mathbb{C} \mid z^n = 1\}$ are called the ***n*^th^ roots of unity.** Using the technique of Examples 1.6 and 1.7, we see that the elements of this set are the numbers

$$\cos\left(m\frac{2\pi}{n}\right) + i\sin\left(m\frac{2\pi}{n}\right) \qquad \text{for} \qquad m = 0, 1, 2, \cdots, n-1.$$

They all have absolute value 1, so $U_n \subset U$. If we let $\zeta = \cos\frac{2\pi}{n} + i\sin\frac{2\pi}{n}$, then these n^{th} roots of unity can be written as

$$1 = \zeta^0, \zeta^1, \zeta^2, \zeta^3, \cdots, \zeta^{n-1}. \tag{10}$$

Because $\zeta^n = 1$, these n powers of ζ are closed under multiplication. For example, with $n = 10$, we have

$$\zeta^6\zeta^8 = \zeta^{14} = \zeta^{10}\zeta^4 = 1 \cdot \zeta^4 = \zeta^4.$$

Thus we see that we can compute $\zeta^i\zeta^j$ by computing $i +_n j$, viewing i and j as elements of $\mathbb{R}_n$.

Let $\mathbb{Z}_n = \{0, 1, 2, 3, \cdots, n-1\}$. We see that $\mathbb{Z}_n \subset \mathbb{R}_n$ and clearly addition modulo n is closed on $\mathbb{Z}_n$.

1.14 Example The solution of the equation $x + 5 = 3$ in $\mathbb{Z}_8$ is $x = 6$, because $5 +_8 6 = 11 - 8 = 3$. ▲

If we rename each of the nth roots of unity in (10) by its exponent, we use for names all the elements of $\mathbb{Z}_n$. This gives a one-to-one correspondence between U_n and $\mathbb{Z}_n$. Clearly,

$$\text{if} \quad \zeta^i \leftrightarrow i \quad \text{and} \quad \zeta^j \leftrightarrow j, \quad \text{then} \quad (\zeta^i \cdot \zeta^j) \leftrightarrow (i +_n j). \tag{11}$$

isomorphism

Thus U_n with complex number multiplication and $\mathbb{Z}_n$ with addition $+_n$ have the same algebraic properties.

1.15 Example It can be shown that there is an isomorphism of U_8 with $\mathbb{Z}_8$ in which $\zeta = e^{i2\pi/8} \leftrightarrow 5$. Under this isomorphism, we must then have $\zeta^2 = \zeta \cdot \zeta \leftrightarrow 5 +_8 5 = 2$. ▲

Exercise 35 asks you to continue the computation in Example 1.15, finding the elements of $\mathbb{Z}_8$ to which each of the remaining six elements of U_8 correspond.

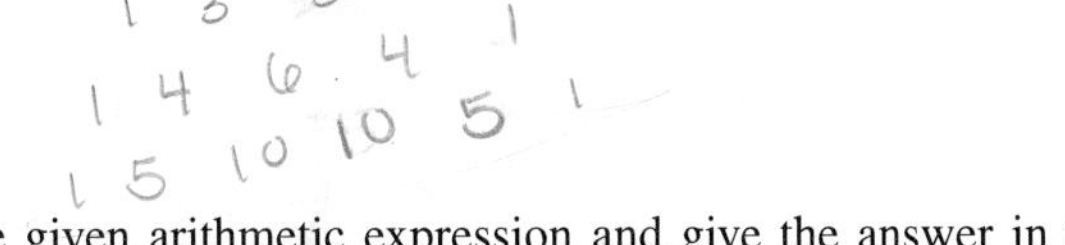

EXERCISES 1

In Exercises 1 through 9 compute the given arithmetic expression and give the answer in the form $a + bi$ for $a, b \in \mathbb{R}$.

1. i^3 **2.** i^4 **3.** i^{23}

4. $(-i)^{35}$ **5.** $(4 - i)(5 + 3i)$ **6.** $(8 + 2i)(3 - i)$

7. $(2 - 3i)(4 + i) + (6 - 5i)$ **8.** $(1 + i)^3$ **9.** $(1 - i)^5$ (Use the binomial theorem.)

10. Find $|3 - 4i|$. **11.** Find $|6 + 4i|$.

In Exercises 12 through 15 write the given complex number z in the polar form $|z|(p + qi)$ where $|p + qi| = 1$.

12. $3 - 4i$ **13.** $-1 + i$ **14.** $12 + 5i$ **15.** $-3 + 5i$

In Exercises 16 through 21, find all solutions in $\mathbb{C}$ of the given equation.

16. $z^4 = 1$ **17.** $z^4 = -1$ **18.** $z^3 = -8$ **19.** $z^3 = -27i$

20. $z^6 = 1$ **21.** $z^6 = -64$

In Exercises 22 through 27, compute the given expression using the indicated modular addition.

22. $10 +_{17} 16$ **23.** $8 +_{10} 6$ **24.** $20.5 +_{25} 19.3$

25. $\frac{1}{2} +_1 \frac{7}{8}$ **26.** $\frac{3\pi}{4} +_{2\pi} \frac{3\pi}{2}$ **27.** $2\sqrt{2} +_{\sqrt{32}} 3\sqrt{2}$

28. Explain why the expression $5 +_6 8$ in $\mathbb{R}_6$ makes no sense.

In Exercises 29 through 34, find *all* solutions x of the given equation.

29. $x +_{15} 7 = 3$ in $\mathbb{Z}_{15}$ **30.** $x +_{2\pi} \frac{3\pi}{2} = \frac{3\pi}{4}$ in $\mathbb{R}_{2\pi}$

31. $x +_7 x = 3$ in $\mathbb{Z}_7$ **32.** $x +_7 x +_7 x = 5$ in $\mathbb{Z}_7$

33. $x +_{12} x = 2$ in $\mathbb{Z}_{12}$ **34.** $x +_4 x +_4 x +_4 x = 0$ in $\mathbb{Z}_4$

35. Example 1.15 asserts that there is an isomorphism of U_8 with $\mathbb{Z}_8$ in which $\zeta = e^{i(\pi/4)} \leftrightarrow 5$ and $\zeta^2 \leftrightarrow 2$. Find the element of $\mathbb{Z}_8$ that corresponds to each of the remaining six elements ζ^m in U_8 for $m = 0, 3, 4, 5, 6,$ and 7.

36. There is an isomorphism of U_7 with $\mathbb{Z}_7$ in which $\zeta = e^{i(2\pi/7)} \leftrightarrow 4$. Find the element in $\mathbb{Z}_7$ to which ζ^m must correspond for $m = 0, 2, 3, 4, 5,$ and 6.

37. Why can there be no isomorphism of U_6 with $\mathbb{Z}_6$ in which $\zeta = e^{i(\pi/3)}$ corresponds to 4?

38. Derive the formulas

$$\sin(a + b) = \sin a \cos b + \cos a \sin b$$

and

$$\cos(a + b) = \cos a \cos b - \sin a \sin b$$

by using Euler's formula and computing $e^{ia}e^{ib}$.

39. Let $z_1 = |z_1|(\cos\theta_1 + i\sin\theta_1)$ and $z_2 = |z_2|(\cos\theta_2 + i\sin\theta_2)$. Use the trigonometric identities in Exercise 38 to derive $z_1z_2 = |z_1||z_2|[\cos(\theta_1 + \theta_2) + i\sin(\theta_1 + \theta_2)]$.

40. **a.** Derive a formula for $\cos 3\theta$ in terms of $\sin\theta$ and $\cos\theta$ using Euler's formula.

b. Derive the formula $\cos 3\theta = 4\cos^3\theta - 3\cos\theta$ from part (a) and the identity $\sin^2\theta + \cos^2\theta = 1$. (We will have use for this identity in Section 32.)

41. Recall the power series expansions

$$e^x = 1 + x + \frac{x^2}{2!} + \frac{x^3}{3!} + \frac{x^4}{4!} + \cdots + \frac{x^n}{n!} + \cdots,$$

$$\sin x = x - \frac{x^3}{3!} + \frac{x^5}{5!} - \frac{x^7}{7!} + \cdots + (-1)^{n-1}\frac{x^{2n-1}}{(2n-1)!} + \cdots, \text{ and}$$

$$\cos x = 1 - \frac{x^2}{2!} + \frac{x^4}{4!} - \frac{x^6}{6!} + \cdots + (-1)^n\frac{x^{2n}}{(2n)!} + \cdots$$

from calculus. Derive Euler's formula $e^{i\theta} = \cos\theta + i\sin\theta$ formally from these three series expansions.

SECTION 2 BINARY OPERATIONS

Suppose that we are visitors to a strange civilization in a strange world and are observing one of the creatures of this world drilling a class of fellow creatures in addition of numbers. Suppose also that we have not been told that the class is learning to add, but were just placed as observers in the room where this was going on. We are asked to give a report on exactly what happens. The teacher makes noises that sound to us approximately like *gloop, poyt.* The class responds with *bimt.* The teacher then gives *ompt, gaft,* and the class responds with *poyt.* What are they doing? We cannot report that they are adding numbers, for we do not even know that the sounds are representing numbers. Of course, we do realize that there is communication going on. All we can say with any certainty is that these creatures know some rule, so that when certain pairs of things are designated in their language, one after another, like *gloop, poyt,* they are able to agree on a response, *bimt.* This same procedure goes on in addition drill in our first grade classes where a teacher may say *four, seven,* and the class responds with *eleven.*

In our attempt to analyze addition and multiplication of numbers, we are thus led to the idea that addition is basically just a rule that people learn, enabling them to associate, with two numbers in a given order, some number as the answer. Multiplication is also such a rule, but a different rule. Note finally that in playing this game with students, teachers have to be a little careful of what two things they give to the class. If a first grade teacher suddenly inserts *ten, sky,* the class will be very confused. The rule is only defined for pairs of things from some specified set.

Definitions and Examples

As mathematicians, let us attempt to collect the core of these basic ideas in a useful definition, generalizing the notions of addition and multiplication of numbers. As we remarked in Section 0, we do not attempt to define a set. However, we can attempt to be somewhat mathematically precise, and we describe our generalizations as *functions* (see Definition 0.10 and Example 0.11) rather than as *rules*. Recall from Definition 0.4 that for any set S, the set $S \times S$ consists of all ordered pairs (a, b) for elements a and b of S.

2.1 Definition A **binary operation** $*$ on a set S is a function mapping $S \times S$ into S. For each $(a, b) \in S \times S$, we will denote the element $*((a, b))$ of S by $a * b$. ■

Intuitively, we may regard a binary operation $*$ on S as assigning, to each ordered pair (a, b) of elements of S, an element $a * b$ of S. We proceed with examples.

2.2 Example Our usual addition $+$ is a binary operation on the set $\mathbb{R}$. Our usual multiplication $\cdot$ is a different binary operation on $\mathbb{R}$. In this example, we could replace $\mathbb{R}$ by any of the sets $\mathbb{C}$, $\mathbb{Z}$, $\mathbb{R}^+$, or $\mathbb{Z}^+$. ▲

Note that we require a binary operation on a set S to be defined for *every* ordered pair (a, b) of elements from S.

2.3 Example Let $M(\mathbb{R})$ be the set of all matrices[†] with real entries. The usual matrix addition $+$ is *not* a binary operation on this set since $A + B$ is not defined for an ordered pair (A, B) of matrices having different numbers of rows or of columns. ▲

Sometimes a binary operation on S provides a binary operation on a subset H of S also. We make a formal definition.

2.4 Definition Let $*$ be a binary operation on S and let H be a subset of S. The subset H is **closed under** $*$ if for all $a, b \in H$ we also have $a * b \in H$. In this case, the binary operation on H given by restricting $*$ to H is the **induced operation** of $*$ on H. ■

By our very definition of a binary operation $*$ on S, the set S is closed under $*$, but a subset may not be, as the following example shows.

2.5 Example Our usual addition $+$ on the set $\mathbb{R}$ of real numbers does not induce a binary operation on the set $\mathbb{R}^*$ of nonzero real numbers because $2 \in \mathbb{R}^*$ and $-2 \in \mathbb{R}^*$, but $2 + (-2) = 0$ and $0 \notin \mathbb{R}^*$. Thus $\mathbb{R}^*$ is not closed under $*$. ▲

In our text, we will often have occasion to decide whether a subset H of S is closed under a binary operation $*$ on S. To arrive at a correct conclusion, *we have to know what it means for an element to be in H*, and to use this fact. Students have trouble here. Be sure you understand the next example.

2.6 Example Let $+$ and $\cdot$ be the usual binary operations of addition and multiplication on the set $\mathbb{Z}$, and let $H = \{n^2 | n \in \mathbb{Z}^+\}$. Determine whether H is closed under (a) addition and (b) multiplication.

For part (a), we need only observe that $1^2 = 1$ and $2^2 = 4$ are in H, but that $1 + 4 = 5$ and $5 \notin H$. Thus H is not closed under addition.

For part (b), suppose that $r \in H$ and $s \in H$. Using what it means for r and s to be in H, we see that there must be integers n and m in $\mathbb{Z}^+$ such that $r = n^2$ and $s = m^2$. Consequently, $rs = n^2m^2 = (nm)^2$. By the characterization of elements in H and the fact that $nm \in \mathbb{Z}^+$, this means that $rs \in H$, so H is closed under multiplication. ▲

[†] Most students of abstract algebra have studied linear algebra and are familiar with matrices and matrix operations. For the benefit of those students, examples involving matrices are often given. The reader who is not familiar with matrices can either skip all references to them or turn to the Appendix at the back of the text, where there is a short summary.

2.7 Example Let F be the set of all real-valued functions f having as domain the set $\mathbb{R}$ of real numbers. We are familiar from calculus with the binary operations $+$, $-$, $\cdot$, and $\circ$ on F. Namely, for each ordered pair (f, g) of functions in F, we define for each $x \in \mathbb{R}$

$$\begin{aligned} f+g \text{ by } (f+g)(x) &= f(x)+g(x) && \text{addition,} \\ f-g \text{ by } (f-g)(x) &= f(x)-g(x) && \text{subtraction,} \\ f\cdot g \text{ by } (f\cdot g)(x) &= f(x)g(x) && \text{multiplication,} \end{aligned}$$

and

$$f \circ g \text{ by } (f\circ g)(x) = f(g(x)) \qquad \text{composition.}$$

All four of these functions are again real valued with domain $\mathbb{R}$, so F is closed under all four operations $+$, $-$, $\cdot$, and $\circ$. ▲

The binary operations described in the examples above are very familiar to you. In this text, we want to *abstract* basic structural concepts from our familiar algebra. To emphasize this concept of *abstraction* from the familiar, we should illustrate these structural concepts with unfamiliar examples. We presented the binary operations of complex number multiplication on U and U_n, addition $+_n$ on $\mathbb{Z}_n$, and addition $+_c$ on $\mathbb{R}_c$ in Section 1.

The most important method of describing a particular binary operation $*$ on a given set is to characterize the element $a * b$ assigned to each pair (a, b) by some property defined in terms of a and b.

2.8 Example On $\mathbb{Z}^+$, we define a binary operation $*$ by $a * b$ equals the smaller of a and b, or the common value if $a = b$. Thus $2 * 11 = 2$; $15 * 10 = 10$; and $3 * 3 = 3$. ▲

2.9 Example On $\mathbb{Z}^+$, we define a binary operation $*'$ by $a *' b = a$. Thus $2 *' 3 = 2$, $25 *' 10 = 25$, and $5 *' 5 = 5$. ▲

2.10 Example On $\mathbb{Z}^+$, we define a binary operation $*''$ by $a *'' b = (a * b) + 2$, where $*$ is defined in Example 2.8. Thus $4 *'' 7 = 6$; $25 *'' 9 = 11$; and $6 *'' 6 = 8$. ▲

It may seem that these examples are of no importance, but consider for a moment. Suppose we go into a store to buy a large, delicious chocolate bar. Suppose we see two identical bars side by side, the wrapper of one stamped \$1.67 and the wrapper of the other stamped \$1.79. Of course we pick up the one stamped \$1.67. Our knowledge of which one we want depends on the fact that at some time we learned the binary operation $*$ of Example 2.8. It is a *very important operation*. Likewise, the binary operation $*'$ of Example 2.9 is defined using our ability to distinguish order. Think what a problem we would have if we tried to put on our shoes first, and then our socks! Thus we should not be hasty about dismissing some binary operation as being of little significance. Of course, our usual operations of addition and multiplication of numbers have a practical importance well known to us.

Examples 2.8 and 2.9 were chosen to demonstrate that a binary operation may or may not depend on the order of the given pair. Thus in Example 2.8, $a * b = b * a$ for all $a, b \in \mathbb{Z}^{+'}$, and in Example 2.9 this is not the case, for $5 *' 7 = 5$ but $7 *' 5 = 7$.

2.11 Definition A binary operation $*$ on a set S is **commutative** if (and only if) $a * b = b * a$ for all $a, b \in S$. ■

As was pointed out in Section 0, it is customary in mathematics to omit the words *and only if* from a definition. Definitions are always understood to be if and only if statements. *Theorems are not always if and only if statements, and no such convention is ever used for theorems.*

Now suppose we wish to consider an expression of the form $a * b * c$. A binary operation $*$ enables us to combine only two elements, and here we have three. The obvious attempts to combine the three elements are to form either $(a * b) * c$ or $a * (b * c)$. With $*$ defined as in Example 2.8, $(2 * 5) * 9$ is computed by $2 * 5 = 2$ and then $2 * 9 = 2$. Likewise, $2 * (5 * 9)$ is computed by $5 * 9 = 5$ and then $2 * 5 = 2$. Hence $(2 * 5) * 9 = 2 * (5 * 9)$, and it is not hard to see that for this $*$,

$$(a * b) * c = a * (b * c),$$

so there is no ambiguity in writing $a * b * c$. But for $*''$ of Example 2.10,

$$(2 *'' 5) *'' 9 = 4 *'' 9 = 6,$$

while

$$2 *'' (5 *'' 9) = 2 *'' 7 = 4.$$

Thus $(a *'' b) *'' c$ need not equal $a *'' (b *'' c)$, and an expression $a *'' b *'' c$ may be ambiguous.

2.12 Definition A binary operation on a set S is **associative** if $(a * b) * c = a * (b * c)$ for all $a, b, c \in S$. ■

It can be shown that if $*$ is associative, then longer expressions such as $a * b * c * d$ are not ambiguous. Parentheses may be inserted in any fashion for purposes of computation; the final results of two such computations will be the same.

Composition of functions mapping $\mathbb{R}$ into $\mathbb{R}$ was reviewed in Example 2.7. For any set S and any functions f and g mapping S into S, we similarly define the composition $f \circ g$ of g followed by f as the function mapping S into S such that $(f \circ g)(x) = f(g(x))$ for all $x \in S$. Some of the most important binary operations we consider are defined using composition of functions. It is important to know that this composition is always associative whenever it is defined.

2.13 Theorem **(Associativity of Composition)** Let S be a set and let f, g, and h be functions mapping S into S. Then $f \circ (g \circ h) = (f \circ g) \circ h$.

Proof To show these two functions are equal, we must show that they give the same assignment to each $x \in S$. Computing we find that

$$(f \circ (g \circ h))(x) = f((g \circ h)(x)) = f(g(h(x)))$$

and

$$((f \circ g) \circ h)(x) = (f \circ g)(h(x)) = f(g(h(x))),$$

so the same element $f(g(h(x)))$ of S is indeed obtained. ◆

As an example of using Theorem 2.13 to save work, recall that it is a fairly painful exercise in summation notation to show that multiplication of $n \times n$ matrices is an associative binary operation. If, in a linear algebra course, we first show that there is a one-to-one correspondence between matrices and linear transformations and that multiplication of matrices corresponds to the composition of the linear transformations (functions), we obtain this associativity at once from Theorem 2.13.

Tables

For a finite set, a binary operation on the set can be defined by means of a table in which the elements of the set are listed across the top as heads of columns and at the left side as heads of rows. We always require that the elements of the set be listed as heads across the top in the same order as heads down the left side. The next example illustrates the use of a table to define a binary operation.

2.14 Example Table 2.15 defines the binary operation $*$ on $S = \{a, b, c\}$ by the following rule:

$$(i\text{th } \textit{entry on the left}) * (j\text{th } \textit{entry on the top})$$
$$= (\textit{entry in the } i\text{th } \textit{row and } j\text{th } \textit{column of the table body}).$$

2.15 Table

$*$	a	b	c
a	b	c	b
b	a	c	b
c	c	b	a

Thus $a * b = c$ and $b * a = a$, so $*$ is not commutative. ▲

We can easily see that *a binary operation defined by a table is commutative if and only if the entries in the table are symmetric with respect to the diagonal that starts at the upper left corner of the table and terminates at the lower right corner.*

2.16 Example Complete Table 2.17 so that $*$ is a commutative binary operation on the set $S = \{a, b, c, d\}$.

2.17 Table

$*$	a	b	c	d
a	b			
b	d	a		
c	a	c	d	
d	a	b	b	c

Solution From Table 2.17, we see that $b * a = d$. For $*$ to be commutative, we must have $a * b = d$ also. Thus we place d in the appropriate square defining $a * b$, which is located symmetrically across the diagonal in Table 2.18 from the square defining $b * a$. We obtain the rest of Table 2.18 in this fashion to give our solution. ▲

2.18 Table

$*$	a	b	c	d
a	b	d	a	a
b	d	a	c	b
c	a	c	d	b
d	a	b	b	c

Some Words of Warning

Classroom experience shows the chaos that may result if a student is given a set and asked to define some binary operation on it. Remember that in an attempt to define a binary operation $*$ on a set S we must be sure that

1. *exactly one element is assigned to each possible ordered pair of elements of S,*
2. *for each ordered pair of elements of S, the element assigned to it is again in S.*

Regarding Condition 1, a student will often make an attempt that assigns an element of S to "most" ordered pairs, but for a few pairs, determines no element. In this event, $*$ is **not everywhere defined** on S. It may also happen that for some pairs, the attempt could assign any of several elements of S, that is, there is ambiguity. In any case of ambiguity, $*$ is **not well defined**. If Condition 2 is violated, then S is **not closed under** $*$.

Following are several illustrations of attempts to define binary operations on sets. Some of them are worthless. The symbol $*$ is used for the attempted operation in all these examples.

2.19 Example On $\mathbb{Q}$, let $a * b = a/b$. Here $*$ is *not everywhere defined* on $\mathbb{Q}$, for no rational number is assigned by this rule to the pair (2, 0). ▲

2.20 Example On $\mathbb{Q}^+$, let $a * b = a/b$. Here both Conditions 1 and 2 are satisfied, and $*$ is a binary operation on $\mathbb{Q}^+$. ▲

2.21 Example On $\mathbb{Z}^+$, let $a * b = a/b$. Here Condition 2 fails, for $1 * 3$ is not in $\mathbb{Z}^+$. Thus $*$ is not a binary operation on $\mathbb{Z}^+$, since $\mathbb{Z}^+$ is *not closed under* $*$. ▲

2.22 Example Let F be the set of all real-valued functions with domain $\mathbb{R}$ as in Example 2.7. Suppose we "define" $*$ to give the usual quotient of f by g, that is, $f * g = h$, where $h(x) = f(x)/g(x)$. Here Condition 2 is violated, for the functions in F were to be defined for *all* real numbers, and for some $g \in F$, $g(x)$ will be zero for some values of x in $\mathbb{R}$ and $h(x)$ would not be defined at those numbers in $\mathbb{R}$. For example, if $f(x) = \cos x$ and $g(x) = x^2$, then $h(0)$ is undefined, so $h \notin F$. ▲

2.23 Example Let F be as in Example 2.22 and let $f * g = h$, where h is the function greater than both f and g. This "definition" is completely worthless. In the first place, we have not defined what it means for one function to be greater than another. Even if we had, any sensible definition would result in there being many functions greater than both f and g, and $*$ would still be *not well defined.* ▲

2.24 Example Let S be a set consisting of 20 people, no two of whom are of the same height. Define $*$ by $a * b = c$, where c is the tallest person among the 20 in S. This is a perfectly good binary operation on the set, although not a particularly interesting one. ▲

2.25 Example Let S be as in Example 2.24 and let $a * b = c$, where c is the shortest person in S who is taller than both a and b. This $*$ is *not everywhere defined,* since if either a or b is the tallest person in the set, $a * b$ is not determined. ▲

EXERCISES 2

Computations

Exercises 1 through 4 concern the binary operation $*$ defined on $S = \{a, b, c, d, e\}$ by means of Table 2.26.

1. Compute $b * d$, $c * c$, and $[(a * c) * e] * a$.
2. Compute $(a * b) * c$ and $a * (b * c)$. Can you say on the basis of this computations whether $*$ is associative?
3. Compute $(b * d) * c$ and $b * (d * c)$. Can you say on the basis of this computation whether $*$ is associative?

2.26 Table

$*$	a	b	c	d	e
a	a	b	c	b	d
b	b	c	a	e	c
c	c	a	b	b	a
d	b	e	b	e	d
e	d	b	a	d	c

2.27 Table

$*$	a	b	c	d
a	a	b	c	
b	b	d		c
c	c	a	d	b
d	d			a

2.28 Table

$*$	a	b	c	d
a	a	b	c	d
b	b	a	c	d
c	c	d	c	d
d				

4. Is $*$ commutative? Why?

5. Complete Table 2.27 so as to define a commutative binary operation $*$ on $S = \{a, b, c, d\}$.

6. Table 2.28 can be completed to define an associative binary operation $*$ on $S = \{a, b, c, d\}$. Assume this is possible and compute the missing entries.

In Exercises 7 through 11, determine whether the binary operation $*$ defined is commutative and whether $*$ is associative.

7. $*$ defined on $\mathbb{Z}$ by letting $a * b = a - b$

8. $*$ defined on $\mathbb{Q}$ by letting $a * b = ab + 1$

9. $*$ defined on $\mathbb{Q}$ by letting $a * b = ab/2$

10. $*$ defined on $\mathbb{Z}^+$ by letting $a * b = 2^{ab}$

11. $*$ defined on $\mathbb{Z}^+$ by letting $a * b = a^b$

12. Let S be a set having exactly one element. How many different binary operations can be defined on S? Answer the question if S has exactly 2 elements; exactly 3 elements; exactly n elements.

13. How many different commutative binary operations can be defined on a set of 2 elements? on a set of 3 elements? on a set of n elements?

Concepts

In Exercises 14 through 16, correct the definition of the italicized term without reference to the text, if correction is needed, so that it is in a form acceptable for publication.

14. A binary operation $*$ is *commutative* if and only if $a * b = b * a$.

15. A binary operation $*$ on a set S is *associative* if and only if, for all $a, b, c \in S$, we have $(b * c) * a = b * (c * a)$.

16. A subset H of a set S is *closed* under a binary operation $*$ on S if and only if $(a * b) \in H$ for all $a, b \in S$.

In Exercises 17 through 22, determine whether the definition of $*$ does give a binary operation on the set. In the event that $*$ is not a binary operation, state whether Condition 1, Condition 2, or both of these conditions on page 24 are violated.

17. On $\mathbb{Z}^+$, define $*$ by letting $a * b = a - b$.

18. On $\mathbb{Z}^+$, define $*$ by letting $a * b = a^b$.

19. On $\mathbb{R}$, define $*$ by letting $a * b = a - b$.

20. On $\mathbb{Z}^+$, define $*$ by letting $a * b = c$, where c is the smallest integer greater than both a and b.

21. On $\mathbb{Z}^+$, define $*$ by letting $a * b = c$, where c is at least 5 more than $a + b$.

22. On $\mathbb{Z}^+$, define $*$ by letting $a * b = c$, where c is the largest integer less than the product of a and b.

23. Let H be the subset of $M_2(\mathbb{R})$ consisting of all matrices of the form $\begin{bmatrix} a & -b \\ b & a \end{bmatrix}$ for $a, b \in \mathbb{R}$. Is H closed under

a matrix addition? **b** matrix multiplication?

24. Mark each of the following true or false.

______ **a.** If $*$ is any binary operation on any set S, then $a * a = a$ for all $a \in S$.

______ **b.** If $*$ is any commutative binary operation on any set S, then $a * (b * c) = (b * c) * a$ for all $a, b, c \in S$.

______ **c.** If $*$ is any associative binary operation on any set S, then $a * (b * c) = (b * c) * a$ for all $a, b, c \in S$.

______ **d.** The only binary operations of any importance are those defined on sets of numbers.

______ **e.** A binary operation $*$ on a set S is commutative if there exist $a, b \in S$ such that $a * b = b * a$.

______ **f.** Every binary operation defined on a set having exactly one element is both commutative and associative.

______ **g.** A binary operation on a set S assigns at least one element of S to each ordered pair of elements of S.

______ **h.** A binary operation on a set S assigns at most one element of S to each ordered pair of elements of S.

______ **i.** A binary operation on a set S assigns exactly one element of S to each ordered pair of elements of S.

______ **j.** A binary operation on a set S may assign more than one element of S to some ordered pair of elements of S.

25. Give a set different from any of those described in the examples of the text and not a set of numbers. Define two different binary operations $*$ and $*'$ on this set. Be sure that your set is *well defined*.

Theory

26. Prove that if $*$ is an associative and commutative binary operation on a set S, then

$$(a * b) * (c * d) = [(d * c) * a] * b$$

for all $a, b, c, d \in S$. Assume the associative law only for triples as in the definition, that is, assume only

$$(x * y) * z = x * (y * z)$$

for all $x, y, z \in S$.

In Exercises 27 and 28, either prove the statement or give a counterexample.

27. Every binary operation on a set consisting of a single element in both commutative and associative.

28. Every commutative binary operation on a set having just two elements is associative.

Let F be the set of all real-valued functions having as domain the set $\mathbb{R}$ of all real numbers. Example 2.7 defined the binary operations $+, -, \cdot$, and $\circ$ on F. In Exercises 29 through 35, either prove the given statement or give a counterexample.

29. Function addition $+$ on F is associative.

30. Function subtraction $-$ on F is commutative

31. Function subtraction $-$ on F is associative.

32. Function multiplication $\cdot$ on F is commutative.

33. Function multiplication $\cdot$ on F is associative.

34. Function composition $\circ$ on F is commutative.

35. If $*$ and $*'$ are any two binary operations on a set S, then

$$a * (b *' c) = (a * b) *' (a * c) \quad \text{for all} \quad a, b, c \in S.$$

36. Suppose that $*$ is an *associative binary* operation on a set S. Let $H = \{a \in S \mid a * x = x * a \text{ for all } x \in S\}$. Show that H is closed under $*$. (We think of H as consisting of all elements of S that *commute* with every element in S.)

37. Suppose that $*$ is an associative and commutative binary operation on a set S. Show that $H = \{a \in S \mid a * a = a\}$ is closed under $*$. (The elements of H are **idempotents** of the binary operation $*$.)

SECTION 3 Isomorphic Binary Structures

Compare Table 3.1 for the binary operation $*$ on the set $S = \{a, b, c\}$ with Table 3.2 for the binary operation $*'$ on the set $T = \{\#, \$, \&\}$.

Notice that if, in Table 3.1, we replace all occurrences of a by #, every b by $, and every c by & using the one-to-one correspondence

$$a \leftrightarrow \# \qquad b \leftrightarrow \$ \qquad c \leftrightarrow \&$$

we obtain precisely Table 3.2. The two tables differ only in the symbols (or names) denoting the elements and the symbols $*$ and $*'$ for the operations. If we rewrite Table 3.3 with elements in the order y, x, z, we obtain Table 3.4. (Here we did not set up any one-one-correpondence; we just listed the same elements in different order outside the heavy bars of the table.) Replacing, in Table 3.1, all occurrences of a by y, every b by x, and every c by z using the one-to-one correspondence

$$a \leftrightarrow y \qquad b \leftrightarrow x \qquad c \leftrightarrow z$$

we obtain Table 3.4. We think of Tables 3.1, 3.2, 3.3, and 3.4 as being *structurally alike*. These four tables differ only in the names (or symbols) for their elements and in the order that those elements are listed as heads in the tables. However, Table 3.5 for binary operation $\bar{*}$ and Table 3.6 for binary operation $\hat{*}$ on the set $S = \{a, b, c\}$ are *structurally different* from each other and from Table 3.1. In Table 3.1, each element appears three times in the body of the table, while the body of Table 3.5 contains the single element b. In Table 3.6, for all $s \in S$ we get the same value c for $s \mathbin{\hat{*}} s$ along the upper-left to lower-right diagonal, while we get three different values in Table 3.1. Thus Tables 3.1 through 3.6 give just three structurally different binary operations on a set of three elements, provided we disregard the names of the elements and the order in which they appear as heads in the tables.

The situation we have just discussed is somewhat akin to children in France and in Germany learning the operation of addition on the set $\mathbb{Z}^+$. The children have different

3.1 Table

$*$	a	b	c
a	c	a	b
b	a	b	c
c	b	c	a

3.2 Table

$*'$	#	$	&
#	&	#	$
$	#	$	&
&	$	&	#

3.3 Table

$*''$	x	y	z
x	x	y	z
y	y	z	x
z	z	x	y

3.4 Table

$*''$	y	x	z
y	z	y	x
x	y	x	z
z	x	z	y

3.5 Table

$\bar{*}$	a	b	c
a	b	b	b
b	b	b	b
c	b	b	b

3.6 Table

$\hat{*}$	a	b	c
a	c	a	b
b	b	c	a
c	a	b	c

names (un, deux, trois, $\cdots$ versus ein, zwei, drei $\cdots$) for the numbers, but they are learning the same binary structure. (In this case, they are also using the same symbols for the numbers, so their addition tables would appear the same if they list the numbers in the same order.)

We are interested in studying the different types of *structures* that binary operations can provide on sets having the same number of elements, as typified by Tables 3.4, 3.5, and 3.6. Let us consider a **binary algebraic structure**[†] $\langle S, *\rangle$ to be a set S together with a binary operation $*$ on S. In order for two such binary structures $\langle S, *\rangle$ and $\langle S', *'\rangle$ to be structurally alike in the sense we have described, we would have to have a one-to-one correspondence between the elements x of S and the elements x' of S' such that

$$\text{if} \quad x \leftrightarrow x' \quad \text{and} \quad y \leftrightarrow y', \quad \text{then} \quad x * y \leftrightarrow x' *' y'. \tag{1}$$

A one-to-one correspondence exists if the sets S and S' have the same number of elements. It is customary to describe a one-to-one correspondence by giving a *one-to-one* function ϕ mapping S *onto* S' (see Definition 0.12). For such a function ϕ, we regard the equation $\phi(x) = x'$ as reading the one-to-one pairing $x \leftrightarrow x$ in left-to-right order. In terms of ϕ, the final $\leftrightarrow$ correspondence in (1), which asserts the algebraic structure in S' is the same as in S, can be expressed as

$$\phi(x * y) = \phi(x) *' \phi(y).$$

Such, a function showing that two algebraic systems are structurally alike is known as an *isomorphism*. We give a formal definition.

3.7 Definition Let $\langle S, *\rangle$ and $\langle S', *'\rangle$ be binary algebraic structures. An **isomorphism of** S with S' is a one-to-one function ϕ mapping S onto S' such that

$$\phi(x * y) = \phi(x) *' \phi(y) \text{ for all } x, y \in S. \tag{2}$$

homomorphism property

[†] Remember that boldface type indicates that a term is being defined.

If such a map ϕ exists, then S and S' are **isomorphic binary structures**, which we denote by $S \simeq S'$, omitting the $*$ and $*'$ from the notation. ■

You may wonder why we labeled the displayed condition in Definition 3.7 the *homomorphism property* rather than the *isomorphism property*. The notion of isomorphism includes the idea of one-to-one correspondence, which appeared in the definition via the words *one-to-one* and *onto* before the display. In Chapter 13, we will discuss the relation between S and S' when $\phi : S \to S'$ satisfies the displayed homomorphism property, but ϕ is not necessarily one to one; ϕ is then called a *homomorphism* rather than an *isomorphism*.

It is apparent that in Section 1, we showed that the binary structures $\langle U, \cdot \rangle$ and $\langle \mathbb{R}_c, +_c \rangle$ are isomorphic for all $c \in \mathbb{R}^+$. Also, $\langle U_n, \cdot \rangle$ and $\langle \mathbb{Z}_n, +_n \rangle$ are isomorphic for each $n \in \mathbb{Z}^+$.

Exercise 27 asks us to show that for a collection of binary algebraic structures, the relation $\simeq$ in Definition 3.7 is an equivalence relation on the collection. Our discussion leading to the preceding definition shows that the binary structures defined by Tables 3.1 through 3.4 are in the same equivalence class, while those given by Tables 3.5 and 3.6 are in different equivalence classes. We proceed to discuss how to try to determine whether binary structures are isomorphic.

How to Show That Binary Structures Are Isomorphic

We now give an outline showing how to proceed from Definition 3.7 to show that two binary structures $\langle S, * \rangle$ and $\langle S', *' \rangle$ are isomorphic.

Step 1 *Define the function ϕ that gives the isomorphism of S with S'.* Now this means that we have to describe, in some fashion, what $\phi(s)$ is to be for every $s \in S$.

Step 2 *Show that ϕ is a one-to-one function.* That is, suppose that $\phi(x) = \phi(y)$ in S' and deduce from this that $x = y$ in S.

Step 3 *Show that ϕ is onto S'.* That is, suppose that $s' \in S'$ is given and show that there does exist $s \in S$ such that $\phi(s) = s'$.

Step 4 *Show that $\phi(x * y) = \phi(x) *' \phi(y)$ for all $x, y \in S$.* This is just a question of computation. Compute both sides of the equation and see whether they are the same.

3.8 Example Let us show that the binary structure $\langle \mathbb{R}, + \rangle$ with operation the usual addition is isomorphic to the structure $\langle \mathbb{R}^+, \cdot \rangle$ where $\cdot$ is the usual multiplication.

Step 1 We have to somehow convert an operation of addition to multiplication. Recall from $a^{b+c} = (a^b)(a^c)$ that addition of exponents corresponds to multiplication of two quantities. Thus we try defining $\phi : \mathbb{R} \to \mathbb{R}^+$ by $\phi(x) = e^x$ for $x \in \mathbb{R}$. Note that $e^x > 0$ for all $x \in \mathbb{R}$, so indeed, $\phi(x) \in \mathbb{R}^+$.

Step 2 If $\phi(x) = \phi(y)$, then $e^x = e^y$. Taking the natural logarithm, we see that $x = y$, so ϕ is indeed one to one.

Step 3 If $r \in \mathbb{R}^+$, then $\ln(r) \in \mathbb{R}$ and $\phi(\ln r) = e^{\ln r} = r$. Thus ϕ is onto $\mathbb{R}^+$.

Step 4 For $x, y \in \mathbb{R}$, we have $\phi(x + y) = e^{x+y} = e^x \cdot e^y = \phi(x) \cdot \phi(y)$. Thus we see that ϕ is indeed an isomorphism. ▲

3.9 Example Let $2\mathbb{Z} = \{2n \mid n \in \mathbb{Z}\}$, so that $2\mathbb{Z}$ is the set of all even integers, positive, negative, and zero. We claim that $\langle \mathbb{Z}, + \rangle$ is isomorphic to $\langle 2\mathbb{Z}, + \rangle$, where $+$ is the usual addition. This will give an example of a binary structure $\langle \mathbb{Z}, + \rangle$ that is actually isomorphic to a structure consisting of a proper subset under the *induced* operation, in contrast to Example 3.8, where the operations were totally different.

Step 1 The obvious function $\phi : \mathbb{Z} \to 2\mathbb{Z}$ to try is given by $\phi(n) = 2n$ for $n \in \mathbb{Z}$.

Step 2 If $\phi(m) = \phi(n)$, then $2m = 2n$ so $m = n$. Thus ϕ is one to one.

Step 3 If $n \in 2\mathbb{Z}$, then n is even so $n = 2m$ for $m = n/2 \in \mathbb{Z}$. Hence $\phi(m) = 2(n/2) = n$ so ϕ is onto $2\mathbb{Z}$.

Step 4 Let $m, n \in \mathbb{Z}$. The equation

$$\phi(m + n) = 2(m + n) = 2m + 2n = \phi(m) + \phi(n)$$

then shows that ϕ is an isomorphism. ▲

How to Show That Binary Structures Are Not Isomorphic

We now turn to the reverse question, namely:

How do we demonstrate that two binary structures $\langle S, * \rangle$ and $\langle S', *' \rangle$ are not isomorphic, if this is the case?

This would mean that there is no one-to-one function ϕ from S onto S' with the property $\phi(x * y) = \phi(x) *' \phi(y)$ for all $x, y \in S$. In general, it is clearly not feasible to try every possible one-to-one function mapping S onto S' and test whether it has this property, except in the case where there are *no* such functions. This is the case precisely when S and S' do not have the same cardinality. (See Definition 0.13.)

3.10 Example The binary structures $\langle \mathbb{Q}, + \rangle$ and $\langle \mathbb{R}, + \rangle$ are not isomorphic because $\mathbb{Q}$ has cardinality $\aleph_0$ while $|\mathbb{R}| \neq \aleph_0$. (See the discussion following Example 0.13.) Note that it is not enough to say that $\mathbb{Q}$ is a proper subset of $\mathbb{R}$. Example 3.9 shows that a proper subset with the induced operation can indeed be isomorphic to the entire binary structure. ▲

A **structural property** of a binary structure is one that must be shared by any isomorphic structure. It is not concerned with names or some other nonstructural characteristics of the elements. For example, the binary structures defined by Tables 3.1 and 3.2 are isomorphic, although the elements are totally different. Also, a structural property is not concerned with what we consider to be the "name" of the binary operation. Example 3.8 showed that a binary structure whose operation is our usual addition can be isomorphic to one whose operation is our usual multiplication. The number of elements in the set S is a structural property of $\langle S, * \rangle$.

> In the event that there are one-to-one mappings of S onto S', we usually show that $\langle S, *\rangle$ is not isomorphic to $\langle S', *'\rangle$ (if this is the case) by showing that one has some structural property that the other does not possess.

3.11 Example The sets $\mathbb{Z}$ and $\mathbb{Z}^+$ both have cardinality $\aleph_0$, and there are lots of one-to-one functions mapping $\mathbb{Z}$ onto $\mathbb{Z}^+$. However, the binary structures $\langle \mathbb{Z}, \cdot\rangle$ and $\langle \mathbb{Z}^+, \cdot\rangle$, where $\cdot$ is the usual multiplication, are not isomorphic. In $\langle \mathbb{Z}, \cdot\rangle$ there are two elements x such that $x \cdot x = x$, namely, 0 and 1. However, in $\langle \mathbb{Z}^+, \cdot\rangle$, there is only the single element 1. ▲

We list a few examples of possible structural properties and nonstructural properties of a binary structure $\langle S, *\rangle$ to get you thinking along the right line.

Possible Structural Properties

1. The set has 4 elements.
2. The operation is commutative.
3. $x * x = x$ for all $x \in S$.
4. The equation $a * x = b$ has a solution x in S for all $a, b \in S$.

Possible Nonstructural Properties

a. The number 4 is an element.
b. The operation is called "addition."
c. The elements of S are matrices.
d. S is a subset of $\mathbb{C}$.

We introduced the algebraic notions of commutativity and associativity in Section 2. One other structural notion that will be of interest to us is illustrated by Table 3.3, where for the binary operation $*''$ on the set $\{x, y, z\}$, we have $x *'' u = u *'' x = u$ for all choices possible choices, x, y, and z for u. Thus x plays the same role as 0 in $\langle \mathbb{R}, +\rangle$ where $0 + u = u + 0 = u$ for all $u \in \mathbb{R}$, and the same role as 1 in $\langle \mathbb{R}, \cdot\rangle$ where $1 \cdot u = u \cdot 1 = u$ for all $u \in \mathbb{R}$. Because Tables 3.1 and 3.2 give structures isomorphic to the one in Table 3.3, they must exhibit an element with a similar property. We see that $b * u = u * b = u$ for all elements u appearing in Table 3.1 and that $\$ *' u = u *' \$ = u$ for all elements u in Table 3.2. We give a formal definition of this structural notion and prove a little theorem.

3.12 Definition Let $\langle S, *\rangle$ be a binary structure. An element e of S is an **identity element for** $*$ if $e * s = s * e = s$ for all $s \in S$. ■

3.13 Theorem **(Uniqueness of Identity Element)** A binary structure $\langle S, *\rangle$ has at most one identity element. That is, if there is an identity element, it is unique.

Proof Proceeding in the standard way to show uniqueness, suppose that both e and $\bar{e}$ are elements of S serving as identity elements. We let them compete with each other. Regarding e as an identity element, we must have $e * \bar{e} = \bar{e}$. However, regarding $\bar{e}$ as an identity element, we must have $e * \bar{e} = e$. We thus obtain $e = \bar{e}$, showing that an identity element must be unique. ◆

If you now have a good grasp of the notion of isomorphic binary structures, it should be evident that having an identity element for $*$ is indeed a structural property of a structure $\langle S, *\rangle$. However, we know from experience that many readers will be unable to see the forest because of all the trees that have appeared. For them, we now supply a careful proof, skipping along to touch those trees that are involved.

3.14 Theorem Suppose $\langle S, *\rangle$ has an identity element e for $*$. If $\phi : S \to S'$ is an isomorphism of $\langle S, *\rangle$ with $\langle S', *'\rangle$, then $\phi(e)$ is an identity element for the binary operation $*'$ on S'.

Proof Let $s' \in S'$. We must show that $\phi(e) *' s' = s' *' \phi(e) = s'$. Because ϕ is an isomorphism, it is a one-to-one map of S onto S'. In particular, there exists $s \in S$ such that $\phi(s) = s'$. Now e is an identity element for $*$ so that we know that $e * s = s * e = s$. Because ϕ is a function, we then obtain

$$\phi(e * s) = \phi(s * e) = \phi(s).$$

Using Definition 3.7 of an isomorphism, we can rewrite this as

$$\phi(e) *' \phi(s) = \phi(s) *' \phi(e) = \phi(s).$$

Remembering that we chose $s \in S$ such that $\phi(s) = s'$, we obtain the desired relation $\phi(e) *' s' = s' *' \phi(e) = s'$. ◆

We conclude with three more examples showing via structural properties that certain binary structures are not isomorphic. In the exercises we ask you to show, as in Theorem 3.14, that the properties we use to distinguish the structures in these examples are indeed structural. That is, they must be shared by any isomorphic structure.

3.15 Example We show that the binary structures $\langle \mathbb{Q}, +\rangle$ and $\langle \mathbb{Z}, +\rangle$ under the usual addition are not isomorphic. (Both $\mathbb{Q}$ and $\mathbb{Z}$ have cardinality $\aleph_0$, so there are lots of one-to-one functions mapping $\mathbb{Q}$ onto $\mathbb{Z}$.) The equation $x + x = c$ has a solution x for all $c \in \mathbb{Q}$, but this is not the case in $\mathbb{Z}$. For example, the equation $x + x = 3$ has no solution in $\mathbb{Z}$. We have exhibited a structural property that *distinguishes* these two structures. ▲

3.16 Example The binary structures $\langle \mathbb{C}, \cdot\rangle$ and $\langle \mathbb{R}, \cdot\rangle$ under the usual multiplication are not isomorphic. (It can be shown that $\mathbb{C}$ and $\mathbb{R}$ have the same cardinality.) The equation $x \cdot x = c$ has a solution x for all $c \in \mathbb{C}$, but $x \cdot x = -1$ has no solution in $\mathbb{R}$. ▲

3.17 Example The binary structure $\langle M_2(\mathbb{R}), \cdot\rangle$ of 2×2 real matrices with the usual matrix multiplication is not isomorphic to $\langle \mathbb{R}, \cdot\rangle$ with the usual number multiplication. (It can be shown that both sets have cardinality $|\mathbb{R}|$.) Multiplication of numbers is commutative, but multiplication of matrices is not. ▲

EXERCISES 3

In all the exercises, $+$ is the usual addition on the set where it is specified, and $\cdot$ is the usual multiplication.

Computations

1. What three things must we check to determine whether a function $\phi: S \to S'$ is an isomorphism of a binary structure $\langle S, *\rangle$ with $\langle S', *'\rangle$?

In Exercises 2 through 10, determine whether the given map ϕ is an isomorphism of the first binary structure with the second. (See Exercise 1.) If it is not an isomorphism, why not?

2. $\langle \mathbb{Z}, +\rangle$ with $\langle \mathbb{Z}, +\rangle$ where $\phi(n) = -n$ for $n \in \mathbb{Z}$

3. $\langle \mathbb{Z}, +\rangle$ with $\langle \mathbb{Z}, +\rangle$ where $\phi(n) = 2n$ for $n \in \mathbb{Z}$

4. $\langle \mathbb{Z}, +\rangle$ with $\langle \mathbb{Z}, +\rangle$ where $\phi(n) = n + 1$ for $n \in \mathbb{Z}$

5. $\langle \mathbb{Q}, +\rangle$ with $\langle \mathbb{Q}, +\rangle$ where $\phi(x) = x/2$ for $x \in \mathbb{Q}$

6. $\langle \mathbb{Q}, \cdot\rangle$ with $\langle \mathbb{Q}, \cdot\rangle$ where $\phi(x) = x^2$ for $x \in \mathbb{Q}$

7. $\langle \mathbb{R}, \cdot\rangle$ with $\langle \mathbb{R}, \cdot\rangle$ where $\phi(x) = x^3$ for $x \in \mathbb{R}$

8. $\langle M_2(\mathbb{R}), \cdot\rangle$ with $\langle \mathbb{R}, \cdot\rangle$ where $\phi(A)$ is the determinant of matrix A

9. $\langle M_1(\mathbb{R}), \cdot\rangle$ with $\langle \mathbb{R}, \cdot\rangle$ where $\phi(A)$ is the determinant of matrix A

10. $\langle \mathbb{R}, +\rangle$ with $\langle \mathbb{R}^+, \cdot\rangle$ where $\phi(r) = 0.5^r$ for $r \in \mathbb{R}$

In Exercises 11 through 15, let F be the set of all functions f mapping $\mathbb{R}$ into $\mathbb{R}$ that have derivatives of all orders. Follow the instructions for Exercises 2 through 10.

11. $\langle F, +\rangle$ with $\langle F, +\rangle$ where $\phi(f) = f'$, the derivative of f

12. $\langle F, +\rangle$ with $\langle \mathbb{R}, +\rangle$ where $\phi(f) = f'(0)$

13. $\langle F, +\rangle$ with $\langle F, +\rangle$ where $\phi(f)(x) = \int_0^x f(t)dt$

14. $\langle F, +\rangle$ with $\langle F, +\rangle$ where $\phi(f)(x) = \frac{d}{dx}[\int_0^x f(t)dt]$

15. $\langle F, \cdot\rangle$ with $\langle F, \cdot\rangle$ where $\phi(f)(x) = x \cdot f(x)$

16. The map $\phi : \mathbb{Z} \to \mathbb{Z}$ defined by $\phi(n) = n + 1$ for $n \in \mathbb{Z}$ is one to one and onto $\mathbb{Z}$. Give the definition of a binary operation $*$ on $\mathbb{Z}$ such that ϕ is an isomorphism mapping

a. $\langle \mathbb{Z}, +\rangle$ onto $\langle \mathbb{Z}, *\rangle$, **b.** $\langle \mathbb{Z}, *\rangle$ onto $\langle \mathbb{Z}, +\rangle$.

In each case, give the identity element for $*$ on $\mathbb{Z}$.

17. The map $\phi : \mathbb{Z} \to \mathbb{Z}$ defined by $\phi(n) = n + 1$ for $n \in \mathbb{Z}$ is one to one and onto $\mathbb{Z}$. Give the definition of a binary operation $*$ on $\mathbb{Z}$ such that ϕ is an isomorphism mapping

a. $\langle \mathbb{Z}, \cdot\rangle$ onto $\langle \mathbb{Z}, *\rangle$, **b.** $\langle \mathbb{Z}, *\rangle$ onto $\langle \mathbb{Z}, \cdot\rangle$.

In each case, give the identity element for $*$ on $\mathbb{Z}$.

18. The map $\phi : \mathbb{Q} \to \mathbb{Q}$ defined by $\phi(x) = 3x - 1$ for $x \in \mathbb{Q}$ is one to one and onto $\mathbb{Q}$. Give the definition of a binary operation $*$ on $\mathbb{Q}$ such that ϕ is an isomorphism mapping

a. $\langle \mathbb{Q}, +\rangle$ onto $\langle \mathbb{Q}, *\rangle$, **b.** $\langle \mathbb{Q}, *\rangle$ onto $\langle \mathbb{Q}, +\rangle$.

In each case, give the identity element for $*$ on $\mathbb{Q}$.

19. The map $\phi : \mathbb{Q} \to \mathbb{Q}$ defined by $\phi(x) = 3x - 1$ for $x \in \mathbb{Q}$ is one to one and onto $\mathbb{Q}$. Give the definition of a binary operation $*$ on $\mathbb{Q}$ such that ϕ is an isomorphism mapping

a. $\langle \mathbb{Q}, \cdot \rangle$ onto $\langle \mathbb{Q}, * \rangle$,

b. $\langle \mathbb{Q}, * \rangle$ onto $\langle \mathbb{Q}, \cdot \rangle$.

In each case, give the identity element for $*$ on $\mathbb{Q}$.

Concepts

20. The displayed homomorphism condition for an isomorphism ϕ in Definition 3.7 is sometimes summarized by saying, "ϕ must commute with the binary operation(s)." Explain how that condition can be viewed in this manner.

In Exercises 21 and 22, correct the definition of the italicized term without reference to the text, if correction is needed, so that it is in a form acceptable for publication.

21. A function $\phi : S \to S'$ is an *isomorphism* if and only if $\phi(a * b) = \phi(a) *' \phi(b)$.

22. Let $*$ be a binary operation on a set S. An element e of S with the property $s * e = s = e * s$ is an *identity element for* $*$ for all $s \in S$.

Proof Synopsis

A good test of your understanding of a proof is your ability to give a one or two sentence synopsis of it, explaining the idea of the proof without all the details and computations. Note that we said "sentence" and not "equation." From now on, some of our exercise sets may contain one or two problems asking for a synopsis of a proof in the text. It should rarely exceed three sentences. We should illustrate for you what we mean by a synopsis. Here is our one-sentence synopsis of Theorem 3.14. Read the statement of the theorem now, and then our synopsis.

> Representing an element of S' as $\phi(s)$ for some $s \in S$, use the homomorphism property of ϕ to carry the computation of $\phi(e) *' \phi(s)$ back to a computation in S.

That is the kind of explanation that one mathematician might give another if asked, "How does the proof go?" We did not make the computation or explain why we could represent an element of S' as $\phi(s)$. To supply every detail would result in a completely written proof. We just gave the guts of the argument in our synopsis.

23. Give a proof synopsis of Theorem 3.13.

Theory

24. An identity element for a binary operation $*$ as described by Definition 3.12 is sometimes referred to as "a two-sided identity element." Using complete sentences, give analogous definitions for

a. a *left identity element* e_L for $*$, and

b. a *right identity element* e_R for $*$.

Theorem 3.13 shows that if a two-sided identity element for $*$ exists, it is unique. Is the same true for a one-sided identity element you just defined? If so, prove it. If not, give a counterexample $\langle S, * \rangle$ for a finite set S and find the first place where the proof of Theorem 3.13 breaks down.

25. Continuing the ideas of Exercise 24 can a binary structure have a left identity element e_L and a right identity element e_R where $e_L \neq e_R$? If so, give an example, using an operation on a finite set S. If not, prove that it is impossible.

26. Recall that if $f : A \to B$ is a one-to-one function mapping A onto B, then $f^{-1}(b)$ is the unique $a \in A$ such that $f(a) = b$. Prove that if $\phi : S \to S'$ is an isomorphism of $\langle S, *\rangle$ with $\langle S', *'\rangle$, then ϕ^{-1} is an isomorphism of $\langle S', *'\rangle$ with $\langle S, *\rangle$.

27. Prove that if $\phi : S \to S'$ is an isomorphism of $\langle S, *\rangle$ with $\langle S', *'\rangle$ and $\psi : S' \to S''$ is an isomorphism of $\langle S', *'\rangle$ with $\langle S'', *''\rangle$, then the composite function $\psi \circ \phi$ is an isomorphism of $\langle S, *\rangle$ with $\langle S'', *''\rangle$.

28. Prove that the relation $\simeq$ of being isomorphic, described in Definition 3.7, is an equivalence relation on any set of binary structures. You may simply quote the results you were asked to prove in the preceding two exercises at appropriate places in your proof.

In Exercises 29 through 32, give a careful proof for a skeptic that the indicated property of a binary structure $\langle S, *\rangle$ is indeed a structural property. (In Theorem 3.14, we did this for the property, "There is an identity element for $*$.")

29. The operation $*$ is commutative.

30. The operation $*$ is associative.

31. For each $c \in S$, the equation $x * x = c$ has a solution x in S.

32. There exists an element b in S such that $b * b = b$.

33. Let H be the subset of $M_2(\mathbb{R})$ consisting of all matrices of the form $\begin{bmatrix} a & -b \\ b & a \end{bmatrix}$ for $a, b \in \mathbb{R}$. Exercise 23 of Section 2 shows that H is closed under both matrix addition and matrix multiplication.

a. Show that $\langle \mathbb{C}, +\rangle$ is isomorphic to $\langle H, +\rangle$.

b. Show that $\langle \mathbb{C}, \cdot\rangle$ is isomorphic to $\langle H, \cdot\rangle$.

(We say that H is a *matrix representation* of the complex numbers $\mathbb{C}$.)

34. There are 16 possible binary structures on the set $\{a, b\}$ of two elements. How many nonisomorphic (that is, structurally different) structures are there among these 16? Phrased more precisely in terms of the isomorphism equivalence relation $\simeq$ on this set of 16 structures, how many equivalence classes are there? Write down one structure from each equivalence class. [*Hint:* Interchanging a and b everywhere in a table and then rewriting the table with elements listed in the original order does not always yield a table different from the one we started with.]

SECTION 4 GROUPS

Let us continue the analysis of our past experience with algebra. Once we had mastered the computational problems of addition and multiplication of numbers, we were ready to apply these binary operations to the solution of problems. Often problems lead to equations involving some unknown number x, which is to be determined. The simplest equations are the linear ones of the forms $a + x = b$ for the operation of addition, and $ax = b$ for multiplication. The additive linear equation always has a numerical solution, and so has the multiplicative one, provided $a \neq 0$. Indeed, the need for solutions of additive linear equations such as $5 + x = 2$ is a very good motivation for the negative numbers. Similarly, the need for rational numbers is shown by equations such as $2x = 3$.

It is desirable for us to be able to solve linear equations involving our binary operations. This is not possible for every binary operation, however. For example, the equation $a * x = a$ has no solution in $S = \{a, b, c\}$ for the operation $*$ of Example 2.14. Let us abstract from familiar algebra those properties of addition that enable us to solve the equation $5 + x = 2$ in $\mathbb{Z}$. We must not refer to subtraction, for we are concerned with the solution phrased in terms of a single binary operation, in this case addition. The steps in the solution are as follows:

$$\begin{aligned} 5 + x &= 2, && \text{given,} \\ -5 + (5 + x) &= -5 + 2, && \text{adding } -5, \\ (-5 + 5) + x &= -5 + 2, && \text{associative law,} \\ 0 + x &= -5 + 2, && \text{computing } -5 + 5, \\ x &= -5 + 2, && \text{property of } 0, \\ x &= -3, && \text{computing } -5 + 2. \end{aligned}$$

Strictly speaking, we have not shown here that -3 is a solution, but rather that it is the only possibility for a solution. To show that -3 is a solution, one merely computes $5 + (-3)$. A similar analysis could be made for the equation $2x = 3$ in the rational numbers with the operation of multiplication:

$$\begin{aligned} 2x &= 3, && \text{given,} \\ \tfrac{1}{2}(2x) &= \tfrac{1}{2}(3), && \text{multiplying by } \tfrac{1}{2}, \\ (\tfrac{1}{2} \cdot 2)x &= \tfrac{1}{2}3, && \text{associative law,} \\ 1 \cdot x &= \tfrac{1}{2}3, && \text{computing } \tfrac{1}{2}2, \\ x &= \tfrac{1}{2}3, && \text{property of } 1, \\ x &= \tfrac{3}{2}, && \text{computing } \tfrac{1}{2}3. \end{aligned}$$

We can now see what properties a set S and a binary operation $*$ on S would have to have to permit imitation of this procedure for an equation $a * x = b$ for $a, b \in S$. Basic to the procedure is the existence of an element e in S with the property that $e * x = x$ for all $x \in S$. For our additive example, 0 played the role of e, and 1 played the role for our multiplicative example. Then we need an element a' in S that has the property that $a' * a = e$. For our additive example with $a = 5$, -5 played the role of a', and $\frac{1}{2}$ played the role for our multiplicative example with $a = 2$. Finally we need the associative law. The remainder is just computation. A similar analysis shows that in order to solve the equation $x * a = b$ (remember that $a * x$ need not equal $x * a$), we would like to have an element e in S such that $x * e = x$ for all $x \in S$ and an a' in S such that $a * a' = e$. With all of these properties of $*$ on S, we could be sure of being able to solve linear equations. Thus we need an associative binary structure $\langle S, * \rangle$ with an identity element e such that for each $a \in S$, there exists $a' \in S$ such that $a * a' = a' * a = e$. This is precisely the notion of a *group*, which we now define.

Definition and Examples

Rather than describe a *group* using terms defined in Sections 2 and 3 as we did at the end of the preceding paragraph, we give a self-contained definition. This enables a person who picks up this text to discover what a group is without having to look up more terms.

4.1 Definition A **group** $\langle G, * \rangle$ is a set G, closed under a binary operation $*$, such that the following axioms are satisfied:

$\mathscr{G}_1$: For all $a, b, c \in G$, we have

$$(a * b) * c = a * (b * c). \quad \textbf{associativity of } *$$

$\mathscr{G}_2$: There is an element e in G such that for all $x \in G$,

$$e * x = x * e = x. \quad \textbf{identity element } e \text{ for } *$$

$\mathscr{G}_3$: Corresponding to each $a \in G$, there is an element a' in G such that

$$a * a' = a' * a = e. \quad \textbf{inverse } a' \text{ of } a$$ ■

4.2 Example We easily see that $\langle U, \cdot \rangle$ and $\langle U_n, \cdot \rangle$ are groups. Multiplication of complex numbers is associative and both U and U_n contain 1, which is an identity for multiplication. For $e^{i\theta} \in U$, the computation

$$e^{i\theta} \cdot e^{i(2\pi - \theta)} = e^{2\pi i} = 1$$

shows that every element of U has an inverse. For $z \in U_n$, the computation

$$z \cdot z^{n-1} = z^n = 1$$

shows that every element of U_n has an inverse. Thus $\langle U, \cdot \rangle$ and $\langle U_n, \cdot \rangle$ are groups. Because $\langle \mathbb{R}_c, +_c \rangle$ is isomorphic to $\langle U, \cdot \rangle$, we see that $\langle \mathbb{R}_c, +_c \rangle$ is a group for all $c \in \mathbb{R}^+$. Similarly, the fact that $\langle \mathbb{Z}_n, +_n \rangle$ is isomorphic to $\langle U_n, \cdot \rangle$ shows that $\langle \mathbb{Z}_n, +_n \rangle$ is a group for all $n \in \mathbb{Z}^+$. ▲

We point out now that we will sometimes be sloppy in notation. Rather than use the binary structure notation $\langle G, * \rangle$ constantly, we often refer to a group G, with the understanding that there is of course a binary operation on the set G. In the event that clarity demands that we specify an operation $*$ on G, we use the phrase "the group G

■ Historical Note

There are three historical roots of the development of abstract group theory evident in the mathematical literature of the nineteenth century: the theory of algebraic equations, number theory, and geometry. All three of these areas used group-theoretic methods of reasoning, although the methods were considerably more explicit in the first area than in the other two.

One of the central themes of geometry in the nineteenth century was the search for invariants under various types of geometric transformations. Gradually attention became focused on the transformations themselves, which in many cases can be thought of as elements of groups.

In number theory, already in the eighteenth century Leonhard Euler had considered the remainders on division of powers a^n by a fixed prime p. These remainders have "group" properties. Similarly, Carl F. Gauss, in his *Disquisitiones Arithmeticae* (1800), dealt extensively with quadratic forms $ax^2 + 2bxy + cy^2$, and in particular showed that equivalence classes of these forms under composition possessed what amounted to group properties.

Finally, the theory of algebraic equations provided the most explicit prefiguring of the group concept. Joseph-Louis Lagrange (1736–1813) in fact initiated the study of permutations of the roots of an equation as a tool for solving it. These permutations, of course, were ultimately considered as elements of a group.

It was Walter von Dyck (1856–1934) and Heinrich Weber (1842–1913) who in 1882 were able independently to combine the three historical roots and give clear definitions of the notion of an abstract group.

under $*$." For example, we may refer to the *groups* $\mathbb{Z}$, $\mathbb{Q}$, and $\mathbb{R}$ *under addition* rather than write the more tedious $\langle\mathbb{Z}, +\rangle$, $\langle\mathbb{Q}, +\rangle$, and $\langle\mathbb{R}, +\rangle$. However, we feel free to refer to the group $\mathbb{Z}_8$ without specifying the operation.

4.3 Definition A group G is **abelian** if its binary operation is commutative. ■

Historical Note

Commutative groups are called *abelian* in honor of the Norwegian mathematician Niels Henrik Abel (1802–1829). Abel was interested in the question of solvability of polynomial equations. In a paper written in 1828, he proved that if all the roots of such an equation can be expressed as rational functions $f, g, \ldots, h$ of one of them, say x, and if for any two of these roots, $f(x)$ and $g(x)$, the relation $f(g(x)) = g(f(x))$ always holds, then the equation is solvable by radicals. Abel showed that each of these functions in fact permutes the roots of the equation; hence, these functions are elements of the group of permutations of the roots. It was this property of commutativity in these permutation groups associated with solvable equations that led Camille Jordan in his 1870 treatise on algebra to name such groups *abelian;* the name since then has been applied to commutative groups in general.

Abel was attracted to mathematics as a teenager and soon surpassed all his teachers in Norway. He finally received a government travel grant to study elsewhere in 1825 and proceeded to Berlin, where he befriended August Crelle, the founder of the most influential German mathematical journal. Abel contributed numerous papers to Crelle's *Journal* during the next several years, including many in the field of elliptic functions, whose theory he created virtually single-handedly. Abel returned to Norway in 1827 with no position and an abundance of debts. He nevertheless continued to write brilliant papers, but died of tuberculosis at the age of 26, two days before Crelle succeeded in finding a university position for him in Berlin.

Let us give some examples of some sets with binary operations that give groups and also of some that do not give groups.

4.4 Example The set $\mathbb{Z}^+$ under addition is *not* a group. There is no identity element for $+$ in $\mathbb{Z}^+$. ▲

4.5 Example The set of all nonnegative integers (including 0) under addition is still *not* a group. There is an identity element 0, but no inverse for 2. ▲

4.6 Example The familiar additive properties of integers and of rational, real, and complex numbers show that $\mathbb{Z}$, $\mathbb{Q}$, $\mathbb{R}$, and $\mathbb{C}$ under addition are abelian groups. ▲

4.7 Example The set $\mathbb{Z}^+$ under multiplication is *not* a group. There is an identity 1, but no inverse of 3. ▲

4.8 Example The familiar multiplicative properties of rational, real, and complex numbers show that the sets $\mathbb{Q}^+$ and $\mathbb{R}^+$ of positive numbers and the sets $\mathbb{Q}^*$, $\mathbb{R}^*$, and $\mathbb{C}^*$ of nonzero numbers under multiplication are abelian groups. ▲

4.9 Example The set of all real-valued functions with domain $\mathbb{R}$ under function addition is a group. This group is abelian. ▲

4.10 Example **(Linear Algebra)** Those who have studied vector spaces should note that the axioms for a vector space V pertaining just to vector addition can be summarized by asserting that V under vector addition is an abelian group. ▲

4.11 Example The set $M_{m \times n}(\mathbb{R})$ of all $m \times n$ matrices under matrix addition is a group. The $m \times n$ matrix with all entries 0 is the identity matrix. This group is abelian. ▲

4.12 Example The set $M_n(\mathbb{R})$ of all $n \times n$ matrices under matrix multiplication is *not* a group. The $n \times n$ matrix with all entries 0 has no inverse. ▲

4.13 Example Show that the subset S of $M_n(\mathbb{R})$ consisting of all *invertible* $n \times n$ matrices under matrix multiplication is a group.

Solution We start by showing that S is closed under matrix multiplication. Let A and B be in S, so that both A^{-1} and B^{-1} exist and $AA^{-1} = BB^{-1} = I_n$. Then

$$(AB)(B^{-1}A^{-1}) = A(BB^{-1})A^{-1} = AI_nA^{-1} = I_n,$$

so that AB is invertible and consequently is also in S.

Since matrix multiplication is associative and I_n acts as the identity element, and since each element of S has an inverse by definition of S, we see that S is indeed a group. This group is *not* commutative. It is our first example of a *nonabelian group*. ▲

The group of invertible $n \times n$ matrices described in the preceding example is of fundamental importance in linear algebra. It is the **general linear group of degree** n, and is usually denoted by $GL(n, \mathbb{R})$. Those of you who have studied linear algebra know that a matrix A in $GL(n, \mathbb{R})$ gives rise to an invertible linear transformation $T : \mathbb{R}^n \to \mathbb{R}^n$, defined by $T(\mathbf{x}) = A\mathbf{x}$, and that conversely, every invertible linear transformation of $\mathbb{R}^n$ into itself is defined in this fashion by some matrix in $GL(n, \mathbb{R})$. Also, matrix multiplication corresponds to composition of linear transformations. Thus all invertible linear transformations of $\mathbb{R}^n$ into itself form a group under function composition; this group is usually denoted by $GL(\mathbb{R}^n)$. Of course, $GL(n, \mathbb{R}) \simeq GL(\mathbb{R}^n)$.

4.14 Example Let $*$ be defined on $\mathbb{Q}^+$ by $a * b = ab/2$. Then

$$(a * b) * c = \frac{ab}{2} * c = \frac{abc}{4},$$

and likewise

$$a * (b * c) = a * \frac{bc}{2} = \frac{abc}{4}.$$

Thus $*$ is associative. Computation shows that

$$2 * a = a * 2 = a$$

for all $a \in \mathbb{Q}^+$, so 2 is an identity element for $*$. Finally,

$$a * \frac{4}{a} = \frac{4}{a} * a = 2,$$

so $a' = 4/a$ is an inverse for a. Hence $\mathbb{Q}^+$ with the operation $*$ is a group. ▲

Elementary Properties of Groups

As we proceed to prove our first theorem about groups, we must use Definition 4.1, which is the only thing we know about groups at the moment. The proof of a second theorem can employ both Definition 4.1 and the first theorem; the proof of a third theorem can use the definition and the first two theorems, and so on.

Our first theorem will establish cancellation laws. In real arithmetic, we know that $2a = 2b$ implies that $a = b$. We need only divide both sides of the equation $2a = 2b$ by 2, or equivalently, multiply both sides by $\frac{1}{2}$, which is the multiplicative inverse of 2. We parrot this proof to establish cancellation laws for any group. Note that we will also use the associative law.

4.15 Theorem If G is a group with binary operation $*$, then the **left and right cancellation laws** hold in G, that is, $a * b = a * c$ implies $b = c$, and $b * a = c * a$ implies $b = c$ for all $a, b, c \in G$.

Proof Suppose $a * b = a * c$. Then by $\mathscr{G}_3$, there exists a', and

$$a' * (a * b) = a' * (a * c).$$

By the associative law,

$$(a' * a) * b = (a' * a) * c.$$

By the definition of a' in $\mathscr{G}_3$, $a' * a = e$, so

$$e * b = e * c.$$

By the definition of e in $\mathscr{G}_2$,

$$b = c.$$

Similarly, from $b * a = c * a$ one can deduce that $b = c$ upon multiplication on the right by a' and use of the axioms for a group. ◆

Our next proof can make use of Theorem 4.15. We show that a "linear equation" in a group has a *unique* solution. Recall that we chose our group properties to allow us to find solutions of such equations.

4.16 Theorem If G is a group with binary operation $*$, and if a and b are any elements of G, then the linear equations $a * x = b$ and $y * a = b$ have unique solutions x and y in G.

Proof First we show the existence of *at least* one solution by just computing that $a' * b$ is a solution of $a * x = b$. Note that

$$\begin{aligned} a * (a' * b) &= (a * a') * b, && \text{associative law,} \\ &= e * b, && \text{definition of } a', \\ &= b, && \text{property of } e. \end{aligned}$$

Thus $x = a' * b$ is a solution of $a * x = b$. In a similar fashion, $y = b * a'$ is a solution of $y * a = b$.

To show uniqueness of y, we use the standard method of assuming that we have two solutions, y_1 and y_2, so that $y_1 * a = b$ and $y_2 * a = b$. Then $y_1 * a = y_2 * a$, and by Theorem 4.15, $y_1 = y_2$. The uniqueness of x follows similarly. ◆

Of course, to prove the uniqueness in the last theorem, we could have followed the procedure we used in motivating the definition of a group, showing that if $a * x = b$, then $x = a' * b$. However, we chose to illustrate the standard way to prove an object is unique; namely, suppose you have two such objects, and then prove they must be the same. Note that the solutions $x = a' * b$ and $y = b * a'$ need not be the same unless $*$ is commutative.

Because a group is a special type of binary structure, we know from Theorem 3.13 that the identity e in a group is unique. We state this again as part of the next theorem for easy reference.

4.17 Theorem In a group G with binary operation $*$, there is only one element e in G such that

$$e * x = x * e = x$$

for all $x \in G$. Likewise for each $a \in G$, there is only one element a' in G such that

$$a' * a = a * a' = e.$$

In summary, the identity element and inverse of each element are unique in a group.

Proof Theorem 3.13 shows that an identity element for any binary structure is unique. No use of the group axioms was required to show this.

Turning to the uniqueness of an inverse, suppose that $a \in G$ has inverses a' and a'' so that $a' * a = a * a' = e$ and $a'' * a = a * a'' = e$. Then

$$a * a'' = a * a' = e$$

and, by Theorem 4.15,

$$a'' = a',$$

so the inverse of a in a group is unique. ◆

Note that in a group G, we have

$$(a * b) * (b' * a') = a * (b * b') * a' = (a * e) * a' = a * a' = e.$$

This equation and Theorem 4.17 show that $b' * a'$ is the unique inverse of $a * b$. That is, $(a * b)' = b' * a'$. We state this as a corollary.

4.18 Corollary Let G be a group. For all $a, b \in G$, we have $(a * b)' = b' * a'$.

For your information, we remark that binary algebraic structures with weaker axioms than those for a group have also been studied quite extensively. Of these weaker structures, the **semigroup,** a set with an associative binary operation, has perhaps had the most attention. A **monoid** is a semigroup that has an identity element for the binary operation. Note that every group is both a semigroup and a monoid.

Finally, it is possible to give axioms for a group $\langle G, *\rangle$ that seem at first glance to be weaker, namely:

1. The binary operation $*$ on G is associative.
2. There exists a **left identity element** e in G such that $e * x = x$ for all $x \in G$.
3. For each $a \in G$, there exists **a left inverse** a' in G such that $a' * a = e$.

From this *one-sided definition*, one can prove that the left identity element is also a right identity element, and a left inverse is also a right inverse for the same element. Thus these axioms should not be called *weaker*, since they result in exactly the same structures being called groups. It is conceivable that it might be easier in some cases to check these *left axioms* than to check our *two-sided axioms*. Of course, by symmetry it is clear that there are also *right axioms* for a group.

Finite Groups and Group Tables

All our examples after Example 4.2 have been of infinite groups, that is, groups where the set G has an infinite number of elements. We turn to finite groups, starting with the smallest finite sets.

Since a group has to have at least one element, namely, the identity, a minimal set that might give rise to a group is a one-element set $\{e\}$. The only possible binary operation $*$ on $\{e\}$ is defined by $e * e = e$. The three group axioms hold. The identity element is always its own inverse in every group.

Let us try to put a group structure on a set of two elements. Since one of the elements must play the role of identity element, we may as well let the set be $\{e, a\}$. Let us attempt to find a table for a binary operation $*$ on $\{e, a\}$ that gives a group structure on $\{e, a\}$. When giving a table for a group operation, we shall always list the identity first, as in the following table.

$*$	e	a
e		
a		

Since e is to be the identity, so

$$e * x = x * e = x$$

for all $x \in \{e, a\}$, we are forced to fill in the table as follows, if $*$ is to give a group:

$*$	e	a
e	e	a
a	a	

Also, a must have an inverse a' such that

$$a * a' = a' * a = e.$$

In our case, a' must be either e or a. Since $a' = e$ obviously does not work, we must have $a' = a$, so we have to complete the table as follows:

$*$	e	a
e	e	a
a	a	e

All the group axioms are now satisfied, except possibly the associative property. Checking associativity on a case-by-case basis from a table defining an operation can be a very tedious process. However, we know that $\mathbb{Z}_2 = \{0, 1\}$ under addition modulo 2 is a group, and by our arguments, its table must be the one above with e replaced by 0 and a by 1. Thus the associative property must be satisfied for our table containing e and a.

With this example as background, we should be able to list some necessary conditions that a table giving a binary operation on a finite set must satisfy for the operation to give a group structure on the set. There must be one element of the set, which we may as well denote by e, that acts as the identity element. The condition $e * x = x$ means that the row of the table opposite e at the extreme left must contain exactly the elements appearing across the very top of the table in the same order. Similarly, the condition $x * e = x$ means that the column of the table under e at the very top must contain exactly the elements appearing at the extreme left in the same order. The fact that every element a has a right and a left inverse means that in the row having a at the extreme left, the element e must appear, and in the column under a at the very top, the e must appear. Thus e must appear in each row and in each column. We can do even better than this, however. By Theorem 4.16, not only the equations $a * x = e$ and $y * a = e$ have unique solutions, but also the equations $a * x = b$ and $y * a = b$. By a similar argument, this means that *each element b of the group must appear once and only once in each row and each column of the table.*

Suppose conversely that a table for a binary operation on a finite set is such that there is an element acting as identity and that in each row and each column, each element of the set appears exactly once. Then it can be seen that the structure is a group structure if and only if the associative law holds. If a binary operation $*$ is given by a table, the associative law is usually messy to check. If the operation $*$ is defined by some characterizing property of $a * b$, the associative law is often easy to check. Fortunately, this second case turns out to be the one usually encountered.

We saw that there was essentially only one group of two elements in the sense that if the elements are denoted by e and a with the identity element e appearing first, the table must be shown in Table 4.19. Suppose that a set has three elements. As before, we may as well let the set be $\{e, a, b\}$. For e to be an identity element, a binary operation $*$ on this set has to have a table of the form shown in Table 4.20. This leaves four places to be filled in. You can quickly see that Table 4.20 must be completed as shown in Table 4.21 if each row and each column are to contain each element exactly once. Because there was only one way to complete the table and $\mathbb{Z}_3 = \{0, 1, 2\}$ under addition modulo 3 is a group, the associative property must hold for our table containing e, a, and b.

Now suppose that G' is any other group of three elements and imagine a table for G' with identity element appearing first. Since our filling out of the table for $G = \{e, a, b\}$ could be done in only one way, we see that if we take the table for G' and rename the identity e, the next element listed a, and the last element b, the resulting table for G' must be the same as the one we had for G. As explained in Section 3, this renaming gives an isomorphism of the group G' with the group G. Definition 3.7 defined the notion of *isomorphism* and of *isomorphic binary structures*. Groups are just certain types of binary structures, so the same definition pertains to them. Thus our work above can be summarized by saying that all groups with a single element are isomorphic, all groups with just two elements are isomorphic, and all groups with just three elements are isomorphic. We use the phrase *up to isomorphism* to express this identification using the equivalence relation $\simeq$. Thus we may say, "There is only one group of three elements, up to isomorphism."

4.19 Table

$*$	e	a
e	e	a
a	a	e

4.20 Table

$*$	e	a	b
e	e	a	b
a	a		
b	b		

4.21 Table

$*$	e	a	b
e	e	a	b
a	a	b	e
b	b	e	a

EXERCISES 4

Computations

In Exercises 1 through 6, determine whether the binary operation $*$ gives a group structure on the given set. If no group results, give the first axiom in the order $\mathscr{G}_1$, $\mathscr{G}_2$, $\mathscr{G}_3$ from Definition 4.1 that does not hold.

1. Let $*$ be defined on $\mathbb{Z}$ by letting $a * b = ab$.

2. Let $*$ be defined on $2\mathbb{Z} = \{2n \mid n \in \mathbb{Z}\}$ by letting $a * b = a + b$.

3. Let $*$ be defined on $\mathbb{R}^+$ by letting $a * b = \sqrt{ab}$.

4. Let $*$ be defined on $\mathbb{Q}$ by letting $a * b = ab$.

5. Let $*$ be defined on the set $\mathbb{R}^*$ of nonzero real numbers by letting $a * b = a/b$.

6. Let $*$ be defined on $\mathbb{C}$ by letting $a * b = |ab|$.

7. Give an example of an abelian group G where G has exactly 1000 elements.

8. We can also consider multiplication $\cdot_n$ modulo n in $\mathbb{Z}_n$. For example, $5 \cdot_7 6 = 2$ in $\mathbb{Z}_7$ because $5 \cdot 6 = 30 = 4(7) + 2$. The set $\{1, 3, 5, 7\}$ with multiplication $\cdot_8$ modulo 8 is a group. Give the table for this group.

9. Show that the group $\langle U, \cdot\rangle$ is not isomorphic to either $\langle \mathbb{R}, +\rangle$ or $\langle \mathbb{R}^*, \cdot\rangle$. (All three groups have cardinality $|\mathbb{R}|$.)

10. Let n be a positive integer and let $n\mathbb{Z} = \{nm \mid m \in \mathbb{Z}\}$.

a. Show that $\langle n\mathbb{Z}, +\rangle$ is a group.

b. Show that $\langle n\mathbb{Z}, +\rangle \simeq \langle \mathbb{Z}, +\rangle$.

In Exercises 11 through 18, determine whether the given set of matrices under the specified operation, matrix addition or multiplication, is a group. Recall that a **diagonal matrix** is a square matrix whose only nonzero entries lie on the **main diagonal,** from the upper left to the lower right corner. An **upper-triangular matrix** is a square matrix with only zero entries below the main diagonal. Associated with each $n \times n$ matrix A is a number called the determinant of A, denoted by $\det(A)$. If A and B are both $n \times n$ matrices, then $\det(AB) = \det(A)\det(B)$. Also, $\det(I_n) = 1$ and A is invertible if and only if $\det(A) \neq 0$.

11. All $n \times n$ diagonal matrices under matrix addition.

12. All $n \times n$ diagonal matrices under matrix multiplication.

13. All $n \times n$ diagonal matrices with no zero diagonal entry under matrix multiplication.

14. All $n \times n$ diagonal matrices with all diagonal entries 1 or -1 under matrix multiplication.

15. All $n \times n$ upper-triangular matrices under matrix multiplication.

16. All $n \times n$ upper-triangular matrices under matrix addition.

17. All $n \times n$ upper-triangular matrices with determinant 1 under matrix multiplication.

18. All $n \times n$ matrices with determinant either 1 or -1 under matrix multiplication.

19. Let S be the set of all real numbers except -1. Define $*$ on S by

$$a * b = a + b + ab.$$

a. Show that $*$ gives a binary operation on S.

b. Show that $\langle S, *\rangle$ is a group.

c. Find the solution of the equation $2 * x * 3 = 7$ in S.

20. This exercise shows that there are two nonisomorphic group structures on a set of 4 elements.

Let the set be $\{e, a, b, c\}$, with e the identity element for the group operation. A group table would then have to start in the manner shown in Table 4.22. The square indicated by the question mark cannot be filled in with a. It must be filled in either with the identity element e or with an element different from both e and a. In this latter case, it is no loss of generality to assume that this element is b. If this square is filled in with e, the table can then be completed in two ways to give a group. Find these two tables. (You need not check the associative law.) If this square is filled in with b, then the table can only be completed in one way to give a group. Find this table. (Again, you need not check the associative law.) Of the three tables you now have, two give isomorphic groups. Determine which two tables these are, and give the one-to-one onto renaming function which is an isomorphism.

a. Are all groups of 4 elements commutative?

b. Which table gives a group isomorphic to the group U_4, so that we know the binary operation defined by the table is associative?

c. Show that the group given by one of the other tables is structurally the same as the group in Exercise 14 for one particular value of n, so that we know that the operation defined by that table is associative also.

21. According to Exercise 12 of Section 2, there are 16 possible binary operations on a set of 2 elements. How many of these give a structure of a group? How many of the 19,683 possible binary operations on a set of 3 elements give a group structure?

Concepts

22. Consider our axioms $\mathscr{G}_1$, $\mathscr{G}_2$, and $\mathscr{G}_3$ for a group. We gave them in the order $\mathscr{G}_1\mathscr{G}_2\mathscr{G}_3$. Conceivable other orders to state the axioms are $\mathscr{G}_1\mathscr{G}_3\mathscr{G}_2$, $\mathscr{G}_2\mathscr{G}_1\mathscr{G}_3$, $\mathscr{G}_2\mathscr{G}_3\mathscr{G}_1$, $\mathscr{G}_3\mathscr{G}_1\mathscr{G}_2$, and $\mathscr{G}_3\mathscr{G}_2\mathscr{G}_1$. Of these six possible orders, exactly three are acceptable for a definition. Which orders are not acceptable, and why? (Remember this. Most instructors ask the student to define a group on at least one test.)

4.22 Table

$*$	e	a	b	c
e	e	a	b	c
a	a	?		
b	b			
c	c			

23. The following "definitions" of a group are taken verbatim, including spelling and punctuation, from papers of students who wrote a bit too quickly and carelessly. Criticize them.

a. A group G is a set of elements together with a binary operation $*$ such that the following conditions are satisfied

$*$ is associative

There exists $e \in G$ such that

$$e * x = x * e = x = \text{identity}.$$

For every $a \in G$ there exists an a' (inverse) such that

$$a \cdot a' = a' \cdot a = e$$

b. A group is a set G such that

The operation on G is associative.

there is an identity element (e) in G.

for every $a \in G$, there is an a' (inverse for each element)

c. A group is a set with a binary operation such

the binary operation is defined

an inverse exists

an identity element exists

d. A set G is called a group over the binery operation $*$ such that for all $a, b \in G$

Binary operation $*$ is associative under addition

there exist an element $\{e\}$ such that

$$a * e = e * a = e$$

Fore every element a there exists an element a' such that

$$a * a' = a' * a = e$$

24. Give a table for a binary operation on the set $\{e, a, b\}$ of three elements satisfying axioms $\mathscr{G}_2$ and $\mathscr{G}_3$ for a group but not axiom $\mathscr{G}_1$.

25. Mark each of the following true or false.

____ **a.** A group may have more than one identity element.

____ **b.** Any two groups of three elements are isomorphic.

____ **c.** In a group, each linear equation has a solution.

_______ **d.** The proper attitude toward a definition is to memorize it so that you can reproduce it word for word as in the text.

_______ **e.** Any definition a person gives for a group is correct provided that everything that is a group by that person's definition is also a group by the definition in the text.

_______ **f.** Any definition a person gives for a group is correct provided he or she can show that everything that satisfies the definition satisfies the one in the text and conversely.

_______ **g.** Every finite group of at most three elements is abelian.

_______ **h.** An equation of the form $a * x * b = c$ always has a unique solution in a group.

_______ **i.** The empty set can be considered a group.

_______ **j.** Every group is a binary algebraic structure.

Proof synopsis

We give an example of a proof synopsis. Here is a one-sentence synopsis of the proof that the inverse of an element a in a group $\langle G, *\rangle$ is unique.

> Assuming that $a * a' = e$ and $a * a'' = e$, apply the left cancellation law to the equation $a * a' = a * a''$.

Note that we said "the left cancellation law" and not "Theorem 4.15." We always suppose that our synopsis was given as an explanation given during a conversation at lunch, with no reference to text numbering and as little notation as is practical.

26. Give a one-sentence synopsis of the proof of the left cancellation law in Theorem 4.15.

27. Give at most a two-sentence synopsis of the proof in Theorem 4.16 that an equation $ax = b$ has a unique solution in a group.

Theory

28. From our intuitive grasp of the notion of isomorphic groups, it should be clear that if $\phi : G \to G'$ is a group isomorphism, then $\phi(e)$ is the identity e' of G'. Recall that Theorem 3.14 gave a proof of this for isomorphic binary structures $\langle S, *\rangle$ and $\langle S', *'\rangle$. Of course, this covers the case of groups.

It should also be intuitively clear that if a and a' are inverse pairs in G, then $\phi(a)$ and $\phi(a')$ are inverse pairs in G', that is, that $\phi(a)' = \phi(a')$. Give a careful proof of this for a skeptic who can't see the forest for all the trees.

29. Show that if G is a finite group with identity e and with an even number of elements, then there is $a \neq e$ in G such that $a * a = e$.

30. Let $\mathbb{R}^*$ be the set of all real numbers except 0. Define $*$ on $\mathbb{R}^*$ by letting $a * b = |a|b$.

a. Show that $*$ gives an associative binary operation on $\mathbb{R}^*$.

b. Show that there is a left identity for $*$ and a right inverse for each element in $\mathbb{R}^*$.

c. Is $\mathbb{R}^*$ with this binary operation a group?

d. Explain the significance of this exercise.

31. If $*$ is a binary operation on a set S, an element x of S is an **idempotent for** $*$ if $x * x = x$. Prove that a group has exactly one idempotent element. (You may use any theorems proved so far in the text.)

32. Show that every group G with identity e and such that $x * x = e$ for all $x \in G$ is abelian. [*Hint:* Consider $(a * b) * (a * b)$.]

33. Let G be an abelian group and let $c^n = c * c * \cdots * c$ for n factors c, where $c \in G$ and $n \in \mathbb{Z}^+$. Give a mathematical induction proof that $(a * b)^n = (a^n) * (b^n)$ for all $a, b \in G$.

34. Let G be a group with a finite number of elements. Show that for any $a \in G$, there exists an $n \in \mathbb{Z}^+$ such that $a^n = e$. See Exercise 33 for the meaning of a^n. [*Hint:* Consider $e, a, a^2, a^3, \ldots, a^m$, where m is the number of elements in G, and use the cancellation laws.]

35. Show that if $(a * b)^2 = a^2 * b^2$ for a and b in a group G, then $a * b = b * a$. See Exercise 33 for the meaning of a^2.

36. Let G be a group and let $a, b \in G$. Show that $(a * b)' = a' * b'$ if and only if $a * b = b * a$.

37. Let G be a group and suppose that $a * b * c = e$ for $a, b, c \in G$. Show that $b * c * a = e$ also.

38. Prove that a set G, together with a binary operation $*$ on G satisfying the left axioms 1, 2, and 3 given on page 43, is a group.

39. Prove that a nonempty set G, together with an associative binary operation $*$ on G such that

$$a * x = b \text{ and } y * a = b \text{ have solutions in } G \text{ for all } a, b \in G,$$

is a group. [*Hint:* Use Exercise 38.]

40. Let $\langle G, \cdot\rangle$ be a group. Consider the binary operation $*$ on the set G defined by

$$a * b = b \cdot a$$

for $a, b \in G$. Show that $\langle G, *\rangle$ is a group and that $\langle G, *\rangle$ is actually isomorphic to $\langle G, \cdot\rangle$. [*Hint:* Consider the map ϕ with $\phi(a) = a'$ for $a \in G$.]

41. Let G be a group and let g be one fixed element of G. Show that the map i_g, such that $i_g(x) = gxg'$ for $x \in G$, is an isomorphism of G with itself.

SECTION 5 SUBGROUPS

Notation and Terminology

It is time to explain some conventional notation and terminology used in group theory. Algebraists as a rule do not use a special symbol $*$ to denote a binary operation different from the usual addition and multiplication. They stick with the conventional additive or multiplicative notation and even call the operation *addition* or *multiplication,* depending on the symbol used. The symbol for addition is, of course, $+$, and usually multiplication is denoted by juxtaposition without a dot, if no confusion results. Thus in place of the notation $a * b$, we shall be using either $a + b$ to be read "the *sum* of a and b," or ab to be read "the *product* of a and b." There is a sort of unwritten agreement that the symbol $+$ should be used only to designate commutative operations. Algebraists feel very uncomfortable when they see $a + b \neq b + a$. For this reason, when developing our theory in a general situation where the operation may or may not be commutative, we shall always use multiplicative notation.

Algebraists frequently use the symbol 0 to denote an additive identity element and the symbol 1 to denote a multiplicative identity element, even though they may not be actually denoting the integers 0 and 1. Of course, if they are also talking about numbers at the same time, so that confusion would result, symbols such as e or u are used as

5.1 Table

	1	a	b
1	1	a	b
a	a	b	1
b	b	1	a

5.2 Table

+	0	a	b
0	0	a	b
a	a	b	0
b	b	0	a

identity elements. Thus a table for a group of three elements might be one like Table 5.1 or, since such a group is commutative, the table might look like Table 5.2. In general situations we shall continue to use e to denote the identity element of a group.

It is customary to denote the inverse of an element a in a group by a^{-1} in multiplicative notation and by $-a$ in additive notation. From now on, we shall be using these notations in place of the symbol a'.

Let n be a positive integer. If a is an element of a group G, written multiplicatively, we denote the product $aaa \dots a$ for n factors a by a^n. We let a^0 be the identity element e, and denote the product $a^{-1}a^{-1}a^{-1} \dots a^{-1}$ for n factors by a^{-n}. It is easy to see that our usual law of exponents, $a^m a^n = a^{m+n}$ for $m, n \in \mathbb{Z}$, holds. For $m, n \in \mathbb{Z}^+$, it is clear. We illustrate another type of case by an example:

$$\begin{aligned} a^{-2}a^5 &= a^{-1}a^{-1}aaaaa = a^{-1}(a^{-1}a)aaaa = a^{-1}eaaaa = a^{-1}(ea)aaa \\ &= a^{-1}aaaa = (a^{-1}a)aaa = eaaa = (ea)aa = aaa = a^3. \end{aligned}$$

In additive notation, we denote $a + a + a + \cdots + a$ for n summands by na, denote $(-a) + (-a) + (-a) + \cdots + (-a)$ for n summands by $-na$, and let $0a$ be the identity element. Be careful: In the notation na, the number n is in $\mathbb{Z}$, not in G. One reason we prefer to present group theory using multiplicative notation, even if G is abelian, is the confusion caused by regarding n as being in G in this notation na. No one ever misinterprets the n when it appears in an exponent.

Let us explain one more term that is used so often it merits a special definition.

5.3 Definition If G is a group, then the **order** $|G|$ of G is the number of elements in G. (Recall from Section 0 that, for any set S, $|S|$ is the cardinality of S.) ■

Subsets and Subgroups

You may have noticed that we sometimes have had groups contained within larger groups. For example, the group $\mathbb{Z}$ under addition is contained within the group $\mathbb{Q}$ under addition, which in turn is contained in the group $\mathbb{R}$ under addition. When we view the group $\langle \mathbb{Z}, + \rangle$ as contained in the group $\langle \mathbb{R}, + \rangle$, it is very important to notice that the operation $+$ on integers n and m as elements of $\langle \mathbb{Z}, + \rangle$ produces the same element $n + m$ as would result if you were to think of n and m as elements in $\langle \mathbb{R}, + \rangle$. Thus we should *not* regard the group $\langle \mathbb{Q}^+, \cdot \rangle$ as contained in $\langle \mathbb{R}, + \rangle$, even though $\mathbb{Q}^+$ is contained in $\mathbb{R}$ as a set. In this instance, $2 \cdot 3 = 6$ in $\langle \mathbb{Q}^+, \cdot \rangle$, while $2 + 3 = 5$ in $\langle \mathbb{R}, + \rangle$. We are requiring not only that the set of one group be a subset of the set of the other, but also that the group operation on the subset be the *induced operation* that assigns the same element to each ordered pair from this subset as is assigned by the group operation on the whole set.

5.4 Definition If a subset H of a group G is closed under the binary operation of G and if H with the induced operation from G is itself a group, then H is a **subgroup of** G. We shall let $H \leq G$ or $G \geq H$ denote that H is a subgroup of G, and $H < G$ or $G > H$ shall mean $H \leq G$ but $H \neq G$. ■

Thus $\langle \mathbb{Z}, + \rangle < \langle \mathbb{R}, + \rangle$ but $\langle \mathbb{Q}^+, \cdot \rangle$ is *not* a subgroup of $\langle \mathbb{R}, + \rangle$, even though as sets, $\mathbb{Q}^+ \subset \mathbb{R}$. Every group G has as subgroups G itself and $\{e\}$, where e is the identity element of G.

5.5 Definition If G is a group, then the subgroup consisting of G itself is the **improper subgroup** of G. All other subgroups are **proper subgroups**. The subgroup $\{e\}$ is the **trivial subgroup** of G. All other subgroups are **nontrivial**. ■

We turn to some illustrations.

5.6 Example Let $\mathbb{R}^n$ be the additive group of all n-component row vectors with real number entries. The subset consisting of all of these vectors having 0 as entry in the first component is a subgroup of $\mathbb{R}^n$. ▲

5.7 Example $\mathbb{Q}^+$ under multiplication is a proper subgroup of $\mathbb{R}^+$ under multiplication. ▲

5.8 Example The nth roots of unity in $\mathbb{C}$ form a subgroup U_n of the group $\mathbb{C}^*$ of nonzero complex numbers under multiplication. ▲

5.9 Example There are two different types of group structures of order 4 (see Exercise 20 of Section 4). We describe them by their group tables (Tables 5.10 and 5.11). The group V is the **Klein 4-group,** and the notation V comes from the German word *Vier* for four. The group $\mathbb{Z}_4$ is isomorphic to the group $U_4 = \{1, i, -1, -i\}$ of fourth roots of unity under multiplication.

The only nontrivial proper subgroup of $\mathbb{Z}_4$ is $\{0, 2\}$. Note that $\{0, 3\}$ is *not* a subgroup of $\mathbb{Z}_4$, since $\{0, 3\}$ is *not closed* under $+$. For example, $3 + 3 = 2$, and $2 \notin \{0, 3\}$. However, the group V has three nontrivial proper subgroups, $\{e, a\}$, $\{e, b\}$, and $\{e, c\}$. Here $\{e, a, b\}$ is *not* a subgroup, since $\{e, a, b\}$ is not closed under the operation of V because $ab = c$, and $c \notin \{e, a, b\}$. ▲

5.10 Table

$\mathbb{Z}_4$:

$+$	0	1	2	3
0	0	1	2	3
1	1	2	3	0
2	2	3	0	1
3	3	0	1	2

5.11 Table

V:

	e	a	b	c
e	e	a	b	c
a	a	e	c	b
b	b	c	e	a
c	c	b	a	e

It is often useful to draw a *subgroup diagram* of the subgroups of a group. In such a diagram, a line running downward from a group G to a group H means that H is a subgroup of G. Thus the larger group is placed nearer the top of the diagram. Figure 5.12 contains the subgroup diagrams for the groups $\mathbb{Z}_4$ and V of Example 5.9.

Note that if $H \leq G$ and $a \in H$, then by Theorem 4.16, the equation $ax = a$ must have a unique solution, namely the identity element of H. But this equation can also be viewed as one in G, and we see that this unique solution must also be the identity element e of G. A similar argument then applied to the equation $ax = e$, viewed in both H and G, shows that the inverse a^{-1} of a in G is also the inverse of a in the subgroup H.

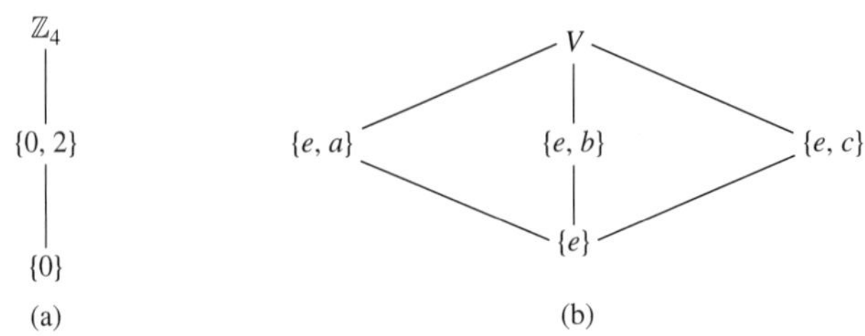

5.12 Figure (a) Subgroup diagram for $\mathbb{Z}_4$. (b) Subgroup diagram for V.

5.13 Example Let F be the group of all real-valued functions with domain $\mathbb{R}$ under addition. The subset of F consisting of those functions that are continuous is a subgroup of F, for the sum of continuous functions is continuous, the function f where $f(x) = 0$ for all x is continuous and is the additive identity element, and if f is continuous, then $-f$ is continuous. ▲

It is convenient to have routine steps for determining whether a subset of a group G is a subgroup of G. Example 5.13 indicates such a routine, and in the next theorem, we demonstrate carefully its validity. While more compact criteria are available, involving only one condition, we prefer this more transparent theorem for a first course.

5.14 Theorem A subset H of a group G is a subgroup of G if and only if

1. H is closed under the binary operation of G,
2. the identity element e of G is in H,
3. for all $a \in H$ it is true that $a^{-1} \in H$ also.

Proof The fact that if $H \leq G$ then Conditions 1, 2, and 3 must hold follows at once from the definition of a subgroup and from the remarks preceding Example 5.13.

Conversely, suppose H is a subset of a group G such that Conditions 1, 2, and 3 hold. By 2 we have at once that $\mathscr{G}_2$ is satisfied. Also $\mathscr{G}_3$ is satisfied by 3. It remains to check the associative axiom, $\mathscr{G}_1$. But surely for all $a, b, c \in H$ it is true that $(ab)c = a(bc)$ in H, for we may actually view this as an equation in G, where the associative law holds. Hence $H \leq G$. ◆

5.15 Example Let F be as in Example 5.13. The subset of F consisting of those functions that are differentiable is a subgroup of F, for the sum of differentiable functions is differentiable, the constant function 0 is differentiable, and if f is differentiable, then $-f$ is differentiable. ▲

5.16 Example Recall from linear algebra that every square matrix A has associated with it a number $\det(A)$ called its determinant, and that A is invertible if and only if $\det(A) \neq 0$. If A and B are square matrices of the same size, then it can be shown that $\det(AB) = \det(A) \cdot \det(B)$. Let G be the multiplicative group of all invertible $n \times n$ matrices with entries in $\mathbb{C}$ and let T be the subset of G consisting of those matrices with determinant 1. The equation $\det(AB) = \det(A) \cdot \det(B)$ shows that T is closed under matrix multiplication. Recall that the identity matrix I_n has determinant 1. From the equation $\det(A) \cdot \det(A^{-1}) = \det(AA^{-1}) = \det(I_n) = 1$, we see that if $\det(A) = 1$, then $\det(A^{-1}) = 1$. Theorem 5.14 then shows that T is a subgroup of G. ▲

Cyclic Subgroups

Let us see how large a subgroup H of $\mathbb{Z}_{12}$ would have to be if it contains 3. It would have to contain the identity element 0 and $3 + 3$, which is 6. Then it has to contain $6 + 3$, which is 9. Note that the inverse of 3 is 9 and the inverse of 6 is 6. It is easily checked that $H = \{0, 3, 6, 9\}$ is a subgroup of $\mathbb{Z}_{12}$, and it is the smallest subgroup containing 3.

Let us imitate this reasoning in a general situation. As we remarked before, for a general argument we always use multiplicative notation. Let G be a group and let $a \in G$. A subgroup of G containing a must, by Theorem 5.14, contain a^n, the result of computing products of a and itself for n factors for every positive integer n. These positive integral powers of a do give a set closed under multiplication. It is possible, however, that the inverse of a is not in this set. Of course, a subgroup containing a must also contain a^{-1}, and, in general, it must contain a^{-m} for all $m \in \mathbb{Z}^+$. It must contain the identity element $e = a^0$. Summarizing, *a subgroup of G containing the element a must contain all elements a^n (or na for additive groups) for all $n \in \mathbb{Z}$*. That is, a subgroup containing a must contain $\{a^n | n \in \mathbb{Z}\}$. Observe that these powers a^n of a need not be distinct. For example, in the group V of Example 5.9,

$$a^2 = e, \quad a^3 = a, \quad a^4 = e, \quad a^{-1} = a, \qquad \text{and so on.}$$

We have almost proved the next theorem.

5.17 Theorem Let G be a group and let $a \in G$. Then

$$H = \{a^n \mid n \in \mathbb{Z}\}$$

is a subgroup of G and is the smallest[†] subgroup of G that contains a, that is, every subgroup containing a contains H.

[†] We may find occasion to distinguish between the terms *minimal* and *smallest* as applied to subsets of a set S that have some property. A subset H of S is minimal with respect to the property if H has the property, and no subset $K \subset H$, $K \neq H$, has the property. If H has the property and $H \subseteq K$ for every subset K with the property, then H is the smallest subset with the property. There may be many minimal subsets, but there can be only one smallest subset. To illustrate, $\{e, a\}$, $\{e, b\}$, and $\{e, c\}$ are all minimal nontrivial subgroups of the group V. (See Fig. 5.12.) However, V contains no smallest nontrivial subgroup.

Proof We check the three conditions given in Theorem 5.14 for a subset of a group to give a subgroup. Since $a^r a^s = a^{r+s}$ for $r, s \in \mathbb{Z}$, we see that the product in G of two elements of H is again in H. Thus H is closed under the group operation of G. Also $a^0 = e$, so $e \in H$, and for $a^r \in H$, $a^{-r} \in H$ and $a^{-r}a^r = e$. Hence all the conditions are satisfied, and $H \leq G$.

Our arguments prior to the statement of the theorem showed that any subgroup of G containing a must contain H, so H is the smallest subgroup of G containing a. ◆

5.18 Definition Let G be a group and let $a \in G$. Then the subgroup $\{a^n \mid n \in \mathbb{Z}\}$ of G, characterized in Theorem 5.17, is called the **cyclic subgroup of G generated by a**, and denoted by $\langle a \rangle$. ■

5.19 Definition An element a of a group G **generates** G and is a **generator for** G if $\langle a \rangle = G$. A group G is **cyclic** if there is some element a in G that generates G. ■

5.20 Example Let $\mathbb{Z}_4$ and V be the groups of Example 5.9. Then $\mathbb{Z}_4$ is cyclic and both 1 and 3 are generators, that is,

$$\langle 1 \rangle = \langle 3 \rangle = \mathbb{Z}_4.$$

However, V is *not* cyclic, for $\langle a \rangle$, $\langle b \rangle$, and $\langle c \rangle$ are proper subgroups of two elements. Of course, $\langle e \rangle$ is the trivial subgroup of one element. ▲

5.21 Example The group $\mathbb{Z}$ under addition is a cyclic group. Both 1 and -1 are generators for this group, and they are the only generators. Also, for $n \in \mathbb{Z}^+$, the group $\mathbb{Z}_n$ under addition modulo n is cyclic. If $n > 1$, then both 1 and $n - 1$ are generators, but there may be others. ▲

5.22 Example Consider the group $\mathbb{Z}$ under addition. Let us find $\langle 3 \rangle$. Here the notation is additive, and $\langle 3 \rangle$ must contain

$$\begin{array}{lllll} & 3, & 3+3=6, & 3+3+3=9, & \text{and so on,} \\ 0, & -3, & -3+-3=-6, & -3+-3+-3=-9, & \text{and so on.} \end{array}$$

In other words, the cyclic subgroup generated by 3 consists of all multiples of 3, positive, negative, and zero. We denote this subgroup by $3\mathbb{Z}$ as well as $\langle 3 \rangle$. In a similar way, we shall let $n\mathbb{Z}$ be the cyclic subgroup $\langle n \rangle$ of $\mathbb{Z}$. Note that $6\mathbb{Z} < 3\mathbb{Z}$. ▲

5.23 Example For each positive integer n, let U_n be the multiplicative group of the nth roots of unity in $\mathbb{C}$. These elements of U_n can be represented geometrically by equally spaced points on a circle about the origin, as illustrated in Fig. 5.24. The heavy point represents the number

$$\zeta = \cos\frac{2\pi}{n} + i\sin\frac{2\pi}{n}.$$

The geometric interpretation of multiplication of complex numbers, explained in Section 1, shows at once that as ζ is raised to powers, it works its way counterclockwise around the circle, landing on each of the elements of U_n in turn. Thus U_n under multiplication is a cyclic group, and ζ is a generator. The group U_n is the cyclic subgroup $\langle \zeta \rangle$ of the group U of all complex numbers z, where $|z| = 1$, under multiplication. ▲

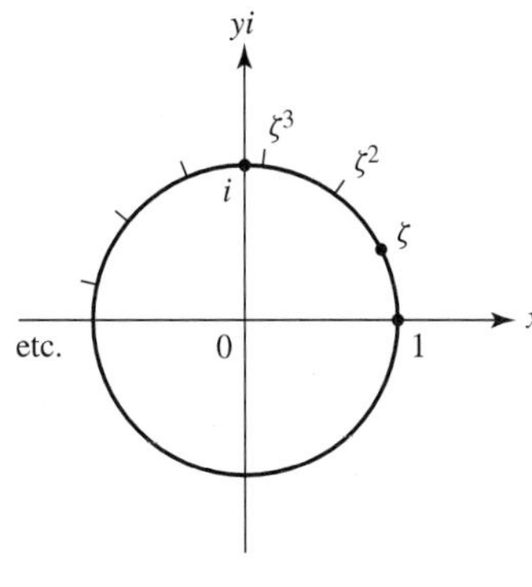

5.24 Figure

EXERCISES 5

Computations

In Exercises 1 through 6, determine whether the given subset of the complex numbers is a subgroup of the group $\mathbb{C}$ of complex numbers under addition.

1. $\mathbb{R}$ **2.** $\mathbb{Q}^+$ **3.** $7\mathbb{Z}$

4. The set $i\mathbb{R}$ of pure imaginary numbers including 0

5. The set $\pi\mathbb{Q}$ of rational multiples of π **6.** The set $\{\pi^n \mid n \in \mathbb{Z}\}$

7. Which of the sets in Exercises 1 through 6 are subgroups of the group $\mathbb{C}^*$ of nonzero complex numbers under multiplication?

In Exercises 8 through 13, determine whether the given set of invertible $n \times n$ matrices with real number entries is a subgroup of $GL(n, \mathbb{R})$.

8. The $n \times n$ matrices with determinant 2

9. The diagonal $n \times n$ matrices with no zeros on the diagonal

10. The upper-triangular $n \times n$ matrices with no zeros on the diagonal

11. The $n \times n$ matrices with determinant -1

12. The $n \times n$ matrices with determinant -1 or 1

13. The set of all $n \times n$ matrices A such that $(A^T)A = I_n$. [These matrices are called **orthogonal.** Recall that A^T, the *transpose* of A, is the matrix whose jth column is the jth row of A for $1 \le j \le n$, and that the transpose operation has the property $(AB)^T = (B^T)(A^T)$.]

Let F be the set of all real-valued functions with domain $\mathbb{R}$ and let $\tilde{F}$ be the subset of F consisting of those functions that have a nonzero value at every point in $\mathbb{R}$. In Exercises 14 through 19, determine whether the given subset of F with the induced operation is (a) a subgroup of the group F under addition, (b) a subgroup of the group $\tilde{F}$ under multiplication.

14. The subset $\tilde{F}$

15. The subset of all $f \in F$ such that $f(1) = 0$

16. The subset of all $f \in \tilde{F}$ such that $f(1) = 1$

17. The subset of all $f \in \tilde{F}$ such that $f(0) = 1$

18. The subset of all $f \in \tilde{F}$ such that $f(0) = -1$

19. The subset of all constant functions in F.

20. Nine groups are given below. Give a *complete* list of all subgroup relations, of the form $G_i \leq G_j$, that exist between these given groups $G_1, G_2, \cdots, G_9$.
$G_1 = \mathbb{Z}$ under addition
$G_2 = 12\mathbb{Z}$ under addition
$G_3 = \mathbb{Q}^+$ under multiplication
$G_4 = \mathbb{R}$ under addition
$G_5 = \mathbb{R}^+$ under multiplication
$G_6 = \{\pi^n \mid n \in \mathbb{Z}\}$ under multiplication
$G_7 = 3\mathbb{Z}$ under addition
$G_8 =$ the set of all integral multiples of 6 under addition
$G_9 = \{6^n \mid n \in \mathbb{Z}\}$ under multiplication

21. Write at least 5 elements of each of the following cyclic groups.

a. $25\mathbb{Z}$ under addition

b. $\{(\frac{1}{2})^n \mid n \in \mathbb{Z}\}$ under multiplication

c. $\{\pi^n \mid n \in \mathbb{Z}\}$ under multiplication

In Exercises 22 through 25, describe all the elements in the cyclic subgroup of $GL(2, \mathbb{R})$ generated by the given 2×2 matrix.

22. $\begin{bmatrix} 0 & -1 \\ -1 & 0 \end{bmatrix}$ **23.** $\begin{bmatrix} 1 & 1 \\ 0 & 1 \end{bmatrix}$ **24.** $\begin{bmatrix} 3 & 0 \\ 0 & 2 \end{bmatrix}$ **25.** $\begin{bmatrix} 0 & -2 \\ -2 & 0 \end{bmatrix}$

26. Which of the following groups are cyclic? For each cyclic group, list all the generators of the group.

$$G_1 = \langle \mathbb{Z}, + \rangle \quad G_2 = \langle \mathbb{Q}, + \rangle \quad G_3 = \langle \mathbb{Q}^+, \cdot \rangle \quad G_4 = \langle 6\mathbb{Z}, + \rangle$$
$$G_5 = \{6^n \mid n \in \mathbb{Z}\} \text{ under multiplication}$$
$$G_6 = \{a + b\sqrt{2} \mid a, b \in \mathbb{Z}\} \text{ under addition}$$

In Exercises 27 through 35, find the order of the cyclic subgroup of the given group generated by the indicated element.

27. The subgroup of $\mathbb{Z}_4$ generated by 3

28. The subgroup of V generated by c (see Table 5.11)

29. The subgroup of U_6 generated by $\cos \frac{2\pi}{3} + i \sin \frac{2\pi}{3}$

30. The subgroup of U_5 generated by $\cos \frac{4\pi}{5} + i \sin \frac{4\pi}{5}$

31. The subgroup of U_8 generated by $\cos \frac{3\pi}{2} + i \sin \frac{3\pi}{2}$

32. The subgroup of U_8 generated by $\cos\frac{5\pi}{4} + i\sin\frac{5\pi}{4}$

33. The subgroup of the multiplicative group G of invertible 4×4 matrices generated by

$$\begin{bmatrix} 0 & 0 & 1 & 0 \\ 0 & 0 & 0 & 1 \\ 1 & 0 & 0 & 0 \\ 0 & 1 & 0 & 0 \end{bmatrix}$$

34. The subgroup of the multiplicative group G of invertible 4×4 matrices generated by

$$\begin{bmatrix} 0 & 0 & 0 & 1 \\ 0 & 0 & 1 & 0 \\ 1 & 0 & 0 & 0 \\ 0 & 1 & 0 & 0 \end{bmatrix}$$

35. The subgroup of the multiplicative group G of invertible 4×4 matrices generated by

$$\begin{bmatrix} 0 & 1 & 0 & 0 \\ 0 & 0 & 0 & 1 \\ 0 & 0 & 1 & 0 \\ 1 & 0 & 0 & 0 \end{bmatrix}$$

36. a. Complete Table 5.25 to give the group $\mathbb{Z}_6$ of 6 elements.

b. Compute the subgroups $\langle 0 \rangle$, $\langle 1 \rangle$, $\langle 2 \rangle$, $\langle 3 \rangle$, $\langle 4 \rangle$, and $\langle 5 \rangle$ of the group $\mathbb{Z}_6$ given in part (a).

c. Which elements are generators for the group $\mathbb{Z}_6$ of part (a)?

d. Give the subgroup diagram for the part (b) subgroups of $\mathbb{Z}_6$. (We will see later that these are all the subgroups of $\mathbb{Z}_6$.)

5.25 Table

$\mathbb{Z}_6$:

+	0	1	2	3	4	5
0	0	1	2	3	4	5
1	1	2	3	4	5	0
2	2					
3	3					
4	4					
5	5					

Concepts

In Exercises 37 and 38, correct the definition of the italicized term without reference to the text, if correction is needed, so that it is in a form acceptable for publication.

37. A *subgroup* of a group G is a subset H of G that contains the identity element e of G and also contains the inverse of each of its elements.

38. A group G is *cyclic* if and only if there exists $a \in G$ such that $G = \{a^n \mid n \in \mathbb{Z}\}$.

39. Mark each of the following true or false.

_______ **a.** The associative law holds in every group.

_______ **b.** There may be a group in which the cancellation law fails.

_______ **c.** Every group is a subgroup of itself.

_______ **d.** Every group has exactly two improper subgroups.

_______ **e.** In every cyclic group, every element is a generator.

_______ **f.** A cyclic group has a unique generator.

_______ **g.** Every set of numbers that is a group under addition is also a group under multiplication.

_______ **h.** A subgroup may be defined as a subset of a group.

_______ **i.** $\mathbb{Z}_4$ is a cyclic group.

_______ **j.** Every subset of every group is a subgroup under the induced operation.

40. Show by means of an example that it is possible for the quadratic equation $x^2 = e$ to have more than two solutions in some group G with identity e.

Theory

In Exercises 41 and 42, let $\phi : G \to G'$ be an isomorphism of a group $\langle G, *\rangle$ with a group $\langle G', *'\rangle$. Write out a proof to convince a skeptic of the intuitively clear statement.

41. If H is a subgroup of G, then $\phi[H] = \{\phi(h) \mid h \in H\}$ is a subgroup of G'. That is, an isomorphism carries subgroups into subgroups.

42. If G is cyclic, then G' is cyclic.

43. Show that if H and K are subgroups of an abelian group G, then

$$\{hk \mid h \in H \text{ and } k \in K\}$$

is a subgroup of G.

44. Find the flaw in the following argument: "Condition 2 of Theorem 5.14 is redundant, since it can be derived from 1 and 3, for let $a \in H$. Then $a^{-1} \in H$ by 3, and by 1, $aa^{-1} = e$ is an element of H, proving 2."

45. Show that a nonempty subset H of a group G is a subgroup of G if and only if $ab^{-1} \in H$ for all $a, b \in H$. (This is one of the *more compact criteria* referred to prior to Theorem 5.14)

46. Prove that a cyclic group with only one generator can have at most 2 elements.

47. Prove that if G is an abelian group, written multiplicatively, with identity element e, then all elements x of G satisfying the equation $x^2 = e$ form a subgroup H of G.

48. Repeat Exercise 47 for the general situation of the set H of all solutions x of the equation $x^n = e$ for a fixed integer $n \geq 1$ in an abelian group G with identity e.

49. Show that if $a \in G$, where G is a finite group with identity e, then there exists $n \in \mathbb{Z}^+$ such that $a^n = e$.

50. Let a nonempty finite subset H of a group G be closed under the binary operation of G. Show that H is a subgroup of G.

51. Let G be a group and let a be one fixed element of G. Show that

$$H_a = \{x \in G \mid xa = ax\}$$

is a subgroup of G.

52. Generalizing Exercise 51, let S be any subset of a group G.

a. Show that $H_S = \{x \in G \mid xs = sx \text{ for all } s \in S\}$ is a subgroup of G.

b. In reference to part (a), the subgroup H_G is the **center of** G. Show that H_G is an abelian group.

53. Let H be a subgroup of a group G. For $a, b \in G$, let $a \sim b$ if and only if $ab^{-1} \in H$. Show that $\sim$ is an equivalence relation on G.

54. For sets H and K, we define the **intersection** $H \cap K$ by

$$H \cap K = \{x \mid x \in H \text{ and } x \in K\}.$$

Show that if $H \leq G$ and $K \leq G$, then $H \cap K \leq G$. (Remember: $\leq$ denotes "is a subgroup of," not "is a subset of.")

55. Prove that every cyclic group is abelian.

56. Let G be a group and let $G_n = \{g^n \mid g \in G\}$. Under what hypothesis about G can we show that G_n is a subgroup of G?

57. Show that a group with no proper nontrivial subgroups is cyclic.

SECTION 6 Cyclic Groups

Recall the following facts and notations from Section 5. If G is a group and $a \in G$, then

$$H = \{a^n \mid n \in \mathbb{Z}\}$$

is a subgroup of G (Theorem 5.17). This group is the **cyclic subgroup $\langle a \rangle$ of G generated by a.** Also, given a group G and an element a in G, if

$$G = \{a^n \mid n \in \mathbb{Z}\},$$

then a is a **generator of** G and the group $G = \langle a \rangle$ is **cyclic.** We introduce one new bit of terminology. Let a be an element of a group G. If the cyclic subgroup $\langle a \rangle$ of G is finite, then the **order of** a is the order $|\langle a \rangle|$ of this cyclic subgroup. Otherwise, we say that a is of **infinite order.** We will see in this section that if $a \in G$ is of finite order m, then m is the smallest positive integer such that $a^m = e$.

The first goal of this section is to describe all cyclic groups and all subgroups of cyclic groups. This is not an idle exercise. We will see later that cyclic groups serve as building blocks for all sufficiently small abelian groups, in particular, for all finite abelian groups. Cyclic groups are fundamental to the understanding of groups.

Elementary Properties of Cyclic Groups

We start with a demonstration that cyclic groups are abelian.

6.1 Theorem Every cyclic group is abelian.

Proof Let G be a cyclic group and let a be a generator of G so that

$$G = \langle a \rangle = \{a^n \mid n \in \mathbb{Z}\}.$$

If g_1 and g_2 are any two elements of G, there exist integers r and s such that $g_1 = a^r$ and $g_2 = a^s$. Then

$$g_1 g_2 = a^r a^s = a^{r+s} = a^{s+r} = a^s a^r = g_2 g_1,$$

so G is abelian. ◆

We shall continue to use multiplicative notation for our general work on cyclic groups, even though they are abelian.

The *division algorithm* that follows is a seemingly trivial, but very fundamental tool for the study of cyclic groups.

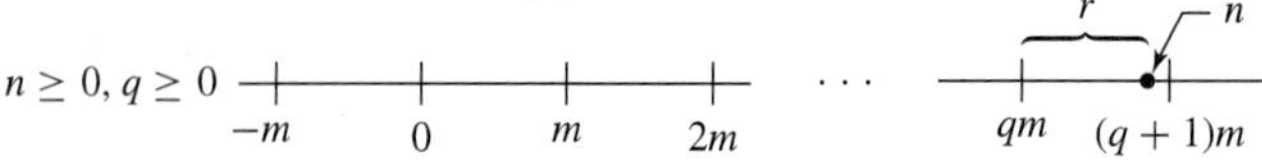

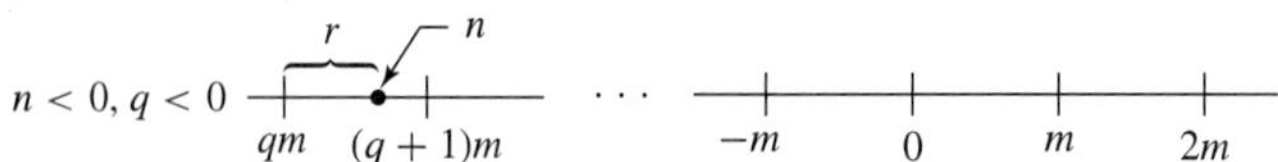

6.2 Figure

6.3 Division Algorithm for $\mathbb{Z}$ If m is a positive integer and n is any integer, then there exist unique integers q and r such that

$$n = mq + r \qquad \text{and} \qquad 0 \leq r < m.$$

Proof We give an intuitive diagrammatic explanation, using Fig. 6.2. On the real x-axis of analytic geometry, mark off the multiples of m and the position of n. Now n falls either on a multiple qm of m and r can be taken as 0, or n falls between two multiples of m. If the latter is the case, let qm be the first multiple of m to the left of n. Then r is as shown in Fig. 6.2. Note that $0 \leq r < m$. Uniqueness of q and r follows since if n is not a multiple of m so that we can take $r = 0$, then there is a unique multiple qm of m to the left of n and at distance less than m from n, as illustrated in Fig. 6.2. ◆

In the notation of the division algorithm, we regard q as the **quotient** and r as the nonnegative **remainder** when n is divided by m.

6.4 Example Find the quotient q and remainder r when 38 is divided by 7 according to the division algorithm.

Solution The positive multiples of 7 are 7, 14, 21, 28, 35, 42, $\cdots$. Choosing the multiple to leave a nonnegative remainder less than 7, we write

$$38 = 35 + 3 = 7(5) + 3$$

so the quotient is $q = 5$ and the remainder is $r = 3$. ▲

6.5 Example Find the quotient q and remainder r when -38 is divided by 7 according to the division algorithm.

Solution The negative multiples of 7 are $-7, -14, -21, -28, -35, -42, \cdots$. Choosing the multiple to leave a nonnegative remainder less than 7, we write

$$-38 = -42 + 4 = 7(-6) + 4$$

so the quotient is $q = -6$ and the remainder is $r = 4$. ▲

We will use the division algorithm to show that a subgroup H of a cyclic group G is also cyclic. Think for a moment what we will have to do to prove this. We will have to use the *definition* of a cyclic group since we have proved little about cyclic groups yet. That is, we will have to use the fact that G has a generating element a. We must then exhibit, in terms of this generator a, some generator $c = a^m$ for H in order to show that H is cyclic. There is really only one natural choice for the power m of a to try. Can you guess what it is before you read the proof of the theorem?

6.6 Theorem A subgroup of a cyclic group is cyclic.

Proof Let G be a cyclic group generated by a and let H be a subgroup of G. If $H = \{e\}$, then $H = \langle e \rangle$ is cyclic. If $H \neq \{e\}$, then $a^n \in H$ for some $n \in \mathbb{Z}^+$. Let m be the smallest integer in $\mathbb{Z}^+$ such that $a^m \in H$.

We claim that $c = a^m$ generates H; that is,

$$H = \langle a^m \rangle = \langle c \rangle.$$

We must show that every $b \in H$ is a power of c. Since $b \in H$ and $H \leq G$, we have $b = a^n$ for some n. Find q and r such that

$$n = mq + r \qquad \text{for} \qquad 0 \leq r < m$$

in accord with the division algorithm. Then

$$a^n = a^{mq+r} = (a^m)^q a^r,$$

so

$$a^r = (a^m)^{-q} a^n.$$

Now since $a^n \in H$, $a^m \in H$, and H is a group, both $(a^m)^{-q}$ and a^n are in H. Thus

$$(a^m)^{-q} a^n \in H; \qquad \text{that is,} \qquad a^r \in H.$$

Since m was the smallest positive integer such that $a^m \in H$ and $0 \leq r < m$, we must have $r = 0$. Thus $n = qm$ and

$$b = a^n = (a^m)^q = c^q,$$

so b is a power of c. ◆

As noted in Examples 5.21 and 5.22, $\mathbb{Z}$ under addition is cyclic and for a positive integer n, the set $n\mathbb{Z}$ of all multiples of n is a subgroup of $\mathbb{Z}$ under addition, the cyclic subgroup generated by n. Theorem 6.6 shows that these cyclic subgroups are the only subgroups of $\mathbb{Z}$ under addition. We state this as a corollary.

6.7 Corollary The subgroups of $\mathbb{Z}$ under addition are precisely the groups $n\mathbb{Z}$ under addition for $n \in \mathbb{Z}$.

This corollary gives us an elegant way to define the *greatest common divisor* of two positive integers r and s. Exercise 45 shows that $H = \{nr + ms \mid n, m \in \mathbb{Z}\}$ is a subgroup of the group $\mathbb{Z}$ under addition. Thus H must be cyclic and have a generator d, which we may choose to be positive.

6.8 Definition Let r and s be two positive integers. The positive generator d of the cyclic group

$$H = \{nr + ms \mid n, m \in \mathbb{Z}\}$$

under addition is the **greatest common divisor** (abbreviated gcd) of r and s. We write $d = \gcd(r, s)$. ■

Note from the definition that d is a divisor of both r and s since both $r = 1r + 0s$ and $s = 0r + 1s$ are in H. Since $d \in H$, we can write

$$d = nr + ms$$

for some integers n and m. We see that every integer dividing both r and s divides the right-hand side of the equation, and hence must be a divisor of d also. Thus d must be the largest number dividing both r and s; this accounts for the name given to d in Definition 6.8.

6.9 Example Find the gcd of 42 and 72.

Solution The positive divisors of 42 are 1, 2, 3, 6, 7, 14, 21, and 42. The positive divisors of 72 are 1, 2, 3, 4, 6, 8, 9, 12, 18, 24, 36, and 72. The greatest common divisor is 6. Note that $6 = (3)(72) + (-5)(42)$. There is an algorithm for expressing the greatest common divisor d of r and s in the form $d = nr + ms$, but we will not need to make use of it here. ▲

Two positive integers are **relatively prime** if their gcd is 1. For example, 12 and 25 are relatively prime. Note that they have no prime factors in common. In our discussion of subgroups of cyclic groups, we will need to know the following:

If r and s are relatively prime and if r divides sm, then r must divide m.	(1)

Let's prove this. If r and s are relatively prime, then we may write

$$1 = ar + bs \qquad \text{for some} \qquad a, b \in \mathbb{Z}.$$

Multiplying by m, we obtain

$$m = arm + bsm.$$

Now r divides both arm and bsm since r divides sm. Thus r is a divisor of the right-hand side of this equation, so r must divide m.

The Structure of Cyclic Groups

We can now describe all cyclic groups, up to an isomorphism.

6.10 Theorem Let G be a cyclic group with generator a. If the order of G is infinite, then G is isomorphic to $\langle \mathbb{Z}, + \rangle$. If G has finite order n, then G is isomorphic to $\langle \mathbb{Z}_n, +_n \rangle$.

Proof **Case I** *For all positive integers m, $a^m \neq e$.* In this case we claim that no two distinct exponents h and k can give equal elements a^h and a^k of G. Suppose that $a^h = a^k$ and say $h > k$. Then

$$a^h a^{-k} = a^{h-k} = e,$$

contrary to our Case I assumption. Hence every element of G can be expressed as a^m for a unique $m \in \mathbb{Z}$. The map $\phi : G \to \mathbb{Z}$ given by $\phi(a^i) = i$ is thus well defined, one to one, and onto $\mathbb{Z}$. Also,

$$\phi(a^i a^j) = \phi(a^{i+j}) = i + j = \phi(a^i) + \phi(a^j),$$

so the homomorphism property is satisfied and ϕ is an isomorphism.

Case II *$a^m = e$ for some positive integer m.* Let n be the smallest positive integer such that $a^n = e$. If $s \in \mathbb{Z}$ and $s = nq + r$ for $0 \leq r < n$, then $a^s = a^{nq+r} = (a^n)^q a^r = e^q a^r = a^r$. As in Case 1, if $0 < k < h < n$ and $a^h = a^k$, then $a^{h-k} = e$ and $0 < h - k < n$, contradicting our choice of n. Thus the elements

$$a^0 = e, a, a^2, a^3, \cdots, a^{n-1}$$

are all distinct and comprise all elements of G. The map $\psi : G \to \mathbb{Z}_n$ given by $\psi(a^i) = i$ for $i = 0, 1, 2, \cdots, n-1$ is thus well defined, one to one, and onto $\mathbb{Z}_n$. Because $a^n = e$, we see that $a^i a^j = a^k$ where $k = i +_n j$. Thus

$$\psi(a^i a^j) = i +_n j = \psi(a^i) +_n \psi(a^j),$$

so the homomorphism property is satisfied and ψ is an isomorphism. ◆

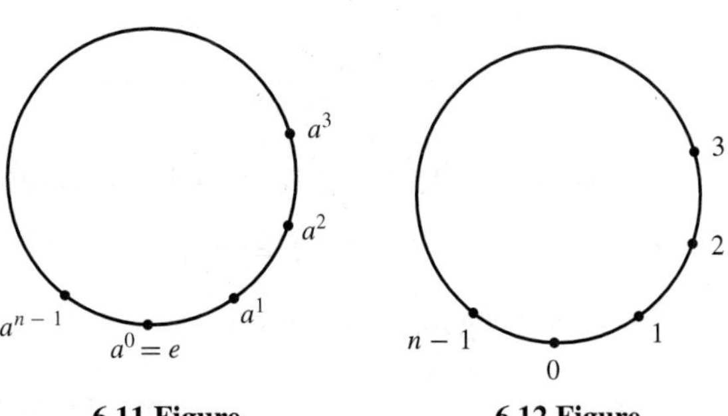

6.11 Figure **6.12 Figure**

6.13 Example Motivated by our work with U_n, it is nice to visualize the elements $e = a^0, a^1, a^2, \cdots, a^{n-1}$ of a cyclic group of order n as being distributed evenly on a circle (see Fig. 6.11). The element a^h is located h of these equal units counterclockwise along the circle, measured from the bottom where $e = a^0$ is located. To multiply a^h and a^k diagrammatically, we start from a^h and go k additional units around counterclockwise. To see arithmetically where we end up, find q and r such that

$$h + k = nq + r \qquad \text{for} \qquad 0 \leq r < n.$$

The nq takes us all the way around the circle q times, and we then wind up at a^r. ▲

Figure 6.12 is essentially the same as Fig. 6.11 but with the points labeled with the exponents on the generator. The operation on these exponents is *addition modulo n*.

Subgroups of Finite Cyclic Groups

We have completed our description of cyclic groups and turn to their subgroups. Corollary 6.7 gives us complete information about subgroups of infinite cyclic groups. Let us give the basic theorem regarding generators of subgroups for the finite cyclic groups.

6.14 Theorem Let G be a cyclic group with n elements and generated by a. Let $b \in G$ and let $b = a^s$. Then b generates a cyclic subgroup H of G containing n/d elements, where d is the greatest common divisor of n and s. Also, $\langle a^s \rangle = \langle a^t \rangle$ if and only if $\gcd(s, n) = gcd(t, n)$.

Proof That b generates a cyclic subgroup H of G is known from Theorem 5.17. We need show only that H has n/d elements. Following the argument of Case II of Theorem 6.10, we see that H has as many elements as the smallest positive power m of b that gives the identity. Now $b = a^s$, and $b^m = e$ if and only if $(a^s)^m = e$, or if and only if n divides ms. What is the smallest positive integer m such that n divides ms? Let d be the gcd of n and s. Then there exists integers u and v such that

$$d = un + vs.$$

Since d divides both n and s, we may write

$$1 = u(n/d) + v(s/d)$$

where both n/d and s/d are integers. This last equation shows that n/d and s/d are relatively prime, for any integer dividing both of them must also divide 1. We wish to find the smallest positive m such that

$$\frac{ms}{n} = \frac{m(s/d)}{(n/d)} \text{ is an integer.}$$

From the boxed division property (1), we conclude that n/d must divide m, so the smallest such m is n/d. Thus the order of H is n/d.

Taking for the moment $\mathbb{Z}_n$ as a model for a cyclic group of order n, we see that if d is a divisor of n, then the cyclic subgroup $\langle d \rangle$ of $\mathbb{Z}_n$ had n/d elements, and contains all the positive integers m less than n such that $\gcd(m, n) = d$. Thus there is only one subgroup of $\mathbb{Z}_n$ of order n/d. Taken with the preceding paragraph, this shows at once that if a is a generator of the cyclic group G, then $\langle a^s \rangle = \langle a^t \rangle$ if and only if $\gcd(s, n) = \gcd(t, n)$. ◆

6.15 Example For an example using additive notation, consider $\mathbb{Z}_{12}$, with the generator $a = 1$. Since the greatest common divisor of 3 and 12 is 3, $3 = 3 \cdot 1$ generates a subgroup of $\frac{12}{3} = 4$ elements, namely

$$\langle 3 \rangle = \{0, 3, 6, 9\}.$$

Since the gcd of 8 and 12 is 4, 8 generates a subgroup of $\frac{12}{4} = 3$ elements, namely,

$$\langle 8 \rangle = \{0, 4, 8\}.$$

Since the gcd of 12 and 5 is 1, 5 generates a subgroup of $\frac{12}{1} = 12$ elements; that is, 5 is a generator of the whole group $\mathbb{Z}_{12}$. ▲

The following corollary follows immediately from Theorem 6.14.

6.16 Corollary If a is a generator of a finite cyclic group G of order n, then the other generators of G are the elements of the form a^r, where r is relatively prime to n.

6.17 Example Let us find all subgroups of $\mathbb{Z}_{18}$ and give their subgroup diagram. All subgroups are cyclic. By Corollary 6.16, the elements 1, 5, 7, 11, 13, and 17 are all generators of $\mathbb{Z}_{18}$. Starting with 2,

$$\langle 2 \rangle = \{0, 2, 4, 6, 8, 10, 12, 14, 16\}.$$

is of order 9 and has as generators elements of the form $h2$, where h is relatively prime to 9, namely, $h = 1, 2, 4, 5, 7$, and 8, so $h2 = 2, 4, 8, 10, 14$, and 16. The element 6 of $\langle 2 \rangle$ generates $\{0, 6, 12\}$, and 12 also is a generator of this subgroup.

We have thus far found all subgroups generated by 0, 1, 2, 4, 5, 6, 7, 8, 10, 11, 12, 13, 14, 16, and 17. This leaves just 3, 9, and 15 to consider.

$$\langle 3 \rangle = \{0, 3, 6, 9, 12, 15\},$$

and 15 also generates this group of order 6, since $15 = 5 \cdot 3$, and the gcd of 5 and 6 is 1. Finally,

$$\langle 9 \rangle = \{0, 9\}.$$

The subgroup diagram for these subgroups of $\mathbb{Z}_{18}$ is given in Fig. 6.18.

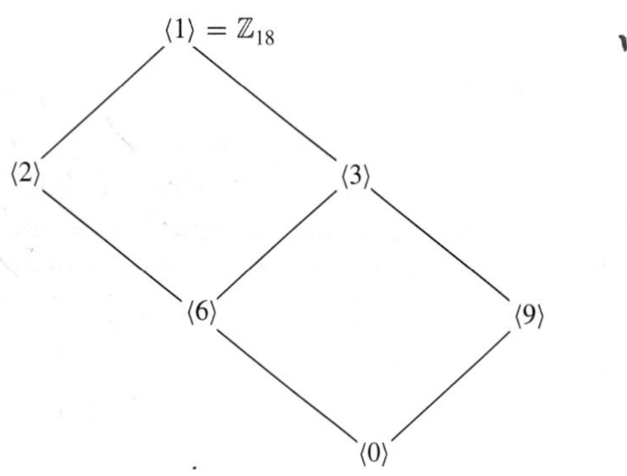

6.18 Figure Subgroup diagram for $\mathbb{Z}_{18}$.

This example is straightforward; we are afraid we wrote it out in such detail that it may look complicated. The exercises give some practice along these lines. ▲

EXERCISES 6

Computations

In Exercises 1 through 4, find the quotient and remainder, according to the division algorithm, when n is divided by m.

1. $n = 42, m = 9$

2. $n = -42, m = 9$

3. $n = -50, m = 8$

4. $n = 50, m = 8$

In Exercises 5 through 7, find the greatest common divisor of the two integers.

5. 32 and 24

6. 48 and 88

7. 360 and 420

In Exercises 8 through 11, find the number of generators of a cyclic group having the given order.

8. 5

9. 8

10. 12

11. 60

An isomorphism of a group with itself is an **automorphism of the group.** In Exercises 12 through 16, find the number of automorphisms of the given group.
[*Hint:* Make use of Exercise 44. What must be the image of a generator under an automorphism?]

12. $\mathbb{Z}_2$

13. $\mathbb{Z}_6$

14. $\mathbb{Z}_8$

15. $\mathbb{Z}$

16. $\mathbb{Z}_{12}$

In Exercises 17 through 21, find the number of elements in the indicated cyclic group.

17. The cyclic subgroup of $\mathbb{Z}_{30}$ generated by 25

18. The cyclic subgroup of $\mathbb{Z}_{42}$ generated by 30

19. The cyclic subgroup $\langle i \rangle$ of the group $\mathbb{C}^*$ of nonzero complex numbers under multiplication

20. The cyclic subgroup of the group $\mathbb{C}^*$ of Exercise 19 generated by $(1 + i)/\sqrt{2}$

21. The cyclic subgroup of the group $\mathbb{C}^*$ of Exercise 19 generated by $1 + i$

In Exercises 22 through 24, find all subgroups of the given group, and draw the subgroup diagram for the subgroups.

22. $\mathbb{Z}_{12}$

23. $\mathbb{Z}_{36}$

24. $\mathbb{Z}_8$

In Exercises 25 through 29, find all orders of subgroups of the given group.

25. $\mathbb{Z}_6$

26. $\mathbb{Z}_8$

27. $\mathbb{Z}_{12}$

28. $\mathbb{Z}_{20}$

29. $\mathbb{Z}_{17}$

Concepts

In Exercises 30 and 31, correct the definition of the italicized term without reference to the text, if correction is needed, so that it is in a form acceptable for publication.

30. An element a of a group G has *order* $n \in \mathbb{Z}^+$ if and only if $a^n = e$.

31. The *greatest common divisor* of two positive integers is the largest positive integer that divides both of them.

32. Mark each of the following true or false.

_____ **a.** Every cyclic group is abelian.
_____ **b.** Every abelian group is cyclic.
_____ **c.** $\mathbb{Q}$ under addition is a cyclic group.
_____ **d.** Every element of every cyclic group generates the group.
_____ **e.** There is at least one abelian group of every finite order >0.
_____ **f.** Every group of order ≤ 4 is cyclic.

______ **g.** All generators of $\mathbb{Z}_{20}$ are prime numbers.

______ **h.** If G and G' are groups, then $G \cap G'$ is a group.

______ **i.** If H and K are subgroups of a group G, then $H \cap K$ is a group.

______ **j.** Every cyclic group of order >2 has at least two distinct generators.

In Exercises 33 through 37, either give an example of a group with the property described, or explain why no example exists.

33. A finite group that is not cyclic

34. An infinite group that is not cyclic

35. A cyclic group having only one generator

36. An infinite cyclic group having four generators

37. A finite cyclic group having four generators

The generators of the cyclic multiplicative group U_n of all nth roots of unity in $\mathbb{C}$ are the **primitive** nth **roots of unity**. In Exercises 38 through 41, find the primitive nth roots of unity for the given value of n.

38. $n = 4$

39. $n = 6$

40. $n = 8$

41. $n = 12$

Proof Synopsis

42. Give a one-sentence synopsis of the proof of Theorem 6.1.

43. Give at most a three-sentence synopsis of the proof of Theorem 6.6.

Theory

44. Let G be a cyclic group with generator a, and let G' be a group isomorphic to G. If $\phi : G \to G'$ is an isomorphism, show that, for every $x \in G$, $\phi(x)$ is completely determined by the value $\phi(a)$. That is, if $\phi : G \to G'$ and $\psi : G \to G'$ are two isomophisms such that $\phi(a) = \psi(a)$, then $\phi(x) = \psi(x)$ for all $x \in G$.

45. Let r and s be positive integers. Show that $\{nr + ms \mid n, m \in \mathbb{Z}\}$ is a subgroup of $\mathbb{Z}$.

46. Let a and b be elements of a group G. Show that if ab has finite order n, then ba also has order n.

47. Let r and s be positive integers.

a. Define the **least common multiple** of r and s as a generator of a certain cyclic group.

b. Under what condition is the least common multiple of r and s their product, rs?

c. Generalizing part (b), show that the product of the greatest common divisor and of the least common multiple of r and s is rs.

48. Show that a group that has only a finite number of subgroups must be a finite group.

49. Show by a counterexample that the following "converse" of Theorem 6.6 is not a theorem: "If a group G is such that every proper subgroup is cyclic, then G is cyclic."

50. Let G be a group and suppose $a \in G$ generates a cyclic subgroup of order 2 and is the *unique* such element. Show that $ax = xa$ for all $x \in G$. [*Hint:* Consider $(xax^{-1})^2$.]

51. Let p and q be distinct prime numbers. Find the number of generators of the cyclic group $\mathbb{Z}_{pq}$.

52. Let p be a prime number. Find the number of generators of the cyclic group $\mathbb{Z}_{p^r}$, where r is an integer ≥ 1.

53. Show that in a finite cyclic group G of order n, written multiplicatively, the equation $x^m = e$ has exactly m solutions x in G for each positive integer m that divides n.

54. With reference to Exercise 53, what is the situation if $1 < m < n$ and m does not divide n?

55. Show that $\mathbb{Z}_p$ has no proper nontrivial subgroups if p is a prime number.

56. Let G be an abelian group and let H and K be finite cyclic subgroups with $|H| = r$ and $|K| = s$.

a. Show that if r and s are relatively prime, then G contains a cyclic subgroup of order rs.

b. Generalizing part (a), show that G contains a cyclic subgroup of order the least common multiple of r and s.

SECTION 7 GENERATING SETS AND CAYLEY DIGRAPHS

Let G be a group, and let $a \in G$. We have described the cyclic subgroup $\langle a \rangle$ of G, which is the smallest subgroup of G that contains the element a. Suppose we want to find as small a subgroup as possible that contains both a and b for another element b in G. By Theorem 5.17, we see that any subgroup containing a and b must contain a^n and b^m for all $m, n \in \mathbb{Z}$, and consequently must contain all finite products of such powers of a and b. For example, such an expression might be $a^2b^4a^{-3}b^2a^5$. Note that we cannot "simplify" this expression by writing first all powers of a followed by the powers of b, since G may not be abelian. However, products of such expressions are again expressions of the same type. Furthermore, $e = a^0$ and the inverse of such an expression is again of the same type. For example, the inverse of $a^2b^4a^{-3}b^2a^5$ is $a^{-5}b^{-2}a^3b^{-4}a^{-2}$. By Theorem 5.14, this shows that all such products of integral powers of a and b form a subgroup of G, which surely must be the smallest subgroup containing both a and b. We call a and b **generators** of this subgroup. If this subgroup should be all of G, then we say that $\{a, b\}$ **generates** G. Of course, there is nothing sacred about taking just two elements $a, b \in G$. We could have made similar arguments for three, four, or any number of elements of G, as long as we take only finite products of their integral powers.

7.1 Example The Klein 4-group $V = \{e, a, b, c\}$ of Example 5.9 is generated by $\{a, b\}$ since $ab = c$. It is also generated by $\{a, c\}$, $\{b, c\}$, and $\{a, b, c\}$. If a group G is generated by a subset S, then every subset of G containing S generates G. ▲

7.2 Example The group $\mathbb{Z}_6$ is generated by $\{1\}$ and $\{5\}$. It is also generated by $\{2, 3\}$ since $2 + 3 = 5$, so that any subgroup containing 2 and 3 must contain 5 and must therefore be $\mathbb{Z}_6$. It is also generated by $\{3, 4\}$, $\{2, 3, 4\}$, $\{1, 3\}$, and $\{3, 5\}$, but it is not generated by $\{2, 4\}$ since $\langle 2 \rangle = \{0, 2, 4\}$ contains 2 and 4. ▲

We have given an intuitive explanation of the subgroup of a group G generated by a subset of G. What follows is a detailed exposition of the same idea approached in another way, namely via intersections of subgroups. After we get an intuitive grasp of a concept, it is nice to try to write it up as neatly as possible. We give a set-theoretic definition and generalize a theorem that was in Exercise 54 of Section 5.

7.3 Definition Let $\{S_i \mid i \in I\}$ be a collection of sets. Here I may be any set of indices. The **intersection** $\bigcap_{i\in I} S_i$ **of the sets** S_i is the set of all elements that are in all the sets S_i; that is,

$$\bigcap_{i\in I} S_i = \{x \mid x \in S_i \text{ for all } i \in I\}.$$

If I is finite, $I = \{1, 2, \ldots, n\}$, we may denote $\bigcap_{i\in I} S_i$ by

$$S_1 \cap S_2 \cap \cdots \cap S_n. \quad \blacksquare$$

7.4 Theorem The intersection of some subgroups H_i of a group G for $i \in I$ is again a subgroup of G.

Proof Let us show closure. Let $a \in \bigcap_{i\in I} H_i$ and $b \in \bigcap_{i\in I} H_i$, so that $a \in H_i$ for all $i \in I$ and $b \in H_i$ for all $i \in I$. Then $ab \in H_i$ for all $i \in I$, since H_i is a group. Thus $ab \in \bigcap_{i\in I} H_i$.

Since H_i is a subgroup for all $i \in I$, we have $e \in H_i$ for all $i \in I$, and hence $e \in \bigcap_{i\in I} H_i$.

Finally, for $a \in \bigcap_{i\in I} H_i$, we have $a \in H_i$ for all $i \in I$, so $a^{-1} \in H_i$ for all $i \in I$, which implies that $a^{-1} \in \bigcap_{i\in I} H_i$. ◆

Let G be a group and let $a_i \in G$ for $i \in I$. There is at least one subgroup of G containing all the elements a_i for $i \in I$, namely G is itself. Theorem 7.4 assures us that if we take the intersection of all subgroups of G containing all a_i for $i \in I$, we will obtain a subgroup H of G. This subgroup H is the smallest subgroup of G containing all the a_i for $i \in I$.

7.5 Definition Let G be a group and let $a_i \in G$ for $i \in I$. The smallest subgroup of G containing $\{a_i \mid i \in I\}$ is the **subgroup generated by** $\{a_i \mid i \in I\}$. If this subgroup is all of G, then $\{a_i \mid i \in I\}$ **generates** G and the a_i are **generators of** G. If there is a finite set $\{a_i \mid i \in I\}$ that generates G, then G is **finitely generated.** ■

Note that this definition is consistent with our previous definition of a generator for a cyclic group. Note also that the statement *a is a generator of G* may mean either that $G = \langle a \rangle$ or that a is a member of a subset of G that generates G. The context in which the statement is made should indicate which is intended. Our next theorem gives the structural insight into the subgroup of G generated by $\{a_i \mid i \in I\}$ that we discussed for two generators before Example 7.1.

7.6 Theorem If G is a group and $a_i \in G$ for $i \in I$, then the subgroup H of G generated by $\{a_i \mid i \in I\}$ has as elements precisely those elements of G that are finite products of integral powers of the a_i, where powers of a fixed a_i may occur several times in the product.

Proof Let K denote the set of all finite products of integral powers of the a_i. Then $K \subseteq H$. We need only observe that K is a subgroup and then, since H is the smallest subgroup containing a_i for $i \in I$, we will be done. Observe that a product of elements in K is again in K. Since $(a_i)^0 = e$, we have $e \in K$. For every element k in K, if we form from the product giving k a new product with the order of the a_i reversed and the opposite sign on all exponents, we have k^{-1}, which is thus in K. For example,

$$[(a_1)^3(a_2)^2(a_1)^{-7}]^{-1} = (a_1)^7(a_2)^{-2}(a_1)^{-3},$$

which is again in K. ◆

Cayley Digraphs

For each generating set S of a finite group G, there is a directed graph representing the group in terms of the generators in S. The term *directed graph* is usually abbreviated as *digraph*. These visual representations of groups were devised by Cayley, and are also referred to as *Cayley diagrams* in the literature.

Intuitively, a **digraph** consists of a finite number of points, called **vertices** of the digraph, and some **arcs** (each with a direction denoted by an arrowhead) joining vertices. In a digraph for a group G using a generating set S we have one vertex, represented by a dot, for each element of G. Each generator in S is denoted by one type of arc. We could use different colors for different arc types in pencil and paperwork. Since different colors are not available in our text, we use different style arcs, like solid, dashed, and dotted, to denote different generators. Thus if $S = \{a, b, c\}$ we might denote

a by ——➤——, b by - - -➤- - -, and c by ·······➤·······.

With this notation, an occurrence of x•——➤——•y in a Cayley digraph means that $xa = y$. That is, traveling an arc in the direction of the arrow indicates that multiplication of the group element at the start of the arc *on the right* by the generator corresponding to that type of arc yields the group element at the end of the arc. Of course, since we are in a group, we know immediately that $ya^{-1} = x$. Thus traveling an arc in the direction opposite to the arrow corresponds to multiplication on the right by the inverse of the corresponding generator. If a generator in S is its own inverse, it is customary to denote this by omitting the arrowhead from the arc, rather than using a double arrow. For example, if $b^2 = e$, we might denote b by - - - - - - - -.

7.7 Example Both of the digraphs shown in Fig. 7.8 represent the group $\mathbb{Z}_6$ with generating set $S = \{1\}$. Neither the length and shape of an arc nor the angle between arcs has any significance. ▲

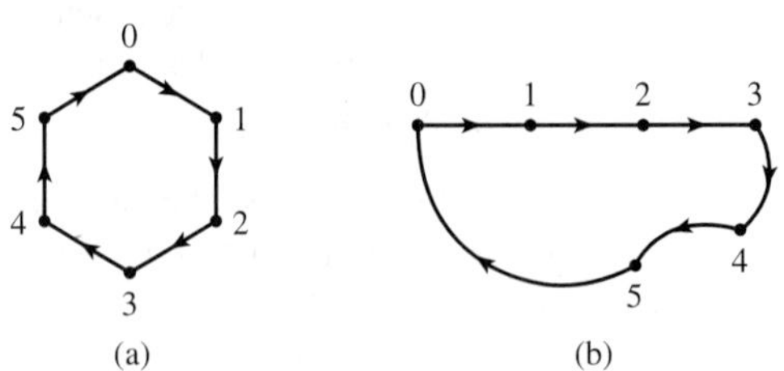

7.8 Figure Two digraphs for $\mathbb{Z}_6$ with $S = \{1\}$ using ——➤—— (1).

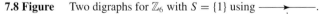

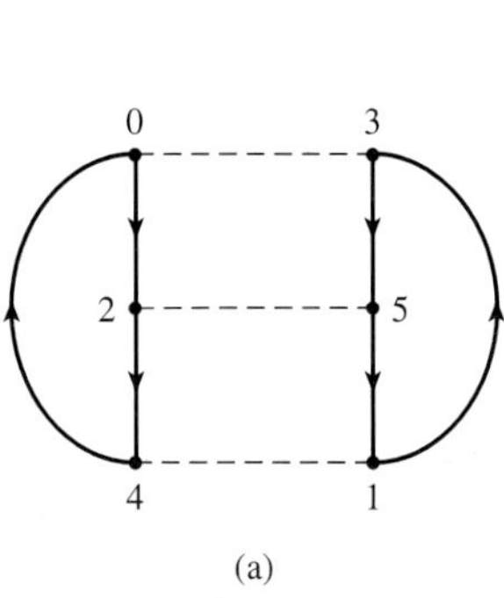

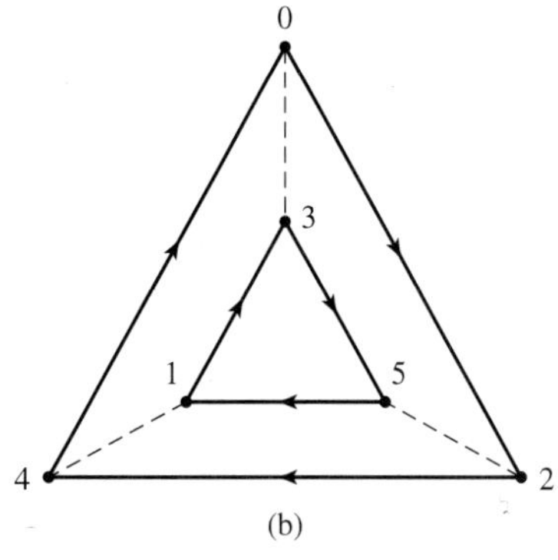

7.9 Figure Two digraphs for $\mathbb{Z}_6$ with $S = \{2, 3\}$ using $\xrightarrow[2]{}$ and - - - - - - - (3).

7.10 Example Both of the digraphs shown in Fig. 7.9 represent the group $\mathbb{Z}_6$ with generating set $S = \{2, 3\}$. Since 3 is its own inverse, there is no arrowhead on the dashed arcs representing 3. Notice how different these Cayley diagrams look from those in Fig. 7.8 for the same group. The difference is due to the different choice for the set of generators. ▲

Every digraph for a group must satisfy these four properties for the reasons indicated.

Property	***Reason***
1. The digraph is connected, that is, we can get from any vertex g to any vertex h by traveling along consecutive arcs, starting at g and ending at h.	Every equation $gx = h$ has a solution in a group.
2. At most one arc goes from a vertex g to a vertex h.	The solution of $gx = h$ is unique.
3. Each vertex g has exactly one arc of each type starting at g, and one of each type ending at g.	For $g \in G$ and each generator b we can compute gb, and $(gb^{-1})b = g$.
4. If two different sequences of arc types starting from vertex g lead to the same vertex h, then those same sequences of arc types starting from any vertex u will lead to the same vertex v.	If $gq = h$ and $gr = h$, then $uq = ug^{-1}h = ur$.

It can be shown that, conversely, every digraph satisfying these four properties is a Cayley digraph for some group. Due to the symmetry of such a digraph, we can choose labels like a, b, c for the various arc types, name any vertex e to represent the identity, and name each other vertex by a product of arc labels and their inverses that we can travel to attain that vertex starting from the one that we named e. Some finite groups were first constructed (found) using digraphs.

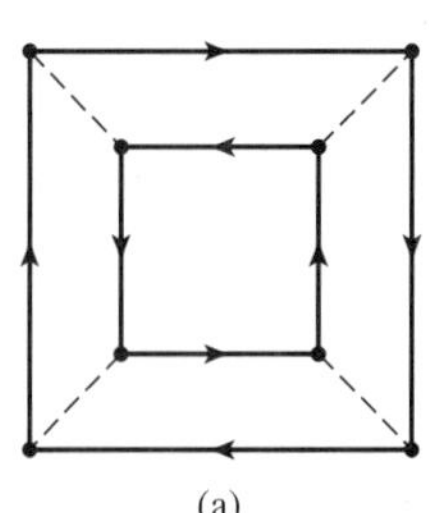

(a)

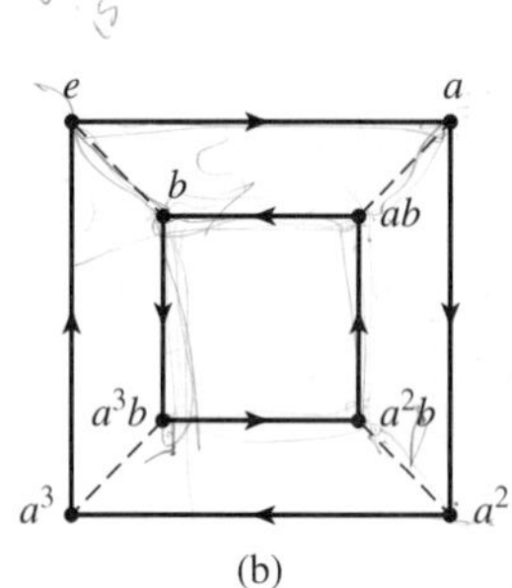

(b)

7.11 Figure

7.12 Example A digraph satisfying the four properties on page 71 is shown in Fig. 7.11 (a). To obtain Fig. 7.11 (b), we selected the labels

$$\xrightarrow[a]{} \quad \text{and} \quad \text{-----}_{b} \text{---},$$

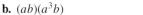

named a vertex e, and then named the other vertices as shown. We have a group $\{e, a, a^2, a^3, b, ab, a^2b, a^3b\}$ of eight elements. Note that the vertex that we named ab could equally well be named ba^{-1}, the vertex that we named a^3 could be named a^{-1}, etc. It is not hard to compute products of elements in this group. To compute $(a^3b)(a^2b)$, we just start at the vertex labeled a^3b and then travel in succession two solid arcs and one dashed arc, arriving at the vertex a, so $(a^3b)(a^2b) = a$. In this fashion, we could write out the table for this eight-element group. ▲

EXERCISES 7

Computations

In Exercises 1 through 6, list the elements of the subgroup generated by the given subset.

1. The subset $\{2, 3\}$ of $\mathbb{Z}_{12}$
2. The subset $\{4, 6\}$ of $\mathbb{Z}_{12}$
3. The subset $\{8, 10\}$ of $\mathbb{Z}_{18}$
4. The subset $\{12, 30\}$ of $\mathbb{Z}_{36}$
5. The subset $\{12, 42\}$ of $\mathbb{Z}$
6. The subset $\{18, 24, 39\}$ of $\mathbb{Z}$
7. For the group described in Example 7.12 compute these products, using Fig. 7.11(b).

   a. $(a^2b)a^3$ **b.** $(ab)(a^3b)$ **c.** $b(a^2b)$

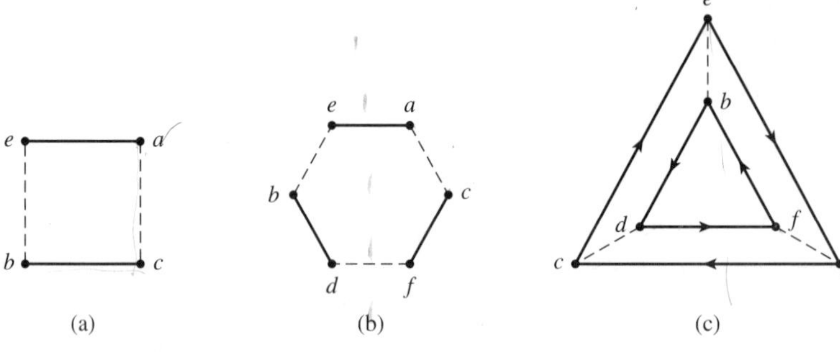

7.13 Figure

In Exercises 8 through 10, give the table for the group having the indicated digraph. In each digraph, take e as identity element. List the identity e first in your table, and list the remaining elements alphabetically, so that your answers will be easy to check.

8. The digraph in Fig. 7.13(a)

9. The digraph in Fig. 7.13(b)

10. The digraph in Fig. 7.13(c)

Concepts

11. How can we tell from a Cayley digraph whether or not the corresponding group is commutative?

12. Referring to Exercise 11, determine whether the group corresponding to the Cayley digraph in Fig. 7.11(b) is commutative.

13. Is it obvious from a Cayley digraph of a group whether or not the group is cyclic? [*Hint:* Look at Fig. 7.9(b).]

14. The large outside triangle in Fig. 7.9(b) exhibits the cyclic subgroup $\{0, 2, 4\}$ of $\mathbb{Z}_6$. Does the smaller inside triangle similarly exhibit a cyclic subgroup of $\mathbb{Z}_6$? Why or why not?

15. The generating set $S = \{1, 2\}$ for $\mathbb{Z}_6$ contains more generators than necessary, since 1 is a generator for the group. Nevertheless, we can draw a Cayley digraph for $\mathbb{Z}_6$ with this generating set S. Draw such a Cayley digraph.

16. Draw a Cayley digraph for $\mathbb{Z}_8$ taking as generating set $S = \{2, 5\}$.

17. A **relation** on a set S of generators of a group G is an equation that equates some product of generators and their inverses to the identity e of G. For example, if $S = \{a, b\}$ and G is commutative so that $ab = ba$, then one relation is $aba^{-1}b^{-1} = e$. If, moreover, b is its own inverse, then another relation is $b^2 = e$.

   a. Explain how we can find some relations on S from a Cayley digraph of G.

   b. Find three relations on the set $S = \{a, b\}$ of generators for the group described by Fig. 7.11(b).

18. Draw digraphs of the two possible structurally different groups of order 4, taking as small a generating set as possible in each case. You need not label vertices.

Theory

19. Show that for $n \geq 3$, there exists a nonabelian group with $2n$ elements that is generated by two elements of order 2.

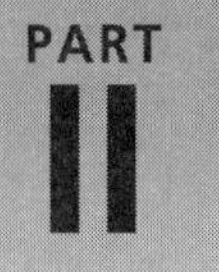

PART II

Permutations, Cosets, and Direct Products

SECTION 8 Groups of Permutations

We have seen examples of groups of numbers, like the groups $\mathbb{Z}$, $\mathbb{Q}$, and $\mathbb{R}$ under addition. We have also introduced groups of matrices, like the group $GL(2, \mathbb{R})$. Each element A of $GL(2, \mathbb{R})$ yields a transformation of the plane $\mathbb{R}^2$ into itself; namely, if we regard $\mathbf{x}$ as a 2-component column vector, then $A\mathbf{x}$ is also a 2-component column vector. The group $GL(2, \mathbb{R})$ is typical of many of the most useful groups in that its elements *act on things* to transform them. Often, an action produced by a group element can be regarded as a *function*, and the binary operation of the group can be regarded as *function composition*. In this section, we construct some finite groups whose elements, called *permutations*, act on finite sets. These groups will provide us with examples of finite nonabelian groups. We shall show that any finite group is structurally the same as some group of permutations. Unfortunately, this result, which sounds very powerful, does not turn out to be particularly useful to us.

You may be familiar with the notion of a permutation of a set as a rearrangement of the elements of the set. Thus for the set $\{1, 2, 3, 4, 5\}$, a rearrangement of the elements could be given schematically as in Fig. 8.1, resulting in the new arrangement $\{4, 2, 5, 3, 1\}$. Let us think of this schematic diagram in Fig. 8.1 as a function mapping of each element listed in the left column into a single (not necessarily different) element from the same set listed at the right. Thus 1 is carried into 4, 2 is mapped into 2, and so on. Furthermore, to be a permutation of the set, this mapping must be such that each element appears in the right column once and only once. For example, the diagram in Fig. 8.2 does *not* give a permutation, for 3 appears twice while 1 does not appear at all in the right column. We now define a permutation to be such a mapping.

† Section 12 is not used in the remainder of the text.

8.1 Figure	8.2 Figure
$1 \to 4$	$1 \to 3$
$2 \to 2$	$2 \to 2$
$3 \to 5$	$3 \to 4$
$4 \to 3$	$4 \to 5$
$5 \to 1$	$5 \to 3$

8.1 Figure **8.2 Figure**

8.3 Definition A **permutation of a set** A is a function $\phi : A \to A$ that is both one to one and onto. ■

Permutation Groups

We now show that function composition $\circ$ is a binary operation on the collection of all permutations of a set A. We call this operation *permutation multiplication*. Let A be a set, and let σ and τ be permutations of A so that σ and τ are both one-to-one functions mapping A onto A. The composite function $\sigma \circ \tau$ defined schematically by

$$A \xrightarrow{\tau} A \xrightarrow{\sigma} A,$$

gives a mapping of A into A. Rather than keep the symbol $\circ$ for permutation multiplication, we will denote $\sigma \circ \tau$ by the juxtaposition $\sigma\tau$, as we have done for general groups. Now $\sigma\tau$ will be a permutation if it is one to one and onto A. *Remember that the action of $\sigma\tau$ on A must be read in right-to-left order: first apply τ and then σ.* Let us show that $\sigma\tau$ is one to one. If

$$(\sigma\tau)(a_1) = (\sigma\tau)(a_2),$$

then

$$\sigma(\tau(a_1)) = \sigma(\tau(a_2)),$$

and since σ is given to be one to one, we know that $\tau(a_1) = \tau(a_2)$. But then, since τ is one to one, this gives $a_1 = a_2$. Hence $\sigma\tau$ is one to one. To show that $\sigma\tau$ is onto A, let $a \in A$. Since σ is onto A, there exists $a' \in A$ such that $\sigma(a') = a$. Since τ is onto A, there exists $a'' \in A$ such that $\tau(a'') = a'$. Thus

$$a = \sigma(a') = \sigma(\tau(a'')) = (\sigma\tau)(a''),$$

so $\sigma\tau$ is onto A.

8.4 Example Suppose that

$$A = \{1, 2, 3, 4, 5\}$$

and that σ is the permutation given by Fig. 8.1. We write σ in a more standard notation, changing the columns to rows in parentheses and omitting the arrows, as

$$\sigma = \begin{pmatrix} 1 & 2 & 3 & 4 & 5 \\ 4 & 2 & 5 & 3 & 1 \end{pmatrix},$$

■ Historical Note

One of the earliest recorded studies of permutations occurs in the *Sefer Yetsirah,* or *Book of Creation,* written by an unknown Jewish author sometime before the eighth century. The author was interested in counting the various ways in which the letters of the Hebrew alphabet can be arranged. The question was in some sense a mystical one. It was believed that the letters had magical powers; therefore, suitable arrangements could subjugate the forces of nature. The actual text of the *Sefer Yetsirah* is very sparse: "Two letters build two words, three build six words, four build 24 words, five build 120, six build 720, seven build 5040." Interestingly enough, the idea of counting the arrangements of the letters of the alphabet also occurred in Islamic mathematics in the eighth and ninth centuries. By the thirteenth century, in both the Islamic and Hebrew cultures, the abstract idea of a permutation had taken root so that both Abu-l-' Abbas ibn al-Banna (1256–1321), a mathematician from Marrakech in what is now Morocco, and Levi ben Gerson, a French rabbi, philosopher, and mathematician, were able to give rigorous proofs that the number of permutations of any set of n elements is $n!$, as well as prove various results about counting combinations.

Levi and his predecessors, however, were concerned with permutations as simply arrangements of a given finite set. It was the search for solutions of polynomial equations that led Lagrange and others in the late eighteenth century to think of permutations as functions from a finite set to itself, the set being that of the roots of a given equation. And it was Augustin-Louis Cauchy (1789–1857) who developed in detail the basic theorems of permutation theory and who introduced the standard notation used in this text.

so that $\sigma(1) = 4$, $\sigma(2) = 2$, and so on. Let

$$\tau = \begin{pmatrix} 1 & 2 & 3 & 4 & 5 \\ 3 & 5 & 4 & 2 & 1 \end{pmatrix}.$$

Then

$$\sigma\tau = \begin{pmatrix} 1 & 2 & 3 & 4 & 5 \\ 4 & 2 & 5 & 3 & 1 \end{pmatrix}\begin{pmatrix} 1 & 2 & 3 & 4 & 5 \\ 3 & 5 & 4 & 2 & 1 \end{pmatrix} = \begin{pmatrix} 1 & 2 & 3 & 4 & 5 \\ 5 & 1 & 3 & 2 & 4 \end{pmatrix}.$$

For example, multiplying in right-to-left order,

$$(\sigma\tau)(1) = \sigma(\tau(1)) = \sigma(3) = 5.$$ ▲

We now show that the collection of all permutations of a nonempty set A forms a group under this permutation multiplication.

8.5 Theorem Let A be a nonempty set, and let S_A be the collection of all permutations of A. Then S_A is a group under permutation multiplication.

Proof We have shown that composition of two permutations of A yields a permutation of A, so S_A is closed under permutation multiplication.

Now permutation multiplication is defined as function composition, and in Section 2, we showed that *function composition is associative*. Hence $\mathscr{G}_1$ is satisfied.

The permutation ι such that $\iota(a) = a$, for all $a \in A$ acts as identity. Therefore $\mathscr{G}_2$ is satisfied.

For a permutation σ, the inverse function, σ^{-1}, is the permutation that reverses the direction of the mapping σ, that is, $\sigma^{-1}(a)$ is the element a' of A such that $a = \sigma(a')$. The existence of exactly one such element a' is a consequence of the fact that, as a function, σ is both one to one and onto. For each $a \in A$ we have

$$\iota(a) = a = \sigma(a') = \sigma(\sigma^{-1}(a)) = (\sigma\sigma^{-1})(a)$$

and also

$$\iota(a') = a' = \sigma^{-1}(a) = \sigma^{-1}(\sigma(a')) = (\sigma^{-1}\sigma)(a'),$$

so that $\sigma^{-1}\sigma$ and $\sigma\sigma^{-1}$ are both the permutation ι. Thus $\mathcal{G}_3$ is satisfied. ◆

Warning: Some texts compute a product $\sigma\mu$ of permutations in left-to-right order, so that $(\sigma\mu)(a) = \mu(\sigma(a))$. Thus the permutation they get for $\sigma\mu$ is the one we would get by computing $\mu\sigma$. Exercise 51 asks us to check in two ways that we still get a group. If you refer to another text on this material, be sure to check its order for permutation multiplication.

There was nothing in our definition of a permutation to require that the set A be finite. However, most of our examples of permutation groups will be concerned with permutations of finite sets. Note that the *structure* of the group S_A is concerned only with the number of elements in the set A, and not what the elements in A are. If sets A and B have the same cardinality, then $S_A \simeq S_B$. To define an isomorphism $\phi : S_A \to S_B$, we let $f : A \to B$ be a one-to-one function mapping A onto B, which establishes that A and B have the same cardinality. For $\sigma \in S_A$, we let $\phi(\sigma)$ be the permutation $\bar{\sigma} \in S_B$ such that $\bar{\sigma}(f(a)) = f(\sigma(a))$ for all $a \in A$. To illustrate this for $A = \{1, 2, 3\}$ and $B = \{\#, \$, \%\}$ and the function $f : A \to B$ defined as

$$f(1) = \#, \quad f(2) = \$, \quad f(3) = \%,$$

ϕ maps

$$\begin{pmatrix} 1 & 2 & 3 \\ 3 & 2 & 1 \end{pmatrix} \text{ into } \begin{pmatrix} \# & \$ & \% \\ \% & \$ & \# \end{pmatrix}.$$

We simply rename the elements of A in our two-row notation by elements in B using the renaming function f, thus renaming elements of S_A to be those of S_B. We can take $\{1, 2, 3, \cdots, n\}$ to be a prototype for a finite set A of n elements.

8.6 Definition Let A be the finite set $\{1, 2, \cdots, n\}$. The group of all permutations of A is the **symmetric group on n letters**, and is denoted by S_n. ■

Note that S_n has $n!$ elements, where

$$n! = n(n-1)(n-2)\cdots(3)(2)(1).$$

Two Important Examples

8.7 Example An interesting example for us is the group S_3 of $3! = 6$ elements. Let the set A be $\{1, 2, 3\}$. We list the permutations of A and assign to each a subscripted Greek letter for a name.

The reasons for the choice of names will be clear later. Let

$$\rho_0 = \begin{pmatrix} 1 & 2 & 3 \\ 1 & 2 & 3 \end{pmatrix}, \qquad \mu_1 = \begin{pmatrix} 1 & 2 & 3 \\ 1 & 3 & 2 \end{pmatrix},$$

$$\rho_1 = \begin{pmatrix} 1 & 2 & 3 \\ 2 & 3 & 1 \end{pmatrix}, \qquad \mu_2 = \begin{pmatrix} 1 & 2 & 3 \\ 3 & 2 & 1 \end{pmatrix},$$

$$\rho_2 = \begin{pmatrix} 1 & 2 & 3 \\ 3 & 1 & 2 \end{pmatrix}, \qquad \mu_3 = \begin{pmatrix} 1 & 2 & 3 \\ 2 & 1 & 3 \end{pmatrix}.$$

8.8 Table

	ρ_0	ρ_1	ρ_2	μ_1	μ_2	μ_3
ρ_0	ρ_0	ρ_1	ρ_2	μ_1	μ_2	μ_3
ρ_1	ρ_1	ρ_2	ρ_0	μ_3	μ_1	μ_2
ρ_2	ρ_2	ρ_0	ρ_1	μ_2	μ_3	μ_1
μ_1	μ_1	μ_2	μ_3	ρ_0	ρ_1	ρ_2
μ_2	μ_2	μ_3	μ_1	ρ_2	ρ_0	ρ_1
μ_3	μ_3	μ_1	μ_2	ρ_1	ρ_2	ρ_0

The multiplication table for S_3 is shown in Table 8.8. Note that this group is not abelian! We have seen that any group of at most 4 elements is abelian. Later we will see that a group of 5 elements is also abelian. Thus S_3 has minimum order for any nonabelian group. ▲

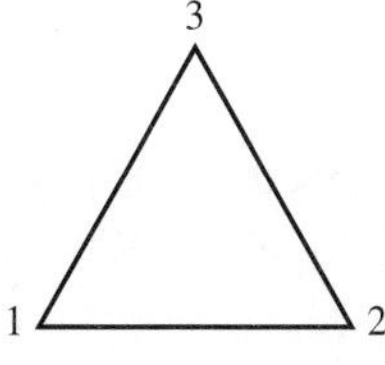

8.9 Figure

There is a natural correspondence between the elements of S_3 in Example 8.7 and the ways in which two copies of an equilateral triangle with vertices 1, 2, and 3 (see Fig. 8.9 can be placed, one covering the other with vertices on top of vertices. For this reason, S_3 is also the **group D_3 of symmetries of an equilateral triangle**. Naively, we used ρ_i for *rotations* and μ_i for *mirror images* in bisectors of angles. The notation D_3 stands for the third dihedral group. The **nth dihedral group D_n** is the group of symmetries of the regular n-gon. See Exercise 44.†

Note that we can consider the elements of S_3 to *act* on the triangle in Fig. 8.9. See the discussion at the start of this section.

8.10 Example Let us form the dihedral group D_4 of permutations corresponding to the ways that two copies of a square with vertices 1, 2, 3, and 4 can be placed, one covering the other with vertices on top of vertices (see Fig. 8.11). D_4 will then be the **group of symmetries of the square**. It is also called the **octic group**. Again, we choose seemingly arbitrary

† Many people denote the nth dihedral group by D_{2n} rather than by D_n since the order of the group is $2n$.

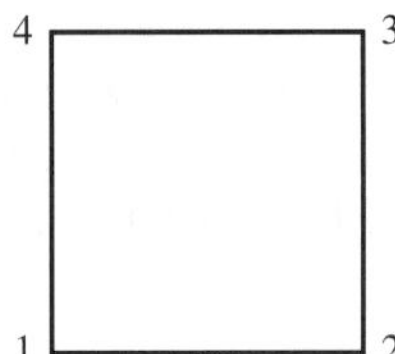

8.11 Figure

notation that we shall explain later. Naively, we are using ρ_i for *rotations*, μ_i for *mirror images* in perpendicular bisectors of sides, and δ_i for *diagonal flips*. There are eight permutations involved here. Let

$$\rho_0 = \begin{pmatrix} 1 & 2 & 3 & 4 \\ 1 & 2 & 3 & 4 \end{pmatrix}, \qquad \mu_1 = \begin{pmatrix} 1 & 2 & 3 & 4 \\ 2 & 1 & 4 & 3 \end{pmatrix},$$

$$\rho_1 = \begin{pmatrix} 1 & 2 & 3 & 4 \\ 2 & 3 & 4 & 1 \end{pmatrix}, \qquad \mu_2 = \begin{pmatrix} 1 & 2 & 3 & 4 \\ 4 & 3 & 2 & 1 \end{pmatrix},$$

$$\rho_2 = \begin{pmatrix} 1 & 2 & 3 & 4 \\ 3 & 4 & 1 & 2 \end{pmatrix}, \qquad \delta_1 = \begin{pmatrix} 1 & 2 & 3 & 4 \\ 3 & 2 & 1 & 4 \end{pmatrix},$$

$$\rho_3 = \begin{pmatrix} 1 & 2 & 3 & 4 \\ 4 & 1 & 2 & 3 \end{pmatrix}, \qquad \delta_2 = \begin{pmatrix} 1 & 2 & 3 & 4 \\ 1 & 4 & 3 & 2 \end{pmatrix}.$$

8.12 Table

	ρ_0	ρ_1	ρ_2	ρ_3	μ_1	μ_2	δ_1	δ_2
ρ_0	ρ_0	ρ_1	ρ_2	ρ_3	μ_1	μ_2	δ_1	δ_2
ρ_1	ρ_1	ρ_2	ρ_3	ρ_0	δ_1	δ_2	μ_2	μ_1
ρ_2	ρ_2	ρ_3	ρ_0	ρ_1	μ_2	μ_1	δ_2	δ_1
ρ_3	ρ_3	ρ_0	ρ_1	ρ_2	δ_2	δ_1	μ_1	μ_2
μ_1	μ_1	δ_2	μ_2	δ_1	ρ_0	ρ_2	ρ_3	ρ_1
μ_2	μ_2	δ_1	μ_1	δ_2	ρ_2	ρ_0	ρ_1	ρ_3
δ_1	δ_1	μ_1	δ_2	μ_2	ρ_1	ρ_3	ρ_0	ρ_2
δ_2	δ_2	μ_2	δ_1	μ_1	ρ_3	ρ_1	ρ_2	ρ_0

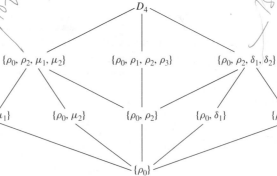

8.13 Figure Subgroup diagram for D_4.

The table for D_4 is given in Table 8.12. Note that D_4 is again nonabelian. This group is simply beautiful. It will provide us with nice examples for many concepts we will introduce in group theory. Look at the lovely symmetries in that table! Finally, we give in Fig. 8.13 the subgroup diagram for the subgroups of D_4. Look at the lovely symmetries in that diagram! ▲

Cayley's Theorem

Look at any group table in the text. Note how each row of the table gives a permutation of the set of elements of the group, as listed at the top of the table. Similarly, each column of the table gives a permutation of the group set, as listed at the left of the table. In view of these observations, it is not surprising that at least every finite group G is isomorphic to a subgroup of the group S_G of all permutations of G. The same is true for infinite groups; Cayley's theorem states that *every* group is isomorphic to some group consisting of permutations under permutation multiplication. This is a nice and intriguing result, and is a classic of group theory. At first glance, the theorem might seem to be a tool to answer *all* questions about groups. What it really shows is the generality of groups of permutations. Examining subgroups of all permutation groups S_A for sets A of all sizes would be a tremendous task. Cayley's theorem does show that if a counterexample exists to some conjecture we have made about groups, then some group of permutations will provide the counterexample.

We now proceed to the proof of Cayley's theorem, starting with a definition and then a lemma that is important in its own right.

Historical Note

Arthur Cayley (1821–1895) gave an abstract-sounding definition of a group in a paper of 1854: "A set of symbols, $1, \alpha, \beta, \cdots$, all of them different and such that the product of any two of them (no matter in what order) or the product of any one of them into itself, belongs to the set, is said to be a group." He then proceeded to define a group table and note that every line and column of the table "will contain all the symbols $1, \alpha, \beta, \cdots$." Cayley's symbols, however, always represented operations on sets; it does not seem that he was aware of any other kind of group. He noted, for instance, that the four matrix operations $1, \alpha =$ inversion, $\beta =$ transposition, and $\gamma = \alpha\beta$, form, abstractly, the non-cyclic group of four elements. In any case, his definition went unnoticed for a quarter of a century.

This paper of 1854 was one of about 300 written during the 14 years Cayley was practicing law, being unable to find a suitable teaching post. In 1863, he finally became a professor at Cambridge. In 1878, he returned to the theory of groups by publishing four papers, in one of which he stated Theorem 8.16 of this text; his "proof" was simply to notice from the group table that multiplication by any group element permuted the group elements. However, he wrote, "this does not in any wise show that the best or the easiest mode of treating the general problem [of finding all groups of a given order] is thus to regard it as a problem of [permutations]. It seems clear that the better course is to consider the general problem in itself."

The papers of 1878, unlike the earlier one, found a receptive audience; in fact, they were an important influence on Walter Van Dyck's 1882 axiomatic definition of an abstract group, the definition that led to the development of abstract group theory.

8.14 Definition Let $f : A \to B$ be a function and let H be a subset of A. The **image of H under f** is $\{f(h) \mid h \in H\}$ and is denoted by $f[H]$. ■

8.15 Lemma Let G and G' be groups and let $\phi : G \to G'$ be a one-to-one function such that $\phi(xy) = \phi(x)\phi(y)$ for all $x, y \in G$. Then $\phi[G]$ is a subgroup of G' and ϕ provides an isomorphism of G with $\phi[G]$.

Proof We show the conditions for a subgroup given in Theorem 8.14 are satisfied by $\phi[G]$. Let $x', y' \in \phi[G]$. Then there exist $x, y \in G$ such that $\phi(x) = x'$ and $\phi(y) = y'$. By hypothesis, $\phi(xy) = \phi(x)\phi(y) = x'y'$, showing that $x'y' \in \phi[G]$. We have shown that $\phi[G]$ is closed under the operation of G'.

Let e' be the identity of G'. Then

$$e'\phi(e) = \phi(e) = \phi(ee) = \phi(e)\phi(e).$$

Cancellation in G' shows that $e' = \phi(e)$ so $e' \in \phi[G]$.

For $x' \in \phi[G]$ where $x' = \phi(x)$, we have

$$e' = \phi(e) = \phi(xx^{-1}) = \phi(x)\phi(x^{-1}) = x'\phi(x^{-1}),$$

which shows that $x'^{-1} = \phi(x^{-1}) \in \phi[G]$. This completes the demonstration that $\phi[G]$ is a subgroup of G'.

That ϕ provides an isomorphism of G with $\phi[G]$ now follows at once because ϕ provides a one-to-one map of G *onto* $\phi[G]$ such that $\phi(xy) = \phi(x)\phi(y)$ for all $x, y \in G$. ◆

8.16 Theorem **(Cayley's Theorem)** Every group is isomorphic to a group of permutations.

Proof Let G be a group. We show that G is isomorphic to a subgroup of S_G. By Lemma 8.15, we need only to define a one-to-one function $\phi : G \to S_G$ such that $\phi(xy) = \phi(x)\phi(y)$ for all $x, y \in G$. For $x \in G$, let $\lambda_x : G \to G$ be defined by $\lambda_x(g) = xg$ for all $g \in G$. (We think of λ_x as performing *left multiplication* by x.) The equation $\lambda_x(x^{-1}c) = x(x^{-1}c) = c$ for all $c \in G$ shows that λ_x maps G onto G. If $\lambda_x(a) = \lambda_x(b)$, then $xa = xb$ so $a = b$ by cancellation. Thus λ_x is also one to one, and is a permutation of G. We now define $\phi : G \to S_G$ by defining $\phi(x) = \lambda_x$ for all $x \in G$.

To show that ϕ is one to one, suppose that $\phi(x) = \phi(y)$. Then $\lambda_x = \lambda_y$ as functions mapping G into G. In particular $\lambda_x(e) = \lambda_y(e)$, so $xe = ye$ and $x = y$. Thus ϕ is one to one. It only remains to show that $\phi(xy) = \phi(x)\phi(y)$, that is, that $\lambda_{xy} = \lambda_x\lambda_y$. Now for any $g \in G$, we have $\lambda_{xy}(g) = (xy)g$. Permutation multiplication is function composition, so $(\lambda_x\lambda_y)(g) = \lambda_x(\lambda_y(g)) = \lambda_x(yg) = x(yg)$. Thus by associativity, $\lambda_{xy} = \lambda_x\lambda_y$. ◆

For the proof of the theorem, we could have considered equally well the permutations ρ_x of G defined by

$$\rho_x(g) = gx$$

for $g \in G$. (We can think of ρ_x as meaning *right multiplication* by x.) Exercise 52 shows that these permutations form a subgroup of S_G, again isomorphic to G, but provided by a map $\mu : G \to S_G$ defined by

$$\mu(x) = \rho_{x^{-1}}.$$

8.17 Definition The map ϕ in the proof of Theorem 8.16 is the **left regular representation** of G, and the map μ in the preceding comment is the **right regular representation** of G. ■

8.18 Example Let us compute the left regular representation of the group given by the group table, Table 8.19. By "compute" we mean give the elements for the left regular representation and the group table. Here the elements are

$$\lambda_e = \begin{pmatrix} e & a & b \\ e & a & b \end{pmatrix}, \quad \lambda_a = \begin{pmatrix} e & a & b \\ a & b & e \end{pmatrix}, \quad \text{and} \quad \lambda_b = \begin{pmatrix} e & a & b \\ b & e & a \end{pmatrix}.$$

The table for this representation is just like the original table with x renamed λ_x, as seen in Table 8.20. For example,

$$\lambda_a\lambda_b = \begin{pmatrix} e & a & b \\ a & b & e \end{pmatrix}\begin{pmatrix} e & a & b \\ b & e & a \end{pmatrix} = \begin{pmatrix} e & a & b \\ e & a & b \end{pmatrix} = \lambda_e.$$ ▲

8.19 Table

	e	a	b
e	e	a	b
a	a	b	e
b	b	e	a

8.20 Table

	λ_e	λ_a	λ_b
λ_e	λ_e	λ_a	λ_b
λ_a	λ_a	λ_b	λ_e
λ_b	λ_b	λ_e	λ_a

For a finite group given by a group table, ρ_a is the permutation of the elements corresponding to their order in the column under a at the very top, and λ_a is the permutation corresponding to the order of the elements in the row opposite a at the extreme left. The notations ρ_a and λ_a were chosen to suggest right and left multiplication by a, respectively.

EXERCISES 8

Computation

In Exercises 1 through 5, compute the indicated product involving the following permutations in S_6:

$$\sigma = \begin{pmatrix} 1 & 2 & 3 & 4 & 5 & 6 \\ 3 & 1 & 4 & 5 & 6 & 2 \end{pmatrix}, \quad \tau = \begin{pmatrix} 1 & 2 & 3 & 4 & 5 & 6 \\ 2 & 4 & 1 & 3 & 6 & 5 \end{pmatrix}, \quad \mu = \begin{pmatrix} 1 & 2 & 3 & 4 & 5 & 6 \\ 5 & 2 & 4 & 3 & 1 & 6 \end{pmatrix}.$$

1. $\tau\sigma$ **2.** $\tau^2\sigma$ 4 3 2 **3.** $\mu\sigma^2$ **4.** $\sigma^{-2}\tau$ **5.** $\sigma^{-1}\tau\sigma$

In Exercises 6 through 9, compute the expressions shown for the permutations σ, τ and μ defined prior to Exercise 1.

6. $|\langle\sigma\rangle|$ **7.** $|\langle\tau^2\rangle|$ **8.** σ^{100} **9.** μ^{100}

10. Partition the following collection of groups into subcollections of isomorphic groups. Here a * superscript means all nonzero elements of the set.

$\mathbb{Z}$ under addition	S_2
$\mathbb{Z}_6$	$\mathbb{R}^*$ under multiplication
$\mathbb{Z}_2$	$\mathbb{R}^+$ under multiplication
S_6	$\mathbb{Q}^*$ under multiplication
$17\mathbb{Z}$ under addition	$\mathbb{C}^*$ under multiplication
$\mathbb{Q}$ under addition	The subgroup $\langle\pi\rangle$ of $\mathbb{R}^*$ under multiplication
$3\mathbb{Z}$ under addition	The subgroup G of S_5 generated by $\begin{pmatrix}1 & 2 & 3 & 4 & 5\\ 3 & 5 & 4 & 1 & 2\end{pmatrix}$
$\mathbb{R}$ under addition	

Let A be a set and let $\sigma \in S_A$. For a fixed $a \in A$, the set

$$\mathcal{O}_{a,\sigma} = \{\sigma^n(a) \mid n \in \mathbb{Z}\}$$

is the **orbit** of a **under** σ. In Exercises 11 through 13, find the orbit of 1 under the permutation defined prior to Exercise 1.

11. σ **12.** τ **13.** μ

14. In Table 8.8, we used $\rho_0, \rho_1, \rho_2, \mu_1, \mu_2, \mu_3$ as the names of the 6 elements of S_3. Some authors use the notations $\epsilon, \rho, \rho^2, \phi, \rho\phi, \rho^2\phi$ for these elements, where their ϵ is our identity ρ_0, their ρ is our ρ_1, and their ϕ is our μ_1. Verify *geometrically* that their six expressions do give all of S_3.

15. With reference to Exercise 14, give a similar alternative labeling for the 8 elements of D_4 in Table 8.12.

16. Find the number of elements in the set $\{\sigma \in S_4 \mid \sigma(3) = 3\}$.

17. Find the number of elements in the set $\{\sigma \in S_5 \mid \sigma(2) = 5\}$.

18. Consider the group S_3 of Example 8.7

a. Find the cyclic subgroups $\langle\rho_1\rangle$, $\langle\rho_2\rangle$, and $\langle\mu_1\rangle$ of S_3.

b. Find *all* subgroups, proper and improper, of S_3 and give the subgroup diagram for them.

19. Verify that the subgroup diagram for D_4 shown in Fig. 8.13 is correct by finding all (cyclic) subgroups generated by one element, then all subgroups generated by two elements, etc.

20. Give the multiplication table for the cyclic subgroup of S_5 generated by

$$\rho = \begin{pmatrix}1 & 2 & 3 & 4 & 5\\ 2 & 4 & 5 & 1 & 3\end{pmatrix}.$$

There will be six elements. Let them be $\rho, \rho^2, \rho^3, \rho^4, \rho^5$, and $\rho^0 = \rho^6$. Is this group isomorphic to S_3?

21. a. Verify that the six matrices

$$\begin{bmatrix}1&0&0\\0&1&0\\0&0&1\end{bmatrix}, \begin{bmatrix}0&1&0\\0&0&1\\1&0&0\end{bmatrix}, \begin{bmatrix}0&0&1\\1&0&0\\0&1&0\end{bmatrix}, \begin{bmatrix}1&0&0\\0&0&1\\0&1&0\end{bmatrix}, \begin{bmatrix}0&0&1\\0&1&0\\1&0&0\end{bmatrix}, \begin{bmatrix}0&1&0\\1&0&0\\0&0&1\end{bmatrix}$$

form a group under matrix multiplication. [*Hint*: Don't try to compute all products of these matrices. Instead, think how the column vector $\begin{bmatrix}1\\2\\3\end{bmatrix}$ is transformed by multiplying it on the left by each of the matrices.]

b. What group discussed in this section is isomorphic to this group of six matrices?

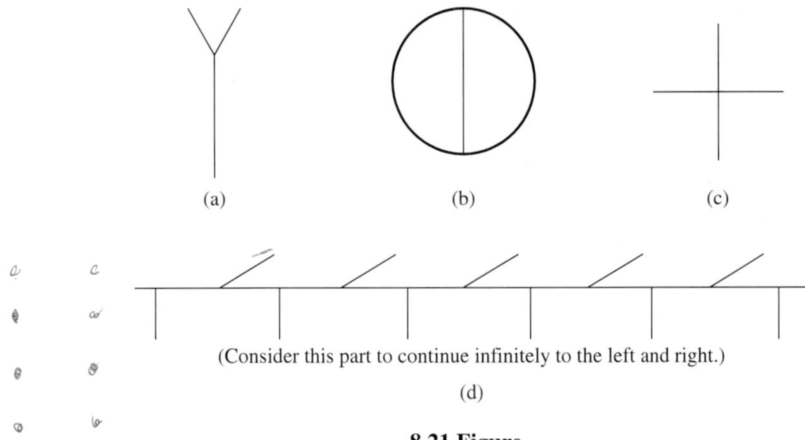

8.21 Figure

22. After working Exercise 21, write down eight matrices that form a group under matrix multiplication that is isomorphic to D_4.

In this section we discussed the group of symmetries of an equilateral triangle and of a square. In Exercises 23 through 26, give a group that we have discussed in the text that is isomorphic to the group of symmetries of the indicated figure. You may want to label some special points on the figure, write some permutations corresponding to symmetries, and compute some products of permutations.

23. The figure in Fig. 8.21 (a)

24. The figure in Fig. 8.21 (b)

25. The figure in Fig. 8.21 (c)

26. The figure in Fig. 8.21 (d)

27. Compute the left regular representation of $\mathbb{Z}_4$. Compute the right regular representation of S_3 using the notation of Example 8.7.

Concepts

In Exercises 28 and 29, correct the definition of the italicized term without reference to the text, if correction is needed, so that it is in a form acceptable for publication.

28. A *permutation* of a set S is a one-to-one map from S to S.

29. The *left regular representation* of a group G is the map of G into S_G whose value at $g \in G$ is the permutation of G that carries each $x \in G$ into gx.

In Exercises 30 through 34, determine whether the given function is a permutation of $\mathbb{R}$.

30. $f_1 : \mathbb{R} \to \mathbb{R}$ defined by $f_1(x) = x + 1$

31. $f_2 : \mathbb{R} \to \mathbb{R}$ defined by $f_2(x) = x^2$

32. $f_3 : \mathbb{R} \to \mathbb{R}$ defined by $f_3(x) = -x^3$

33. $f_4 : \mathbb{R} \to \mathbb{R}$ defined by $f_4(x) = e^x$

34. $f_5 : \mathbb{R} \to \mathbb{R}$ defined by $f_5(x) = x^3 - x^2 - 2x$

35. Mark each of the following true or false.

_____ **a.** Every permutation is a one-to-one function.

_____ **b.** Every function is a permutation if and only if it is one to one.

_____ **c.** Every function from a finite set onto itself must be one to one.

_____ **d.** Every group G is isomorphic to a subgroup of S_G.

_____ **e.** Every subgroup of an abelian group is abelian.
_____ **f.** Every element of a group generates a cyclic subgroup of the group.
_____ **g.** The symmetric group S_{10} has 10 elements.
_____ **h.** The symmetric group S_3 is cyclic.
_____ **i.** S_n is not cyclic for any n.
_____ **j.** Every group is isomorphic to some group of permutations.

36. Show by an example that every proper subgroup of a nonabelian group may be abelian.

37. Let A be a nonempty set. What type of algebraic structure mentioned previously in the text is given by the set of *all* functions mapping A into itself under function composition?

38. Indicate schematically a Cayley digraph for D_n using a generating set consisting of a rotation through $2\pi/n$ radians and a reflection (mirror image). See Exercise 44.

Proof Synopsis

39. Give a two-sentence synopsis of the proof of Cayley's theorem.

Theory

In Exercises 40 through 43, let A be a set, B a subset of A, and let b be one particular element of B. Determine whether the given set is sure to be a subgroup of S_A under the induced operation. Here $\sigma[B] = \{\sigma(x) \mid x \in B\}$.

40. $\{\sigma \in S_A \mid \sigma(b) = b\}$

41. $\{\sigma \in S_A \mid \sigma(b) \in B\}$

42. $\{\sigma \in S_A \mid \sigma[B] \subseteq B\}$

43. $\{\sigma \in S_A \mid \sigma[B] = B\}$

44. In analogy with Examples 8.7 and 8.10, consider a regular plane n-gon for $n \geq 3$. Each way that two copies of such an n-gon can be placed, with one covering the other, corresponds to a certain permutation of the vertices. The set of these permutations in a group, the **nth dihedral group** D_n, under permutation multiplication. Find the order of this group D_n. Argue *geometrically* that this group has a subgroup having just half as many elements as the whole group has.

45. Consider a cube that exactly fills a certain cubical box. As in Examples 8.7 and 8.10, the ways in which the cube can be placed into the box correspond to a certain group of permutations of the vertices of the cube. This group is the **group of rigid motions** (or **rotations**) **of the cube**. (It should not be confused with the *group of symmetries of the figure*, which will be discussed in the exercises of Section 12.) How many elements does this group have? Argue *geometrically* that this group has at least three different subgroups of order 4 and at least four different subgroups of order 3.

46. Show that S_n is a nonabelian group for $n \geq 3$.

47. Strengthening Exercise 46, show that if $n \geq 3$, then the only element of σ of S_n satisfying $\sigma\gamma = \gamma\sigma$ for all $\gamma \in S_n$ is $\sigma = \iota$, the identity permutation.

48. Orbits were defined before Exercise 11. Let $a, b \in A$ and $\sigma \in S_A$. Show that if $\mathcal{O}_{a,\sigma}$ and $\mathcal{O}_{b,\sigma}$ have an element in common, then $\mathcal{O}_{a,\sigma} = \mathcal{O}_{b,\sigma}$.

49. If A is a set, then a subgroup H of S_A is **transitive on** A if for each $a, b \in A$ there exists $\sigma \in H$ such that $\sigma(a) = b$. Show that if A is a nonempty finite set, then there exists a finite cyclic subgroup H of S_A with $|H| = |A|$ that is transitive on A.

50. Referring to the definition before Exercise 11 and to Exercise 49, show that for $\sigma \in S_A$, $\langle\sigma\rangle$ is transitive on A if and only if $\mathcal{O}_{a,\sigma} = A$ for some $a \in A$.

51. (See the warning on page 78). Let G be a group with binary operation $*$. Let G' be the same set as G, and define a binary operation $*'$ on G' by $x *' y = y * x$ for all $x, y \in G'$.

a. (Intuitive argument that G' under $*'$ is a group.) Suppose the front wall of your class room were made of transparent glass, and that all possible products $a * b = c$ and all possible instances $a * (b * c) =$

$(a * b) * c$ of the associative property for G under $*$ were written on the wall with a magic marker. What would a person see when looking at the other side of the wall from the next room in front of yours?

b. Show from the mathematical definition of $*'$ that G' is a group under $*'$.

52. Let G be a group. Prove that the permutations $\rho_a : G \to G$, where $\rho_a(x) = xa$ for $a \in G$ and $x \in G$, do form a group isomorphic to G. $\phi(g) = \rho_{g^{-1}}$

53. A **permutation matrix** is one that can be obtained from an identity matrix by reordering its rows. If P is an $n \times n$ permutation matrix and A is any $n \times n$ matrix and $C = PA$, then C can be obtained from A by making precisely the same reordering of the rows of A as the reordering of the rows which produced P from I_n.

a. Show that every finite group of order n is isomorphic to a group consisting of $n \times n$ permutation matrices under matrix multiplication.

b. For each of the four elements e, a, b, and c in the Table 5.11 for the group V, give a specific 4×4 matrix that corresponds to it under such an isomorphism.

SECTION 9 ORBITS, CYCLES, AND THE ALTERNATING GROUPS

Orbits

Each permutation σ of a set A determines a natural partition of A into cells with the property that $a, b \in A$ are in the same cell if and only if $b = \sigma^n(a)$ for some $n \in \mathbb{Z}$. We establish this partition using an appropriate equivalence relation:

$$\text{For } a, b \in A, \text{ let } a \sim b \text{ if and only if } b = \sigma^n(a) \text{ for some } n \in \mathbb{Z}. \quad \textbf{(1)}$$

We now check that $\sim$ defined by Condition (1) is indeed an equivalence relation.

Reflexive Clearly $a \sim a$ since $a = \iota(a) = \sigma^0(a)$.

Symmetric If $a \sim b$, then $b = \sigma^n(a)$ for some $n \in \mathbb{Z}$. But then $a = \sigma^{-n}(b)$ and $-n \in \mathbb{Z}$, so $b \sim a$.

Transitive Suppose $a \sim b$ and $b \sim c$, then $b = \sigma^n(a)$ and $c = \sigma^m(b)$ for some $n, m \in \mathbb{Z}$. Substituting, we find that $c = \sigma^m(\sigma^n(a)) = \sigma^{n+m}(a)$, so $a \sim c$.

9.1 Definition Let σ be a permutation of a set A. The equivalence classes in A determined by the equivalence relation (1) are the **orbits of σ**. ■

9.2 Example Since the identity permutation ι of A leaves each element of A fixed, the orbits of ι are the one-element subsets of A. ▲

9.3 Example Find the orbits of the permutation

$$\sigma = \begin{pmatrix} 1 & 2 & 3 & 4 & 5 & 6 & 7 & 8 \\ 3 & 8 & 6 & 7 & 4 & 1 & 5 & 2 \end{pmatrix}$$

in S_8.

Solution To find the orbit containing 1, we apply σ repeatedly, obtaining symbolically

$$1 \xrightarrow{\sigma} 3 \xrightarrow{\sigma} 6 \xrightarrow{\sigma} 1 \xrightarrow{\sigma} 3 \xrightarrow{\sigma} 6 \xrightarrow{\sigma} 1 \xrightarrow{\sigma} 3 \xrightarrow{\sigma} \cdots .$$

Since σ^{-1} would simply reverse the directions of the arrows in this chain, we see that the orbit containing 1 is $\{1, 3, 6\}$. We now choose an integer from 1 to 8 not in $\{1, 3, 6\}$, say 2, and similarly find that the orbit containing 2 is $\{2, 8\}$. Finally, we find that the orbit containing 4 is $\{4, 7, 5\}$. Since these three orbits include all integers from 1 to 8, we see that the complete list of orbits of σ is

$$\{1, 3, 6\}, \quad \{2, 8\}, \quad \{4, 5, 7\}.$$ ▲

Cycles

For the remainder of this section, we consider just permutations of a finite set A of n elements. We may as well suppose that $A = \{1, 2, 3, \cdots, n\}$ and that we are dealing with elements of the symmetric group S_n.

Refer back to Example 9.3. The orbits of

$$\sigma = \begin{pmatrix} 1 & 2 & 3 & 4 & 5 & 6 & 7 & 8 \\ 3 & 8 & 6 & 7 & 4 & 1 & 5 & 2 \end{pmatrix} \quad \text{(2)}$$

are indicated graphically in Fig. 9.4. That is, σ acts on each integer from 1 to 8 on one of the circles by carrying it into the next integer on the circle traveled counterclockwise, in the direction of the arrows. For example, the leftmost circle indicates that $\sigma(1) = 3$, $\sigma(3) = 6$, and $\sigma(6) = 1$. Figure 9.4 is a nice way to visualize the structure of the permutation σ.

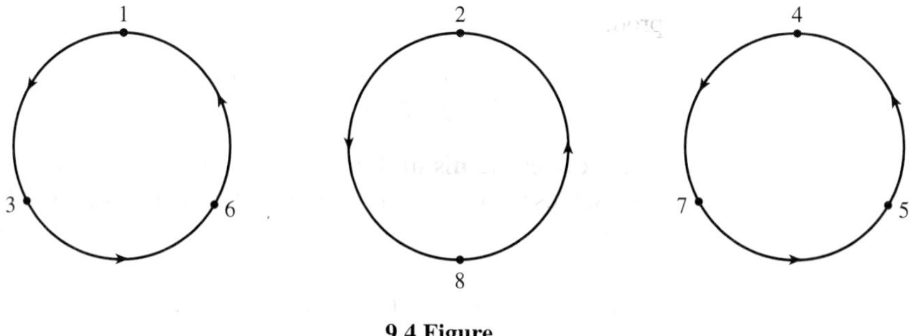

9.4 Figure

Each individual circle in Fig. 9.4 also defines, by itself, a permutation in S_8. For example, the leftmost circle corresponds to the permutation

$$\mu = \begin{pmatrix} 1 & 2 & 3 & 4 & 5 & 6 & 7 & 8 \\ 3 & 2 & 6 & 4 & 5 & 1 & 7 & 8 \end{pmatrix} \quad \text{(3)}$$

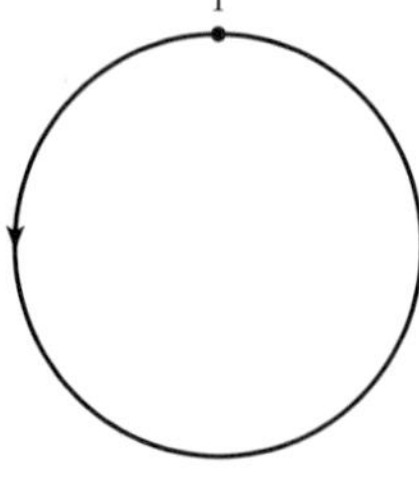

9.5 Figure

that acts on 1, 3, and 6 just as σ does, but leaves the remaining integers 2, 4, 5, 7, and 8 fixed. In summary, μ has one three-element orbit $\{1, 3, 6\}$ and five one-element orbits $\{2\}, \{4\}, \{5\}, \{7\}$, and $\{8\}$. Such a permutation, described graphically by a single circle, is called a *cycle* (for circle). We consider the identity permutation to be a cycle since it can be represented by a circle having only the integer 1, as shown in Fig. 9.5. We now define the term *cycle* in a mathematically precise way.

9.6 Definition A permutation $\sigma \in S_n$ is a **cycle** if it has at most one orbit containing more than one element. The **length** of a cycle is the number of elements in its largest orbit. ■

To avoid the cumbersome notation, as in Eq. (3), for a cycle, we introduce a single-row *cyclic notation*. In cyclic notation, the cycle in Eq. (3) becomes

$$\mu = (1, 3, 6).$$

We understand by this notation that μ carries the first number 1 into the second number 3, the second number 3 into the next number 6, etc., until finally the last number 6 is carried into the first number 1. An integer not appearing in this notation for μ is understood to be left fixed by μ. Of course, the set on which μ acts, which is $\{1, 2, 3, 4, 5, 6, 7, 8\}$ in our example, must be made clear by the context.

9.7 Example Working within S_5, we see that

$$(1, 3, 5, 4) = \begin{pmatrix} 1 & 2 & 3 & 4 & 5 \\ 3 & 2 & 5 & 1 & 4 \end{pmatrix}.$$

Observe that

$$(1, 3, 5, 4) = (3, 5, 4, 1) = (5, 4, 1, 3) = (4, 1, 3, 5).$$ ▲

Of course, since cycles are special types of permutations, they can be multiplied just as any two permutations. The product of two cycles need not again be a cycle, however.

Using cyclic notation, we see that the permutation σ in Eq. (2) can be written as a product of cycles:

$$\sigma = \begin{pmatrix} 1 & 2 & 3 & 4 & 5 & 6 & 7 & 8 \\ 3 & 8 & 6 & 7 & 4 & 1 & 5 & 2 \end{pmatrix} = (1, 3, 6)(2, 8)(4, 7, 5). \quad (4)$$

These cycles are **disjoint**, meaning that any integer is moved by at most one of these cycles; thus no one number appears in the notations of two different cycles. Equation (4) exhibits σ in terms of its orbits, and is a one-line description of Fig. 9.4. Every permutation in S_n can be expressed in a similar fashion as a product of the disjoint cycles corresponding to its orbits. We state this as a theorem and write out the proof.

9.8 Theorem Every permutation σ of a finite set is a product of disjoint cycles.

Proof Let $B_1, B_2, \cdots, B_r$ be the orbits of σ, and let μ_i be the cycle defined by

$$\mu_i(x) = \begin{cases} \sigma(x) & \text{for } x \in B_i \\ x & \text{otherwise.} \end{cases}$$

Clearly $\sigma = \mu_1\mu_2 \cdots \mu_r$. Since the equivalence-class orbits $B_1, B_2, \cdots, B$, being distinct equivalence classes, are disjoint, the cycles $\mu_1, \mu_2, \cdots, \mu_r$ are disjoint also. ◆

While permutation multiplication in general is not commutative, it is readily seen that *multiplication of disjoint cycles is commutative*. Since the orbits of a permutation are unique, the representation of a permutation as a product of disjoint cycles, none of which is the identity permutation, is unique up to the order of the factors.

9.9 Example Consider the permutation

$$\begin{pmatrix} 1 & 2 & 3 & 4 & 5 & 6 \\ 6 & 5 & 2 & 4 & 3 & 1 \end{pmatrix}.$$

Let us write it as a product of disjoint cycles. First, 1 is moved to 6 and then 6 to 1, giving the cycle $(1, 6)$. Then 2 is moved to 5, which is moved to 3, which is moved to 2, or $(2, 5, 3)$. This takes care of all elements but 4, which is left fixed. Thus

$$\begin{pmatrix} 1 & 2 & 3 & 4 & 5 & 6 \\ 6 & 5 & 2 & 4 & 3 & 1 \end{pmatrix} = (1, 6)(2, 5, 3).$$

Multiplication of *disjoint* cycles is commutative, so the order of the factors $(1, 6)$ and $(2, 5, 3)$ is not important. ▲

You should practice multiplying permutations in cyclic notation where the cycles may or may not be disjoint. We give an example and provide further practice in Exercises 7 through 9.

9.10 Example Consider the cycles (1,4,5,6) and (2,1,5) in S_6. Multiplying, we find that

$$(1, 4, 5, 6)(2, 1, 5) = \begin{pmatrix} 1 & 2 & 3 & 4 & 5 & 6 \\ 6 & 4 & 3 & 5 & 2 & 1 \end{pmatrix}$$

and

$$(2, 1, 5)(1, 4, 5, 6) = \begin{pmatrix} 1 & 2 & 3 & 4 & 5 & 6 \\ 4 & 1 & 3 & 2 & 6 & 5 \end{pmatrix}.$$

Neither of these permutations is a cycle. ▲

Even and Odd Permutations

It seems reasonable that every reordering of the sequence $1, 2, \ldots, n$ can be achieved by repeated interchange of positions of pairs of numbers. We discuss this a bit more formally.

9.11 Definition A cycle of length 2 is a **transposition**. ■

Thus a transposition leaves all elements but two fixed, and maps each of these onto the other. A computation shows that

$$(a_1, a_2, \cdots, a_n) = (a_1, a_n)(a_1, a_{n-1}) \cdots (a_1, a_3)(a_1, a_2).$$

Therefore any cycle is a product of transpositions. We then have the following as a corollary to Theorem 9.8.

9.12 Corollary Any permutation of a finite set of at least two elements is a product of transpositions.

Naively, this corollary just states that any rearrangement of n objects can be achieved by successively interchanging pairs of them.

9.13 Example Following the remarks prior to the corollary, we see that (1, 6) (2, 5, 3) is the product (1, 6) (2, 3) (2, 5) of transpositions. ▲

9.14 Example In S_n for $n \geq 2$, the identity permutation is the product (1, 2) (1, 2) of transpositions. ▲

We have seen that every permutation of a finite set with at least two elements is a product of transpositions. The transpositions may not be disjoint, and a representation of the permutation in this way is not unique. For example, we can always insert at the beginning the transposition (1, 2) twice, because (1, 2) (1, 2) is the identity permutation. What is true is that the number of transpositions used to represent a given permutation must either always be even or always be odd. This is an important fact. We will give two proofs. The first uses a property of determinants from linear algebra. The second involves counting orbits and was suggested by David M. Bloom.

9.15 Theorem No permutation in S_n can be expressed both as a product of an even number of transpositions and as a product of an odd number of transpositions.

Proof 1 (From linear algebra) We remarked in Section 8 that $S_A \simeq S_B$ if A and B have the same cardinality. We work with permutations of the n rows of the $n \times n$ identity matrix I_n, rather than of the numbers $1, 2, \ldots, n$. The identity matrix has determinant 1. Interchanging any two rows of a square matrix changes the sign of the determinant. Let C be a matrix obtained by a permutation σ of the rows of I_n. If C could be obtained from I_n by both an even number and an odd number of transpositions of rows, its determinant would have to be both 1 and -1, which is impossible. Thus σ cannot be expressed both as a product of an even number and an odd number of transpositions.

Proof 2 (Counting orbits) Let $\sigma \in S_n$ and let $\tau = (i, j)$ be a transposition in S_n. We claim that the number of orbits of σ and of $\tau\sigma$ differ by 1.

Case I Suppose i and j are in different orbits of σ. Write σ as a product of disjoint cycles, the first of which contains j and the second of which contains i, symbolized by the two circles in Fig. 9.16. We may write the product of these two cycles symbolically as

$$(b, j, \times, \times, \times)(a, i, \times, \times)$$

where the symbols $\times$ denote possible other elements in these orbits.

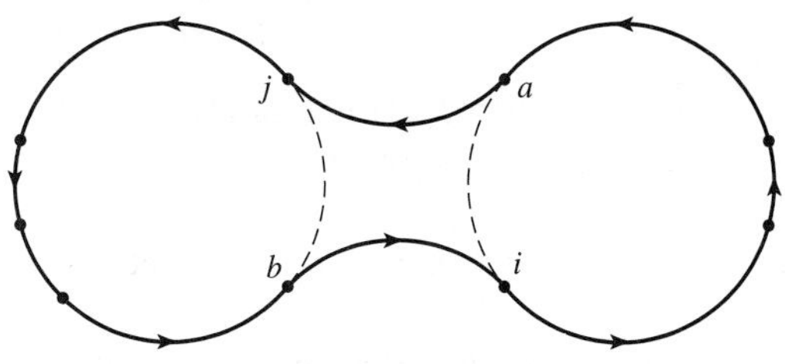

9.16 Figure

Computing the product of the first three cycles in $\tau\sigma = (i, j)\sigma$, we obtain

$$(i, j)(b, j, \times, \times, \times)(a, i, \times, \times) = (a, j, \times, \times, \times, b, i, \times, \times).$$

The original 2 orbits have been joined to form just one in $\tau\sigma$ as symbolized in Fig. 9.16. Exercise 28 asks us to repeat the computation to show that the same thing happens if either one or both of i and j should be only element of their orbit in σ.

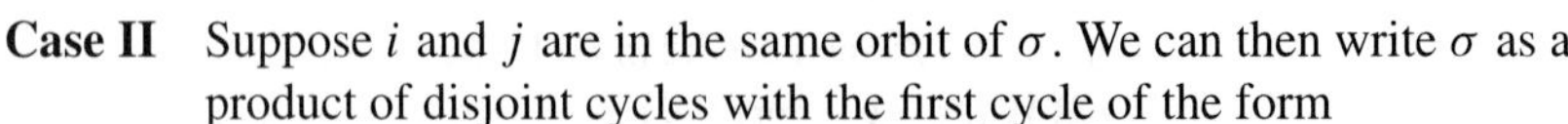

Case II Suppose i and j are in the same orbit of σ. We can then write σ as a product of disjoint cycles with the first cycle of the form

$$(a, i, \times, \times, \times, b, j, \times, \times)$$

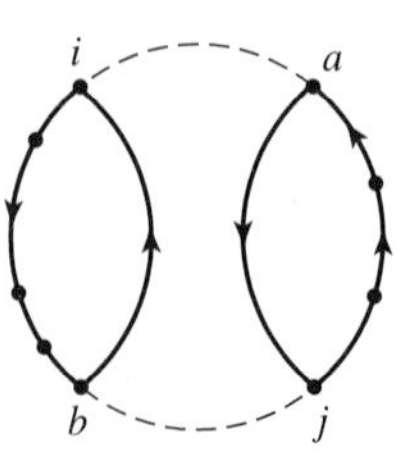

9.17 Figure

shown symbolically by the circle in Fig. 9.17. Computing the product of the first two cycles in $\tau\sigma = (i, j)\sigma$, we obtain

$$(i, j)(a, i, \times, \times, \times, b, j, \times, \times) = (a, j, \times, \times)(b, i, \times, \times, \times).$$

The original single orbit has been split into two as symbolized in Fig. 9.17.

We have shown that the number of orbits of $\tau\sigma$ differs from the number of orbits of σ by 1. The identity permutation ι has n orbits, because each element is the only member of its orbit. Now the number of orbits of a given permutation $\sigma \in S_n$ differs from n by either an even or an odd number, but not both. Thus it is impossible to write

$$\sigma = \tau_1\tau_2\tau_3 \cdots \tau_m\iota$$

where the τ_k are transpositions in two ways, once with m even and once with m odd. ◆

9.18 Definition A permutation of a finite set is **even** or **odd** according to whether it can be expressed as a product of an even number of transpositions or the product of an odd number of transpositions, respectively. ■

9.19 Example The identity permutation ι in S_n is an even permutation since we have $\iota = (1, 2)(1, 2)$. If $n = 1$ so that we cannot form this product, we define ι to be even. On the other hand, the permutation (1, 4, 5, 6) (2, 1, 5) in S_6 can be written as

$$(1, 4, 5, 6)(2, 1, 5) = (1, 6)(1, 5)(1, 4)(2, 5)(2, 1)$$

which has five transpositions, so this is an odd permutation. ▲

The Alternating Groups

We claim that for $n \geq 2$, the number of even permutations in S_n is the same as the number of odd permutation; that is, S_n is split equally and both numbers are $(n!)/2$. To show this, let A_n be the set of even permutations in S_n and let B_n be the set of odd permutations for $n \geq 2$. We proceed to define a one-to-one function from A_n onto B_n. This is exactly what is needed to show that A_n and B_n have the same number of elements.

Let τ be any fixed transposition in S_n; it exists since $n \geq 2$. We may as well suppose that $\tau = (1, 2)$. We define a function

$$\lambda_\tau : A_n \to B_n$$

by

$$\lambda_\tau(\sigma) = \tau\sigma,$$

that is, $\sigma \in A_n$ is mapped into $(1, 2)\sigma$ by λ_τ. Observe that since σ is even, the permutation $(1, 2)\sigma$ can be expressed as a product of a (1 + even number), or odd number, of transpositions, so $(1, 2)\sigma$ is indeed in B_n. If for σ and μ in A_n it is true that $\lambda_\tau(\sigma) = \lambda_\tau(\mu)$, then

$$(1, 2)\sigma = (1, 2)\mu,$$

and since S_n is a group, we have $\sigma = \mu$. Thus λ_τ is a one-to-one function. Finally,

$$\tau = (1, 2) = \tau^{-1},$$

so if $\rho \in B_n$, then

$$\tau^{-1}\rho \in A_n,$$

and

$$\lambda_\tau(\tau^{-1}\rho) = \tau(\tau^{-1}\rho) = \rho.$$

Thus λ_τ is onto B_n. Hence the number of elements in A_n is the same as the number in B_n since there is a one-to-one correspondence between the elements of the sets.

Note that the product of two even permutations is again even. Also since $n \geq 2$, S_n has the transposition $(1, 2)$ and $\iota = (1, 2)(1, 2)$ is an even permutation. Finally, note that if σ is expressed as a product of transpositions, the product of the same transpositions taken in just the opposite order is σ^{-1}. Thus if σ is an even permutation, σ^{-1} must also be even. Referring to Theorem 5.14, we see that we have proved the following statement.

9.20 Theorem If $n \geq 2$, then the collection of all even permutations of $\{1, 2, 3, \cdots, n\}$ forms a subgroup of order $n!/2$ of the symmetric group S_n.

9.21 Definition The subgroup of S_n consisting of the even permutations of n letters is the **alternating group A_n on n letters.** ■

Both S_n and A_n are very important groups. Cayley's theorem shows that every finite group G is structurally identical to some subgroup of S_n for $n = |G|$. It can be shown that there are no formulas involving just radicals for solution of polynomial equations of degree n for $n \geq 5$. This fact is actually due to the structure of A_n, surprising as that may seem!

EXERCISES 9

Computations

In Exercises 1 through 6, find all orbits of the given permutation.

1. $\begin{pmatrix} 1 & 2 & 3 & 4 & 5 & 6 \\ 5 & 1 & 3 & 6 & 2 & 4 \end{pmatrix}$

2. $\begin{pmatrix} 1 & 2 & 3 & 4 & 5 & 6 & 7 & 8 \\ 5 & 6 & 2 & 4 & 8 & 3 & 1 & 7 \end{pmatrix}$

3. $\begin{pmatrix} 1 & 2 & 3 & 4 & 5 & 6 & 7 & 8 \\ 2 & 3 & 5 & 1 & 4 & 6 & 8 & 7 \end{pmatrix}$

4. $\sigma : \mathbb{Z} \to \mathbb{Z}$ where $\sigma(n) = n + 1$

5. $\sigma : \mathbb{Z} \to \mathbb{Z}$ where $\sigma(n) = n + 2$

6. $\sigma : \mathbb{Z} \to \mathbb{Z}$ where $\sigma(n) = n - 3$

In Exercises 7 through 9, compute the indicated product of cycles that are permutations of $\{1, 2, 3, 4, 5, 6, 7, 8\}$.

7. $(1, 4, 5)(7, 8)(2, 5, 7)$

8. $(1, 3, 2, 7)(4, 8, 6)$

9. $(1, 2)(4, 7, 8)(2, 1)(7, 2, 8, 1, 5)$

In Exercises 10 through 12, express the permutation of $\{1, 2, 3, 4, 5, 6, 7, 8\}$ as a product of disjoint cycles, and then as a product of transpositions.

10. $\begin{pmatrix} 1 & 2 & 3 & 4 & 5 & 6 & 7 & 8 \\ 8 & 2 & 6 & 3 & 7 & 4 & 5 & 1 \end{pmatrix}$

11. $\begin{pmatrix} 1 & 2 & 3 & 4 & 5 & 6 & 7 & 8 \\ 3 & 6 & 4 & 1 & 8 & 2 & 5 & 7 \end{pmatrix}$

12. $\begin{pmatrix} 1 & 2 & 3 & 4 & 5 & 6 & 7 & 8 \\ 3 & 1 & 4 & 7 & 2 & 5 & 8 & 6 \end{pmatrix}$

13. Recall that element a of a group G with identity element e has order $r > 0$ if $a^r = e$ and no smaller positive power of a is the identity. Consider the group S_8.

a. What is the order of the cycle $(1, 4, 5, 7)$?

b. State a theorem suggested by part (a).

c. What is the order of $\sigma = (4, 5)(2, 3, 7)$? of $\tau = (1, 4)(3, 5, 7, 8)$?

d. Find the order of each of the permutations given in Exercises 10 through 12 by looking at its decomposition into a product of disjoint cycles.

e. State a theorem suggested by parts (c) and (d). [*Hint:* The important words you are looking for are *least common multiple*.]

In Exercises 14 through 18, find the maximum possible order for an element of S_n for the given value of n.

14. $n = 5$ **15.** $n = 6$ **16.** $n = 7$ **17.** $n = 10$ **18.** $n = 15$

19. Figure 9.22 shows a Cayley digraph for the alternating group A_4 using the generating set $S = \{(1, 2, 3), (1, 2)(3, 4)\}$. Continue labeling the other nine vertices with the elements of A_4, expressed as a product of disjoint cycles.

Concepts

In Exercises 20 through 22, correct the definition of the italicized term without reference to the text, if correction is needed, so that it is in a form acceptable for publication.

20. For a permutation σ of a set A, an *orbit* of σ is a nonempty minimal subset of A that is mapped onto itself by σ.

21. A *cycle* is a permutation having only one orbit.

22. The *alternating group* is the group of all even permutations.

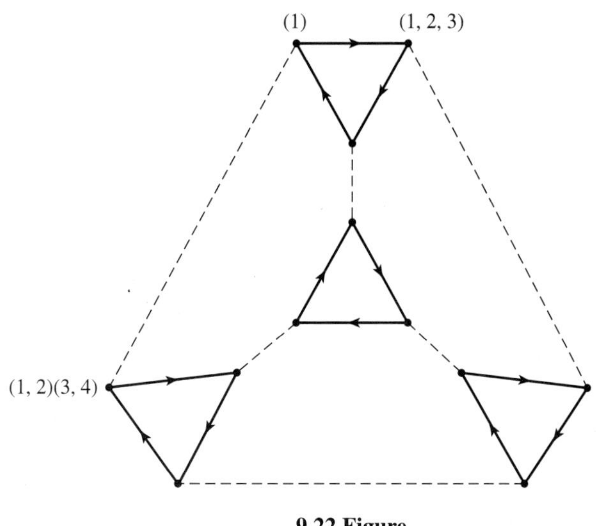

9.22 Figure

23. Mark each of the following true or false.

_____ **a.** Every permutation is a cycle.
_____ **b.** Every cycle is a permutation.
_____ **c.** The definition of even and odd permutations could have been given equally well before Theorem 9.15.
_____ **d.** Every nontrivial subgroup H of S_9 containing some odd permutation contains a transposition.
_____ **e.** A_5 has 120 elements.
_____ **f.** S_n is not cyclic for any $n \geq 1$.
_____ **g.** A_3 is a commutative group.
_____ **h.** S_7 is isomorphic to the subgroup of all those elements of S_8 that leave the number 8 fixed.
_____ **i.** S_7 is isomorphic to the subgroup of all those elements of S_8 that leave the number 5 fixed.
_____ **j.** The odd permutations in S_8 form a subgroup of S_8.

24. Which of the permutations in S_3 of Example 8.7 are even permutations? Give the table for the alternating group A_3.

Proof Synopsis

25. Give a one-sentence synopsis of Proof 1 of Theorem 9.15.

26. Give a two-sentence synopsis of Proof 2 of Theorem 9.15.

Theory

27. Prove the following about S_n if $n \geq 3$.

a. Every permutation in S_n can be written as a product of at most $n - 1$ transpositions.

b. Every permutation in S_n that is not a cycle can be written as a product of at most $n - 2$ transpositions.

c. Every odd permutation in S_n can be written as a product of $2n + 3$ transpositions, and every even permutation as a product of $2n + 8$ transpositions.

28. **a.** Draw a figure like Fig. 9.16 to illustrate that if i and j are in different orbits of σ and $\sigma(i) = i$, then the number of orbits of $(i, j)\sigma$ is one less than the number of orbits of σ.

b. Repeat part (a) if $\sigma(j) = j$ also.

29. Show that for every subgroup H of S_n for $n \geq 2$, either all the permutations in H are even or exactly half of them are even.

30. Let σ be a permutation of a set A. We shall say "σ **moves** $a \in A$" if $\sigma(a) \neq a$. If A is a finite set, how many elements are moved by a cycle $\sigma \in S_A$ of length n?

31. Let A be an infinite set. Let H be the set of all $\sigma \in S_A$ such that the number of elements moved by σ (see Exercise 30) is finite. Show that H is a subgroup of S_A.

32. Let A be an infinite set. Let K be the set of all $\sigma \in S_A$ that move (see Exercise 30) at most 50 elements of A. Is K a subgroup of S_A? Why?

33. Consider S_n for a fixed $n \geq 2$ and let σ be a fixed odd permutation. Show that every odd permutation in S_n is a product of σ and some permutation in A_n.

34. Show that if σ is a cycle of odd length, then σ^2 is a cycle.

35. Following the line of thought opened by Exercise 34, complete the following with a condition involving n and r so that the resulting statement is a theorem:

If σ is a cycle of length n, then σ^r is also a cycle if and only if . . .

36. Let G be a group and let a be a fixed element of G. Show that the map $\lambda_a : G \to G$, given by $\lambda_a(g) = ag$ for $g \in G$, is a permutation of the set G.

37. Referring to Exercise 36, show that $H = \{\lambda_a \mid a \in G\}$ is a subgroup of S_G, the group of all permutations of G.

38. Referring to Exercise 49 of Section 8, show that H of Exercise 37 is transitive on the set G. [*Hint:* This is an immediate corollary of one of the theorems in Section 4.]

39. Show that S_n is generated by $\{(1, 2), (1, 2, 3, \cdots, n)\}$. [*Hint:* Show that as r varies, $(1, 2, 3, \cdots, n)^r(1, 2)(1, 2, 3, \cdots, n)^{n-r}$ gives all the transpositions $(1, 2), (2, 3), (3, 4), \cdots, (n-1, n), (n, 1)$. Then show that any transposition is a product of some of these transpositions and use Corollary 9.12]

SECTION 10 Cosets and the Theorem of Lagrange

You may have noticed that the order of a subgroup H of a finite group G seems always to be a divisor of the order of G. This is the theorem of Lagrange. We shall prove it by exhibiting a partition of G into cells, all having the same size as H. Thus if there are r such cells, we will have

$$r(\text{order of } H) = (\text{order of } G)$$

from which the theorem follows immediately. The cells in the partition will be called *cosets of H*, and they are important in their own right. In Section 14, we will see that if H satisfies a certain property, then each coset can be regarded as an element of a group in a very natural way. We give some indication of such *coset groups* in this section to help you develop a feel for the topic.

Cosets

Let H be a subgroup of a group G, which may be of finite or infinite order. We exhibit two partitions of G by defining two equivalence relations, $\sim_L$ and $\sim_R$ on G.

10.1 Theorem Let H be a subgroup of G. Let the relation $\sim_L$ be defined on G by

$$a \sim_L b \quad \text{if and only if} \quad a^{-1}b \in H.$$

Let $\sim_R$ be defined by

$$a \sim_R b \quad \text{if and only if} \quad ab^{-1} \in H.$$

Then $\sim_L$ and $\sim_R$ are both equivalence relations on G.

Proof We show that $\sim_L$ is an equivalence relation, and leave the proof for $\sim_R$ to Exercise 26. When reading the proof, notice how we must constantly make use of the fact that H is a *subgroup* of G.

Reflexive Let $a \in G$. Then $a^{-1}a = e$ and $e \in H$ since H is a subgroup. Thus $a \sim_L a$.

Symmetric Suppose $a \sim_L b$. Then $a^{-1}b \in H$. Since H is a subgroup, $(a^{-1}b)^{-1}$ is in H and $(a^{-1}b)^{-1} = b^{-1}a$, so $b^{-1}a$ is in H and $b \sim_L a$.

Transitive Let $a \sim_L b$ and $b \sim_L c$. Then $a^{-1}b \in H$ and $b^{-1}c \in H$. Since H is a subgroup, $(a^{-1}b)(b^{-1}c) = a^{-1}c$ is in H, so $a \sim_L c$. ◆

The equivalence relation $\sim_L$ in Theorem 10.1 defines a partition of G, as described in Theorem 0.22. Let's see what the cells in this partition look like. Suppose $a \in G$. The cell containing a consists of all $x \in G$ such that $a \sim_L x$, which means all $x \in G$ such that $a^{-1}x \in H$. Now $a^{-1}x \in H$ if and only if $a^{-1}x = h$ for some $h \in H$, or equivalently, if and only if $x = ah$ for some $h \in H$. Therefore the cell containing a is $\{ah \mid h \in H\}$, which we denote by aH. If we go through the same reasoning for the equivalence relation $\sim_R$ defined by H, we find the cell in this partition containing $a \in G$ is $Ha = \{ha \mid h \in H\}$. Since G need not be abelian, we have no reason to expect aH and Ha to be the same subset of G. We give a formal definition.

10.2 Definition Let H be a subgroup of a group G. The subset $aH = \{ah \mid h \in H\}$ of G is the **left coset** of H containing a, while the subset $Ha = \{ha \mid h \in H\}$ is the **right coset** of H containing a. ■

10.3 Example Exhibit the left cosets and the right cosets of the subgroup $3\mathbb{Z}$ of $\mathbb{Z}$.

Solution Our notation here is additive, so the left coset of $3\mathbb{Z}$ containing m is $m + 3\mathbb{Z}$. Taking $m = 0$, we see that

$$3\mathbb{Z} = \{\cdots, -9, -6, -3, 0, 3, 6, 9, \cdots\}$$

is itself one of its left cosets, the coset containing 0. To find another left coset, we select an element of $\mathbb{Z}$ not in $3\mathbb{Z}$, say 1, and find the left coset containing it. We have

$$1 + 3\mathbb{Z} = \{\cdots, -8, -5, -2, 1, 4, 7, 10, \cdots\}.$$

These two left cosets, $3\mathbb{Z}$ and $1 + 3\mathbb{Z}$, do not yet exhaust $\mathbb{Z}$. For example, 2 is in neither of them. The left coset containing 2 is

$$2 + 3\mathbb{Z} = \{\cdots, -7, -4, -1, 2, 5, 8, 11, \cdots\}.$$

It is clear that these three left cosets we have found do exhaust $\mathbb{Z}$, so they constitute the partition of $\mathbb{Z}$ into left cosets of $3\mathbb{Z}$.

Since $\mathbb{Z}$ is abelian, the left coset $m + 3\mathbb{Z}$ and the right coset $3\mathbb{Z} + m$ are the same, so the partition of $\mathbb{Z}$ into right cosets is the same. ▲

We observe two things from Example 10.3.

> For a subgroup H of an abelian group G, the partition of G into left cosets of H and the partition into right cosets are the same.

Also, looking back at Examples 0.17 and 0.20, we see that the equivalence relation $\sim_R$ for the subgroup $n\mathbb{Z}$ of $\mathbb{Z}$ is the same as the relation of congruence modulo n. Recall that $h \equiv k \pmod{n}$ in $\mathbb{Z}$ if $h - k$ is divisible by n. This is the same as saying that $h + (-k)$ is in $n\mathbb{Z}$, which is relation $\sim_R$ of Theorem 10.1 in additive notation. Thus the partition of $\mathbb{Z}$ into cosets of $n\mathbb{Z}$ is the partition of $\mathbb{Z}$ into residue classes modulo n. For that reason, we often refer to the cells of this partition as *cosets modulo* $n\mathbb{Z}$. Note that we do not have to specify *left* or *right* cosets since they are the same for this abelian group $\mathbb{Z}$.

10.4 Example The group $\mathbb{Z}_6$ is abelian. Find the partition of $\mathbb{Z}_6$ into cosets of the subgroup $H = \{0, 3\}$.

Solution One coset is $\{0, 3\}$ itself. The coset containing 1 is $1 + \{0, 3\} = \{1, 4\}$. The coset containing 2 is $2 + \{0, 3\} = \{2, 5\}$. Since $\{0, 3\}$, $\{1, 4\}$, and $\{2, 5\}$ exhaust all of $\mathbb{Z}_6$, these are all the cosets. ▲

We point out a fascinating thing that we will develop in detail in Section 14. Referring back to Example 10.4, Table 10.5 gives the binary operation for $\mathbb{Z}_6$ but with elements listed in the order they appear in the cosets $\{0, 3\}$, $\{1, 4\}$, $(2, 5\}$. We shaded the table according to these cosets.

Suppose we denote these cosets by LT(light), MD(medium), and DK(dark) according to their shading. Table 10.5 then defines a binary operation on these shadings, as shown in Table 10.6. Note that if we replace LT by 0, MD by 1, and DK by 2 in Table 10.6, we obtain the table for $\mathbb{Z}_3$. Thus the table of shadings forms a group! We will see in

10.5 Table

$+_6$	0	3	1	4	2	5
0	0	3	1	4	2	5
3	3	0	4	1	5	2
1	1	4	2	5	3	0
4	4	1	5	2	0	3
2	2	5	3	0	4	1
5	5	2	0	3	1	4

10.6 Table

	LT	MD	DK
LT	LT	MD	DK
MD	MD	DK	LT
DK	DK	LT	MD

Section 14 that for a partition of an *abelian* group into cosets of a subgroup, reordering the group table according to the elements in the cosets always gives rise to such a *coset group*.

10.7 Example Table 10.8 again shows Table 8.8 for the symmetric group S_3 on three letters. Let H be the subgroup $\langle \mu_1 \rangle = \{\rho_0, \mu_1\}$ of S_3. Find the partitions of S_3 into left cosets of H, and the partition into right cosets of H.

Solution For the partition into left cosets, we have

$$\begin{aligned} H &= \{\rho_0, \mu_1\}, \\ \rho_1 H &= \{\rho_1\rho_0, \rho_1\mu_1\} = \{\rho_1, \mu_3\}, \\ \rho_2 H &= \{\rho_2\rho_0, \rho_2\mu_1\} = \{\rho_2, \mu_2\}. \end{aligned}$$

The partition into right cosets is

$$\begin{aligned} H &= \{\rho_0, \mu_1\}, \\ H\rho_1 &= \{\rho_0\rho_1, \mu_1\rho_1\} = \{\rho_1, \mu_2\}, \\ H\rho_2 &= \{\rho_0\rho_2, \mu_1\rho_2\} = \{\rho_2, \mu_3\}. \end{aligned}$$

The partition into left cosets of H is different from the partition into right cosets. For example, the left coset containing ρ_1 is $\{\rho_1, \mu_3\}$, while the right coset containing ρ_1 is $\{\rho_1, \mu_2\}$. This does not surprise us since the group S_3 is not abelian. ▲

Referring to Example 10.7, Table 10.9 gives permutation multiplication in S_3. The elements are listed in the order they appear in the left cosets $\{\rho_0, \mu_1\}, \{\rho_1, \mu_3\}, \{\rho_2, \mu_2\}$ found in that example. Again, we have shaded the table light, medium, and dark according to the coset to which the element belongs. Note the difference between this table and Table 10.5. This time, the body of the table does not split up into 2×2 blocks opposite and under the shaded cosets at the left and the top, as in Table 10.5 and we don't get a coset group. The product of a light element and a dark one may be either dark or medium.

Table 10.8 is shaded according to the two left cosets of the subgroup $\langle \rho_1 \rangle = \{\rho_0, \rho_1, \rho_2\}$ of S_3. These are also the two right cosets, even though S_3 is not abelian.

10.8 Table

	ρ_0	ρ_1	ρ_2	μ_1	μ_2	μ_3
ρ_0	ρ_0	ρ_1	ρ_2	μ_1	μ_2	μ_3
ρ_1	ρ_1	ρ_2	ρ_0	μ_3	μ_1	μ_2
ρ_2	ρ_2	ρ_0	ρ_1	μ_2	μ_3	μ_1
μ_1	μ_1	μ_2	μ_3	ρ_0	ρ_1	ρ_2
μ_2	μ_2	μ_3	μ_1	ρ_2	ρ_0	ρ_1
μ_3	μ_3	μ_1	μ_2	ρ_1	ρ_2	ρ_0

10.9 Table

	ρ_0	μ_1	ρ_1	μ_3	ρ_2	μ_2
ρ_0	ρ_0	μ_1	ρ_1	μ_3	ρ_2	μ_2
μ_1	μ_1	ρ_0	μ_2	ρ_2	μ_3	ρ_1
ρ_1	ρ_1	μ_3	ρ_2	μ_2	ρ_0	μ_1
μ_3	μ_3	ρ_1	μ_1	ρ_0	μ_2	ρ_2
ρ_2	ρ_2	μ_2	ρ_0	μ_1	ρ_1	μ_3
μ_2	μ_2	ρ_2	μ_3	ρ_1	μ_1	ρ_0

From Table 10.8 it is clear that we do have a coset group isomorphic to $\mathbb{Z}_2$ in this case. We will see in Section 14 that the left cosets of a subgroup H of a group G give rise to a coset group precisely when the partition of G into left cosets of H is the same as the partition into right cosets of H. In such a case, we may simply speak of the *cosets of* H, omitting the adjective left or right. We discuss coset groups in detail in Section 14, but we think it will be easier for you to understand them then if you experiment a bit with them now. Some of the exercises in this section are designed for such experimentation.

The Theorem of Lagrange

Let H be a subgroup of a group G. We claim that every left coset and every right coset of H have the same number of elements as H. We show this by exhibiting a *one-to-one* map of H *onto* a left coset gH of H for a fixed element g of G. If H is of finite order, this will show that gH has the same number of elements as H. If H is infinite, the existence of such a map is taken as the *definition* for equality of the size of H and the size of gH. (See Definition 0.13.)

Our choice for a one-to-one map $\phi : H \to gH$ is the natural one. Let $\phi(h) = gh$ for each $h \in H$. This map is onto gH by the definition of gH as $\{gh \mid h \in H\}$. To show that it is one to one, suppose that $\phi(h_1) = \phi(h_2)$ for h_1 and h_2 in H. Then $gh_1 = gh_2$ and by the cancellation law in the group G, we have $h_1 = h_2$. Thus ϕ is one to one.

Of course, a similar one-to-one map of H onto the right coset Hg can be constructed. (See Exercise 27.) We summarize as follows:

> Every coset (left or right) of a subgroup H of a group G has the same number of elements as H.

We can now prove the theorem of Lagrange.

10.10 Theorem **(Theorem of Lagrange)** Let H be a subgroup of a finite group G. Then the order of H is a divisor of the order of G.

Proof Let n be the order of G, and let H have order m. The preceding boxed statement shows that every coset of H also has m elements. Let r be the number of cells in the partition of G into left cosets of H. Then $n = rm$, so m is indeed a divisor of n. ◆

Note that this elegant and important theorem comes from the simple counting of cosets and the number of elements in each coset. *Never underestimate results that count something*! We continue to derive consequences of Theorem 10.10, which should be regarded as a counting theorem.

10.11 Corollary Every group of prime order is cyclic.

Proof Let G be of prime order p, and let a be an element of G different from the identity. Then the cyclic subgroup $\langle a \rangle$ of G generated by a has at least two elements, a and e. But by

Theorem 10.10, the order $m \geq 2$ of $\langle a \rangle$ must divide the prime p. Thus we must have $m = p$ and $\langle a \rangle = G$, so G is cyclic. ◆

Since every cyclic group of order p is isomorphic to $\mathbb{Z}_p$, we see that *there is only one group structure, up to isomorphism, of a given prime order p*. Now doesn't this elegant result follow easily from the theorem of Lagrange, a *counting* theorem? *Never underestimate a theorem that counts something.* Proving the preceding corollary is a favorite examination question.

10.12 Theorem The order of an element of a finite group divides the order of the group.

Proof Remembering that the order of an element is the same as the order of the cyclic subgroup generated by the element, we see that this theorem follows directly from Theorem 10.10. ◆

10.13 Definition Let H be a subgroup of a group G. The number of left cosets of H in G is the **index** $(G : H)$ **of H in G**. ■

The index $(G : H)$ just defined may be finite or infinite. If G is finite, then obviously $(G : H)$ is finite and $(G : H) = |G|/|H|$, since every coset of H contains $|H|$ elements. Exercise 35 shows the index $(G : H)$ could be equally well defined as the number of right cosets of H in G. We state a basic theorem concerning indices of subgroups, and leave the proof to the exercises (see Exercise 38).

10.14 Theorem Suppose H and K are subgroups of a group G such that $K \leq H \leq G$, and suppose $(H : K)$ and $(G : H)$ are both finite. Then $(G : K)$ is finite, and $(G : K) = (G : H)(H : K)$.

Theorem 10.10 shows that if there is a subgroup H of a finite group G, then the order of H divides the order of G. Is the converse true? That is, if G is a group of order n, and m divides n, is there always a subgroup of order m? We will see in the next section that this is true for abelian groups. However, A_4 can be shown to have no subgroup of order 6, which gives a counterexample for nonabelian groups.

EXERCISES 10

Computations

1. Find all cosets of the subgroup $4\mathbb{Z}$ of $\mathbb{Z}$.

2. Find all cosets of the subgroup $4\mathbb{Z}$ of $2\mathbb{Z}$.

3. Find all cosets of the subgroup $\langle 2 \rangle$ of $\mathbb{Z}_{12}$.

4. Find all cosets of the subgroup $\langle 4 \rangle$ of $\mathbb{Z}_{12}$.

5. Find all cosets of the subgroup $\langle 18 \rangle$ of $\mathbb{Z}_{36}$.

6. Find all left cosets of the subgroup $\{\rho_0, \mu_2\}$ of the group D_4 given by Table 8.12.

7. Repeat the preceding exercise, but find the right cosets this time. Are they the same as the left cosets?

8. Rewrite Table 8.12 in the order exhibited by the left cosets in Exercise 6. Do you seem to get a coset group of order 4? If so, is it isomorphic to $\mathbb{Z}_4$ or to the Klein 4-group V?

9. Repeat Exercise 6 for the subgroup $\{\rho_0, \rho_2\}$ of D_4.

10. Repeat the preceding exercise, but find the right cosets this time. Are they the same as the left coset?

11. Rewrite Table 8.12 in the order exhibited by the left cosets in Exercise 9. Do you seem to get a coset group of order 4? If so, is it isomorphic to $\mathbb{Z}_4$ or to the Klein 4-group V?

12. Find the index of $\langle 3 \rangle$ in the group $\mathbb{Z}_{24}$.

13. Find the index of $\langle \mu_1 \rangle$ in the group S_3, using the notation of Example 10.7

14. Find the index of $\langle \mu_2 \rangle$ in the group D_4 given in Table 8.12

15. Let $\sigma = (1, 2, 5, 4)(2, 3)$ in S_5. Find the index of $\langle \sigma \rangle$ in S_5.

16. Let $\mu = (1, 2, 4, 5)(3, 6)$ in S_6. Find the index of $\langle \mu \rangle$ in S_6.

Concepts

In Exercises 17 and 18, correct the definition of the italicized term without reference to the text, if correction is needed, so that it is in a form acceptable for publication.

17. Let G be a group and let $H \subseteq G$. The *left coset of H containing a* is $aH = \{ah \mid h \in H\}$.

18. Let G be a group and let $H \leq G$. The *index of H in G* is the number of right cosets of H in G.

19. Mark each of the following true or false.

______ **a.** Every subgroup of every group has left cosets.

______ **b.** The number of left cosets of a subgroup of a finite group divides the order of the group.

______ **c.** Every group of prime order is abelian.

______ **d.** One cannot have left cosets of a finite subgroup of an infinite group.

______ **e.** A subgroup of a group is a left coset of itself.

______ **f.** Only subgroups of finite groups can have left cosets.

______ **g.** A_n is of index 2 in S_n for $n > 1$.

______ **h.** The theorem of Lagrange is a nice result.

______ **i.** Every finite group contains an element of every order that divides the order of the group.

______ **j.** Every finite cyclic group contains an element of every order that divides the order of the group.

In Exercises 20 through 24, give an example of the desired subgroup and group if possible. If impossible, say why it is impossible.

20. A subgroup of an abelian group G whose left cosets and right cosets give different partitions of G

21. A subgroup of a group G whose left cosets give a partition of G into just one cell

22. A subgroup of a group of order 6 whose left cosets give a partition of the group into 6 cells

23. A subgroup of a group of order 6 whose left cosets give a partition of the group into 12 cells

24. A subgroup of a group of order 6 whose left cosets give a partition of the group into 4 cells

Proof Synopsis

25. Give a one-sentence synopsis of the proof of Theorem 10.10.

Theory

26. Prove that the relation $\sim_R$ of Theorem 10.1 is an equivalence relation.

27. Let H be a subgroup of a group G and let $g \in G$. Define a one-to-one map of H onto Hg. Prove that your map is one to one and is onto Hg.

28. Let H be a subgroup of a group G such that $g^{-1}hg \in H$ for all $g \in G$ and all $h \in H$. Show that every left coset gH is the same as the right coset Hg.

29. Let H be a subgroup of a group G. Prove that if the partition of G into left cosets of H is the same as the partition into right cosets of H, then $g^{-1}hg \in H$ for all $g \in G$ and all $h \in H$. (Note that this is the converse of Exercise 28.)

Let H be a subgroup of a group G and let $a, b \in G$. In Exercises 30 through 33 prove the statement or give a counterexample.

30. If $aH = bH$, then $Ha = Hb$.

31. If $Ha = Hb$, then $b \in Ha$.

32. If $aH = bH$, then $Ha^{-1} = Hb^{-1}$.

33. If $aH = bH$, then $a^2H = b^2H$.

34. Let G be a group of order pq, where p and q are prime numbers. Show that every proper subgroup of G is cyclic.

35. Show that there are the same number of left as right cosets of a subgroup H of a group G; that is, exhibit a one-to-one map of the collection of left cosets onto the collection of right cosets. (Note that this result is obvious by counting for finite groups. Your proof must hold for any group.)

36. Exercise 29 of Section 4 showed that every finite group of even order $2n$ contains an element of order 2. Using the theorem of Lagrange, show that if n is odd, then an abelian group of order $2n$ contains precisely one element of order 2.

37. Show that a group with at least two elements but with no proper nontrivial subgroups must be finite and of prime order.

38. Prove Theorem 10.14 [*Hint:* Let $\{a_i H \mid i = 1, \cdots, r\}$ be the collection of distinct left cosets of H in G and $\{b_j K \mid j = 1, \cdots, s\}$ be the collection of distinct left cosets of K in H. Show that

$$\{(a_i b_j)K \mid i = 1, \cdots, r; j = 1, \cdots, s\}$$

is the collection of distinct left cosets of K in G.]

39. Show that if H is a subgroup of index 2 in a finite group G, then every left coset of H is also a right coset of H.

40. Show that if a group G with identity e has finite order n, then $a^n = e$ for all $a \in G$.

41. Show that every left coset of the subgroup $\mathbb{Z}$ of the additive group of real numbers contains exactly one element x such that $0 \leq x < 1$.

42. Show that the function *sine* assigns the same value to each element of any fixed left coset of the subgroup $\langle 2\pi \rangle$ of the additive group $\mathbb{R}$ of real numbers. (Thus *sine* induces a well-defined function on the set of cosets; the value of the function on a coset is obtained when we choose an element x of the coset and compute $\sin x$.)

43. Let H and K be subgroups of a group G. Define $\sim$ on G by $a \sim b$ if and only if $a = hbk$ for some $h \in H$ and some $k \in K$.

a. Prove that $\sim$ is an equivalence relation on G.

b. Describe the elements in the equivalence class containing $a \in G$. (These equivalence classes are called **double cosets.**)

44. Let S_A be the group of all permutations of the set A, and let c be one particular element of A.

a. Show that $\{\sigma \in S_A \mid \sigma(c) = c\}$ is a subgroup $S_{c,c}$ of S_A.

b. Let $d \neq c$ be another particular element of A. Is $S_{c,d} = \{\sigma \in S_A \mid \sigma(c) = d\}$ a subgroup of S_A? Why or why not?

c. Characterize the set $S_{c,d}$ of part (b) in terms of the subgroup $S_{c,c}$ of part (a).

45. Show that a finite cyclic group of order n has exactly one subgroup of each order d dividing n, and that these are all the subgroups it has.

46. The **Euler phi-function** is defined for positive integers n by $\varphi(n) = s$, where s is the number of positive integers less than or equal to n that are relatively prime to n. Use Exercise 45 to show that

$$n = \sum_{d \mid n} \varphi(d),$$

the sum being taken over all positive integers d dividing n. [*Hint:* Note that the number of generators of $\mathbb{Z}_d$ is $\varphi(d)$ by Corollary 6.16.]

47. Let G be a finite group. Show that if for each positive integer m the number of solutions x of the equation $x^m = e$ in G is at most m, then G is cyclic. [*Hint:* Use Theorem 10.12 and Exercise 46 to show that G must contain an element of order $n = |G|$.]

SECTION 11 DIRECT PRODUCTS AND FINITELY GENERATED ABELIAN GROUPS

Direct Products

Let us take a moment to review our present stockpile of groups. Starting with finite groups, we have the cyclic group $\mathbb{Z}_n$, the symmetric group S_n, and the alternating group A_n for each positive integer n. We also have the dihedral groups D_n of Section 8, and the Klein 4-group V. Of course we know that subgroups of these groups exist. Turning to infinite groups, we have groups consisting of sets of numbers under the usual addition or multiplication, as, for example, $\mathbb{Z}$, $\mathbb{R}$, and $\mathbb{C}$ under addition, and their nonzero elements under multiplication. We have the group U of complex numbers of magnitude 1 under multiplication, which is isomorphic to each of the groups $\mathbb{R}_c$ under addition modulo c, where $c \in \mathbb{R}^+$. We also have the group S_A of all permutations of an infinite set A, as well as various groups formed from matrices.

One purpose of this section is to show a way to use known groups as building blocks to form more groups. The Klein 4-group will be recovered in this way from the cyclic groups. Employing this procedure with the cyclic groups gives us a large class of abelian groups that can be shown to include all possible structure types for a finite abelian group. We start by generalizing Definition 0.4.

11.1 Definition The **Cartesian product of sets** $S_1, S_2, \cdots, S_n$ is the set of all ordered n-tuples $(a_1, a_2, \cdots, a_n)$, where $a_i \in S_i$ for $i = 1, 2, \cdots, n$. The Cartesian product is denoted by either

$$S_1 \times S_2 \times \cdots \times S_n$$

or by

$$\prod_{i=1}^{n} S_i. \quad \blacksquare$$

We could also define the Cartesian product of an infinite number of sets, but the definition is considerably more sophisticated and we shall not need it.

Now let $G_1, G_2, \cdots, G_n$ be groups, and let us use multiplicative notation for all the group operations. Regarding the G_i as sets, we can form $\prod_{i=1}^{n} G_i$. Let us show that we can make $\prod_{i=1}^{n} G_i$ into a group by means of a binary operation of *multiplication by components*. Note again that we are being sloppy when we use the same notation for a group as for the set of elements of the group.

11.2 Theorem Let $G_1, G_2, \cdots, G_n$ be groups. For $(a_1, a_2, \cdots, a_n)$ and $(b_1, b_2, \cdots, b_n)$ in $\prod_{i=1}^n G_i$, define $(a_1, a_2, \cdots, a_n)(b_1, b_2, \cdots, b_n)$ to be the element $(a_1b_1, a_2b_2, \cdots, a_nb_n)$. Then $\prod_{i=1}^n G_i$ is a group, the **direct product of the groups** G_i, under this binary operation.

Proof Note that since $a_i \in G_i$, $b_i \in G_i$, and G_i is a group, we have $a_ib_i \in G_i$. Thus the definition of the binary operation on $\prod_{i=1}^n G_i$ given in the statement of the theorem makes sense; that is, $\prod_{i=1}^n G_i$ is closed under the binary operation.

The associative law in $\prod_{i=1}^n G_i$ is thrown back onto the associative law in each component as follows:

$$\begin{aligned}
&(a_1, a_2, \cdots, a_n)[(b_1, b_2, \cdots, b_n)(c_1, c_2, \cdots, c_n)]\\
&\quad = (a_1, a_2, \cdots, a_n)(b_1c_1, b_2c_2, \cdots, b_nc_n)\\
&\quad = (a_1(b_1c_1), a_2(b_2c_2), \cdots, a_n(b_nc_n))\\
&\quad = ((a_1b_1)c_1, (a_2b_2)c_2, \cdots, (a_nb_n)c_n)\\
&\quad = (a_1b_1, a_2b_2, \cdots, a_nb_n)(c_1, c_2, \cdots, c_n)\\
&\quad = [(a_1, a_2, \cdots, a_n)(b_1, b_2, \cdots, b_n)](c_1, c_2, \cdots, c_n).
\end{aligned}$$

If e_i is the identity element in G_i, then clearly, with multiplication by components, $(e_1, e_2, \cdots, e_n)$ is an identity in $\prod_{i=1}^n G_i$. Finally, an inverse of $(a_1, a_2, \cdots, a_n)$ is $(a_1^{-1}, a_2^{-1}, \cdots, a_n^{-1})$; compute the product by components. Hence $\prod_{i=1}^n G_i$ is a group. ◆

In the event that the operation of each G_i is commutative, we sometimes use additive notation in $\prod_{i=1}^n G_i$ and refer to $\prod_{i=1}^n G_i$ as the **direct sum of the groups** G_i. The notation $\oplus_{i=1}^n G_i$ is sometimes used in this case in place of $\prod_{i=1}^n G_i$, especially with abelian groups with operation $+$. The direct sum of abelian groups $G_1, G_2, \cdots, G_n$ may be written $G_1 \oplus G_2 \oplus \cdots \oplus G_n$. We leave to Exercise 46 the proof that a direct product of abelian groups is again abelian.

It is quickly seen that if the S_i has r_i elements for $i = 1, \cdots, n$, then $\prod_{i=1}^n S_i$ has $r_1r_2 \cdots r_n$ elements, for in an n-tuple, there are r_1 choices for the first component from S_1, and for each of these there are r_2 choices for the next component from S_2, and so on.

11.3 Example Consider the group $\mathbb{Z}_2 \times \mathbb{Z}_3$, which has $2 \cdot 3 = 6$ elements, namely (0, 0), (0, 1), (0, 2), (1, 0), (1, 1), and (1, 2). We claim that $\mathbb{Z}_2 \times \mathbb{Z}_3$ is cyclic. It is only necessary to find a generator. Let us try (1, 1). Here the operations in $\mathbb{Z}_2$ and $\mathbb{Z}_3$ are written additively, so we do the same in the direct product $\mathbb{Z}_2 \times \mathbb{Z}_3$.

$$\begin{aligned}
(1, 1) &= (1, 1)\\
2(1, 1) &= (1, 1) + (1, 1) = (0, 2)\\
3(1, 1) &= (1, 1) + (1, 1) + (1, 1) = (1, 0)\\
4(1, 1) &= 3(1, 1) + (1, 1) = (1, 0) + (1, 1) = (0, 1)\\
5(1, 1) &= 4(1, 1) + (1, 1) = (0, 1) + (1, 1) = (1, 2)\\
6(1, 1) &= 5(1, 1) + (1, 1) = (1, 2) + (1, 1) = (0, 0)
\end{aligned}$$

Thus (1, 1) generates all of $\mathbb{Z}_2 \times \mathbb{Z}_3$. Since there is, up to isomorphism, only one cyclic group structure of a given order, we see that $\mathbb{Z}_2 \times \mathbb{Z}_3$ is isomorphic to $\mathbb{Z}_6$. ▲

11.4 Example Consider $\mathbb{Z}_3 \times \mathbb{Z}_3$. This is a group of nine elements. We claim that $\mathbb{Z}_3 \times \mathbb{Z}_3$ is *not* cyclic. Since the addition is by components, and since in $\mathbb{Z}_3$ every element added to itself three times gives the identity, the same is true in $\mathbb{Z}_3 \times \mathbb{Z}_3$. Thus no element can generate the group, for a generator added to itself successively could only give the identity after nine summands. We have found another group structure of order 9. A similar argument shows that $\mathbb{Z}_2 \times \mathbb{Z}_2$ is not cyclic. Thus $\mathbb{Z}_2 \times \mathbb{Z}_2$ must be isomorphic to the Klein 4-group. ▲

The preceding examples illustrate the following theorem:

11.5 Theorem The group $\mathbb{Z}_m \times \mathbb{Z}_n$ is cyclic and is isomorphic to $\mathbb{Z}_{mn}$ if and only if m and n are relatively prime, that is, the gcd of m and n is 1.

Proof Consider the cyclic subgroup of $\mathbb{Z}_m \times \mathbb{Z}_n$ generated by (1, 1) as described by Theorem 5.17. As our previous work has shown, the order of this cyclic subgroup is the smallest power of (1, 1) that gives the identity (0, 0). Here taking a power of (1, 1) in our additive notation will involve adding (1, 1) to itself repeatedly. Under addition by components, the first component $1 \in \mathbb{Z}_m$ yields 0 only after m summands, $2m$ summands, and so on, and the second component $1 \in \mathbb{Z}_n$ yields 0 only after n summands, $2n$ summands, and so on. For them to yield 0 simultaneously, the number of summands must be a multiple of both m and n. The smallest number that is a multiple of both m and n will be mn if and only if the gcd of m and n is 1; in this case, (1, 1) generates a cyclic subgroup of order mn, which is the order of the whole group. This shows that $\mathbb{Z}_m \times \mathbb{Z}_n$ is cyclic of order mn, and hence isomorphic to $\mathbb{Z}_{mn}$ if m and n are relatively prime.

For the converse, suppose that the gcd of m and n is $d > 1$. The mn/d is divisible by both m and n. Consequently, for any (r, s) in $\mathbb{Z}_m \times \mathbb{Z}_n$, we have

$$\underbrace{(r, s) + (r, s) + \cdots + (r, s)}_{mn/d \text{ summands}} = (0, 0).$$

Hence no element (r, s) in $\mathbb{Z}_m \times \mathbb{Z}_n$ can generate the entire group, so $\mathbb{Z}_m \times \mathbb{Z}_n$ is not cyclic and therefore not isomorphic to $\mathbb{Z}_{mn}$. ◆

This theorem can be extended to a product of more than two factors by similar arguments. We state this as a corollary without going through the details of the proof.

11.6 Corollary The group $\prod_{i=1}^{n} \mathbb{Z}_{m_i}$ is cyclic and isomorphic to $\mathbb{Z}_{m_1 m_2 \cdots m_n}$ if and only if the numbers m_i for $i = 1, \cdots, n$ are such that the gcd of any two of them is 1.

11.7 Example The preceding corollary shows that if n is written as a product of powers of distinct prime numbers, as in

$$n = (p_1)^{n_1}(p_2)^{n_2} \cdots (p_r)^{n_r},$$

then $\mathbb{Z}_n$ is isomorphic to

$$\mathbb{Z}_{(p_1)^{n_1}} \times \mathbb{Z}_{(p_2)^{n_2}} \times \cdots \times \mathbb{Z}_{(p_r)^{n_r}}.$$

In particular, $\mathbb{Z}_{72}$ is isomorphic to $\mathbb{Z}_8 \times \mathbb{Z}_9$. ▲

We remark that changing the order of the factors in a direct product yields a group isomorphic to the original one. The names of elements have simply been changed via a permutation of the components in the n-tuples.

Exercise 47 of Section 6 asked you to define the least common multiple of two positive integers r and s as a generator of a certain cyclic group. It is straightforward to prove that the subset of $\mathbb{Z}$ consisting of all integers that are multiples of both r and s is a subgroup of $\mathbb{Z}$, and hence is a cyclic group. Likewise, the set of all common multiples of n positive integers $r_1, r_2, \cdots, r_n$ is a subgroup of $\mathbb{Z}$, and hence is cyclic.

11.8 Definition Let $r_1, r_2, \cdots, r_n$ be positive integers. Their **least common multiple** (abbreviated lcm) is the positive generator of the cyclic group of all common multiples of the r_i, that is, the cyclic group of all integers divisible by each r_i for $i = 1, 2, \cdots, n$. ■

From Definition 11.8 and our work on cyclic groups, we see that the lcm of $r_1, r_2, \cdots, r_n$ is the smallest positive integer that is a multiple of each r_i for $i = 1, 2, \cdots, n$, hence the name *least common multiple*.

11.9 Theorem Let $(a_1, a_2, \cdots, a_n) \in \prod_{i=1}^n G_i$. If a_i is of finite order r_i in G_i, then the order of $(a_1, a_2, \cdots, a_n)$ in $\prod_{i=1}^n G_i$ is equal to the least common multiple of all the r_i.

Proof This follows by a repetition of the argument used in the proof of Theorem 11.5. For a power of $(a_1, a_2, \cdots, a_n)$ to give $(e_1, e_2, \cdots, e_n)$, the power must simultaneously be a multiple of r_1 so that this power of the first component a_1 will yield e_1, a multiple of r_2, so that this power of the second component a_2 will yield e_2, and so on. ◆

11.10 Example Find the order of (8, 4, 10) in the group $\mathbb{Z}_{12} \times \mathbb{Z}_{60} \times \mathbb{Z}_{24}$.

Solution Since the gcd of 8 and 12 is 4, we see that 8 is of order $\frac{12}{4} = 3$ in $\mathbb{Z}_{12}$. (See Theorem 6.14.) Similarly, we find that 4 is of order 15 in $\mathbb{Z}_{60}$ and 10 is of order 12 in $\mathbb{Z}_{24}$. The lcm of 3, 15, and 12 is $3 \cdot 5 \cdot 4 = 60$, so (8, 4, 10) is of order 60 in the group $\mathbb{Z}_{12} \times \mathbb{Z}_{60} \times \mathbb{Z}_{24}$. ▲

11.11 Example The group $\mathbb{Z} \times \mathbb{Z}_2$ is generated by the elements (1, 0) and (0, 1). More generally, the direct product of n cyclic groups, each of which is either $\mathbb{Z}$ or $\mathbb{Z}_m$ for some positive integer m, is generated by the n n-tuples

$$(1, 0, 0, \cdots, 0), \quad (0, 1, 0, \cdots, 0), \quad (0, 0, 1, \cdots, 0), \quad \cdots, \quad (0, 0, 0, \cdots, 1).$$

Such a direct product might also be generated by fewer elements. For example, $\mathbb{Z}_3 \times \mathbb{Z}_4 \times \mathbb{Z}_{35}$ is generated by the single element (1, 1, 1). ▲

Note that if $\prod_{i=1}^n G_i$ is the direct product of groups G_i, then the subset

$$\overline{G}_i = \{(e_1, e_2, \cdots, e_{i-1}, a_i, e_{i+1}, \cdots, e_n) \mid a_i \in G_i\},$$

that is, the set of all n-tuples with the identity elements in all places but the ith, is a subgroup of $\prod_{i=1}^n G_i$. It is also clear that this subgroup $\overline{G}_i$ is naturally isomorphic to G_i; just rename

$$(e_1, e_2, \cdots, e_{i-1}, a_i, e_{i+1}, \cdots, e_n) \text{ by } a_i.$$

The group G_i is mirrored in the ith component of the elements of $\overline{G}_i$, and the e_j in the other components just ride along. We consider $\prod_{i=1}^{n} G_i$ to be the *internal direct product* of these subgroups $\overline{G}_i$. The direct product given by Theorem 11.2 is called the *external direct product* of the groups G_i. The terms *internal* and *external,* as applied to a direct product of groups, just reflect whether or not (respectively) we are regarding the component groups as subgroups of the product group. We shall usually omit the words *external* and *internal* and just say *direct product*. Which term we mean will be clear from the context.

Historical Note

In his *Disquisitiones Arithmeticae,* Carl Gauss demonstrated various results in what is today the theory of abelian groups in the context of number theory. Not only did he deal extensively with equivalence classes of quadratic forms, but he also considered residue classes modulo a given integer. Although he noted that results in these two areas were similar, he did not attempt to develop an abstract theory of abelian groups.

In the 1840s, Ernst Kummer in dealing with ideal complex numbers noted that his results were in many respects analogous to those of Gauss. (See the Historical Note in Section 26.) But it was Kummer's student Leopold Kronecker (see the Historical Note in Section 29) who finally realized that an abstract theory could be developed out of the analogies. As he wrote in 1870, "these principles [from the work of Gauss and Kummer] belong to a more general, abstract realm of ideas. It is therefore appropriate to free their development from all unimportant restrictions, so that one can spare oneself from the necessity of repeating the same argument in different cases. This advantage already appears in the development itself, and the presentation gains in simplicity, if it is given in the most general admissible manner, since the most important features stand out with clarity." Kronecker then proceeded to develop the basic principles of the theory of finite abelian groups and was able to state and prove a version of Theorem 11.12 restricted to finite groups.

The Structure of Finitely Generated Abelian Groups

Some theorems of abstract algebra are easy to understand and use, although their proofs may be quite technical and time-consuming to present. This is one section in the text where we explain the meaning and significance of a theorem but omit its proof. The meaning of any theorem whose proof we omit is well within our understanding, and we feel we should be acquainted with it. It would be impossible for us to meet some of these fascinating facts in a one-semester course if we were to insist on wading through complete proofs of all theorems. The theorem that we now state gives us complete structural information about all sufficiently small abelian groups, in particular, about all finite abelian groups.

11.12 Theorem **(Fundamental Theorem of Finitely Generated Abelian Groups)** Every finitely generated abelian group G is isomorphic to a direct product of cyclic groups in the form

$$\mathbb{Z}_{(p_1)^{r_1}} \times \mathbb{Z}_{(p_2)^{r_2}} \times \cdots \times \mathbb{Z}_{(p_n)^{r_n}} \times \mathbb{Z} \times \mathbb{Z} \times \cdots \times \mathbb{Z},$$

where the p_i are primes, not necessarily distinct, and the r_i are positive integers. The direct product is unique except for possible rearrangement of the factors; that is, the number (**Betti number** of G) of factors $\mathbb{Z}$ is unique and the prime powers $(p_i)^{r_i}$ are unique.

Proof The proof is omitted here. ◆

11.13 Example Find all abelian groups, up to isomorphism, of order 360. The phrase *up to isomorphism* signifies that any abelian group of order 360 should be structurally identical (isomorphic) to one of the groups of order 360 exhibited.

Solution We make use of Theorem 11.12. Since our groups are to be of the finite order 360, no factors $\mathbb{Z}$ will appear in the direct product shown in the statement of the theorem.

First we express 360 as a product of prime powers $2^3 3^2 5$. Then using Theorem 11.12, we get as possibilities

1. $\mathbb{Z}_2 \times \mathbb{Z}_2 \times \mathbb{Z}_2 \times \mathbb{Z}_3 \times \mathbb{Z}_3 \times \mathbb{Z}_5$
2. $\mathbb{Z}_2 \times \mathbb{Z}_4 \times \mathbb{Z}_3 \times \mathbb{Z}_3 \times \mathbb{Z}_5$
3. $\mathbb{Z}_2 \times \mathbb{Z}_2 \times \mathbb{Z}_2 \times \mathbb{Z}_9 \times \mathbb{Z}_5$
4. $\mathbb{Z}_2 \times \mathbb{Z}_4 \times \mathbb{Z}_9 \times \mathbb{Z}_5$
5. $\mathbb{Z}_8 \times \mathbb{Z}_3 \times \mathbb{Z}_3 \times \mathbb{Z}_5$
6. $\mathbb{Z}_8 \times \mathbb{Z}_9 \times \mathbb{Z}_5$

Thus there are six different abelian groups (up to isomorphism) of order 360. ▲

Applications

We conclude this section with a sampling of the many theorems we could now prove regarding abelian groups.

11.14 Definition A group G is **decomposable** if it is isomorphic to a direct product of two proper nontrivial subgroups. Otherwise G is **indecomposable**. ■

11.15 Theorem The finite indecomposable abelian groups are exactly the cyclic groups with order a power of a prime.

Proof Let G be a finite indecomposable abelian group. Then by Theorem 11.12, G is isomorphic to a direct product of cyclic groups of prime power order. Since G is indecomposable, this direct product must consist of just one cyclic group whose order is a power of a prime number.

Conversely, let p be a prime. Then $\mathbb{Z}_{p^r}$ is indecomposable, for if $\mathbb{Z}_{p^r}$ were isomorphic to $\mathbb{Z}_{p^i} \times \mathbb{Z}_{p^j}$, where $i + j = r$, then every element would have an order at most $p^{\max(i,j)} < p^r$. ◆

11.16 Theorem If m divides the order of a finite abelian group G, then G has a subgroup of order m.

Proof By Theorem 11.12, we can think of G as being

$$\mathbb{Z}_{(p_1)^{r_1}} \times \mathbb{Z}_{(p_2)^{r_2}} \times \cdots \times \mathbb{Z}_{(p_n)^{r_n}},$$

where not all primes p_i need be distinct. Since $(p_1)^{r_1}(p_2)^{r_2}\cdots(p_n)^{r_n}$ is the order of G, then m must be of the form $(p_1)^{s_1}(p_2)^{s_2}\cdots(p_n)^{s_n}$, where $0 \le s_i \le r_i$. By Theorem 6.14, $(p_i)^{r_i-s_i}$ generates a cyclic subgroup of $\mathbb{Z}_{(p_i)^{r_i}}$ of order equal to the quotient of $(p_i)^{r_i}$ by the gcd of $(p_i)^{r_i}$ and $(p_i)^{r_i-s_i}$. But the gcd of $(p_i)^{r_i}$ and $(p_i)^{r_i-s_i}$ is $(p_i)^{r_i-s_i}$. Thus $(p_i)^{r_i-s_i}$ generates a cyclic subgroup of $\mathbb{Z}_{(p_i)^{r_i}}$ of order

$$[(p_i)^{r_i}]/[(p_i)^{r_i-s_i}] = (p_i)^{s_i}.$$

Recalling that $\langle a\rangle$ denotes the cyclic subgroup generated by a, we see that

$$\langle (p_1)^{r_1-s_1}\rangle \times \langle (p_2)^{r_2-s_2}\rangle \times \cdots \times \langle (p_n)^{r_n-s_n}\rangle$$

is the required subgroup of order m. ◆

11.17 Theorem If m is a square free integer, that is, m is not divisible by the square of any prime, then every abelian group of order m is cyclic.

Proof Let G be an abelian group of square free order m. Then by Theorem 11.12, G is isomorphic to

$$\mathbb{Z}_{(p_1)^{r_1}} \times \mathbb{Z}_{(p_2)^{r_2}} \times \cdots \times \mathbb{Z}_{(p_n)^{r_n}},$$

where $m = (p_1)^{r_1}(p_2)^{r_2}\cdots(p_n)^{r_n}$. Since m is square free, we must have all $r_i = 1$ and all p_i distinct primes. Corollary 11.6 then shows that G is isomorphic to $\mathbb{Z}_{p_1p_2\cdots p_n}$, so G is cyclic. ◆

EXERCISES 11

Computations

1. List the elements of $\mathbb{Z}_2 \times \mathbb{Z}_4$. Find the order of each of the elements. Is this group cyclic?

2. Repeat Exercise 1 for the group $\mathbb{Z}_3 \times \mathbb{Z}_4$.

In Exercises 3 through 7, find the order of the given element of the direct product.

3. (2, 6) in $\mathbb{Z}_4 \times \mathbb{Z}_{12}$ **4.** (2, 3) in $\mathbb{Z}_6 \times \mathbb{Z}_{15}$ **5.** (8, 10) in $\mathbb{Z}_{12} \times \mathbb{Z}_{18}$

6. (3, 10, 9) in $\mathbb{Z}_4 \times \mathbb{Z}_{12} \times \mathbb{Z}_{15}$ **7.** (3, 6, 12, 16) in $\mathbb{Z}_4 \times \mathbb{Z}_{12} \times \mathbb{Z}_{20} \times \mathbb{Z}_{24}$

8. What is the largest order among the orders of all the cyclic subgroups of $\mathbb{Z}_6 \times \mathbb{Z}_8$? of $\mathbb{Z}_{12} \times \mathbb{Z}_{15}$?

9. Find all proper nontrivial subgroups of $\mathbb{Z}_2 \times \mathbb{Z}_2$.

10. Find all proper nontrivial subgroups of $\mathbb{Z}_2 \times \mathbb{Z}_2 \times \mathbb{Z}_2$.

11. Find all subgroups of $\mathbb{Z}_2 \times \mathbb{Z}_4$ of order 4.

12. Find all subgroups of $\mathbb{Z}_2 \times \mathbb{Z}_2 \times \mathbb{Z}_4$ that are isomorphic to the Klein 4-group.

13. Disregarding the order of the factors, write direct products of two or more groups of the form $\mathbb{Z}_n$ so that the resulting product is isomorphic to $\mathbb{Z}_{60}$ in as many ways as possible.

14. Fill in the blanks.

a. The cyclic subgroup of $\mathbb{Z}_{24}$ generated by 18 has order___.

b. $\mathbb{Z}_3 \times \mathbb{Z}_4$ is of order___.

c. The element (4, 2) of $\mathbb{Z}_{12} \times \mathbb{Z}_8$ has order__.

d. The Klein 4-group is isomorphic to $\mathbb{Z}_{\_\_} \times \mathbb{Z}_{\_\_}$.

e. $\mathbb{Z}_2 \times \mathbb{Z} \times \mathbb{Z}_4$ has__elements of finite order.

15. Find the maximum possible order for some element of $\mathbb{Z}_4 \times \mathbb{Z}_6$.

16. Are the groups $\mathbb{Z}_2 \times \mathbb{Z}_{12}$ and $\mathbb{Z}_4 \times \mathbb{Z}_6$ isomorphic? Why or why not?

17. Find the maximum possible order for some element of $\mathbb{Z}_8 \times \mathbb{Z}_{10} \times \mathbb{Z}_{24}$.

18. Are the groups $\mathbb{Z}_8 \times \mathbb{Z}_{10} \times \mathbb{Z}_{24}$ and $\mathbb{Z}_4 \times \mathbb{Z}_{12} \times \mathbb{Z}_{40}$ isomorphic? Why or why not?

19. Find the maximum possible order for some element of $\mathbb{Z}_4 \times \mathbb{Z}_{18} \times \mathbb{Z}_{15}$.

20. Are the groups $\mathbb{Z}_4 \times \mathbb{Z}_{18} \times \mathbb{Z}_{15}$ and $\mathbb{Z}_3 \times \mathbb{Z}_{36} \times \mathbb{Z}_{10}$ isomorphic? Why or why not?

In Exercises 21 through 25, proceed as in Example 11.13 to find all abelian groups, up to isomorphism, of the given order.

21. Order 8 **22.** Order 16 **23.** Order 32

24. Order 720 **25.** Order 1089

26. How many abelian groups (up to isomorphism) are there of order 24? of order 25? of order (24)(25)?

27. Following the idea suggested in Exercise 26, let m and n be relatively prime positive integers. Show that if there are (up to isomorphism) r abelian groups of order m and s of order n, then there are (up to isomorphism) rs abelian groups of order mn.

28. Use Exercise 27 to determine the number of abelian groups (up to isomorphism) of order $(10)^5$.

29. a. Let p be a prime number. Fill in the second row of the table to give the number of abelian groups of order p^n, up to isomorphism.

n	2	3	4	5	6	7	8
number of groups							

b. Let p, q, and r be distinct prime numbers. Use the table you created to find the number of abelian groups, up to isomorphism, of the given order.

i. $p^3q^4r^7$ **ii.** $(qr)^7$ **iii.** $q^5r^4q^3$

30. Indicate schematically a Cayley digraph for $\mathbb{Z}_m \times \mathbb{Z}_n$ for the generating set $S = \{(1, 0), (0, 1)\}$.

31. Consider Cayley digraphs with two arc types, a solid one with an arrow and a dashed one with no arrow, and consisting of two regular n-gons, for $n \geq 3$, with solid arc sides, one inside the other, with dashed arcs joining the vertices of the outer n-gon to the inner one. Figure 7.9(b) shows such a Cayley digraph with $n = 3$, and Figure 7.11(b) shows one with $n = 4$. The arrows on the outer n-gon may have the same (clockwise or counterclockwise) direction as those on the inner n-gon, or they may have the opposite direction. Let G be a group with such a Cayley digraph.

a. Under what circumstances will G be abelian?

b. If G is abelian, to what familiar group is it isomorphic?

c. If G is abelian, under what circumstances is it cyclic?

d. If G is not abelian, to what group we have discussed is it isomorphic?

Concepts

32. Mark each of the following true or false.

_____ **a.** If G_1 and G_2 are any groups, then $G_1 \times G_2$ is always isomorphic to $G_2 \times G_1$.
_____ **b.** Computation in an external direct product of groups is easy if you know how to compute in each component group.
_____ **c.** Groups of finite order must be used to form an external direct product.
_____ **d.** A group of prime order could not be the internal direct product of two proper nontrivial subgroups.
_____ **e.** $\mathbb{Z}_2 \times \mathbb{Z}_4$ is isomorphic to $\mathbb{Z}_8$.
_____ **f.** $\mathbb{Z}_2 \times \mathbb{Z}_4$ is isomorphic to S_8.
_____ **g.** $\mathbb{Z}_3 \times \mathbb{Z}_8$ is isomorphic to S_4.
_____ **h.** Every element in $\mathbb{Z}_4 \times \mathbb{Z}_8$ has order 8.
_____ **i.** The order of $\mathbb{Z}_{12} \times \mathbb{Z}_{15}$ is 60.
_____ **j.** $\mathbb{Z}_m \times \mathbb{Z}_n$ has mn elements whether m and n are relatively prime or not.

33. Give an example illustrating that not every nontrivial abelian group is the internal direct product of two proper nontrivial subgroups.

34. a. How many subgroups of $\mathbb{Z}_5 \times \mathbb{Z}_6$ are isomorphic to $\mathbb{Z}_5 \times \mathbb{Z}_6$?
b. How many subgroups of $\mathbb{Z} \times \mathbb{Z}$ are isomorphic to $\mathbb{Z} \times \mathbb{Z}$?

35. Give an example of a nontrivial group that is not of prime order and is not the internal direct product of two nontrivial subgroups.

36. Mark each of the following true or false.

_____ **a.** Every abelian group of prime order is cyclic.
_____ **b.** Every abelian group of prime power order is cyclic.
_____ **c.** $\mathbb{Z}_8$ is generated by $\{4, 6\}$.
_____ **d.** $\mathbb{Z}_8$ is generated by $\{4, 5, 6\}$.
_____ **e.** All finite abelian groups are classified up to isomorphism by Theorem 11.12.
_____ **f.** Any two finitely generated abelian groups with the same Betti number are isomorphic.
_____ **g.** Every abelian group of order divisible by 5 contains a cyclic subgroup of order 5.
_____ **h.** Every abelian group of order divisible by 4 contains a cyclic subgroup of order 4.
_____ **i.** Every abelian group of order divisible by 6 contains a cyclic subgroup of order 6.
_____ **j.** Every finite abelian group has a Betti number of 0.

37. Let p and q be distinct prime numbers. How does the number (up to isomorphism) of abelian groups of order p^r compare with the number (up to isomorphism) of abelian groups of order q^r?

38. Let G be an abelian group of order 72.

a. Can you say how many subgroups of order 8 G has? Why, or why not?
b. Can you say how many subgroups of order 4 G has? Why, or why not?

39. Let G be an abelian group. Show that the elements of finite order in G form a subgroup. This subgroup is called the **torsion subgroup** of G.

Exercises 40 through 43 deal with the concept of the torsion subgroup just defined.

40. Find the order of the torsion subgroup of $\mathbb{Z}_4 \times \mathbb{Z} \times \mathbb{Z}_3$; of $\mathbb{Z}_{12} \times \mathbb{Z} \times \mathbb{Z}_{12}$.

41. Find the torsion subgroup of the multiplicative group $\mathbb{R}^*$ of nonzero real numbers.

42. Find the torsion subgroup T of the multiplicative group $\mathbb{C}^*$ of nonzero complex numbers.

43. An abelian group is **torsion free** if e is the only element of finite order. Use Theorem 11.12 to show that every finitely generated abelian group is the internal direct product of its torsion subgroup and of a torsion-free subgroup. (Note that $\{e\}$ may be the torsion subgroup, and is also torsion free.)

44. The part of the decomposition of G in Theorem 11.12 corresponding to the subgroups of prime-power order can also be written in the form $\mathbb{Z}_{m_1} \times \mathbb{Z}_{m_2} \times \cdots \times \mathbb{Z}_{m_r}$, where m_i divides m_{i+1} for $i = 1, 2, \cdots, r-1$. The numbers m_i can be shown to be unique, and are the **torsion coefficients** of G.

a. Find the torsion coefficients of $\mathbb{Z}_4 \times \mathbb{Z}_9$.

b. Find the torsion coefficients of $\mathbb{Z}_6 \times \mathbb{Z}_{12} \times \mathbb{Z}_{20}$.

c. Describe an algorithm to find the torsion coefficients of a direct product of cyclic groups.

Proof Synopsis

45. Give a two-sentence synopsis of the proof of Theorem 11.5.

Theory

46. Prove that a direct product of abelian groups is abelian.

47. Let G be an abelian group. Let H be the subset of G consisting of the identity e together with all elements of G of order 2. Show that H is a subgroup of G.

48. Following up the idea of Exercise 47 determine whether H will always be a subgroup for every abelian group G if H consists of the identity e together with all elements of G of order 3; of order 4. For what positive integers n will H always be a subgroup for every abelian group G, if H consists of the identity e together with all elements of G of order n? Compare with Exercise 48 of Section 5.

49. Find a counterexample of Exercise 47 with the hypothesis that G is abelian omitted.

Let H and K be subgroups of a group G. Exercises 50 and 51 ask you to establish necessary and sufficient criteria for G to appear as the internal direct product of H and K.

50. Let H and K be groups and let $G = H \times K$. Recall that both H and K appear as subgroups of G in a natural way. Show that these subgroups H (actually $H \times \{e\}$) and K (actually $\{e\} \times K$) have the following properties.

a. Every element of G is of the form hk for some $h \in H$ and $k \in K$.

b. $hk = kh$ for all $h \in H$ and $k \in K$.

c. $H \cap K = \{e\}$.

51. Let H and K be subgroups of a group G satisfying the three properties listed in the preceding exercise. Show that for each $g \in G$, the expression $g = hk$ for $h \in H$ and $k \in K$ is unique. Then let each g be renamed (h, k). Show that, under this renaming, G becomes structurally identical (isomorphic) to $H \times K$.

52. Show that a finite abelian group is not cyclic if and only if it contains a subgroup isomorphic to $\mathbb{Z}_p \times \mathbb{Z}_p$ for some prime p.

53. Prove that if a finite abelian group has order a power of a prime p, then the order of every element in the group is a power of p. Can the hypothesis of commutativity be dropped? Why, or why not?

54. Let G, H, and K be finitely generated abelian groups. Show that if $G \times K$ is isomorphic to $H \times K$, then $G \simeq H$.

SECTION 12 †PLANE ISOMETRIES

Consider the Euclidean plane $\mathbb{R}^2$. An **isometry of** $\mathbb{R}^2$ is a permutation $\phi : \mathbb{R}^2 \to \mathbb{R}^2$ that preserves distance, so that the distance between points P and Q is the same as the distance between the points $\phi(P)$ and $\phi(Q)$ for all points P and Q in $\mathbb{R}^2$. If ψ is also an isometry of $\mathbb{R}^2$, then the distance between $\psi(\phi(P))$ and $\psi(\phi(Q))$ must be the same as the distance between $\phi(P)$ and $\phi(Q)$, which in turn is the distance between P and Q, showing that the composition of two isometries is again an isometry. Since the identity map is an isometry and the inverse of an isometry is an isometry, we see that the isometries of $\mathbb{R}^2$ form a subgroup of the group of all permutations of $\mathbb{R}^2$.

Given any subset S of $\mathbb{R}^2$, the isometries of $\mathbb{R}^2$ that carry S onto itself form a subgroup of the group of isometries. This subgroup is the **group of symmetries of** S **in** $\mathbb{R}^2$. In Section 8 we gave tables for the group of symmetries of an equilateral triangle and for the group of symmetries of a square in $\mathbb{R}^2$.

Everything we have defined in the two preceding paragraphs could equally well have been done for n-dimensional Euclidean space $\mathbb{R}^n$, but we will concern ourselves chiefly with plane isometries here.

It can be proved that every isometry of the plane is one of just four types (see Artin [5]). We will list the types and show, for each type, a labeled figure that can be carried into itself by an isometry of that type. In each of Figs. 12.1, 12.3, and 12.4, consider the line with spikes shown to be extended infinitely to the left and to the right. We also give an example of each type in terms of coordinates.

translation τ: Slide every point the same distance in the same direction. See Fig. 12.1. (*Example:* $\tau(x, y) = (x, y) + (2, -3) = (x + 2, y - 3)$.)

rotation ρ: Rotate the plane about a point P through an angle θ. See Fig. 12.2. (*Example:* $\rho(x, y) = (-y, x)$ is a rotation through $90°$ counterclockwise about the origin $(0, 0)$.)

reflection μ: Map each point into its mirror image (μ for mirror) across a line L, each point of which is left fixed by μ. See Fig. 12.3. The line L is the *axis of reflection*. (*Example:* $\mu(x, y) = (y, x)$ is a reflection across the line $y = x$.)

glide reflection γ: The product of a translation and a reflection across a line mapped into itself by the translation. See Fig. 12.4. (*Example:* $\gamma(x, y) = (x + 4, -y)$ is a glide reflection along the x-axis.)

Notice the little curved arrow that is carried into another curved arrow in each of Figs. 12.1 through 12.4. For the translation and rotation, the counterclockwise directions of the curved arrows remain the same, but for the reflection and glide reflection, the counterclockwise arrow is mapped into a clockwise arrow. We say that translations and rotations *preserve orientation,* while the reflection and glide reflection *reverse orientation.* We do not classify the identity isometry as any definite one of the four types listed; it could equally well be considered to be a translation by the zero vector or a rotation about any point through an angle of $0°$. We always consider a glide reflection to be the product of a reflection and a translation that is different from the identity isometry.

† This section is not used in the remainder of the text.

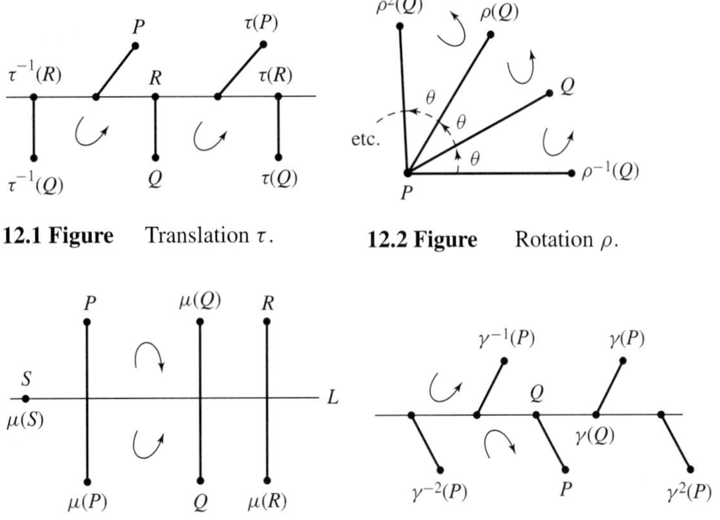

12.1 Figure Translation τ.

12.2 Figure Rotation ρ.

12.3 Figure Reflection μ.

12.4 Figure Glide reflection γ.

The theorem that follows describes the possible structures of finite subgroups of the full isometry group.

12.5 Theorem Every finite group G of isometries of the plane is isomorphic to either $\mathbb{Z}_n$ or to a dihedral group D_n for some positive integer n.

Proof Outline First we show that there is a point P in the plane that is left fixed by every isometry in G. This can be done in the following way, using coordinates in the plane. Suppose $G = \{\phi_1, \phi_2, \cdots, \phi_m\}$ and let

$$(x_i, y_i) = \phi_i(0, 0).$$

Then the point

$$P = (\overline{x}, \overline{y}) = \left(\frac{x_1 + x_2 + \cdots + x_m}{m}, \frac{y_1 + y_2 + \cdots + y_m}{m} \right)$$

is the *centroid* of the set $S = \{(x_i, y_i) \mid i = 1, 2, \cdots, m\}$. The isometries in G permute the points in S among themselves, since if $\phi_i \phi_j = \phi_k$ then $\phi_i(x_j, y_j) = \phi_i\{\phi_j(0, 0)] = \phi_k(0, 0) = (x_k, y_k)$. It can be shown that the centroid of a set of points is uniquely determined by its distances from the points, and since each isometry in G just permutes the set S, it must leave the centroid $(\overline{x}, \overline{y})$ fixed. Thus G consists of the identity, rotations about P, and reflections across a line through P.

The orientation-preserving isometries in G form a subgroup H of G which is either all of G or of order $m/2$. This can be shown in the same way that we showed that the even permutations are a subgroup of S_n containing just half the elements of S_n. (See Exercise 22.) Of course H consists of the identity and the rotations in G. If we choose a rotation in G that rotates the plane through as small an angle $\theta > 0$ as possible, it can be shown to generate the subgroup H. (See Exercise 23.) This shows that if $H = G$, then G is cyclic of order m and thus isomorphic to $\mathbb{Z}_m$. Suppose $H \neq G$ so that G contains some reflections. Let $H = \{\iota, \rho_1, \cdots, \rho_{n-1}\}$ with $n = m/2$. If μ is a reflection in G, then the coset $H\mu$ consists of all n of the reflections in G.

Consider now a regular n-gon in the plane having P as its center and with a vertex lying on the line through P left fixed by μ. The elements of H rotate this n-gon through all positions, and the elements of $H\mu$ first reflect in an axis through a vertex, effectively turning the n-gon over, and then rotate through all positions. Thus the action of G on this n-gon is the action of D_n, so G is isomorphic to D_n. ◆

The preceding theorem gives the complete story about finite plane isometry groups. We turn now to some infinite groups of plane isometries that arise naturally in decorating and art. Among these are the *discrete frieze groups*. A discrete frieze consists of a pattern of finite width and height that is repeated endlessly in both directions along its baseline to form a strip of infinite length but finite height; think of it as a decorative border strip that goes around a room next to the ceiling on wallpaper. We consider those isometries that carry each basic pattern onto itself or onto another instance of the pattern in the frieze. The set of all such isometries is called the **"frieze group."** All discrete frieze groups are infinite and have a subgroup isomorphic to $\mathbb{Z}$ generated by the translation that slides the frieze lengthwise until the basic pattern is superimposed on the position of its next neighbor pattern in that direction. As a simple example of a discrete frieze, consider integral signs spaced equal distances apart and continuing infinitely to the left and right, indicated schematically like this.

Let us consider the integral signs to be one unit apart. The symmetry group of this frieze is generated by a translation τ sliding the plane one unit to the right, and by a rotation ρ of 180° about a point in the center of some integral sign. There are no horizontal or vertical reflections, and no glide reflections. This frieze group is nonabelian; we can check that $\tau\rho = \rho\tau^{-1}$. The n-th dihedral group D_n is generated by two elements that do not commute, a rotation ρ_1 through $360/n°$ of order n and a reflection μ of order 2 satisfying $\rho_1\mu = \mu\rho_1^{-1}$. Thus it is natural to use the notation D_∞ for this nonabelian frieze group generated by τ of infinite order and ρ of order 2.

As another example, consider the frieze given by an infinite string of D's.

$$\cdots\mathbf{D\,D\,D\,D\,D\,D\,D\,D\,D\,D\,D}\cdots$$

Its group is generated by a translation τ one step to the right and by a vertical reflection μ across a horizontal line cutting through the middle of all the D's. We can check that these group generators commute this time, that is, $\tau\mu = \mu\tau$, so this frieze group is abelian and is isomorphic to $\mathbb{Z} \times \mathbb{Z}_2$.

It can be shown that if we classify such discrete friezes only by whether or not their groups contain a

rotation	horizontal axis reflection
vertical axis reflection	nontrivial glide reflection

then there are a total of seven possibilities. A *nontrivial glide reflection* in a symmetry group is one that is not equal to a product of a translation in that group and a reflection in that group. The group for the string of D's above contains glide reflections across the horizontal line through the centers of the D's, but the translation component of each glide reflection is also in the group so they are all considered trivial glide reflections in that group. The frieze group for

. . . D D D D D . . .
. . . D D D D D . . .

contains a nontrivial glide reflection whose translation component is not an element of the group. The exercises exhibit the seven possible cases, and ask you to tell, for each case, which of the four types of isometries displayed above appear in the symmetry group. We do not obtain seven different group structures. Each of the groups obtained can be shown to be isomorphic to one of

$$\mathbb{Z}, \quad D_\infty, \quad \mathbb{Z} \times \mathbb{Z}_2, \quad \text{or} \quad D_\infty \times \mathbb{Z}_2.$$

Equally interesting is the study of symmetries when a pattern in the shape of a square, parallelogram, rhombus, or hexagon is repeated by translations along *two nonparallel vector directions* to fill the entire plane, like patterns that appear on wallpaper. These groups are called the *wallpaper groups* or the *plane crystallographic groups*. While a frieze could not be carried into itself by a rotation through a positive angle less than 180°, it is possible to have rotations of 60°, 90°, 120°, and 180° for some of these plane-filling patterns. Figure 12.6 provides an illustration where the pattern consists of a square. We are interested in the group of plane isometries that carry this square onto itself or onto another square. Generators for this group are given by two translations (one sliding a square to the next neighbor to the right and one to the next above), by a rotation through 90° about the center of a square, and by a reflection in a vertical (or horizontal) line along the edges of the square. The one reflection is all that is needed to "turn the plane over"; a diagonal reflection can also be used. After being turned over, the translations and rotations can be used again. The isometry group for this *periodic pattern* in the plane surely contains a subgroup isomorphic to $\mathbb{Z} \times \mathbb{Z}$ generated by the unit translations to the right and upward, and a subgroup isomorphic to D_4 generated by those isometries that carry one square (it can be any square) into itself.

If we consider the plane to be filled with parallelograms as in Fig. 12.7, we do not get all the types of isometries that we did for Fig. 12.6. The symmetry group this time is

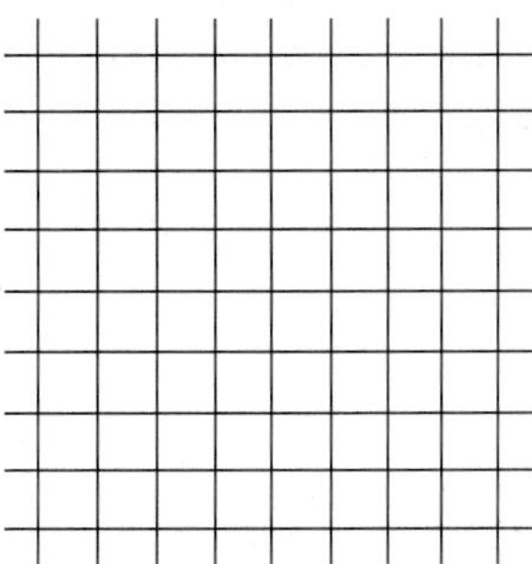

12.6 Figure

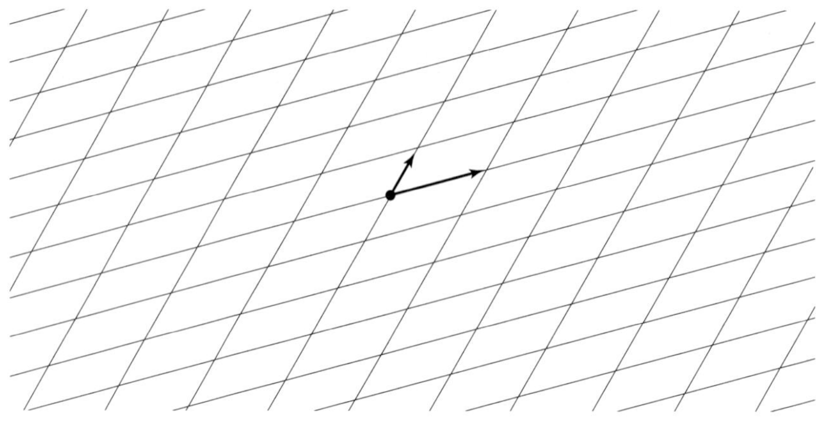

12.7 Figure

generated by the translations indicated by the arrows and a rotation through 180° about any vertex of a parallelogram.

It can be shown that there are 17 different types of wallpaper patterns when they are classified according to the types of rotations, reflections, and nontrivial glide reflections that they admit. We refer you to Gallian [8] for pictures of these 17 possibilities and a chart to help you identify them. The exercises illustrate a few of them. The situation in space is more complicated; it can be shown that there are 230 three-dimensional crystallographic groups. The final exercise we give involves rotations in space.

M. C. Escher (1898–1973) was an artist whose work included plane-filling patterns. The exercises include reproductions of four of his works of this type.

EXERCISES 12

1. This exercise shows that the group of symmetries of a certain type of geometric figure may depend on the dimension of the space in which we consider the figure to lie.

a. Describe all symmetries of a point in the real line $\mathbb{R}$; that is, describe all isometries of $\mathbb{R}$ that leave one point fixed.

b. Describe all symmetries (translations, reflections, etc.) of a point in the plane $\mathbb{R}^2$.

c. Describe all symmetries of a line segment in $\mathbb{R}$.

d. Describe all symmetries of a line segment in $\mathbb{R}^2$.

e. Describe some symmetries of a line segment in $\mathbb{R}^3$.

2. Let P stand for an orientation preserving plane isometry and R for an orientation reversing one. Fill in the table with P or R to denote the orientation preserving or reversing property of a product.

	P	R
P		
R		

3. Fill in the table to give *all* possible types of plane isometries given by a product of two types. For example, a product of two rotations may be a rotation, or it may be another type. Fill in the box corresponding to $\rho\rho$ with both letters. Use your answer to Exercise 2 to eliminate some types. Eliminate the identity from consideration.

	τ	ρ	μ	γ
τ				
ρ				
μ				
γ				

4. Draw a plane figure that has a one-element group as its group of symmetries in $\mathbb{R}^2$.

5. Draw a plane figure that has a two-element group as its group of symmetries in $\mathbb{R}^2$.

6. Draw a plane figure that has a three-element group as its group of symmetries in $\mathbb{R}^2$.

7. Draw a plane figure that has a four-element group isomorphic to $\mathbb{Z}_4$ as its group of symmetries in $\mathbb{R}^2$.

8. Draw a plane figure that has a four-element group isomorphic to the Klein 4-group V as its group of symmetries in $\mathbb{R}^2$.

9. For each of the four types of plane isometries (other than the identity), give the possibilities for the order of an isometry of that type in the group of plane isometries.

10. A plane isometry ϕ has a *fixed point* if there exists a point P in the plane such that $\phi(P) = P$. Which of the four types of plane isometries (other than the identity) can have a fixed point?

11. Referring to Exercise 10, which types of plane isometries, if any, have exactly one fixed point?

12. Referring to Exercise 10, which types of plane isometries, if any, have exactly two fixed points?

13. Referring to Exercise 10, which types of plane isometries, if any, have an infinite number of fixed points?

14. Argue geometrically that a plane isometry that leaves three noncolinear points fixed must be the identity map.

15. Using Exercise 14, show algebraically that if two plane isometries ϕ and ψ agree on three noncolinear points, that is, if $\phi(P_i) = \psi(P_i)$ for noncolinear points P_1, P_2, and P_3, then ϕ and ψ are the same map.

16. Do the rotations, together with the identity map, form a subgroup of the group of plane isometries? Why or why not?

17. Do the translations, together with the identity map, form a subgroup of the group of plane isometries? Why or why not?

18. Do the rotations about one particular point P, together with the identity map, form a subgroup of the group of plane isometries? Why or why not?

19. Does the reflection across one particular line L, together with the identity map, form a subgroup of the group of plane isometries? Why or why not?

20. Do the glide reflections, together with the identity map, form a subgroup of the group of plane isometries? Why or why not?

21. Which of the four types of plane isometries can be elements of a *finite* subgroup of the group of plane isometries?

22. Completing a detail of the proof of Theorem 12.5, let G be a finite group of plane isometries. Show that the rotations in G, together with the identity isometry, form a subgroup H of G, and that either $H = G$ or $|G| = 2|H|$. [*Hint:* Use the same method that we used to show that $|S_n| = 2|A_n|$.]

23. Completing a detail in the proof of Theorem 12.5, let G be a finite group consisting of the identity isometry and rotations about one point P in the plane. Show that G is cyclic, generated by the rotation in G that turns the plane counterclockwise about P through the smallest angle $\theta > 0$. [*Hint:* Follow the idea of the proof that a subgroup of a cyclic group is cyclic.]

Exercises 24 through 30 illustrate the seven different types of friezes when they are classified according to their symmetries. Imagine the figure shown to be continued infinitely to the right and left. The symmetry group of a frieze always contains translations. For each of these exercises answer these questions about the symmetry group of the frieze.

- **a.** Does the group contain a rotation?
- **b.** Does the group contain a reflection across a horizontal line?
- **c.** Does the group contain a reflection across a vertical line?
- **d.** Does the group contain a nontrivial glide reflection?
- **e.** To which of the possible groups $\mathbb{Z}$, D_∞, $\mathbb{Z} \times \mathbb{Z}_2$, or $D_\infty \times \mathbb{Z}_2$ do you think the symmetry group of the frieze is isomorphic?

24. F F F F F F F F F F F F F F F

25. T T T T T T T T T T T

26. E E E E E E E E E E E E

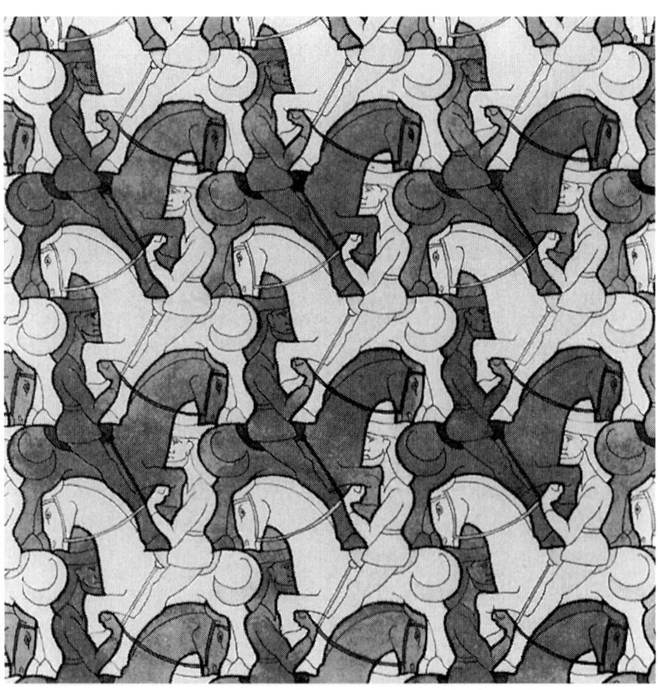

12.8 Figure The Study of Regular Division of the Plane with Horsemen (© 1946 M. C. Escher Foundation–Baarn–Holland. All rights reserved.)

27. Z Z Z Z Z Z Z Z Z Z Z

28. H H H H H H H H H H

29. ⌋ ⌉ ⌋ ⌉ ⌋ ⌉ ⌋ ⌉ ⌋ ⌉

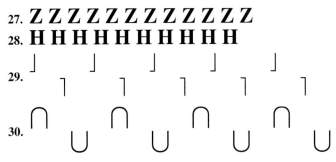

30. ∩ ∪ ∩ ∪ ∩ ∪ ∩ ∪

Exercises 31 through 37 describe a pattern to be used to fill the plane by translation in the two directions given by the specified vectors. Answer these questions in each case.

a. Does the symmetry group contain any rotations? If so, through what possible angles θ where $0 < \theta \leq 180°$?

12.9 Figure The Study of Regular Division of the Plane with Imaginary Human Figures (© 1936 M. C. Escher Foundation–Baarn–Holland. All rights reserved.)

12.10 Figure The Study of Regular Division of the Plane with Reptiles (© 1939 M. C. Escher Foundation–Baarn–Holland. All rights reserved.)

b. Does the symmetry group contain any reflections?

c. Does the symmetry group contain any nontrivial glide reflections?

31. A square with horizontal and vertical edges using translation directions given by vectors (1, 0) and (0, 1).

32. A square as in Exercise 31 using translation directions given by vectors (1, 1/2) and (0, 1).

33. A square as in Exercise 31 with the letter L at its center using translation directions given by vectors (1, 0) and (0, 1).

34. A square as in Exercise 31 with the letter E at its center using translation directions given by vectors (1, 0) and (0, 1).

35. A square as in Exercise 31 with the letter H at its center using translation directions given by vectors (1, 0) and (0, 1).

36. A regular hexagon with a vertex at the top using translation directions given by vectors (1, 0) and $(1, \sqrt{3})$.

37. A regular hexagon with a vertex at the top containing an equilateral triangle with vertex at the top and centroid at the center of the hexagon, using translation directions given by vectors (1, 0) and $(1, \sqrt{3})$.

Exercises 38 through 41 are concerned with art works of M. C. Escher. Neglect the shading in the figures and assume the markings in each human figure, reptile, or horseman are the same, even though they may be invisible due to shading. Answer the same questions (a), (b), and (c) that were asked for Exercises 31 through 36, and also answer this part (d).

d. Assuming horizontal and vertical coordinate axes with equal scales as usual, give vectors in the two nonparallel directions of vectors that generate the translation subgroup. Do not concern yourself with the length of these vectors.

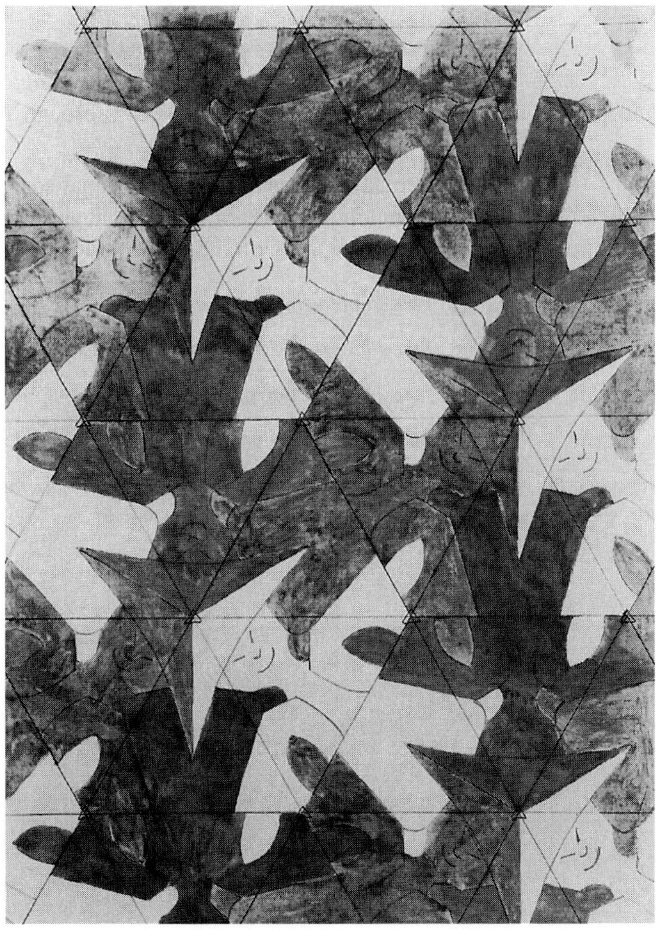

12.11 Figure The Study of Regular Division of the Plane with Human Figures ()

38. *The Study of Regular Division of the Plane with Horsemen* in Fig. 12.8.

39. *The Study of Regular Division of the Plane with Imaginary Human Figures* in Fig. 12.9.

40. *The Study of Regular Division of the Plane with Reptiles* in Fig. 12.10.

41. *The Study of Regular Division of the Plane with Human Figures* in Fig. 12.11.

42. Show that the rotations of a cube in space form a group isomorphic to S_4. [*Hint:* A rotation of the cube permutes the diagonals through the center of the cube.]

Homomorphisms and Factor Groups

SECTION 13 HOMOMORPHISMS

Structure-Relating Maps

Let G and G' be groups. We are interested in maps from G to G' that relate the group structure of G to the group structure of G'. Such a map often gives us information about one of the groups from known structural properties of the other. An isomorphism $\phi : G \to G'$, if one exists, is an example of such a structure-relating map. If we know all about the group G and know that ϕ is an isomorphism, we immediately know all about the group structure of G', for it is structurally just a copy of G. We now consider more general structure-relating maps, weakening the conditions from those of an isomorphism by no longer requiring that the maps be one to one and onto. You see, those conditions are the purely *set-theoretic portion* of our definition of an isomorphism, and have nothing to do with the binary operations of G and of G'. The binary operations are what give us the *algebra* which is the focus of our study in this text. We keep just the homomorphism property of an isomorphism related to the binary operations for the definition we now make.

13.1 Definition A map ϕ of a group G into a group G' is a **homomorphism** if the homomorphism property

$$\phi(ab) = \phi(a)\phi(b) \qquad (1)$$

holds for all $a, b \in G$. ■

‡ Section 16 is a prerequisite only for Sections 17 and 36.

† Section 17 is not required for the remainder of the text.

Let us now examine the idea behind the requirement (1) for a homomorphism $\phi : G \to G'$. In Eq. (1), the product ab on the left-hand side takes place in G, while the product $\phi(a)\phi(b)$ on the right-hand side takes place in G'. Thus Eq. (1) gives a relation between these binary operations, and hence between the two group structures.

For any groups G and G', there is always at least one homomorphism $\phi : G \to G'$, namely the **trivial homomorphism** defined by $\phi(g) = e'$ for all $g \in G$, where e' is the identity in G'. Equation (1) then reduces to the true equation $e' = e'e'$. No information about the structure of G or G' can be gained from the other group using this trivial homomorphism. We give an example illustrating how a homomorphism ϕ mapping G onto G' may give structural information about G'.

13.2 Example Let $\phi : G \to G'$ be a group homomorphism of G *onto* G'. We claim that if G is abelian, then G' must be abelian. Let $a', b' \in G'$. We must show that $a'b' = b'a'$. Since ϕ is onto G', there exist $a, b \in G$ such that $\phi(a) = a'$ and $\phi(b) = b'$. Since G is abelian, we have $ab = ba$. Using property (1), we have $a'b' = \phi(a)\phi(b) = \phi(ab) = \phi(ba) = \phi(b)\phi(a) = b'a'$, so G' is indeed abelian? ▲

Example 13.16 will give an illustration showing how information about G' may give information about G via a homomorphism $\phi : G \to G'$. We now give examples of homomorphisms for specific groups.

13.3 Example Let S_n be the symmetric group on n letters, and let $\phi : S_n \to \mathbb{Z}_2$ be defined by

$$\phi(\sigma) = \begin{cases} 0 & \text{if } \sigma \text{ is an even permutation,} \\ 1 & \text{if } \sigma \text{ is an odd permutation.} \end{cases}$$

Show that ϕ is a homomorphism.

Solution We must show that $\phi(\sigma\mu) = \phi(\sigma) + \phi(\mu)$ for all choices of $\sigma, \mu \in S_n$. Note that the operation on the right-hand side of this equation is written additively since it takes place in the group $\mathbb{Z}_2$. Verifying this equation amounts to checking just four cases:

σ odd and μ odd,

σ odd and μ even,

σ even and μ odd,

σ even and μ even.

Checking the first case, if σ and μ can both be written as a product of an odd number of transpositions, then $\sigma\mu$ can be written as the product of an even number of transpositions. Thus $\phi(\sigma\mu) = 0$ and $\phi(\sigma) + \phi(\mu) = 1 + 1 = 0$ in $\mathbb{Z}_2$. The other cases can be checked similarly. ▲

13.4 Example **(Evaluation Homomorphism)** Let F be the additive group of all functions mapping $\mathbb{R}$ into $\mathbb{R}$, let $\mathbb{R}$ be the additive group of real numbers, and let c be any real number. Let $\phi_c : F \to \mathbb{R}$ be the **evaluation homomorphism** defined by $\phi_c(f) = f(c)$ for $f \in F$. Recall that, by definition, the sum of two functions f and g is the function $f + g$ whose value at x is $f(x) + g(x)$. Thus we have

$$\phi_c(f + g) = (f + g)(c) = f(c) + g(c) = \phi_c(f) + \phi_c(g),$$

and Eq. (1) is satisfied, so we have a homomorphism. ▲

13.5 Example Let $\mathbb{R}^n$ be the additive group of column vectors with n real-number components. (This group is of course isomorphic to the direct product of $\mathbb{R}$ under addition with itself for n factors.) Let A be an $m \times n$ matrix of real numbers. Let $\phi : \mathbb{R}^n \to \mathbb{R}^m$ be defined by $\phi(\mathbf{v}) = A\mathbf{v}$ for each column vector $\mathbf{v} \in \mathbb{R}^n$. Then ϕ is a homomorphism, since for $\mathbf{v}, \mathbf{w} \in \mathbb{R}^n$, matrix algebra shows that $\phi(\mathbf{v}+\mathbf{w}) = A(\mathbf{v}+\mathbf{w}) = A\mathbf{v} + A\mathbf{w} = \phi(\mathbf{v}) + \phi(\mathbf{w})$. In linear algebra, such a map computed by multiplying a column vector on the left by a matrix A is known as a **linear transformation**. ▲

13.6 Example Let $GL(n, \mathbb{R})$ be the multiplicative group of all invertible $n \times n$ matrices. Recall that a matrix A is invertible if and only if its determinant, $\det(A)$, is nonzero. Recall also that for matrices $A, B \in GL(n, \mathbb{R})$ we have

$$\det(AB) = \det(A)\det(B).$$

This means that det is a homomorphism mapping $GL(n, \mathbb{R})$ into the multiplicative group $\mathbb{R}^*$ of nonzero real numbers. ▲

Homomorphisms of a group G into itself are often useful for studying the structure of G. Our next example gives a nontrivial homomorphism of a group into itself.

13.7 Example Let $r \in \mathbb{Z}$ and let $\phi_r : \mathbb{Z} \to \mathbb{Z}$ be defined by $\phi_r(n) = rn$ for all $n \in \mathbb{Z}$. For all $m, n \in \mathbb{Z}$, we have $\phi_r(m+n) = r(m+n) = rm + rn = \phi_r(m) + \phi_r(n)$ so ϕ_r is a homomorphism. Note that ϕ_0 is the trivial homomorphism, ϕ_1 is the identity map, and ϕ_{-1} maps $\mathbb{Z}$ *onto* $\mathbb{Z}$. For all other r in $\mathbb{Z}$, the map ϕ_r is not onto $\mathbb{Z}$. ▲

13.8 Example Let $G = G_1 \times G_2 \times \cdots \times G_i \times \cdots \times G_n$ be a direct product of groups. The **projection map** $\pi_i : G \to G_i$ where $\pi_i(g_1, g_2, \cdots, g_i, \cdots, g_n) = g_i$ is a homomorphism for each $i = 1, 2, \cdots, n$. This follows immediately from the fact that the binary operation of G coincides in the ith component with the binary operation in G_i. ▲

13.9 Example Let F be the additive group of continuous functions with domain [0, 1] and let $\mathbb{R}$ be the additive group of real numbers. The map $\sigma : F \to \mathbb{R}$ defined by $\sigma(f) = \int_0^1 f(x)dx$ for $f \in F$ is a homomorphism, for

$$\begin{aligned}\sigma(f+g) &= \int_0^1 (f+g)(x)dx = \int_0^1 [f(x)+g(x)]dx \\ &= \int_0^1 f(x)dx + \int_0^1 g(x)dx = \sigma(f) + \sigma(g)\end{aligned}$$

for all $f, g \in F$. ▲

13.10 Example **(Reduction Modulo *n*)** Let γ be the natural map of $\mathbb{Z}$ into $\mathbb{Z}_n$ given by $\gamma(m) = r$, where r is the remainder given by the division algorithm when m is divided by n. Show that γ is a homomorphism.

Solution We need to show that

$$\gamma(s+t) = \gamma(s) + \gamma(t)$$

for $s, t \in \mathbb{Z}$. Using the division algorithm, we let

$$s = q_1 n + r_1 \qquad \textbf{(2)}$$

and

$$t = q_2 n + r_2 \tag{3}$$

where $0 \leq r_i < n$ for $i = 1, 2$. If

$$r_1 + r_2 = q_3 n + r_3 \tag{4}$$

for $0 \leq r_3 < n$, then adding Eqs. (2) and (3) we see that

$$s + t = (q_1 + q_2 + q_3)n + r_3,$$

so that $\gamma(s + t) = r_3$.

From Eqs. (2) and (3) we see that $\gamma(s) = r_1$ and $\gamma(t) = r_2$. Equation (4) shows that the sum $r_1 + r_2$ in $\mathbb{Z}_n$ is equal to r_3 also.

Consequently $\gamma(s + t) = \gamma(s) + \gamma(t)$, so we do indeed have a homomorphism. ▲

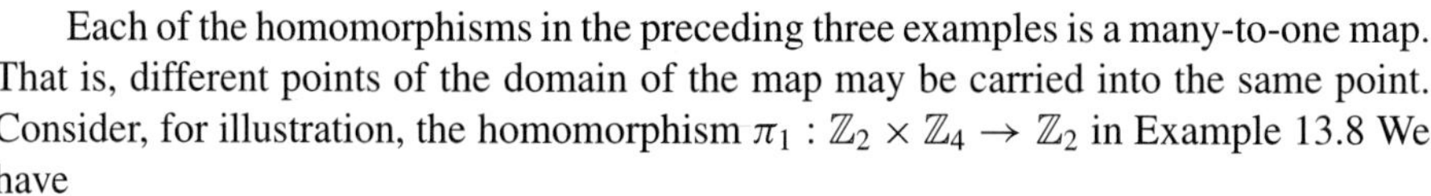

Each of the homomorphisms in the preceding three examples is a many-to-one map. That is, different points of the domain of the map may be carried into the same point. Consider, for illustration, the homomorphism $\pi_1 : \mathbb{Z}_2 \times \mathbb{Z}_4 \to \mathbb{Z}_2$ in Example 13.8 We have

$$\pi_1(0, 0) = \pi_1(0, 1) = \pi_1(0, 2) = \pi_1(0, 3) = 0,$$

so four elements in $\mathbb{Z}_2 \times \mathbb{Z}_4$ are mapped into 0 in $\mathbb{Z}_2$ by π_1.

Composition of group homomorphisms is again a group homomorphism. That is, if $\phi : G \to G'$ and $\gamma : G' \to G''$ are both group homomorphisms then their composition $(\gamma \circ \phi) : G \to G''$, where $(\gamma \circ \phi)(g) = \gamma(\phi(g))$ for $g \in G$, is also a homomorphism. (See Exercise 49.)

Properties of Homomorphisms

We turn to some structural features of G and G' that are *preserved* by a homomorphism $\phi : G \to G'$. First we review set-theoretic definitions. Note the use of *square brackets* when we apply a function to a *subset* of its domain.

13.11 Definition Let ϕ be a mapping of a set X into a set Y, and let $A \subseteq X$ and $B \subseteq Y$. The **image** $\phi[A]$ **of** A **in** Y **under** ϕ is $\{\phi(a) \mid a \in A\}$. The set $\phi[X]$ is the **range of** ϕ. The **inverse image** $\phi^{-1}[B]$ **of** B **in** X is $\{x \in X \mid \phi(x) \in B\}$. ■

The first three properties of a homomorphism stated in the theorem that follows have already been encountered for the special case of an isomorphism; namely, in Theorem 3.14, Exercise 28 of Section 4, and Exercise 41 of Section 5. There they were really obvious because the structures of G and G' were identical. We will now see that they hold for structure-relating maps of groups, even if the maps are not one to one and onto. We do not consider them obvious in this new context.

13.12 Theorem Let ϕ be a homomorphism of a group G into a group G'.

1. If e is the identity element in G, then $\phi(e)$ is the identity element e' in G'.
2. If $a \in G$, then $\phi(a^{-1}) = \phi(a)^{-1}$.

3. If H is a subgroup of G, then $\phi[H]$ is a subgroup of G'.
4. If K' is a subgroup of G', then $\phi^{-1}[K']$ is a subgroup of G.

Loosely speaking, ϕ preserves the identity element, inverses, and subgroups.

Proof Let ϕ be a homomorphism of G into G'. Then

$$\phi(a) = \phi(ae) = \phi(a)\phi(e).$$

Multiplying on the left by $\phi(a)^{-1}$, we see that $e' = \phi(e)$. Thus $\phi(e)$ must be the identity element e' in G'. The equation

$$e' = \phi(e) = \phi(aa^{-1}) = \phi(a)\phi(a^{-1})$$

shows that $\phi(a^{-1}) = \phi(a)^{-1}$.

Turning to Statement (3), let H be a subgroup of G, and let $\phi(a)$ and $\phi(b)$ be any two elements in $\phi[H]$. Then $\phi(a)\phi(b) = \phi(ab)$, so we see that $\phi(a)\phi(b) \in \phi[H]$; thus, $\phi[H]$ is closed under the operation of G'. The fact that $e' = \phi(e)$ and $\phi(a^{-1}) = \phi(a)^{-1}$ completes the proof that $\phi[H]$ is a subgroup of G'.

Going the other way for Statement (4), let K' be a subgroup of G'. Suppose a and b are in $\phi^{-1}[K']$. Then $\phi(a)\phi(b) \in K'$ since K' is a subgroup. The equation $\phi(ab) = \phi(a)\phi(b)$ shows that $ab \in \phi^{-1}[K']$. Thus $\phi^{-1}[K']$ is closed under the binary operation in G. Also, K' must contain the identity element $e' = \phi(e)$, so $e \in \phi^{-1}[K']$. If $a \in \phi^{-1}[K']$, then $\phi(a) \in K'$, so $\phi(a)^{-1} \in K'$. But $\phi(a)^{-1} = \phi(a^{-1})$, so we must have $a^{-1} \in \phi^{-1}[K']$. Hence $\phi^{-1}[K']$ is a subgroup of G. ◆

Let $\phi : G \to G'$ be a homomorphism and let e' be the identity element of G'. Now $\{e'\}$ is a subgroup of G', so $\phi^{-1}[\{e'\}]$ is a subgroup H of G by Statement (4) in Theorem 13.12. This subgroup is critical to the study of homomorphisms.

13.13 Definition Let $\phi : G \to G'$ be a homomorphism of groups. The subgroup $\phi^{-1}[\{e'\}] = \{x \in G \mid \phi(x) = e'\}$ is the **kernel of** ϕ, denoted by Ker(ϕ). ■

Example 13.5 discussed the homomorphism $\phi : \mathbb{R}^n \to \mathbb{R}^m$ given by $\phi(\mathbf{v}) = A\mathbf{v}$ where A is an $m \times n$ matrix. In this context, Ker(ϕ) is called the *null space* of A. It consists of all $\mathbf{v} \in \mathbb{R}^n$ such that $A\mathbf{v} = \mathbf{0}$, the zero vector.

Let $H = \text{Ker}(\phi)$ for a homomorphism $\phi : G \to G'$. We think of ϕ as "collapsing" H down onto e'. Theorem 13.15 that follows shows that for $g \in G$, the cosets gH and Hg are the same, and are collapsed onto the single element $\phi(g)$ by ϕ. That is $\phi^{-1}[\{\phi(g)\}] = gH = Hg$. (Be sure that you understand the reason for the uses of (), [], and {} in $\phi^{-1}[\{\phi(g)\}]$.) We have attempted to symbolize this collapsing in Fig. 13.14, where the shaded rectangle represents G, the solid vertical line segments represent the cosets of $H = \text{Ker}(\phi)$, and the horizontal line at the bottom represents G'. We view ϕ as projecting the elements of G, which are in the shaded rectangle, straight down onto elements of G', which are on the horizontal line segment at the bottom. Notice the downward arrow labeled ϕ at the left, starting at G and ending at G'. Elements of $H = \text{Ker}(\phi)$ thus lie on the solid vertical line segment in the shaded box lying over e', as labeled at the top of the figure.

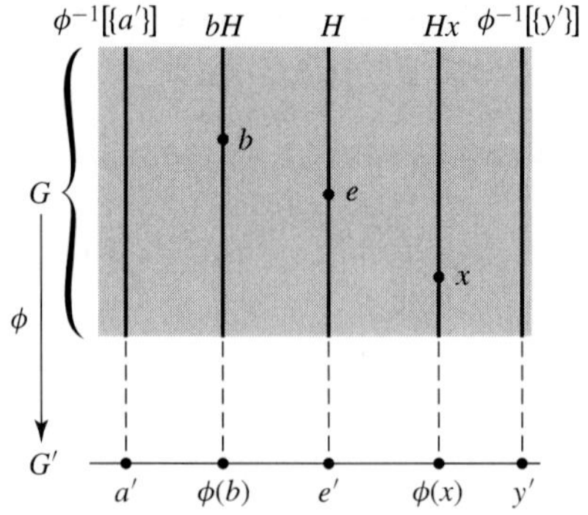

$\phi^{-1}[\{a'\}]$ bH H Hx $\phi^{-1}[\{y'\}]$

G b e x ϕ G'

a' $\phi(b)$ e' $\phi(x)$ y'

13.14 Figure Cosets of H collapsed by ϕ.

13.15 Theorem Let $\phi : G \to G'$ be a group homomorphism, and let $H = \text{Ker}(\phi)$. Let $a \in G$. Then the set

$$\phi^{-1}[\{\phi(a)\}] = \{x \in G \mid \phi(x) = \phi(a)\}$$

is the left coset aH of H, and is also the right coset Ha of H. Consequently, the two partitions of G into left cosets and into right cosets of H are the same.

→ ***Proof*** We want to show that

$$\{x \in G \mid \phi(x) = \phi(a)\} = aH.$$

There is a standard way to show that two sets are equal; show that each is a subset of the other.

Suppose that $\phi(x) = \phi(a)$. Then

$$\phi(a)^{-1}\phi(x) = e',$$

where e' is the identity of G'. By Theorem 13.12, we know that $\phi(a)^{-1} = \phi(a^{-1})$, so we have

$$\phi(a^{-1})\phi(x) = e'.$$

Since ϕ is a homomorphism, we have

$$\phi(a^{-1})\phi(x) = \phi(a^{-1}x), \qquad \text{so} \qquad \phi(a^{-1}x) = e'.$$

But this shows that $a^{-1}x$ is in $H = \text{Ker}(\phi)$, so $a^{-1}x = h$ for some $h \in H$, and $x = ah \in aH$. This shows that

$$\{x \in G \mid \phi(x) = \phi(a)\} \subseteq aH.$$

To show containment in the other direction, let $y \in aH$, so that $y = ah$ for some $h \in H$. Then

$$\phi(y) = \phi(ah) = \phi(a)\phi(h) = \phi(a)e' = \phi(a),$$

so that $y \in \{x \in G \mid \phi(x) = \phi(a)\}$.

We leave the similar demonstration that $\{x \in G \mid \phi(x) = \phi(a)\} = Ha$ to Exercise 52. ◆

13.16 Example Equation 5 of Section 1 shows that $|z_1 z_2| = |z_1||z_2|$ for complex numbers z_1 and z_2. This means that the absolute value function $|\ \ |$ is a homomorphism of the group $\mathbb{C}^*$ of nonzero complex numbers under multiplication onto the group $\mathbb{R}^+$ of positive real numbers under multiplication. Since $\{1\}$ is a subgroup of $\mathbb{R}^+$, Theorem 13.12 shows again that the complex numbers of magnitude 1 form a subgroup U of $\mathbb{C}^*$. Recall that the complex numbers can be viewed as filling the coordinate plane, and that the magnitude of a complex number is its distance from the origin. Consequently, the cosets of U are circles with center at the origin. Each circle is collapsed by this homomorphism onto its point of intersection with the positive real axis. ▲

We give an illustration of Theorem 13.15 from calculus.

13.17 Example Let D be the additive group of all differentiable functions mapping $\mathbb{R}$ into $\mathbb{R}$, and let F be the additive group of all functions mapping $\mathbb{R}$ into $\mathbb{R}$. Then differentiation gives us a map $\phi : D \to F$, where $\phi(f) = f'$ for $f \in F$. We easily see that ϕ is a homomorphism, for $\phi(f + g) = (f + g)' = f' + g' = \phi(f) + \phi(g)$; the derivative of a sum is the sum of the derivatives.

Now Ker(ϕ) consists of all functions f such that $f' = 0$, the zero constant function. Thus Ker(ϕ) consists of all constant functions, which form a subgroup C of F. Let us find all functions in G mapped into x^2 by ϕ, that is, all functions whose derivative is x^2. Now we know that $x^3/3$ is one such function. By Theorem 13.15, all such functions form the coset $x^3/3 + C$. Doesn't this look familiar? ▲

We will often use the following corollary of Theorem 13.15.

13.18 Corollary A group homomorphism $\phi : G \to G'$ is a one-to-one map if and only if Ker(ϕ) $= \{e\}$.

Proof If Ker(ϕ) $= \{e\}$, then for every $a \in G$, the elements mapped into $\phi(a)$ are precisely the elements of the left coset $a\{e\} = \{a\}$, which shows that ϕ is one to one.

Conversely, suppose ϕ is one to one. Now by Theorem 13.12, we know that $\phi(e) = e'$, the identity element of G'. Since ϕ is one to one, we see that e is the only element mapped into e' by ϕ, so Ker(ϕ) $= \{e\}$. ◆

In view of Corollary 13.18, we modify the outline given prior to Example 3.8 for showing that a map ϕ is an isomorphism of binary structures when the structures are groups G and G'.

To Show $\phi : G \to G'$ Is an Isomorphism

Step 1 Show ϕ is a homomorphism.

Step 2 Show $\mathrm{Ker}(\phi) = \{e\}$.

Step 3 Show ϕ maps G onto G'.

Theorem 13.15 shows that the kernel of a group homomorphism $\phi : G \to G'$ is a subgroup H of G whose left and right cosets coincide, so that $gH = Hg$ for all $g \in G$. We will see in Section 14 that when left and right cosets coincide, we can form a coset group, as discussed intuitively in Section 10. Furthermore, we will see that H then appears as the kernel of a homomorphism of G onto this coset group in a very natural way. Such subgroups H whose left and right cosets coincide are very useful in studying a group, and are given a special name. We will work with them a lot in Section 14.

Historical Note

Normal subgroups were introduced by Evariste Galois in 1831 as a tool for deciding whether a given polynomial equation was solvable by radicals. Galois noted that a subgroup H of a group G of permutations induced two decompositions of G into what we call *left cosets* and *right cosets*. If the two decompositions coincide, that is, if the left cosets are the same as the right cosets, Galois called the decomposition *proper*. Thus a subgroup giving a proper decomposition is what we call a *normal subgroup*. Galois stated that if the group of permutations of the roots of an equation has a proper decomposition, then one can solve the given equation if one can first solve an equation corresponding to the subgroup H and then an equation corresponding to the cosets.

Camille Jordan, in his commentaries on Galois's work in 1865 and 1869, elaborated on these ideas considerably. He also defined normal subgroups, although without using the term, essentially as on this page and likewise gave the first definition of a simple group (page 149).

13.19 Definition A subgroup H of a group G is **normal** if its left and right cosets coincide, that is, if $gH = Hg$ for all $g \in G$. ■

Note that all subgroups of abelian groups are normal.

13.20 Corollary If $\phi : G \to G'$ is a group homomorphism, then $\mathrm{Ker}(\phi)$ is a normal subgroup of G.

Proof This follows immediately from the last sentence in the statement of Theorem 13.15 and Definition 13.19. ◆

For any group homomorphism $\phi : G \to G'$, two things are of primary importance: the *kernel of* ϕ, and the *image* $\phi[G]$ *of* G *in* G'. We have indicated the importance of

Ker(ϕ). Section 14 will indicate the importance of the image $\phi[G]$. Exercise 44 asks us to show that if $|G|$ is finite, then $|\phi[G]|$ is finite and is a divisor of $|G|$.

EXERCISES 13

Computations

In Exercises 1 through 15, determine whether the given map ϕ is a homomorphism. [*Hint:* The straightforward way to proceed is to check whether $\phi(ab) = \phi(a)\phi(b)$ for all a and b in the domain of ϕ. However, if we should happen to notice that $\phi^{-1}[\{e'\}]$ is not a subgroup whose left and right cosets coincide, or that ϕ does not satisfy the properties given in Exercise 44 or 45 for finite groups, then we can say at once that ϕ is not a homomorphism.]

1. Let $\phi : \mathbb{Z} \to \mathbb{R}$ under addition be given by $\phi(n) = n$.

2. Let $\phi : \mathbb{R} \to \mathbb{Z}$ under addition be given by $\phi(x) =$ the greatest integer $\leq x$.

3. Let $\phi : \mathbb{R}^* \to \mathbb{R}^*$ under multiplication be given by $\phi(x) = |x|$.

4. Let $\phi : \mathbb{Z}_6 \to \mathbb{Z}_2$ be given by $\phi(x) =$ the remainder of x when divided by 2, as in the division algorithm.

5. Let $\phi : \mathbb{Z}_9 \to \mathbb{Z}_2$ be given by $\phi(x) =$ the remainder of x when divided by 2, as in the division algorithm.

6. Let $\phi : \mathbb{R} \to \mathbb{R}^*$, where $\mathbb{R}$ is additive and $\mathbb{R}^*$ is multiplicative, be given by $\phi(x) = 2^x$.

7. Let $\phi_i : G_i \to G_1 \times G_2 \times \cdots \times G_i \times \cdots \times G_r$ be given by $\phi_i(g_i) = (e_1, e_2, \ldots, g_i, \ldots, e_r)$, where $g_i \in G_i$ and e_j is the identity element of G_j. This is an **injection map**. Compare with Example 13.8.

8. Let G be any group and let $\phi : G \to G$ be given by $\phi(g) = g^{-1}$ for $g \in G$.

9. Let F be the additive group of functions mapping $\mathbb{R}$ into $\mathbb{R}$ having derivatives of all orders. Let $\phi : F \to F$ be given by $\phi(f) = f''$, the second derivative of f.

10. Let F be the additive group of all continuous functions mapping $\mathbb{R}$ into $\mathbb{R}$. Let $\mathbb{R}$ be the additive group of real numbers, and let $\phi : F \to \mathbb{R}$ be given by

$$\phi(f) = \int_0^4 f(x)dx.$$

11. Let F be the additive group of all functions mapping $\mathbb{R}$ into $\mathbb{R}$, and let $\phi : F \to F$ be given by $\phi(f) = 3f$.

12. Let M_n be the additive group of all $n \times n$ matrices with real entries, and let $\mathbb{R}$ be the additive group of real numbers. Let $\phi(A) = \det(A)$, the determinant of A, for $A \in M_n$.

13. Let M_n and $\mathbb{R}$ be as in Exercise 12. Let $\phi(A) = \text{tr}(A)$ for $A \in M_n$, where the **trace** $\text{tr}(A)$ is the sum of the elements on the main diagonal of A, from the upper-left to the lower-right corner.

14. Let $GL(n, \mathbb{R})$ be the multiplicative group of invertible $n \times n$ matrices, and let $\mathbb{R}$ be the additive group of real numbers. Let $\phi : GL(n, \mathbb{R}) \to \mathbb{R}$ be given by $\phi(A) = \text{tr}(A)$, where $\text{tr}(A)$ is defined in Exercise 13.

15. Let F be the multiplicative group of all continuous functions mapping $\mathbb{R}$ into $\mathbb{R}$ that are nonzero at every $x \in \mathbb{R}$. Let $\mathbb{R}^*$ be the multiplicative group of nonzero real numbers. Let $\phi : F \to \mathbb{R}^*$ be given by $\phi(f) = \int_0^1 f(x)dx$.

In Exercises 16 through 24, compute the indicated quantities for the given homomorphism ϕ. (See Exercise 46.)

16. Ker(ϕ) for $\phi : S_3 \to \mathbb{Z}_2$ in Example 13.3

17. Ker(ϕ) and $\phi(25)$ for $\phi : \mathbb{Z} \to \mathbb{Z}_7$ such that $\phi(1) = 4$

18. Ker(ϕ) and $\phi(18)$ for $\phi : \mathbb{Z} \to \mathbb{Z}_{10}$ such that $\phi(1) = 6$

19. Ker(ϕ) and $\phi(20)$ for $\phi : \mathbb{Z} \to S_8$ such that $\phi(1) = (1, 4, 2, 6)(2, 5, 7)$

20. Ker(ϕ) and $\phi(3)$ for $\phi : \mathbb{Z}_{10} \to \mathbb{Z}_{20}$ such that $\phi(1) = 8$

21. Ker(ϕ) and $\phi(14)$ for $\phi : \mathbb{Z}_{24} \to S_8$ where $\phi(1) = (2, 5)(1, 4, 6, 7)$

22. Ker(ϕ) and $\phi(-3, 2)$ for $\phi : \mathbb{Z} \times \mathbb{Z} \to \mathbb{Z}$ where $\phi(1, 0) = 3$ and $\phi(0, 1) = -5$

23. Ker(ϕ) and $\phi(4, 6)$ for $\phi : \mathbb{Z} \times \mathbb{Z} \to \mathbb{Z} \times \mathbb{Z}$ where $\phi(1, 0) = (2, -3)$ and $\phi(0, 1) = (-1, 5)$

24. Ker(ϕ) and $\phi(3, 10)$ for $\phi : \mathbb{Z} \times \mathbb{Z} \to S_{10}$ where $\phi(1, 0) = (3, 5)(2, 4)$ and $\phi(0, 1) = (1, 7)(6, 10, 8, 9)$

25. How many homomorphisms are there of $\mathbb{Z}$ onto $\mathbb{Z}$?

26. How many homomorphisms are there of $\mathbb{Z}$ into $\mathbb{Z}$?

27. How many homomorphisms are there of $\mathbb{Z}$ into $\mathbb{Z}_2$?

28. Let G be a group, and let $g \in G$. Let $\phi_g : G \to G$ be defined by $\phi_g(x) = gx$ for $x \in G$. For which $g \in G$ is ϕ_g a homomorphism?

29. Let G be a group, and let $g \in G$. Let $\phi_g : G \to G$ be defined by $\phi_g(x) = gxg^{-1}$ for $x \in G$. For which $g \in G$ is ϕ_g a homomorphism?

Concepts

In Exercises 30 and 31, correct the definition of the italicized term without reference to the text, if correction is needed, so that it is in a form acceptable for publication.

30. A *homomorphism* is a map such that $\phi(xy) = \phi(x)\phi(y)$.

31. Let $\phi : G \to G'$ be a homomorphism of groups. The *kernel of* ϕ is $\{x \in G \mid \phi(x) = e'\}$ where e' is the identity in G'.

32. Mark each of the following true or false.

_____ **a.** A_n is a normal subgroup of S_n.

_____ **b.** For any two groups G and G', there exists a homomorphism of G into G'.

_____ **c.** Every homomorphism is a one-to-one map.

_____ **d.** A homomorphism is one to one if and only if the kernel consists of the identity element alone.

_____ **e.** The image of a group of 6 elements under some homomorphism may have 4 elements. (See Exercise 44.)

_____ **f.** The image of a group of 6 elements under a homomorphism may have 12 elements.

_____ **g.** There is a homomorphism of some group of 6 elements into some group of 12 elements.

_____ **h.** There is a homomorphism of some groups of 6 elements into some group of 10 elements.

_____ **i.** A homomorphism may have an empty kernel.

_____ **j.** It is not possible to have a nontrivial homomorphism of some finite group into some infinite group.

In Exercises 33 through 43, give an example of a nontrivial homomorphism ϕ for the given groups, if an example exists. If no such homomorphism exists, explain why that is so. You may use Exercises 44 and 45.

33. $\phi : \mathbb{Z}_{12} \to \mathbb{Z}_5$

34. $\phi : \mathbb{Z}_{12} \to \mathbb{Z}_4$

35. $\phi : \mathbb{Z}_2 \times \mathbb{Z}_4 \to \mathbb{Z}_2 \times \mathbb{Z}_5$

36. $\phi : \mathbb{Z}_3 \to \mathbb{Z}$

37. $\phi : \mathbb{Z}_3 \to S_3$

38. $\phi : \mathbb{Z} \to S_3$

39. $\phi : \mathbb{Z} \times \mathbb{Z} \to 2\mathbb{Z}$

40. $\phi : 2\mathbb{Z} \to \mathbb{Z} \times \mathbb{Z}$

41. $\phi : D_4 \to S_3$

42. $\phi : S_3 \to S_4$

43. $\phi : S_4 \to S_3$

Theory

44. Let $\phi : G \to G'$ be a group homomorphism. Show that if $|G|$ is finite, then $|\phi[G]|$ is finite and is a divisor of $|G|$.

45. Let $\phi : G \to G'$ be a group homomorphism. Show that if $|G'|$ is finite, then, $|\phi[G]|$ is finite and is a divisor of $|G'|$.

46. Let a group G be generated by $\{a_i \mid i \in I\}$, where I is some indexing set and $a_i \in G$ for all $i \in I$. Let $\phi : G \to G'$ and $\mu : G \to G'$ be two homomorphisms from G into a group G', such that $\phi(a_i) = \mu(a_i)$ for every $i \in I$. Prove that $\phi = \mu$. [Thus, for example, a homomorphism of a cyclic group is completely determined by its value on a generator of the group.] [*Hint*: Use Theorem 7.6 and, of course, Definition 13.1.]

47. Show that any group homomorphism $\phi : G \to G'$ where $|G|$ is a prime must either be the trivial homomorphism or a one-to-one map.

48. The **sign of an even permutation** is $+1$ and the **sign of an odd permutation** is -1. Observe that the map $\text{sgn}_n : S_n \to \{1, -1\}$ defined by

$$\text{sgn}_n(\sigma) = \text{sign of } \sigma$$

is a homomorphism of S_n onto the multiplicative group $\{1, -1\}$. What is the kernel? Compare with Example 13.3.

49. Show that if G, G', and G'' are groups and if $\phi : G \to G'$ and $\gamma : G' \to G''$ are homomorphisms, then the composite map $\gamma\phi : G \to G''$ is a homomorphism.

50. Let $\phi : G \to H$ be a group homomorphism. Show that $\phi[G]$ is abelian if and only if for all $x, y \in G$, we have $xyx^{-1}y^{-1} \in \text{Ker}(\phi)$.

51. Let G be any group and let a be any element of G. Let $\phi : \mathbb{Z} \to G$ be defined by $\phi(n) = a^n$. Show that ϕ is a homomorphism. Describe the image and the possibilities for the kernel of ϕ.

52. Let $\phi : G \to G'$ be a homomorphism with kernel H and let $a \in G$. Prove the set equality $\{x \in G \mid \phi(x) = \phi(a)\} = Ha$.

53. Let G be a group, Let $h, k \in G$ and let $\phi : \mathbb{Z} \times \mathbb{Z} \to G$ be defined by $\phi(m, n) = h^m k^n$. Give a necessary and sufficient condition, involving h and k, for ϕ to be a homomorphism. Prove your condition.

54. Find a necessary and sufficient condition on G such that the map ϕ described in the preceding exercise is a homomorphism for *all* choices of $h, k \in G$.

55. Let G be a group, h an element of G, and n a positive integer. Let $\phi : \mathbb{Z}_n \to G$ be defined by $\phi(i) = h^i$ for $0 \le i \le n$. Give a necessary and sufficient condition (in terms of h and n) for ϕ to be a homomorphism. Prove your assertion.

SECTION 14 FACTOR GROUPS

Let H be a subgroup of a finite group G. Suppose we write a table for the group operation of G, listing element heads at the top and at the left as they occur in the left cosets of H. We illustrated this in Section 10. The body of the table may break up into blocks corresponding to the cosets (Table 10.5), giving a group operation on the cosets, or they may not break up that way (Table 10.9). We start this section by showing that if H is the kernel of a group homomorphism $\phi : G \to G'$, then the cosets of H (remember that left and right cosets then coincide) are indeed elements of a group whose binary operation is derived from the group operation of G.

Factor Groups from Homomorphisms

Let G be a group and let S be a set having the same cardinality as G. Then there is a one-to-one correspondence $\leftrightarrow$ between S and G. We can use $\leftrightarrow$ to define a binary operation on S, making S into a group isomorphic to G. Naively, we simply use the correspondence to rename each element of G by the name of its corresponding (under $\leftrightarrow$) element in S. We can describe explicitly the computation of xy for $x, y \in S$ as follows:

$$\text{if } x \leftrightarrow g_1 \quad \text{and} \quad y \leftrightarrow g_2 \quad \text{and} \quad z \leftrightarrow g_1g_2, \quad \text{then} \quad xy = z. \tag{1}$$

The direction $\rightarrow$ of the one-to-one correspondence $s \leftrightarrow g$ between $s \in S$ and $g \in G$ gives us a one-to-one function μ mapping S onto G. (Of course, the direction $\leftarrow$ of $\leftrightarrow$ gives us the inverse function μ^{-1}). Expressed in terms of μ, the computation (1) of xy for $x, y \in S$ becomes

$$\text{if } \mu(x) = g_1 \quad \text{and} \quad \mu(y) = g_2 \quad \text{and} \quad \mu(z) = g_1g_2, \quad \text{then} \quad xy = z. \tag{2}$$

The map $\mu : S \to G$ now becomes an isomorphism mapping the group S onto the group G. Notice that from (2), we obtain $\mu(xy) = \mu(z) = g_1g_2 = \mu(x)\mu(y)$, the required homomorphism property.

Let G and G' be groups, let $\phi : G \to G'$ be a homomorphism, and let $H = \text{Ker}(\phi)$. Theorem 13.15 shows that for $a \in G$, we have $\phi^{-1}[\{\phi(a)\}] = aH = Ha$. We have a one-to-one correspondence $aH \leftrightarrow \phi(a)$ between cosets of H in G and elements of the subgroup $\phi[G]$ of G'. Remember that if $x \in aH$, so that $x = ah$ for some $h \in H$, then $\phi(x) = \phi(ah) = \phi(a)\phi(h) = \phi(a)e' = \phi(a)$, so the computation of the element of $\phi[G]$ corresponding to the coset $aH = xH$ is the same whether we compute it as $\phi(a)$ or as $\phi(x)$. Let us denote the set of all cosets of H by G/H. (We read G/H as "G over H" or as "G modulo H" or as "G mod H," but *never* as "G divided by H.")

In the preceding paragraph, we started with a homomorphism $\phi : G \to G'$ having kernel H, and we finished with the set G/H of cosets in one-to-one correspondence with the elements of the group $\phi[G]$. In our work above that, we had a set S with elements in one-to-one correspondence with a those of a group G, and we made S into a group isomorphic to G with an isomorphism μ. Replacing S by G/H and replacing G by $\phi[G]$ in that construction, we can consider G/H to be a group isomorphic to $\phi[G]$ with that isomorphism μ. In terms of G/H and $\phi[G]$, the computation (2) of the product $(xH)(yH)$ for $xH, yH \in G/H$ becomes

$$\begin{aligned}&\text{if } \mu(xH) = \phi(x) \quad \text{and} \quad \mu(yH) = \phi(y) \quad \text{and} \quad \mu(zH) = \phi(x)\phi(y),\\ &\text{then} \quad (xH)(yH) = zH.\end{aligned} \tag{3}$$

But because ϕ is a homomorphism, we can easily find $z \in G$ such that $\mu(zH) = \phi(x)\phi(y)$; namely, we take $z = xy$ in G, and find that

$$\mu(zH) = \mu(xyH) = \phi(xy) = \phi(x)\phi(y).$$

This shows that the product $(xH)(yH)$ of two cosets is the coset $(xy)H$ that contains the product xy of x and y in G. While this computation of $(xH)(yH)$ may seem to depend on our choices x from xH and y from yH, our work above shows it does not. We demonstrate it again here because it is such an important point. If $h_1, h_2 \in H$ so that xh_1 is an element of xH and yh_2 is an element of yH, then there exists $h_3 \in H$ such that $h_1 y = yh_3$ because $Hy = yH$ by Theorem 13.15. Thus we have

$$(xh_1)(yh_2) = x(h_1 y)h_2 = x(yh_3)h_2 = (xy)(h_3 h_2) \in (xy)H,$$

so we obtain the same coset. Computation of the product of two cosets is accomplished by *choosing* an element from each coset and taking, as product of the cosets, the coset that contains the product in G of the choices. Any time we define something (like a product) in terms of choices, it is important to show that it is **well defined,** which means that it is independent of the choices made. This is precisely what we have just done. We summarize this work in a theorem.

14.1 Theorem Let $\phi : G \to G'$ be a group homomorphism with kernel H. Then the cosets of H form a **factor group**, G/H, where $(aH)(bH) = (ab)H$. Also, the map $\mu : G/H \to \phi[G]$ defined by $\mu(aH) = \phi(a)$ is an isomorphism. Both coset multiplication and μ are well defined, independent of the choices a and b from the cosets.

14.2 Example Example 13.10 considered the map $\gamma : \mathbb{Z} \to \mathbb{Z}_n$, where $\gamma(m)$ is the remainder when m is divided by n in accordance with the division algorithm. We know that γ is a homomorphism. Of course, $\text{Ker}(\gamma) = n\mathbb{Z}$. By Theorem 14.1, we see that the factor group $\mathbb{Z}/n\mathbb{Z}$ is isomorphic to $\mathbb{Z}_n$. The cosets of $n\mathbb{Z}$ are the *residue classes modulo* n. For example, taking $n = 5$, we see the cosets of $5\mathbb{Z}$ are

$$\begin{aligned} 5\mathbb{Z} &= \{\cdots, -10, -5, 0, 5, 10, \cdots\}, \\ 1 + 5\mathbb{Z} &= \{\cdots, -9, -4, 1, 6, 11, \cdots\}, \\ 2 + 5\mathbb{Z} &= \{\cdots, -8, -3, 2, 7, 12, \cdots\}, \\ 3 + 5\mathbb{Z} &= \{\cdots, -7, -2, 3, 8, 13, \cdots\}, \\ 4 + 5\mathbb{Z} &= \{\cdots, -6, -1, 4, 9, 14, \cdots\}. \end{aligned}$$

Note that the isomorphism $\mu : \mathbb{Z}/5\mathbb{Z} \to \mathbb{Z}_5$ of Theorem 14.1 assigns to each coset of $5\mathbb{Z}$ its smallest nonnegative element. That is, $\mu(5\mathbb{Z}) = 0$, $\mu(1 + 5\mathbb{Z}) = 1$, etc. ▲

It is very important that we learn how to compute in a factor group. We can multiply (add) two cosets by choosing *any* two representative elements, multiplying (adding) them and finding the coset in which the resulting product (sum) lies.

14.3 Example Consider the factor group $\mathbb{Z}/5\mathbb{Z}$ with the cosets shown above. We can add $(2 + 5\mathbb{Z}) + (4 + 5\mathbb{Z})$ by choosing 2 and 4, finding $2 + 4 = 6$, and noticing that 6 is in the coset $1 + 5\mathbb{Z}$. We could equally well add these two cosets by choosing 27 in $2 + 5\mathbb{Z}$ and -16 in $4 + 5\mathbb{Z}$; the sum $27 + (-16) = 11$ is also in the coset $1 + 5\mathbb{Z}$. ▲

The factor groups $\mathbb{Z}/n\mathbb{Z}$ in the preceding example are classics. Recall that we refer to the cosets of $n\mathbb{Z}$ as *residue classes modulo* n. Two integers in the same coset are *congruent modulo* n. This terminology is carried over to other factor groups. A factor group G/H is often called the **factor group of** G **modulo** H. Elements in the same coset of H are often said to be **congruent modulo** H. By abuse of notation, we may sometimes write $\mathbb{Z}/n\mathbb{Z} = \mathbb{Z}_n$ and think of $\mathbb{Z}_n$ as the additive group of residue classes of $\mathbb{Z}$ modulo $\langle n \rangle$, or abusing notation further, modulo n.

Factor Groups from Normal Subgroups

So far, we have obtained factor groups only from homomorphisms. Let G be a group and let H be a subgroup of G. Now H has both left cosets and right cosets, and in general, a left coset aH need not be the same set as the right coset Ha. Suppose we try to define a binary operation on left cosets by defining

$$(aH)(bH) = (ab)H \tag{4}$$

as in the statement of Theorem 14.1 Equation 4 attempts to define left coset multiplication by choosing representatives a and b from the cosets. Equation 4 is meaningless unless it gives a *well-defined* operation, independent of the representative elements a and b chosen from the cosets. The theorem that follows shows that Eq. 4 gives a well-defined binary operation if and only if H is a normal subgroup of G.

14.4 Theorem Let H be a subgroup of a group G. Then left coset multiplication is well defined by the equation

$$(aH)(bH) = (ab)H$$

if and only if H is a normal subgroup of G.

Proof Suppose first that $(aH)(bH) = (ab)H$ does give a well-defined binary operation on left cosets. Let $a \in G$. We want to show that aH and Ha are the same set. We use the standard technique of showing that each is a subset of the other.

Let $x \in aH$. Choosing representatives $x \in aH$ and $a^{-1} \in a^{-1}H$, we have $(xH)(a^{-1}H) = (xa^{-1})H$. On the other hand, choosing representatives $a \in aH$ and $a^{-1} \in a^{-1}H$, we see that $(aH)(a^{-1}H) = eH = H$. Using our assumption that left coset multiplication by representatives is well defined, we must have $xa^{-1} = h \in H$. Then $x = ha$, so $x \in Ha$ and $aH \subseteq Ha$. We leave the symmetric proof that $Ha \subseteq aH$ to Exercise 25.

We turn now to the converse: If H is a normal subgroup, then left coset multiplication by representatives is well-defined. Due to our hypothesis, we can simply say *cosets,* omitting *left* and *right*. Suppose we wish to compute $(aH)(bH)$. Choosing $a \in aH$ and $b \in bH$, we obtain the coset $(ab)H$. Choosing different representatives $ah_1 \in aH$ and $bh_2 \in bH$, we obtain the coset ah_1bh_2H. We must show that these are the same coset. Now $h_1b \in Hb = bH$, so $h_1b = bh_3$ for some $h_3 \in H$. Thus

$$(ah_1)(bh_2) = a(h_1b)h_2 = a(bh_3)h_2 = (ab)(h_3h_2)$$

and $(ab)(h_3h_2) \in (ab)H$. Therefore, ah_1bh_2 is in $(ab)H$. ◆

Theorem 14.4 shows that if left and right cosets of H coincide, then Eq. 4 gives a well-defined binary operation on cosets. We wonder whether the cosets do form a group with such coset multiplication. This is indeed true.

14.5 Corollary Let H be a normal subgroup of G. Then the cosets of H form a group G/H under the binary operation $(aH)(bH) = (ab)H$. ▲

Proof Computing, $(aH)[(bH)(cH)] = (aH)[(bc)H] = [a(bc)]H$, and similarly, we have $[(aH)(bH)](cH) = [(ab)c]H$, so associativity in G/H follows from associativity in G. Because $(aH)(eH) = (ae)H = aH = (ea)H = (eH)(aH)$, we see that $eH = H$ is the identity element in G/H. Finally, $(a^{-1}H)(aH) = (a^{-1}a)H = eH = (aa^{-1})H = (aH)(a^{-1}H)$ shows that $a^{-1}H = (aH)^{-1}$. ◆

14.6 Definition The group G/H in the preceding corollary is the **factor group** (or **quotient group**) of G by H. ■

14.7 Example Since $\mathbb{Z}$ is an abelian group, $n\mathbb{Z}$ is a normal subgroup. Corollary 14.5 allows us to construct the factor group $\mathbb{Z}/n\mathbb{Z}$ with no reference to a homomorphism. As we observed in Example 14.2, $\mathbb{Z}/n\mathbb{Z}$ is isomorphic to $\mathbb{Z}_n$. ▲

14.8 Example Consider the abelian group $\mathbb{R}$ under addition, and let $c \in \mathbb{R}^+$. The cyclic subgroup $\langle c\rangle$ of $\mathbb{R}$ contains as elements

$$\cdots - 3c, -2c, -c, 0, c, 2c, 3c, \cdots.$$

Every coset of $\langle c\rangle$ contains just one element of x such that $0 \le x < c$. If we choose these elements as representatives of the cosets when computing in $\mathbb{R}/\langle c\rangle$, we find that we are computing their sum modulo c as discussed for the computation in $\mathbb{R}_c$ in Section 1. For example, if $c = 5.37$, then the sum of the cosets $4.65 + \langle 5.37\rangle$ and $3.42 + \langle 5.37\rangle$ is the coset $8.07 + \langle 5.37\rangle$, which contains $8.07 - 5.37 = 2.7$, which is $4.65 +_{5.37} 3.42$. Working with these coset elements x where $0 \le x < c$, we thus see that the group $\mathbb{R}_c$ of Example 4.2 is isomorphic to $\mathbb{R}/\langle c\rangle$ under an isomorphism ψ where $\psi(x) = x + \langle c\rangle$ for all $x \in \mathbb{R}_c$. Of course, $\mathbb{R}/\langle c\rangle$ is then also isomorphic to the circle group U of complex numbers of magnitude 1 under multiplication. ▲

We have seen that the group $\mathbb{Z}/\langle n\rangle$ is isomorphic to the group $\mathbb{Z}_n$, and as a set, $\mathbb{Z}_n = \{0, 1, 3, 4, \cdots, n-1\}$, the set of nonnegative integers less than n. Example 14.8 shows that the group $\mathbb{R}/\langle c\rangle$ is isomorphic to the group $\mathbb{R}_c$. In Section 1, we choose the notation $\mathbb{R}_c$ rather than the conventional $[0, c)$ for the half-open interval of nonnegative real numbers less than c. We did that to bring out now the comparison of these factor groups of $\mathbb{Z}$ with these factor groups of $\mathbb{R}$.

The Fundamental Homomorphism Theorem

We have seen that every homomorphism $\phi : G \to G'$ gives rise to a natural factor group (Theorem 14.1), namely, $G/\text{Ker}(\phi)$. We now show that each factor group G/H gives rise to a natural homomorphism having H as kernel.

14.9 Theorem Let H be a normal subgroup of G. Then $\gamma : G \to G/H$ given by $\gamma(x) = xH$ is a homomorphism with kernel H.

Proof Let $x, y \in G$. Then

$$\gamma(xy) = (xy)H = (xH)(yH) = \gamma(x)\gamma(y),$$

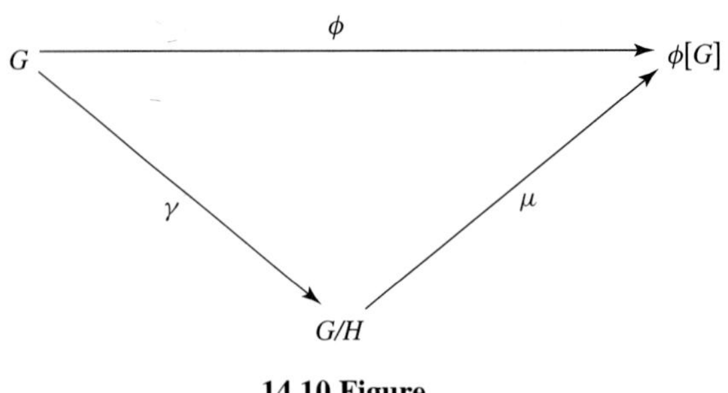

14.10 Figure

so γ is a homomorphism. Since $xH = H$ if an only if $x \in H$, we see that the kernel of γ is indeed H. ◆

We have seen in Theorem 14.1 that if $\phi : G \to G'$ is a homomorphism with kernel H, then $\mu : G/H \to \phi[G]$ where $\mu(gH) = \phi(g)$ is an isomorphism. Theorem 14.9 shows that $\gamma : G \to G/H$ defined by $\gamma(g) = gH$ is a homomorphism. Figure 14.10 shows these groups and maps. We see that the homomorphism ϕ can be *factored,* $\phi = \mu\gamma$, where γ is a homomorphism and μ is an isomorphism of G/H with $\phi[G]$. We state this as a theorem.

14.11 Theorem **(The Fundamental Homomorphism Theorem)** Let $\phi : G \to G'$ be a group homomorphism with kernel H. Then $\phi[G]$ is a group, and $\mu : G/H \to \phi[G]$ given by $\mu(gH) = \phi(g)$ is an isomorphism. If $\gamma : G \to G/H$ is the homomorphism given by $\gamma(g) = gH$, then $\phi(g) = \mu\gamma(g)$ for each $g \in G$.

The isomorphism μ in Theorem 14.11 is referred to as a *natural* or *canonical* isomorphism, and the same adjectives are used to describe the homomorphism γ. There may be other isomorphisms and homomorphisms for these same groups, but the maps μ and γ have a special status with ϕ and are uniquely determined by Theorem 14.11.

In summary, every homomorphism with domain G gives rise to a factor group G/H, and every factor group G/H gives rise to a homomorphism mapping G into G/H. Homomorphisms and factor groups are closely related. We give an example indicating how useful this relationship can be.

14.12 Example Classify the group $(\mathbb{Z}_4 \times \mathbb{Z}_2)/(\{0\} \times \mathbb{Z}_2)$ according to the fundamental theorem of finitely generated abelian groups (Theorem 11.12).

Solution The projection map $\pi_1 : \mathbb{Z}_4 \times \mathbb{Z}_2 \to \mathbb{Z}_4$ given by $\pi_1(x, y) = x$ is a homomorphism of $\mathbb{Z}_4 \times \mathbb{Z}_2$ onto $\mathbb{Z}_4$ with kernel $\{0\} \times \mathbb{Z}_2$. By Theorem 14.11, we know that the given factor group is isomorphic to $\mathbb{Z}_4$. ▲

Normal Subgroups and Inner Automorphisms

We derive some alternative characterizations of normal subgroups, which often provide us with an easier way to check normality than finding both the left and the right coset decompositions.

Suppose that H is a subgroup of G such that $ghg^{-1} \in H$ for all $g \in G$ and all $h \in H$. Then $gHg^{-1} = \{ghg^{-1} \mid h \in H\} \subseteq H$ for all $g \in G$. We claim that actually $gHg^{-1} = H$. We must show that $H \subseteq gHg^{-1}$ for all $g \in G$. Let $h \in H$. Replacing g by g^{-1} in the relation $ghg^{-1} \in H$, we obtain $g^{-1}h(g^{-1})^{-1} = g^{-1}hg = h_1$ where $h_1 \in H$. Consequently, $h = gh_1g^{-1} \in gHg^{-1}$, and we are done.

Suppose that $gH = Hg$ for all $g \in G$. Then $gh = h_1g$, so $ghg^{-1} \in H$ for all $g \in G$ and all $h \in H$. By the preceding paragraph, this means that $gHg^{-1} = H$ for all $g \in G$. Conversely, if $gHg^{-1} = H$ for all $g \in G$, then $ghg^{-1} = h_1$ so $gh = h_1g \in Hg$, and $gH \subseteq Hg$. But also, $g^{-1}Hg = H$ giving $g^{-1}hg = h_2$, so that $hg = gh_2$ and $Hg \subseteq gH$.

We summarize our work as a theorem.

14.13 Theorem The following are three equivalent conditions for a subgroup H of a group G to be a *normal* subgroup of G.

1. $ghg^{-1} \in H$ for all $g \in G$ and $h \in H$.
2. $gHg^{-1} = H$ for all $g \in G$.
3. $gH = Hg$ for all $g \in G$.

Condition (2) of Theorem 14.13 is often taken as the definition of a normal subgroup H of a group G.

14.14 Example Every subgroup H of an abelian group G is normal. We need only note that $gh = hg$ for all $h \in H$ and all $g \in G$, so, of course, $ghg^{-1} = h \in H$ for all $g \in G$ and all $h \in H$. ▲

Exercise 29 of Section 13 shows that the map $i_g : G \to G$ defined by $i_g(x) = gxg^{-1}$ is a homomorphism of G into itself. We see that $gag^{-1} = gbg^{-1}$ if and only if $a = b$, so i_g is one to one. Since $g(g^{-1}yg)g^{-1} = y$, we see that i_g is onto G, so it is an isomorphism of G with itself.

14.15 Definition An isomorphism $\phi : G \to G$ of a group G with itself is an **automorphism** of G. The automorphism $i_g : G \to G$, where $i_g(x) = gxg^{-1}$ for all $x \in G$, is the **inner automorphism of** G **by** g. Performing i_g on x is called **conjugation of** x **by** g. ■

The equivalence of conditions (1) and (2) in Theorem 14.13 shows that $gH = Hg$ for all $g \in G$ if and only if $i_g[H] = H$ for all $g \in G$, that is, if and only if H is **invariant** under all inner automorphisms of G. It is important to realize that $i_g[H] = H$ is an equation in *sets;* we need not have $i_g(h) = h$ for all $h \in H$. That is i_g may perform a nontrivial *permutation* of the set H. We see that the normal subgroups of a group G are precisely those that are invariant under all inner automorphisms. A subgroup K of G is a **conjugate subgroup** of H if $K = i_g[H]$ for some $g \in G$.

EXERCISES 14

Computations

In Exercises 1 through 8, find the order of the given factor group.

1. $\mathbb{Z}_6/\langle 3\rangle$

2. $(\mathbb{Z}_4 \times \mathbb{Z}_{12})/(\langle 2\rangle \times \langle 2\rangle)$

3. $(\mathbb{Z}_4 \times \mathbb{Z}_2)/\langle(2, 1)\rangle$

4. $(\mathbb{Z}_3 \times \mathbb{Z}_5)/(\{0\} \times \mathbb{Z}_5)$

5. $(\mathbb{Z}_2 \times \mathbb{Z}_4)/\langle(1, 1)\rangle$

6. $(\mathbb{Z}_{12} \times \mathbb{Z}_{18})/\langle(4, 3)\rangle$

7. $(\mathbb{Z}_2 \times S_3)/\langle(1, \rho_1)\rangle$

8. $(\mathbb{Z}_{11} \times \mathbb{Z}_{15})/\langle(1, 1)\rangle$

In Exercises 9 through 15, give the order of the element in the factor group.

9. $5 + \langle 4\rangle$ in $\mathbb{Z}_{12}/\langle 4\rangle$

10. $26 + \langle 12\rangle$ in $\mathbb{Z}_{60}/\langle 12\rangle$

11. $(2, 1) + \langle(1, 1)\rangle$ in $(\mathbb{Z}_3 \times \mathbb{Z}_6)/\langle(1, 1)\rangle$

12. $(3, 1) + \langle(1, 1)\rangle$ in $(\mathbb{Z}_4 \times \mathbb{Z}_4)/\langle(1, 1)\rangle$

13. $(3, 1) + \langle(0, 2)\rangle$ in $(\mathbb{Z}_4 \times \mathbb{Z}_8)/\langle(0, 2)\rangle$

14. $(3, 3) + \langle(1, 2)\rangle$ in $(\mathbb{Z}_4 \times \mathbb{Z}_8)/\langle(1, 2)\rangle$

15. $(2, 0) + \langle(4, 4)\rangle$ in $(\mathbb{Z}_6 \times \mathbb{Z}_8)/\langle(4, 4)\rangle$

16. Compute $i_{\rho_1}[H]$ for the subgroup $H = \{\rho_0, \mu_1\}$ of the group S_3 of Example 8.7.

Concepts

In Exercises 17 through 19, correct the definition of the italicized term without reference to the text, if correction is needed, so that it is in a form acceptable for publication.

17. A *normal subgroup* H of G is one satisfying $hG = Gh$ for all $h \in H$.

18. A *normal subgroup* H of G is one satisfying $g^{-1}hg \in H$ for all $h \in H$ and all $g \in G$.

19. An *automorphism* of a group G is a homomorphism mapping G into G.

20. What is the importance of a *normal* subgroup of a group G?

Students often write nonsense when first proving theorems about factor groups. The next two exercises are designed to call attention to one basic type of error.

21. A student is asked to show that if H is a normal subgroup of an abelian group G, then G/H is abelian. The student's proof starts as follows:

We must show that G/H is abelian. Let a and b be two elements of G/H.

a. Why does the instructor reading this proof expect to find nonsense from here on in the student's paper?

b. What should the student have written?

c. Complete the proof.

22. A **torsion group** is a group all of whose elements have finite order. A group is **torsion free** if the identity is the only element of finite order. A student is asked to prove that if G is a torsion group, then so is G/H for every normal subgroup H of G. The student writes

We must show that each element of G/H is of finite order. Let $x \in G/H$.

Answer the same questions as in Exercise 21.

23. Mark each of the following true or false.

_____ **a.** It makes sense to speak of the factor group G/N if and only if N is a normal subgroup of the group G.

_____ **b.** Every subgroup of an abelian group G is a normal subgroup of G.

_____ **c.** An inner automorphism of an abelian group must be just the identity map.

_______ **d.** Every factor group of a finite group is again of finite order.

_______ **e.** Every factor group of a torsion group is a torsion group. (See Exercise 22.)

_______ **f.** Every factor group of a torsion-free group is torsion free. (See Exercise 22.)

_______ **g.** Every factor group of an abelian group is abelian.

_______ **h.** Every factor group of a nonabelian group is nonabelian.

_______ **i.** $\mathbb{Z}/n\mathbb{Z}$ is cyclic of order n.

_______ **j.** $\mathbb{R}/n\mathbb{R}$ is cyclic of order n, where $n\mathbb{R} = \{nr \mid r \in \mathbb{R}\}$ and $\mathbb{R}$ is under addition.

Theory

24. Show that A_n is a normal subgroup of S_n and compute S_n/A_n; that is, find a known group to which S_n/A_n is isomorphic.

25. Complete the proof of Theorem 14.4 by showing that if H is a subgroup of a group G and if left coset multiplication $(aH)(bH) = (ab)H$ is well defined, then $Ha \subseteq aH$.

26. Prove that the torsion subgroup T of an abelian group G is a normal subgroup of G, and that G/T is torsion free. (See Exercise 22.)

27. A subgroup H is **conjugate to a subgroup** K of a group G if there exists an inner automorphism i_g of G such that $i_g[H] = K$. Show that conjugacy is an equivalence relation on the collection of subgroups of G.

28. Characterize the normal subgroups of a group G in terms of the cells where they appear in the partition given by the conjugacy relation in the preceding exercise.

29. Referring to Exercise 27, find all subgroups of S_3 (Example 8.7) that are conjugate to $\{\rho_0, \mu_2\}$.

30. Let H be a normal subgroup of a group G, and let $m = (G : H)$. Show that $a^m \in H$ for every $a \in G$.

31. Show that an intersection of normal subgroups of a group G is again a normal subgroup of G.

32. Given any subset S of a group G, show that it makes sense to speak of the smallest normal subgroup that contains S. [*Hint:* Use Exercise 31.]

33. Let G be a group. An element of G that can be expressed in the form $aba^{-1}b^{-1}$ for some $a, b \in G$ is a **commutator** in G. The preceding exercise shows that there is a smallest normal subgroup C of a group G containing all commutators in G; the subgroup C is the **commutator subgroup** of G. Show that G/C is an abelian group.

34. Show that if a finite group G has exactly one subgroup H of a given order, then H is a normal subgroup of G.

35. Show that if H and N are subgroups of a group G, and N is normal in G, then $H \cap N$ is normal in H. Show by an example that $H \cap N$ need not be normal in G.

36. Let G be a group containing at least one subgroup of a fixed finite order s. Show that the intersection of all subgroups of G of order s is a normal subgroup of G. [*Hint:* Use the fact that if H has order s, then so does $x^{-1}Hx$ for all $x \in G$.]

37. **a.** Show that all automorphisms of a group G form a group under function composition.

b. Show that the inner automorphisms of a group G form a normal subgroup of the group of all automorphisms of G under function composition. [*Warning:* Be sure to show that the inner automorphisms do form a subgroup.]

38. Show that the set of all $g \in G$ such that $i_g : G \to G$ is the identity inner automorphism i_e is a normal subgroup of a group G.

39. Let G and G' be groups, and let H and H' be normal subgroups of G and G', respectively. Let ϕ be a homomorphism of G into G'. Show that ϕ induces a natural homomorphism $\phi_* : (G/H) \to (G'/H')$ if $\phi[H] \subseteq H'$. (This fact is used constantly in algebraic topology.)

40. Use the properties $\det(AB) = \det(A) \cdot \det(B)$ and $\det(I_n) = 1$ for $n \times n$ matrices to show the following:

a. The $n \times n$ matrices with determinant 1 form a normal subgroup of $GL(n, \mathbb{R})$.

b. The $n \times n$ matrices with determinant ± 1 form a normal subgroup of $GL(n, \mathbb{R})$.

41. Let G be a group, and let $\mathscr{P}(G)$ be the set of all *subsets* of G. For any $A, B \in \mathscr{P}(G)$, let us define the product subset $AB = \{ab \mid a \in A, b \in B\}$.

a. Show that this multiplication of subsets is associative and has an identity element, but that $\mathscr{P}(G)$ is not a group under this operation.

b. Show that if N is a normal subgroup of G, then the set of cosets of N is closed under the above operation on $\mathscr{P}(G)$, and that this operation agrees with the multiplication given by the formula in Corollary 14.5.

c. Show (without using Corollary 14.5) that the cosets of N in G form a group under the above operation. Is its identity element the same as the identity element of $\mathscr{P}(G)$?

SECTION 15 FACTOR-GROUP COMPUTATIONS AND SIMPLE GROUPS

Factor groups can be a tough topic for students to grasp. There is nothing like a bit of computation to strengthen understanding in mathematics. We start by attempting to improve our intuition concerning factor groups. Since we will be dealing with normal subgroups throughout this section, we often denote a subgroup of a group G by N rather than by H.

Let N be a normal subgroup of G. In the factor group G/N, the subgroup N acts as identity element. We may regard N as being *collapsed* to a single element, either to 0 in additive notation or to e in multiplicative notation. This collapsing of N together with the algebraic structure of G require that other subsets of G, namely, the cosets of N, also collapse into a single element in the factor group. A visualization of this collapsing is provided by Fig. 15.1. Recall from Theorem 14.9 that $\gamma : G \to G/N$ defined by $\gamma(a) = aN$ for $a \in G$ is a homomorphism of G onto G/N. Figure 15.1 is very similar to Fig. 13.14, but in Fig. 15.1 the image group under the homomorphism is actually formed from G. We can view the "line" G/N at the bottom of the figure as obtained by collapsing to a point each coset of N in another copy of G. Each point of G/N thus corresponds to a whole vertical line segment in the shaded portion, representing a coset of N in G. It is crucial to remember that multiplication of cosets in G/N can be computed by multiplying in G, using any representative elements of the cosets as shown in the figure.

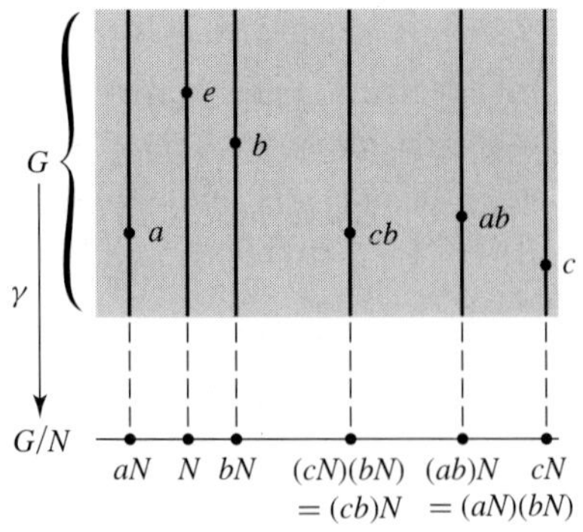

15.1 Figure

Additively, two elements of G will collapse into the same element of G/N if they differ by an element of N. Multiplicatively, a and b collapse together if ab^{-1} is in N. The degree of collapsing can vary from nonexistent to catastrophic. We illustrate the two extreme cases by examples.

15.2 Example The trivial subgroup $N = \{0\}$ of $\mathbb{Z}$ is, of course, a normal subgroup. Compute $\mathbb{Z}/\{0\}$.

Solution Since $N = \{0\}$ has only one element, every coset of N has only one element. That is, the cosets are of the form $\{m\}$ for $m \in \mathbb{Z}$. There is no collapsing at all, and consequently, $\mathbb{Z}/\{0\} \simeq \mathbb{Z}$. Each $m \in \mathbb{Z}$ is simply renamed $\{m\}$ in $\mathbb{Z}/\{0\}$. ▲

15.3 Example Let n be a positive integer. The set $n\mathbb{R} = \{nr \mid r \in \mathbb{R}\}$ is a subgroup of $\mathbb{R}$ under addition, and it is normal since $\mathbb{R}$ is abelian. Compute $\mathbb{R}/n\mathbb{R}$.

Solution A bit of thought shows that actually $n\mathbb{R} = \mathbb{R}$, because each $x \in \mathbb{R}$ is of the form $n(x/n)$ and $x/n \in \mathbb{R}$. Thus $\mathbb{R}/n\mathbb{R}$ has only one element, the subgroup $n\mathbb{R}$. The factor group is a trivial group consisting only of the identity element. ▲

As illustrated in Examples 15.2 and 15.3 for any group G, we have $G/\{e\} \simeq G$ and $G/G \simeq \{e\}$, where $\{e\}$ is the trivial group consisting only of the identity element e. These two extremes of factor groups are of little importance. We would like knowledge of a factor group G/N to give some information about the structure of G. If $N = \{e\}$, the factor group has the same structure as G and we might as well have tried to study G directly. If $N = G$, the factor group has no significant structure to supply information about G. If G is a finite group and $N \neq \{e\}$ is a normal subgroup of G, then G/N is a smaller group than G, and consequently may have a more simple structure than G. The multiplication of cosets in G/N reflects the multiplication in G, since products of cosets can be computed by multiplying in G representative elements of the cosets.

We give two examples showing that even when G/N has order 2, we may be able to deduce some useful results. If G is a finite group and G/N has just two elements, then we must have $|G| = 2|N|$. Note that every subgroup H containing just half the elements of a finite group G must be a normal subgroup, since for each element a in G but not in H, both the left coset aH and the right coset Ha must consist of all elements in G that are not in H. Thus the left and right cosets of H coincide and H is a normal subgroup of G.

15.4 Example Because $|S_n| = 2|A_n|$, we see that A_n is a normal subgroup of S_n, and S_n/A_n has order 2. Let σ be an odd permutation in S_n, so that $S_n/A_n = \{A_n, \sigma A_n\}$. Renaming the element A_n "even" and the element σA_n "odd," the multiplication in S_n/A_n shown in Table 15.5 becomes

15.5 Table

	A_n	σA_n
A_n	A_n	σA_n
σA_n	σA_n	A_n

$$\text{(even)(even)} = \text{even} \qquad \text{(odd)(even)} = \text{odd}$$
$$\text{(even)(odd)} = \text{odd} \qquad \text{(odd)(odd)} = \text{even}.$$

Thus the factor group reflects these multiplicative properties for all the permutations in S_n. ▲

Example 15.4 illustrates that while knowing the product of two cosets in G/N does not tell us what the product of two elements of G is, it may tell us that the product in G of two *types* of elements is itself of a certain type.

15.6 Example **(Falsity of the Converse of the Theorem of Lagrange)** The theorem of Lagrange states if H is a subgroup of a finite group G, then the order of H divides the order of G. We show that it is false that if d divides the order of G, then there must exist a subgroup H of G having order d. Namely, we show that A_4, which has order 12, contains no subgroup of order 6.

Suppose that H were a subgroup of A_4 having order 6. As observed before in Example 15.4, it would follow that H would be a normal subgroup of A_4. Then A_4/H would have only two elements, H and σH for some $\sigma \in A_4$ not in H. Since in a group of order 2, the square of each element is the identity, we would have $HH = H$ and $(\sigma H)(\sigma H) = H$. Now computation in a factor group can be achieved by computing with representatives in the original group. Thus, computing in A_4, we find that for each $\alpha \in H$ we must have $\alpha^2 \in H$ and for each $\beta \in \sigma H$ we must have $\beta^2 \in H$. That is, the square of every element in A_4 must be in H. But in A_4, we have

$$(1, 2, 3) = (1, 3, 2)^2 \quad \text{and} \quad (1, 3, 2) = (1, 2, 3)^2$$

so $(1, 2, 3)$ and $(1, 3, 2)$ are in H. A similar computation shows that $(1, 2, 4)$, $(1, 4, 2)$, $(1, 3, 4)$, $(1, 4, 3)$, $(2, 3, 4)$, and $(2, 4, 3)$ are all in H. This shows that there must be at least 8 elements in H, contradicting the fact that H was supposed to have order 6. ▲

We now turn to several examples that *compute* factor groups. If the group we start with is finitely generated and abelian, then its factor group will be also. *Computing* such a factor group means classifying it according to the fundamental theorem (Theorem 11.12).

15.7 Example Let us compute the factor group $(\mathbb{Z}_4 \times \mathbb{Z}_6)/\langle(0, 1)\rangle$. Here $\langle(0, 1)\rangle$ is the cyclic subgroup H of $\mathbb{Z}_4 \times \mathbb{Z}_6$ generated by $(0, 1)$. Thus

$$H = \{(0, 0), (0, 1), (0, 2), (0, 3), (0, 4), (0, 5)\}.$$

Since $\mathbb{Z}_4 \times \mathbb{Z}_6$ has 24 elements and H has 6 elements, all cosets of H must have 6 elements, and $(\mathbb{Z}_4 \times \mathbb{Z}_6)/H$ must have order 4. Since $\mathbb{Z}_4 \times \mathbb{Z}_6$ is abelian, so is $(\mathbb{Z}_4 \times \mathbb{Z}_6)/H$ (remember, we compute in a factor group by means of representatives from the original group). In additive notation, the cosets are

$$H = (0, 0) + H, \qquad (1, 0) + H, \qquad (2, 0) + H, \qquad (3, 0) + H.$$

Since we can compute by choosing the representatives $(0, 0)$, $(1, 0)$, $(2, 0)$, and $(3, 0)$, it is clear that $(\mathbb{Z}_4 \times \mathbb{Z}_6)/H$ is isomorphic to $\mathbb{Z}_4$. Note that this is what we would expect, since in a factor group modulo H, everything in H becomes the identity element; that is, we are essentially setting everything in H equal to zero. Thus the whole second factor $\mathbb{Z}_6$ of $\mathbb{Z}_4 \times \mathbb{Z}_6$ is collapsed, leaving just the first factor $\mathbb{Z}_4$. ▲

Example 15.7 is a special case of a general theorem that we now state and prove. We should acquire an intuitive feeling for this theorem in terms of *collapsing one of the factors to the identity element.*

15.8 Theorem Let $G = H \times K$ be the direct product of groups H and K. Then $\overline{H} = \{(h, e) \mid h \in H\}$ is a normal subgroup of G. Also $G/\overline{H}$ is isomorphic to K in a natural way. Similarly, $G/\overline{K} \simeq H$ in a natural way.

Proof Consider the homomorphism $\pi_2 : H \times K \to K$, where $\pi_2(h, k) = k$. (See Example 13.8). Because $\text{Ker}(\pi_2) = \overline{H}$, we see that $\overline{H}$ is a normal subgroup of $H \times K$. Because π_2 is onto K, Theorem 14.11 tells us that $(H \times K)/\overline{H} \simeq K$. ◆

We continue with additional computations of abelian factor groups. To illustrate how easy it is to compute in a factor group if we can compute in the whole group, we prove the following theorem.

15.9 Theorem A factor group of a cyclic group is cyclic.

Proof Let G be cyclic with generator a, and let N be a normal subgroup of G. We claim the coset aN generates G/N. We must compute all powers of aN. But this amounts to computing, in G, all powers of the representative a and all these powers give all elements in G. Hence the powers of aN certainly give all cosets of N and G/N is cyclic. ◆

15.10 Example Let us compute the factor group $(\mathbb{Z}_4 \times \mathbb{Z}_6)/\langle(0, 2)\rangle$. Now $(0, 2)$ generates the subgroup

$$H = \{(0, 0), (0, 2), (0, 4)\}$$

of $\mathbb{Z}_4 \times \mathbb{Z}_6$ of order 3. Here the first factor $\mathbb{Z}_4$ of $\mathbb{Z}_4 \times \mathbb{Z}_6$ is left alone. The $\mathbb{Z}_6$ factor, on the other hand, is essentially collapsed by a subgroup of order 3, giving a factor group in the second factor of order 2 that must be isomorphic to $\mathbb{Z}_2$. Thus $(\mathbb{Z}_4 \times \mathbb{Z}_6)/\langle(0, 2)\rangle$ is isomorphic to $\mathbb{Z}_4 \times \mathbb{Z}_2$. ▲

15.11 Example Let us compute the factor group $(\mathbb{Z}_4 \times \mathbb{Z}_6)/\langle(2, 3)\rangle$. *Be careful*! There is a great temptation to say that we are setting the 2 of $\mathbb{Z}_4$ and the 3 of $\mathbb{Z}_6$ both equal to zero, so that $\mathbb{Z}_4$ is collapsed to a factor group isomorphic to $\mathbb{Z}_2$ and $\mathbb{Z}_6$ to one isomorphic to $\mathbb{Z}_3$, giving a total factor group isomorphic to $\mathbb{Z}_2 \times \mathbb{Z}_3$. *This is wrong*! Note that

$$H = \langle(2, 3)\rangle = \{(0, 0), (2, 3)\}$$

is of order 2, so $(\mathbb{Z}_4 \times \mathbb{Z}_6)/\langle(2, 3)\rangle$ has order 12, not 6. Setting $(2, 3)$ equal to zero does not make $(2, 0)$ and $(0, 3)$ equal to zero individually, so the factors do not collapse separately.

The possible abelian groups of order 12 are $\mathbb{Z}_4 \times \mathbb{Z}_3$ and $\mathbb{Z}_2 \times \mathbb{Z}_2 \times \mathbb{Z}_3$, and we must decide to which one our factor group is isomorphic. These two groups are most easily distinguished in that $\mathbb{Z}_4 \times \mathbb{Z}_3$ has an element of order 4, and $\mathbb{Z}_2 \times \mathbb{Z}_2 \times \mathbb{Z}_3$ does not. We claim that the coset $(1, 0) + H$ is of order 4 in the factor group $(\mathbb{Z}_4 \times \mathbb{Z}_6)/H$. To find the smallest power of a coset giving the identity in a factor group modulo H, we must, by choosing representatives, find the smallest power of a representative that is in the subgroup H. Now,

$$4(1, 0) = (1, 0) + (1, 0) + (1, 0) + (1, 0) = (0, 0)$$

is the first time that $(1, 0)$ added to itself gives an element of H. Thus $(\mathbb{Z}_4 \times \mathbb{Z}_6)/\langle(2, 3)\rangle$ has an element of order 4 and is isomorphic to $\mathbb{Z}_4 \times \mathbb{Z}_3$ or $\mathbb{Z}_{12}$. ▲

15.12 Example Let us compute (that is, classify as in Theorem 11.12 the group $(\mathbb{Z} \times \mathbb{Z})/\langle(1, 1)\rangle$. We may visualize $\mathbb{Z} \times \mathbb{Z}$ as the points in the plane with both coordinates integers, as indicated by the dots in Fig. 15.13. The subgroup $\langle(1, 1)\rangle$ consists of those points that lie on the $45°$ line through the origin, indicated in the figure. The coset $(1, 0) + \langle(1, 1)\rangle$ consists of those dots on the $45°$ line through the point $(1, 0)$, also shown in the figure. Continuing, we see that each coset consists of those dots lying on one of the $45°$ lines in the figure. We may choose the representatives

$$\cdots, (-3, 0), (-2, 0), (-1, 0), (0, 0), (1, 0), (2, 0), (3, 0), \cdots$$

of these cosets to compute in the factor group. Since these representatives correspond precisely to the points of $\mathbb{Z}$ on the x-axis, we see that the factor group $(\mathbb{Z} \times \mathbb{Z})/\langle(1, 1)\rangle$ is isomorphic to $\mathbb{Z}$. ▲

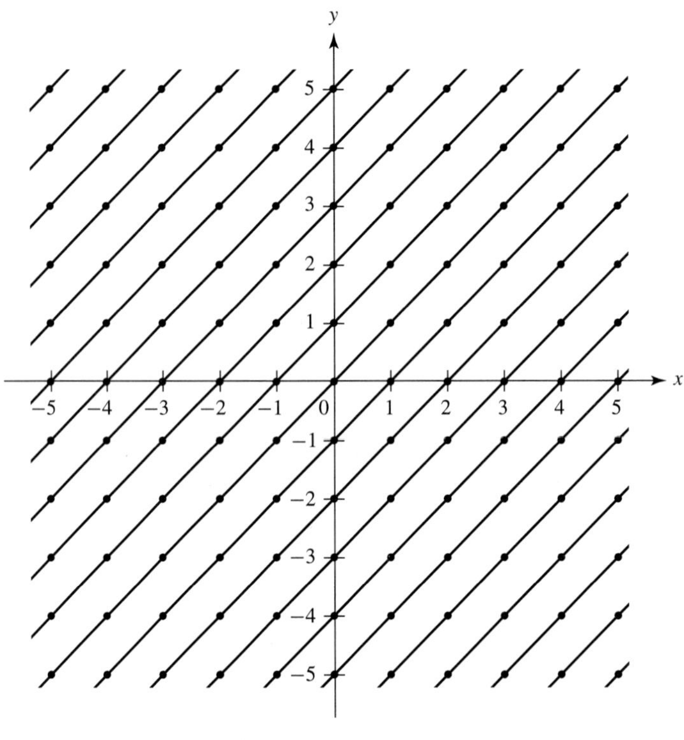

15.13 Figure

Simple Groups

As we mentioned in the preceding section, one feature of a factor group is that it gives crude information about the structure of the whole group. Of course, sometimes there may be no nontrivial proper normal subgroups. For example, Theorem 10.10 shows that a group of prime order can have no nontrivial proper subgroups of any sort.

15.14 Definition A group is **simple** if it is nontrivial and has no proper nontrivial normal subgroups. ■

15.15 Theorem The alternating group A_n is simple for $n \geq 5$.

Proof See Exercise 39. ◆

There are many simple groups other than those given above. For example, A_5 is of order 60 and A_6 is of order 360, and there is a simple group of nonprime order, namely 168, between these orders.

The complete determination and classification of all finite simple groups were recently completed. Hundreds of mathematicians worked on this task from 1950 to 1980. It can be shown that a finite group has a sort of factorization into simple groups, where the factors are unique up to order. The situation is similar to the factorization of positive integers into primes. The new knowledge of all finite simple groups can now be used to solve some problems of finite group theory.

We have seen in this text that a finite simple abelian group is isomorphic to $\mathbb{Z}_p$ for some prime p. In 1963, Thompson and Feit [21] published their proof of a longstanding conjecture of Burnside, showing that every finite nonabelian simple group is of even order. Further great strides toward the complete classification were made by Aschbacher in the 1970s. Early in 1980, Griess announced that he had constructed a predicted "monster" simple group of order

$$808,017,424,794,512,875,886,459,904,961,710,757,005,754,368,\\000,000,000.$$

Aschbacher added the final details of the classification in August 1980. The research papers contributing to the entire classification fill roughly 5000 journal pages.

We turn to the characterization of those normal subgroups N of a group G for which G/N is a simple group. First we state an addendum to Theorem 13.12 on properties of a group homomorphism. The proof is left to Exercises 35 and 36.

15.16 Theorem Let $\phi : G \to G'$ be a group homomorphism. If N is a normal subgroup of G, then $\phi[N]$ is a normal subgroup of $\phi[G]$. Also, if N' is a normal subgroup of $\phi[G]$, then $\phi^{-1}[N']$ is a normal subgroup of G.

Theorem 15.16 should be viewed as saying that a homomorphism $\phi : G \to G'$ preserves normal subgroups between G and $\phi[G]$. It is important to note that $\phi[N]$ may not be normal in G', even though N is normal in G. For example, $\phi : \mathbb{Z}_2 \to S_3$, where $\phi(0) = \rho_0$ and $\phi(1) = \mu_1$ is a homomorphism, and $\mathbb{Z}_2$ is a normal subgroup of itself, but $\{\rho_0, \mu_1\}$ is not a normal subgroup of S_3.

We can now characterize when G/N is a simple group.

15.17 Definition A **maximal normal subgroup of a group** G is a normal subgroup M not equal to G such that there is no proper normal subgroup N of G properly containing M. ■

15.18 Theorem M is a maximal normal subgroup of G if and only if G/M is simple.

Proof Let M be a maximal normal subgroup of G. Consider the canonical homomorphism $\gamma : G \to G/M$ given by Theorem 14.9. Now γ^{-1} of any nontrivial proper normal subgroup of G/M is a proper normal subgroup of G properly containing M. But M is maximal, so this can not happen. Thus G/M is simple.

Conversely, Theorem 15.16 shows that if N is a normal subgroup of G properly containing M, then $\gamma[N]$ is normal in G/M. If also $N \neq G$, then

$$\gamma[N] \neq G/M \qquad \text{and} \qquad \gamma[N] \neq \{M\}.$$

Thus, if G/M is simple so that no such $\gamma[N]$ can exist, no such N can exist, and M is maximal. ◆

The Center and Commutator Subgroups

Every nonabelian group G has two important normal subgroups, the *center* $Z(G)$ of G and the *commutator subgroup* C of G. (The letter Z comes from the German word *zentrum,* meaning center.) The center $Z(G)$ is defined by

$$Z(G) = \{z \in G \mid zg = gz \text{ for all } g \in G\}.$$

Exercise 52 of Section 5 shows that $Z(G)$ is an abelian subgroup of G. Since for each $g \in G$ and $z \in Z(G)$ we have $gzg^{-1} = zgg^{-1} = ze = z$, we see at once that $Z(G)$ is a normal subgroup of G. If G is abelian, then $Z(G) = G$; in this case, the center is not useful.

15.19 Example The center of a group G always contains the identity element e. It may be that $Z(G) = \{e\}$, in which case we say that **the center of G is trivial**. For example, examination of Table 8.8 for the group S_3 shows us that $Z(S_3) = \{\rho_0\}$, so the center of S_3 is trivial. (This is a special case of Exercise 38, which shows that the center of every nonabelian group of order pq for primes p and q is trivial.) Consequently, the center of $S_3 \times \mathbb{Z}_5$ must be $\{\rho_0\} \times \mathbb{Z}_5$, which is isomorphic to $\mathbb{Z}_5$. ▲

Turning to the commutator subgroup, recall that in forming a factor group of G modulo a normal subgroup N, we are essentially putting every element in G that is in N equal to e, for N forms our new identity in the factor group. This indicates another use for factor groups. Suppose, for example, that we are studying the structure of a nonabelian group G. Since Theorem 11.12 gives complete information about the structure of all sufficiently small abelian groups, it might be of interest to try to form an abelian group as much like G as possible, an *abelianized version* of G, by starting with G and then requiring that $ab = ba$ for all a and b in our new group structure. To require that $ab = ba$ is to say that $aba^{-1}b^{-1} = e$ in our new group. An element $aba^{-1}b^{-1}$ in a group is a **commutator of the group.** Thus we wish to attempt to form an abelianized version of G by replacing every commutator of G by e. By the first observation of this paragraph, we should then attempt to form the factor group of G modulo the smallest normal subgroup we can find that contains all commutators of G.

15.20 Theorem Let G be a group. The set of all commutators $aba^{-1}b^{-1}$ for $a, b \in G$ generates a subgroup C (the **commutator subgroup**) of G. This subgroup C is a normal subgroup of G. Furthermore, if N is a normal subgroup of G, then G/N is abelian if and only if $C \leq N$.

Proof The commutators certainly generate a subgroup C; we must show that it is normal in G. Note that the inverse $(aba^{-1}b^{-1})^{-1}$ of a commutator is again a commutator, namely, $bab^{-1}a^{-1}$. Also $e = eee^{-1}e^{-1}$ is a commutator. Theorem 7.6 then shows that C consists precisely of all finite products of commutators. For $x \in C$, we must show that $g^{-1}xg \in C$ for all $g \in G$, or that if x is a product of commutators, so is $g^{-1}xg$ for all $g \in G$. By inserting $e = gg^{-1}$ between each product of commutators occurring in x, we see that it is sufficient to show for each commutator $cdc^{-1}d^{-1}$ that $g^{-1}(cdc^{-1}d^{-1})g$ is in C. But

$$\begin{aligned} g^{-1}(cdc^{-1}d^{-1})g &= (g^{-1}cdc^{-1})(e)(d^{-1}g) \\ &= (g^{-1}cdc^{-1})(gd^{-1}dg^{-1})(d^{-1}g) \\ &= [(g^{-1}c)d(g^{-1}c)^{-1}d^{-1}][dg^{-1}d^{-1}g], \end{aligned}$$

which is in C. Thus C is normal in G.

The rest of the theorem is obvious if we have acquired the proper feeling for factor groups. One doesn't visualize in this way, but writing out that G/C is abelian follows from

$$\begin{aligned} (aC)(bC) &= abC = ab(b^{-1}a^{-1}ba)C \\ &= (abb^{-1}a^{-1})baC = baC = (bC)(aC). \end{aligned}$$

Furthermore, if N is a normal subgroup of G and G/N is abelian, then $(a^{-1}N)(b^{-1}N) = (b^{-1}N)(a^{-1}N)$; that is, $aba^{-1}b^{-1}N = N$, so $aba^{-1}b^{-1} \in N$, and $C \leq N$. Finally, if $C \leq N$, then

$$\begin{aligned} (aN)(bN) &= abN = ab(b^{-1}a^{-1}ba)N \\ &= (abb^{-1}a^{-1})baN = baN = (bN)(aN). \end{aligned}$$

◆

15.21 Example For the group S_3 in Table 8.8, we find that one commutator is $\rho_1\mu_1\rho_1^{-1}\mu_1^{-1} = \rho_1\mu_1\rho_2\mu_1 = \mu_3\mu_2 = \rho_2$. We similarly find that $\rho_2\mu_1\rho_2^{-1}\mu_1^{-1} = \rho_2\mu_1\rho_1\mu_1 = \mu_2\mu_3 = \rho_1$. Thus the commutator subgroup C of S_3 contains A_3. Since A_3 is a normal subgroup of S_3 and S_3/A_3 is abelian, Theorem 15.20 shows that $C = A_3$. ▲

EXERCISES 15

Computations

In Exercises 1 through 12, classify the given group according to the fundamental theorem of finitely generated abelian groups.

1. $(\mathbb{Z}_2 \times \mathbb{Z}_4)/\langle(0, 1)\rangle$
2. $(\mathbb{Z}_2 \times \mathbb{Z}_4)/\langle(0, 2)\rangle$
3. $(\mathbb{Z}_2 \times \mathbb{Z}_4)/\langle(1, 2)\rangle$
4. $(\mathbb{Z}_4 \times \mathbb{Z}_8)/\langle(1, 2)\rangle$
5. $(\mathbb{Z}_4 \times \mathbb{Z}_4 \times \mathbb{Z}_8)/\langle(1, 2, 4)\rangle$
6. $(\mathbb{Z} \times \mathbb{Z})/\langle(0, 1)\rangle$
7. $(\mathbb{Z} \times \mathbb{Z})/\langle(1, 2)\rangle$
8. $(\mathbb{Z} \times \mathbb{Z} \times \mathbb{Z})/\langle(1, 1, 1)\rangle$
9. $(\mathbb{Z} \times \mathbb{Z} \times \mathbb{Z}_4)/\langle(3, 0, 0)\rangle$
10. $(\mathbb{Z} \times \mathbb{Z} \times \mathbb{Z}_8)/\langle(0, 4, 0)\rangle$
11. $(\mathbb{Z} \times \mathbb{Z})/\langle(2, 2)\rangle$
12. $(\mathbb{Z} \times \mathbb{Z} \times \mathbb{Z})/\langle(3, 3, 3)\rangle$

13. Find both the center $Z(D_4)$ and the commutator subgroup C of the group D_4 of symmetries of the square in Table 8.12.

14. Find both the center and the commutator subgroup of $\mathbb{Z}_3 \times S_3$.

15. Find both the center and the commutator subgroup of $S_3 \times D_4$.

16. Describe all subgroups of order ≤ 4 of $\mathbb{Z}_4 \times \mathbb{Z}_4$, and in each case classify the factor group of $\mathbb{Z}_4 \times \mathbb{Z}_4$ modulo the subgroup by Theorem 11.12. That is, describe the subgroup and say that the factor group of $\mathbb{Z}_4 \times \mathbb{Z}_4$ modulo the subgroup is isomorphic to $\mathbb{Z}_2 \times \mathbb{Z}_4$, or whatever the case may be. [*Hint:* $\mathbb{Z}_4 \times \mathbb{Z}_4$ has six different cyclic subgroups of order 4. Describe them by giving a generator, such as the subgroup $\langle(1, 0)\rangle$. There is one subgroup of order 4 that is isomorphic to the Klein 4-group. There are three subgroups of order 2.]

Concepts

In Exercises 17 and 18, correct the definition of the italicized term without reference to the text, if correction is needed, so that it is in a form acceptable for publication.

17. The *center* of a group G contains all elements of G that commute with every element of G.

18. The *commutator subgroup* of a group G is $\{a^{-1}b^{-1}ab \mid a, b \in G\}$.

19. Mark each of the following true or false.

_______ **a.** Every factor group of a cyclic group is cyclic.
_______ **b.** A factor group of a noncyclic group is again noncyclic.
_______ **c.** $\mathbb{R}/\mathbb{Z}$ under addition has no element of order 2.
_______ **d.** $\mathbb{R}/\mathbb{Z}$ under addition has elements of order n for all $n \in \mathbb{Z}^+$.
_______ **e.** $\mathbb{R}/\mathbb{Z}$ under addition has an infinite number of elements of order 4.
_______ **f.** If the commutator subgroup C of a group G is $\{e\}$, then G is abelian.
_______ **g.** If G/H is abelian, then the commutator subgroup of C of G contains H.
_______ **h.** The commutator subgroup of a simple group G must be G itself.
_______ **i.** The commutator subgroup of a nonabelian simple group G must be G itself.
_______ **j.** All nontrivial finite simple groups have prime order.

In Exercises 20 through 23, let F be the additive group of all functions mapping $\mathbb{R}$ into $\mathbb{R}$, and let F^* be the multiplicative group of all elements of F that do not assume the value 0 at any point of $\mathbb{R}$.

20. Let K be the subgroup of F consisting of the constant functions. Find a subgroup of F to which F/K is isomorphic.

21. Let K^* be the subgroup of F^* consisting of the nonzero constant functions. Find a subgroup of F^* to which F^*/K^* is isomorphic.

22. Let K be the subgroup of continuous functions in F. Can you find an element of F/K having order 2? Why or why not?

23. Let K^* be the subgroup of F^* consisting of the continuous functions in F^*. Can you find an element of F^*/K^* having order 2? Why or why not?

In Exercises 24 through 26, let U be the multiplicative group $\{z \in \mathbb{C} \mid |z| = 1\}$.

24. Let $z_0 \in U$. Show that $z_0U = \{z_0z \mid z \in U\}$ is a subgroup of U, and compute U/z_0U.

25. To what group we have mentioned in the text is $U/\langle -1\rangle$ isomorphic?

26. Let $\zeta_n = \cos(2\pi/n) + i\sin(2\pi/n)$ where $n \in \mathbb{Z}^+$. To what group we have mentioned is $U/\langle\zeta_n\rangle$ isomorphic?

27. To what group mentioned in the text is the additive group $\mathbb{R}/\mathbb{Z}$ isomorphic?

28. Give an example of a group G having no elements of finite order > 1 but having a factor group G/H, all of whose elements are of finite order.

29. Let H and K be normal subgroups of a group G. Give an example showing that we may have $H \simeq K$ while G/H is not isomorphic to G/K.

30. Describe the center of every simple

a. abelian group

b. nonabelian group.

31. Describe the commutator subgroup of every simple

a. abelian group

b. nonabelian group.

Proof Synopsis

32. Give a one-sentence synopsis of the proof of Theorem 15.9.

33. Give at most a two-sentence synopsis of the proof of Theorem 15.18.

Theory

34. Show that if a finite group G contains a nontrivial subgroup of index 2 in G, then G is not simple.

35. Let $\phi : G \to G'$ be a group homomorphism, and let N be a normal subgroup of G. Show that $\phi[N]$ is normal subgroup of $\phi[G]$.

36. Let $\phi : G \to G'$ be a group homomorphism, and let N' be a normal subgroup of G'. Show that $\phi^{-1}[N']$ is a normal subgroup of G.

37. Show that if G is nonabelian, then the factor group $G/Z(G)$ is not cyclic. [*Hint:* Show the equivalent contrapositive, namely, that if $G/Z(G)$ is cyclic then G is abelian (and hence $Z(G) = G$).]

38. Using Exercise 37, show that a nonabelian group G of order pq where p and q are primes has a trivial center.

39. Prove that A_n is simple for $n \geq 5$, following the steps and hints given.

a. Show A_n contains every 3-cycle if $n \geq 3$.

b. Show A_n is generated by the 3-cycles for $n \geq 3$. [*Hint:* Note that $(a, b)(c, d) = (a, c, b)(a, c, d)$ and $(a, c)(a, b) = (a, b, c)$.]

c. Let r and s be fixed elements of $\{1, 2, \cdots, n\}$ for $n \geq 3$. Show that A_n is generated by the n "special" 3-cycles of the form (r, s, i) for $1 \leq i \leq n$ [*Hint:* Show every 3-cycle is the product of "special" 3-cycles by computing

$$(r, s, i)^2, \quad (r, s, j)(r, s, i)^2, \quad (r, s, j)^2(r, s, i),$$

and

$$(r, s, i)^2(r, s, k)(r, s, j)^2(r, s, i).$$

Observe that these products give all possible types of 3-cycles.]

d. Let N be a normal subgroup of A_n for $n \geq 3$. Show that if N contains a 3-cycle, then $N = A_n$. [*Hint:* Show that $(r, s, i) \in N$ implies that $(r, s, j) \in N$ for $j = 1, 2, \cdots, n$ by computing

$$((r, s)(i, j))(r, s, i)^2((r, s)(i, j))^{-1}.]$$

e. Let N be a nontrivial normal subgroup of A_n for $n \geq 5$. Show that one of the following cases must hold, and conclude in each case that $N = A_n$.

Case I N contains a 3-cycle.

Case II N contains a product of disjoint cycles, at least one of which has length greater than 3. [*Hint:* Suppose N contains the disjoint product $\sigma = \mu(a_1, a_2, \cdots, a_r)$. Show $\sigma^{-1}(a_1, a_2, a_3)\sigma(a_1, a_2, a_3)^{-1}$ is in N, and compute it.]

Case III N contains a disjoint product of the form $\sigma = \mu(a_4, a_5, a_6)(a_1, a_2, a_3)$. [*Hint:* Show $\sigma^{-1}(a_1, a_2, a_4)$ $\sigma(a_1, a_2, a_4)^{-1}$ is in N, and compute it.]

Case IV N contains a disjoint product of the form $\sigma = \mu(a_1, a_2, a_3)$ where μ is a product of disjoint 2-cycles. [*Hint:* Show $\sigma^2 \in N$ and compute it.]

Case V N contains a disjoint product σ of the form $\sigma = \mu(a_3, a_4)(a_1, a_2)$, where μ is a product of an even number of disjoint 2-cycles. [*Hint:* Show that $\sigma^{-1}(a_1, a_2, a_3)\sigma(a_1, a_2, a_3)^{-1}$ is in N, and compute it to deduce that $\alpha = (a_2, a_4)(a_1, a_3)$ is in N. Using $n \geq 5$ for the first time, find $i \neq a_1, a_2, a_3, a_4$ in $\{1, 2, \cdots, n\}$. Let $\beta = (a_1, a_3, i)$. Show that $\beta^{-1}\alpha\beta\alpha \in N$, and compute it.]

40. Let N be a normal subgroup of G and let H be any subgroup of G. Let $HN = \{hn \mid h \in H, n \in N\}$. Show that HN is a subgroup of G, and is the smallest subgroup containing both N and H.

41. With reference to the preceding exercise, let M also be a normal subgroup of G. Show that NM is again a normal subgroup of G.

42. Show that if H and K are normal subgroups of a group G such that $H \cap K = \{e\}$, then $hk = kh$ for all $h \in H$ and $k \in K$. [*Hint:* Consider the commutator $hkh^{-1}k^{-1} = (hkh^{-1})k^{-1} = h(kh^{-1}k^{-1})$.]

SECTION 16 †Group Action on a Set

We have seen examples of how groups may *act on things*, like the group of symmetries of a triangle or of a square, the group of rotations of a cube, the general linear group acting on $\mathbb{R}^n$, and so on. In this section, we give the general notion of group action on a set. The next section will give an application to counting.

The Notion of a Group Action

Definition 2.1 defines a binary operation $*$ on a set S to be a function mapping $S \times S$ into S. The function $*$ gives us a rule for "multiplying" an element s_1 in S and an element s_2 in S to yield an element $s_1 * s_2$ in S.

More generally, for any sets A, B, and C, we can view a map $* : A \times B \to C$ as defining a "multiplication," where any element a of A times any element b of B has as value some element c of C. Of course, we write $a * b = c$, or simply $ab = c$. In this section, we will be concerned with the case where X is a set, G is a group, and we have a map $* : G \times X \to X$. We shall write $*(g, x)$ as $g * x$ or gx.

16.1 Definition Let X be a set and G a group. An **action of G on X** is a map $* : G \times X \to X$ such that ■

1. $ex = x$ for all $x \in X$,
2. $(g_1g_2)(x) = g_1(g_2x)$ for all $x \in X$ and all $g_1, g_2 \in G$.

Under these conditions, X is a ***G*-set.**

---

† This section is a prerequisite only for Sections 17 and 36.

16.2 Example Let X be any set, and let H be a subgroup of the group S_X of all permutations of X. Then X is an H-set, where the action of $\sigma \in H$ on X is its action as an element of S_X, so that $\sigma x = \sigma(x)$ for all $x \in X$. Condition 2 is a consequence of the definition of permutation multiplication as function composition, and Condition 1 is immediate from the definition of the identity permutation as the identity function. Note that, in particular, $\{1, 2, 3, \cdots, n\}$ is an S_n set. ▲

Our next theorem will show that for every G-set X and each $g \in G$, the map $\sigma_g : X \to X$ defined by $\sigma_g(x) = gx$ is a permutation of X, and that there is a homomorphism $\phi : G \to S_X$ such that the action of G on X is essentially the Example 16.2 action of the image subgroup $H = \phi[G]$ of S_X on X. So actions of subgroups of S_X on X describe all possible group actions on X. When studying the set X, actions using subgroups of S_X suffice. However, sometimes a set X is used to study G via a group action of G on X. Thus we need the more general concept given by Definition 16.1.

16.3 Theorem Let X be a G-set. For each $g \in G$, the function $\sigma_g : X \to X$ defined by $\sigma_g(x) = gx$ for $x \in X$ is a permutation of X. Also, the map $\phi : G \to S_X$ defined by $\phi(g) = \sigma_g$ is a homomorphism with the property that $\phi(g)(x) = gx$.

Proof To show that σ_g is a permutation of X, we must show that it is a one-to-one map of X onto itself. Suppose that $\sigma_g(x_1) = \sigma_g(x_2)$ for $x_1, x_2 \in X$. Then $gx_1 = gx_2$. Consequently, $g^{-1}(gx_1) = g^{-1}(gx_2)$. Using Condition 2 in Definition 16.1, we see that $(g^{-1}g)x_1 = (g^{-1}g)x_2$, so $ex_1 = ex_2$. Condition 1 of the definition then yields $x_1 = x_2$, so σ_g is one to one. The two conditions of the definition show that for $x \in X$, we have $\sigma_g(g^{-1}x) = g(g^{-1})x = (gg^{-1})x = ex = x$, so σ_g maps X onto X. Thus σ_g is indeed a permutation.

To show that $\phi : G \to S_X$ defined by $\phi(g) = \sigma_g$ is a homomorphism, we must show that $\phi(g_1g_2) = \phi(g_1)\phi(g_2)$ for all $g_1, g_2 \in G$. We show the equality of these two permutations in S_X by showing they both carry an $x \in X$ into the same element. Using the two conditions in Definition 16.1 and the rule for function composition, we obtain

$$\begin{aligned}\phi(g_1g_2)(x) &= \sigma_{g_1g_2}(x) = (g_1g_2)x = g_1(g_2x) = g_1\sigma_{g_2}(x) = \sigma_{g_1}(\sigma_{g_2}(x))\\ &= (\sigma_{g_1} \circ \sigma_{g_2})(x) = (\sigma_{g_1}\sigma_{g_2})(x) = (\phi(g_1)\phi(g_2))(x).\end{aligned}$$

Thus ϕ is a homomorphism. The stated property of ϕ follows at once since by our definitions, we have $\phi(g)(x) = \sigma_g(x) = gx$. ◆

It follows from the preceding theorem and Theorem 13.15 that if X is G-set, then the subset of G leaving every element of X fixed is a normal subgroup N of G, and we can regard X as a G/N-set where the action of a coset gN on X is given by $(gN)x = gx$ for each $x \in X$. If $N = \{e\}$, then the identity element of G is the only element that leaves every $x \in X$ fixed; we then say that G **acts faithfully** on X. A group G is **transitive** on a G-set X if for each $x_1, x_2 \in X$, there exists $g \in G$ such that $gx_1 = x_2$. Note that G is transitive on X if and only if the subgroup $\phi[G]$ of S_X is transitive on X, as defined in Exercise 49 of Section 8.

We continue with more examples of G-sets.

16.4 Example Every group G is itself a G-set, where the action on $g_2 \in G$ by $g_1 \in G$ is given by left multiplication. That is, $*(g_1, g_2) = g_1 g_2$. If H is a subgroup of G, we can also regard G as an H-set, where $*(h, g) = hg$. ▲

16.5 Example Let H be a subgroup of G. Then G is an H-set under conjugation where $*(h, g) = hgh^{-1}$ for $g \in G$ and $h \in H$. Condition 1 is obvious, and for Condition 2 note that

$$*(h_1h_2, g) = (h_1h_2)g(h_1h_2)^{-1} = h_1(h_2gh_2^{-1})h_1^{-1} = *(h_1, *(h_2, g)).$$

We always write this action of H on G by conjugation as hgh^{-1}. The abbreviation hg described before the definition would cause terrible confusion with the group operation of G. ▲

16.6 Example For students who have studied vector spaces with real (or complex) scalars, we mention that the axioms $(rs)\mathbf{v} = r(s\mathbf{v})$ and $1\mathbf{v} = \mathbf{v}$ for scalars r and s and a vector $\mathbf{v}$ show that the set of vectors is an $\mathbb{R}^*$-set (or a $\mathbb{C}^*$-set) for the multiplicative group of nonzero scalars. ▲

16.7 Example Let H be a subgroup of G, and let L_H be the set of all left cosets of H. Then L_H is a G-set, where the action of $g \in G$ on the left coset xH is given by $g(xH) = (gx)H$. Observe that this action is well defined: if $yH = xH$, then $y = xh$ for some $h \in H$, and $g(yH) = (gy)H = (gxh)H = (gx)(hH) = (gx)H = g(xH)$. A series of exercises shows that every G-set is isomorphic to one that may be formed using these left coset G-sets as building blocks. (See Exercises 14 through 17.) ▲

16.8 Example Let G be the group $D_4 = \{\rho_0, \rho_1, \rho_2, \rho_3, \mu_1, \mu_2, \delta_1, \delta_2\}$ of symmetries of the square, described in Example 8.10. In Fig. 16.9 we show the square with vertices 1, 2, 3, 4 as in Fig. 8.11. We also label the sides s_1, s_2, s_3, s_4, the diagonals d_1 and d_2, vertical and horizontal axes m_1 and m_2, the center point C, and midpoints P_i of the sides s_i. Recall that ρ_i corresponds to rotating the square counterclockwise through $\pi i/2$ radians, μ_i

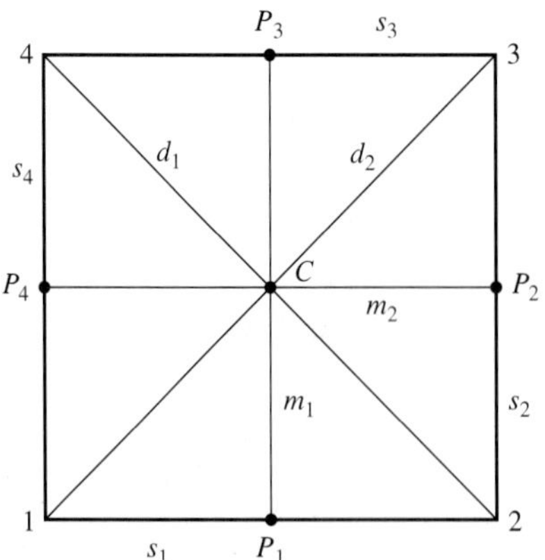

16.9 Figure

16.10 Table

	1	2	3	4	s_1	s_2	s_3	s_4	m_1	m_2	d_1	d_2	C	P_1	P_2	P_3	P_4
ρ_0	1	2	3	4	s_1	s_2	s_3	s_4	m_1	m_2	d_1	d_2	C	P_1	P_2	P_3	P_4
ρ_1	2	3	4	1	s_2	s_3	s_4	s_1	m_2	m_1	d_2	d_1	C	P_2	P_3	P_4	P_1
ρ_2	3	4	1	2	s_3	s_4	s_1	s_2	m_1	m_2	d_1	d_2	C	P_3	P_4	P_1	P_2
ρ_3	4	1	2	3	s_4	s_1	s_2	s_3	m_2	m_1	d_2	d_1	C	P_4	P_1	P_2	P_3
μ_1	2	1	4	3	s_1	s_4	s_3	s_2	m_1	m_2	d_2	d_1	C	P_1	P_4	P_3	P_2
μ_2	4	3	2	1	s_3	s_2	s_1	s_4	m_1	m_2	d_2	d_1	C	P_3	P_2	P_1	P_4
δ_1	3	2	1	4	s_2	s_1	s_4	s_3	m_2	m_1	d_1	d_2	C	P_2	P_1	P_4	P_3
δ_2	1	4	3	2	s_4	s_3	s_2	s_1	m_2	m_1	d_1	d_2	C	P_4	P_3	P_2	P_1

corresponds to flipping on the axis m_i, and δ_i to flipping on the diagonal d_i. We let

$$X = \{1, 2, 3, 4, s_1, s_2, s_3, s_4, m_1, m_2, d_1, d_2, C, P_1, P_2, P_3, P_4\}.$$

Then X can be regarded as a D_4-set in a natural way. Table 16.10 describes completely the action of D_4 on X and is given to provide geometric illustrations of ideas to be introduced. We should be sure that we understand how this table is formed before continuing. ▲

Isotropy Subgroups

Let X be a G-set. Let $x \in X$ and $g \in G$. It will be important to know when $gx = x$. We let

$$X_g = \{x \in X \mid gx = x\} \qquad \text{and} \qquad G_x = \{g \in G \mid gx = x\}.$$

16.11 Example For the D_4-set X in Example 16.8, we have

$$X_{\rho_0} = X, \qquad X_{\rho_1} = \{C\}, \qquad X_{\mu_1} = \{s_1, s_3, m_1, m_2, C, P_1, P_3\}$$

Also, with $G = D_4$,

$$G_1 = \{\rho_0, \delta_2\}, \qquad G_{s_3} = \{\rho_0, \mu_1\}, \qquad G_{d_1} = \{\rho_0, \rho_2, \delta_1, \delta_2\}.$$

We leave the computation of the other X_σ and G_x to Exercises 1 and 2. ▲

Note that the subsets G_x given in the preceding example were, in each case, subgroups of G. This is true in general.

16.12 Theorem Let X be a G-set. Then G_x is a subgroup of G for each $x \in X$.

Proof Let $x \in X$ and let $g_1, g_2 \in G_x$. Then $g_1 x = x$ and $g_2 x = x$. Consequently, $(g_1 g_2)x = g_1(g_2 x) = g_1 x = x$, so $g_1 g_2 \in G_x$, and G_x is closed under the induced operation of G. Of course $ex = x$, so $e \in G_x$. If $g \in G_x$, then $gx = x$, so $x = ex = (g^{-1}g)x = g^{-1}(gx) = g^{-1}x$, and consequently $g^{-1} \in G_x$. Thus G_x is a subgroup of G. ◆

16.13 Definition Let X be a G-set and let $x \in X$. The subgroup G_x is the **isotropy subgroup of** x. ■

Orbits

For the D_4-set X of Example 16.8 with action table in Table 16.10, the elements in the subset $\{1, 2, 3, 4\}$ are carried into elements of this same subset under action by D_4. Furthermore, each of the elements 1, 2, 3, and 4 is carried into all the other elements of the subset by the various elements of D_4. We proceed to show that every G-set X can be partitioned into subsets of this type.

16.14 Theorem Let X be a G-set. For $x_1, x_2 \in X$, let $x_1 \sim x_2$ if and only if there exists $g \in G$ such that $gx_1 = x_2$. Then $\sim$ is an equivalence relation on X.

Proof For each $x \in X$, we have $ex = x$, so $x \sim x$ and $\sim$ is reflexive.

Suppose $x_1 \sim x_2$, so $gx_1 = x_2$ for some $g \in G$. Then $g^{-1}x_2 = g^{-1}(gx_1) = (g^{-1}g)x_1 = ex_1 = x_1$, so $x_2 \sim x_1$, and $\sim$ is symmetric.

Finally, if $x_1 \sim x_2$ and $x_2 \sim x_3$, then $g_1x_1 = x_2$ and $g_2x_2 = x_3$ for some $g_1, g_2 \in G$. Then $(g_2g_1)x_1 = g_2(g_1x_1) = g_2x_2 = x_3$, so $x_1 \sim x_3$ and $\sim$ is transitive. ◆

16.15 Definition Let X be a G-set. Each cell in the partition of the equivalence relation described in Theorem 16.14 is an **orbit in X under G**. If $x \in X$, the cell containing x is the **orbit of** x. We let this cell be Gx. ■

The relationship between the orbits in X and the group structure of G lies at the heart of the applications that appear in Section 17. The following theorem gives this relationship. Recall that for a set X, we use $|X|$ for the number of elements in X, and $(G : H)$ is the index of a subgroup H in a group G.

16.16 Theorem Let X be a G-set and let $x \in X$. Then $|Gx| = (G : G_x)$. If $|G|$ is finite, then $|Gx|$ is a divisor of $|G|$.

Proof We define a one-to-one map ψ from Gx onto the collection of left cosets of G_x in G. Let $x_1 \in Gx$. Then there exists $g_1 \in G$ such that $g_1x = x_1$. We define $\psi(x_1)$ to be the left coset g_1G_x of G_x. We must show that this map ψ is well defined, independent of the choice of $g_1 \in G$ such that $g_1x = x_1$. Suppose also that $g_1'x = x_1$. Then, $g_1x = g_1'x$, so $g_1^{-1}(g_1x) = g_1^{-1}(g_1'x)$, from which we deduce $x = (g_1^{-1}g_1')x$. Therefore $g_1^{-1}g_1' \in G_x$, so $g_1' \in g_1G_x$, and $g_1G_x = g_1'G_x$. Thus the map ψ is well defined.

To show the map ψ is one to one, suppose $x_1, x_2 \in Gx$, and $\psi(x_1) = \psi(x_2)$. Then there exist $g_1, g_2 \in G$ such that $x_1 = g_1x$, $x_2 = g_2x$, and $g_2 \in g_1G_x$. Then $g_2 = g_1g$ for some $g \in G_x$, so $x_2 = g_2x = g_1(gx) = g_1x = x_1$. Thus ψ is one to one.

Finally, we show that each left coset of G_x in G is of the form $\psi(x_1)$ for some $x_1 \in Gx$. Let g_1G_x be a left coset. Then if $g_1x = x_1$, we have $g_1G_x = \psi(x_1)$. Thus ψ maps Gx one to one onto the collection of left cosets so $|Gx| = (G : G_x)$.

If $|G|$ is finite, then the equation $|G| = |G_x|(G : G_x)$ shows that $|Gx| = (G : G_x)$ is a divisor of $|G|$. ◆

16.17 Example Let X be the D_4-set in Example 16.8, with action table given by Table 16.10. With $G = D_4$, we have $G1 = \{1, 2, 3, 4\}$ and $G_1 = \{\rho_0, \delta_2\}$. Since $|G| = 8$, we have $|G1| = (G : G_1) = 4$. ▲

We should remember not only the cardinality equation in Theorem 16.16 but also that the *elements of G carrying x into g_1x are precisely the elements of the left coset g_1G_x*. Namely, if $g \in G_x$, then $(g_1g)x = g_1(gx) = g_1x$. On the other hand, if $g_2x = g_1x$, then $g_1^{-1}(g_2x) = x$ so $(g_1^{-1}g_2)x = x$. Thus $g_1^{-1}g_2 \in G_x$ so $g_2 \in g_1G_x$.

EXERCISES 16

Computations

In Exercises 1 through 3, let

$$X = \{1,\ 2,\ 3,\ 4,\ s_1, s_2, s_3, s_4, m_1, m_2, d_1, d_2, C, P_1, P_2, P_3, P_4\}$$

be the D_4-set of Example 16.8 with action table in Table 16.10. Find the following, where $G = D_4$.

1. The fixed sets X_σ for each $\sigma \in D_4$, that is, $X_{\rho_0}, X_{\rho_1}, \cdots, X_{\delta_2}$

2. The isotropy subgroups G_x for each $x \in X$, that is, $G_1, G_2, \cdots, G_{P_3}, G_{P_4}$

3. The orbits in X under D_4

Concepts

In Exercises 4 and 5, correct the definition of the italicized term without reference to the text, if correction is needed, so that it is in a form acceptable for publication.

4. A group G *acts faithfully* on X if and only if $gx = x$ implies that $g = e$.

5. A group G is *transitive* on a G-set X if and only if, for some $g \in G$, gx can be every other x.

6. Let X be a G-set and let $S \subseteq X$. If $Gs \subseteq S$ for all $s \in S$, then S is a **sub-G-set.** Characterize a sub-G-set of a G-set X in terms of orbits in X and G.

7. Characterize a transitive G-set in terms of its orbits.

8. Mark each of the following true or false.

_____ **a.** Every G-set is also a group.

_____ **b.** Each element of a G-set is left fixed by the identity of G.

_____ **c.** If every element of a G-set is left fixed by the same element g of G, then g must be the identity e.

_____ **d.** Let X be a G-set with $x_1, x_2 \in X$ and $g \in G$. If $gx_1 = gx_2$, then $x_1 = x_2$.

_____ **e.** Let X be a G-set with $x \in X$ and $g_1, g_2 \in G$. If $g_1x = g_2x$, then $g_1 = g_2$.

_____ **f.** Each orbit of a G-set X is a transitive sub-G-set.

_____ **g.** Let X be a G-set and let $H \leq G$. Then X can be regarded in a natural way as an H-set.

_____ **h.** With reference to (g), the orbits in X under H are the same as the orbits in X under G.

_____ **i.** If X is a G-set, then each element of G acts as a permutation of X.

_____ **j.** Let X be a G-set and let $x \in X$. If G is finite, then $|G| = |Gx| \cdot |G_x|$.

9. Let X and Y be G-sets with the *same* group G. An **isomorphism** between G-sets X and Y is a map $\phi : X \to Y$ that is one to one, onto Y, and satisfies $g\phi(x) = \phi(gx)$ for all $x \in X$ and $g \in G$. Two G-sets are **isomorphic** if such an isomorphism between them exists. Let X be the D_4-set of Example 16.8.

a. Find two distinct orbits of X that are isomorphic sub-D_4-sets.

b. Show that the orbits $\{1, 2, 3, 4\}$ and $\{s_1, s_2, s_3, s_4\}$ are not isomorphic sub-D_4-sets. [*Hint:* Find an element of G that acts in an essentially different fashion on the two orbits.]

c. Are the orbits you gave for your answer to part (a) the only two different isomorphic sub-D_4-sets of X?

10. Let X be the D_4-set in Example 16.8.

a. Does D_4 act faithfully on X?

b. Find all orbits in X on which D_4 acts faithfully as a sub-D_4-set.

Theory

11. Let X be a G-set. Show that G acts faithfully on X if and only if no two distinct elements of G have the same action on each element of X.

12. Let X be a G-set and let $Y \subseteq X$. Let $G_Y = \{g \in G \mid gy = y \text{ for all } y \in Y\}$. Show G_Y is a subgroup of G, generalizing Theorem 16.12.

13. Let G be the additive group of real numbers. Let the action of $\theta \in G$ on the real plane $\mathbb{R}^2$ be given by rotating the plane counterclockwise about the origin through θ radians. Let P be a point other than the origin in the plane.

a. Show $\mathbb{R}^2$ is a G-set.

b. Describe geometrically the orbit containing P.

c. Find the group G_P.

Exercises 14 through 17 show how all possible G-sets, up to isomorphism (see Exercise 9), can be formed from the group G.

14. Let $\{X_i \mid i \in I\}$ be a disjoint collection of sets, so $X_i \cap X_j = \varnothing$ for $i \neq j$. Let each X_i be a G-set for the same group G.

a. Show that $\bigcup_{i \in I} X_i$ can be viewed in a natural way as a G-set, the **union** of the G-sets X_i.

b. Show that every G-set X is the union of its orbits.

15. Let X be a transitive G-set, and let $x_0 \in X$. Show that X is isomorphic (see Exercise 9) to the G-set L of all left cosets of G_{x_0}, described in Example 16.7. [*Hint:* For $x \in X$, suppose $x = gx_0$, and define $\phi : X \to L$ by $\phi(x) = gG_{x_0}$. Be sure to show ϕ is well defined!]

16. Let X_i for $i \in I$ be G-sets for the same group G, and suppose the sets X_i are not necessarily disjoint. Let $X_i' = \{(x, i) \mid x \in X_i\}$ for each $i \in I$. Then the sets X_i' are disjoint, and each can still be regarded as a G-set in an obvious way. (The elements of X_i have simply been tagged by i to distinguish them from the elements of X_j for $i \neq j$.) The G-set $\bigcup_{i \in I} X_i'$ is the **disjoint union** of the G-sets X_i. Using Exercises 14 and 15, show that every G-set is isomorphic to a disjoint union of left coset G-sets, as described in Example 16.7.

17. The preceding exercises show that every G-set X is isomorphic to a disjoint union of left coset G-sets. The question then arises whether left coset G-sets of distinct subgroups H and K of G can themselves be isomorphic. Note that the map defined in the hint of Exercise 15 depends on the choice of x_0 as "base point." If x_0 is replaced by g_0x_0 and if $G_{x_0} \neq G_{g_0x_0}$, then the collections L_H of left cosets of $H = G_{x_0}$ and L_K of left cosets of $K = G_{g_0x_0}$ form distinct G-sets that must be isomorphic, since both L_H and L_K are isomorphic to X.

a. Let X be a transitive G-set and let $x_0 \in X$ and $g_0 \in G$. If $H = G_{x_0}$ describe $K = G_{g_0x_0}$ in terms of H and g_0.

b. Based on part (a), conjecture conditions on subgroups H and K of G such that the left coset G-sets of H and K are isomorphic.

c. Prove your conjecture in part (b).

18. Up to isomorphism, how many transitive $\mathbb{Z}_4$ sets X are there? (Use the preceding exercises.) Give an example of each isomorphism type, listing an action table of each as in Table 16.10. Take lowercase names a, b, c, and so on for the elements in the set X.

19. Repeat Exercise 18 for the group $\mathbb{Z}_6$.

20. Repeat Exercise 18 for the group S_3. List the elements of S_3 in the order ι, $(1, 2, 3)$, $(1, 3, 2)$, $(2, 3)$, $(1, 3)$, $(1, 2)$.

SECTION 17 †APPLICATIONS OF G-SETS TO COUNTING

This section presents an application of our work with G-sets to counting. Suppose, for example, we wish to count how many distinguishable ways the six faces of a cube can be marked with from one to six dots to form a die. The standard die is marked so that when placed on a table with the 1 on the bottom and the 2 toward the front, the 6 is on top, the 3 on the left, the 4 on the right, and the 5 on the back. Of course, other ways of marking the cube to give a distinguishably different die are possible.

Let us distinguish between the faces of the cube for the moment and call them the bottom, top, left, right, front, and back. Then the bottom can have any one of six marks from one dot to six dots, the top any one of the five remaining marks, and so on. There are $6! = 720$ ways the cube faces can be marked in all. Some markings yield the same die as others, in the sense that one marking can be carried into another by a rotation of the marked cube. For example, if the standard die described above is rotated 90° counterclockwise as we look down on it, then 3 will be on the front face rather than 2, but it is the same die.

There are 24 possible positions of a cube on a table, for any one of six faces can be placed down, and then any one of four to the front, giving $6 \cdot 4 = 24$ possible positions. Any position can be achieved from any other by a rotation of the die. These rotations form a group G, which is isomorphic to a subgroup of S_8 (see Exercise 45 of Section 8). We let X be the 720 possible ways of marking the cube and let G act on X by rotation of the cube. We consider two markings to give the same die if one can be carried into the other under action by an element of G, that is, by rotating the cube. In other words, we consider each *orbit* in X under G to correspond to a single die, and different orbits to give different dice. The determination of the number of distinguishable dice thus leads to the question of determining the number of orbits under G in a G-set X.

The following theorem gives a tool for determining the number of orbits in a G-set X under G. Recall that for each $g \in G$ we let X_g be the set of elements of X left fixed by g, so that $X_g = \{x \in X \mid gx = x\}$. Recall also that for each $x \in X$, we let $G_x = \{g \in G \mid gx = x\}$, and Gx is the orbit of x under G.

17.1 Theorem **(Burnside's Formula)** Let G be a finite group and X a finite G-set. If r is the number of orbits in X under G, then

$$r \cdot |G| = \sum_{g \in G} |X_g|. \tag{1}$$

† This section is not used in the remainder of the text.

Proof We consider all pairs (g, x) where $gx = x$, and let N be the number of such pairs. For each $g \in G$ there are $|X_g|$ pairs having g as first member. Thus,

$$N = \sum_{g \in G} |X_g|. \tag{2}$$

On the other hand, for each $x \in X$ there are $|G_x|$ pairs having x as second member. Thus we also have

$$N = \sum_{x \in X} |G_x|.$$

By Theorem 16.16 we have $|Gx| = (G : G_x)$. But we know that $(G : G_x) = |G|/|G_x|$, so we obtain $|G_x| = |G|/|Gx|$. Then

$$N = \sum_{x \in X} \frac{|G|}{|Gx|} = |G| \left(\sum_{x \in X} \frac{1}{|Gx|} \right). \tag{3}$$

Now $1/|Gx|$ has the same value for all x in the same orbit, and if we let $\mathcal{O}$ be any orbit, then

$$\sum_{x \in \mathcal{O}} \frac{1}{|Gx|} = \sum_{x \in \mathcal{O}} \frac{1}{|\mathcal{O}|} = 1. \tag{4}$$

Substituting (4) in (3), we obtain

$$N = |G| \text{ (number of orbits in } X \text{ under } G) = |G| \cdot r. \tag{5}$$

Comparison of Eq. 2 and Eq. 5 gives Eq. 1. ◆

17.2 Corollary If G is a finite group and X is a finite G-set, then

$$\text{(number of orbits in } X \text{ under } G) = \frac{1}{|G|} \cdot \sum_{g \in G} |X_g|.$$

Proof The proof of this corollary follows immediately from the preceding theorem. ◆

Let us continue our computation of the number of distinguishable dice as our first example.

17.3 Example We let X be the set of 720 different markings of faces of a cube using from one to six dots. Let G be the group of 24 rotations of the cube as discussed above. We saw that the number of distinguishable dice is the number of orbits in X under G. Now $|G| = 24$. For $g \in G$ where $g \neq e$, we have $|X_g| = 0$, because any rotation other than the identity element changes any one of the 720 markings into a different one. However, $|X_e| = 720$ since the identity element leaves all 720 markings fixed. Then by Corollary 17.2,

$$\text{(number of orbits)} = \frac{1}{24} \cdot 720 = 30,$$

so there are 30 distinguishable dice. ▲

Of course the number of distinguishable dice could be counted without using the machinery of the preceding corollary, but by using elementary combinatorics as often taught in a freshman finite math course. In marking a cube to make a die, we can, by rotation if necessary, assume the face marked 1 is down. There are five choices for the top (opposite) face. By rotating the die as we look down on it, any one of the remaining four faces could be brought to the front position, so there are no different choices involved for the front face. But with respect to the number on the front face, there are $3 \cdot 2 \cdot 1$ possibilities for the remaining three side faces. Thus there are $5 \cdot 3 \cdot 2 \cdot 1 = 30$ possibilities in all.

The next two examples appear in some finite math texts and are easy to solve by elementary means. We use Corollary 17.2 so that we have more practice thinking in terms of orbits.

17.4 Example How many distinguishable ways can seven people be seated at a round table, where there is no distinguishable "head" to the table? Of course there are 7! ways to assign people to the different chairs. We take X to be the 7! possible assignments. A rotation of people achieved by asking each person to move one place to the right results in the same arrangement. Such a rotation generates a cyclic group G of order 7, which we consider to act on X in the obvious way. Again, only the identity e leaves any arrangement fixed, and it leaves all 7! arrangements fixed. By Corollary 17.2

$$(\text{number of orbits}) = \frac{1}{7} \cdot 7! = 6! = 720. \qquad \blacktriangle$$

17.5 Example How many distinguishable necklaces (with no clasp) can be made using seven different-colored beads of the same size? Unlike the table in Example 17.4, the necklace can be turned over as well as rotated. Thus we consider the full dihedral group D_7 of order $2 \cdot 7 = 14$ as acting on the set X of 7! possibilities. Then the number of distinguishable necklaces is

$$(\text{number of orbits}) = \frac{1}{14} \cdot 7! = 360. \qquad \blacktriangle$$

In using Corollary 17.2, we have to compute $|G|$ and $|X_g|$ for each $g \in G$. In the examples and the exercises, $|G|$ will pose no real problem. Let us give an example where $|X_g|$ is not as trivial to compute as in the preceding examples. We will continue to assume knowledge of very elementary combinatorics.

17.6 Example Let us find the number of distinguishable ways the edges of an equilateral triangle can be painted if four different colors of paint are available, assuming only one color is used on each edge, and the same color may be used on different edges.

Of course there are $4^3 = 64$ ways of painting the edges in all, since each of the three edges may be any one of four colors. We consider X to be the set of these 64 possible painted triangles. The group G acting on X is the group of symmetries of the triangle, which is isomorphic to S_3 and which we consider to be S_3. We use the notation for elements in S_3 given in Section 8. We need to compute $|X_g|$ for each of the six elements g in S_3.

$\lvert X_{\rho_0}\rvert = 64$	Every painted triangle is left fixed by ρ_0.
$\lvert X_{\rho_1}\rvert = 4$	To be invariant under ρ_1, all edges must be the same color, and there are 4 possible colors.
$\lvert X_{\rho_2}\rvert = 4$	Same reason as for ρ_1.
$\lvert X_{\mu_1}\rvert = 16$	The edges that are interchanged must be the same color (4 possibilities) and the other edge may also be any of the colors (times 4 possibilities).
$\lvert X_{\mu_2}\rvert = \lvert X_{\mu_3}\rvert = 16$	Same reason as for μ_1.

Then

$$\sum_{g\in S_3} |X_g| = 64 + 4 + 4 + 16 + 16 + 16 = 120.$$

Thus

$$(\text{number of orbits}) = \frac{1}{6} \cdot 120 = 20,$$

and there are 20 distinguishable painted triangles. ▲

17.7 Example We repeat Example 17.6 with the assumption that a different color is used on each edge. The number of possible ways of painting the edges is then $4 \cdot 3 \cdot 2 = 24$, and we let X be the set of 24 possible painted triangles. Again, the group acting on X can be considered to be S_3. Since all edges are a different color, we see $|X_{\rho_0}| = 24$ while $|X_g| = 0$ for $g \neq \rho_0$. Thus

$$(\text{number of orbits}) = \frac{1}{6} \cdot 24 = 4,$$

so there are four distinguishable triangles. ▲

EXERCISES 17

Computations

In each of the following exercises use Corollary 17.2 to work the problem, even though the answer might be obtained by more elementary methods.

1. Find the number of orbits in $\{1, 2, 3, 4, 5, 6, 7, 8\}$ under the cyclic subgroup $\langle(1, 3, 5, 6)\rangle$ of S_8.

2. Find the number of orbits in $\{1, 2, 3, 4, 5, 6, 7, 8\}$ under the subgroup of S_8 generated by $(1, 3)$ and $(2, 4, 7)$.

3. Find the number of distinguishable tetrahedral dice that can be made using one, two, three, and four dots on the faces of a regular tetrahedron, rather than a cube.

4. Wooden cubes of the same size are to be painted a different color on each face to make children's blocks. How many distinguishable blocks can be made if eight colors of paint are available?

5. Answer Exercise 4 if colors may be repeated on different faces at will. [*Hint:* The 24 rotations of a cube consist of the identity, 9 that leave a pair of opposite faces invariant, 8 that leave a pair of opposite vertices invariant, and 6 leaving a pair of opposite edges invariant.]

6. Each of the eight corners of a cube is to be tipped with one of four colors, each of which may be used on from one to all eight corners. Find the number of distinguishable markings possible. (See the hint in Exercise 5.)

7. Find the number of distinguishable ways the edges of a square of cardboard can be painted if six colors of paint are available and

a. no color is used more than once.

b. the same color can be used on any number of edges.

8. Consider six straight wires of equal lengths with ends soldered together to form edges of a regular tetrahedron. Either a 50-ohm or 100-ohm resistor is to be inserted in the middle of each wire. Assume there are at least six of each type of resistor available. How many essentially different wirings are possible?

9. A rectangular prism 2 ft long with 1-ft square ends is to have each of its six faces painted with one of six possible colors. How many distinguishable painted prisms are possible if

a. no color is to be repeated on different faces,

b. each color may be used on any number of faces?

PART IV

Rings and Fields

SECTION 18 RINGS AND FIELDS

All our work thus far has been concerned with sets on which a single binary operation has been defined. Our years of work with the integers and real numbers show that a study of sets on which two binary operations have been defined should be of great importance. Algebraic structures of this type are introduced in this section. In one sense, this section seems more intutive than those that precede it, for the structures studied are closely related to those we have worked with for many years. However, we will be continuing with our axiomatic approach. So, from another viewpoint this study is more complicated than group theory, for we now have two binary operations and more axioms to deal with.

Definitions and Basic Properties

The most general algebraic structure with two binary operations that we shall study is called a *ring*. As Example 18.2 following Definition 18.1 indicates, we have all worked with rings since grade school.

18.1 Definition A **ring** $\langle R, +, \cdot \rangle$ is a set R together with two binary operations $+$ and $\cdot$, which we call *addition* and *multiplication*, defined on R such that the following axioms are satisfied:

$\mathscr{R}_1$. $\langle R, + \rangle$ is an abelian group.

$\mathscr{R}_2$. Multiplication is associative.

$\mathscr{R}_3$. For all $a, b, c \in R$, the **left distributive law**, $a \cdot (b + c) = (a \cdot b) + (a \cdot c)$ and the **right distributive law** $(a + b) \cdot c = (a \cdot c) + (b \cdot c)$ hold. ■

† Sections 24 and 25 are not required for the remainder of the text.

18.2 Example We are well aware that axioms $\mathscr{R}_1$, $\mathscr{R}_2$, and $\mathscr{R}_3$ for a ring hold in any subset of the complex numbers that is a group under addition and that is closed under multiplication. For example, $\langle \mathbb{Z}, +, \cdot \rangle$, $\langle \mathbb{Q}, +, \cdot \rangle$, $\langle \mathbb{R}, +, \cdot \rangle$, and $\langle \mathbb{C}, +, \cdot \rangle$ are rings. ▲

Historical Note

The theory of rings grew out of the study of two particular classes of rings, polynomial rings in n variables over the real or complex numbers (Section 22) and the "integers" of an algebraic number field. It was David Hilbert (1862–1943) who first introduced the term *ring,* in connection with the latter example, but it was not until the second decade of the twentieth century that a fully abstract definition appeared. The theory of commutative rings was given a firm axiomatic foundation by Emmy Noether (1882–1935) in her monumental paper "Ideal Theory in Rings," which appeared in 1921. A major concept of this paper is the ascending chain condition for ideals. Noether proved that in any ring in which every ascending chain of ideals has a maximal element, every ideal is finitely generated.

Emmy Noether received her doctorate from the University of Erlangen, Germany, in 1907. Hilbert invited her to Göttingen in 1915, but his efforts to secure her a paid position were blocked because of her sex. Hilbert complained, "I do not see that the sex of the candidate is an argument against her admission [to the faculty]. After all, we are a university, not a bathing establishment." Noether was, however, able to lecture under Hilbert's name. Ultimately, after the political changes accompanying the end of the First World War reached Göttingen, she was given in 1923 a paid position at the University. For the next decade, she was very influential in the development of the basic concepts of modern algebra. Along with other Jewish faculty members, however, she was forced to leave Göttingen in 1933. She spent the final two years of her life at Bryn Mawr College near Philadelphia.

It is customary to denote multiplication in a ring by juxtaposition, using ab in place of $a \cdot b$. We shall also observe the usual convention that multiplication is performed before addition in the absence of parentheses, so the left distributive law, for example, becomes

$$a(b + c) = ab + ac,$$

without the parentheses on the right side of the equation. Also, as a convenience analogous to our notation in group theory, we shall somewhat incorrectly refer to a *ring* R in place of a *ring* $\langle R, +, \cdot \rangle$, provided that no confusion will result. In particular, from now on $\mathbb{Z}$ will always be $\langle \mathbb{Z}, +, \cdot \rangle$, and $\mathbb{Q}$, $\mathbb{R}$, and $\mathbb{C}$ will also be the rings in Example 18.2. We may on occasion refer to $\langle R, + \rangle$ as *the additive group of the ring* R.

18.3 Example Let R be any ring and let $M_n(R)$ be the collection of all $n \times n$ matrices having elements of R as entries. The operations of addition and multiplication in R allow us to add and multiply matrices in the usual fashion, explained in the appendix. We can quickly check that $\langle M_n(R), + \rangle$ is an abelian group. The associativity of matrix multiplication and the two distributive laws in $M_n(R)$ are more tedious to demonstrate, but straightforward calculations indicate that they follow from the same properties in R. We will assume from now on that we know that $M_n(R)$ is a ring. In particular, we have the rings $M_n(\mathbb{Z})$, $M_n(\mathbb{Q})$, $M_n(\mathbb{R})$, and $M_n(\mathbb{C})$. Note that multiplication is not a commutative operation in any of these rings for $n \geq 2$. ▲

18.4 Example Let F be the set of all functions $f : \mathbb{R} \to \mathbb{R}$. We know that $\langle F, +\rangle$ is an abelian group under the usual function addition,

$$(f + g)(x) = f(x) + g(x).$$

We define multiplication on F by

$$(fg)(x) = f(x)g(x).$$

That is, fg is the function whose value at x is $f(x)g(x)$. It is readily checked that F is a ring; we leave the demonstration to Exercise 34. We have used this juxtaposition notation $\sigma\mu$ for the composite function $\sigma(\mu(x))$ when discussing permutation multiplication. If we were to use both function multiplication and function composition in F, we would use the notation $f \circ g$ for the composite function. However, we will be using composition of functions almost exclusively with homomorphisms, which we will denote by Greek letters, and the usual product defined in this example chiefly when multiplying polynomial function $f(x)g(x)$, so no confusion should result. ▲

18.5 Example Recall that in group theory, $n\mathbb{Z}$ is the cyclic subgroup of $\mathbb{Z}$ under addition consisting of all integer multiples of the integer n. Since $(nr)(ns) = n(nrs)$, we see that $n\mathbb{Z}$, is closed under multiplication. The associative and distributive laws which hold in $\mathbb{Z}$ then assure us that $\langle n\mathbb{Z}, +, \cdot\rangle$ is as ring. From now on in the text, we will consider $n\mathbb{Z}$ to be this ring. ▲

18.6 Example Consider the cyclic group $\langle \mathbb{Z}_n, +\rangle$. If we define for $a, b \in \mathbb{Z}_n$ the product ab as the remainder of the usual product of integers when divided by n, it can be shown that $\langle \mathbb{Z}_n, +, \cdot\rangle$ is a ring. We shall feel free to use this fact. For example, in $\mathbb{Z}_{10}$ we have $(3)(7) = 1$. This operation on $\mathbb{Z}_n$ is **multiplication modulo** n. We do not check the ring axioms here, for they will follow in Section 26 from some of the theory we develop there. From now on, $\mathbb{Z}_n$ will always be the ring $\langle \mathbb{Z}_n, +, \cdot\rangle$. ▲

18.7 Example If $R_1, R_2, \cdots, R_n$ are rings, we can form the set $R_1 \times R_2 \times \cdots \times R_n$ of all ordered n-tuples $(r_1, r_2, \cdots, r_n)$, where $r_i \in R_i$. Defining addition and multiplication of n-tuples by components (just as for groups), we see at once from the ring axioms in each component that the set of all these n-tuples forms a ring under addition and multiplication by components. The ring $R_1 \times R_2 \times \cdots \times R_n$ is the **direct product** of the rings R_i. ▲

Continuing matters of notation, we shall always let 0 be the additive identity of a ring. The additive inverse of an element a of a ring is $-a$. We shall frequently have occasion to refer to a sum

$$a + a + \cdots + a$$

having n summands. We shall let this sum be $n \cdot a$, always using the dot. However, $n \cdot a$ *is not to be constructed as a multiplication of n and a in the ring, for the integer n may not be in the ring at all.* If $n < 0$, we let

$$n \cdot a = (-a) + (-a) + \cdots + (-a)$$

for $|n|$ summands. Finally, we define

$$0 \cdot a = 0$$

for $0 \in \mathbb{Z}$ on the left side of the equations and $0 \in R$ on the right side. Actually, the equation $0a = 0$ holds also for $0 \in R$ on both sides. The following theorem proves this and various other elementary but important facts. Note the strong use of the distributive laws in the proof of this theorem. Axiom $\mathscr{R}_1$ for a ring concerns only addition, and axiom $\mathscr{R}_2$ concerns only multiplication. This shows that in order to prove anything that gives a relationship between these two operations, we are going to have to use axiom $\mathscr{R}_3$. For example, the first thing that we will show in Theorem 18.8 is that $0a = 0$ for any element a in a ring R. Now this relation involves both addition and multiplication. The multiplication $0a$ stares us in the face, and 0 is an *additive* concept. Thus we will have to come up with an argument that uses a distributive law to prove this.

18.8 Theorem If R is a ring with additive identity 0, then for any $a, b \in R$ we have

1. $0a = a0 = 0$,
2. $a(-b) = (-a)b = -(ab)$,
3. $(-a)(-b) = ab$.

Proof For Property 1, note that by axioms $\mathscr{R}_1$ and $\mathscr{R}_2$,

$$a0 + a0 = a(0 + 0) = a0 = 0 + a0.$$

Then by the cancellation law for the additive group $\langle R, +\rangle$, we have $a0 = 0$. Likewise,

$$0a + 0a = (0 + 0)a = 0a = 0 + 0a$$

implies that $0a = 0$. This proves Property 1.

In order to understand the proof of Property 2, we must remember that, by *definition*, $-(ab)$ is the element that when added to ab gives 0. Thus to show that $a(-b) = -(ab)$, we must show precisely that $a(-b) + ab = 0$. By the left distributive law,

$$a(-b) + ab = a(-b + b) = a0 = 0,$$

since a0 = 0 by Property 1. Likewise,

$$(-a)b + ab = (-a + a)b = 0b = 0.$$

For Property 3, note that

$$(-a)(-b) = -(a(-b))$$

by Property 2. Again by Property 2,

$$-(a(-b)) = -(-(ab)),$$

and $-(-(ab))$ is the element that when added to $-(ab)$ gives 0. This is ab by definition of $-(ab)$ and by the uniqueness of an inverse in a group. Thus, $(-a)(-b) = ab$. ◆

It is important that you *understand* the preceding proof. The theorem allows us to use our usual rules for signs.

Homomorphisms and Isomorphisms

From our work in group theory, it is quite clear how a structure-relating map of a ring R into a ring R' should be defined.

18.9 Definition For rings R and R', a map $\phi : R \to R'$ is a **homomorphism** if the following two conditions are satisfied for all $a, b \in R$:

1. $\phi(a + b) = \phi(a) + \phi(b)$,
2. $\phi(ab) = \phi(a)\phi(b)$. ■

In the preceding definition, Condition 1 is the statement that ϕ is a homomorphism mapping the abelian group $\langle R, +\rangle$ into $\langle R', +\rangle$. Condition 2 requires that ϕ relate the multiplicative structures of the rings R and R' in the same way. Since ϕ is also a group homomorphism, all the results concerning group homomorphisms are valid for the additive structure of the rings. In particular, ϕ is one to one if and only if its **kernel** $\text{Ker}(\phi) = \{a \in R \mid \phi(a) = 0'\}$ is just the subset $\{0\}$ of R. The homomorphism ϕ of the group $\langle R, +\rangle$ gives rise to a factor group. We expect that a ring homomorphism will give rise to a factor ring. This is indeed the case. We delay discussion of this to Section 26, where the treatment will parallel our treatment of factor groups in Section 14.

18.10 Example Let F be the ring of all functions mapping $\mathbb{R}$ into $\mathbb{R}$ defined in Example 18.4. For each $a \in \mathbb{R}$, we have the **evaluation homomorphism** $\phi_a : F \to \mathbb{R}$, where $\phi_a(f) = f(a)$ for $f \in F$. We defined this homomorphism for the group $\langle F, +\rangle$ in Example 13.4, but we did not do much with it in group theory. We will be working a great deal with it in the rest of this text, for finding a real solution of a polynomial equation $p(x) = 0$ amounts precisely to finding $a \in \mathbb{R}$ such that $\phi_a(p) = 0$. Much of the remainder of this text deals with solving polynomial equations. We leave the demonstration of the multiplicative homomorphism property 2 for ϕ_a to Exercise 35. ▲

18.11 Example The map $\phi : \mathbb{Z} \to \mathbb{Z}_n$ where $\phi(a)$ is the remainder of a modulo n is a ring homomorphism for each positive integer n. We know $\phi(a + b) = \phi(a) + \phi(b)$ by group theory. To show the multiplicative property, write $a = q_1n + r_1$ and $b = q_2n + r_2$ according to the division algorithm. Then $ab = n(q_1q_2n + r_1q_2 + q_1r_2) + r_1r_2$. Thus $\phi(ab)$ is the remainder of r_1r_2 when divided by n. Since $\phi(a) = r_1$ and $\phi(b) = r_2$, Example 18.6 indicates that $\phi(a)\phi(b)$ is also this same remainder, so $\phi(ab) = \phi(a)\phi(b)$. From group theory, we anticipate that the ring $\mathbb{Z}_n$ might be isomorphic to a factor ring $\mathbb{Z}/n\mathbb{Z}$. This is indeed the case; factor rings will be discussed in Section 26. ▲

We realize that in the study of any sort of mathematical structure, an idea of basic importance is the concept of two systems being *structurally identical*, that is, one being just like the other except for names. In algebra this concept is always called *isomorphism*.

The concept of two things being just alike except for names of elements leads us, just as it did for groups, to the following definition.

18.12 Definition An **isomorphism** $\phi : R \to R'$ from a ring R to a ring R' is a homomorphism that is one to one and onto R'. The rings R and R' are then **isomorphic**. ■

From our work in group theory, we expect that isomorphism gives an equivalence relation on any collection of rings. We need to check that the multiplicative property of an isomorphism is satisfied for the inverse map $\phi^{-1} : R' \to R$ (to complete the symmetry argument). Similarly, we check that if $\mu : R' \to R''$ is also a ring ismorphism, then the multiplicative requirement holds for the composite map $\mu\phi : R \to R''$ (to complete the transitivity argument). We ask you to do this in Exercise 36.

18.13 Example As abelian groups, $\langle \mathbb{Z}, + \rangle$ and $\langle 2\mathbb{Z}, + \rangle$ are isomorphic under the map $\phi : \mathbb{Z} \to \mathbb{Z}$, with $\phi(x) = 2x$ for $x \in \mathbb{Z}$. Here ϕ is *not* a ring isomorphism, for $\phi(xy) = 2xy$, while $\phi(x)\phi(y) = 2x2y = 4xy$. ▲

Multiplicative Questions: Fields

Many of the rings we have mentioned, such as $\mathbb{Z}$, $\mathbb{Q}$, and $\mathbb{R}$, have a multiplicative identity element 1. However, $2\mathbb{Z}$ does not have an identity element for multiplication. Note also that multiplication is not commutative in the matrix rings described in Example 18.3.

It is evident that $\{0\}$, with $0 + 0 = 0$ and $(0)(0) = 0$, gives a ring, the **zero ring.** Here 0 acts as multiplicative as well as additive identity element. By Theorem 18.8, this is the only case in which 0 could act as a multiplicative identity element, for from $0a = 0$, we can then deduce that $a = 0$. Theorem 3.13 shows that if a ring has a multiplicative identity element, it is unique. We denote a multiplicative identity element in a ring by 1.

18.14 Definition A ring in which the multiplication is commutative is a **commutative ring.** A ring with a multiplicative identity element is a **ring with unity;** the multiplicative identity element 1 is called "**unity.**" ■

In a ring with unity 1 the distributive laws show that

$$\underbrace{(1 + 1 + \cdots + 1)}_{n \text{ summands}} \underbrace{(1 + 1 + \cdots + 1)}_{m \text{ summands}} = \underbrace{(1 + 1 + \cdots + 1)}_{nm \text{ summands}},$$

that is, $(n \cdot 1)(m \cdot 1) = (nm) \cdot 1$. The next example gives an application of this observation.

18.15 Example We claim that for integers r and s where $\gcd(r, s) = 1$, the rings $\mathbb{Z}_{rs}$ and $\mathbb{Z}_r \times \mathbb{Z}_s$ are isomorphic. Additively, they are both cyclic abelian groups of order rs with generators 1 and $(1, 1)$ respectively. Thus $\phi : \mathbb{Z}_{rs} \to \mathbb{Z}_r \times \mathbb{Z}_s$ defined by $\phi(n \cdot 1) = n \cdot (1, 1)$ is an additive group isomorphism. To check the multiplicative Condition 2 of Definition 18.9, we use the observation preceding this example for the unity (1, 1) in the ring $\mathbb{Z}_r \times \mathbb{Z}_s$, and compute.

$$\phi(nm) = (nm) \cdot (1, 1) = [n \cdot (1, 1)][m \cdot (1, 1)] = \phi(n)\phi(m). \quad \blacktriangle$$

Note that a direct product $R_1 \times R_2 \times \cdots \times R_n$ of rings is commutative or has unity if and only if each R_i is commutative or has unity, respectively.

In a ring R with unity $1 \neq 0$, the set R^* of nonzero elements, if closed under the ring multiplication, will be a multiplicative group if multiplicative inverses exist. A **multiplicative inverse** of an element a in a ring R with unity $1 \neq 0$ is an element $a^{-1} \in R$ such that $aa^{-1} = a^{-1}a = 1$. Precisely as for groups, a multiplicative inverse for an element a in R is unique, if it exists at all (see Exercise 43). Theorem 18.8 shows that it would be hopeless to have a multiplicative inverse for 0 except for the ring $\{0\}$, where $0 + 0 = 0$ and $(0)(0) = 0$, with 0 as both additive and multiplicative identity element. We are thus led to discuss the existence of multiplicative inverses for nonzero elements in a ring with nonzero unity. There is unavoidably a lot of terminology to be defined in this introductory section on rings. We are almost done.

18.16 Definition Let R be a ring with unity $1 \neq 0$. An element u in R is a **unit** of R if it has a multiplicative inverse in R. If every nonzero element of R is a unit, then R is a **division ring** (or **skew field**). A **field** is a commutative division ring. A noncommutative division ring is called a "**strictly skew field.**" ■

18.17 Example Let us find the units in $\mathbb{Z}_{14}$. Of course, 1 and $-1 = 13$ are units. Since $(3)(5) = 1$ we see that 3 and 5 are units; therefore $-3 = 11$ and $-5 = 9$ are also units. None of the remaining elements of $\mathbb{Z}_{14}$ can be units, since no multiple of 2, 4, 6, 7, 8, or 10 can be one more than a multiple of 14; they all have a common factor, either 2 or 7, with 14. Section 20 will show that the units in $\mathbb{Z}_n$ are precisely those $m \in \mathbb{Z}_n$ such that $\gcd(m, n) = 1$. ▲

18.18 Example $\mathbb{Z}$ is not a field, because 2, for example, has no multiplicative inverse, so 2 is not a unit in $\mathbb{Z}$. The only units in $\mathbb{Z}$ are 1 and -1. However, $\mathbb{Q}$ and $\mathbb{R}$ are fields. An example of a strictly skew field is given in Section 24. ▲

We have the natural concepts of a subring of a ring and subfield of a field. A **subring of a ring** is a subset of the ring that is a ring under induced operations from the whole ring; a **subfield** is defined similarly for a subset of a field. In fact, let us say here once and for all that if we have a set, together with a certain specified type of *algebraic structure* (group, ring, field, integral domain, vector space, and so on), then any subset of this set, together with a natural induced algebraic structure *that yields an algebraic structure of the same type*, is a *substructure*. If K and L are both structures, we shall let $K \leq L$ denote that K is a substructure of L and $K < L$ denote that $K \leq L$ but $K \neq L$. Exercise 48 gives criteria for a subset S of a ring R to form a subring of R.

Finally, be careful not to confuse our use of the words *unit* and *unity*. *Unity* is the multiplicative identity element, while a *unit* is any element having a multiplicative inverse. Thus the multiplicative identity element or unity is a unit, but not every unit is unity. For example, -1 is a unit in $\mathbb{Z}$, but -1 is not unity, that is, $-1 \neq 1$.

Historical Note

Although fields were implict in the early work on the solvability of equations by Abel and Galois, it was Leopold Kronecker (1823–1891) who in connection with his own work on this subject first published in 1881 a definition of what he called a "domain of rationality": "The domain of rationality $(R', R'', R''', \cdots)$ contains $\cdots$ every one of those quantities which are rational functions of the quantities $R', R'', R''', \cdots$ with integral coefficients." Kronecker, however, who insisted that any mathematical subject must be constructible in finitely many steps, did not view the domain of rationality as a complete entity, but merely as a region in which took place various operations on its elements.

Richard Dedekind (1831–1916), the inventor of the Dedekind cut definition of a real number, considered a field as a completed entity. In 1871, he published the following definition in his supplement to the second edition of Dirichlet's text on number theory: "By a field we mean any system of infinitely many real or complex numbers, which in itself is so closed and complete, that the addition, subtraction, multiplication, and division of any two numbers always produces a number of the same system." Both Kronecker and Dedekind had, however, dealt with their varying ideas of this notion as early as the 1850s in their university lectures.

A more abstract definition of a field, similar to the one in the text, was given by Heinrich Weber (1842–1913) in a paper of 1893. Weber's definition, unlike that of Dedekind, specifically included fields with finitely many elements as well as other fields, such as function fields, which were not subfields of the field of complex numbers.

EXERCISES 18

Computations

In Exercises 1 through 6, compute the product in the given ring.

1. $(12)(16)$ in $\mathbb{Z}_{24}$

2. $(16)(3)$ in $\mathbb{Z}_{32}$

3. $(11)(-4)$ in $\mathbb{Z}_{15}$

4. $(20)(-8)$ in $\mathbb{Z}_{26}$

5. $(2,3)(3,5)$ in $\mathbb{Z}_5 \times \mathbb{Z}_9$

6. $(-3,5)(2,-4)$ in $\mathbb{Z}_4 \times \mathbb{Z}_{11}$

In Exercises 7 through 13, decide whether the indicated operations of addition and multiplication are defined (closed) on the set, and give a ring structure. If a ring is not formed, tell why this is the case. If a ring is formed, state whether the ring is commutative, whether it has unity, and whether it is a field.

7. $n\mathbb{Z}$ with the usual addition and multiplication

8. $\mathbb{Z}^+$ with the usual addition and multiplication

9. $\mathbb{Z} \times \mathbb{Z}$ with addition and multiplication by components

10. $2\mathbb{Z} \times \mathbb{Z}$ with addition and multiplication by components

11. $\{a + b\sqrt{2} \mid a, b \in \mathbb{Z}\}$ with the usual addition and multiplication

12. $\{a + b\sqrt{2} \mid a, b \in \mathbb{Q}\}$ with the usual addition and multiplication

13. The set of all pure imaginary complex numbers ri for $r \in \mathbb{R}$ with the usual addition and multiplication

In Exercises 14 through 19, describe all units in the given ring

14. $\mathbb{Z}$ **15.** $\mathbb{Z} \times \mathbb{Z}$ **16.** $\mathbb{Z}_5$

17. $\mathbb{Q}$ **18.** $\mathbb{Z} \times \mathbb{Q} \times \mathbb{Z}$ **19.** $\mathbb{Z}_4$

20. Consider the matrix ring $M_2(\mathbb{Z}_2)$.

a. Find the **order** of the ring, that is, the number of elements in it.

b. List all units in the ring.

21. If possible, give an example of a homomorphism $\phi : R \to R'$ where R and R' are rings with unity $1 \neq 0$ and $1' \neq 0'$, and where $\phi(1) \neq 0'$ and $\phi(1) \neq 1'$.

22. (Linear algebra) Consider the map det of $M_n(\mathbb{R})$ into $\mathbb{R}$ where $\det(A)$ is the determinant of the matrix A for $A \in M_n(\mathbb{R})$. Is det a ring homomorphism? Why or why not?

23. Describe all ring homomorphisms of $\mathbb{Z}$ into $\mathbb{Z}$.

24. Describe all ring homomorphisms of $\mathbb{Z}$ into $\mathbb{Z} \times \mathbb{Z}$.

25. Describe all ring homomorphisms of $\mathbb{Z} \times \mathbb{Z}$ into $\mathbb{Z}$.

26. How many homomorphisms are there of $\mathbb{Z} \times \mathbb{Z} \times \mathbb{Z}$ into $\mathbb{Z}$?

27. Consider this solution of the equation $X^2 = I_3$ in the ring $M_3(\mathbb{R})$.

$X^2 = I_3$ implies $X^2 - I_3 = 0$, the zero matrix, so factoring, we have $(X - I_3)(X + I_3) = 0$ whence either $X = I_3$ or $X = -I_3$.

Is this reasoning correct? If not, point out the error, and if possible, give a counterexample to the conclusion.

28. Find all solutions of the equation $x^2 + x - 6 = 0$ in the ring $\mathbb{Z}_{14}$ by factoring the quadratic polynomial. Compare with Exercise 27.

Concepts

In Exercises 29 and 30, correct the definition of the italicized term without reference to the text, if correction is needed, so that it is in a from acceptable for publication.

29. A *field* F is a ring with nonzero unity such that the set of nonzero elements of F is a group under multiplication.

30. A *unit* in a ring is an element of magnitude 1.

31. Give an example of a ring having two elements a and b such that $ab = 0$ but neither a nor b is zero.

32. Give an example of a ring with unity $1 \neq 0$ that has a subring with nonzero unity $1' \neq 1$. [*Hint:* Consider a direct product, or a subring of $\mathbb{Z}_6$.]

33. Mark each of the following true or false.

_____ **a.** Every field is also a ring.

_____ **b.** Every ring has a multiplicative identity.

_____ **c.** Every ring with unity has at least two units.

_____ **d.** Every ring with unity has at most two units.

_____ **e.** It is possible for a subset of some field to be a ring but not a subfield, under the induced operations.
_____ **f.** The distributive laws for a ring are not very important.
_____ **g.** Multiplication in a field is commutative.
_____ **h.** The nonzero elements of a field form a group under the multiplication in the field.
_____ **i.** Addition in every ring is commutative.
_____ **j.** Every element in a ring has an additive inverse.

Theory

34. Show that the multiplication defined on the set F of functions in Example 18.4 satisfies axioms $\mathscr{R}_2$ and $\mathscr{R}_3$ for a ring.

35. Show that the evaluation map ϕ_a of Example 18.10 satisfies the multiplicative requirement for a homomorphism.

36. Complete the argument outlined after Definitions 18.12 to show that isomorphism gives an equivalence relation on a collection of rings.

37. Show that if U is the collection of all units in a ring $\langle R, +, \cdot \rangle$ with unity, then $\langle U, \cdot \rangle$ is a group. [*Warning:* Be sure to show that U is closed under multiplication.]

38. Show that $a^2 - b^2 = (a+b)(a-b)$ for all a and b in a ring R if and only if R is commutative.

39. Let $(R, +)$ be an abelian group. Show that $(R, +, \cdot)$ is a ring if we define $ab = 0$ for all $a, b \in R$.

40. Show that the rings $2\mathbb{Z}$ and $3\mathbb{Z}$ are not isomorphic. Show that the fields $\mathbb{R}$ and $\mathbb{C}$ are not isomorphic.

41. (Freshman exponentiation) Let p be a prime. Show that in the ring $\mathbb{Z}_p$ we have $(a+b)^p = a^p + b^p$ for all $a, b \in \mathbb{Z}_p$. [*Hint:* Observe that the usual binomial expansion for $(a+b)^n$ is valid in a *commutative* ring.]

42. Show that the unity element in a subfield of a field must be the unity of the whole field, in contrast to Exercise 32 for rings.

43. Show that the multiplicative inverse of a unit a ring with unity is unique.

44. An element a of a ring R is **idempotent** if $a^2 = a$.

a. Show that the set of all idempotent elements of a commutative ring is closed under multiplication.
b. Find all idempotents in the ring $\mathbb{Z}_6 \times \mathbb{Z}_{12}$.

45. (Linear algebra) Recall that for an $m \times n$ matrix A, the *transpose* A^T of A is the matrix whose jth column is the jth row of A. Show that if A is an $m \times n$ matrix such that A^TA is invertible, then the *projection matrix* $P = A(A^TA)^{-1}A^T$ is an idempotent in the ring of $n \times n$ matrices.

46. An element a of a ring R is **nilpotent** if $a^n = 0$ for some $n \in \mathbb{Z}^+$. Show that if a and b are nilpotent elements of a *commutative* ring, then $a + b$ is also nilpotent.

47. Show that a ring R has no nonzero nilpotent element if and only if 0 is the only solution of $x^2 = 0$ in R.

48. Show that a subset S of a ring R gives a subring of R if and only if the following hold:

$$0 \in S;$$
$$(a - b) \in S \text{ for all } a, b \in S;$$
$$ab \in S \text{ for all } a, b \in S.$$

49. a. Show that an intersection of subrings of a ring R is again a subring of R.
b. Show that an intersection of subfields of a field F is again a subfield of F.

50. Let R be a ring, and let a be a fixed element of R. Let $I_a = \{x \in R \mid ax = 0\}$. Show that I_a is a subring of R.

51. Let R be a ring, and let a be a fixed element of R. Let R_a be the subring of R that is the intersection of all subrings of R containing a (see Exercise 49). The ring R_a is the **subring of R generated by** a. Show that the abelian group $\langle R_a, +\rangle$ is generated (in the sense of Section 7) by $\{a^n \mid n \in \mathbb{Z}^+\}$.

52. (Chinese Remainder Theorem for two congruences) Let r and s be positive integers such that $\gcd(r, s) = 1$. Use the isomorphism in Example 18.15 to show that for $m, n \in \mathbb{Z}$, there exists an integer x such that $x \equiv m \pmod{r}$ and $x \equiv n \pmod{s}$.

53. **a.** State and prove the generalization of Example 18.15 for a direct product with n factors.

b. Prove the Chinese Remainder Theorem: Let $a_i, b_i \in \mathbb{Z}^+$ for $i = 1, 2, \cdots, n$ and let $\gcd(b_i, b_j) = 1$ for $i \neq j$. Then there exists $x \in \mathbb{Z}^+$ such that $x \equiv a_i \pmod{b_i}$ for $i = 1, 2, \cdots, n$.

54. Consider $\langle S, +, \cdot\rangle$, where S is a set and $+$ and $\cdot$ are binary operations on S such that

$\langle S, +\rangle$ is a group,

$\langle S^*, \cdot\rangle$ is a group where S^* consists of all elements of S except the additive identity element,

$a(b + c) = (ab) + (ac)$ and $(a + b)c = (ac) + (bc)$ for all $a, b, c \in S$.

Show that $\langle S, +, \cdot\rangle$ is a division ring. [*Hint:* Apply the distributive laws to $(1 + 1)(a + b)$ to prove the commutativity of addition.]

55. A ring R is a **Boolean ring** if $a^2 = a$ for all $a \in R$, so that every element is idempotent. Show that every Boolean ring is commutative.

56. (For students having some knowledge of the laws of set theory) For a set S, let $\mathscr{P}(S)$ be the collection of all subsets of S. Let binary operations $+$ and $\cdot$ on $\mathscr{P}(S)$ be defined by

$$A + B = (A \cup B) - (A \cap B) = \{x \mid x \in A \text{ or } x \in B \text{ but } x \notin (A \cap B)\}$$

and

$$A \cdot B = A \cap B$$

for $A, B \in \mathscr{P}(S)$.

a. Give the tables for $+$ and $\cdot$ for $\mathscr{P}(S)$, where $S = \{a, b\}$. [*Hint:* $\mathscr{P}(S)$ has four elements.]

b. Show that for *any* set S, $\langle \mathscr{P}(S), +, \cdot\rangle$ is a Boolean ring (see Exercise 55).

SECTION 19 Integral Domains

While a careful treatment of polynomials is not given until Section 22, for purposes of motivation we shall make intuitive use of them in this section.

Divisors of Zero and Cancellation

One of the most important algebraic properties of our usual number system is that a product of two numbers can only be 0 if at least one of the factors is 0. We have used this fact many times in solving equations, perhaps without realizing that we were using it. Suppose, for example, we are asked to solve the equation

$$x^2 - 5x + 6 = 0.$$

The first thing we do is to factor the left side:

$$x^2 - 5x + 6 = (x - 2)(x - 3).$$

Then we conclude that the only possible values for x are 2 and 3. Why? The reason is that if x is replaced by any number a, the product $(a-2)(a-3)$ of the resulting numbers is 0 if and only if either $a-2=0$ or $a-3=0$.

19.1 Example Solve the equation $x^2 - 5x + 6 = 0$ in $\mathbb{Z}_{12}$.

Solution The factorization $x^2 - 5x + 6 = (x-2)(x-3)$ is still valid if we think of x as standing for any number in $\mathbb{Z}_{12}$. But in $\mathbb{Z}_{12}$, not only is $0a = a0 = 0$ for all $a \in \mathbb{Z}_{12}$, but also

$$\begin{aligned}(2)(6) &= (6)(2) = (3)(4) = (4)(3) = (3)(8) = (8)(3)\\ &= (4)(6) = (6)(4) = (4)(9) = (9)(4) = (6)(6) = (6)(8)\\ &= (8)(6) = (6)(10) = (10)(6) = (8)(9) = (9)(8) = 0.\end{aligned}$$

We find, in fact, that our equation has not only 2 and 3 as solutions, but also 6 and 11, for $(6-2)(6-3) = (4)(3) = 0$ and $(11-2)(11-3) = (9)(8) = 0$ in $\mathbb{Z}_{12}$. ▲

These ideas are of such importance that we formalize them in a definition.

19.2 Definition If a and b are two nonzero elements of a ring R such that $ab = 0$, then a and b are **divisors of 0** (or **0 divisors**). ■

Example 19.1 shows that in $\mathbb{Z}_{12}$ the elements 2, 3, 4, 6, 8, 9, and 10 are divisors of 0. Note that these are exactly the numbers in $\mathbb{Z}_{12}$ that are not relatively prime to 12, that is, whose gcd with 12 is not 1. Our next theorem shows that this is an example of a general situation.

19.3 Theorem In the ring $\mathbb{Z}_n$, the divisors of 0 are precisely those nonzero elements that are not relatively prime to n.

Proof Let $m \in \mathbb{Z}_n$, where $m \neq 0$, and let the gcd of m and n be $d \neq 1$. Then

$$m\left(\frac{n}{d}\right) = \left(\frac{m}{d}\right)n,$$

and $(m/d)n$ gives 0 as a multiple of n. Thus $m(n/d) = 0$ in $\mathbb{Z}_n$, while neither m nor n/d is 0, so m is a divisor of 0.

On the other hand, suppose $m \in \mathbb{Z}_n$ is relatively prime to n. If for $s \in \mathbb{Z}_n$ we have $ms = 0$, then n divides the product ms of m and s as elements in the ring $\mathbb{Z}$. Since n is relatively prime to m, boxed Property 1 following Example 6.9 shows that n divides s, so $s = 0$ in $\mathbb{Z}_n$. ◆

19.4 Corollary If p is a prime, then $\mathbb{Z}_p$ has no divisors of 0.

Proof This corollary is immediate from Theorem 19.3. ◆

Another indication of the importance of the concept of 0 divisors is shown in the following theorem. Let R be a ring, and let $a, b, c \in R$. The **cancellation laws** hold in R if $ab = ac$ with $a \neq 0$ implies $b = c$, and $ba = ca$ with $a \neq 0$ implies $b = c$. These are multiplicative cancellation laws. Of course, the additive cancellation laws hold in R, since $\langle R, +\rangle$ is a group.

19.5 Theorem The cancellation laws hold in a ring R if and only if R has no divisors of 0.

Proof Let R be a ring in which the cancellation laws hold, and suppose $ab = 0$ for some $a, b \in R$. We must show that either a or b is 0. If $a \neq 0$, then $ab = a0$ implies that $b = 0$ by cancellation laws. Similarly, $b \neq 0$ implies that $a = 0$, so there can be no divisors of 0 if the cancellation laws hold.

Conversely, suppose that R has no divisors of 0, and suppose that $ab = ac$ with $a \neq 0$. Then

$$ab - ac = a(b - c) = 0.$$

Since $a \neq 0$, and since R has no divisors of 0, we must have $b - c = 0$, so $b = c$. A similar argument shows that $ba = ca$ with $a \neq 0$ implies $b = c$. ◆

Suppose that R is a ring with no divisors of 0. Then an equation $ax = b$, with $a \neq 0$, in R can have at most one solution x in R, for if $ax_1 = b$ and $ax_2 = b$, then $ax_1 = ax_2$, and by Theorem 19.5 $x_1 = x_2$, since R has no divisors of 0. If R has unity $1 \neq 0$ and a is a unit in R with multiplicative inverse a^{-1}, then the solution x of $ax = b$ is $a^{-1}b$. In the case that R is commutative, in particular if R is a field, it is customary to denote $a^{-1}b$ and ba^{-1} (they are equal by commutativity) by the formal quotient b/a. This quotient notation must not be used in the event that R is not commutative, for then we do not know whether b/a denotes $a^{-1}b$ or ba^{-1}. In particular, the multiplicative inverse a^{-1} of a nonzero element a of a field may be written $1/a$.

Integral Domains

The integers are really our most familiar number system. In terms of the algebraic properties we are discussing, $\mathbb{Z}$ is a commutative ring with unity and no divisors of 0. Surely this is responsible for the name that the next definition gives to such a structure.

19.6 Definition An **integral domain** D is a commutative ring with unity $1 \neq 0$ and containing no divisors of 0. ■

Thus, if the coefficients of a polynomial are from an integral domain, one can solve a polynomial equation in which the polynomial can be factored into linear factors in the usual fashion by setting each factor equal to 0.

In our hierarchy of algebraic structures, an integral domain belongs between a commutative ring with unity and a field, as we shall show. Theorem 19.5 shows that the cancellation laws for multiplication hold in an integral domain.

19.7 Example We have seen that $\mathbb{Z}$ and $\mathbb{Z}_p$ for any prime p are integral domains, but $\mathbb{Z}_n$ is not an integral domain if n is not prime. A moment of thought shows that the direct product $R \times S$ of two nonzero rings R and S is not an integral domain. Just observe that for $r \in R$ and $s \in S$ both nonzero, we have $(r, 0)(0, s) = (0, 0)$. ▲

19.8 Example Show that although $\mathbb{Z}_2$ is an integral domain, the matrix ring $M_2(\mathbb{Z}_2)$ has divisors of zero.

Solution We need only observe that

$$\begin{pmatrix} 1 & 0 \\ 0 & 0 \end{pmatrix}\begin{pmatrix} 0 & 0 \\ 1 & 0 \end{pmatrix} = \begin{pmatrix} 0 & 0 \\ 0 & 0 \end{pmatrix}.$$

▲

Our next theorem shows that the structure of a field is still the most restrictive (that is, the richest) one we have defined.

19.9 Theorem Every field F is an integral domain.

Proof Let $a, b \in F$, and suppose that $a \neq 0$. Then if $ab = 0$, we have

$$\left(\frac{1}{a}\right)(ab) = \left(\frac{1}{a}\right)0 = 0.$$

But then

$$0 = \left(\frac{1}{a}\right)(ab) = \left[\left(\frac{1}{a}\right)a\right]b = 1b = b.$$

We have shown that $ab = 0$ with $a \neq 0$ implies that $b = 0$ in F, so there are no divisors of 0 in F. Of course, F is a commutative ring with unity, so our theorem is proved. ◆

Figure 19.10 gives a Venn diagram view of containment for the algebraic structures having two binary operations with which we will be chiefly concerned. In Exercise 20 we ask you to redraw this figure to include strictly skew fields as well.

Thus far the only fields we know are $\mathbb{Q}$, $\mathbb{R}$, and $\mathbb{C}$. The corollary of the next theorem will exhibit some fields of finite order! The proof of this theorem is a personal favorite. It is done by counting. Counting is one of the most powerful techniques in mathematics.

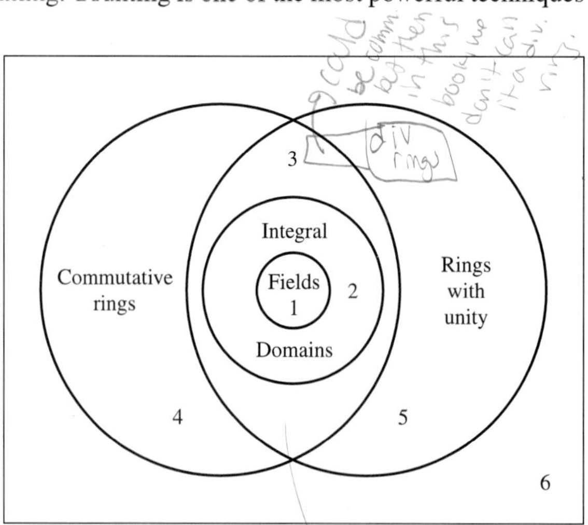

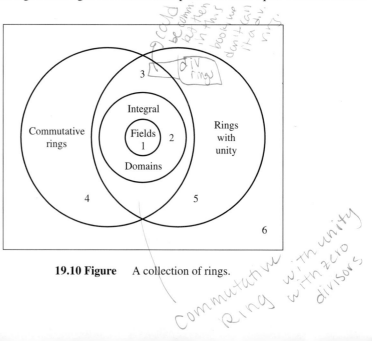

19.10 Figure A collection of rings.

19.11 Theorem Every finite integral domain is a field.

Proof Let

$$0, 1, a_1, \cdots, a_n$$

be all the elements of a finite domain D. We need to show that for $a \in D$, where $a \neq 0$, there exists $b \in D$ such that $ab = 1$. Now consider

$$a1, aa_1, \cdots, aa_n.$$

We claim that all these elements of D are distinct, for $aa_i = aa_j$ implies that $a_i = a_j$, by the cancellation laws that hold in an integral domain. Also, since D has no 0 divisors, none of these elements is 0. Hence by *counting,* we find that $a1, aa_1, \cdots, aa_n$ are elements $1, a_1, \cdots, a_n$ in some order, so that either $a1 = 1$, that is, $a = 1$, or $aa_i = 1$ for some i. Thus a has a multiplicative inverse. ◆

19.12 Corollary If p is a prime, then $\mathbb{Z}_p$ is a field.

Proof This corollary follows immediately from the fact that $\mathbb{Z}_p$ is an integral domain and from Theorem 19.11. ◆

The preceding corollary shows that when we consider the ring $M_n(\mathbb{Z}_p)$, we are talking about a ring of matrices over a *field*. In the typical undergraduate linear algebra course, only the field properties of the real or complex numbers are used in much of the work. Such notions as matrix reduction to solve linear systems, determinants, Cramer's rule, eigenvalues and eigenvectors, and similarity transformations to try to diagonalize a matrix are valid using matrices over any field; they depend only on the arithmetic properties of a field. Considerations of linear algebra involving notions of magnitude, such as least-squares approximate solutions or orthonormal bases, only make sense using fields where we have an idea of magnitude. The relation

$$p \cdot 1 = \underbrace{1 + 1 + \cdots + 1}_{p \text{ summands}} = 0$$

indicates that there can be no very natural notion of magnitude in the field $\mathbb{Z}_p$.

The Characteristic of a Ring

Let R be any ring. We might ask whether there is a positive integer n such that $n \cdot a = 0$ for all $a \in R$, where $n \cdot a$ means $a + a + \cdots + a$ for n summands, as explained in Section 18. For example, the integer m has this property for the ring $\mathbb{Z}_m$.

19.13 Definition If for a ring R a positive integer n exists such that $n \cdot a = 0$ for all $a \in R$, then the least such positive integer is the **characteristic of the ring** R. If no such positive integer exists, then R is of **characteristic** 0. ■

We shall be using the concept of a characteristic chiefly for fields. Exercise 29 asks us to show that the characteristic of an integral domain is either 0 or a prime p.

19.14 Example The ring $\mathbb{Z}_n$ is of characteristic n, while $\mathbb{Z}$, $\mathbb{Q}$, $\mathbb{R}$, and $\mathbb{C}$ all have characteristic 0. ▲

At first glance, determination of the characteristic of a ring seems to be a tough job, unless the ring is obviously of characteristic 0. Do we have to examine *every* element a of the ring in accordance with Definition 19.13? Our final theorem of this section shows that if the ring has unity, it suffices to examine only $a = 1$.

19.15 Theorem Let R be a ring with unity. If $n \cdot 1 \neq 0$ for all $n \in \mathbb{Z}^+$, then R has characteristic 0. If $n \cdot 1 = 0$ for some $n \in \mathbb{Z}^+$, then the smallest such integer n is the characteristic of R.

Proof If $n \cdot 1 \neq 0$ for all $n \in \mathbb{Z}^+$, then surely we cannot have $n \cdot a = 0$ for all $a \in R$ for some positive integer n, so by Definition 19.13, R has characteristic 0.

Suppose that n is a positive integer such that $n \cdot 1 = 0$. Then for any $a \in R$, we have

$$n \cdot a = a + a + \cdots + a = a(1 + 1 + \cdots + 1) = a(n \cdot 1) = a0 = 0.$$

Our theorem follows directly. ◆

EXERCISES 19

Computations

1. Find all solutions of the equation $x^3 - 2x^2 - 3x = 0$ in $\mathbb{Z}_{12}$.
2. Solve the equation $3x = 2$ in the field $\mathbb{Z}_7$; in the field $\mathbb{Z}_{23}$.
3. Find all solutions of the equation $x^2 + 2x + 2 = 0$ in $\mathbb{Z}_6$.
4. Find all solutions of $x^2 + 2x + 4 = 0$ in $\mathbb{Z}_6$.

In Exercises 5 through 10, find the characteristic of the given ring.

5. $2\mathbb{Z}$
6. $\mathbb{Z} \times \mathbb{Z}$
7. $\mathbb{Z}_3 \times 3\mathbb{Z}$
8. $\mathbb{Z}_3 \times \mathbb{Z}_3$
9. $\mathbb{Z}_3 \times \mathbb{Z}_4$
10. $\mathbb{Z}_6 \times \mathbb{Z}_{15}$
11. Let R be a commutative ring with unity of characteristic 4. Compute and simplify $(a + b)^4$ for $a, b \in R$.
12. Let R be a commutative ring with unity of characteristic 3. Compute and simplify $(a + b)^9$ for $a, b \in R$.
13. Let R be a commutative ring with unity of characteristic 3. Compute and simplify $(a + b)^6$ for $a, b \in R$.
14. Show that the matrix $\begin{bmatrix} 1 & 2 \\ 2 & 4 \end{bmatrix}$ is a divisor of zero in $M_2(\mathbb{Z})$.

Concepts

In Exercises 15 and 16, correct the definition of the italicized term without reference to the text, if correction is needed, so that it is in a form acceptable for publication.

15. If $ab = 0$, then a and b are *divisors of zero*.
16. If $n \cdot a = 0$ for all elements a in a ring R, then n is the *characteristic of* R.
17. Mark each of the following true or false.

______ **a.** $n\mathbb{Z}$ has zero divisors if n is not prime.
______ **b.** Every field is an integral domain.
______ **c.** The characteristic of $n\mathbb{Z}$ is n.

______ **d.** As a ring, $\mathbb{Z}$ is isomorphic to $n\mathbb{Z}$ for all $n \geq 1$.

______ **e.** The cancellation law holds in any ring that is isomorphic to an integral domain.

______ **f.** Every integral domain of characteristic 0 is infinite.

______ **g.** The direct product of two integral domains is again an integral domain.

______ **h.** A divisor of zero in a commutative ring with unity can have no multiplicative inverse.

______ **i.** $n\mathbb{Z}$ is a subdomain of $\mathbb{Z}$.

______ **j.** $\mathbb{Z}$ is a subfield of $\mathbb{Q}$.

18. Each of the six numbered regions in Fig. 19.10 corresponds to a certain type of a ring. Give an example of a ring in each of the six cells. For example, a ring in the region numbered 3 must be commutative (it is inside the commutative circle), have unity, but not be an integral domain.

19. (For students who have had a semester of linear algebra) Let F be a field. Give five different characterizations of the elements A of $M_n(F)$ that are divisors of 0.

20. Redraw Fig. 19.10 to include a subset corresponding to strictly skew fields.

Proof Synopsis

21. Give a one-sentence synopsis of the proof of the "if" part of Theorem 19.5.

22. Give a one-sentence synopsis of the proof of Theorem 19.11.

Theory

23. An element a of a ring R is **idempotent** if $a^2 = a$. Show that a division ring contains exactly two idempotent elements.

24. Show that an intersection of subdomains of an integral domain D is again a subdomain of D.

25. Show that a finite ring R with unity $1 \neq 0$ and no divisors of 0 is a division ring. (It is actually a field, although commutativity is not easy to prove. See Theorem 24.10.) [*Note:* In your proof, to show that $a \neq 0$ is a unit, you must show that a "left multiplicative inverse" of $a \neq 0$ in R is also a "right multiplicative inverse."]

26. Let R be a ring that contains at least two elements. Suppose for each nonzero $a \in R$, there exists a unique $b \in R$ such that $aba = a$.

a. Show that R has no divisors of 0.

b. Show that $bab = b$.

c. Show that R has unity.

d. Show that R is a division ring.

27. Show that the characteristic of a subdomain of an integral domain D is equal to the characteristic of D.

28. Show that if D is an integral domain, then $\{n \cdot 1 \mid n \in \mathbb{Z}\}$ is a subdomain of D contained in every subdomain of D.

29. Show that the characteristic of an integral domain D must be either 0 or a prime p. [*Hint:* If the characteristic of D is mn, consider $(m \cdot 1)(n \cdot 1)$ in D.]

30. This exercise shows that every ring R can be enlarged (if necessary) to a ring S with unity, having the same characteristic as R. Let $S = R \times Z$ if R has characteristic 0, and $R \times \mathbb{Z}_n$ if R has characteristic n. Let addition in S be the usual addition by components, and let multiplication be defined by

$$(r_1, n_1)(r_2, n_2) = (r_1 r_2 + n_1 \cdot r_2 + n_2 \cdot r_1, n_1 n_2)$$

where $n \cdot r$ has the meaning explained in Section 18.

a. Show that S is a ring.
b. Show that S has unity.
c. Show that S and R have the same characteristic.
d. Show that the map $\phi : R \to S$ given by $\phi(r) = (r, 0)$ for $r \in R$ maps R isomorphically onto a subring of S.

SECTION 20 Fermat's and Euler's Theorems

Fermat's Theorem

We know that as additive groups, $\mathbb{Z}_n$ and $\mathbb{Z}/n\mathbb{Z}$ are naturally isomorphic, with the coset $a + n\mathbb{Z}$ corresponding to a for each $a \in \mathbb{Z}_n$. Furthermore, addition of cosets in $\mathbb{Z}/n\mathbb{Z}$ may be performed by choosing any representatives, adding them in $\mathbb{Z}$, and finding the coset of $n\mathbb{Z}$ containing their sum. It is easy to see that $\mathbb{Z}/n\mathbb{Z}$ can be made into a ring by multiplying cosets in the same fashion, that is, by multiplying any chosen representatives. While we will be showing this later in a more general situation, we do this special case now. We need only show that such coset multiplication is well defined, because the associativity of multiplication and the distributive laws will follow immediately from those properties of the chosen representatives in $\mathbb{Z}$. To this end, choose representatives $a + rn$ and $b + sn$, rather than a and b, from the cosets $a + n\mathbb{Z}$ and $b + n\mathbb{Z}$. Then

$$(a + rn)(b + sn) = ab + (as + rb + rsn)n,$$

which is also an element of $ab + n\mathbb{Z}$. Thus the multiplication is well-defined, and our cosets form a ring isomorphic to the ring $\mathbb{Z}_n$.

The following is a special case of Exercise 37 in Section 18.

For any field, the nonzero elements form a group under the field multiplication.

In particular, for $\mathbb{Z}_p$, the elements

$$1, 2, 3, \cdots, p - 1$$

form a group of order $p - 1$ under multiplication modulo p. Since the order of any element in a group divides the order of the group, we see that for $b \neq 0$ and $b \in \mathbb{Z}_p$, we have $b^{p-1} = 1$ in $\mathbb{Z}_p$. Using the fact that $\mathbb{Z}_p$ is isomorphic to the ring of cosets of the form $a + p\mathbb{Z}$ described above, we see at once that for any $a \in \mathbb{Z}$ not in the coset $0 + p\mathbb{Z}$, we must have

$$a^{p-1} \equiv 1 \pmod{p}.$$

This gives us at once the so-called Little Theorem of Fermat.

20.1 Theorem **(Little Theorem of Fermat)** If $a \in \mathbb{Z}$ and p is a prime not dividing a, then p divides $a^{p-1} - 1$, that is, $a^{p-1} \equiv 1 \pmod{p}$ for $a \not\equiv 0 \pmod{p}$.

20.2 Corollary If $a \in \mathbb{Z}$, then $a^p \equiv a \pmod{p}$ for any prime p.

Proof The corollary follows from Theorem 20.1 if $a \not\equiv 0 \pmod{p}$. If $a \equiv 0 \pmod{p}$, then both sides reduce to 0 modulo p. ◆

Historical Note

The statement of Theorem 20.1 occurs in a letter from Pierre de Fermat (1601–1665) to Bernard Frenicle de Bessy, dated 18 October 1640. Fermat's version of the theorem was that for any prime p and any geometric progression $a, a^2, \cdots, a^t, \cdots$, there is a least number a^T of the progression such that p divides $a^T - 1$. Furthermore, T divides $p - 1$ and p also divides all numbers of the form $a^{KT} - 1$. (It is curious that Fermat failed to note the condition that p not divide a; perhaps he felt that it was obvious that the result fails in that case.)

Fermat did not in the letter or elsewhere indicate a proof of the result and, in fact, never mentioned it again. But we can infer from other parts of this correspondence that Fermat's interest in this result came from his study of perfect numbers. (A perfect number is a positive integer m that is the sum of all of its divisors less than m; for example, $6 = 1 + 2 + 3$ is a perfect number.) Euclid had shown that $2^{n-1}(2^n - 1)$ is perfect if $2^n - 1$ is prime. The question then was to find methods for determining whether $2^n - 1$ was prime. Fermat noted that $2^n - 1$ was composite if n is composite, and then derived from his theorem the result that if n is prime, the only possible divisors of $2^n - 1$ are those of the form $2kn + 1$. From this result he was able quickly to show, for example, that $2^{37} - 1$ was divisible by $223 = 2 \cdot 3 \cdot 37 + 1$.

20.3 Example Let us compute the remainder of 8^{103} when divided by 13. Using Fermat's theorem, we have

$$\begin{aligned} 8^{103} &\equiv (8^{12})^8(8^7) \equiv (1^8)(8^7) \equiv 8^7 \equiv (-5)^7 \\ &\equiv (25)^3(-5) \equiv (-1)^3(-5) \equiv 5 \pmod{13}. \end{aligned}$$ ▲

20.4 Example Show that $2^{11,213} - 1$ is not divisible by 11.

Solution By Fermat's theorem, $2^{10} \equiv 1 \pmod{11}$, so

$$\begin{aligned} 2^{11,213} - 1 &\equiv [(2^{10})^{1,121} \cdot 2^3] - 1 \equiv [1^{1,121} \cdot 2^3] - 1 \\ &\equiv 2^3 - 1 \equiv 8 - 1 \equiv 7 \pmod{11}. \end{aligned}$$

Thus the remainder of $2^{11,213} - 1$ when divided by 11 is 7, not 0. (The number 11,213 is prime, and it has been shown that $2^{11,213} - 1$ is a prime number. Primes of the form $2^p - 1$ where p is prime are known as **Mersenne primes.**) ▲

20.5 Example Show that for every integer n, the number $n^{33} - n$ is divisible by 15.

Solution This seems like an incredible result. It means that 15 divides $2^{33} - 2$, $3^{33} - 3$, $4^{33} - 4$, etc.

Now $15 = 3 \cdot 5$, and we shall use Fermat's theorem to show that $n^{33} - n$ is divisible by both 3 and 5 for every n. Note that $n^{33} - n = n(n^{32} - 1)$.

If 3 divides n, then surely 3 divides $n(n^{32} - 1)$. If 3 does not divide n, then by Fermat's theorem, $n^2 \equiv 1 \pmod 3$ so

$$n^{32} - 1 \equiv (n^2)^{16} - 1 \equiv 1^{16} - 1 \equiv 0 \pmod 3,$$

and hence 3 divides $n^{32} - 1$.

If $n \equiv 0 \pmod 5$, then $n^{33} - n \equiv 0 \pmod 5$. If $n \not\equiv 0 \pmod 5$, then by Fermat's theorem, $n^4 \equiv 1 \pmod 5$, so

$$n^{32} - 1 \equiv (n^4)^8 - 1 \equiv 1^8 - 1 \equiv 0 \pmod 5.$$

Thus $n^{33} - n \equiv 0 \pmod 5$ for every n also. ▲

Euler's Generalization

Euler gave a generalization of Fermat's theorem. His generalization will follow at once from our next theorem, which is proved by *counting*, using essentially the same argument as in Theorem 19.11.

20.6 Theorem The set G_n of nonzero elements of $\mathbb{Z}_n$ that are not 0 divisors forms a group under multiplication modulo n.

Proof First we must show that G_n is closed under multiplication modulo n. Let $a, b \in G_n$. If $ab \notin G_n$, then there would exist $c \neq 0$ in $\mathbb{Z}_n$ such that $(ab)c = 0$. Now $(ab)c = 0$ implies that $a(bc) = 0$. Since $b \in G_n$ and $c \neq 0$, we have $bc \neq 0$ by definition of G_n. But then $a(bc) = 0$ would imply that $a \notin G_n$ contrary to assumption. *Note that we have shown that for any ring the set of elements that are not divisors of* 0 *is closed under multiplication.* No structure of $\mathbb{Z}_n$ other than ring structure has been involved so far.

We now show that G_n is a group. Of course, multiplication modulo n is associative, and $1 \in G_n$. It remains to show that for $a \in G_n$, there is $b \in G_n$ such that $ab = 1$. Let

$$1, a_1, \cdots, a_r$$

be the elements of G_n. The elements

$$a1, aa_1, \cdots, aa_r$$

are all different, for if $aa_i = aa_j$, then $a(a_i - a_j) = 0$, and since $a \in G_n$ and thus is not a divisor of 0, we must have $a_i - a_j = 0$ or $a_i = a_j$. Therefore by counting, we find that either $a1 = 1$, or some aa_i must be 1, so a has a multiplicative inverse. ◆

Note that the only property of $\mathbb{Z}_n$ used in this last theorem, other than the fact that it was a ring with unity, was that it was finite. In both Theorem 19.11 and Theorem 20.6 we have (in essentially the same construction) employed a counting argument. *Counting arguments are often simple, but they are among the most powerful tools of mathematics.*

Let n be a positive integer. Let $\varphi(n)$ be defined as the number of positive integers less than or equal to n and relatively prime to n. Note that $\varphi(1) = 1$.

20.7 Example Let $n = 12$. The positive integers less than or equal to 12 and relatively prime to 12 are 1, 5, 7, and 11, so $\varphi(12) = 4$. ▲

By Theorem 19.3, $\varphi(n)$ is the number of nonzero elements of $\mathbb{Z}_n$ that are not divisors of 0. This function $\varphi : \mathbb{Z}^+ \to \mathbb{Z}^+$ is the **Euler phi-function.** We can now describe Euler's generalization of Fermat's theorem.

20.8 Theorem **(Euler's Theorem)** If a is an integer relatively prime to n, then $a^{\varphi(n)} - 1$ is divisible by n, that is, $a^{\varphi(n)} \equiv 1 \pmod{n}$.

Proof If a is relatively prime to n, then the coset $a + n\mathbb{Z}$ of $n\mathbb{Z}$ containing a contains an integer $b < n$ and relatively prime to n. Using the fact that multiplication of these cosets by multiplication modulo n of representatives is well-defined, we have

$$a^{\varphi(n)} \equiv b^{\varphi(n)} \pmod{n}.$$

But by Theorems 19.3 and 20.6, b can be viewed as an element of the multiplicative group G_n of order $\varphi(n)$ consisting of the $\varphi(n)$ elements of $\mathbb{Z}_n$ relatively prime to n. Thus

$$b^{\varphi(n)} \equiv 1 \pmod{n},$$

and our theorem follows. ◆

20.9 Example Let $n = 12$. We saw in Example 20.7 that $\varphi(12) = 4$. Thus if we take any integer a relatively prime to 12, then $a^4 \equiv 1 \pmod{12}$. For example, with $a = 7$, we have $7^4 = (49)^2 = 2,401 = 12(200) + 1$, so $7^4 \equiv 1 \pmod{12}$. Of course, the easy way to compute $7^4 \pmod{12}$, without using Euler's theorem, is to compute it in $\mathbb{Z}_{12}$. In $\mathbb{Z}_{12}$, we have $7 = -5$ so

$$7^2 = (-5)^2 = (5)^2 = 1 \qquad \text{and} \qquad 7^4 = 1^2 = 1.$$ ▲

Application to $ax \equiv b \pmod{m}$

Using Theorem 20.6, we can find all solutions of a linear congruence $ax \equiv b \pmod{m}$. We prefer to work with an equation in $\mathbb{Z}_m$ and interpret the results for congruences.

20.10 Theorem Let m be a positive integer and let $a \in \mathbb{Z}_m$ be relatively prime to m. For each $b \in \mathbb{Z}_m$, the equation $ax = b$ has a unique solution in $\mathbb{Z}_m$.

Proof By Theorem 20.6, a is a unit in $\mathbb{Z}_m$ and $s = a^{-1}b$ is certainly a solution of the equation. Multiplying both sides of $ax = b$ on the left by a^{-1}, we see this is the only solution. ◆

Interpreting this theorem for congruences, we obtain at once the following corollary.

20.11 Corollary If a and m are relatively prime intergers, then for any integer b, the congruence $ax \equiv b \pmod{m}$ has as solutions all integers in precisely one residue class modulo m.

Theorem 20.10 serves as a lemma for the general case.

20.12 Theorem Let m be a positive integer and let $a, b \in \mathbb{Z}_m$. Let d be the gcd of a and m. The equation $ax = b$ has a solution in $\mathbb{Z}_m$ if and only if d divides b. When d divides b, the equation has exactly d solutions in $\mathbb{Z}_m$.

Proof First we show there is no solution of $ax = b$ in $\mathbb{Z}_m$ unless d divides b. Suppose $s \in \mathbb{Z}_m$ is a solution. Then $as - b = qm$ in $\mathbb{Z}$, so $b = as - qm$. Since d divides both a and m, we see that d divides the right-hand side of the equation $b = as - qm$, and hence divides b. Thus a solution s can exist only if d divides b.

Suppose now that d does divide b. Let

$$a = a_1 d, \quad b = b_1 d, \quad \text{and} \quad m = m_1 d.$$

Then the equation $as - b = qm$ in $\mathbb{Z}$ can be rewritten as $d(a_1 s - b_1) = dqm_1$. We see that $as - b$ is a multiple of m if and only if $a_1 s - b_1$ is a multiple of m_1. Thus the solutions s of $ax = b$ in $\mathbb{Z}_m$ are precisely the elements that, read modulo m_1, yield solutions of $a_1 x = b_1$ in $\mathbb{Z}_{m_1}$. Now let $s \in \mathbb{Z}_{m_1}$ be the unique solution of $a_1 x = b_1$ in $\mathbb{Z}_{m_1}$ given by Theorem 20.10. The numbers in $\mathbb{Z}_m$ that reduce to s modulo m_1 are precisely those that can be computed in $\mathbb{Z}_m$ as

$$s, s + m_1, s + 2m_1, s + 3m_1, \cdots, s + (d-1)m_1.$$

Thus there are exactly d solutions of the equation in $\mathbb{Z}_m$. ◆

Theorem 20.12 gives us at once this classical result on the solutions of a linear congruence.

20.13 Corollary Let d be the gcd of positive integers a and m. The congruence $ax \equiv b \pmod{m}$ has a solution if and only if d divides b. When this is the case, the solutions are the integers in exactly d distinct residue classes modulo m.

Actually, our proof of Theorem 20.12 shows a bit more about the solutions of $ax \equiv b \pmod{m}$ than we stated in this corollary; namely, it shows that if any solution s is found, then the solutions are precisely all elements of the residue classes $(s + km_1) + (m\mathbb{Z})$ where $m_1 = m/d$ and k runs through the integers from 0 to $d - 1$. It also tells us that we can find such an s by finding $a_1 = a/d$ and $b_1 = b/d$, and solving $a_1 x \equiv b_1 \pmod{m_1}$. To solve this congruence, we may consider a_1 and b_1 to be replaced by their remainders modulo m_1 and solve the equation $a_1 x = b_1$ in $\mathbb{Z}_{m_1}$.

20.14 Example Find all solutions of the congruence $12x \equiv 27 \pmod{18}$.

Solution The gcd of 12 and 18 is 6, and 6 is not a divisor of 27. Thus by the preceding corollary, there are no solutions. ▲

20.15 Example Find all solutions of the congruence $15x \equiv 27 \pmod{18}$.

Solution The gcd of 15 and 18 is 3, and 3 does divide 27. Proceeding as explained before Example 20.14, we divide everything by 3 and consider the congruence $5x \equiv 9 \pmod{6}$, which amounts to solving the equation $5x = 3$ in $\mathbb{Z}_6$. Now the units in $\mathbb{Z}_6$ are 1 and 5, and 5 is clearly its own inverse in this group of units. Thus the solution in $\mathbb{Z}_6$ is $x = (5^{-1})(3) = (5)(3) = 3$. Consequently, the solutions of $15x \equiv 27 \pmod{18}$ are the integers in the three residue classes.

$$3 + 18\mathbb{Z} = \{\cdots, -33, -15, 3, 21, 39, \cdots\},$$

$$9 + 18\mathbb{Z} = \{\cdots, -27, -9, 9, 27, 45, \cdots\}.$$

$$15 + 18\mathbb{Z} = \{\cdots, -21, -3, 15, 33, 51, \cdots\},$$

illustrating Corollary 20.13. Note the $d = 3$ solutions 3, 9, and 15 in $\mathbb{Z}_{18}$. All the solutions in the three displayed residue classes modulo 18 can be collected in the one residue class $3 + 6\mathbb{Z}$ modulo 6, for they came from the solution $x = 3$ of $5x = 3$ in $\mathbb{Z}_6$. ▲

EXERCISES 20

Computations

We will see later that the multiplicative group of nonzero elements of a finite field is cyclic. Illustrate this by finding a generator for this group for the given finite field.

1. $\mathbb{Z}_7$ **2.** $\mathbb{Z}_{11}$ **3.** $\mathbb{Z}_{17}$

4. Using Fermat's theorem, find the remainder of 3^{47} when it is divided by 23.

5. Use Fermat's theorem to find the remainder of 37^{49} when it is divided by 7.

6. Compute the remainder of $2^{(2^{17})} + 1$ when divided by 19. [*Hint:* You will need to compute the remainder of 2^{17} modulo 18.]

7. Make a table of values of $\varphi(n)$ for $n \leq 30$.

8. Compute $\varphi(p^2)$ where p is a prime.

9. Compute $\varphi(pq)$ where both p and q are primes.

10. Use Euler's generalization of Fermat's theorem to find the remainder of 7^{1000} when divided by 24.

In Exercises 11 through 18, describe all solutions of the given congruence, as we did in Examples 20.14 and 20.15.

11. $2x \equiv 6 \pmod{4}$ **12.** $22x \equiv 5 \pmod{15}$

13. $36x \equiv 15 \pmod{24}$ **14.** $45x \equiv 15 \pmod{24}$

15. $39x \equiv 125 \pmod{9}$ **16.** $41x \equiv 125 \pmod{9}$

17. $155x \equiv 75 \pmod{65}$ **18.** $39x \equiv 52 \pmod{130}$

19. Let p be a prime ≥ 3. Use Exercise 28 below to find the remainder of $(p-2)!$ modulo p.

20. Using Exercise 28 below, find the remainder of 34! modulo 37.

21. Using Exercise 28 below, find the remainder of 49! modulo 53.

22. Using Exercise 28 below, find the remainder of 24! modulo 29.

Concepts

23. Mark each of the following true or false.

_____ **a.** $a^{p-1} \equiv 1 \pmod{p}$ for all integers a and primes p.

_____ **b.** $a^{p-1} \equiv 1 \pmod{p}$ for all integers a such that $a \not\equiv 0 \pmod{p}$ for a prime p.

_____ **c.** $\varphi(n) \leq n$ for all $n \in \mathbb{Z}^+$.

_____ **d.** $\varphi(n) \leq n - 1$ for all $n \in \mathbb{Z}^+$.

_____ **e.** The units in $\mathbb{Z}_n$ are the positive integers less than n and relatively prime to n.

_____ **f.** The product of two units in $\mathbb{Z}_n$ is always a unit.

_____ **g.** The product of two nonunits in $\mathbb{Z}_n$ may be a unit.

_____ **h.** The product of a unit and a nonunit in $\mathbb{Z}_n$ is never a unit.

_______ **i.** Every congruence $ax \equiv b \pmod{p}$, where p is a prime, has a solution.

_______ **j.** Let d be the gcd of positive integers a and m. If d divides b, then the congruence $ax \equiv b \pmod{m}$ has exactly d incongruent solutions.

24. Give the group multiplication table for the multiplicative group of units in $\mathbb{Z}_{12}$. To which group of order 4 is it isomorphic?

Proof Synopsis

25. Give a one-sentence synopsis of the proof of Theorem 20.1.

26. Give a one-sentence synopsis of the proof of Theorem 20.8.

Theory

27. Show that 1 and $p - 1$ are the only elements of the field $\mathbb{Z}_p$ that are their own multiplicative inverse. [*Hint:* Consider the equation $x^2 - 1 = 0$.]

28. Using Exercise 27, deduce the half of *Wilson's theorem* that states that if p is a prime, then $(p-1)! \equiv -1 \pmod{p}$. [The other half states that if n is an integer >1 such that $(n-1)! \equiv -1 \pmod{n}$, then n is a prime. Just think what the remainder of $(n-1)!$ would be modulo n if n is not a prime.]

29. Use Fermat's theorem to show that for any positive integer n, the integer $n^{37} - n$ is divisible by 383838. [*Hint:* $383838 = (37)(19)(13)(7)(3)(2)$.]

30. Referring to Exercise 29, find a number larger than 383838 that divides $n^{37} - n$ for all positive integers n.

SECTION 21 The Field of Quotients of an Integral Domain

If an integral domain is such that every nonzero element has a multiplicative inverse, then it is a field. However, many integral domains, such as the integers $\mathbb{Z}$, do not form a field. This dilemma is not too serious. It is the purpose of this section to show that every integral domain can be regarded as being contained in a certain field, *a field of quotients of the integral domain*. This field will be a minimal field containing the integral domain in a sense that we shall describe. For example, the integers are contained in the field $\mathbb{Q}$, whose elements can all be expressed as quotients of integers. Our construction of a field of quotients of an integral domain is exactly the same as the construction of the rational numbers from the integers, which often appears in a course in foundations or advanced calculus. To follow this construction through is such a good exercise in the use of definitions and the concept of isomorphism that we discuss it in some detail, although to write out, or to read, every last detail would be tedious. We can be motivated at every step by the way $\mathbb{Q}$ can be formed from $\mathbb{Z}$.

The Construction

Let D be an integral domain that we desire to enlarge to a field of quotients F. A coarse outline of the steps we take is as follows:

1. Define what the elements of F are to be.
2. Define the binary operations of addition and multiplication on F.

3. Check all the field axioms to show that F is a field under these operations.
4. Show that F can be viewed as containing D as an integral subdomain.

Steps 1, 2, and 4 are very interesting, and Step 3 is largely a mechanical chore. We proceed with the construction.

Step 1 Let D be a given integral domain, and form the Cartesian product

$$D \times D = \{(a, b) \mid a, b \in D\}$$

We are going to think of an ordered pair (a, b) as representing a *formal quotient* a/b, that is, if $D = \mathbb{Z}$, the pair (2, 3) will eventually represent the number $\frac{2}{3}$ for us. The pair (2, 0) represents no element of $\mathbb{Q}$ and suggests that we cut the set $D \times D$ down a bit. Let S be the subset of $D \times D$ given by

$$S = \{(a, b) \mid a, b \in D, b \neq 0\}.$$

Now S is still not going to be our field as is indicated by the fact that, with $D = \mathbb{Z}$, *different* pairs of integers such as (2, 3) and (4, 6) can represent the *same* rational number. We next define when two elements of S will eventually represent the same element of F, or, as we shall say, when two elements of S are *equivalent*.

21.1 Definition Two elements (a, b) and (c, d) in S are **equivalent,** denoted by $(a, b) \sim (c, d)$, if and only if $ad = bc$. ■

Observe that this definition is reasonable, since the criterion for $(a, b) \sim (c, d)$ is an equation $ad = bc$ involving elements in D and concerning the known multiplication in D. Note also that for $D = \mathbb{Z}$, the criterion gives us our usual definition of *equality* of $\frac{a}{b}$ with $\frac{c}{d}$, for example, $\frac{2}{3} = \frac{4}{6}$ since (2)(6) = (3)(4). The rational number that we usually denote by $\frac{2}{3}$ can be thought of as the collection of *all* quotients of integers that reduce to, or are equivalent to, $\frac{2}{3}$.

21.2 Lemma The relation $\sim$ between elements of the set S as just described is an equivalence relation.

Proof We must check the three properties of an equivalence relation.

Reflexive $(a, b) \sim (a, b)$ since $ab = ba$, for multiplication in D is commutative.

Symmetric If $(a, b) \sim (c, d)$, then $ad = bc$. Since multiplication in D is commutative, we deduce that $cb = da$, and consequently $(c, d) \sim (a, b)$.

Transitive If $(a, b) \sim (c, d)$ and $(c, d) \sim (r, s)$, then $ad = bc$ and $cs = dr$. Using these relations and the fact that multiplication in D is commutative, we have

$$asd = sad = sbc = bcs = bdr = brd.$$

Now $d \neq 0$, and D is an integral domain, so cancellation is valid; this is a crucial step in the argument. Hence from $asd = brd$ we obtain $as = br$, so that $(a, b) \sim (r, s)$.

◆

We now know, in view of Theorem 0.22, that $\sim$ gives a partition of S into equivalence classes. To avoid long bars over extended expressions, we shall let $[(a, b)]$, rather than $\overline{(a, b)}$, be the equivalence class of (a, b) in S under the relation $\sim$. We now finish Step 1 by defining F to be the set of all equivalence classes $[(a, b)]$ for $(a, b) \in S$.

Step 2 The next lemma serves to define addition and multiplication in F. Observe that if $D = \mathbb{Z}$ and $[(a, b)]$ is viewed as $(a/b) \in \mathbb{Q}$, these definitions applied to $\mathbb{Q}$ give the usual operations.

21.3 Lemma For $[(a, b)]$ and $[(c, d)]$ in F, the equations

$$[(a, b)] + [(c, d)] = [(ad + bc, bd)]$$

and

$$[(a, b)][(c, d)] = [(ac, bd)]$$

give well-defined operations of addition and multiplication on F.

Proof Observe first that if $[(a, b)]$ and $[(c, d)]$ are in F, then (a, b) and (c, d) are in S, so $b \neq 0$ and $d \neq 0$. Because D is an integral domain, $bd \neq 0$, so both $(ad + bc, bd)$ and (ac, bd) are in S. (Note the crucial use here of the fact that D has no divisors of 0.) This shows that the right-hand sides of the defining equations are at least in F.

It remains for us to show that these operations of addition and multiplication are well defined. That is, they were defined by means of representatives in S of elements of F, so we must show that if different representatives in S are chosen, the same element of F will result. To this end, suppose that $(a_1, b_1) \in [(a, b)]$ and $(c_1, d_1) \in [(c, d)]$. We must show that

$$(a_1 d_1 + b_1 c_1, b_1 d_1) \in [(ad + bc, bd)]$$

and

$$(a_1 c_1, b_1 d_1) \in [(ac, bd)].$$

Now $(a_1, b_1) \in [(a, b)]$ means that $(a_1, b_1) \sim (a, b)$; that is,

$$a_1 b = b_1 a.$$

Similarly, $(c_1, d_1) \in [(c, d)]$ implies that

$$c_1 d = d_1 c.$$

To get a "common denominator" (common second member) for the four pairs (a, b), (a_1, b_1), (c, d), and (c_1, d_1), we multiply the first equation by $d_1 d$ and the second equation by $b_1 b$. Adding the resulting equations, we obtain the following equation in D:

$$a_1 b d_1 d + c_1 d b_1 b = b_1 a d_1 d + d_1 c b_1 b.$$

Using various axioms for an integral domain, we see that

$$(a_1 d_1 + b_1 c_1) bd = b_1 d_1 (ad + bc),$$

so

$$(a_1d_1 + b_1c_1, b_1d_1) \sim (ad + bc, bd),$$

giving $(a_1d_1 + b_1c_1, b_1d_1) \in [(ad + bc, bd)]$. This takes care of addition in F. For multiplication in F, on multiplying the equations $a_1b = b_1a$ and $c_1d = d_1c$, we obtain

$$a_1bc_1d = b_1ad_1c,$$

so, using axioms of D, we get

$$a_1c_1bd = b_1d_1ac,$$

which implies that

$$(a_1c_1, b_1d_1) \sim (ac, bd).$$

Thus $(a_1c_1, b_1d_1) \in [(ac, bd)]$, which completes the proof. ◆

It is important to *understand* the meaning of the last lemma and the necessity for proving it. This completes our Step 2.

Step 3 Step 3 is routine, but it is good for us to work through a few of these details. The reason for this is that we cannot work through them unless we *understand* what we have done. Thus working through them will contribute to our understanding of this construction. We list the things that must be proved and prove a couple of them. The rest are left to the exercises.

1. Addition in F is commutative.

Proof Now $[(a, b)] + [(c, d)]$ is by definition $[(ad + bc, bd)]$. Also $[(c, d)] + [(a, b)]$ is by definition $[(cb + da, db)]$. We need to show that $(ad + bc, bd) \sim (cb + da, db)$. This is true, since $ad + bc = cb + da$ and $bd = db$, by the axioms of D. ◆

2. Addition is associative.

3. $[(0, 1)]$ is an identity element for addition in F.

4. $[(-a, b)]$ is an additive inverse for $[(a, b)]$ in F.

5. Multiplication in F is associative.

6. Multiplication in F is commutative.

7. The distributive laws hold in F.

8. $[(1, 1)]$ is a multiplicative identity element in F.

9. If $[(a, b)] \in F$ is not the additive identity element, then $a \neq 0$ in D and $[(b, a)]$ is a multiplicative inverse for $[(a, b)]$.

Proof Let $[(a, b)] \in F$. If $a = 0$, then

$$a1 = b0 = 0,$$

so

$$(a, b) \sim (0, 1),$$

that is, $[(a, b)] = [(0, 1)]$. But $[(0, 1)]$ is the additive identity by Part 3. Thus if $[(a, b)]$ is not the additive identity in F, we have $a \neq 0$, so it makes sense to talk about $[(b, a)]$ in F. Now $[(a, b)][(b, a)] = [(ab, ba)]$. But in D we have $ab = ba$, so $(ab)1 = (ba)1$, and

$$(ab, ba) \sim (1, 1).$$

Thus

$$[(a, b)][(b, a)] = [(1, 1)],$$

and $[(1, 1)]$ is the multiplicative identity by Part 8. ◆

This completes Step 3.

Step 4 It remains for us to show that F can be regarded as containing D. To do this, we show that there is an isomorphism i of D with a subdomain of F. Then if we rename the image of D under i using the names of the elements of D, we will be done. The next lemma gives us this isomorphism. We use the letter i for this isomorphism to suggest *injection* (see the footnote on page 4); we will inject D into F.

21.4 Lemma The map $i : D \to F$ given by $i(a) = [(a, 1)]$ is an isomorphism of D with a subring of F.

Proof For a and b in D, we have

$$i(a + b) = [(a + b, 1)].$$

Also,

$$i(a) + i(b) = [(a, 1)] + [(b, 1)] = [(a1 + 1b, 1)] = [(a + b, 1)].$$

so $i(a + b) = i(a) + i(b)$. Furthermore,

$$i(ab) = [(ab, 1)],$$

while

$$i(a)i(b) = [(a, 1)][(b, 1)] = [(ab, 1)],$$

so $i(ab) = i(a)i(b)$.

It remains for us to show only that i is one to one. If $i(a) = i(b)$, then

$$[(a, 1)] = [(b, 1)],$$

so $(a, 1) \sim (b, 1)$ giving $a1 = 1b$; that is,

$$a = b.$$

Thus i is an isomorphism of D with $i[D]$, and, of course, $i[D]$ is then a subdomain of F. ◆

Since $[(a, b)] = [(a, 1)][(1, b)] = [(a, 1)]/[(b, 1)] = i(a)/i(b)$ clearly holds in F, we have now proved the following theorem.

21.5 Theorem Any integral domain D can be enlarged to (or embedded in) a field F such that every element of F can be expressed as a quotient of two elements of D. (Such a field F is a **field of quotients of** D.)

Uniqueness

We said in the beginning that F could be regarded in some sense as a minimal field containing D. This is intuitively evident, since every field containing D must contain all elements a/b for every $a, b \in D$ with $b \neq 0$. The next theorem will show that every field containing D contains a subfield which is a field of quotients of D, and that any two fields of quotients of D are isomorphic.

21.6 Theorem Let F be a field of quotients of D and let L be any field containing D. Then there exists a map $\psi : F \to L$ that gives an isomorphism of F with a subfield of L such that $\psi(a) = a$ for $a \in D$.

Proof The subring and mapping diagram in Fig. 21.7 may help you to visualize the situation for this theorem.

An element of F is of the form $a /_F b$ where $/_F$ denotes the quotient of $a \in D$ by $b \in D$ regarded as elements of F. We of course want to map $a /_F b$ onto $a /_L b$ where $/_L$ denotes the quotient of elements in L. The main job will be to show that such a map is well defined.

We must define $\psi : F \to L$, and we start by defining

$$\psi(a) = a \qquad \text{for} \qquad a \in D.$$

Every $x \in F$ is a quotient $a /_F b$ of some two elements a and b, $b \neq 0$, of D. Let us attempt to define ψ by

$$\psi(a /_F b) = \psi(a) /_L \psi(b).$$

We must first show that this map ψ is sensible and well-defined. Since ψ is the identity on D, for $b \neq 0$ we have $\psi(b) \neq 0$, so our definition of $\psi(a /_F b)$ as $\psi(a) /_L \psi(b)$ makes sense. If $a /_F b = c /_F d$ in F, then $ad = bc$ in D, so $\psi(ad) = \psi(bc)$. But since ψ is the identity on D,

$$\psi(ad) = \psi(a)\psi(d) \qquad \text{and} \qquad \psi(bc) = \psi(b)\psi(c).$$

Thus

$$\psi(a) /_L \psi(b) = \psi(c) /_L \psi(d)$$

in L, so ψ is well-defined.

The equations

$$\psi(xy) = \psi(x)\psi(y)$$

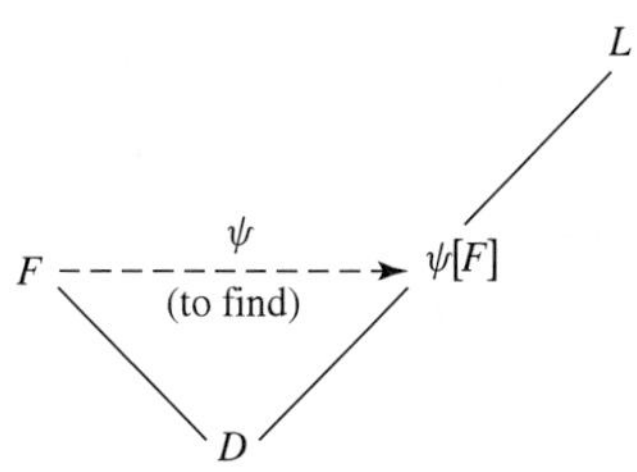

21.7 Figure

and

$$\psi(x+y) = \psi(x) + \psi(y)$$

follow easily from the definition of ψ on F and from the fact that ψ is the identity on D.

If $\psi(a /_F b) = \psi(c /_F d)$, we have

$$\psi(a) /_L \psi(b) = \psi(c) /_L \psi(d)$$

so

$$\psi(a)\psi(d) = \psi(b)\psi(c).$$

Since ψ is the identity on D, we then deduce that $ad = bc$, so $a /_F b = c /_F d$. Thus ψ is one to one.

By definition, $\psi(a) = a$ for $a \in D$. ◆

21.8 Corollary Every field L containing an integral domain D contains a field of quotients of D.

Proof In the proof of Theorem 21.6 every element of the subfield $\psi[F]$ of L is a quotient in L of elements of D. ◆

21.9 Corollary Any two fields of quotients of an integral domain D are isomorphic.

Proof Suppose in Theorem 21.6 that L is a field of quotients of D, so that every element x of L can be expressed in the form $a /_L b$ for $a, b \in D$. Then L is the field $\psi[F]$ of the proof of Theorem 21.6 and is thus isomorphic to F. ◆

EXERCISES 21

Computations

1. Describe the field F of quotients of the integral subdomain

$$D = \{n + mi \mid n, m \in \mathbb{Z}\}$$

of $\mathbb{C}$. "Describe" means give the elements of $\mathbb{C}$ that make up the field of quotients of D in $\mathbb{C}$. (The elements of D are the **Gaussian integers.**)

2. Describe (in the sense of Exercise 1) the field F of quotients of the integral subdomain $D = \{n + m\sqrt{2} \mid n, m \in \mathbb{Z}\}$ of $\mathbb{R}$.

Concepts

3. Correct the definition of the italicized term without reference to the text, if correction is needed, so that it is in a form acceptable for publication.

A *field of quotients* of an integral domain D is a field F in which D can be embedded so that every nonzero element of D is a unit in F.

4. Mark each of the following true or false.

_____ **a.** $\mathbb{Q}$ is a field of quotients of $\mathbb{Z}$.

_____ **b.** $\mathbb{R}$ is a field of quotients of $\mathbb{Z}$.

_____ **c.** $\mathbb{R}$ is a field of quotients of $\mathbb{R}$.

_____ **d.** $\mathbb{C}$ is a field of quotients of $\mathbb{R}$.

_____ **e.** If D is a field, then any field of quotients of D is isomorphic to D.

_____ **f.** The fact that D has no divisors of 0 was used strongly several times in the construction of a field F of quotients of the integral domain D.

_____ **g.** Every element of an integral domain D is a unit in a field F of quotients of D.

_____ **h.** Every nonzero element of an integral domain D is a unit in a field F of quotients of D.

_____ **i.** A field of quotients F' of a subdomain D' of an integral domain D can be regarded as a subfield of some field of quotients of D.

_____ **j.** Every field of quotients of $\mathbb{Z}$ is isomorphic to $\mathbb{Q}$.

5. Show by an example that a field F' of quotients of a proper subdomain D' of an integral domain D may also be a field of quotients for D.

Theory

6. Prove Part 2 of Step 3. You may assume any preceding part of Step 3.

7. Prove Part 3 of Step 3. You may assume any preceding part of Step 3.

8. Prove Part 4 of Step 3. You may assume any preceding part of Step 3.

9. Prove Part 5 of Step 3. You may assume any preceding part of Step 3.

10. Prove Part 6 of Step 3. You may assume any preceding part of Step 3.

11. Prove Part 7 of Step 3. You may assume any preceding part of Step 3.

12. Let R be a commutative ring, and let T be a nonempty subset of R closed under multiplication and containing neither 0 nor divisors of 0. Starting with $R \times T$ and otherwise exactly following the construction in this section, we can show that the ring R can be enlarged to a *partial ring of quotients* $Q(R, T)$. Think about this for 15 minutes or so; look back over the construction and see why things still work. In particular, show the following:

a. $Q(R, T)$ has unity even if R does not.

b. In $Q(R, T)$, every nonzero element of T is a unit.

13. Prove from Exercise 12 that every nonzero commutative ring containing an element a that is not a divisor of 0 can be enlarged to a commutative ring with unity. Compare with Exercise 30 of Section 19.

14. With reference to Exercise 12, how many elements are there in the ring $Q(\mathbb{Z}_4, \{1, 3\})$?

15. With reference to Exercise 12, describe the ring $Q(\mathbb{Z}, \{2^n \mid n \in \mathbb{Z}^+\})$, by describing a subring of $\mathbb{R}$ to which it is isomorphic.

16. With reference to Exercise 12, describe the ring $Q(3\mathbb{Z}, \{6^n \mid n \in \mathbb{Z}^+\})$ by describing a subring of $\mathbb{R}$ to which it is isomorphic.

17. With reference to Exercise 12, suppose we drop the condition that T have no divisors of zero and just require that nonempty T not containing 0 be closed under multiplication. The attempt to enlarge R to a commutative ring with unity in which every nonzero element of T is a unit must fail if T contains an element a that is a divisor of 0, for a divisor of 0 cannot also be a unit. Try to discover where a construction parallel to that in the text but starting with $R \times T$ first runs into trouble. In particular, for $R = \mathbb{Z}_6$ and $T = \{1, 2, 4\}$, illustrate the first difficulty encountered. [*Hint:* It is in Step 1.]

SECTION 22 RINGS OF POLYNOMIALS

Polynomials in an Indeterminate

We all have a pretty workable idea of what constitutes a *polynomial in x with coefficients in a ring R*. We can guess how to add and multiply such polynomials and know what is meant by the *degree* of a polynomial. We expect that the set $R[x]$ of all polynomials with coefficients in the ring R is itself a ring with the usual operations of polynomial addition and multiplication, and that R is a subring of $R[x]$. However, we will be working with polynomials from a slightly different viewpoint than the approach in high school algebra or calculus, and there are a few things that we want to say.

In the first place, we will call x an **indeterminate** rather than a variable. Suppose, for example that our ring of coefficients is $\mathbb{Z}$. One of the polynomials in the ring $\mathbb{Z}[x]$ is $1x$, which we shall write simply as x. Now x is not 1 or 2 or any of the other elements of $\mathbb{Z}[x]$. Thus from now on we will never write such things as "$x = 1$" or "$x = 2$," as we have done in other courses. We call x an indeterminate rather than a variable to emphasize this change. Also, we will never write an expression such as "$x^2 - 4 = 0$," simply because $x^2 - 4$ is not the zero polynomial in our ring $\mathbb{Z}[x]$. We are accustomed to speaking of "solving a polynomial equation," and will be spending a lot of time in the remainder of our text discussing this, but we will always refer to it as "finding a zero of a polynomial." In summary, we try to be careful in our discussion of algebraic structures not to say in one context that things are equal and in another context that they are not equal.

HISTORICAL NOTE

The use of x and other letters near the end of the alphabet to represent an "indeterminate" is due to René Descartes (1596–1650). Earlier, François Viete (1540–1603) had used vowels for indeterminates and consonants for known quantities. Descartes is also responsible for the first publication of the factor theorem (Corollary 23.3) in his work *The Geometry,* which appeared as an appendix to his *Discourse on Method* (1637). This work also contained the first publication of the basic concepts of analytic geometry; Descartes showed how geometric curves can be described algebraically.

Descartes was born to a wealthy family in La Haye, France; since he was always of delicate health, he formed the habit of spending his mornings in bed. It was at these times that he accomplished his most productive work. The *Discourse on Method* was Descartes' attempt to show the proper procedures for "searching for truth in the sciences." The first step in this process was to reject as absolutely false everything of which he had the least doubt; but, since it was necessary that he who was thinking was "something," he conceived his first principle of philosophy: "I think, therefore I am." The most enlightening parts of the *Discourse on Method,* however, are the three appendices: *The Optics, The Geometry,* and *The Meteorology*. It was here that Descartes provided examples of how he actually applied his method. Among the important ideas Descartes discovered and published in these works were the sine law of refraction of light, the basics of the theory of equations, and a geometric explanation of the rainbow.

In 1649, Descartes was invited by Queen Christina of Sweden to come to Stockholm to tutor her. Unfortunately, the Queen required him, contrary to his long-established habits, to rise at an early hour. He soon contracted a lung disease and died in 1650.

If a person knows nothing about polynomials, it is not an easy task to describe precisely the nature of a polynomial in x with coefficients in a ring R. If we just define such a polynomial to be a *finite formal sum*

$$\sum_{i=0}^{n} a_i x^i = a_0 + a_1 x + \cdots + a_n x^n,$$

where $a_i \in R$, we get ourselves into a bit of trouble. For surely $0 + a_1 x$ and $0 + a_1 x + 0x^2$ are different as formal sums, but we want to regard them as the same polynomial. A practical solution to this problem is to define a polynomial as an *infinite formal sum*

$$\sum_{i=0}^{\infty} a_i x^i = a_0 + a_1 x + \cdots + a_n x^n + \cdots,$$

where $a_i = 0$ for all but a finite number of values of i. Now there is no problem of having more than one formal sum represent what we wish to consider a single polynomial.

22.1 Definition Let R be a ring. A **polynomial** $f(x)$ **with coefficients in** R is an infinite formal sum

$$\sum_{i=0}^{\infty} a_i x^i = a_0 + a_1 x + \cdots + a_n x^n + \cdots,$$

where $a_i \in R$ and $a_i = 0$ for all but a finite number of values of i. The a_i are **coefficients of** $f(x)$. If for some $i \geq 0$ it is true that $a_i \neq 0$, the largest such value of i is the **degree of** $f(x)$. If all $a_i = 0$, then the degree of $f(x)$ is undefined.[†] ■

To simplify working with polynomials, let us agree that if $f(x) = a_0 + a_1 x + \cdots + a_n x^n + \cdots$ has $a_i = 0$ for $i > n$, then we may denote $f(x)$ by $a_0 + a_1 x + \cdots + a_n x^n$. Also, if R has unity $1 \neq 0$, we will write a term $1x^k$ in such a sum as x^k. For example, in $\mathbb{Z}[x]$, we will write the polynomial $2 + 1x$ as $2 + x$. Finally, we shall agree that we may omit altogether from the formal sum any term $0x^i$, or a_0 if $a_0 = 0$ but not all $a_i = 0$. Thus 0, 2, x, and $2 + x^2$ are polynomials with coefficients in $\mathbb{Z}$. An element of R is a **constant polynomial.**

Addition and multiplication of polynomials with coefficients in a ring R are defined in a way familiar to us. If

$$f(x) = a_0 + a_1 x + \cdots + a_n x^n + \cdots$$

and

$$g(x) = b_0 + b_1 x + \cdots + b_n x^n + \cdots,$$

then for polynomial addition, we have

$$f(x) + g(x) = c_0 + c_1 x + \cdots + c_n x^n + \cdots \quad \text{where } c_n = a_n + b_n,$$

[†] The degree of the zero polynomial is sometimes defined to be -1, which is the first integer less than 0, or defined to be $-\infty$ so that the degree of $f(x)g(x)$ will be the sum of the degrees of $f(x)$ and $g(x)$ if one of them is zero.

and for polynomial multiplication, we have

$$f(x)g(x) = d_0 + d_1x + \cdots + d_nx^n + \cdots \text{ where } d_n = \sum_{i=0}^{n} a_ib_{n-i}$$

Observe that both c_i and d_i are 0 for all but a finite number of values of i, so these definitions make sense. Note that $\sum_{i=0}^{n} a_ib_{n-i}$ need not equal $\sum_{i=0}^{n} b_ia_{n-i}$ if R is not commutative. With these definitions of addition and multiplication, we have the following theorem.

22.2 Theorem The set $R[x]$ of all polynomials in an indeterminate x with coefficients in a ring R is a ring under polynomial addition and multiplication. If R is commutative, then so is $R[x]$, and if R has unity $1 \neq 0$, then 1 is also unity for $R[x]$.

Proof That $\langle R[x], +\rangle$ is an abelian group is apparent. The associative law for multiplication and the distributive laws are straightforward, but slightly cumbersome, computations. We illustrate by proving the associative law.

Applying ring axioms to $a_i, b_j, c_k \in R$, we obtain

$$\begin{aligned}
\left[\left(\sum_{i=0}^{\infty} a_ix^i\right)\left(\sum_{j=0}^{\infty} b_jx^j\right)\right]\left(\sum_{k=0}^{\infty} c_kx^k\right) &= \left[\sum_{n=0}^{\infty}\left(\sum_{i=0}^{n} a_ib_{n-i}\right)x^n\right]\left(\sum_{k=0}^{\infty} c_kx^k\right) \\
&= \sum_{s=0}^{\infty}\left[\sum_{n=0}^{s}\left(\sum_{i=0}^{n} a_ib_{n-i}\right)c_{s-n}\right]x^s \\
&= \sum_{s=0}^{\infty}\left(\sum_{i+j+k=s} a_ib_jc_k\right)x^s \\
&= \sum_{s=0}^{\infty}\left[\sum_{m=0}^{s} a_{s-m}\left(\sum_{j=0}^{m} b_jc_{m-j}\right)\right]x^s \\
&= \left(\sum_{i=0}^{\infty} a_ix^i\right)\left[\sum_{m=0}^{\infty}\left(\sum_{j=0}^{m} b_jc_{m-j}\right)x^m\right] \\
&= \left(\sum_{i=0}^{\infty} a_ix^i\right)\left[\left(\sum_{j=0}^{\infty} b_jx^j\right)\left(\sum_{k=0}^{\infty} c_kx^k\right)\right].
\end{aligned}$$

Whew!! In this computation, the fourth expression, having just two summation signs, should be viewed as the value of the triple product $f(x)g(x)h(x)$ of these polynomials under this associative multiplication. (In a similar fashion, we view $f(g(h(x)))$ as the value of the associative composition $(f \circ g \circ h)(x)$ of three functions f, g, and h.)

The distributive laws are similarly proved. (See Exercise 26.)

The comments prior to the statement of the theorem show that $R[x]$ is a commutative ring if R is commutative, and a unity $1 \neq 0$ in R is also unity for $R[x]$, in view of the definition of multiplication in $R[x]$. ◆

Thus $\mathbb{Z}[x]$ is the ring of polynomials in the indeterminate x with integral coefficients, $Q[x]$ the ring of polynomials in x with rational coefficients, and so on.

22.3 Example In $\mathbb{Z}_2[x]$, we have

$$(x+1)^2 = (x+1)(x+1) = x^2 + (1+1)x + 1 = x^2 + 1.$$

Still working in $\mathbb{Z}_2[x]$, we obtain

$$(x+1) + (x+1) = (1+1)x + (1+1) = 0x + 0 = 0. \qquad \blacktriangle$$

If R is a ring and x and y are two indeterminates, then we can form the ring $(R[x])[y]$, that is, the ring of polynomials in y with coefficients that are polynomials in x. Every polynomial in y with coefficients that are polynomials in x can be rewritten in a natural way as a polynomial in x with coefficients that are polynomials in y as illustrated by Exercise 20. This indicates that $(R[x])[y]$ is naturally isomorphic to $(R[y])[x]$, although a careful proof is tedious. We shall identify these rings by means of this natural isomorphism, and shall consider this ring $R[x, y]$ the **ring of polynomials in two indeterminates x and y with coefficients in R**. The **ring $R[x_1, \cdots, x_n]$ of polynomials in the n indeterminates x_i with coefficients in R** is similarly defined.

We leave as Exercise 24 the proof that if D is an integral domain then so is $D[x]$. In particular, if F is a field, then $F[x]$ is an integral domain. Note that $F[x]$ is not a field, for x is not a unit in $F[x]$. That is, there is no polynomial $f(x) \in F[x]$ such that $xf(x) = 1$. By Theorem 21.5, one can construct the field of quotients $F(x)$ of $F[x]$. Any element in $F(x)$ can be represented as a quotient $f(x)/g(x)$ of two polynomials in $F[x]$ with $g(x) \neq 0$. We similarly define $F(x_1, \cdots, x_n)$ to be the field of quotients of $F[x_1, \cdots, x_n]$. This field $F(x_1, \cdots, x_n)$ is the **field of rational functions in n indeterminates over F**. These fields play a very important role in algebraic geometry.

The Evaluation Homomorphisms

We are now ready to proceed to show how homomorphisms can be used to study what we have always referred to as "solving a polynomial equation." Let E and F be fields, with F a subfield of E, that is, $F \leq E$. The next theorem asserts the existence of very important homomorphisms of $F[x]$ into E. *These homomorphisms will be the fundamental tools for much of the rest of our work.*

22.4 Theorem **(The Evaluation Homomorphisms for Field Theory)** Let F be a subfield of a field E, let α be any element of E, and let x be an indeterminate. The map $\phi_\alpha : F[x] \to E$ defined by

$$\phi_\alpha(a_0 + a_1x + \cdots + a_nx^n) = a_0 + a_1\alpha + \cdots + a_n\alpha^n$$

for $(a_0 + a_1x + \cdots + a_nx^n) \in F[x]$ is a homomorphism of $F[x]$ into E. Also, $\phi_\alpha(x) = \alpha$, and ϕ_α maps F isomorphically by the identity map; that is, $\phi_\alpha(a) = a$ for $a \in F$. The homomorphism ϕ_α is **evaluation at α**.

Proof The subfield and mapping diagram in Fig. 22.5 may help us to visualize this situation. The dashed lines indicate an element of the set. The theorem is really an immediate

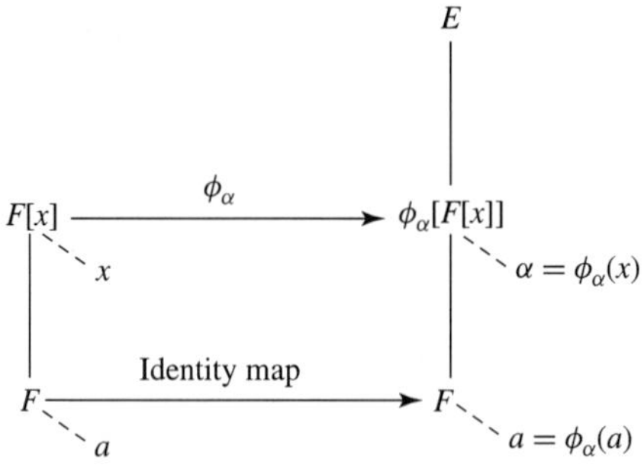

22.5 Figure

consequence of our definitions of addition and multiplication in $F[x]$. The map ϕ_α is well defined, that is, independent of our representation of $f(x) \in F[x]$ as a finite sum

$$a_0 + a_1 x + \cdots + a_n x^n,$$

since such a finite sum representing $f(x)$ can be changed only by insertion or deletion of terms $0x^i$, which does not affect the value of $\phi_\alpha(f(x))$.

If $f(x) = a_0 + a_1 x + \cdots + a_n x^n$, $g(x) = b_0 + b_1 x + \cdots + b_m x^m$, and $h(x) = f(x) + g(x) = c_0 + c_1 x + \cdots + c_r x^r$, then

$$\phi_\alpha(f(x) + g(x)) = \phi_\alpha(h(x)) = c_0 + c_1 \alpha + \cdots + c_r \alpha^r,$$

while

$$\phi_\alpha(f(x)) + \phi_\alpha(g(x)) = \left(a_0 + a_1 \alpha + \cdots + a_n \alpha^n\right) + \left(b_0 + b_1 \alpha + \cdots + b_m \alpha^m\right).$$

Since by definition of polynomial addition we have $c_i = a_i + b_i$, we see that

$$\phi_\alpha(f(x) + g(x)) = \phi_\alpha(f(x)) + \phi_\alpha(g(x)).$$

Turning to multiplication, we see that if

$$f(x)g(x) = d_0 + d_1 x + \cdots + d_s x^s,$$

then

$$\phi_\alpha(f(x)g(x)) = d_0 + d_1 \alpha + \cdots + d_s \alpha^s,$$

while

$$[\phi_\alpha(f(x))][\phi_\alpha(g(x))] = \left(a_0 + a_1 \alpha + \cdots + \alpha_n \alpha^n\right)\left(b_0 + b_1 \alpha + \cdots + b_m \alpha^m\right).$$

Since by definition of polynomial multiplication $d_j = \sum_{i=0}^{j} a_i b_{j-i}$, we see that

$$\phi_\alpha(f(x)g(x)) = [\phi_\alpha(f(x))][\phi_\alpha(g(x))].$$

Thus ϕ_α is a homomorphism.

The very definition of ϕ_α applied to a constant polynomial $a \in F[x]$, where $a \in F$, gives $\phi_\alpha(a) = a$, so ϕ_α maps F isomorphically by the identity map. Again by definition of ϕ_α, we have $\phi_\alpha(x) = \phi_\alpha(1x) = 1\alpha = \alpha$. ◆

We point out that this theorem is valid with the identical proof if F and E are merely commutative rings with unity rather than fields. However, we shall be interested primarily in the case in which they are fields.

It is hard to overemphasize the importance of this simple theorem for us. It is the very foundation for all of our further work in field theory. It is so simple that it could justifiably be called an *observation* rather than a theorem. It was perhaps a little misleading to write out the proof because the polynomial notation makes it look so complicated that you may be fooled into thinking it is a difficult theorem.

22.6 Example Let F be $\mathbb{Q}$ and E be $\mathbb{R}$ in Theorem 22.4, and consider the evaluation homomorphism $\phi_0 : \mathbb{Q}[x] \to \mathbb{R}$. Here

$$\phi_0\left(a_0 + a_1x + \cdots + a_nx^n\right) = a_0 + a_1 0 + \cdots + a_n 0^n = a_0.$$

Thus every polynomial is mapped onto its constant term. ▲

22.7 Example Let F be $\mathbb{Q}$ and E be $\mathbb{R}$ in Theorem 22.4 and consider the evaluation homomorphism $\phi_2 : \mathbb{Q}[x] \to \mathbb{R}$. Here

$$\phi_2\left(a_0 + a_1x + \cdots + a_nx^n\right) = a_0 + a_1 2 + \cdots + a_n 2^n.$$

Note that

$$\phi_2(x^2 + x - 6) = 2^2 + 2 - 6 = 0.$$

Thus $x^2 + x - 6$ is in the kernel N of ϕ_2. Of course,

$$x^2 + x - 6 = (x - 2)(x + 3),$$

and the reason that $\phi_2(x^2 + x - 6) = 0$ is that $\phi_2(x - 2) = 2 - 2 = 0$. ▲

22.8 Example Let F be $\mathbb{Q}$ and E be $\mathbb{C}$ in Theorem 22.4 and consider the evaluation homomorphism $\phi_i : \mathbb{Q}[x] \to \mathbb{C}$. Here

$$\phi_i\left(a_0 + a_1x + \cdots + a_nx^n\right) = a_0 + a_1 i + \cdots + a_n i^n$$

and $\phi_i(x) = i$. Note that

$$\phi_i(x^2 + 1) = i^2 + 1 = 0,$$

so $x^2 + 1$ is in the kernel N of ϕ_i. ▲

22.9 Example Let F be $\mathbb{Q}$ and let E be $\mathbb{R}$ in Theorem 22.4 and consider the evaluation homomorphism $\phi_\pi : \mathbb{Q}[x] \to \mathbb{R}$. Here

$$\phi_\pi(a_0 + a_1x + \cdots + a_nx^n) = a_0 + a_1\pi + \cdots + a_n\pi^n.$$

It can be proved that $a_0 + a_1\pi + \cdots + a_n\pi^n = 0$ if and only if $a_i = 0$ for $i = 0, 1, \cdots, n$. Thus the kernel of ϕ_π is $\{0\}$, and ϕ_π is a one-to-one map. This shows that all *formal polynomials in π with rational coefficients* form a ring isomorphic to $\mathbb{Q}[x]$ in a natural way with $\phi_\pi(x) = \pi$. ▲

The New Approach

We now complete the connection between our new ideas and the classical concept of solving a polynomial equation. Rather than speak of *solving a polynomial equation,* we shall refer to *finding a zero of a polynomial.*

22.10 Definition Let F be a subfield of a field E, and let α be an element of E. Let $f(x) = a_0 + a_1x + \cdots + a_nx^n$ be in $F[x]$, and let $\phi_\alpha : F[x] \to E$ be the evaluation homomorphism of Theorem 22.4. Let $f(\alpha)$ denote

$$\phi_\alpha(f(x)) = a_0 + a_1\alpha + \cdots + a_n\alpha^n.$$

If $f(\alpha) = 0$, then α is a **zero of** $f(x)$. ■

In terms of this definition, we can rephrase the classical problem of finding all real numbers r such that $r^2 + r - 6 = 0$ by letting $F = \mathbb{Q}$ and $E = \mathbb{R}$ and *finding all $\alpha \in \mathbb{R}$ such that*

$$\phi_\alpha(x^2 + x - 6) = 0,$$

that is, finding all zeros of $x^2 + x - 6$ in $\mathbb{R}$. Both problems have the same answer, since

$$\{\alpha \in \mathbb{R} \mid \phi_\alpha(x^2 + x - 6) = 0\} = \{r \in \mathbb{R} \mid r^2 + r - 6 = 0\} = \{2, -3\}.$$

It may seem that we have merely succeeded in making a simple problem seem quite complicated. In fact, *what we have done is to phrase the problem in the language of mappings, and we can now use all the mapping machinery that we have developed and will continue to develop for its solution.*

Our Basic Goal

We continue to attempt to put our future work in perspective. Sections 26 and 27 are concerned with topics in ring theory that are analogous to the material on factor groups and homomorphisms for group theory. However, our aim in developing these analogous concepts for rings will be quite different from our aims in group theory. In group theory we used the concepts of factor groups and homomorphisms to study the structure of a given group and to determine the types of group structures of certain orders that could exist. We will be talking about homomorphisms and factor rings in Section 26 with an eye to finding zeros of polynomials, which is one of the oldest and most fundamental problems in algebra. Let us take a moment to talk about this aim in the light of mathematical history, using the language of "solving polynomial equations" to which we are accustomed.

We start with the Pythagorean school of mathematics of about 525 B.C. The Pythagoreans worked with the assumption that all distances are **commensurable**; that is, given distances a and b, there should exist a unit of distance u and integers n and m such that $a = (n)(u)$ and $b = (m)(u)$. In terms of numbers, then, thinking of u as being one unit of distance, they maintained that all numbers are integers. This idea of commensurability can be rephrased according to our ideas as an assertion that all numbers are rational, for if a and b are rational numbers, then each is an integral multiple of the reciprocal of the least common multiple of their denominators. For example, if $a = \frac{7}{12}$ and $b = \frac{19}{15}$, then $a = (35)(\frac{1}{60})$ and $b = (76)(\frac{1}{60})$.

The Pythagoreans knew, of course, what is now called the *Pythagorean theorem;* that is, for a right triangle with legs of lengths a and b and a hypotenuse of length c,

$$a^2 + b^2 = c^2.$$

They also had to grant the existence of a hypotenuse of a right triangle having two legs of equal length, say one unit each. The hypotenuse of such a right triangle would, as we know, have to have a length of $\sqrt{2}$. Imagine then their consternation and dismay when one of their society—according to some stories it was Pythagoras himself—came up with the embarrassing fact that is stated in our terminology in the following theorem.

22.11 Theorem The polynomial $x^2 - 2$ has no zeros in the rational numbers. Thus $\sqrt{2}$ is not a rational number.

Proof Suppose that m/n for $m, n \in \mathbb{Z}$ is a rational number such that $(m/n)^2 = 2$. We assume that we have canceled any factors common to m and n, so that the fraction m/n is in lowest terms with $\gcd(m, n) = 1$. Then

$$m^2 = 2n^2,$$

where both m^2 and $2n^2$ are integers. Since m^2 and $2n^2$ are the same integer, and since 2 is a factor of $2n^2$, we see that 2 must be one of the factors of m^2. But as a square, m^2 has as factors the factors of m repeated twice. Thus m^2 must have two factors 2. Then $2n^2$ must have two factors 2, so n^2 must have 2 as a factor, and consequently n has 2 as a factor. We have deduced from $m^2 = 2n^2$ that both m and n must be divisible by 2, contradicting the fact that the fraction m/n is in lowest terms. Thus we have $2 \neq (m/n)^2$ for any $m, n \in \mathbb{Z}$. ◆

Thus the Pythagoreans ran right into the question of a solution of a polynomial equation, $x^2 - 2 = 0$. We refer the student to Shanks [36, Chapter 3], for a lively and totally delightful account of this Pythagorean dilemma and its significance in mathematics.

Historical Note

The solution of polynomial equations has been a goal of mathematics for nearly 4000 years. The Babylonians developed versions of the quadratic formula to solve quadratic equations. For example, to solve $x^2 - x = 870$, the Babylonian scribe instructed his students to take half of 1 ($\frac{1}{2}$), square it ($\frac{1}{4}$), and add that to 870. The square root of $870\frac{1}{4}$, namely $29\frac{1}{2}$, is then added to $\frac{1}{2}$ to give 30 as the answer. What the scribes did not discuss, however, was what to do if the square root in this process was not a rational number. Chinese mathematicians, however, from about 200 B.C., discovered a method similar to what is now called *Horner's method* to solve quadratic equations numerically; since they used a decimal system, they were able in principle to carry out the computation to as many places as necessary and could therefore ignore the distinction between rational and irrational solutions. The Chinese, in fact, extended their numerical techniques to polynomial equations of higher degree. In the Arab world, the Persian poet–mathematician Omar Khayyam (1048–1131) developed methods for solving cubic equations geometrically by finding the point(s) of intersection of appropriately chosen conic sections, while Sharaf al-Din al-Tusi (died 1213) used, in effect, techniques of calculus to determine whether or not a cubic equation had a real positive root. It was the Italian Girolamo Cardano (1501–1576) who first published a procedure for solving cubic equations algebraically.

In our motivation of the definition of a group, we commented on the necessity of having negative numbers, so that equations such as $x + 2 = 0$ might have solutions. The introduction of negative numbers caused a certain amount of consternation in some philosophical circles. We can visualize 1 apple, 2 apples, and even $\frac{13}{11}$ apples, but how can we point to anything and say that it is -17 apples? Finally, consideration of the equation $x^2 + 1 = 0$ led to the introduction of the number i. The very name of an "imaginary number" given to i shows how this number was regarded. Even today, many students are led by this name to regard i with some degree of suspicion. The negative numbers were introduced to us at such an early stage in our mathematical development that we accepted them without question.

We first met polynomials in high school freshman algebra. The first problem there was to learn how to add, multiply, and factor polynomials. Then, in both freshman algebra and in the second course in algebra in high school, considerable emphasis was placed on solving polynomial equations. These topics are exactly those with which we shall be concerned. The difference is that while in high school, only polynomials with real number coefficients were considered, *we shall be doing our work for polynomials with coefficients from any field.*

Once we have developed the machinery of homomorphisms and factor rings in Section 26, we will proceed with our **basic goal**: to show that given any polynomial of degree ≥ 1, where the coefficients of the polynomial may be from any field, we can find a zero of this polynomial in some field containing the given field. After the machinery is developed in Sections 26 and 27, the achievement of this goal will be very easy, and is really a very elegant piece of mathematics.

All this fuss may seem ridiculous, but just think back in history. This is the *culmination of more than 2000 years of mathematical endeavor in working with polynomial equations*. After achieving our *basic goal,* we shall spend the rest of our time studying the nature of these solutions of polynomial equations. We need have no fear in approaching this material. *We shall be dealing with familiar topics of high school algebra. This work should seem much more natural than group theory.*

In conclusion, we remark that the machinery of factor rings and ring homomorphisms is not really necessary in order for us to achieve our *basic goal*. For a direct demonstration, see Artin [27, p. 29]. However, factor rings and ring homomorphisms are fundamental ideas that we should grasp, and our *basic goal* will follow very easily once we have mastered them.

EXERCISES 22

Computations

In Exercises 1 through 4, find the sum and the product of the given polynomials in the given polynomial ring.

1. $f(x) = 4x - 5$, $g(x) = 2x^2 - 4x + 2$ in $\mathbb{Z}_8[x]$.

2. $f(x) = x + 1$, $g(x) = x + 1$ in $\mathbb{Z}_2[x]$.

3. $f(x) = 2x^2 + 3x + 4$, $g(x) = 3x^2 + 2x + 3$ in $\mathbb{Z}_6[x]$.

4. $f(x) = 2x^3 + 4x^2 + 3x + 2$, $g(x) = 3x^4 + 2x + 4$ in $\mathbb{Z}_5[x]$.

5. How many polynomials are there of degree ≤ 3 in $\mathbb{Z}_2[x]$? (Include 0.)

6. How many polynomials are there of degree ≤ 2 in $\mathbb{Z}_5[x]$? (Include 0.)

In Exercises 7 and 8, $F = E = \mathbb{C}$ in Theorem 22.4. Compute for the indicated evaluation homomorphism.

7. $\phi_2(x^2 + 3)$

8. $\phi_i(2x^3 - x^2 + 3x + 2)$

In Exercises 9 through 11, $F = E = \mathbb{Z}_7$ in Theorem 22.4. Compute for the indicated evaluation homomorphism.

9. $\phi_3[(x^4 + 2x)(x^3 - 3x^2 + 3)]$

10. $\phi_5[(x^3 + 2)(4x^2 + 3)(x^7 + 3x^2 + 1)]$

11. $\phi_4(3x^{106} + 5x^{99} + 2x^{53})$ [*Hint:* Use Fermat's theorem.]

In Exercises 12 through 15, find all zeros in the indicated finite field of the given polynomial with coefficients in that field. [*Hint:* One way is simply to try all candidates!]

12. $x^2 + 1$ in $\mathbb{Z}_2$

13. $x^3 + 2x + 2$ in $\mathbb{Z}_7$

14. $x^5 + 3x^3 + x^2 + 2x$ in $\mathbb{Z}_5$

15. $f(x)g(x)$ where $f(x) = x^3 + 2x^2 + 5$ and $g(x) = 3x^2 + 2x$ in $\mathbb{Z}_7$

16. Let $\phi_a : \mathbb{Z}_5[x] \to \mathbb{Z}_5$ be an evaluation homomorphism as in Theorem 22.4. Use Fermat's theorem to evaluate $\phi_3(x^{231} + 3x^{117} - 2x^{53} + 1)$.

17. Use Fermat's theorem to find all zeros in $\mathbb{Z}_5$ of $2x^{219} + 3x^{74} + 2x^{57} + 3x^{44}$.

Concepts

In Exercises 18 and 19, correct the definition of the italicized term without reference to the text, if correction is needed, so that it is in a form acceptable for publication.

18. A *polynomial with coefficients in a ring R is an infinite formal sum*

$$\sum_{i=0}^{\infty} a_i x^i = a_0 + a_1 x + a_2 x^2 + \cdots + a_n x^n + \cdots$$

where $a_i \in R$ for $i = 0, 1, 2, \cdots$.

19. Let F be a field and let $f(x) \in F[x]$. A *zero of* $f(x)$ is an $\alpha \in F$ such that $\phi_\alpha(f(x)) = 0$, where $\phi_\alpha : F(x) \rightarrow F$ is the evaluation homomorphism mapping x into α.

20. Consider the element

$$f(x, y) = (3x^3 + 2x)y^3 + (x^2 - 6x + 1)y^2 + (x^4 - 2x)y + (x^4 - 3x^2 + 2)$$

of $(\mathbb{Q}[x])[y]$. Write $f(x, y)$ as it would appear if viewed as an element of $(\mathbb{Q}[y])[x]$.

21. Consider the evaluation homomorphism $\phi_5 : \mathbb{Q}[x] \rightarrow \mathbb{R}$. Find six elements in the kernel of the homomorphism ϕ_5.

22. Find a polynomial of degree >0 in $\mathbb{Z}_4[x]$ that is a unit.

23. Mark each of the following true or false.

______ **a.** The polynomial $(a_n x^n + \cdots + a_1 x + a_0) \in R[x]$ is 0 if and only if $a_i = 0$, for $i = 0, 1, \cdots, n$.

______ **b.** If R is a commutative ring, then $R[x]$ is commutative.

______ **c.** If D is an integral domain, then $D[x]$ is an integral domain.

______ **d.** If R is a ring containing divisors of 0, then $R[x]$ has divisors of 0.

______ **e.** If R is a ring and $f(x)$ and $g(x)$ in $R[x]$ are of degrees 3 and 4, respectively, then $f(x)g(x)$ may be of degree 8 in $R[x]$.

______ **f.** If R is any ring and $f(x)$ and $g(x)$ in $R[x]$ are of degrees 3 and 4, respectively, then $f(x)g(x)$ is always of degree 7.

______ **g.** If F is a subfield E and $\alpha \in E$ is a zero of $f(x) \in F[x]$, then α is a zero of $h(x) = f(x)g(x)$ for all $g(x) \in F[x]$.

______ **h.** If F is a field, then the units in $F[x]$ are precisely the units in F.

______ **i.** If R is a ring, then x is never a divisor of 0 in $R[x]$.

______ **j.** If R is a ring, then the zero divisors in $R[x]$ are precisely the zero divisors in R.

Theory

24. Prove that if D is an integral domain, then $D[x]$ is an integral domain.

25. Let D be an integral domain and x an indeterminate.

a. Describe the units in $D[x]$.

b. Find the units in $\mathbb{Z}[x]$.

c. Find the units in $\mathbb{Z}_7[x]$.

26. Prove the left distributive law for $R[x]$, where R is a ring and x is an indeterminate.

27. Let F be a field of characteristic zero and let D be the formal polynomial differentiation map, so that

$$D(a_0 + a_1 x + a_2 x^2 + \cdots + a_n x^n) = a_1 + 2 \cdot a_2 x + \cdots + n \cdot a_n x^{n-1}.$$

a. Show that $D : F[x] \rightarrow F[x]$ is a group homomorphism of $\langle F[x], + \rangle$ into itself. Is D a ring homomorphism?

b. Find the kernel of D.

c. Find the image of $F[x]$ under D.

28. Let F be a subfield of a field E.

a. Define an *evaluation homomorphism*

$$\phi_{\alpha_1,\cdots,\alpha_n} : F[x_1, \cdots, x_n] \to E \qquad \text{for} \quad \alpha_i \in E,$$

stating the analog of Theorem 22.4.

b. With $E = F = \mathbb{Q}$, compute $\phi_{-3,2}(x_1{}^2 x_2{}^3 + 3x_1{}^4 x_2)$.

c. Define the concept of a *zero of a polynomial* $f(x_1, \cdots, x_n) \in F[x_1, \cdots, x_n]$ in a way analogous to the definition in the text of a zero of $f(x)$.

29. Let R be a ring, and let R^R be the set of all functions mapping R into R. For $\phi, \psi \in R^R$, define the sum $\phi + \psi$ by

$$(\phi + \psi)(r) = \phi(r) + \psi(r)$$

and the product $\phi \cdot \psi$ by

$$(\phi \cdot \psi)(r) = \phi(r)\psi(r)$$

for $r \in R$. Note that $\cdot$ is *not* function composition. Show that $\langle R^R, +, \cdot \rangle$ is a ring.

30. Referring to Exercise 29, let F be a field. An element ϕ of F^F is a **polynomial function on** F, if there exists $f(x) \in F[x]$ such that $\phi(a) = f(a)$ for all $a \in F$.

a. Show that the set P_F of all polynomial functions on F forms a subring of F^F.

b. Show that the ring P_F is not necessarily isomorphic to $F[x]$. [*Hint:* Show that if F is a finite field, P_F and $F[x]$ don't even have the same number of elements.]

31. Refer to Exercises 29 and 30 for the following questions.

a. How many elements are there in $\mathbb{Z}_2{}^{\mathbb{Z}_2}$? in $\mathbb{Z}_3{}^{\mathbb{Z}_3}$?

b. Classify $\langle \mathbb{Z}_2{}^{\mathbb{Z}_2}, + \rangle$ and $\langle \mathbb{Z}_3{}^{\mathbb{Z}_3}, + \rangle$ by Theorem 11.12, the Fundamental Theorem of finitely generated abelian groups.

c. Show that if F is a finite field, then $F^F = P_F$. [*Hint:* Of course, $P_F \subseteq F^F$. Let F have as elements $a_1, \cdots, a_n$. Note that if

$$f_i(x) = c(x - a_1) \cdots (x - a_{i-1})(x - a_{i+1}) \cdots (x - a_n),$$

then $f_i(a_j) = 0$ for $j \neq i$, and the value $f_i(a_i)$ can be controlled by the choice of $c \in F$. Use this to show that every function on F is a polynomial function.]

SECTION 23 Factorization of Polynomials over a Field

Recall that we are concerned with finding zeros of polynomials. Let E and F be fields, with $F \leq E$. Suppose that $f(x) \in F[x]$ factors in $F[x]$, so that $f(x) = g(x)h(x)$ for $g(x), h(x) \in F[x]$ and let $\alpha \in E$. Now for the evaluation homomorphism ϕ_α, we have

$$f(\alpha) = \phi_\alpha(f(x)) = \phi_\alpha(g(x)h(x)) = \phi_\alpha(g(x))\phi_\alpha(h(x)) = g(\alpha)h(\alpha).$$

Thus if $\alpha \in E$, then $f(\alpha) = 0$ if and only if either $g(\alpha) = 0$ or $h(\alpha) = 0$. The attempt to find a zero of $f(x)$ is reduced to the problem of finding a zero of a factor of $f(x)$. This is one reason why it is useful to study factorization of polynomials.

The Division Algorithm in $F[x]$

The following theorem is the basic tool for our work in this section. Note the similarity with the division algorithm for $\mathbb{Z}$ given in Theorem 6.3, the importance of which has been amply demonstrated.

23.1 Theorem **(Division Algorithm for $F[x]$)** Let

$$f(x) = a_n x^n + a_{n-1}x^{n-1} + \cdots + a_0$$

and

$$g(x) = b_m x^m + b_{m-1}x^{m-1} + \cdots + b_0$$

be two elements of $F[x]$, with a_n and b_m both nonzero elements of F and $m > 0$. Then there are unique polynomials $q(x)$ and $r(x)$ in $F[x]$ such that $f(x) = g(x)q(x) + r(x)$, where either $r(x) = 0$ or the degree of $r(x)$ is less than the degree m of $g(x)$.

Proof Consider the set $S = \{f(x) - g(x)s(x) \mid s(x) \in F[x]\}$. If $0 \in S$ then there exists an $s(x)$ such that $f(x) - g(x)s(x) = 0$, so $f(x) = g(x)s(x)$. Taking $q(x) = s(x)$ and $r(x) = 0$, we are done. Otherwise, let $r(x)$ be an element of minimal degree in S. Then

$$f(x) = g(x)q(x) + r(x)$$

for some $q(x) \in F[x]$. We must show that the degree of $r(x)$ is less than m. Suppose that

$$r(x) = c_t x^t + c_{t-1}x^{t-1} + \cdots + c_0,$$

with $c_j \in F$ and $c_t \neq 0$. If $t \geq m$, then

$$f(x) - q(x)g(x) - (c_t/b_m)x^{t-m}g(x) = r(x) - (c_t/b_m)x^{t-m}g(x), \qquad \textbf{(1)}$$

and the latter is of the form

$$r(x) - (c_t x^t + \text{terms of lower degree}),$$

which is a polynomial of degree lower than t, the degree of $r(x)$. However, the polynomial in Eq. (1) can be written in the form

$$f(x) - g(x)\big[q(x) + (c_t/b_m)x^{t-m}\big],$$

so it is in S, contradicting the fact that $r(x)$ was selected to have minimal degree in S. Thus the degree of $r(x)$ is less than the degree m of $g(x)$.

For uniqueness, if

$$f(x) = g(x)q_1(x) + r_1(x)$$

and

$$f(x) = g(x)q_2(x) + r_2(x),$$

then subtracting we have

$$g(x)[q_1(x) - q_2(x)] = r_2(x) - r_1(x).$$

Because either $r_2(x) - r_1(x) = 0$ or the degree of $r_2(x) - r_1(x)$ is less than the degree of $g(x)$, this can hold only if $q_1(x) - q_2(x) = 0$ so $q_1(x) = q_2(x)$. Then we must have $r_2(x) - r_1(x) = 0$ so $r_1(x) = r_2(x)$. ◆

We can compute the polynomials $q(x)$ and $r(x)$ of Theorem 23.1 by long division just as we divided polynomials in $\mathbb{R}[x]$ in high school.

23.2 Example Let us work with polynomials in $\mathbb{Z}_5[x]$ and divide

$$f(x) = x^4 - 3x^3 + 2x^3 + 4x - 1$$

by $g(x) = x^2 - 2x + 3$ to find $q(x)$ and $r(x)$ of Theorem 23.1. The long division should be easy to follow, but remember that we are in $\mathbb{Z}_5[x]$, so, for example, $4x - (-3x) = 2x$.

$$\begin{array}{r|l}
 & x^2 - x - 3 \\
\hline
x^2 - 2x + 3 & x^4 - 3x^3 + 2x^2 + 4x - 1 \\
 & x^4 - 2x^3 + 3x^2 \\
\cline{2-2}
 & \quad - x^3 - x^2 + 4x \\
 & \quad - x^3 + 2x^2 - 3x \\
\cline{2-2}
 & \qquad - 3x^2 + 2x - 1 \\
 & \qquad - 3x^2 + x - 4 \\
\cline{2-2}
 & \qquad\qquad x + 3
\end{array}$$

Thus

$$q(x) = x^2 - x - 3, \quad \text{and} \quad r(x) = x + 3. \qquad \blacktriangle$$

We give three important corollaries of Theorem 23.1. The first one appears in high school algebra for the special case $F[x] = \mathbb{R}[x]$. We phrase our proof in terms of the mapping (homomorphism) approach described in Section 22.

23.3 Corollary **(Factor Theorem)** An element $a \in F$ is a zero of $f(x) \in F[x]$ if and only if $x - a$ is a factor of $f(x)$ in $F[x]$.

Proof Suppose that for $a \in F$ we have $f(a) = 0$. By Theorem 23.1, there exist $q(x)$, $r(x) \in F[x]$ such that

$$f(x) = (x - a)q(x) + r(x),$$

where either $r(x) = 0$ or the degree of $r(x)$ is less than 1. Thus we must have $r(x) = c$ for $c \in F$, so

$$f(x) = (x - a)q(x) + c.$$

Applying our evaluation homomorphism, $\phi_a : F[x] \to F$ of Theorem 22.4, we find

$$0 = f(a) = 0q(a) + c,$$

so it must be that $c = 0$. Then $f(x) = (x - a)q(x)$, so $x - a$ is a factor of $f(x)$.

Conversely, if $x - a$ is a factor of $f(x)$ in $F[x]$, where $a \in F$, then applying our evaluation homomorpohism ϕ_a to $f(x) = (x - a)q(x)$, we have $f(a) = 0q(a) = 0$. ◆

23.4 Example Working again in $\mathbb{Z}_5[x]$, note that 1 is a zero of

$$(x^4 + 3x^3 + 2x + 4) \in \mathbb{Z}_5[x].$$

Thus by Corollary 23.3, we should be able to factor $x^4 + 3x^3 + 2x + 4$ into $(x - 1)q(x)$ in $\mathbb{Z}_5[x]$. Let us find the factorization by long division.

$$\begin{array}{r|l}
 & x^3 + 4x^2 + 4x + 1 \\
\hline
x - 1 & x^4 + 3x^3 + \phantom{4x^2 +{}} 2x + 4 \\
 & x^4 - x^3 \\
 & \phantom{x^4 +{}} 4x^3 \\
 & \phantom{x^4 +{}} 4x^3 - 4x^2 \\
 & \phantom{x^4 + 4x^3 +{}} 4x^2 + 2x \\
 & \phantom{x^4 + 4x^3 +{}} 4x^2 - 4x \\
 & \phantom{x^4 + 4x^3 + 4x^2 +{}} x + 4 \\
 & \phantom{x^4 + 4x^3 + 4x^2 +{}} x - 1 \\
 & \phantom{x^4 + 4x^3 + 4x^2 + x +{}} 0
\end{array}$$

Thus $x^4 + 3x^3 + 2x + 4 = (x - 1)(x^3 + 4x^2 + 4x + 1)$ in $\mathbb{Z}_5[x]$. Since 1 is seen to be a zero of $x^3 + 4x^2 + 4x + 1$ also, we can divide this polynomial by $x - 1$ and get

$$\begin{array}{r|l}
 & x^2 + 4 \\
\hline
x - 1 & x^3 + 4x^2 + 4x + 1 \\
 & x^3 - x^2 \\
 & \phantom{x^3 +{}} 0 \quad + 4x + 1 \\
 & \phantom{x^3 + 4x^2 +{}} 4x - 4 \\
 & \phantom{x^3 + 4x^2 + 4x +{}} 0
\end{array}$$

Since $x^2 + 4$ still has 1 as a zero, we can divide again by $x - 1$ and get

$$\begin{array}{r|l}
 & x + 1 \\
\hline
x - 1 & x^2 + 4 \\
 & x^2 - x \\
 & \phantom{x^2 +{}} x + 4 \\
 & \phantom{x^2 +{}} x - 1 \\
 & \phantom{x^2 + x +{}} 0
\end{array}$$

Thus $x^4 + 3x^3 + 2x + 4 = (x - 1)^3(x + 1)$ in $\mathbb{Z}_5[x]$. ▲

The next corollary should also look familiar.

23.5 Corollary A nonzero polynomial $f(x) \in F[x]$ of degree n can have at most n zeros in a field F.

Proof The preceding corollary shows that if $a_1 \in F$ is a zero of $f(x)$, then

$$f(x) = (x - a_1)q_1(x),$$

where, of course, the degree of $q_1(x)$ is $n - 1$. A zero $a_2 \in F$ of $q_1(x)$ then results in a factorization

$$f(x) = (x - a_1)(x - a_2)q_2(x).$$

Continuing this process, we arrive at

$$f(x) = (x - a_1)\cdots(x - a_r)q_r(x),$$

where $q_r(x)$ has no further zeros in F. Since the degree of $f(x)$ is n, at most n factors $(x - a_i)$ can appear on the right-hand side of the preceding equation, so $r \leq n$. Also, if $b \neq a_i$ for $i = 1, \cdots, r$ and $b \in F$, then

$$f(b) = (b - a_1)\cdots(b - a_r)q_r(b) \neq 0,$$

since F has no divisors of 0 and none of $b - a_i$ or $q_r(b)$ are 0 by construction. Hence the a_i for $i = 1, \cdots, r \leq n$ are all the zeros in F of $f(x)$. ◆

Our final corollary is concerned with the structure of the multiplicative group F^* of nonzero elements of a field F, rather than with factorization in $F[x]$. It may at first seem surprising that such a result follows from the division algorithm in $F[x]$, but recall that the result that a subgroup of a cyclic group is cyclic follows from the division algorithm in $\mathbb{Z}$.

23.6 Corollary If G is a finite subgroup of the multiplicative group $\langle F^*, \cdot\rangle$ of a field F, then G is cyclic. In particular, the multiplicative group of all nonzero elements of a finite field is cyclic.

Proof By Theorem 11.12 as a finite abelian group, G is isomorphic to a direct product $\mathbb{Z}_{d_1} \times \mathbb{Z}_{d_2} \times \cdots \times \mathbb{Z}_{d_r}$, where each d_i is a power of a prime. Let us think of each of the $\mathbb{Z}_{d_i}$ as a cyclic group of order d_i in *multiplicative* notation. Let m be the least common multiple of all the d_i for $i = 1, 2, \cdots, r$; note that $m \leq d_1 d_2 \cdots d_r$. If $a_i \in \mathbb{Z}_{d_i}$, then ${a_i}^{d_i} = 1$, so ${a_i}^m = 1$ since d_i divides m. Thus for all $\alpha \in G$, we have $\alpha^m = 1$, so every element of G is zero of $x^m - 1$. But G has $d_1 d_2 \cdots d_r$ elements, while $x^m - 1$ can have at most m zeros in the field F by Corollary 23.5, so $m \geq d_1 d_2 \cdots d_r$. Hence $m = d_1 d_2 \cdots d_r$, so the primes involved in the prime powers $d_1, d_2, \cdots, d_r$ are distinct, and the group G is isomorphic to the cyclic group $\mathbb{Z}_m$. ◆

Exercises 5 through 8 ask us to find all generators of the cyclic groups of units for some finite fields. The fact that the multiplicative group of units of a finite field is cyclic has been applied in algebraic coding.

Irreducible Polynomials

Our next definition singles out a type of polynomial in $F[x]$ that will be of utmost importance to us. The concept is probably already familiar. We really *are* doing high school algebra in a more general setting.

23.7 Definition A nonconstant polynomial $f(x) \in F[x]$ is **irreducible over** F or is an **irreducible polynomial in** $F[x]$ if $f(x)$ cannot be expressed as a product $g(x)h(x)$ of two polynomials $g(x)$ and $h(x)$ in $F[x]$ both of lower degree than the degree of $f(x)$. If $f(x) \in F[x]$ is a nonconstant polynomial that is not irreducible over F, then $f(x)$ is **reducible over** F. ■

Note that the preceding definition concerns the concept *irreducible over F* and not just the concept *irreducible*. A polynomial $f(x)$ may be irreducible over F, but may not be irreducible if viewed over a larger field E containing F. We illustrate this.

23.8 Example Theorem 22.11 shows that $x^2 - 2$ viewed in $\mathbb{Q}[x]$ has no zeros in $\mathbb{Q}$. This shows that $x^2 - 2$ is irreducible over $\mathbb{Q}$, for a factorization $x^2 - 2 = (ax + b)(cx + d)$ for $a, b, c, d \in \mathbb{Q}$ would give rise to zeros of $x^2 - 2$ in $\mathbb{Q}$. However, $x^2 - 2$ viewed in $\mathbb{R}[x]$ is not irreducible over $\mathbb{R}$, because $x^2 - 2$ factors in $\mathbb{R}[x]$ into $(x - \sqrt{2})(x + \sqrt{2})$. ▲

It is worthwhile to remember that *the units in $F[x]$ are precisely the nonzero elements of F*. Thus we could have defined an irreducible polynomial $f(x)$ as a nonconstant polynomial such that in any factorization $f(x) = g(x)h(x)$ in $F[x]$, either $g(x)$ or $h(x)$ is a unit.

23.9 Example Let us show that $f(x) = x^3 + 3x + 2$ viewed in $\mathbb{Z}_5[x]$ is irreducible over $\mathbb{Z}_5$. If $x^3 + 3x + 2$ factored in $\mathbb{Z}_5[x]$ into polynomials of lower degree then there would exist at least one linear factor of $f(x)$ of the form $x - a$ for some $a \in \mathbb{Z}_5$. But then $f(a)$ would be 0, by Corollary 23.3. However, $f(0) = 2$, $f(1) = 1$, $f(-1) = -2$, $f(2) = 1$, and $f(-2) = -2$, showing that $f(x)$ has no zeros in $\mathbb{Z}_5$. Thus $f(x)$ is irreducible over $\mathbb{Z}_5$. This test for irreducibility by finding zeros works nicely for quadratic and cubic polynomials over a finite field with a small number of elements. ▲

Irreducible polynomials will play a very important role in our work from now on. The problem of determining whether a given $f(x) \in F[x]$ is irreducible over F may be difficult. We now give some criteria for irreducibility that are useful in certain cases. One technique for determining irreducibility of quadratic and cubic polynomials was illustrated in Examples 23.8 and 23.9. We formalize it in a theorem.

23.10 Theorem Let $f(x) \in F[x]$, and let $f(x)$ be of degree 2 or 3. Then $f(x)$ is reducible over F if and only if it has a zero in F.

Proof If $f(x)$ is reducible so that $f(x) = g(x)h(x)$, where the degree of $g(x)$ and the degree of $h(x)$ are both less than the degree of $f(x)$, then since $f(x)$ is either quadratic or cubic, either $g(x)$ or $h(x)$ is of degree 1. If, say, $g(x)$ is of degree 1, then except for a possible factor in F, $g(x)$ is of the form $x - a$. Then $g(a) = 0$, which implies that $f(a) = 0$, so $f(x)$ has a zero in F.

Conversely, Corollary 23.3 shows that if $f(a) = 0$ for $a \in F$, then $x - a$ is a factor of $f(x)$, so $f(x)$ is reducible. ◆

We turn to some conditions for irreducibility over $\mathbb{Q}$ of polynomials in $\mathbb{Q}[x]$. The most important condition that we shall give is contained in the next theorem. We shall not prove this theorem here; it involves clearing denominators and gets a bit messy.

23.11 Theorem If $f(x) \in \mathbb{Z}[x]$, then $f(x)$ factors into a product of two polynomials of lower degrees r and s in $\mathbb{Q}[x]$ if and only if it has such a factorization with polynomials of the same degrees r and s in $\mathbb{Z}[x]$.

Proof The proof is omitted here. ◆

23.12 Corollary If $f(x) = x^n + a_{n-1}x^{n-1} + \cdots + a_0$ is in $\mathbb{Z}[x]$ with $a_0 \neq 0$, and if $f(x)$ has a zero in $\mathbb{Q}$, then it has a zero m in $\mathbb{Z}$, and m must divide a_0.

Proof If $f(x)$ has a zero a in $\mathbb{Q}$, then $f(x)$ has a linear factor $x - a$ in $\mathbb{Q}[x]$ by Corollary 23.3. But then by Theorem 23.11, $f(x)$ has a factorization with a linear factor in $\mathbb{Z}[x]$, so for some $m \in \mathbb{Z}$ we must have

$$f(x) = (x - m)\big(x^{n-1} + \cdots - a_0/m\big).$$

Thus a_0/m is in $\mathbb{Z}$, so m divides a_0. ◆

23.13 Example Corollary 23.12 gives us another proof of the irreducibility of $x^2 - 2$ over $\mathbb{Q}$, for $x^2 - 2$ factors nontrivially in $\mathbb{Q}[x]$ if and only if it has a zero in $\mathbb{Q}$ by Theorem 23.10. By Corollary 23.12, it has a zero in $\mathbb{Q}$ if and only if it has a zero in $\mathbb{Z}$, and moreover the only possibilities are the divisors ± 1 and ± 2 of 2. A check shows that none of these numbers is a zero of $x^2 - 2$. ▲

23.14 Example Let us use Theorem 23.11 to show that

$$f(x) = x^4 - 2x^2 + 8x + 1$$

viewed in $\mathbb{Q}[x]$ is irreducible over $\mathbb{Q}$. If $f(x)$ has a linear factor in $\mathbb{Q}[x]$, then it has a zero in $\mathbb{Z}$, and by Corollary 23.12, this zero would have to be a divisor in $\mathbb{Z}$ of 1, that is, either ± 1. But $f(1) = 8$, and $f(-1) = -8$, so such a factorization is impossible.

If $f(x)$ factors into two quadratic factors in $\mathbb{Q}[x]$, then by Theorem 23.11, it has a factorization.

$$(x^2 + ax + b)(x^2 + cx + d)$$

in $\mathbb{Z}[x]$. Equating coefficients of powers of x, we find that we must have

$$bd = 1, \quad ad + bc = 8, \quad ac + b + d = -2, \quad \text{and} \quad a + c = 0$$

for integers $a, b, c, d \in \mathbb{Z}$. From $bd = 1$, we see that either $b = d = 1$ or $b = d = -1$. In any case, $b = d$ and from $ad + bc = 8$, we deduce that $d(a + c) = 8$. But this is impossible since $a + c = 0$. Thus a factorization into two quadratic polynomials is also impossible and $f(x)$ is irreducible over $\mathbb{Q}$. ▲

We conclude our irreducibility criteria with the famous Eisenstein criterion for irreducibility. An additional very useful criterion is given in Exercise 37.

23.15 Theorem **(Eisenstein Criterion)** Let $p \in \mathbb{Z}$ be a prime. Suppose that $f(x) = a_n x^n + \cdots + a_0$ is in $\mathbb{Z}[x]$, and $a_n \not\equiv 0 \pmod{p}$, but $a_i \equiv 0 \pmod{p}$ for all $i < n$, with $a_0 \not\equiv 0 \pmod{p^2}$. Then $f(x)$ is irreducible over $\mathbb{Q}$.

Proof By Theorem 23.11 we need only show that $f(x)$ does not factor into polynomials of lower degree in $\mathbb{Z}[x]$. If

$$f(x) = (b_r x^r + \cdots + b_0)(c_s x^s + \cdots + c_0)$$

is a factorization in $\mathbb{Z}[x]$, with $b_r \neq 0$, $c_s \neq 0$ and $r, s < n$, then $a_0 \not\equiv 0 \pmod{p^2}$ implies that b_0 and c_0 are not both congruent to 0 modulo p. Suppose that $b_0 \not\equiv 0 \pmod{p}$ and $c_0 \equiv 0 \pmod{p}$. Now $a_n \not\equiv 0 \pmod{p}$ implies that $b_r, c_s \not\equiv 0 \pmod{p}$, since $a_n = b_r c_s$. Let m be the smallest value of k such that $c_k \not\equiv 0 \pmod{p}$. Then

$$a_m = b_0 c_m + b_1 c_{m-1} + \cdots + \begin{cases} b_m c_0 \text{ if } r \geq m, \\ b_r c_{m-r} \text{ if } r < m. \end{cases}$$

The fact that neither b_0 nor c_m are congruent to 0 modulo p while $c_{m-1}, \cdots, c_0$ are all congruent to 0 modulo p implies that $a_m \not\equiv 0$ modulo p, so $m = n$. Consequently, $s = n$, contradicting our assumption that $s < n$; that is, that our factorization was nontrivial. ◆

Note that if we take $p = 2$, the Eisenstein criterion gives us still another proof of the irreducibility of $x^2 - 2$ over $\mathbb{Q}$.

23.16 Example Taking $p = 3$, we see by Theorem 23.15 that

$$25x^5 - 9x^4 - 3x^2 - 12$$

is irreducible over $\mathbb{Q}$. ▲

23.17 Corollary The polynomial

$$\Phi_p(x) = \frac{x^p - 1}{x - 1} = x^{p-1} + x^{p-2} + \cdots + x + 1$$

is irreducible over $\mathbb{Q}$ for any prime p.

Proof Again by Theorem 23.11, we need only consider factorizations in $\mathbb{Z}[x]$. We remarked following Theorem 22.5 that its proof actually shows that evaluation homomorphims can be used for commutative rings. Here we want to use the evaluation homomorphism $\phi_{x+1} : \mathbb{Q}[x] \to \mathbb{Q}[x]$. It is natural for us to denote $\phi_{x+1}(f(x))$ by $f(x+1)$ for $f(x) \in \mathbb{Q}[x]$. Let

$$g(x) = \Phi_p(x+1) = \frac{(x+1)^p - 1}{(x+1) - 1} = \frac{x^p + \binom{p}{1} x^{p-1} + \cdots + px}{x}.$$

The coefficient of x^{p-r} for $0 < r < p$ is the binomial coefficient $p!/[r!(p-r)!]$ which is divisible by p because p divides $p!$ but does not divide either $r!$ or $(p-r)!$ when $0 < r < p$. Thus

$$g(x) = x^{p-1} + \binom{p}{1} x^{p-2} + \cdots + p$$

satisifies the Eisenstein criterion for the prime p and is thus irreducible over $\mathbb{Q}$. But if $\Phi_p(x) = h(x)r(x)$ were a nontrivial factorization of $\Phi_p(x)$ in $\mathbb{Z}[x]$, then

$$\Phi_p(x+1) = g(x) = h(x+1)r(x+1)$$

would give a nontrivial factorization of $g(x)$ in $\mathbb{Z}[x]$. Thus $\Phi_p(x)$ must also be irreducible over $\mathbb{Q}$. ◆

The polynomial $\Phi_p(x)$ in Corollary 23.17 is the pth **cyclotomic polynomial.**

Uniqueness of Factorization in *F*[*x*]

Polynomials in $F[x]$ can be factored into a product of irreducible polynomials in $F[x]$ in an essentially unique way. For $f(x), g(x) \in F[x]$ we say that $g(x)$ **divides** $f(x)$ **in** $F[x]$ if there exists $q(x) \in F[x]$ such that $f(x) = g(x)q(x)$. Note the similarity of the theorem that follows with boxed Property (1) for $\mathbb{Z}$ following Example 6.9.

23.18 Theorem Let $p(x)$ be an irreducible polynomial in $F[x]$. If $p(x)$ divides $r(x)s(x)$ for $r(x), s(x) \in F[x]$, then either $p(x)$ divides $r(x)$ or $p(x)$ divides $s(x)$.

Proof We delay the proof of this theorem to Section 27. (See Theorem 27.27.) ◆

23.19 Corollary If $p(x)$ is irreducible in $F[x]$ and $p(x)$ divides the product $r_1(x)\cdots r_n(x)$ for $r_i(x) \in F[x]$, then $p(x)$ divides $r_i(x)$ for at least one i.

Proof Using mathematical induction, we find that this is immediate from Theorem 23.18. ◆

23.20 Theorem If F is a field, then every nonconstant polynomial $f(x) \in F[x]$ can be factored in $F[x]$ into a product of irreducible polynomials, the irreducible polynomials being unique except for order and for unit (that is, nonzero constant) factors in F.

Proof Let $f(x) \in F[x]$ be a nonconstant polynomial. If $f(x)$ is not irreducible, then $f(x) = g(x)h(x)$, with the degree of $g(x)$ and the degree of $h(x)$ both less than the degree of $f(x)$. If $g(x)$ and $h(x)$ are both irreducible, we stop here. If not, at least one of them factors into polynomials of lower degree. Continuing this process, we arrive at a factorization

$$f(x) = p_1(x)p_2(x)\cdots p_r(x),$$

where $p_i(x)$ is irreducible for $i = 1, 2, \cdots, r$.

It remains for us to show uniqueness. Suppose that

$$f(x) = p_1(x)p_2(x)\cdots p_r(x) = q_1(x)q_2(x)\cdots q_s(x)$$

are two factorizations of $f(x)$ into irreducible polynomials. Then by Corollary 23.19, $p_1(x)$ divides some $q_j(x)$, let us assume $q_1(x)$. Since $q_1(x)$ is irreducible,

$$q_1(x) = u_1 p_1(x),$$

where $u_1 \neq 0$, but u_1 is in F and thus is a unit. Then substituting $u_1 p_1(x)$ for $q_1(x)$ and canceling, we get

$$p_2(x)\cdots p_r(x) = u_1 q_2(x)\cdots q_s(x).$$

By a similar argument, say $q_2(x) = u_2 p_2(x)$, so

$$p_3(x) \cdots p_r(x) = u_1 u_2 q_3(x) \cdots q_s(x).$$

Continuing in this manner, we eventually arrive at

$$1 = u_1 u_2 \cdots u_r q_{r+1}(x) \cdots q_s(x).$$

This is only possible if $s = r$, so that this equation is actually $1 = u_1 u_2 \cdots u_r$. Thus the irreducible factors $p_i(x)$ and $q_j(x)$ were the same except possibly for order and unit factors. ◆

23.21 Example Example 23.4 shows a factorization of $x^4 + 3x^3 + 2x + 4$ in $\mathbb{Z}_5[x]$ is $(x - 1)^3(x + 1)$. These irreducible factors in $\mathbb{Z}_5[x]$ are only unique up to units in $\mathbb{Z}_5[x]$, that is, nonzero constants in $\mathbb{Z}_5$. For example, $(x - 1)^3(x + 1) = (x - 1)^2(2x - 2)(3x + 3)$. ▲

EXERCISES 23

Computations

In Exercises 1 through 4, find $q(x)$ and $r(x)$ as described by the division algorithm so that $f(x) = g(x)q(x) + r(x)$ with $r(x) = 0$ or of degree less than the degree of $g(x)$.

1. $f(x) = x^6 + 3x^5 + 4x^2 - 3x + 2$ and $g(x) = x^2 + 2x - 3$ in $\mathbb{Z}_7[x]$.
2. $f(x) = x^6 + 3x^5 + 4x^2 - 3x + 2$ and $g(x) = 3x^2 + 2x - 3$ in $\mathbb{Z}_7[x]$.
3. $f(x) = x^5 - 2x^4 + 3x - 5$ and $g(x) = 2x + 1$ in $\mathbb{Z}_{11}[x]$.
4. $f(x) = x^4 + 5x^3 - 3x^2$ and $g(x) = 5x^2 - x + 2$ in $\mathbb{Z}_{11}[x]$.

In Exercises 5 through 8, find all generators of the cyclic multiplicative group of units of the given finite field. (Review Corollary 6.16.)

5. $\mathbb{Z}_5$ 6. $\mathbb{Z}_7$ 7. $\mathbb{Z}_{17}$ 8. $\mathbb{Z}_{23}$

9. The polynomial $x^4 + 4$ can be factored into linear factors in $\mathbb{Z}_5[x]$. Find this factorization.
10. The polynomial $x^3 + 2x^2 + 2x + 1$ can be factored into linear factors in $\mathbb{Z}_7[x]$. Find this factorization.
11. The polynomial $2x^3 + 3x^2 - 7x - 5$ can be factored into linear factors in $\mathbb{Z}_{11}[x]$. Find this factorization.
12. Is $x^3 + 2x + 3$ an irreducible polynomial of $\mathbb{Z}_5[x]$? Why? Express it as a product of irreducible polynomials of $\mathbb{Z}_5[x]$.
13. Is $2x^3 + x^2 + 2x + 2$ an irreducible polynomial in $\mathbb{Z}_5[x]$? Why? Express it as a product of irreducible polynomials in $\mathbb{Z}_5[x]$.
14. Show that $f(x) = x^2 + 8x - 2$ is irreducible over $\mathbb{Q}$. Is $f(x)$ irreducible over $\mathbb{R}$? Over $\mathbb{C}$?
15. Repeat Exercise 14 with $g(x) = x^2 + 6x + 12$ in place of $f(x)$.
16. Demonstrate that $x^3 + 3x^2 - 8$ is irreducible over $\mathbb{Q}$.
17. Demonstrate that $x^4 - 22x^2 + 1$ is irreducible over $\mathbb{Q}$.

In Exercises 18 through 21, determine whether the polynomial in $\mathbb{Z}[x]$ satisfies an Eisenstein criterion for irreducibility over $\mathbb{Q}$.

18. $x^2 - 12$

19. $8x^3 + 6x^2 - 9x + 24$

20. $4x^{10} - 9x^3 + 24x - 18$

21. $2x^{10} - 25x^3 + 10x^2 - 30$

22. Find all zeros of $6x^4 + 17x^3 + 7x^2 + x - 10$ in $\mathbb{Q}$. (This is a tedious high school algebra problem. *You* might use a bit of analytic geometry and calculus and make a graph, or use Newton's method to see which are the best candidates for zeros.)

Concepts

In Exercises 23 and 24, correct the definition of the italicized term without reference to the text, if correction is needed, so that it is in a form acceptable for publication.

23. A polynomial $f(x) \in F[x]$ is *irreducible over the field F* if and only if $f(x) \neq g(x)h(x)$ for any polynomials $g(x), h(x) \in F[x]$.

24. A nonconstant polynomial $f(x) \in F[x]$ is *irreducible over the field F* if and only if in any factorization of it in $F[x]$, one of the factors is in F.

25. Mark each of the following true or false.

______ **a.** $x - 2$ is irreducible over $\mathbb{Q}$.

______ **b.** $3x - 6$ is irreducible over $\mathbb{Q}$.

______ **c.** $x^2 - 3$ is irreducible over $\mathbb{Q}$.

______ **d.** $x^2 + 3$ is irreducible over $\mathbb{Z}_7$.

______ **e.** If F is a field, the units of $F[x]$ are precisely the nonzero elements of F.

______ **f.** If F is a field, the units of $F(x)$ are precisely the nonzero elements of F.

______ **g.** A polynomial $f(x)$ of degree n with coefficients in a field F can have at most n zeros in F.

______ **h.** A polynomial $f(x)$ of degree n with coefficients in a field F can have at most n zeros in any given field E such that $F \leq E$.

______ **i.** Every polynomial of degree 1 in $F[x]$ has at least one zero in the field F.

______ **j.** Each polynomial in $F[x]$ can have at most a finite number of zeros in the field F.

26. Find all prime numbers p such that $x + 2$ is a factor of $x^4 + x^3 + x^2 - x + 1$ in $\mathbb{Z}_p[x]$.

In Exercises 27 through 30, find all irreducible polynomials of the indicated degree in the given ring.

27. Degree 2 in $\mathbb{Z}_2[x]$

28. Degree 3 in $\mathbb{Z}_2[x]$

29. Degree 2 in $\mathbb{Z}_3[x]$

30. Degree 3 in $\mathbb{Z}_3[x]$

31. Find the number of irreducible quadratic polynomials in $\mathbb{Z}_p[x]$, where p is a prime. [*Hint:* Find the number of reducible polynomials of the form $x^2 + ax + b$, then the number of reducible quadratics, and subtract this from the total number of quadratics.]

Proof Synopsis

32. Give a synopsis of the proof of Corollary 23.5.

33. Give a synopsis of the proof of Corollary 23.6.

Theory

34. Show that for p a prime, the polynomial $x^p + a$ in $\mathbb{Z}_p[x]$ is not irreducible for any $a \in \mathbb{Z}_p$.

35. If F is a field and $a \neq 0$ is a zero of $f(x) = a_0 + a_1x + \cdots + a_nx^n$ in $F[x]$, show that $1/a$ is a zero of $a_n + a_{n-1}x + \cdots + a_0x^n$.

36. (Remainder Theorem) Let $f(x) \in F[x]$ where F is a field, and let $\alpha \in F$. Show that the remainder $r(x)$ when $f(x)$ is divided by $x - \alpha$, in accordance with the division algorithm, is $f(\alpha)$.

37. Let $\sigma_m : \mathbb{Z} \to \mathbb{Z}_m$ be the natural homomorphism given by $\sigma_m(a) =$ (the remainder of a when divided by m) for $a \in \mathbb{Z}$.

a. Show that $\overline{\sigma_m} : \mathbb{Z}[x] \to \mathbb{Z}_m[x]$ given by

$$\overline{\sigma_m}(a_0 + a_1x + \cdots + a_nx^n) = \sigma_m(a_0) + \sigma_m(a_1)x + \cdots + \sigma_m(a_n)x^n$$

is a homomorphism of $\mathbb{Z}[x]$ onto $\mathbb{Z}_m[x]$.

b. Show that if $f(x) \in \mathbb{Z}[x]$ and $\overline{\sigma_m}(f(x))$ both have degree n and $\overline{\sigma_m}(f(x))$ does not factor in $\mathbb{Z}_m[x]$ into two polynomials of degree less than n, then $f(x)$ is irreducible in $\mathbb{Q}[x]$.

c. Use part (b) to show that $x^3 + 17x + 36$ is irreducible in $\mathbb{Q}[x]$. [*Hint:* Try a prime value of m that simplifies the coefficients.]

SECTION 24 †Noncommutative Examples

Thus far, the only example we have presented of a ring that is not commutative is the ring $M_n(F)$ of all $n \times n$ matrices with entries in a field F. We shall be doing almost nothing with noncommutative rings and strictly skew fields. To show that there are other important noncommutative rings occurring very naturally in algebra, we give several examples of such rings.

Rings of Endomorphisms

Let A be any abelian group. A homomorphism of A into itself is an **endomorphism of** A. Let the set of all endomorphisms of A be End(A). Since the composition of two homomorphisms of A into itself is again such a homomorphism, we define multiplication on End(A) by function composition, and thus multiplication is associative.

To define addition, for $\phi, \psi \in \text{End}(A)$, we have to describe the value of $(\phi + \psi)$ on each $a \in A$. Define

$$(\phi + \psi)(a) = \phi(a) + \psi(a).$$

Since

$$\begin{aligned}(\phi + \psi)(a + b) &= \phi(a + b) + \psi(a + b)\\ &= [\phi(a) + \phi(b)] + [\psi(a) + \psi(b)]\\ &= [\phi(a) + \psi(a)] + [\phi(b) + \psi(b)]\\ &= (\phi + \psi)(a) + (\phi + \psi)(b)\end{aligned}$$

we see that $\phi + \psi$ is again in End(A).

Since A is commutative, we have

$$(\phi + \psi)(a) = \phi(a) + \psi(a) = \psi(a) + \phi(a) = (\psi + \phi)(a)$$

for all $a \in A$, so $\phi + \psi = \psi + \phi$ and addition in End(A) is commutative. The associativity of addition follows from

† This section is not used in the remainder of the text.

$$\begin{aligned}[\phi + (\psi + \theta)](a) &= \phi(a) + [(\psi + \theta)(a)] \\ &= \phi(a) + [\psi(a) + \theta(a)] \\ &= [\phi(a) + \psi(a)] + \theta(a) \\ &= (\phi + \psi)(a) + \theta(a) \\ &= [(\phi + \psi) + \theta](a).\end{aligned}$$

If e is the additive identity of A, then the homomorphism 0 defined by

$$0(a) = e$$

for $a \in A$ is an additive identity in End(A). Finally, for

$$\phi \in \text{End}(A),$$

$-\phi$ defined by

$$(-\phi)(a) = -\phi(a)$$

is in End(A), since

$$\begin{aligned}(-\phi)(a + b) &= -\phi(a + b) = -[\phi(a) + \phi(b)] \\ &= -\phi(a) - \phi(b) = (-\phi)(a) + (-\phi)(b),\end{aligned}$$

and $\phi + (-\phi) = 0$. Thus $\langle \text{End}(A), + \rangle$ is an abelian group.

Note that we have not yet used the fact that our functions are *homomorphisms* except to show that $\phi + \psi$ and $-\phi$ are again *homomorphisms*. Thus the set A^A of *all functions* from A into A is an abelian group under exactly the same definition of addition, and, of course, function composition again gives a nice associative multiplication in A^A. However, we do need the fact that these functions in End(A) are homomorphisms now to prove the left distributive law in End(A). Except for this left distributive law, $\langle A^A, +, \cdot \rangle$ satisfies all the axioms for a ring. Let ϕ, ψ, and θ be in End(A), and let $a \in A$. Then

$$(\theta(\phi + \psi))(a) = \theta((\phi + \psi)(a)) = \theta(\phi(a) + \psi(a)).$$

Since θ is a *homomorphism,*

$$\begin{aligned}\theta(\phi(a) + \psi(a)) &= \theta(\phi(a)) + \theta(\psi(a)) \\ &= (\theta\phi)(a) + (\theta\psi)(a) \\ &= (\theta\phi + \theta\psi)(a).\end{aligned}$$

Thus $\theta(\phi + \psi) = \theta\phi + \theta\psi$. The right distributive law causes no trouble, even in A^A, and follows from

$$\begin{aligned}((\psi + \theta)\phi)(a) &= (\psi + \theta)(\phi(a)) = \psi(\phi(a)) + \theta(\phi(a)) \\ &= (\psi\phi)(a) + (\theta\phi)(a) = (\psi\phi + \theta\phi)(a).\end{aligned}$$

Thus we have proved the following theorem.

24.1 Theorem The set End(A) of all endomorphisms of an abelian group A forms a ring under homomorphism addition and homomorphism multiplication (function composition).

Again, to show relevance to this section, we should give an example showing that End(A) need not be commutative. Since function composition is in general not commutative, this seems reasonable to expect. However, End(A) may be commutative in some cases. Indeed, Exercise 15 asks us to show that End($\langle \mathbb{Z}, + \rangle$) is commutative.

24.2 Example Consider the abelian group $\langle \mathbb{Z} \times \mathbb{Z}, +\rangle$ discussed in Section 11. It is straightforward to verify that two elements of $\text{End}(\langle \mathbb{Z} \times \mathbb{Z}, +\rangle)$ are ϕ and ψ defined by

$$\phi((m,n)) = (m+n, 0) \qquad \text{and} \qquad \psi((m,n)) = (0,n).$$

Note that ϕ maps everything onto the first factor of $\mathbb{Z} \times \mathbb{Z}$, and ψ collapses the first factor. Thus

$$(\psi\phi)(m,n) = \psi(m+n, 0) = (0,0).$$

while

$$(\phi\psi)(m,n) = \phi(0,n) = (n,0).$$

Hence $\phi\psi \neq \psi\phi$. ▲

24.3 Example Let F be a field of characteristic zero, and let $\langle F[x], +\rangle$ be the additive group of the ring $F[x]$ of polynomials with coefficients in F. For this example, let us denote this additive group by $F[x]$, to simplify this notation. We can consider $\text{End}(F[x])$. One element of $\text{End}(F[x])$ acts on each polynomial in $F[x]$ by multiplying it by x. Let this endomorphism be X, so

$$X\left(a_0 + a_1x + a_2x^2 + \cdots + a_nx^n\right) = a_0x + a_1x^2 + a_2x^3 + \cdots + a_nx^{n+1}.$$

Another element of $\text{End}(F[x])$ is formal differentiation with respect to x. (The familiar formula "the derivation of a sum is the sum of the derivatives" guarantees that differentiation is an endomorphism of $F[x]$.) Let Y be this endomorphism, so

$$Y\left(a_0 + a_1x + a_2x^2 + \cdots + a_nx^n\right) = a_1 + 2a_2x + \cdots + na_nx^{n-1}.$$

Exercise 17 asks us to show that $YX - XY = 1$, where 1 is unity (the identity map) in $\text{End}(F[x])$. Thus $XY \neq YX$. Multiplication of polynomials in $F[x]$ by any element of F also gives an element of End $(F[x])$. The subring of $\text{End}(F[x])$ generated by X and Y and multiplications by elements of F is the **Weyl algebra** and is important in quantum mechanics. ▲

Group Rings and Group Algebras

Let $G = \{g_i \mid i \in I\}$ be any group written multiplicatively and let R be any commutative ring with nonzero unity. Let RG be the set of all *formal sums.*

$$\sum_{i \in I} a_i g_i$$

for $a_i \in R$ and $g_i \in G$, *where all but a finite number of the a_i are 0.* Define the sum of two elements of RG by

$$\left(\sum_{i\in I} a_i g_i\right) + \left(\sum_{i\in I} b_i g_i\right) = \sum_{i\in I}(a_i + b_i)g_i.$$

Observe that $(a_i + b_i) = 0$ except for a finite number of indices i, so $\Sigma_{i\in I}(a_i + b_i)g_i$ is again in RG. It is immediate that $\langle RG, +\rangle$ is an abelian group with additive identity $\Sigma_{i\in I}0g_i$.

Multiplication of two elements of RG is defined by the use of the multiplications in G and R as follows:

$$\left(\sum_{i\in I} a_i g_i\right)\left(\sum_{i\in I} b_i g_i\right) = \sum_{i\in I}\left(\sum_{g_j g_k = g_i} a_j b_k\right) g_i.$$

Naively, we formally distribute the sum $\Sigma_{i\in I} a_i g_i$ over the sum $\Sigma_{i\in I} b_i g_i$ and rename a term $a_j g_j b_k g_k$ by $a_j b_k g_i$ where $g_j g_k = g_i$ in G. Since a_i and b_i are 0 for all but a finite number of i, the sum $\Sigma_{g_j g_k = g_i} a_j b_k$ contains only a finite number of nonzero summands $a_j b_k \in R$ and may thus be viewed as an element of R. Again, at most a finite number of such sums $\Sigma_{g_j g_k = g_i} a_j b_k$ are nonzero. Thus multiplication is closed on RG.

The distributive laws follow at once from the definition of addition and the formal way we used distributivity to define multiplication. For the associativity of multiplication

$$\begin{aligned}\left(\sum_{i\in I} a_i g_i\right)\left[\left(\sum_{i\in I} b_i g_i\right)\left(\sum_{i\in I} c_i g_i\right)\right] &= \left(\sum_{i\in I} a_i g_i\right)\left[\sum_{i\in I}\left(\sum_{g_j g_k = g_i} b_j c_k\right) g_i\right] \\ &= \sum_{i\in I}\left(\sum_{g_h g_j g_k = g_i} a_h b_j c_k\right) g_i \\ &= \left[\sum_{i\in I}\left(\sum_{g_h g_j = g_i} a_h b_j\right) g_i\right]\left(\sum_{i\in I} c_i g_i\right) \\ &= \left[\left(\sum_{i\in I} a_i g_i\right)\left(\sum_{i\in I} b_i g_i\right)\right]\left(\sum_{i\in I} c_i g_i\right).\end{aligned}$$

Thus we have proved the following theorem.

24.4 Theorem If G is any group written multiplicatively and R is a commutative ring with nonzero unity, then $\langle RG, +, \cdot\rangle$ is a ring.

Corresponding to each $g \in G$, we have an element $1g$ in RG. If we identify (rename) $1g$ with g, we see that $\langle RG, \cdot\rangle$ can be considered to contain G naturally as a multiplicative subsystem. Thus, if G is not abelian, RG is not a commutative ring.

24.5 Definition The ring RG defined above is the **group ring of G over R**. If F is a field, then FG is the **group algebra of G over F**. ■

24.6 Example Let us give the addition and multiplication tables for the group algebra $\mathbb{Z}_2 G$, where $G = \{e, a\}$ is cyclic of order 2. The elements of $Z_2 G$ are

$$0e + 0a, \quad 0e + 1a, \quad 1e + 0a, \quad \text{and} \quad 1e + 1a.$$

If we denote these elements in the obvious, natural way by

$$0, \quad a, \quad e, \quad \text{and} \quad e + a,$$

24.7 Table

$+$	0	a	e	$e+a$
0	0	a	e	$e+a$
a	a	0	$e+a$	e
e	e	$e+a$	0	a
$e+a$	$e+a$	e	a	0

24.8 Table

	0	a	e	$e+a$
0	0	0	0	0
a	0	e	a	$e+a$
e	0	a	e	$e+a$
$e+a$	0	$e+a$	$e+a$	0

respectively, we get Tables 24.7 and 24.8. For example, to see that $(e+a)(e+a) = 0$, we have

$$(1e + 1a)(1e + 1a) = (1+1)e + (1+1)a = 0e + 0a.$$

This example shows that a group algebra may have 0 divisors. Indeed, this is usually the case. ▲

The Quaternions

We have not yet given an example of a noncommutative division ring. The *quaternions* of Hamilton are the standard example of a strictly skew field; let us describe them.

■ Historical Note

Sir William Rowan Hamilton (1805–1865) discovered quaternions in 1843 while he was searching for a way to multiply number triplets (vectors in $\mathbb{R}^3$). Six years earlier he had developed the complex numbers abstractly as pairs (a, b) of real numbers with addition $(a, b) + (a' + b') = (a + a', b + b')$ and multiplication $(a, b)(a'b') = (aa' - bb', ab' + a'b)$; he was then looking for an analogous multiplication for 3-vectors that was distributive and such that the length of the product vector was the product of the lengths of the factors. After many unsuccessful attempts to multiply vectors of the form $a + bi + cj$ (where $1, i, j$ are mutually perpendicular), he realized while walking along the Royal Canal in Dublin on October 16, 1843, that he needed a new "imaginary symbol" k to be perpendicular to the other three elements. He could not "resist the impulse . . . to cut with a knife on a stone of Brougham Bridge" the fundamental defining formulas on page 225 for multiplying these quaternions.

The quaternions were the first known example of a strictly skew field. Though many others were subsequently discovered, it was eventually noted that none were finite. In 1909 Joseph Henry Maclagan Wedderburn (1882–1948), then a preceptor at Princeton University, gave the first proof of Theorem 24.10.

Let the set $\mathbb{H}$, for Hamilton, be $\mathbb{R} \times \mathbb{R} \times \mathbb{R} \times \mathbb{R}$. Now $\langle \mathbb{R} \times \mathbb{R} \times \mathbb{R} \times \mathbb{R}, +\rangle$ is a group under addition by components, the direct product of $\mathbb{R}$ under addition with itself four times. This gives the operation of addition on $\mathbb{H}$. Let us rename certain elements of $\mathbb{H}$. We shall let

$$1 = (1, 0, 0, 0), \quad i = (0, 1, 0, 0),$$
$$j = (0, 0, 1, 0), \quad \text{and} \quad k = (0, 0, 0, 1).$$

We furthermore agree to let

$$a_1 = (a_1, 0, 0, 0), \quad a_2 i = (0, a_2, 0, 0),$$
$$a_3 j = (0, 0, a_3, 0) \quad \text{and} \quad a_4 k = (0, 0, 0, a_4).$$

In view of our definition of addition, we then have

$$(a_1, a_2, a_3, a_4) = a_1 + a_2 i + a_3 j + a_4 k.$$

Thus

$$\begin{aligned}(a_1 + a_2 i + a_3 j + a_4 k) &+ (b_1 + b_2 i + b_3 j + b_4 k)\\ &= (a_1 + b_1) + (a_2 + b_2)i + (a_3 + b_3)j + (a_4 + b_4)k.\end{aligned}$$

To define multiplication on $\mathbb{H}$, we start by defining

$$1a = a1 = a \quad \text{for} \quad a \in \mathbb{H},$$
$$i^2 = j^2 = k^2 = -1,$$

and

$$ij = k, \quad jk = i, \quad ki = j, \quad ji = -k, \quad kj = -i, \quad \text{and} \quad ik = -j.$$

Note the similarity with the so-called cross product of vectors. These formulas are easy to remember if we think of the sequence

$$i, j, k, i, j, k.$$

The product from left to right of two adjacent elements is the next one to the right. The product from right to left of two adjacent elements is the negative of the next one to the left. We then define a product to be what it must be to make the distributive laws hold, namely,

$$\begin{aligned}(a_1 + a_2 i &+ a_3 j + a_4 k)(b_1 + b_2 i + b_3 j + b_4 k)\\ &= (a_1 b_1 - a_2 b_2 - a_3 b_3 - a_4 b_4) + (a_1 b_2 + a_2 b_1 + a_3 b_4 - a_4 b_3)i\\ &\quad + (a_1 b_3 - a_2 b_4 + a_3 b_1 + a_4 b_2)j\\ &\quad + (a_1 b_4 + a_2 b_3 - a_3 b_2 + a_4 b_1)k.\end{aligned}$$

Exercise 19 shows that the quaternions are isomorphic to a subring of $M_2(\mathbb{C})$, so multiplication is associative. Since $ij = k$ and $ji = -k$, we see that multiplication is not commutative, so $\mathbb{H}$ is definitely not a field. Turning to the existence of multiplicative inverses, let $a = a_1 + a_2 i + a_3 j + a_4 k$, with not all $a_i = 0$. Computation shows that

$$(a_1 + a_2 i + a_3 j + a_4 k)(a_1 - a_2 i - a_3 j - a_4 k) = a_1{}^2 + a_2{}^2 + a_3{}^2 + a_4{}^2.$$

If we let

$$|a|^2 = a_1{}^2 + a_2{}^2 + a_3{}^2 + a_4{}^2 \qquad \text{and} \qquad \bar{a} = a_1 - a_2 i - a_3 j - a_4 k,$$

we see that

$$\frac{\bar{a}}{|a|^2} = \frac{a_1}{|a|^2} - \left(\frac{a_2}{|a|^2}\right)i - \left(\frac{a_3}{|a|^2}\right)j - \left(\frac{a_4}{|a|^2}\right)k$$

is a multiplicative inverse for a. We consider that we have demonstrated the following theorem.

24.9 Theorem The quaternions $\mathbb{H}$ form a strictly skew field under addition and multiplication.

Note that $G = \{\pm 1, \pm i, \pm j, \pm k\}$ is a group of order 8 under quaternion multiplication. This group is generated by i and j, where

$$i^4 = 1, \qquad j^2 = i^2 \qquad \text{and} \qquad ji = i^3 j.$$

There are no finite strictly skew fields. This is the content of a famous theorem of Wedderburn, which we state without proof.

24.10 Theorem **(Wedderburn's Theorem)** Every finite division ring is a field.

Proof See Artin, Nesbitt, and Thrall [24] for proof of Wedderburn's theorem. ◆

EXERCISES 24

Computations

In Exercises 1 through 3, let $G = \{e, a, b\}$ be a cyclic group of order 3 with identity element e. Write the element in the group algebra $\mathbb{Z}_5 G$ in the form

$$re + sa + tb \qquad \text{for} \qquad r, s, t \in \mathbb{Z}_5.$$

1. $(2e + 3a + 0b) + (4e + 2a + 3b)$ **2.** $(2e + 3a + 0b)(4e + 2a + 3b)$ **3.** $(3e + 3a + 3b)^4$

In Exercises 4 through 7, write the element of $\mathbb{H}$ in the form $a_1 + a_2 i + a_3 j + a_4 k$ for $a_i \in \mathbb{R}$.

4. $(i + 3j)(4 + 2j - k)$ **5.** $i^2 j^3 kji^5$

6. $(i + j)^{-1}$ **7.** $[(1 + 3i)(4j + 3k)]^{-1}$

8. Referring to the group S_3 given in Example 8.7, compute the product

$$(0\rho_0 + 1\rho_1 + 0\rho_2 + 0\mu_1 + 1\mu_2 + 1\mu_3)(1\rho_0 + 1\rho_1 + 0\rho_2 + 1\mu_1 + 0\mu_2 + 1\mu_3)$$

in the group algebra $\mathbb{Z}_2 S_3$.

9. Find the center of the group $\langle \mathbb{H}^*, \cdot \rangle$, where $\mathbb{H}^*$ is the set of nonzero quaternions.

Concepts

10. Find two subsets of $\mathbb{H}$ different from $\mathbb{C}$ and from each other, each of which is a field isomorphic to $\mathbb{C}$ under the induced addition and multiplication from $\mathbb{H}$.

11. Mark each of the following true or false.

_____ **a.** $M_n(F)$ has no divisors of 0 for any n and any field F.

_____ **b.** Every nonzero element of $M_2(\mathbb{Z}_2)$ is a unit.

_____ **c.** $\text{End}(A)$ is always a ring with unity $\neq 0$ for every abelian group A.

_____ **d.** $\text{End}(A)$ is never a ring with unity $\neq 0$ for any abelian group A.

_____ **e.** The subset Iso(A) of End(A), consisting of the isomorphisms of A onto A, forms a subring of End(A) for every abelian group A.

_____ **f.** $R\langle\mathbb{Z}, +\rangle$ is isomorphic to $\langle Z, +, \cdot\rangle$ for every commutative ring R with unity.

_____ **g.** The group ring RG of an abelian group G is a commutative ring for any commutative ring R with unity.

_____ **h.** The quaternions are a field.

_____ **i.** $\langle\mathbb{H}^*, \cdot\rangle$ is a group where $\mathbb{H}^*$ is the set of nonzero quaternions.

_____ **j.** No subring of $\mathbb{H}$ is a field.

12. Show each of the following by giving an example.

a. A polynomial of degree n with coefficients in a strictly skew field may have more than n zeros in the skew field.

b. A finite multiplicative subgroup of a strictly skew field need not be cyclic.

Theory

13. Let ϕ be the element of End($\langle\mathbb{Z} \times \mathbb{Z}, +\rangle$) given in Example 24.2. That example showed that ϕ is a right divisor of 0. Show that ϕ is also a left divisor of 0.

14. Show that $M_2(F)$ has at least six units for every field F. Exhibit these units. [Hint: F has at least two elements, 0 and 1.]

15. Show that End ($\langle\mathbb{Z}, +\rangle$) is naturally isomorphic to $\langle\mathbb{Z}, +, \cdot\rangle$ and that End($\langle\mathbb{Z}_n, +\rangle$) is naturally isomorphic to $\langle\mathbb{Z}_n, +, \cdot\rangle$.

16. Show that End($\langle\mathbb{Z}_2 \times \mathbb{Z}_2, +\rangle$) is not isomorphic to $\langle\mathbb{Z}_2 \times \mathbb{Z}_2, +, \cdot\rangle$.

17. Referring to Example 24.3, show that $YX - XY = 1$.

18. If $G = \{e\}$, the group of one element, show that RG is isomorphic to R for any ring R.

19. There exists a matrix $K \in M_2(\mathbb{C})$ such that $\phi : \mathbb{H} \to M_2(\mathbb{C})$ defined by

$$\phi(a + bi + cj + dk) = a\begin{bmatrix} 1 & 0 \\ 0 & 1 \end{bmatrix} + b\begin{bmatrix} 0 & 1 \\ -1 & 0 \end{bmatrix} + c\begin{bmatrix} 0 & i \\ i & 0 \end{bmatrix} + dK,$$

for all $a, b, c, d \in \mathbb{R}$, gives an isomorphism of $\mathbb{H}$ with $\phi[\mathbb{H}]$

a. Find the matrix K.

b. What 8 equations should you check to see that ϕ really is a homomorphism?

c. What other thing should you check to show that ϕ gives an isomorphism of $\mathbb{H}$ with $\phi[\mathbb{H}]$?

SECTION 25 †ORDERED RINGS AND FIELDS

We are familiar with the inequality relation $<$ on the set $\mathbb{R}$ and on any subset of $\mathbb{R}$. (We remind you that relations were discussed in Section 0. See Definition 0.7.) We regard $<$ as providing an *ordering* of the real numbers. In this section, we study orderings of rings and fields. *We assume throughout this section that the rings under discussion have nonzero unity* 1.

In the real numbers, $a < b$ if and only if $b - a$ is positive, so the order relation $<$ on $\mathbb{R}$ is completely determined if we know which real numbers are positive. We use the idea of labeling certain elements as positive to define the notion of order in a ring.

† This section is not used in the remainder of the text.

25.1 Definition An **ordered ring** is a ring R together with a nonempty subset P of R satisfying these two properties.

Closure For all $a, b \in P$, both $a + b$ and ab are in P.

Trichotomy For each $a \in R$, one and only one of the following holds:

$$a \in P, \quad a = 0, \quad -a \in P.$$

Elements of P are called **"positive."** ■

It is easy to see that if R is an ordered ring with set P of positive elements and S is a subring of R, then $P \cap S$ satisfies the requirements for a set of positive elements in the ring S, and thus gives an ordering of S. (See Exercise 26.) This is the **induced ordering** from the given ordering of R.

We observe at once that for each of the rings $\mathbb{Z}$, $\mathbb{Q}$ and $\mathbb{R}$ the set of elements that we have always considered to be positive satisfies the conditions of closure and trichotomy. We will refer to the familiar ordering of these rings and the induced ordering on their subrings as the *natural ordering*. We now give an unfamiliar illustration.

25.2 Example Let R be an ordered ring with set P of positive elements. There are two natural ways to define an ordering of the polynomial ring $R[x]$. We describe two possible sets, P_{low} and P_{high}, of positive elements. A nonzero polynomial in $R[x]$ can be written in the form

$$f(x) = a_r x^r + a_{r+1} x^{r+1} + \cdots + a_n x^n$$

where $a_r \neq 0$ and $a_n \neq 0$, so that $a_r x^r$ and $a_n x^n$ are the nonzero terms of lowest and highest degree, respectively. Let P_{low} be the set of all such $f(x)$ for which $a_r \in P$, and let P_{high} be the set of all such $f(x)$ for which $a_n \in P$. The closure and trichotomy requirements that P_{low} and P_{high} must satisfy to give orderings of $R[x]$ follow at once from those same properties for P and the definition of addition and multiplication in $R[x]$. Illustrating in $\mathbb{Z}[x]$, with ordering given by P_{low}, the polynomial $f(x) = -2x + 3x^4$ would not be positive because -2 is not positive in $\mathbb{Z}$. With ordering given by P_{high}, this same polynomial would be positive because 3 is positive in $\mathbb{Z}$. ▲

Suppose now that P is the set of positive elements in an ordered ring R. Let a be any nonzero element of R. Then either a or $-a$ is in P, so by closure, $a^2 = (-a)^2$ is also in P. Thus all squares of nonzero elements of R are positive. In particular, $1 = 1^2$ is positive. By closure, we see that $1 + 1 + \cdots + 1$ for any finite number of summands is always in P, so it is never zero. Thus an ordered ring has characteristic zero.

Because squares of nonzero elements must be positive, we see that the natural ordering of $\mathbb{R}$ is the only possible ordering. The positive real numbers are precisely the squares of nonzero real numbers and the set could not be enlarged without destroying trichotomy. Because $1 + 1 + \cdots + 1$ must be positive, the only possible ordering of $\mathbb{Z}$ is the natural ordering also. All ordered rings have characteristic zero so we can, by identification (renaming), consider every ordered ring to contain $\mathbb{Z}$ as an ordered subring.

If a and b are nonzero elements of P then either $-a$ or a is in P and either $-b$ or b is in P. Consequently by closure, either ab or $-ab$ is in P. By trichotomy, ab cannot be zero so an ordered ring can have no zero divisors.

We summarize these observations in a theorem and corollary.

25.3 Theorem Let R be an ordered ring. All squares of nonzero elements of R are positive, R has characteristic 0, and there are no zero divisors.

25.4 Corollary We can consider $\mathbb{Z}$ to be embedded in any ordered ring R, and the induced ordering of $\mathbb{Z}$ from R is the natural ordering of $\mathbb{Z}$. The only possible ordering of $\mathbb{R}$ is the natural ordering.

Theorem 25.3 shows that the field $\mathbb{C}$ of complex numbers cannot be ordered, because both $1 = 1^2$ and $-1 = i^2$ are squares. It also shows that no finite ring can be ordered because the characteristic of an ordered ring is zero.

The theorem that follows defines a relation $<$ in an ordered ring, and gives properties of $<$. The definition of $<$ is motivated by the observation that, in the real numbers, $a < b$ if and only if $b - a$ is positive. The theorem also shows that ordering could have been defined in terms of a relation $<$ having the listed properties.

25.5 Theorem Let R be an ordered ring with set P of positive elements. Let $<$, read "is less than," be the relation on R defined by

$$a < b \text{ if and only if } (b - a) \in P \tag{1}$$

for $a, b \in R$. The relation $<$ has these properties for all $a, b, c \in R$.

Trichotomy One and only one of the following holds:

$$a < b, \quad a = b, \quad b < a.$$

Transitivity If $a < b$ and $b < c$, then $a < c$.

Isotonicity If $b < c$, then $a + b < a + c$.
If $b < c$ and $0 < a$, then $ab < ac$ and $ba < ca$.

Conversely, given a relation $<$ on a nonzero ring R satisfying these three conditions, the set $P = \{x \in R \mid 0 < x\}$ satisfies the two criteria for a set of positive elements in Definition 25.1, and the relation $<_P$ defined as in Condition (1) with this P is the given relation $<$.

Proof Let R be an ordered ring with set P of positive elements, and let $a < b$ mean $(b - a) \in P$. We prove the three properties for $<$.

Trichotomy Let $a, b \in R$. By the trichotomy property of P in Definition 25.1 applied to $b - a$, exactly one of

$$(b - a) \in P, \quad b - a = 0, \quad (a - b) \in P$$

holds. These translate in terms of $<$ to

$$a < b, \quad a = b, \quad b < a$$

respectively.

Transitivity Let $a < b$ and $b < c$. Then $(b - a) \in P$ and $(c - b) \in P$. By closure of P under addition, we have

$$(b - a) + (c - b) = (c - a) \in P$$

so $a < c$.

Isotonicity Let $b < c$, so $(c - b) \in P$. Then $(a + c) - (a + b) = (c - b) \in P$ so $a + b < a + c$. Also if $a > 0$, then by closure of P both $a(c - b) = ac - ab$ and $(c - b)a = ca - ba$ are in P, so $ab < ac$ and $ba < ca$.

We leave the "conversely" part of the theorem as an equally easy exercise. (See Exercise 27.) ◆

In view of Theorem 25.5, we will now feel free to use the $<$ notation in an ordered ring. The notations $>$, $\leq$, and $\geq$ are defined as usual in terms of $<$ and $=$. Namely,

$$b > a \text{ means } a < b, \quad a \leq b \text{ means either } a = b \text{ or } a < b,$$
$$a \geq b \text{ means either } b < a \text{ or } b = a.$$

25.6 Example Let R be an ordered ring. It is illustrative to think what the orderings of $R[x]$ given by P_{low} and P_{high} in Example 25.2 mean in terms of the relation $<$ of Theorem 25.5.

Taking P_{low}, we observe, for every $a > 0$ in R, that $a - x$ is positive so $x < a$. Also, $x = x - 0$ is positive, so $0 < x$. Thus $0 < x < a$ for every $a \in R$. We have $(x^i - x^j) \in P_{\text{low}}$ when $i < j$, so $x^j < x^i$ if $i < j$. Our monomials have the ordering

$$0 < \cdots x^6 < x^5 < x^4 < x^3 < x^2 < x < a$$

for any positive $a \in R$. Taking $R = \mathbb{R}$, we see that in this ordering of $\mathbb{R}[x]$ there are infinitely many positive elements that are less than any positive real number!

We leave a similar discussion of $<$ for the ordering of $R[x]$ given by P_{high} to Exercise 1. ▲

The preceding example is of interest because it exhibits an ordering that is *not Archimedian*. We give a definition explaining this terminology. Remember that we can consider $\mathbb{Z}$ to be a subring of every ordered ring.

25.7 Definition An ordering of a ring R with this property:

For each given positive a and b in R, there exists a positive integer n such that $na > b$.

is an **Archimedian ordering.** ■

The natural ordering of $\mathbb{R}$ is Archimedian, but the ordering of $\mathbb{R}[x]$ given by P_{low} discussed in Example 25.6 is not Archimedian because for every positive integer n we have $(17 - nx) \in P_{\text{low}}$, so $nx < 17$ for all $n \in \mathbb{Z}^+$.

We give two examples describing types of ordered rings and fields that are of interest in more advanced work.

25.8 Example **(Formal Power Series Rings)** Let R be a ring. In Section 22 we defined a polynomial in $R[x]$ to be a formal sum $\sum_{i=0}^{\infty} a_i x^i$ where all but a finite number of the a_i are 0. If we do not require any of the a_i to be zero, we obtain a **formal power series** in x with coefficients in the ring R. (The adjective, *formal*, is customarily used because we are not dealing with convergence of series.) Exactly the same formulas are used to define the sum and product of these series as for polynomials in Section 22. Most of us had some practice adding and multiplying series when we studied calculus. These series form a ring which we denote by $R[[x]]$, and which contains $R[x]$ as a subring.

If R is an ordered ring, we can extend the ordering to $R[[x]]$ exactly as we extended the ordering to $R[x]$ using the set P_{low} of positive elements. (We cannot use P_{high}. Why not?) The monomials have the same ordering that we displayed in Example 25.6. ▲

25.9 Example **(Formal Laurent Series Fields)** Continuing with the idea of Example 25.8, we let F be a field and consider formal series of the form $\sum_{i=N}^{\infty} a_i x^i$ where N may be any integer, positive, zero, or negative, and $a_i \in F$. (Equivalently, we could consider $\sum_{i=-\infty}^{\infty} a_i x^i$ where all but a finite number of the a_i are zero for negative values of i. In studying calculus for functions of a complex variable, one encounters series of this form called "*Laurent series*.") With the natural addition and multiplication of these series, we actually have a *field* which we denote by $F((x))$. The inverse of x is the series $x^{-1} + 0 + 0x + 0x^2 + \cdots$. Inverses of elements and quotients can be computed by series division. We compute three terms of $(x^{-1} - 1 + x - x^2 + x^3 + \cdots)/(x^3 + 2x^4 + 3x^5 + \cdots)$ in $\mathbb{R}((x))$ for illustration.

$$
\begin{array}{r|l}
 & x^{-4} - 3x^{-3} + 4x^{-2} + \cdots \\
\hline
x^3 + 2x^4 + 3x^5 + \cdots & x^{-1} - 1 + x - x^2 + x^3 + \cdots \\
 & x^{-1} + 2 + 3x + \cdots \\
\hline
 & \quad -3 - 2x + \cdots \\
 & \quad -3 - 6x - 9x^2 + \cdots \\
\hline
 & \qquad 4x + \cdots
\end{array}
$$

If F is an ordered field, we can use the obvious analog of P_{low} in $R[[x]]$ to define an ordering of $F((x))$. In Exercise 2 we ask you to symbolically order the monomials $\cdots x^{-3}, x^{-2}, x^{-1}, x^0 = 1, x, x^2, x^3, \cdots$ as we did for $R[x]$ in Example 25.6. Note that $F((x))$ contains, as a subfield, a field of quotients of $F[x]$, and thus induces an ordering on this field of quotients. ▲

Let R be an ordered ring and let $\phi : R \to R'$ be a ring isomorphism. It is intuitively clear that by identification (renaming), the map ϕ can be used to carry over the ordering of R to provide an ordering of R'. We state as a theorem what would have to be proved for a skeptic, and leave the proof as Exercise 25.

25.10 Theorem Let R be an ordered ring with set P of positive elements and let $\phi : R \to R'$ be a ring isomorphism. The subset $P' = \phi[P]$ satisfies the requirements of Definition 25.1 for a set of positive elements of R'. Furthermore, in the ordering of R' given by P', we have $\phi(a) <' \phi(b)$ in R' if and only if $a < b$ in R.

We call the ordering of R' described in the preceding theorem the **"ordering induced by"** ϕ from the ordering of R.

25.11 Example Example 22.9 stated that the evaluation homomorphism $\phi_\pi : \mathbb{Q}[x] \to \mathbb{R}$ where

$$\phi\big(a_0 + a_1 x + \cdots + a_n x^n\big) = a_0 + a_1\pi + \cdots + a_n\pi^n$$

is one to one. Thus it provides an isomorphism of $\mathbb{Q}[x]$ with $\phi[\mathbb{Q}[x]]$. We denote this image ring by $\mathbb{Q}[\pi]$. If we provide $\mathbb{Q}[x]$ with the ordering using the set P_{low} of Examples 25.2 and 25.6, the ordering on $\mathbb{Q}[\pi]$ induced by ϕ_π is very different from that induced by the natural (and only) ordering of $\mathbb{R}$. In the P_{low} ordering, π is less than any element of $\mathbb{Q}$! ▲

An isomorphism of a ring R onto itself is called an **automorphism** of R. Theorem 25.10 can be used to exhibit different orderings of an ordered ring R if there exist automorphisms of R that do not carry the set P of positive elements onto itself. We give an example.

25.12 Example Exercise 11 of Section 18 shows that $\{m + n\sqrt{2} \mid m, n \in \mathbb{Z}\}$ is a ring. Let us denote this ring by $\mathbb{Z}[\sqrt{2}]$. This ring has a natural order induced from $\mathbb{R}$ in which $\sqrt{2}$ is positive. However, we claim that $\phi : \mathbb{Z}[\sqrt{2}] \to \mathbb{Z}[\sqrt{2}]$ defined by $\phi(m + n\sqrt{2}) = m - n\sqrt{2}$ is an automorphism. It is clearly one to one and onto $\mathbb{Z}[\sqrt{2}]$. We leave the verification of the homomorphism property to Exercise 17. Because $\phi(\sqrt{2}) = -\sqrt{2}$, we see the ordering induced by ϕ will be one where $-\sqrt{2}$ is positive! In the natural order on $\mathbb{Z}[\sqrt{2}]$, an element $m + n\sqrt{2}$ is positive if m and n are both positive, or if m is positive and $2n^2 < m^2$, or if n is positive and $m^2 < 2n^2$. In Exercise 3, we ask you to give the analogous descriptions for positive elements in the ordering of $\mathbb{Z}[\sqrt{2}]$ induced by ϕ. ▲

In view of Examples 25.11 and 25.12, which exhibit orderings on subrings of $\mathbb{R}$ that are not the induced orderings, we wonder whether $\mathbb{Q}$ can have an ordering other than the natural one. Our final theorem shows that this is not possible.

25.13 Theorem Let D be an ordered integral domain with P as set of positive elements, and let F be a field of quotients of D. The set

$$P' = \{x \in F \mid x = a/b \text{ for } a, b \in D \text{ and } ab \in P\}$$

is well-defined and gives an order on F that induces the given order on D. Furthermore, P' is the only subset of F with this property.

Proof To show that P' is well-defined, suppose that $x = a/b = a'/b'$ for $a, b, a', b' \in D$ and that $ab \in P$. We must show that $a'b' \in P$. From $a/b = a'/b'$ we obtain $ab' = a'b$. Multiplying by b, we have $(ab)b' = a'b^2$. Now $b^2 \in P$ and by assumption, $ab \in P$. Using trichotomy and the properties $a(-b) = (-a)b = -(ab)$ of a ring, we see that either a' and b' are both in P or both not in P. In either case, we have $a'b' \in P$.

We proceed to closure for P'. Let $x = a/b$ and $y = c/d$ be two elements of P', so $ab \in P$ and $cd \in P$. Now $x + y = (ad + bc)/bd$ and $(ad + bc)bd = (ab)d^2 + b^2(cd)$ is in P because squares are also in P and P is closed under addition and multiplication. Thus $(x + y) \in P'$. Also $xy = ac/bd$ is in P' because $acbd = (ab)(cd)$ is a product of two elements of P and thus in P.

For trichotomy, we need only observe that for $x = a/b$, the product ab satisfies just one of

$$ab \in P, \qquad ab = 0, \qquad ab \notin P$$

by trichotomy for P. For P', these translate into $x \in P'$, $x = 0$, and $x \notin P'$, respectively.

We have shown that P' does give an ordering of F. For $a \in D$, we see that $a = a/1$ is in P' if and only if $a1 = a$ is in P, so the given ordering on D is indeed the induced ordering from F by P'.

Finally, suppose that P'' is a set of positive elements of F satisfying the conditions of Definition 25.1 and such that $P'' \cap D = P$. Let $x = a/b \in P''$ where $a, b \in D$. Then $xb^2 = ab$ must be in P'', so $ab \in (P'' \cap D) = P$. Thus $x \in P'$ so $P'' \subseteq P'$. The law of trichotomy shows that we then must have $P' = P''$. Therefore P' gives the only ordering of F that maintains original order for elements of D. ◆

EXERCISES 25

Computations

1. Let R be an ordered ring. Describe the ordering of a positive element a of R and the monomials $x, x^2, x^3, \cdots, x^n, \cdots$ in $R[x]$ as we did in Example 25.6, but using the set P_{high} of Example 25.6 as set of positive elements of $R[x]$.

2. Let F be an ordered field and let $F((x))$ be the field of formal Laurent series with coefficients in F, discussed in Example 25.9. Describe the ordering of the monomials $\cdots x^{-3}, x^{-2}, x^{-1}, x^0 = 1, x, x^2, x^3, \cdots$ in the ordering of $F((x))$ described in that example.

3. Example 25.12 described an ordering of $\mathbb{Z}[\sqrt{2}] = \{m + n\sqrt{2} \mid m, n \in \mathbb{Z}\}$ in which $-\sqrt{2}$ is positive. Describe, in terms of m and n, all positive elements of $\mathbb{Z}[\sqrt{2}]$ in that ordering.

In Exercises 4 through 9, let $\mathbb{R}[x]$ have the ordering given by

i. P_{low} ii. P_{high}

as described in Example 25.2. In each case (i) and (ii), list the labels a, b, c, d, e of the given polynomials in an order corresponding to increasing order of the polynomials as described by the relation $<$ of Theorem 25.5.

4. **a.** $-5 + 3x$ **b.** $5 - 3x$ **c.** $-x + 7x^2$ **d.** $x - 7x^2$ **e.** $2 + 4x^2$

5. **a.** -1 **b.** $3x - 8x^3$ **c.** $-5x + 7x^2 - 11x^4$ **d.** $8x^2 + x^5$ **e.** $-3x^3 - 4x^5$

6. **a.** $-3 + 5x^2$ **b.** $-2x + 5x^2 + x^3$ **c.** -5 **d.** $6x^3 + 8x^4$ **e.** $8x^4 - 5x^5$

7. **a.** $-2x^2 + 5x^3$ **b.** $x^3 + 4x^4$ **c.** $2x - 3x^2$ **d.** $-3x - 4x^2$ **e.** $2x - 2x^2$

8. **a.** $4x - 3x^2$ **b.** $4x + 2x^2$ **c.** $4x - 6x^3$ **d.** $5x - 6x^3$ **e.** $3x - 2x^2$

9. **a.** $x - 3x^2 + 5x^3$ **b.** $2 - 3x^2 + 5x^3$ **c.** $x - 3x^2 + 4x^3$ **d.** $x + 3x^2 + 4x^4$ **e.** $x + 3x^2 - 4x^3$

In Exercises 10 through 13, let $\mathbb{Q}((x))$ have the ordering described in Example 25.9. List the labels a, b, c, d, e of the given elements in an order corresponding to increasing order of the elements as described by the relation $<$ of Theorem 25.5.

10. **a.** $\frac{1}{x}$ **b.** $\frac{-5}{x^2}$ **c.** $\frac{2}{x}$ **d.** $\frac{-3}{x^2}$ **e.** $4x$

11. **a.** $\frac{1}{1-x}$ **b.** $\frac{x^2}{1+x}$ **c.** $\frac{1}{x-x^2}$ **d.** $\frac{-x}{1+x^2}$ **e.** $\frac{3-2x}{x^3+4x}$

12. **a.** $\frac{5-7x}{x^2+3x^3}$ **b.** $\frac{-2+4x}{4-3x}$ **c.** $\frac{7+2x}{4-3x}$ **d.** $\frac{9-3x^2}{2+6x}$ **e.** $\frac{3-5x}{-6+2x}$

13. **a.** $\dfrac{1-x}{1+x}$ **b.** $\dfrac{3-5x}{3+5x}$ **c.** $\dfrac{1}{4x+x^2}$ **d.** $\dfrac{1}{-3x+x^2}$ **e.** $\dfrac{4x+x^2}{1-x}$

Concepts

14. It can be shown that the smallest subfield of $\mathbb{R}$ containing $\sqrt[3]{2}$ is isomorphic to the smallest subfield of $\mathbb{C}$ containing $\sqrt[3]{2}(\frac{-1+i\sqrt{3}}{2})$. Explain why this shows that, although there is no ordering for $\mathbb{C}$, there may be an ordering of a subfield of $\mathbb{C}$ that contains some elements that are not real numbers.

15. Mark each of the following true or false.

______ **a.** There is only one ordering possible for the ring $\mathbb{Z}$.
______ **b.** The field $\mathbb{R}$ can be ordered in only one way.
______ **c.** Any subfield of $\mathbb{R}$ can be ordered in only one way.
______ **d.** The field $\mathbb{Q}$ can be ordered in only one way.
______ **e.** If R is an ordered ring, then $R[x]$ can be ordered in a way that induces the given order on R.
______ **f.** An ordering of a ring R is Archimedian if for each $a, b \in R$, there exists $n \in \mathbb{Z}^+$ such that $b < na$.
______ **g.** An ordering of a ring R is Archimedian if for each $a, b \in R$ such that $0 < a$, there exists $n \in \mathbb{Z}^+$ such that $b < na$.
______ **h.** If R is an ordered ring and $a \in R$, then $-a$ cannot be positive.
______ **i.** If R is an ordered ring and $a \in R$, then either a or $-a$ is positive.
______ **j.** Every ordered ring has an infinite number of elements.

16. Describe an ordering of the ring $\mathbb{Q}[\pi]$, discussed in Example 25.11, in which π is greater than any rational number.

Theory

17. Referring to Example 25.12, show that the map $\phi : \mathbb{Z}[\sqrt{2}] \to \mathbb{R}$ where $\phi(m + n\sqrt{2}) = m - n\sqrt{2}$ is a homomorphism.

In Exercises 18 through 24, let R be an ordered ring with set P of positive elements, and let $<$ be the relation on R defined in Theorem 25.5. Prove the given statement. (All the proofs have to be in terms of Definition 25.1 and Theorem 25.5. For example, you must not say, "We know that negative times positive is negative, so if $a < 0$ and $0 < b$ then $ab < 0$.")

18. If $a \in P$, then $0 < a$.

19. If $a, b \in P$ and $ac = bd$, then either $c = d = 0$ or $cd \in P$.

20. If $a < b$, then $-b < -a$.

21. If $a < 0$ and $0 < b$, then $ab < 0$.

22. If R is a field and a and b are positive, then a/b is positive.

23. If R is a field and $0 < a < 1$, then $1 < 1/a$.

24. If R is a field and $-1 < a < 0$, then $1/a < -1$.

25. Prove Theorem 25.10 of the text.

26. Show that if R is an ordered ring with set P of positive elements and S is a subring of R, then $P \cap S$ satisfies the requirements for a set of positive elements in the ring S, and thus gives an ordering of S.

27. Show that if $<$ is a relation on a ring R satisfying the properties of trichotomy, transitivity, and isotonicity stated in Theorem 25.5, then there exists a subset P of R satisfying the conditions for a set of positive elements in Definition 25.1, and such that the relation $<_P$ defined by $a <_P b$ if and only if $(b - a) \in P$ is the same as the relation $<$.

28. Let R be an ordered integral domain. Show that if $a^{2n+1} = b^{2n+1}$ where $a, b \in R$ and n is a positive integer, then $a = b$.

29. Let R be an ordered ring and consider the ring $R[x, y]$ of polynomials in two variables with coefficients in R. Example 25.2 describes two ways in which we can order $R[x]$, and for each of these, we can continue on and order $(R[x])[y]$ in the analogous two ways, giving four ways of arriving at an ordering of $R[x, y]$. There are another four ways of arriving at an ordering of $R[x, y]$ if we first order $R[y]$ and then $(R[y])[x]$. Show that all eight of these orderings of $R[x, y]$ are different. [*Hint:* You might start by considering whether $x < y$ or $y < x$ in each of these orderings, and continue in this fashion.]

Ideals and Factor Rings

SECTION 26 HOMOMORPHISMS AND FACTOR RINGS

Homomorphisms

We defined the concepts of *homomorphism* and *isomorphism* for rings in Section 18, since we wished to talk about evaluation homomorphisms for polynomials and about isomorphic rings. We repeat some definitions here for easy reference. Recall that a homomorphism is a *structure-relating map*. A homomorphism for rings must relate both their additive structure and their multiplicative structure.

26.1 Definition A map ϕ of a ring R into a ring R' is a **homomorphism** if

$$\phi(a+b) = \phi(a) + \phi(b)$$

and

$$\phi(ab) = \phi(a)\phi(b)$$

for all elements a and b in R. ■

In Example 18.10 we defined evaluation homomorphisms, and Example 18.11 showed that the map $\phi : \mathbb{Z} \to \mathbb{Z}_n$, where $\phi(m)$ is the remainder of m when divided by n, is a homomorphism. We give another simple but very fundamental example of a homomorphism.

26.2 Example **(Projection Homomorphisms)** Let $R_1, R_2, \cdots, R_n$ be rings. For each i, the map $\pi_i : R_1 \times R_2 \times \cdots \times R_n \to R_i$ defined by $\pi_i(r_1, r_2, \cdots, r_n) = r_i$ is a homomorphism, *projection onto the ith component*. The two required properties of a homomorphism hold

† Section 28 is not required for the remainder of the text.

for π_i since both addition and multiplication in the direct product are computed by addition and multiplication in each individual component. ▲

Properties of Homomorphisms

We work our way through the exposition of Section 13 but for ring homomorphisms.

26.3 Theorem **(Analogue of Theorem 13.12)** Let ϕ be a homomorphism of a ring R into a ring R'. If 0 is the additive identity in R, then $\phi(0) = 0'$ is the additive identity in R', and if $a \in R$, then $\phi(-a) = -\phi(a)$. If S is a subring of R, then $\phi[S]$ is a subring of R'. Going the other way, if S' is a subring of R', then $\phi^{-1}[S']$ is a subring of R. Finally, if R has unity 1, then $\phi(1)$ is unity for $\phi[R]$. Loosely speaking, subrings correspond to subrings, and rings with unity correspond to rings with unity under a ring homomorphism.

Proof Let ϕ be a homomorphism of a ring R into a ring R'. Since, in particular, ϕ can be viewed as a group homomorphism of $\langle R, +\rangle$ into $\langle R', +'\rangle$, Theorem 13.12 tells us that $\phi(0) = 0'$ is the additive identity element of R' and that $\phi(-a) = -\phi(a)$.

Theorem 13.12 also tells us that if S is a subring of R, then, considering the additive group $\langle S, +\rangle$, the set $\langle \phi[S], +'\rangle$ gives a subgroup of $\langle R', +'\rangle$. If $\phi(s_1)$ and $\phi(s_2)$ are two elements of $\phi[S]$, then

$$\phi(s_1)\phi(s_2) = \phi(s_1 s_2)$$

and $\phi(s_1 s_2) \in \phi[S]$. Thus $\phi(s_1)\phi(s_2) \in \phi[S]$, so $\phi[S]$ is closed under multiplication. Consequently, $\phi[S]$ is a subring of R'.

Going the other way, Theorem 13.12 also shows that if S' is a subring of R', then $\langle \phi^{-1}[S'], +\rangle$ is a subgroup of $\langle R, +\rangle$. Let $a, b \in \phi^{-1}[S']$, so that $\phi(a) \in S'$ and $\phi(b) \in S'$. Then

$$\phi(ab) = \phi(a)\phi(b).$$

Since $\phi(a)\phi(b) \in S'$, we see that $ab \in \phi^{-1}[S']$ so $\phi^{-1}[S']$ is closed under multiplication and thus is a subring of R.

Finally, if R has unity 1, then for all $r \in R$,

$$\phi(r) = \phi(1r) = \phi(r1) = \phi(1)\phi(r) = \phi(r)\phi(1),$$

so $\phi(1)$ is unity for $\phi[R]$. ◆

Note in Theorem 26.3 that $\phi(1)$ is unity for $\phi[R]$, but not necessarily for R' as we ask you to illustrate in Exercise 9.

26.4 Definition Let a map $\phi : R \to R$ be a homomorphism of rings. The subring

$$\phi^{-1}[0'] = \{r \in R \mid \phi(r) = 0'\}$$

is the **kernel** of ϕ, denoted by Ker(ϕ). ■

Now this Ker(ϕ) is the same as the kernel of the group homomorphism of $\langle R, +\rangle$ into $\langle R', +\rangle$ given by ϕ. Theorem 13.15 and Corollary 13.18 on group homomorphisms give us at once analogous results for ring homomorphisms.

26.5 Theorem **(Analogue of Theorem 13.15)** Let $\phi : R \to R'$ be a ring homomorphism, and let $H = \text{Ker}(\phi)$. Let $a \in R$. Then $\phi^{-1}[\phi(a)] = a + H = H + a$, where $a + H = H + a$ is the coset containing a of the commutative additive group $\langle H, +\rangle$.

26.6 Corollary **(Analogue of Corollary 13.18)** A ring homomorphism $\phi : R \to R'$ is a one-to-one map if and only if $\text{Ker}(\phi) = \{0\}$.

Factor (Quotient) Rings

We are now ready to describe the analogue for rings of Section 14. We start with the analogue of Theorem 14.1.

26.7 Theorem **(Analogue of Theorem 14.1)** Let $\phi : R \to R'$ be a ring homomorphism with kernel H. Then the additive cosets of H form a ring R/H whose binary operations are defined by choosing representatives. That is, the sum of two cosets is defined by

$$(a + H) + (b + H) = (a + b) + H,$$

and the product of the cosets is defined by

$$(a + H)(b + H) = (ab) + H.$$

Also, the map $\mu : R/H \to \phi[R]$ defined by $\mu(a + H) = \phi(a)$ is an isomorphism.

Proof Once again, the additive part of the theory is done for us in Theorem 14.1. We proceed to check the multiplicative aspects.

We must first show that multiplication of cosets by choosing representatives is well defined. To this end, let $h_1, h_2, \in H$ and consider the representatives $a + h_1$ of $a + H$ and $b + h_2$ of $b + H$. Let

$$c = (a + h_1)(b + h_2) = ab + ah_2 + h_1b + h_1h_2.$$

We must show that this element c lies in the coset $ab + H$. Since $ab + H = \phi^{-1}[\phi(ab)]$, we need only show that $\phi(c) = \phi(ab)$. Since ϕ is a homomorphism and $\phi(h) = 0'$ for $h \in H$, we obtain

$$\begin{aligned}\phi(c) &= \phi(ab + ah_2 + h_1b + h_1h_2)\\ &= \phi(ab) + \phi(ah_2) + \phi(h_1b) + \phi(h_1h_2)\\ &= \phi(ab) + \phi(a)0' + 0'\phi(b) + 0'0'\\ &= \phi(ab) + 0' + 0' + 0' = \phi(ab). \end{aligned} \qquad (1)$$

Thus multiplication by choosing representatives is well defined.

To show that R/H is a ring, it remains to show that the associative property for multiplication and the distributive laws hold in R/H. Since addition and multiplication are computed by choosing representatives, these properties follow at once from corresponding properties in R.

Theorem 14.1 shows that the map μ defined in the statement of Theorem 26.4 is well defined, one to one, onto $\phi[R]$, and satisfies the additive property for a homomorphism.

Multiplicatively, we have

$$\mu[(a+H)(b+H)] = \mu(ab+H) = \phi(ab)$$
$$= \phi(a)\phi(b) = \mu(a+H)\mu(b+H).$$

This completes the demonstration that μ is an isomorphism. ◆

26.8 Example Example 18.11 shows that the map $\phi : \mathbb{Z} \to \mathbb{Z}_n$ defined by $\phi(m) = r$, where r is the remainder of m when divided by n, is a homomorphism. Since $\text{Ker}(\phi) = n\mathbb{Z}$, Theorem 26.7 shows that $\mathbb{Z}/n\mathbb{Z}$ is a ring where operations on residue classes can be computed by choosing representatives and performing the corresponding operation in $\mathbb{Z}$. The theorem also shows that this ring $\mathbb{Z}/n\mathbb{Z}$ is isomorphic to $\mathbb{Z}_n$. ▲

It remains only to characterize those subrings H of a ring R such that multiplication of additive cosets of H by choosing representatives is well defined. The coset multiplication in Theorem 26.7 was shown to be well defined in Eq. (1). The success of Eq. (1) is due to the fact that $\phi(ah_2) = \phi(h_1 b) = \phi(h_1 h_2) = 0'$. That is, if $h \in H$ where $H = \text{Ker}(\phi)$, then for every $a, b \in R$ we have $ah \in H$ and $hb \in H$. This suggests Theorem 26.9 below, which is the analogue of Theorem 14.4.

26.9 Theorem **(Analogue of Theorem 14.4)** Let H be a subring of the ring R. Multiplication of additive cosets of H is well defined by the equation

$$(a+H)(b+H) = ab + H$$

if and only if $ah \in H$ and $hb \in H$ for all $a, b \in R$ and $h \in H$.

Proof Suppose first that $ah \in H$ and $hb \in H$ for all $a, b \in R$ and all $h \in H$. Let $h_1, h_2 \in H$ so that $a + h_1$ and $b + h_2$ are also representatives of the cosets $a + H$ and $b + H$ containing a and b. Then

$$(a+h_1)(b+h_2) = ab + ah_2 + h_1 b + h_1 h_2.$$

Since ah_2 and $h_1 b$ and $h_1 h_2$ are all in H by hypothesis, we see that $(a+h_1)(b+h_2) \in ab + H$.

Conversely, suppose that multiplication of additive cosets by representatives is well defined. Let $a \in R$ and consider the coset product $(a+H)H$. Choosing representatives $a \in (a+H)$ and $0 \in H$, we see that $(a+H)H = a0 + H = 0 + H = H$. Since we can also compute $(a+H)H$ by choosing $a \in (a+H)$ and any $h \in H$, we see that $ah \in H$ for any $h \in H$. A similar argument starting with the product $H(b+H)$ shows that $hb \in H$ for any $h \in H$. ◆

In group theory, normal subgroups are precisely the type of substructure of groups required to form a factor group with a well-defined operation on cosets given by operating with chosen representatives. Theorem 26.9 shows that in ring theory, the analogous substructure must be a subring H of a ring R such that $aH \subseteq H$ and $Hb \subseteq H$ for all $a, b \in R$, where $aH = \{ah \mid h \in H\}$ and $Hb = \{hb \mid h \in H\}$. From now on we will usually denote such a substructure by N rather than H. Recall that we started using N to mean a normal subgroup in Section 15.

26.10 Definition An addive subgroup N of a ring R satisfying the properties

$$aN \subseteq N \quad \text{and} \quad Nb \subseteq N \quad \text{for all } a, b \in R$$

is an **ideal.** ■

26.11 Example We see that $n\mathbb{Z}$ is an ideal in the ring $\mathbb{Z}$ since we know it is a subring, and $s(nm) = (nm)s = n(ms) \in n\mathbb{Z}$ for all $s \in \mathbb{Z}$. ▲

26.12 Example Let F be the ring of all functions mapping $\mathbb{R}$ into $\mathbb{R}$, and let C be the subring of F consisting of all the constant functions in F. Is C an ideal in F? Why?

Solution It is not true that the product of a constant function with every function is again a constant function. For example, the product of $\sin x$ and 2 is the function $2 \sin x$. Thus C is not an ideal of F. ▲

Historical Note

It was Ernst Eduard Kummer (1810–1893) who introduced the concept of an "ideal complex number" in 1847 in order to preserve the notion of unique factorization in certain rings of algebraic integers. In particular, Kummer wanted to be able to factor into primes numbers of the form $a_0 + a_1\alpha + a_2\alpha^2 + \cdots + a_{p-1}\alpha^{p-1}$, where α is a complex root of $x^p = 1$ (p prime) and the a_i are ordinary integers. Kummer had noticed that the naive definition of primes as "unfactorable numbers" does not lead to the expected results; the product of two such "unfactorable" numbers may well be divisible by other "unfactorable" numbers. Kummer defined "ideal prime factors" and "ideal numbers" in terms of certain congruence relationships; these "ideal factors" were then used as the divisors necessary to preserve unique factorization. By use of these, Kummer was in fact able to prove certain cases of Fermat's Last Theorem, which states that $x^n + y^n = z^n$ has no solutions $x, y, z \in \mathbb{Z}^+$ if $n > 2$.

It turned out that an "ideal number," which was in general not a "number" at all, was uniquely determined by the set of integers it "divided." Richard Dedekind took advantage of this fact to identify the ideal factor with this set; he therefore called the set itself an ideal, and proceeded to show that it satisfied the definition given in the text. Dedekind was then able to define the notions of prime ideal and product of two ideals and show that any ideal in the ring of integers of any algebraic number field could be written uniquely as a product of prime ideals.

26.13 Example Let F be as in the preceding example, and let N be the subring of all functions f such that $f(2) = 0$. Is N an ideal in F? Why or why not?

Solution Let $f \in N$ and let $g \in F$. Then $(fg)(2) = f(2)g(2) = 0g(2) = 0$, so $fg \in N$. Similarly, we find that $gf \in N$. Therefore N is an ideal of F. We could also have proved this by just observing that N is the kernel of the evaluation homomorphism $\phi_2 : F \to \mathbb{R}$. ▲

Once we know that multiplication by choosing representatives is well defined on additive cosets of a subring N of R, the associative law for multiplication and the distributive laws for these cosets follow at once from the same properties in R. We have at once this corollary of Theorem 26.9.

26.14 Corollary **(Analogue of Corollary 14.5)** Let N be an ideal of a ring R. Then the additive cosets of N form a ring R/N with the binary operations defined by

$$(a+N)+(b+N)=(a+b)+N$$

and

$$(a+N)(b+N)=ab+N.$$

26.15 Definition The ring R/N in the preceding corollary is the **factor ring** (or **quotient ring**) **of** R **by** N. ■

If we use the term *quotient ring*, be sure not to confuse it with the notion of the *field of quotients* of an integral domain, discussed in Section 21.

Fundamental Homomorphism Theorem

To complete our analogy with Sections 13 and 14, we give the analogues of Theorems 14.9 and 14.11.

26.16 Theorem **(Analogue of Theorem 14.9)** Let N be an ideal of a ring R. Then $\gamma : R \to R/N$ given by $\gamma(x) = x + N$ is a ring homomorphism with kernel N.

Proof The additive part is done in Theorem 14.9. Turning to the multiplicative question, we see that

$$\gamma(xy)=(xy)+N=(x+N)(y+N)=\gamma(x)\gamma(y).$$ ◆

26.17 Theorem **(Fundamental Homomorphism Theorem; Analogue of Theorem 14.11)** Let $\phi : R \to R'$ be a ring homomorphism with kernel N. Then $\phi[R]$ is a ring, and the map $\mu : R/N \to \phi[R]$ given by $\mu(x+N) = \phi(x)$ is an isomorphism. If $\gamma : R \to R/N$ is the homomorphism given by $\gamma(x) = x + N$, then for each $x \in R$, we have $\phi(x) = \mu\gamma(x)$.

Proof This follows at once from Theorems 26.7 and 26.16. Figure 26.18 is the analogue of Fig. 14.10. ◆

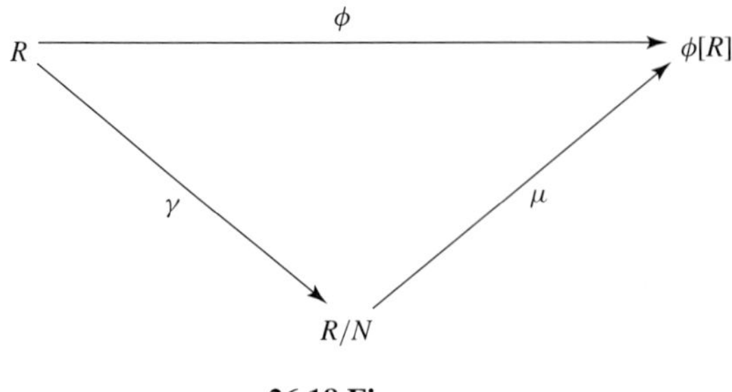

26.18 Figure

26.19 Example Example 26.11 shows that $n\mathbb{Z}$ is an ideal of $\mathbb{Z}$, so we can form the factor ring $\mathbb{Z}/n\mathbb{Z}$. Example 18.11 shows that $\phi : \mathbb{Z} \to \mathbb{Z}_n$ where $\phi(m)$ is the remainder of m modulo n is a homomorphism, and we see that $\text{Ker}(\phi) = n\mathbb{Z}$. Theorem 26.17 then shows that the map $\mu : \mathbb{Z}/n\mathbb{Z} \to \mathbb{Z}_n$ where $\mu(m + n\mathbb{Z})$ is the remainder of m modulo n is well defined and is an isomorphism. ▲

In summary, every ring homomorphism with domain R gives rise to a factor ring R/N, and every factor ring R/N gives rise to a homomorphism mapping R into R/N. An *ideal* in ring theory is analogous to a *normal subgroup* in the group theory. Both are the type of substructure needed to form a factor structure.

We should now add an addendum to Theorem 26.3 on properties of homomorphisms. Let $\phi : R \to R'$ be a homomorphism, and let N be an ideal of R. Then $\phi[N]$ is an ideal of $\phi[R]$, although it need not be an ideal of R'. Also, if N' is an ideal of either $\phi[R]$ or of R', then $\phi^{-1}[N']$ is an ideal of R. We leave the proof of this to Exercise 22.

EXERCISES 26

Computations

1. Describe all ring homomorphisms of $\mathbb{Z} \times \mathbb{Z}$ into $\mathbb{Z} \times \mathbb{Z}$. [*Hint:* Note that if ϕ is such a homomorphism, then $\phi((1, 0)) = \phi((1, 0))\phi((1, 0))$ and $\phi((0, 1)) = \phi((0, 1))\phi((0, 1))$. Consider also $\phi((1, 0)(0, 1))$.]
2. Find all positive integers n such that $\mathbb{Z}_n$ contains a subring isomorphic to $\mathbb{Z}_2$.
3. Find all ideals N of $\mathbb{Z}_{12}$. In each case compute $\mathbb{Z}_{12}/N$; that is, find a known ring to which the quotient ring is isomorphic.
4. Give addition and multiplication tables for $2\mathbb{Z}/8\mathbb{Z}$. Are $2\mathbb{Z}/8\mathbb{Z}$ and $\mathbb{Z}_4$ isomorphic rings?

Concepts

In Exercises 5 through 7, correct the definition of the italicized term without reference to the text, if correction is needed, so that it is in a form acceptable for publication.

5. An *isomorphism of a ring* R with a ring R' is a homomorphism $\phi : R \to R'$ such that $\text{Ker}(\phi) = \{0\}$.
6. An *ideal* N of a ring R is an additive subgroup of $\langle R, +\rangle$ such that for all $r \in R$ and all $n \in N$, we have $rn \in N$ and $nr \in N$.
7. The *kernel of a homomorphism* ϕ mapping a ring R into a ring R' is $\{\phi(r) = 0' \mid r \in R\}$.
8. Let F be the ring of all functions mapping $\mathbb{R}$ into $\mathbb{R}$ and having derivatives of all orders. Differentiation gives a map $\delta : F \to F$ where $\delta(f(x)) = f'(x)$. Is δ a homomorphism? Why? Give the connection between this exercise and Example 26.12.
9. Give an example of a ring homomorphism $\phi : R \to R'$ where R has unity 1 and $\phi(1) \neq 0'$, but $\phi(1)$ is not unity for R'.
10. Mark each of the following true or false.
    _____ **a.** The concept of a ring homomorphism is closely connected with the idea of a factor ring.
    _____ **b.** A ring homomorphism $\phi : R \to R'$ carries ideals of R into ideals of R'.
    _____ **c.** A ring homomorphism is one to one if and only if the kernel is $\{0\}$.
    _____ **d.** $\mathbb{Q}$ is an ideal in $\mathbb{R}$.

_______ **e.** Every ideal in a ring is a subring of the ring.

_______ **f.** Every subring of every ring is an ideal of the ring.

_______ **g.** Every quotient ring of every commutative ring is again a commutative ring.

_______ **h.** The rings $\mathbb{Z}/4\mathbb{Z}$ and $\mathbb{Z}_4$ are isomorphic.

_______ **i.** An ideal N in a ring R with unity 1 is all of R if and only if $1 \in N$.

_______ **j.** The concept of an ideal is to the concept of a ring as the concept of a normal subgroup is to the concept of a group.

11. Let R be a ring. Observe that $\{0\}$ and R are both ideals of R. Are the factor rings R/R and $R/\{0\}$ of real interest? Why?

12. Give an example to show that a factor ring of an integral domain may be a field.

13. Give an example to show that a factor ring of an integral domain may have divisors of 0.

14. Give an example to show that a factor ring of a ring with divisors of 0 may be an integral domain.

15. Find a subring of the ring $\mathbb{Z} \times \mathbb{Z}$ that is not an ideal of $\mathbb{Z} \times \mathbb{Z}$.

16. A student is asked to prove that a quotient ring of a ring R modulo an ideal N is commutative if and only if $(rs - sr) \in N$ for all $r, s \in R$. The student starts out:
Assume R/N is commutative. Then $rs = sr$ for all $r, s \in R/N$.

a. Why does the instructor reading this expect nonsense from there on?

b. What should the student have written?

c. Prove the assertion. (Note the "if and only if.")

Theory

17. Let $R = \{a + b\sqrt{2} \mid a, b \in \mathbb{Z}\}$ and let R' consist of all 2×2 matrices of the form $\begin{bmatrix} a & 2b \\ b & a \end{bmatrix}$ for $a, b \in \mathbb{Z}$. Show that R is a subring of $\mathbb{R}$ and that R' is a subring of $M_2(\mathbb{Z})$. Then show that $\phi : R \to R'$, where $\phi(a + b\sqrt{2}) = \begin{bmatrix} a & 2b \\ b & a \end{bmatrix}$ is an isomorphism.

18. Show that each homomorphism from a field to a ring is either one to one or maps everything onto 0.

19. Show that if R, R', and R'' are rings, and if $\phi : R \to R'$ and $\psi : R' \to R''$ are homomorphisms, then the composite function $\psi\phi : R \to R''$ is a homomorphism. (Use Exercise 49 of Section 13.)

20. Let R be a commutative ring with unity of prime characteristic p. Show that the map $\phi_p : R \to R$ given by $\phi_p(a) = a^p$ is a homomorphism (the **Frobenius homomorphism**).

21. Let R and R' be rings and let $\phi : R \to R'$ be a ring homomorphism such that $\phi[R] \neq \{0'\}$. Show that if R has unity 1 and R' has no 0 divisors, then $\phi(1)$ is unity for R'.

22. Let $\phi : R \to R'$ be a ring homomorphism and let N be an ideal of R.

a. Show that $\phi[N]$ is an ideal of $\phi[R]$.

b. Give an example to show that $\phi[N]$ need not be an ideal of R'.

c. Let N' be an ideal either of $\phi[R]$ or of R'. Show that $\phi^{-1}[N']$ is an ideal of R.

23. Let F be a field, and let S be any subset of $F \times F \times \cdots \times F$ for n factors. Show that the set N_S of all $f(x_1, \cdots, x_n) \in F[x_1, \cdots, x_n]$ that have every element $(a_1, \cdots, a_n)$ of S as a zero (see Exercise 28 of Section 22) is an ideal in $F[x_1, \cdots, x_n]$. This is of importance in algebraic geometry.

24. Show that a factor ring of a field is either the trivial (zero) ring of one element or is isomorphic to the field.

25. Show that if R is a ring with unity and N is an ideal of R such that $N \neq R$, then R/N is a ring with unity.

26. Let R be a commutative ring and let $a \in R$. Show that $I_a = \{x \in R \mid ax = 0\}$ is an ideal of R.

27. Show that an intersection of ideals of a ring R is again an ideal of R.

28. Let R and R' be rings and let N and N' be ideals of R and R', respectively. Let ϕ be a homomorphism of R into R'. Show that ϕ induces a natural homomorphism $\phi_* : R/N \to R'/N'$ if $\phi[N] \subseteq N'$. (Use Exercise 39 of Section 14.)

29. Let ϕ be a homomorphism of a ring R with unity onto a nonzero ring R'. Let u be a unit in R. Show that $\phi(u)$ is a unit in R'.

30. An element a of a ring R is **nilpotent** if $a^n = 0$ for some $n \in \mathbb{Z}^+$. Show that the collection of all nilpotent elements in a commutative ring R is an ideal, the **nilradical of** R.

31. Referring to the definition given in Exercise 30, find the nilradical of the ring $\mathbb{Z}_{12}$ and observe that it is one of the ideals of $\mathbb{Z}_{12}$ found in Exercise 3. What is the nilradical of $\mathbb{Z}$? of $\mathbb{Z}_{32}$?

32. Referring to Exercise 30, show that if N is the nilradical of a commutative ring R, then R/N has as nilradical the trivial ideal $\{0 + N\}$.

33. Let R be a commutative ring and N an ideal of R. Referring to Exercise 30, show that if every element of N is nilpotent and the nilradical of R/N is R/N, then the nilradical of R is R.

34. Let R be a commutative ring and N an ideal of R. Show that the set $\sqrt{N}$ of all $a \in R$, such that $a^n \in N$ for some $n \in \mathbb{Z}^+$, is an ideal of R, the **radical of** N.

35. Referring to Exercise 34, show by examples that for proper ideals N of a commutative ring R,

a. $\sqrt{N}$ need not equal N

b. $\sqrt{N}$ may equal N.

36. What is the relationship of the ideal $\sqrt{N}$ of Exercise 34 to the nilradical of R/N (see Exercise 30)? Word your answer carefully.

37. Show that $\phi : \mathbb{C} \to M_2(\mathbb{R})$ given by

$$\phi(a + bi) = \begin{pmatrix} a & b \\ -b & a \end{pmatrix}$$

for $a, b \in \mathbb{R}$ gives an isomorphism of $\mathbb{C}$ with the subring $\phi[\mathbb{C}]$ of $M_2(\mathbb{R})$.

38. Let R be a ring with unity and let $\text{End}(\langle R, +\rangle)$ be the ring of endomorphisms of $\langle R, +\rangle$ as described in Section 24. Let $a \in R$, and let $\lambda_a : R \to R$ be given by

$$\lambda_a(x) = ax$$

for $x \in R$.

a. Show that λ_a is an endomorphism of $\langle R, +\rangle$.

b. Show that $R' = \{\lambda_a \mid a \in R\}$ is a subring of $\text{End}(\langle R, +\rangle)$.

c. Prove the analogue of Cayley's theorem for R by showing that R' of (b) is isomorphic to R.

SECTION 27 Prime and Maximal Ideals

Exercises 12 through 14 of the preceding section asked us to provide examples of factor rings R/N where R and R/N have very different structural properties. We start with some examples of this situation, and in the process, provide solutions to those exercises.

27.1 Example As was shown in Corollary 19.12, the ring $\mathbb{Z}_p$, which is isomorphic to $\mathbb{Z}/p\mathbb{Z}$, is a field for p a prime. *Thus a factor ring of an integral domain may be a field.* ▲

27.2 Example The ring $\mathbb{Z} \times \mathbb{Z}$ is not an integral domain, for

$$(0, 1)(1, 0) = (0, 0),$$

showing that $(0, 1)$ and $(1, 0)$ are 0 divisors. Let $N = \{(0, n) \mid n \in \mathbb{Z}\}$. Now N is an ideal of $\mathbb{Z} \times \mathbb{Z}$, and $(\mathbb{Z} \times \mathbb{Z})/N$ is isomorphic to $\mathbb{Z}$ under the correspondence $[(m, 0) + N] \leftrightarrow m$, where $m \in \mathbb{Z}$. Thus a *factor* ring of a ring may be an integral domain, even though the original ring is not. ▲

27.3 Example The subset $N = \{0, 3\}$ of $\mathbb{Z}_6$ is easily seen to be an ideal of $\mathbb{Z}_6$, and $\mathbb{Z}_6/N$ has three elements, $0 + N$, $1 + N$, and $2 + N$. These add and multiply in such a fashion as to show that $\mathbb{Z}_6/N \simeq \mathbb{Z}_3$ under the correspondence

$$(0 + N) \leftrightarrow 0, \qquad (1 + N) \leftrightarrow 1, \qquad (2 + N) \leftrightarrow 2.$$

This example shows that *if R is not even an integral domain, that is, if R has zero divisors, it is still possible for R/N to be a field.* ▲

27.4 Example Note that $\mathbb{Z}$ is an integral domain, but $\mathbb{Z}/6\mathbb{Z} \simeq \mathbb{Z}_6$ is not. The preceding examples showed that a factor ring may have a structure that seems *better* than the original ring. This example indicates that the structure of a factor ring may seem *worse* than that of the original ring. ▲

Every nonzero ring R has at least two ideals, the **improper ideal** R and the **trivial ideal** $\{0\}$. For these ideals, the factor rings are R/R, which has only one element, and $R/\{0\}$, which is isomorphic to R. These are uninteresting cases. Just as for a subgroup of a group, a **proper nontrivial ideal** of a ring R is an ideal N of R such that $N \neq R$ and $N \neq \{0\}$.

While factor rings of rings and integral domains may be of great interest, as the above examples indicate, Corollary 27.6, which follows our next theorem, shows that a factor ring of a field is really not useful to us.

27.5 Theorem If R is a ring with unity, and N is an ideal of R containing a unit, then $N = R$.

Proof Let N be an ideal of R, and suppose that $u \in N$ for some unit u in R. Then the condition $rN \subseteq N$ for all $r \in R$ implies, if we take $r = u^{-1}$ and $u \in N$, that $1 = u^{-1}u$ is in N. But then $rN \subseteq N$ for all $r \in R$ implies that $r1 = r$ is in N for all $r \in R$, so $N = R$. ◆

27.6 Corollary A field contains no proper nontrivial ideals.

Proof Since every nonzero element of a field is a unit, it follows at once from Theorem 27.5 that an ideal of a field F is either $\{0\}$ or all of F. ◆

Maximal and Prime Ideals

We now take up the question of when a factor ring is a field and when it is an integral domain. The analogy with groups in Section 15 can be stretched a bit further to cover the case in which the factor ring is a field.

27.7 Definition A **maximal ideal of a ring** R is an ideal M different from R such that there is no proper ideal N of R properly containing M. ■

27.8 Example Let p be a prime positive integer. We know that $\mathbb{Z}/p\mathbb{Z}$ is isomorphic to $\mathbb{Z}_p$. Forgetting about multiplication for the moment and regarding $\mathbb{Z}/p\mathbb{Z}$ and $\mathbb{Z}_p$ as additive groups, we know that $\mathbb{Z}_p$ is a simple group, and consequently $p\mathbb{Z}$ must be a maximal normal subgroup of $\mathbb{Z}$ by Theorem 15.18. Since $\mathbb{Z}$ is an abelian group and every subgroup is a normal subgroup, we see that $p\mathbb{Z}$ is a maximal proper subgroup of $\mathbb{Z}$. Since $p\mathbb{Z}$ is an ideal of the ring $\mathbb{Z}$, it follows that $p\mathbb{Z}$ is a maximal ideal of $\mathbb{Z}$. We know that $\mathbb{Z}/p\mathbb{Z}$ is isomorphic to the ring $\mathbb{Z}_p$, and that $\mathbb{Z}_p$ is actually a field. Thus $\mathbb{Z}/p\mathbb{Z}$ is a field. This illustrates the next theorem. ▲

27.9 Theorem **(Analogue of Theorem 15.18)** Let R be a commutative ring with unity. Then M is a maximal ideal of R if and only if R/M is a field.

Proof Suppose M is a maximal ideal in R. Observe that if R is a commutative ring with unity, then R/M is also a nonzero commutative ring with unity if $M \neq R$, which is the case if M is maximal. Let $(a + M) \in R/M$, with $a \notin M$, so that $a + M$ is not the additive identity element of R/M. Suppose $a + M$ has no multiplicative inverse in R/M. Then the set $(R/M)(a + M) = \{(r + M)(a + M) \mid (r + M) \in R/M\}$ does not contain $1 + M$. We easily see that $(R/M)(a + M)$ is an ideal of R/M. It is nontrivial because $a \notin M$, and it is a proper ideal because it does not contain $1 + M$. By the final paragraph of Section 26, if $\gamma : R \to R/M$ is the canonical homomorphism, then $\gamma^{-1}[(R/M)(a + M)]$ is a proper ideal of R properly containing M. But this contradicts our assumption that M is a maximal ideal, so $a + M$ must have a multiplicative inverse in R/M.

Conversely, suppose that R/M is a field. By the final paragraph of Section 26, if N is any ideal of R such that $M \subset N \subset R$ and γ is the canonical homomorphism of R onto R/M, then $\gamma[N]$ is an ideal of R/M with $\{(0 + M)\} \subset \gamma[N] \subset R/M$. But this is contrary to Corollary 27.6, which states that the field R/M contains no proper nontrivial ideals. Hence if R/M is a field, M is maximal. ◆

27.10 Example Since $\mathbb{Z}/n\mathbb{Z}$ is isomorphic to $\mathbb{Z}_n$ and $\mathbb{Z}_n$ is a field if and only if n is a prime, we see that the maximal ideals of $\mathbb{Z}$ are precisely the ideals $p\mathbb{Z}$ for prime positive integers p. ▲

27.11 Corollary A commutative ring with unity is a field if and only if it has no proper nontrivial ideals.

Proof Corollary 27.6 shows that a field has no proper nontrivial ideals.

Conversely, if a commutative ring R with unity has no proper nontrivial ideals, then $\{0\}$ is a maximal ideal and $R/\{0\}$, which is isomorphic to R, is a field by Theorem 27.9. ◆

We now turn to the question of characterizing, for a commutative ring R with unity, the ideals $N \neq R$ such that R/N is an integral domain. The answer here is rather obvious. The factor ring R/N will be an integral domain if and only if $(a + N)(b + N) = N$ implies that either

$$a + N = N \quad \text{or} \quad b + N = N.$$

This is exactly the statement that R/N has no divisors of 0, since the coset N plays the role of 0 in R/N. Looking at representatives, we see that this condition amounts to saying that $ab \in N$ implies that either $a \in N$ or $b \in N$.

27.12 Example All ideals of $\mathbb{Z}$ are of the form $n\mathbb{Z}$. For $n = 0$, we have $n\mathbb{Z} = \{0\}$, and $\mathbb{Z}/\{0\} \simeq \mathbb{Z}$, which is an integral domain. For $n > 0$, we have $\mathbb{Z}/n\mathbb{Z} \simeq \mathbb{Z}_n$ and $\mathbb{Z}_n$ is an integral domain if and only if n is a prime. Thus the nonzero ideals $n\mathbb{Z}$ such that $\mathbb{Z}/n\mathbb{Z}$ is an integral domain are of the form $p\mathbb{Z}$, where p is a prime. Of course, $\mathbb{Z}/p\mathbb{Z}$ is actually a field, so that $p\mathbb{Z}$ is a maximal ideal of $\mathbb{Z}$. Note that for a product rs of integers to be in $p\mathbb{Z}$, the prime p must divide either r or s. The role of prime integers in this example makes the use of the word *prime* in the next definition more reasonable. ▲

27.13 Definition An ideal $N \neq R$ in a commutative ring R is a **prime ideal** if $ab \in N$ implies that either $a \in N$ or $b \in N$ for $a, b \in R$. ■

Note that $\{0\}$ is a prime ideal in $\mathbb{Z}$, and indeed, in any integral domain.

27.14 Example Note that $\mathbb{Z} \times \{0\}$ is a prime ideal of $\mathbb{Z} \times \mathbb{Z}$, for if $(a, b)(c, d) \in \mathbb{Z} \times \{0\}$, then we must have $bd = 0$ in $\mathbb{Z}$. This implies that either $b = 0$ so $(a, b) \in \mathbb{Z} \times \{0\}$ or $d = 0$ so $(c, d) \in \mathbb{Z} \times \{0\}$. Note that $(\mathbb{Z} \times \mathbb{Z})/(\mathbb{Z} \times \{0\})$ is isomorphic to $\mathbb{Z}$, which is an integral domain. ▲

Our remarks preceding Example 27.12 constitute a proof of the following theorem, which is illustrated by Example 27.14.

27.15 Theorem Let R be a commutative ring with unity, and let $N \neq R$ be an ideal in R. Then R/N is an integral domain if and only if N is a prime ideal in R.

27.16 Corollary Every maximal ideal in a commutative ring R with unity is a prime ideal.

Proof If M is maximal in R, then R/M is a field, hence an integral domain, and therefore M is a prime ideal by Theorem 27.15. ◆

The material that has just been presented regarding maximal and prime ideals is very important and we shall be using it quite a lot. We should keep the main ideas well in mind. We must know and understand the definitions of maximal and prime ideals and must remember the following facts that we have demonstrated.

For a commutative ring R with unity:

1. An ideal M of R is maximal if and only if R/M is a field.

2. An ideal N of R is prime if and only if R/N is an integral domain.

3. Every maximal ideal of R is a prime ideal.

Prime Fields

We now proceed to show that the rings $\mathbb{Z}$ and $\mathbb{Z}_n$ form foundations upon which all rings with unity rest, and that $\mathbb{Q}$ and $\mathbb{Z}_p$ perform a similar service for all fields. Let R be any ring with unity 1. Recall that by $n \cdot 1$ we mean $1 + 1 + \cdots + 1$ for n summands for $n > 0$, and $(-1) + (-1) + \cdots + (-1)$ for $|n|$ summands for $n < 0$, while $n \cdot 1 = 0$ for $n = 0$.

27.17 Theorem If R is a ring with unity 1, then the map $\phi : \mathbb{Z} \to R$ given by

$$\phi(n) = n \cdot 1$$

for $n \in \mathbb{Z}$ is a homomorphism of $\mathbb{Z}$ into R.

Proof Observe that

$$\phi(n + m) = (n + m) \cdot 1 = (n \cdot 1) + (m \cdot 1) = \phi(n) + \phi(m).$$

The distributive laws in R show that

$$\underbrace{(1 + 1 + \cdots + 1)}_{n \text{ summands}}\underbrace{(1 + 1 + \cdots + 1)}_{m \text{ summands}} = \underbrace{(1 + 1 + \cdots + 1)}_{nm \text{ summands}}.$$

Thus $(n \cdot 1)(m \cdot 1) = (nm) \cdot 1$ for $n, m > 0$. Similar arguments with the distributive laws show that for all $n, m \in \mathbb{Z}$, we have

$$(n \cdot 1)(m \cdot 1) = (nm) \cdot 1.$$

Thus

$$\phi(nm) = (nm) \cdot 1 = (n \cdot 1)(m \cdot 1) = \phi(n)\phi(m).$$ ◆

27.18 Corollary If R is a ring with unity and characteristic $n > 1$, then R contains a subring isomorphic to $\mathbb{Z}_n$. If R has characteristic 0, then R contains a subring isomorphic to $\mathbb{Z}$.

Proof The map $\phi : \mathbb{Z} \to R$ given by $\phi(m) = m \cdot 1$ for $m \in \mathbb{Z}$ is a homomorphism by Theorem 27.17. The kernel must be an ideal in $\mathbb{Z}$. All ideals in $\mathbb{Z}$ are of the form $s\mathbb{Z}$ for some $s \in \mathbb{Z}$. By Theorem 19.15 we see that if R has characteristic $n > 0$, then the kernel of ϕ is $n\mathbb{Z}$. Then the image $\phi[\mathbb{Z}] \leq R$ is isomorphic to $\mathbb{Z}/n\mathbb{Z} \simeq \mathbb{Z}_n$. If the characteristic of R is 0, then $m \cdot 1 \neq 0$ for all $m \neq 0$, so the kernel of ϕ is $\{0\}$. Thus, the image $\phi[\mathbb{Z}] \leq R$ is isomorphic to $\mathbb{Z}$. ◆

27.19 Theorem A field F is either of prime characteristic p and contains a subfield isomorphic to $\mathbb{Z}_p$ or of characteristic 0 and contains a subfield isomorphic to $\mathbb{Q}$.

Proof If the characteristic of F is not 0, the above corollary shows that F contains a subring isomorphic to $\mathbb{Z}_n$. Then n must be a prime p, or F would have 0 divisors. If F is of characteristic 0, then F must contain a subring isomorphic to $\mathbb{Z}$. In this case Corollaries

21.8 and 21.9 show that F must contain a field of quotients of this subring and that this field of quotients must be isomorphic to $\mathbb{Q}$. ◆

Thus every field contains either a subfield isomorphic to $\mathbb{Z}_p$ for some prime p or a subfield isomorphic to $\mathbb{Q}$. These fields $\mathbb{Z}_p$ and $\mathbb{Q}$ are the fundamental building blocks on which all fields rest.

27.20 Definition The fields $\mathbb{Z}_p$ and $\mathbb{Q}$ are **prime fields.** ■

Ideal Structure in $F[x]$

Throughout the rest of this section, we assume that F is a field. We give the next definition for a general commutative ring R with unity, although we are only interested in the case $R = F[x]$. Note that for a commutative ring R with unity and $a \in R$, the set $\{ra \mid r \in R\}$ is an ideal in R that contains the element a.

27.21 Definition If R is a commutative ring with unity and $a \in R$, the ideal $\{ra \mid r \in R\}$ of all multiples of a is the **principal ideal generated by** a and is denoted by $\langle a \rangle$. An ideal N of R is a **principal ideal** if $N = \langle a \rangle$ for some $a \in R$. ■

27.22 Example Every ideal of the ring $\mathbb{Z}$ is of the form $n\mathbb{Z}$, which is generated by n, so every ideal of $\mathbb{Z}$ is a principal ideal. ▲

27.23 Example The ideal $\langle x \rangle$ in $F[x]$ consists of all polynomials in $F[x]$ having zero constant term. ▲

The next theorem is another simple but very important application of the division algorithm for $F[x]$. (See Theorem 23.1.) The proof of this theorem is to the division algorithm in $F[x]$ as the proof that a subgroup of a cyclic group is cyclic is to the division algorithm in $\mathbb{Z}$.

27.24 Theorem If F is a field, every ideal in $F[x]$ is principal.

Proof Let N be an ideal of $F[x]$. If $N = \{0\}$, then $N = \langle 0 \rangle$. Suppose that $N \neq \{0\}$, and let $g(x)$ be a nonzero element of N of minimal degree. If the degree of $g(x)$ is 0, then $g(x) \in F$ and is a unit, so $N = F[x] = \langle 1 \rangle$ by Theorem 27.5, so N is principal. If the degree of $g(x)$ is ≥ 1, let $f(x)$ be any element of N. Then by Theorem 23.1, $f(x) = g(x)q(x) + r(x)$, where $r(x) = 0$ or (degree $r(x)$) < (degree $g(x)$). Now $f(x) \in N$ and $g(x) \in N$ imply that $f(x) - g(x)q(x) = r(x)$ is in N by definition of an ideal. Since $g(x)$ is a nonzero element of minimal degree in N, we must have $r(x) = 0$. Thus $f(x) = g(x)q(x)$ and $N = \langle g(x) \rangle$. ◆

We can now characterize the maximal ideals of $F[x]$. This is a crucial step in achieving our **basic goal:** to show that any nonconstant polynomial $f(x)$ in $F[x]$ has a zero in some field E containing F.

27.25 Theorem An ideal $\langle p(x)\rangle \neq \{0\}$ of $F[x]$ is maximal if and only if $p(x)$ is irreducible over F.

Proof Suppose that $\langle p(x)\rangle \neq \{0\}$ is a maximal ideal of $F[x]$. Then $\langle p(x)\rangle \neq F[x]$, so $p(x) \notin F$. Let $p(x) = f(x)g(x)$ be a factorization of $p(x)$ in $F[x]$. Since $\langle p(x)\rangle$ is a maximal ideal and hence also a prime ideal, $(f(x)g(x)) \in \langle p(x)\rangle$ implies that $f(x) \in \langle p(x)\rangle$ or $g(x) \in \langle p(x)\rangle$; that is, either $f(x)$ or $g(x)$ has $p(x)$ as a factor. But then we can't have the degrees of both $f(x)$ and $g(x)$ less than the degree of $p(x)$. This shows that $p(x)$ is irreducible over F.

Conversely, if $p(x)$ is irreducible over F, suppose that N is an ideal such that $\langle p(x)\rangle \subseteq N \subseteq F[x]$. Now N is a principal ideal by Theorem 27.24, so $N = \langle g(x)\rangle$ for some $g(x) \in N$. Then $p(x) \in N$ implies that $p(x) = g(x)q(x)$ for some $q(x) \in F[x]$. But $p(x)$ is irreducible, which implies that either $g(x)$ or $q(x)$ is of degree 0. If $g(x)$ is of degree 0, that is, a nonzero constant in F, then $g(x)$ is a unit in $F[x]$, so $\langle g(x)\rangle = N = F[x]$. If $q(x)$ is of degree 0, then $q(x) = c$, where $c \in F$, and $g(x) = (1/c)p(x)$ is in $\langle p(x)\rangle$, so $N = \langle p(x)\rangle$. Thus $\langle p(x)\rangle \subset N \subset F[x]$ is impossible, so $\langle p(x)\rangle$ is maximal. ◆

27.26 Example Example 23.9 shows that $x^3 + 3x + 2$ is irreducible in $\mathbb{Z}_5[x]$, so $\mathbb{Z}_5[x]/\langle x^3 + 3x + 2\rangle$ is a field. Similarly, Theorem 22.11 shows that $x^2 - 2$ is irreducible in $\mathbb{Q}[x]$, so $\mathbb{Q}[x]/\langle x^2 - 2\rangle$ is a field. We shall examine such fields in more detail later. ▲

Application to Unique Factorization in $F[x]$

In Section 23, we stated without proof Theorem 27.27, which follows. (See Theorem 23.18.) Assuming this theorem, we proved in Section 23 that factorization of polynomials in $F[x]$ into irreducible polynomials is unique, except for order of factors and units in F. We delayed the proof of Theorem 27.27 until now since the machinery we have developed enables us to give such a simple, four-line proof. This proof fills the gap in our proof of unique factorization in $F[x]$.

27.27 Theorem Let $p(x)$ be an irreducible polynomial in $F[x]$. If $p(x)$ divides $r(x)s(x)$ for $r(x), s(x) \in F[x]$, then either $p(x)$ divides $r(x)$ or $p(x)$ divides $s(x)$.

Proof Suppose $p(x)$ divides $r(x)s(x)$. Then $r(x)s(x) \in \langle p(x)\rangle$, which is maximal by Theorem 27.25. Therefore, $\langle p(x)\rangle$ is a prime ideal by Corollary 27.16. Hence $r(x)s(x) \in \langle p(x)\rangle$ implies that either $r(x) \in \langle p(x)\rangle$, giving $p(x)$ divides $r(x)$, or that $s(x) \in \langle p(x)\rangle$, giving $p(x)$ divides $s(x)$. ◆

A Preview of Our Basic Goal

We close this section with an outline of the demonstration in Section 29 of our basic goal. We have all the ideas for the proof at hand now; perhaps you can fill in the details from this outline.

Basic goal: Let F be a field and let $f(x)$ be a nonconstant polynomial in $F[x]$. Show that there exists a field E containing F and containing a zero α of $f(x)$.

Outline of the Proof

1. Let $p(x)$ be an irreducible factor of $f(x)$ in $F[x]$.
2. Let E be the *field* $F[x]/\langle p(x)\rangle$. (See Theorems 27.25 and 27.9.)
3. Show that no two different elements of F are in the same coset of $F[x]/\langle p(x)\rangle$, and deduce that we may consider F to be (isomorphic to) a subfield of E.
4. Let α be the coset $x + \langle p(x)\rangle$ in E. Show that for the evaluation homomorphism $\phi_\alpha : F[x] \to E$, we have $\phi_\alpha(f(x)) = 0$. That is, α is a zero of $f(x)$ in E.

An example of a field constructed according to this outline is given in Section 29. There, we give addition and multiplication tables for the field $\mathbb{Z}_2[x]/\langle x^2 + x + 1\rangle$. We show there that this field has just four elements, the cosets

$$0 + \langle x^2 + x + 1\rangle, \qquad 1 + \langle x^2 + x + 1\rangle, \qquad x + \langle x^2 + x + 1\rangle,$$

and

$$(x + 1) + \langle x^2 + x + 1\rangle.$$

We rename these four cosets 0, 1, α, and $\alpha + 1$ respectively, and obtain Tables 29.20 and 29.21 for addition and multiplication in this 4-element field. To see how these tables are constructed, remember that we are in a field of characteristic 2, so that $\alpha + \alpha = \alpha(1 + 1) = \alpha 0 = 0$. Remember also that α is a zero of $x^2 + x + 1$, so that $\alpha^2 + \alpha + 1 = 0$ and consequently $\alpha^2 = -\alpha - 1 = \alpha + 1$.

EXERCISES 27

Computations

1. Find all prime ideals and all maximal ideals of $\mathbb{Z}_6$.
2. Find all prime ideals and all maximal ideals of $\mathbb{Z}_{12}$.
3. Find all prime ideals and all maximal ideals of $\mathbb{Z}_2 \times \mathbb{Z}_2$.
4. Find all prime ideals and all maximal ideals of $\mathbb{Z}_2 \times \mathbb{Z}_4$.
5. Find all $c \in \mathbb{Z}_3$ such that $\mathbb{Z}_3[x]/\langle x^2 + c\rangle$ is a field.
6. Find all $c \in \mathbb{Z}_3$ such that $\mathbb{Z}_3[x]/\langle x^3 + x^2 + c\rangle$ is a field.
7. Find all $c \in \mathbb{Z}_3$ such that $\mathbb{Z}_3[x]/\langle x^3 + cx^2 + 1\rangle$ is a field.
8. Find all $c \in \mathbb{Z}_5$ such that $\mathbb{Z}_5[x]/\langle x^2 + x + c\rangle$ is a field.
9. Find all $c \in \mathbb{Z}_5$ such that $\mathbb{Z}_5[x]/\langle x^2 + cx + 1\rangle$ is a field.

Concepts

In Exercises 10 through 13, correct the definition of the italicized term without reference to the text, if correction is needed, so that it is in a form acceptable for publication.

10. A *maximal ideal* of a ring R is an ideal that is not contained in any other ideal of R.
11. A *prime ideal* of a commutative ring R is an ideal of the form $pR = \{pr \mid r \in R\}$ for some prime p.

12. A *prime field* is a field that has no proper subfields.

13. A *principal ideal* of a commutative ring with unity is an ideal N with the property that there exists $a \in N$ such that N is the smallest ideal that contains a.

14. Mark each of the following true or false.

_______ **a.** Every prime ideal of every commutative ring with unity is a maximal ideal.
_______ **b.** Every maximal ideal of every commutative ring with unity is a prime ideal.
_______ **c.** $\mathbb{Q}$ is its own prime subfield.
_______ **d.** The prime subfield of $\mathbb{C}$ is $\mathbb{R}$.
_______ **e.** Every field contains a subfield isomorphic to a prime field.
_______ **f.** A ring with zero divisors may contain one of the prime fields as a subring.
_______ **g.** Every field of characteristic zero contains a subfield isomorphic to $\mathbb{Q}$.
_______ **h.** Let F be a field. Since $F[x]$ has no divisors of 0, every ideal of $F[x]$ is a prime ideal.
_______ **i.** Let F be a field. Every ideal of $F[x]$ is a principal ideal.
_______ **j.** Let F be a field. Every principal ideal of $F[x]$ is a maximal ideal.

15. Find a maximal ideal of $\mathbb{Z} \times \mathbb{Z}$.

16. Find a prime ideal of $\mathbb{Z} \times \mathbb{Z}$ that is not maximal.

17. Find a nontrivial proper ideal of $\mathbb{Z} \times \mathbb{Z}$ that is not prime.

18. Is $\mathbb{Q}[x]/\langle x^2 - 5x + 6\rangle$ a field? Why?

19. Is $\mathbb{Q}[x]/\langle x^2 - 6x + 6\rangle$ a field? Why?

Proof Synopsis

20. Give a one- or two-sentence synopsis of "only if" part of Theorem 27.9.

21. Give a one- or two-sentence synopsis of "if" part of Theorem 27.9.

22. Give a one- or two-sentence synopsis of Theorem 27.24.

23. Give a one- or two-sentence synopsis of the "only if" part of Theorem 27.25.

Theory

24. Let R be a finite commutative ring with unity. Show that every prime ideal in R is a maximal ideal.

25. Corollary 27.18 tells us that every ring with unity contains a subring isomorphic to either $\mathbb{Z}$ or some $\mathbb{Z}_n$. Is it possible that a ring with unity may simultaneously contain two subrings isomorphic to $\mathbb{Z}_n$ and $\mathbb{Z}_m$ for $n \neq m$? If it is possible, give an example. If it is impossible, prove it.

26. Continuing Exercise 25, is it possible that a ring with unity may simultaneously contain two subrings isomorphic to the fields $\mathbb{Z}_p$ and $\mathbb{Z}_q$ for two different primes p and q? Give an example or prove it is impossible.

27. Following the idea of Exercise 26, is it possible for an integral domain to contain two subrings isomorphic to $\mathbb{Z}_p$ and $\mathbb{Z}_q$ for $p \neq q$ and p and q both prime? Give reasons or an illustration.

28. Prove directly from the definitions of maximal and prime ideals that every maximal ideal of a commutative ring R with unity is a prime ideal. [*Hint*: Suppose M is maximal in R, $ab \in M$, and $a \notin M$. Argue that the smallest ideal $\{ra + m \mid r \in R, m \in M\}$ containing a and M must contain 1. Express 1 as $ra + m$ and multiply by b.]

29. Show that N is a maximal ideal in a ring R if and only if R/N is a **simple ring,** that is, it is nontrivial and has no proper nontrivial ideals. (Compare with Theorem 15.18.)

30. Prove that if F is a field, every proper nontrivial prime ideal of $F[x]$ is maximal.

31. Let F be a field and $f(x), g(x) \in F[x]$. Show that $f(x)$ divides $g(x)$ if and only if $g(x) \in \langle f(x) \rangle$.

32. Let F be a field and let $f(x), g(x) \in F[x]$. Show that

$$N = \{r(x)f(x) + s(x)g(x) \mid r(x), s(x) \in F[x]\}$$

is an ideal of $F[x]$. Show that if $f(x)$ and $g(x)$ have different degrees and $N \neq F[x]$, then $f(x)$ and $g(x)$ cannot both be irreducible over F.

33. Use Theorem 27.24 to prove the *equivalence* of these two theorems:

Fundamental Theorem of Algebra: Every nonconstant polynomial in $\mathbb{C}[x]$ has a zero in $\mathbb{C}$.

Nullstellensatz for $\mathbb{C}[x]$: Let $f_1(x), \cdots, f_r(x) \in \mathbb{C}[x]$ and suppose that every $\alpha \in \mathbb{C}$ that is a zero of all r of these polynomials is also a zero of a polynomial $g(x)$ in $\mathbb{C}[x]$. Then some power of $g(x)$ is in the smallest ideal of $\mathbb{C}[x]$ that contains the r polynomials $f_1(x), \cdots, f_r(x)$.

There is a sort of arithmetic of ideals in a ring. The next three exercises define sum, product, and quotient of ideals.

34. If A and B are ideals of a ring R, the **sum** $A + B$ **of** A **and** B is defined by

$$A + B = \{a + b \mid a \in A, b \in B\}.$$

a. Show that $A + B$ is an ideal.

b. Show that $A \subseteq A + B$ and $B \subseteq A + B$.

35. Let A and B be ideals of a ring R. The **product** AB **of** A **and** B is defined by

$$AB = \left\{ \sum_{i=1}^{n} a_i b_i \mid a_i \in A, b_i \in B, n \in \mathbb{Z}^+ \right\}.$$

a. Show that AB is an ideal in R.

b. Show that $AB \subseteq (A \cap B)$.

36. Let A and B be ideals of a *commutative* ring R. The **quotient** $A : B$ **of** A **by** B is defined by

$$A : B = \{r \in R \mid rb \in A \text{ for all } b \in B\}.$$

Show that $A : B$ is an ideal of R.

37. Show that for a field F, the set S of all matrices of the form

$$\begin{pmatrix} a & b \\ 0 & 0 \end{pmatrix}$$

for $a, b \in F$ is a **right ideal** but not a **left ideal** of $M_2(F)$. That is, show that S is a subring closed under multiplication on the *right* by any element of $M_2(F)$, but is not closed under *left* multiplication.

38. Show that the matrix ring $M_2(\mathbb{Z}_2)$ is a simple ring; that is, $M_2(\mathbb{Z}_2)$ has no proper nontrivial ideals.

SECTION 28 †GRÖBNER BASES FOR IDEALS

This section gives a brief introduction to algebraic geometry. In particular, we are concerned with the problem of finding as simple a description as we can of the set of common zeros of a finite number of polynomials. In order to accomplish our goal in a single section of this text, we will be stating a few theorems without proof. We recommend the book by Adams and Loustaunau [23] for the proofs and further study.

† This section is not used in the remainder of the text.

Algebraic Varieties and Ideals

Let F be a field. Recall that $F[x_1, x_2, \cdots, x_n]$ is the ring of polynomials in n indeterminants $x_1, x_2, \cdots, x_n$ with coefficients in F. We let F^n be the Cartesian product $F \times F \times \cdots \times F$ for n factors. For ease in writing, we denote an element $(a_1, a_2, \cdots, a_n)$ of F^n by **a**, in bold type. Using similar economy, we let $F[\mathbf{x}] = F[x_1, x_2, \cdots, x_n]$. For each $\mathbf{a} \in F^n$, we have an evaluation homomorphism $\phi_{\mathbf{a}}: F[\mathbf{x}] \to F$ just as in Theorem 22.4. That is, for $f(\mathbf{x}) = f(x_1, x_2, \cdots, x_n) \in F[\mathbf{x}]$, we define $\phi_{\mathbf{a}}(f(\mathbf{x})) = f(\mathbf{a}) = f(a_1, a_2, \cdots, a_n)$. The proof that $\phi_{\mathbf{a}}$ is indeed a homomorphism follows from the associative, commutative, and distributive properties of the operations in $F[\mathbf{x}]$ and F. Just as for the one-indeterminate case, an element **a** of F^n is a **zero of** $f(\mathbf{x}) \in F[\mathbf{x}]$ if $f(\mathbf{a}) = 0$. In what follows, we further abbreviate a polynomial $f(\mathbf{x})$ by "f."

In this section we discuss the problem of finding common zeros in F^n of a finite number of polynomials $f_1, f_2, \cdots, f_r$ in $F[\mathbf{x}]$. Finding and studying geometric properties of the set of all these common zeros is the subject of algebraic geometry.

28.1 Definition Let S be a finite subset of $F[\mathbf{x}]$. The **algebraic variety** $V(S)$ in F^n is the set of all common zeros in F^n of the polynomials in S. ■

In our illustrative examples, which usually involve at most three indeterminates, we use x, y, z in place of x_1, x_2, and x_3.

28.2 Example Let $S = \{2x + y - 2\} \subset \mathbb{R}[x, y]$. The algebraic variety $V(S)$ in $\mathbb{R}^2$ is the line with x-intercept 1 and y-intercept 2. ▲

We leave to Exercise 29 the straightforward proof that for r elements $f_1, f_2, \cdots, f_r$ in a commutative ring R with unity, the set

$$I = \{c_1 f_1 + c_2 f_2 + \cdots + c_r f_r \mid c_i \in R \text{ for } i = 1, \cdots, r\}$$

is an ideal of R. We denote this ideal by $\langle f_1, f_2, \cdots, f_r\rangle$. We are interested in the case $R = F[\mathbf{x}]$ where all the c_i and all the f_i are polynomials in $F[\mathbf{x}]$. We regard the c_i as "coefficient polynomials." By its construction, this ideal I is the smallest ideal containing the polynomials $f_1, f_2, \cdots, f_r$; it can also be described as the intersection of all ideals containing these r polynomials.

28.3 Definition Let I be an ideal in a commutative ring R with unity. A subset $\{b_1, b_2, \cdots, b_r\}$ of I is a **basis** for I if $I = \langle b_1, b_2, \cdots, b_r\rangle$. ■

Unlike the situation in linear algebra, there is no requirement of independence for elements of a basis, or of unique representation of an ideal member in terms of a basis.

28.4 Theorem Let $f_1, f_2, \cdots, f_r \in F[\mathbf{x}]$. The set of common zeros in F^n of the polynomials f_i for $i = 1, 2, \cdots, r$ is the same as the set of common zeros in F^n of all the polynomials in the entire ideal $I = \langle f_1, f_2, \cdots, f_r\rangle$.

Proof Let

$$f = c_1 f_1 + c_2 f_2 + \cdots + c_r f_r \qquad \textbf{(1)}$$

be any element of I, and let $\mathbf{a} \in F^n$ be a common zero of $f_1, f_2, \cdots$, and f_r. Applying the evaluation homomorphism $\phi_{\mathbf{a}}$ to Eq. (1), we obtain

$$\begin{aligned} f(\mathbf{a}) &= c_1(\mathbf{a})f_1(\mathbf{a}) + c_2(\mathbf{a})f_2(\mathbf{a}) + \cdots + c_r(\mathbf{a})f_r(\mathbf{a}) \\ &= c_1(\mathbf{a})0 + c_2(\mathbf{a})0 + \cdots + c_r(\mathbf{a})0 = 0, \end{aligned}$$

showing that $\mathbf{a}$ is also a zero of every polynomial f in I. Of course, a zero of every polynomial in I will be a zero of each f_i because each $f_i \in I$. ◆

For an ideal I in $F[\mathbf{x}]$, we let $V(I)$ be the set of all common zeros of all elements of I. We can summarize Theorem 28.4 as

$$V(\{f_1, f_2, \cdots, f_r\}) = V(\langle f_1, f_2, \cdots, f_r\rangle).$$

We state without proof the Hilbert Basis Theorem. (See Adams and Loustaunau [23].)

28.5 Theorem **(Hilbert Basis Theorem)** Every ideal in $F[x_1, x_2, \cdots, x_n]$ has a finite basis.

> **Our objective**: Given a basis for an ideal I in $F[\mathbf{x}]$, modify it if possible to become a basis that better exhibits the structure of I and the geometry of the associated algebraic variety $V(I)$.

The theorem that follows provides a tool for this task. You should notice that the theorem gives information about the division algorithm that we did not mention in Theorem 23.1. We use the same notation here as in Theorem 23.1, but with $\mathbf{x}$ rather than x. If $f(\mathbf{x}) = g(\mathbf{x})h(\mathbf{x})$ in $F(\mathbf{x})$, then $g(\mathbf{x})$ and $h[\mathbf{x}]$ are called **"divisors"** or **"factors"** of $f(\mathbf{x})$.

28.6 Theorem **(Property of the Division Algorithm)** Let $f(\mathbf{x})$, $g(\mathbf{x})$, $q(\mathbf{x})$ and $r(\mathbf{x})$ be polynomials in $F[\mathbf{x}]$ such that $f(\mathbf{x}) = g(\mathbf{x})q(\mathbf{x}) + r(\mathbf{x})$. The common zeros in F^n of $f(\mathbf{x})$ and $g(\mathbf{x})$ are the same as the common zeros of $g(\mathbf{x})$ and $r(\mathbf{x})$. Also the common divisors in $F[\mathbf{x}]$ of $f(\mathbf{x})$ and $g(\mathbf{x})$ are the same as the common divisors of $g(\mathbf{x})$ and $r(\mathbf{x})$.

If $f(\mathbf{x})$ and $g(\mathbf{x})$ are two members of a basis for an ideal I of $F[\mathbf{x}]$, then replacement of $f(\mathbf{x})$ by $r(\mathbf{x})$ in the basis still yields a basis for I.

Proof If $\mathbf{a} \in F^n$ is a common zero of $g(\mathbf{x})$ and $r(\mathbf{x})$, then applying $\phi_{\mathbf{a}}$ to both sides of the equation $f(\mathbf{x}) = g(\mathbf{x})q(\mathbf{x}) + r(\mathbf{x})$, we obtain $f(\mathbf{a}) = g(\mathbf{a})q(\mathbf{a}) + r(\mathbf{a}) = 0q(\mathbf{a}) + 0 = 0$, so $\mathbf{a}$ is a zero of both $f(\mathbf{x})$ and $g(\mathbf{x})$. If $\mathbf{b} \in F[\mathbf{x}]$ is a common zero of $f(\mathbf{x})$ and $g(\mathbf{x})$, then applying $\phi_{\mathbf{b}}$ yields $f(\mathbf{b}) = g(\mathbf{b})q(\mathbf{b}) + r(\mathbf{b})$ so $0 = 0q(\mathbf{b}) + r(\mathbf{b})$ and we see that $r(\mathbf{b}) = 0$ as well as $g(\mathbf{b})$.

The proof concerning common divisors is essentially the same, and is left as Exercise 30.

Finally, let B be a basis for an ideal I, let $f(\mathbf{x}), g(\mathbf{x}), \in B$ and let $f(\mathbf{x}) = g(\mathbf{x})q(\mathbf{x}) + r(\mathbf{x})$. Let B' be the set obtained by replacing $f(\mathbf{x})$ by $r(\mathbf{x})$ in B, and let I' be the ideal having B' as a basis. Let S be the set obtained from B by adjoining $r(\mathbf{x})$ to B. Note that S can also be obtained by adjoining $f(\mathbf{x})$ to B'. The equation $f(\mathbf{x}) = g(\mathbf{x})q(\mathbf{x}) + r(\mathbf{x})$

shows that $f(\mathbf{x}) \in I'$, so we have $B' \subseteq S \subseteq I'$. Thus S is a basis for I'. The equation $r(\mathbf{x}) = f(\mathbf{x}) - q(\mathbf{x})g(\mathbf{x})$ shows that $r(\mathbf{x}) \in I$, so we have $B \subseteq S \subseteq I$. Thus S is basis for I. Therefore $I = I'$ and B' is a basis for I. ◆

A Familiar Linear Illustration

A basic technique for problem solving in linear algebra is finding all common solutions of a finite number of linear equations. For the moment we abandon our practice of never writing "$f(\mathbf{x}) = 0$" for a nonzero polynomial, and work a typical problem as we do in a linear algebra course.

28.7 Example **(Solution as in a Linear Algebra Course)** Find all solutions in $\mathbb{R}^3$ of the linear system

$$\begin{aligned} x + y - 3z &= 8 \\ 2x + y + z &= -5. \end{aligned}$$

Solution We multiply the first equation by -2 and add it to the second, obtaining the new system

$$\begin{aligned} x + y - 3z &= 8 \\ -y + 7z &= -21 \end{aligned}$$

which has the same solution set in $\mathbb{R}^3$ as the preceding one. For any value z, we can find the corresponding y-value from the second equation and then determine x from the first equation. Keeping z as parameter, we obtain $\{(-4z - 13, 7z + 21, z) \mid z \in \mathbb{R}\}$ as solution set, which is a line in Euclidean 3-space through the point $(-13, 21, 0)$. ▲

In the notation of this section, the problem in the preceding example can be phrased as follows:

$$\text{Describe } V(\langle x + y - 3z - 8, 2x + y + z + 5\rangle) \text{ in } \mathbb{R}^3.$$

We solved it by finding a more useful basis, namely

$$\{x + y - 3z - 8, -y + 7z + 21\}.$$

Notice that the second member, $-y + 7z + 21$, of this new basis can be obtained from the original two basis polynomials as a remainder $r(x, y, z)$ in a division process, namely

$$\begin{array}{r|l} & \quad 2 \\ \hline x + y - 3z - 8 & 2x + y + z + 5 \\ & 2x + 2y - 6z - 16 \\ \cline{2-2} & - y + 7z + 21 \end{array}$$

Thus $2x + y + z + 5 = (x + y - 3z - 8)(2) + (-y + 7z + 21)$, an expression of the form $f(x, y, z) = g(x, y, z)q(x, y, z) + r(x, y, z)$. We replaced the polynomial f by the polynomial r, as in Theorem 28.6, which assures us that $V(\langle f, g\rangle) = V(\langle g, r\rangle)$ and that $\langle f, g\rangle = \langle g, r\rangle$. We chose a very simple, 1-step problem in Example 28.7. However, it is clear that the method introduced in a linear algebra course for solving a linear system can be phrased in terms of applying a division algorithm process repeatedly to change a given ideal basis into one that better illuminates the geometry of the associated algebraic variety.

A Single Indeterminate Illustration

Suppose now that we want to find the variety $V(I)$ in $\mathbb{R}$ associated with an ideal I in $F[x]$, the ring of polynomials in the single indeterminate x. By Theorem 27.24, every ideal in $F[x]$ is principal, so there exists $f(x) \in F[x]$ such that $I = \langle f(x) \rangle$. Thus $V(I)$ consists of the zeros of a single polynomial, and $\{f(x)\}$ is probably as simple a basis for I as we could desire. We give an example illustrating computation of such a single generator $f(x)$ for I in a case where the given basis for I contains more than one polynomial. Because a polynomial in $\mathbb{R}[x]$ has only a finite number of zeros in $\mathbb{R}$, we expect two or more randomly selected polynomials in $\mathbb{R}[x]$ to have no common zeros, but we constructed the basis in our example carefully!

28.8 Example Let us describe the algebraic variety V in $\mathbb{R}$ consisting of common zeros of

$$f(x) = x^4 + x^3 - 3x^2 - 5x - 2 \qquad \text{and} \qquad g(x) = x^3 + 3x^2 - 6x - 8.$$

We want to find a new basis for $\langle f, g \rangle$ having polynomials of as small degree as possible, so we use the division algorithm $f(x) = g(x)q(x) + r(x)$ in Theorem 23.1, where $r(x)$ will have degree at most 2. We then replace the basis $\{f, g\}$ by the basis $\{g, r\}$.

$$\begin{array}{r|l} & x - 2 \\ \hline x^3 + 3x^2 - 6x - 8 & x^4 + x^3 - 3x^2 - 5x - 2 \\ & x^4 + 3x^3 - 6x^2 - 8x \\ \hline & -2x^3 + 3x^2 + 3x - 2 \\ & -2x^3 - 6x^2 + 12x + 16 \\ \hline & 9x^2 - 9x - 18 \end{array}$$

Because zeros of $9x^2 - 9x - 18$ are the same as zeros of $x^2 - x - 2$, we let $r(x) = x^2 - x - 2$, and take as new basis

$$\{g, r\} = (x^3 + 3x^2 - 6x - 8, x^2 - x - 2).$$

By dividing $g(x)$ by $r(x)$ to obtain a remainder $r_1(x)$, we will now be able to find a basis $\{r(x), r_1(x)\}$ consisting of polynomials of degree at most 2.

$$\begin{array}{r|l} & x + 4 \\ \hline x^2 - x - 2 & x^3 + 3x^2 - 6x - 8 \\ & x^3 - x^2 - 2x \\ \hline & 4x^2 - 4x - 8 \\ & 4x^2 - 4x - 8 \\ \hline & 0 \end{array}$$

Our new basis $\{r(x), r_1(x)\}$ now becomes $\{x^2 - x - 2\}$. Thus $I = \langle f(x), g(x) \rangle = \langle x^2 - x - 2 \rangle = \langle (x - 2)(x + 1) \rangle$, and we see that $V = \{-1, 2\}$. ▲

Theorem 28.6 tells us that the common divisors of $f(x)$ and $g(x)$ in the preceding example are the same as the common divisors of $r(x)$ and $r_1(x)$. Because $0 = (0)r(x)$, we see that $r(x)$ itself divides 0, so the common divisors of $f(x)$ and $g(x)$ are just those of $r(x)$, which, of course, include $r(x)$ itself. Thus $r(x)$ is called a "*greatest common divisor*" (abbreviated gcd) of $f(x)$ and $g(x)$.

Gröbner Bases

We tackle the problem of finding a nice basis for an ideal I in $F[\mathbf{x}] = F[x_1, x_2, \cdots, x_n]$. In view of our illustrations for the linear and single indeterminant cases, it seems reasonable to try to replace polynomials in a basis by polynomials of lower degree, or containing fewer indeterminates. It is crucial to have a systematic way to accomplish this. Every instructor in linear algebra has had an occasional student who refuses to master matrix reduction and creates zero entries in columns of a matrix in an almost random fashion, rather than finishing the first column and then proceeding to the second, etc. As a first step in our goal, we tackle this problem of specifying an order for polynomials in a basis.

Our polynomials in $F[\mathbf{x}]$ have terms of the form $a{x_1}^{m_1}{x_2}^{m_2}\cdots{x_n}^{m_n}$ where $a \in F$.

Properties for an Ordering of Power Products

1. $1 < P$ for all power products $P \neq 1$.
2. For any two power products P_i and P_j, exactly one of $P_i < P_j$, $P_i = P_j$, $P_j < P_i$ holds.
3. If $P_i < P_j$ and $P_j < P_k$, then $P_i < P_k$.
4. If $P_i < P_j$, then $PP_i < PP_j$ for any power product P.

Let us consider a **power product** in $F[\mathbf{x}]$ to be an expression

$$P = {x_1}^{m_1}{x_2}^{m_2}\cdots{x_n}^{m_n} \text{ where all the } m_i \geq 0 \text{ in } \mathbb{Z}.$$

Notice that all x_i are present, perhaps some with exponent 0. Thus in $F[x, y, z\}$, we must write xz^2 as xy^0z^2 to be a power product. We want to describe a *total ordering* $<$ on the set of all power products so that we know just what it means to say that $P_i < P_j$ for two power products, providing us with a notion of relative size for power products. We can then try to change an ideal basis in a systematic way to create one with polynomials having terms a_iP_i with as "small" power products P_i as possible. We denote by 1 the power product with all exponents 0, and require that an ordering of the power products has the properties shown in the box. Suppose that such an ordering has been described and that $P_i \neq P_j$ and P_i divides P_j so that $P_j = PP_i$ where $1 < P$. From Property 4 in the box, we then have $1P_i < PP_i = P_j$, so $P_i < P_j$. Thus P_i divides P_j implies that $P_i < P_j$. In Exercise 28, we ask you to show by a counterexample that $P_i < P_j$ does not imply that P_i divides P_j.

It can also be shown that these properties guarantee that any step-by-step process for modifying a finite ideal basis that does not increase the size of any maximal power product in a basis element and replaces at least one by something smaller at each step will terminate in a finite number of steps.

In $F[x]$ with x the only indeterminate, there is only one power product ordering, for by Property 1, we must have $1 < x$. Multiplying repeatedly by x and using Property 4, we have $x < x^2$, $x^2 < x^3$, etc. Property 3 then shows that $1 < x < x^2 < x^3 < \cdots$ is the only possible order. Notice that in Example 28.8, we modified a basis by replacing basis polynomials by polynomials containing smaller power products.

There are a number of possible orderings for power products in $F[\mathbf{x}]$ with n indeterminates. We present just one, the *lexicographical order* (denoted by "lex"). In lex, we define

$$x_1^{s_1}x_2^{s_2}\cdots x_n^{s_n} < x_1^{t_1}x_2^{t_2}\cdots x_n^{t_n} \tag{2}$$

if and only if $s_i < t_i$ for the first subscript i, reading from left to right, such that $s_i \neq t_i$. Thus in $F[x, y]$, if we write power products in the order $x^n y^m$, we have $y = x^0 y^1 < x^1 y^0 = x$ and $xy < xy^2$. Using lex, the order of n indeterminates is given by $1 < x_n < x_{n-1} < \cdots < x_2 < x_1$. Our reduction in Example 28.7, where we first got rid of all "big" x's that we could and then the "smaller" y's, corresponded to the lex order $z < y < x$, that is, to writing all power products in the $x^m y^n z^s$ order. For the two-indeterminate case with $y < x$, the total lex term order schematically is

$$1 < y < y^2 < y^3 \cdots < x < xy < xy^2 < xy^3 < \cdots < x^2 < x^2y < x^2y^2 < \cdots.$$

An ordering of power products P induces an obvious ordering of terms aP of a polynomial in $F[\mathbf{x}]$, which we will refer to as a **term order.** From now on, given an ordering of power products, we consider every polynomial f in $F[\mathbf{x}]$ to be written in decreasing order of terms, so that the leading (first) term has the highest order. We denote by $\mathrm{lt}(f)$ the leading term of f and by $\mathrm{lp}(f)$ the power product of the leading term. If f and g are polynomials in $F[\mathbf{x}]$ such that $\mathrm{lp}(g)$ divides $\mathrm{lp}(f)$, then we can execute a division of f by g, as illustrated in the linear and one-indeterminate cases, to obtain $f(\mathbf{x}) = g(\mathbf{x})q(\mathbf{x}) + r(\mathbf{x})$ where $\mathrm{lp}(r) < \mathrm{lp}(f)$. Note that we did not say that $\mathrm{lp}(r) < \mathrm{lp}(g)$. We illustrate with an example.

28.9 Example By division, reduce the basis $\{xy^2, y^2 - y\}$ for the ideal $I = \langle xy^2, y^2 - y\rangle$ in $\mathbb{R}[x, y]$ to one with smaller maximum term size, assuming the order lex with $y < x$.

Solution We see that y^2 divides xy^2 and compute

$$\begin{array}{r|l} & x \\ \hline y^2 - y & xy^2 \\ & \underline{xy^2 - xy} \\ & \qquad xy \end{array}$$

Because y^2 does not divide xy, we cannot continue the division. Note that $\mathrm{lp}(xy) = xy$ is not less than $\mathrm{lp}(y^2 - y) = y^2$. However, we do have $\mathrm{lp}(xy) < \mathrm{lp}(xy^2)$. Our new basis for I is $\{xy, y^2 - y\}$. ▲

When dealing with more than one indeterminate, it is often easier to perform basis reduction by multiplying a basis polynomial $g(\mathbf{x})$ by a polynomial $-q(\mathbf{x})$ and adding it to a polynomial $f(\mathbf{x})$ to obtain $r(\mathbf{x})$, as we perform matrix reduction in linear algebra, rather than writing out the division display as we did in the preceding example. Starting with basis polynomials xy^2 and $y^2 - y$, we can reduce the xy^2 by multiplying $y^2 - y$ by $-x$ and adding the resulting $-xy^2 + xy$ to xy^2, obtaining the replacement xy for xy^2. We can do that in our head, and write down the result directly.

Referring again to Example 28.9, it will follow from what we state later that given any polynomial $f(x, y) = c_1(x, y)(xy) + c_2(x, y)(y^2 - y)$ in $\langle xy, y^2 - y\rangle$, either xy or

y^2 will divide $\text{lp}(f)$. (See Exercises 31.) This illustrates the defining property of a *Gröbner basis*.

28.10 Definition A set $\{g_1, g_2, \cdots, g_r\}$ of nonzero polynomials in $F[x_1, x_2, \cdots, x_n]$, with term ordering $<$, is a **Gröbner basis** for the ideal $I = \langle g_1, g_2, \cdots, g_r\rangle$ if and only if, for each nonzero $f \in I$, there exists some i where $1 \leq i \leq r$ such that $\text{lp}(g_i)$ divides $\text{lp}(f)$. ■

While we have illustrated the computation of a Gröbner basis from a given basis for an ideal in Examples 28.7, 28.8, and 28.9, we have not given a specific algorithm. We refer the reader to Adams and Loustaunau [23]. The method consists of multiplying some polynomial in the basis by any polynomial in $F[\mathbf{x}]$ and adding the result to another polynomial in the basis in a manner that reduces the size of power products. In our illustrations, we have treated the case involving division of $f(\mathbf{x})$ by $g(\mathbf{x})$ where $\text{lp}(g)$ divides $\text{lp}(f)$, but we can also use the process if $\text{lp}(g)$ only divides some other power product in f. For example, if two elements in a basis are $xy - y^3$ and $y^2 - 1$, we can multiply $y^2 - 1$ by y and add it to $xy - y^3$, reducing $xy - y^3$ to $xy - y$. Theorem 28.6 shows that this is a valid computation.

You may wonder how any basis $\{g_1, g_2, \cdots, g_r\}$ can fail to be a Gröbner basis for $I = \langle g_1, g_2, \cdots, g_r\rangle$ because, when we form an element $c_1g_1 + c_2g_2 + \cdots + c_rg_r$ in I, we see that $\text{lp}(g_i)$ is a divisor of $\text{lp}(c_ig_i)$ for $i = 1, 2, \cdots, r$. However, cancellation of power products can occur in the addition. We illustrate with an example.

28.11 Example Consider the ideal $I = \langle x^2y - 2, xy^2 - y\rangle$ in $\mathbb{R}[x, y\}$. The polynomials in the basis shown cannot be reduced further. However, the ideal I contains $y(x^2y - 2) - x(xy^2 - y) = xy - 2y$, whose leading power product xy is not divisible by either of the leading power products x^2y or xy^2 of the given basis. Thus $\{x^2y - 2, xy^2 - y\}$ is not a Gröbner basis for I, according to Definition 28.10. ▲

When we run into a situation like that in Example 28.11, we realize that a Gröbner basis must contain some polynomial with a smaller leading power product than those in the given basis. Let f and g be polynomials in the given basis. Just as we did in Example 28.11, we can multiply f and g by as small power products as possible so that the resulting two leading power products will be the same, the *least common multiple* (lcm) of $\text{lp}(f)$ and $\text{lp}(g)$, and then subtract or add with suitable coefficients from F so cancellation results. We denote a polynomial formed in this fashion by $S(f, g)$. We state without proof a theorem that can be used to test whether a basis is a Gröbner basis.

28.12 Theorem A basis $G = \{g_1, g_2, \cdots, g_r\}$ is a Gröbner basis for the ideal $\langle g_1, g_2, \cdots, g_r\rangle$ if and only if, for all $i \neq j$, the polynomial $S(g_i, g_j)$ can be reduced to zero by repeatedly dividing remainders by elements of G, as in the division algorithm.

As we mentioned before, we may prefer to think of reducing $S(g_i, g_j)$ by a sequence of operations consisting of adding (or subtracting) multiples of polynomials in G, rather than writing out division.

We can now indicate how we can obtain a Gröbner basis from a given basis. First, reduce the polynomials in the basis as far as possible among themselves. Then choose polynomials g_i and g_j in the basis, and form the polynomial $S(g_i, g_j)$. See if $S(g_i, g_j)$ can be reduced to zero as just described. If so, choose a different pair of polynomials, and repeat the procedure with them. If $S(g_i, g_j)$ cannot be reduced to zero as described above, augment the given basis with this $S(g_i, g_j)$, and start all over, reducing this basis as much as possible. By Theorem 28.12, when every polynomial $S(g_i, g_j)$ for all $i \neq j$ can be reduced to zero using polynomials from the latest basis, we have arrived at a Gröbner basis. We conclude with a continuation of Example 28.11.

28.13 Example Continuing Example 2.8.11, let $g_1 = x^2y - 2$, $g_2 = xy^2 - y$, and $I = \langle g_1, g_2 \rangle$ in $\mathbb{R}^2$. In Example 28.11, we obtained the polynomial $S(g_1, g_2) = xy - 2y$, which cannot be reduced to zero using g_1 and g_2. We now reduce the basis $\{x^2y - 2, xy^2 - y, xy - 2y\}$, indicating each step.

$\{x^2y - 2, xy^2 - y, xy - 2y\}$	augmented basis
$\{2xy - 2, xy^2 - y, xy - 2y\}$	by adding $(-x)$ (third) to first
$\{2xy - 2, 2y^2 - y, xy - 2y\}$	by adding $(-y)$ (third) to second
$\{4y - 2, 2y^2 - y, xy - 2y\}$	by adding (-2) (third) to first
$\{4y - 2, 0, xy - 2y\}$	by adding $(-\frac{y}{2})$ (first) to second
$\{4y - 2, 0, \frac{1}{2}x - 2y\}$	by adding $(-\frac{x}{4})$ (first) to third
$\{4y - 2, 0, \frac{1}{2}x - 1\}$	by adding $(\frac{1}{2})$ (first) to third

Clearly, $\{y - \frac{1}{2}, x - 2\}$ is a Gröbner basis. Note that if $f = y - \frac{1}{2}$ and $g = x - 2$, then $S(f, g) = xf - yg = (xy - \frac{x}{2}) - (xy - 2y) = -\frac{x}{2} + 2y$, which can readily be reduced to zero by adding $\frac{1}{2}(x - 2)$ and $-2(y - \frac{1}{2})$.

From the Gröbner basis, we see that the algebraic variety $V(I)$ contains only one point, $(2, \frac{1}{2})$, in $\mathbb{R}^2$. ▲

The importance of Gröbner bases in applications is due to the fact that they are *machine computable*. They have applications to engineering and computer science as well as to mathematics.

EXERCISES 28

In Exercises 1 through 4, write the polynomials in $\mathbb{R}[x, y, z]$ in decreasing term order, using the order lex for power products $x^m y^n z^s$ where $z < y < x$.

1. $2xy^3z^5 - 5x^2yz^3 + 7x^2y^2z - 3x^3$

2. $3y^2z^5 - 4x + 5y^3z^3 - 8z^7$

3. $3y - 7x + 10z^3 - 2xy^2z^2 + 2x^2yz^2$

4. $38 - 4xz + 2yz - 8xy + 3yz^3$

In Exercises 5 through 8, write the polynomials in $\mathbb{R}[x, y, z]$ in decreasing term order, using the order lex for power products $z^m y^n x^s$ where $x < y < z$.

5. The polynomial in Exercise 1.

6. The polynomial in Exercise 2.

7. The polynomial in Exercise 3.

8. The polynomial in Exercise 4.

Another ordering, deglex, for power products in $F[\mathbf{x}]$ is defined as follows:

$$x_1^{s_1} x_2^{s_2} \cdots x_n^{s_n} < x_1^{t_1} x_2^{t_2} \cdots x_n^{t_n}$$

if and only if either $\sum_{i=1}^{n} s_i < \sum_{i=1}^{n} t_i$, or these two sums are equal and $s_i < t_i$ for the smallest value of i such that $s_i \neq t_i$. Exercises 9 through 13 are concerned with the order deglex.

9. List, in increasing order, the smallest 20 power products in $\mathbb{R}[x, y, z]$ for the order deglex with power products $x^m y^n z^s$ where $z < y < x$.

In Exercises 10 through 13, write the polynomials in order of decreasing terms using the order deglex with power products $x^m y^n z^s$ where $z < y < x$.

10. The polynomial in Exercise 1.

11. The polynomial in Exercise 2.

12. The polynomial in Exercise 3.

13. The polynomial in Exercise 4.

For Exercises 14 through 17, let power products in $\mathbb{R}[x, y, z]$ have order lex where $z < y < x$. If possible, perform a single-step division algorithm reduction that changes the given ideal basis to one having smaller maximum term order.

14. $\langle xy^2 - 2x, x^2y + 4xy, xy - y^2 \rangle$

15. $\langle xy + y^3, y^3 + z, x - y^4 \rangle$

16. $\langle xyz - 3z^2, x^3 + y^2z^3, x^2yz^3 + 4 \rangle$

17. $\langle y^2z^3 + 3, y^3z^2 - 2z, y^2z^2 + 3 \rangle$

In Exercises 18 and 19, let the order of power products in $\mathbb{R}[w, x, y, z]$ be lex with $z < y < x < w$. Find a Gröbner basis for the given ideal.

18. $\langle w + x - y + 4z - 3, 2w + x + y - 2z + 4, w + 3x - 3y + z - 5 \rangle$

19. $\langle w - 4x + 3y - z + 2, 2w - 2x + y - 2z + 5, w - 10x + 8y - z - 1 \rangle$

In Exercises 20 through 22, find a Gröbner basis for the indicated ideal in $\mathbb{R}[x]$.

20. $\langle x^4 + x^3 - 3x^2 - 4x - 4, x^3 + x^2 - 4x - 4 \rangle$

21. $\langle x^4 - 4x^3 + 5x^2 - 2x, x^3 - x^2 - 4x + 4, x^3 - 3x + 2 \rangle$

22. $\langle x^5 + x^2 + 2x - 5, x^3 - x^2 + x - 1 \rangle$

In Exercises 23 through 26, find a Gröbner basis for the given ideal in $\mathbb{R}[x, y]$. Consider the order of power products to be lex with $y < x$. If you can, describe the corresponding algebraic variety in $\mathbb{R}[x, y]$.

23. $\langle x^2y - x - 2, xy + 2y - 9 \rangle$

24. $\langle x^2y + x, xy^2 - y \rangle$

25. $\langle x^2y + x + 1, xy^2 + y - 1 \rangle$

26. $\langle x^2y + xy^2, xy - x \rangle$

Concepts

27. Let F be a field. Mark each of the following true or false.

_______ **a.** Every ideal in $F[\mathbf{x}]$ has a finite basis.

_______ **b.** Every subset of $\mathbb{R}^2$ is an algebraic variety.

_______ **c.** The empty subset of $\mathbb{R}^2$ is an algebraic variety.

_______ **d.** Every finite subset of $\mathbb{R}^2$ is an algebraic variety.

_______ **e.** Every line in $\mathbb{R}^2$ is an algebraic variety.

_______ **f.** Every finite collection of lines in $\mathbb{R}^2$ is an algebraic variety.

_______ **g.** A greatest common divisor of a finite number of polynomials in $\mathbb{R}[x]$ (one indeterminate) can be computed using the division algorithm repeatedly.

_______ **h.** I have computed Gröbner bases before I knew what they were.

_______ **i.** Any ideal in $F[\mathbf{x}]$ has a unique Gröbner basis.

_______ **j.** The ideals $\langle x, y \rangle$ and $\langle x^2, y^2 \rangle$ are equal because they both yield the same algebraic variety, namely $\{(0, 0)\}$, in $\mathbb{R}^2$.

28. Let $\mathbb{R}[x, y]$ be ordered by lex. Give an example to show that $P_i < P_j$ does not imply that P_i divides P_j.

Theory

29. Show that if $f_1, f_2, \cdots, f_r$ are elements of a commutative ring R with unity, then $I = \{c_1 f_1 + c_2 f_2 + \cdots + c_r f_r \mid c_i \in I \text{ for } i = 1, \cdots, r\}$ is an ideal of R.

30. Show that if $f(\mathbf{x}) = g(\mathbf{x})q(\mathbf{x}) + r(\mathbf{x})$ in $F[\mathbf{x}]$, then the common divisors in $F[\mathbf{x}]$ of $f(\mathbf{x})$ and $g(\mathbf{x})$ are the same as the common divisors in $F[\mathbf{x}]$ of $g(\mathbf{x})$ and $r(\mathbf{x})$.

31. Show that $\{xy, y^2 - y)$ is a Gröbner basis for $\langle xy, y^2 - y\rangle$, as asserted after Example 28.9.

32. Let F be a field. Show that if S is a nonempty subset of F^n, then

$$I(S) = \{f(\mathbf{x}) \in F[\mathbf{x}] \mid f(\mathbf{s}) = 0 \text{ for all } \mathbf{s} \in S\}$$

is an ideal of $F[\mathbf{x}]$.

33. Referring to Exercise 32, show that $S \subseteq V(I(S))$.

34. Referring to Exercise 32, give an example of a subset S of $\mathbb{R}^2$ such that $V(I(S)) \neq S$.

35. Referring to Exercise 32, show that if N is an ideal of $F[\mathbf{x}]$, then $N \subseteq I(V(N))$.

36. Referring to Exercise 32, give an example of an ideal N in $\mathbb{R}[x, y]$ such that $I(V(N)) \neq N$.

PART VI

Extension Fields

SECTION 29 INTRODUCTION TO EXTENSION FIELDS

Our Basic Goal Achieved

We are now in a position to achieve our **basic goal,** which, loosely stated, is to show that every nonconstant polynomial has a zero. This will be stated more precisely and proved in Theorem 29.3. We first introduce some new terminology for some old ideas.

29.1 Definition A field E is an **extension field of a field** F if $F \leq E$. ■

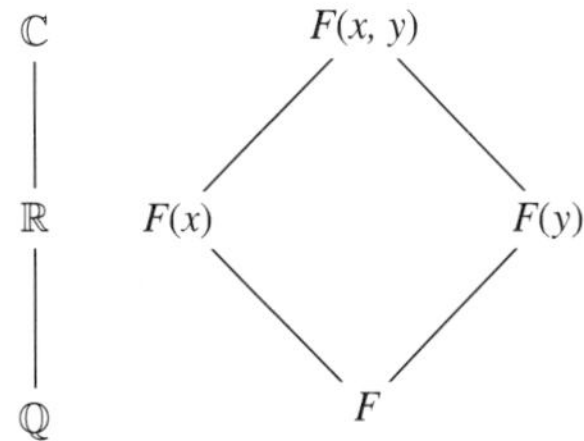

29.2 Figure

Thus $\mathbb{R}$ is an extension field of $\mathbb{Q}$, and $\mathbb{C}$ is an extension field of both $\mathbb{R}$ and $\mathbb{Q}$. As in the study of groups, it will often be convenient to use subfield diagrams to picture extension fields, the larger field being on top. We illustrate this in Fig. 29.2. A configuration where there is just one single column of fields, as at the left-hand side of Fig. 29.2, is often referred to, without any precise definition, as a **tower of fields.**

[†] Section 32 is not required for the remainder of the text.

Now for our *basic goal!* This great and important result follows quickly and elegantly from the techniques we now have at our disposal.

29.3 Theorem **(Kronecker's Theorem) (Basic Goal)** Let F be a field and let $f(x)$ be a nonconstant polynomial in $F[x]$. Then there exists an extension field E of F and an $\alpha \in E$ such that $f(\alpha) = 0$.

Proof By Theorem 23.20, $f(x)$ has a factorization in $F[x]$ into polynomials that are irreducible over F. Let $p(x)$ be an irreducible polynomial in such a factorization. It is clearly sufficient to find an extension field E of F containing an element α such that $p(\alpha) = 0$.

By Theorem 27.25, $\langle p(x)\rangle$ is a maximal ideal in $F[x]$, so $F[x]/\langle p(x)\rangle$ is a field. We claim that F can be identified with a subfield of $F[x]/\langle p(x)\rangle$ in a natural way by use of the map $\psi : F \to F[x]/\langle p(x)\rangle$ given by

$$\psi(a) = a + \langle p(x)\rangle$$

for $a \in F$. This map is one to one, for if $\psi(a) = \psi(b)$, that is, if $a + \langle p(x)\rangle = b + \langle p(x)\rangle$ for some $a, b \in F$, then $(a - b) \in \langle p(x)\rangle$, so $a - b$ must be a multiple of the polynomial $p(x)$, which is of degree ≥ 1. Now $a, b \in F$ implies that $a - b$ is in F. Thus we must have $a - b = 0$, so $a = b$. We defined addition and multiplication in $F[x]/\langle p(x)\rangle$ by choosing any representatives, so we may choose $a \in (a + \langle p(x)\rangle)$. Thus ψ is a homomorphism that maps F one-to-one onto a subfield of $F[x]/\langle p(x)\rangle$. We identify F with $\{a + \langle p(x)\rangle \mid a \in F\}$ by means of this map ψ. Thus we shall view $E = F[x]/\langle p(x)\rangle$ as an extension field of F. We have now manufactured our desired extension field E of F. It remains for us to show that E contains a zero of $p(x)$.

Let us set

$$\alpha = x + \langle p(x)\rangle,$$

so $\alpha \in E$. Consider the evaluation homomorphism $\phi_\alpha : F[x] \to E$, given by Theorem 22.4. If $p(x) = a_0 + a_1x + \cdots + a_nx^n$, where $a_i \in F$, then we have

$$\phi_\alpha(p(x)) = a_0 + a_1(x + \langle p(x)\rangle) + \cdots + a_n(x + \langle p(x)\rangle)^n$$

■ Historical Note

Leopold Kronecker is known for his insistence on constructibility of mathematical objects. As he noted, "God made the integers; all else is the work of man." Thus, he wanted to be able to construct new "domains of rationality" (fields) by using only the existence of integers and indeterminates. He did not believe in starting with the real or complex numbers, because as far as he was concerned, those fields could not be determined in a constructive way. Hence in an 1881 paper, Kronecker created an extension field by simply adjoining to a given field a root α of an irreducible nth degree polynomial $p(x)$; that is, his new field consisted of expressions rational in the original field elements and his new root α with the condition that $p(\alpha) = 0$. The proof of the theorem presented in the text (Theorem 29.3) dates from the twentieth century.

Kronecker completed his dissertation in 1845 at the University of Berlin. For many years thereafter, he managed the family business, ultimately becoming financially independent. He then returned to Berlin, where he was elected to the Academy of Sciences and thus permitted to lecture at the university. On the retirement of Kummer, he became a professor at Berlin, and with Karl Weierstrass (1815–1897) directed the influential mathematics seminar.

in $E = F[x]/\langle p(x)\rangle$. *But we can compute in* $F[x]/\langle p(x)\rangle$ *by choosing representatives, and* x *is a representative of the coset* $\alpha = x + \langle p(x)\rangle$. Therefore,

$$\begin{aligned} p(\alpha) &= \left(a_0 + a_1x + \cdots + a_nx^n\right) + \langle p(x)\rangle \\ &= p(x) + \langle p(x)\rangle = \langle p(x)\rangle = 0 \end{aligned}$$

in $F[x]/\langle p(x)\rangle$. We have found an element α in $E = F[x]/\langle p(x)\rangle$ such that $p(\alpha) = 0$, and therefore $f(\alpha) = 0$. ◆

We illustrate the construction involved in the proof of Theorem 29.3 by two examples.

29.4 Example Let $F = \mathbb{R}$, and let $f(x) = x^2 + 1$, which is well known to have no zeros in $\mathbb{R}$ and thus is irreducible over $\mathbb{R}$ by Theorem 23.10. Then $\langle x^2 + 1\rangle$ is a maximal ideal in $\mathbb{R}[x]$, so $\mathbb{R}[x]/\langle x^2 + 1\rangle$ is a field. Identifying $r \in \mathbb{R}$ with $r + \langle x^2 + 1\rangle$ in $\mathbb{R}[x]/\langle x^2 + 1\rangle$, we can view $\mathbb{R}$ as a subfield of $E = \mathbb{R}[x]/\langle x^2 + 1\rangle$. Let

$$\alpha = x + \langle x^2 + 1\rangle.$$

Computing in $\mathbb{R}[x]/\langle x^2 + 1\rangle$, we find

$$\begin{aligned} \alpha^2 + 1 &= (x + \langle x^2 + 1\rangle)^2 + (1 + \langle x^2 + 1\rangle) \\ &= (x^2 + 1) + \langle x^2 + 1\rangle = 0. \end{aligned}$$

Thus α is a zero of $x^2 + 1$. We shall identify $\mathbb{R}[x]/\langle x^2 + 1\rangle$ with $\mathbb{C}$ at the close of this section. ▲

29.5 Example Let $F = \mathbb{Q}$, and consider $f(x) = x^4 - 5x^2 + 6$. This time $f(x)$ factors in $\mathbb{Q}[x]$ into $(x^2 - 2)(x^2 - 3)$, both factors being irreducible over $\mathbb{Q}$, as we have seen. We can start with $x^2 - 2$ and construct an extension field E of $\mathbb{Q}$ containing α such that $\alpha^2 - 2 = 0$, or we can construct an extension field K of $\mathbb{Q}$ containing an element β such that $\beta^2 - 3 = 0$. The construction in either case is just as in Example 29.4. ▲

Algebraic and Transcendental Elements

As we said before, most of the rest of this text is devoted to the study of zeros of polynomials. We commence this study by putting an element of an extension field E of a field F into one of two categories.

29.6 Definition An element α of an extension field E of a field F is **algebraic over** F if $f(\alpha) = 0$ for some nonzero $f(x) \in F[x]$. If α is not algebraic over F, then α is **transcendental over** F. ■

29.7 Example $\mathbb{C}$ is an extension field of $\mathbb{Q}$. Since $\sqrt{2}$ is a zero of $x^2 - 2$, we see that $\sqrt{2}$ is an algebraic element over $\mathbb{Q}$. Also, i is an algebraic element over $\mathbb{Q}$, being a zero of $x^2 + 1$. ▲

29.8 Example It is well known (but not easy to prove) that the real numbers π and e are transcendental over $\mathbb{Q}$. Here e is the base for the natural logarithms. ▲

Just as we do not speak simply of an *irreducible polynomial,* but rather of an *irreducible polynomial over F*, similarly we don't speak simply of an *algebraic element,* but rather of an *element algebraic over F*. The following illustration shows the reason for this.

29.9 Example The real number π is transcendental over $\mathbb{Q}$, as we stated in Example 29.8. However, π is algebraic over $\mathbb{R}$, for it is a zero of $(x - \pi) \in \mathbb{R}[x]$. ▲

29.10 Example It is easy to see that the real number $\sqrt{1 + \sqrt{3}}$ is algebraic over $\mathbb{Q}$. For if $\alpha = \sqrt{1 + \sqrt{3}}$, then $\alpha^2 = 1 + \sqrt{3}$, so $\alpha^2 - 1 = \sqrt{3}$ and $(\alpha^2 - 1)^2 = 3$. Therefore $\alpha^4 - 2\alpha^2 - 2 = 0$, so α is a zero of $x^4 - 2x^2 - 2$, which is in $\mathbb{Q}[x]$. ▲

To connect these ideas with those of number theory, we give the following definition.

29.11 Definition An element of $\mathbb{C}$ that is algebraic over $\mathbb{Q}$ is an **algebraic number.** A **transcendental number** is an element of $\mathbb{C}$ that is transcendental over $\mathbb{Q}$. ■

There is an extensive and elegant theory of algebraic numbers. (See the Bibliography.)

The next theorem gives a useful characterization of algebraic and transcendental elements over F in an extension field E of F. It also illustrates the importance of our evaluation homomorphisms ϕ_α. *Note that once more we are describing our concepts in terms of mappings.*

29.12 Theorem Let E be an extension field of a field F and let $\alpha \in E$. Let $\phi_\alpha : F[x] \to E$ be the evaluation homomorphism of $F[x]$ into E such that $\phi_\alpha(a) = a$ for $a \in F$ and $\phi_\alpha(x) = \alpha$. Then α is transcendental over F if and only if ϕ_α gives an isomorphism of $F[x]$ with a subdomain of E, that is, if and only if ϕ_α is a one-to-one map.

Proof The element α is transcendental over F if and only if $f(\alpha) \neq 0$ for all nonzero $f(x) \in F[x]$, which is true (by definition) if and only if $\phi_\alpha(f(x)) \neq 0$ for all nonzero $f(x) \in F[x]$, which is true if and only if the kernel of ϕ_α is $\{0\}$, that is, if and only if ϕ_α is a one-to-one map. ◆

The Irreducible Polynomial for α over F

Consider the extension field $\mathbb{R}$ of $\mathbb{Q}$. We know that $\sqrt{2}$ is algebraic over $\mathbb{Q}$, being a zero of $x^2 - 2$. Of course, $\sqrt{2}$ is also a zero of $x^3 - 2x$ and of $x^4 - 3x^2 + 2 = (x^2 - 2)(x^2 - 1)$. Both these other polynomials having $\sqrt{2}$ as a zero were multiples of $x^2 - 2$. The next theorem shows that this is an illustration of a general situation. This theorem plays a central role in our later work.

29.13 Theorem Let E be an extension field of F, and let $\alpha \in E$, where α is algebraic over F. Then there is an irreducible polynomial $p(x) \in F[x]$ such that $p(\alpha) = 0$. This irreducible polynomial $p(x)$ is uniquely determined up to a constant factor in F and is a polynomial of minimal degree ≥ 1 in $F[x]$ having α as a zero. If $f(\alpha) = 0$ for $f(x) \in F[x]$, with $f(x) \neq 0$, then $p(x)$ divides $f(x)$.

Proof Let ϕ_α be the evaluation homomorphism of $F[x]$ into E, given by Theorem 22.4. The kernel of ϕ_α is an ideal and by Theorem 27.24 it must be a principal ideal generated by some $p(x) \in F[x]$. Now $\langle p(x) \rangle$ consists precisely of those elements of $F[x]$ having α as a zero. Thus, if $f(\alpha) = 0$ for $f(x) \neq 0$, then $f(x) \in \langle p(x) \rangle$, so $p(x)$ divides $f(x)$. Thus $p(x)$ is a polynomial of minimal degree ≥ 1 having α as a zero, and any other such polynomial of the same degree as $p(x)$ must be of the form $(a)p(x)$ for some $a \in F$.

It only remains for us to show that $p(x)$ is irreducible. If $p(x) = r(x)s(x)$ were a factorization of $p(x)$ into polynomials of lower degree, then $p(\alpha) = 0$ would imply that $r(\alpha)s(\alpha) = 0$, so either $r(\alpha) = 0$ or $s(\alpha) = 0$, since E is a field. This would contradict the fact that $p(x)$ is of minimal degree ≥ 1 such that $p(\alpha) = 0$. Thus $p(x)$ is irreducible. ◆

By multiplying by a suitable constant in F, we can assume that the coefficient of the highest power of x appearing in $p(x)$ of Theorem 29.13 is 1. Such a polynomial having 1 as the coefficient of the highest power of x appearing is a **monic polynomial.**

29.14 Definition Let E be an extension field of a field F, and let $\alpha \in E$ be algebraic over F. The unique monic polynomial $p(x)$ having the property described in Theorem 29.13 is the **irreducible polynomial for α over F** and will be denoted by $\operatorname{irr}(\alpha, F)$. The degree of $\operatorname{irr}(\alpha, F)$ is the **degree of α over F**, denoted by $\deg(\alpha, F)$. ■

29.15 Example We know that $\operatorname{irr}(\sqrt{2}, \mathbb{Q}) = x^2 - 2$. Referring to Example 29.10, we see that for $\alpha = \sqrt{1 + \sqrt{3}}$ in $\mathbb{R}$, α is a zero of $x^4 - 2x^2 - 2$, which is in $\mathbb{Q}[x]$. Since $x^4 - 2x^2 - 2$ is irreducible over $\mathbb{Q}$ (by Eisenstein with $p = 2$, or by application of the technique of Example 23.14), we see that

$$\operatorname{irr}(\sqrt{1 + \sqrt{3}}, \mathbb{Q}) = x^4 - 2x^2 - 2.$$

Thus $\sqrt{1 + \sqrt{3}}$ is algebraic of degree 4 over $\mathbb{Q}$. ▲

Just as we must speak of an element α as *algebraic over F* rather than simply as *algebraic,* we must speak of the *degree of α over F* rather than the *degree of α*. To take a trivial illustration, $\sqrt{2} \in \mathbb{R}$ is algebraic of degree 2 over $\mathbb{Q}$ but algebraic of degree 1 over $\mathbb{R}$, for $\operatorname{irr}(\sqrt{2}, \mathbb{R}) = x - \sqrt{2}$.

The quick development of the theory here is due to the machinery of homomorphisms and ideal theory that we now have at our disposal. Note especially our constant use of the evaluation homomorphisms ϕ_α.

Simple Extensions

Let E be an extension field of a field F, and let $\alpha \in E$. Let ϕ_α be the evaluation homomorphism of $F[x]$ into E with $\phi_\alpha(a) = a$ for $a \in F$ and $\phi_\alpha(x) = \alpha$, as in Theorem 22.4. We consider two cases.

Case I *Suppose α is algebraic over F.* Then as in Theorem 29.13, the kernel of ϕ_α is $\langle \text{irr}(\alpha, F)\rangle$ and by Theorem 27.25, $\langle \text{irr}(\alpha, F)\rangle$ is a maximal ideal of $F[x]$. Therefore, $F[x]/\langle \text{irr}(\alpha, F)\rangle$ is a field and is isomorphic to the image $\phi_\alpha[F[x]]$ in E. This subfield $\phi_\alpha[F[x]]$ of E is then the smallest subfield of E containing F and α. We shall denote this field by $F(\alpha)$.

Case II *Suppose α is transcendental over F.* Then by Theorem 29.12, ϕ_α gives an isomorphism of $F[x]$ with a subdomain of E. Thus in this case $\phi_\alpha[F[x]]$ is *not* a field but an integral domain that we shall denote by $F[\alpha]$. By Corollary 21.8, E contains a field of quotients of $F[\alpha]$, which is thus the smallest subfield of E containing F and α. As in Case I, we denote this field by $F(\alpha)$.

29.16 Example Since π is transcendental over $\mathbb{Q}$, the field $\mathbb{Q}(\pi)$ is isomorphic to the field $\mathbb{Q}(x)$ of rational functions over $\mathbb{Q}$ in the indeterminate x. Thus from a structural viewpoint, an element that is transcendental over a field F behaves as though it were an indeterminate over F. ▲

29.17 Definition An extension field E of a field F is a **simple extension of** F if $E = F(\alpha)$ for some $\alpha \in E$. ■

Many important results appear throughout this section. We have now developed so much machinery that results are starting to pour out of our efficient plant at an alarming rate. The next theorem gives us insight into the nature of the field $F(\alpha)$ in the case where α is algebraic over F.

29.18 Theorem Let E be a simple extension $F(\alpha)$ of a field F, and let α be algebraic over F. Let the degree of $\text{irr}(\alpha, F)$ be $n \geq 1$. Then every element β of $E = F(\alpha)$ can be uniquely expressed in the form

$$\beta = b_0 + b_1\alpha + \cdots + b_{n-1}\alpha^{n-1},$$

where the b_i are in F.

Proof For the usual evaluation homomorphism ϕ_α, every element of

$$F(\alpha) = \phi_\alpha[F[x]]$$

is of the form $\phi_\alpha(f(x)) = f(\alpha)$, a formal polynomial in α with coefficients in F. Let

$$\text{irr}(\alpha, F) = p(x) = x^n + a_{n-1}x^{n-1} + \cdots + a_0.$$

Then $p(\alpha) = 0$, so

$$\alpha^n = -a_{n-1}\alpha^{n-1} - \cdots - a_0.$$

This equation in $F(\alpha)$ can be used to express every monomial α^m for $m \geq n$ in terms of powers of α that are less than n. For example,

$$\begin{aligned}\alpha^{n+1} &= \alpha\alpha^n = -a_{n-1}\alpha^n - a_{n-2}\alpha^{n-1} - \cdots - a_0\alpha \\ &= -a_{n-1}\left(-a_{n-1}\alpha^{n-1} - \cdots - a_0\right) - a_{n-2}\alpha^{n-1} - \cdots - a_0\alpha.\end{aligned}$$

Thus, if $\beta \in F(\alpha)$, β can be expressed in the required form

$$\beta = b_0 + b_1\alpha + \cdots + b_{n-1}\alpha^{n-1}.$$

For uniqueness, if

$$b_0 + b_1\alpha + \cdots + b_{n-1}\alpha^{n-1} = b_0' + b_1'\alpha + \cdots + b_{n-1}'\alpha^{n-1}$$

for $b_i' \in F$, then

$$(b_0 - b_0') + (b_1 - b_1')x + \cdots + (b_{n-1} - b_{n-1}')x^{n-1} = g(x)$$

is in $F[x]$ and $g(\alpha) = 0$. Also, the degree of $g(x)$ is less than the degree of $\mathrm{irr}(\alpha, F)$. Since $\mathrm{irr}(\alpha, F)$ is a nonzero polynomial of minimal degree in $F[x]$ having α as a zero, we must have $g(x) = 0$. Therefore, $b_i - b_i' = 0$, so

$$b_i = b_i',$$

and the uniqueness of the b_i is established. ◆

We give an impressive example illustrating Theorem 29.18.

29.19 Example The polynomial $p(x) = x^2 + x + 1$ in $\mathbb{Z}_2[x]$ is irreducible over $\mathbb{Z}_2$ by Theorem 23.10, since neither element 0 nor element 1 of $\mathbb{Z}_2$ is a zero of $p(x)$. By Theorem 29.3, we know that there is an extension field E of $\mathbb{Z}_2$ containing a zero α of $x^2 + x + 1$. By Theorem 29.18, $\mathbb{Z}_2(\alpha)$ has as elements $0 + 0\alpha$, $1 + 0\alpha$, $0 + 1\alpha$, and $1 + 1\alpha$, that is, 0, 1, α, and $1 + \alpha$. *This gives us a new finite field, of four elements!* The addition and multiplication tables for this field are shown in Tables 29.20 and 29.21. For example, to compute $(1 + \alpha)(1 + \alpha)$ in $\mathbb{Z}_2(\alpha)$, we observe that since $p(\alpha) = \alpha^2 + \alpha + 1 = 0$, then

$$\alpha^2 = -\alpha - 1 = \alpha + 1.$$

Therefore,

$$(1 + \alpha)(1 + \alpha) = 1 + \alpha + \alpha + \alpha^2 = 1 + \alpha^2 = 1 + \alpha + 1 = \alpha. \quad ▲$$

Finally, we can use Theorem 29.18 to fulfill our promise of Example 29.4 and show that $\mathbb{R}[x]/\langle x^2 + 1\rangle$ is isomorphic to the field $\mathbb{C}$ of complex numbers. We saw in Example 29.4 that we can view $\mathbb{R}[x]/\langle x^2 + 1\rangle$ as an extension field of $\mathbb{R}$. Let

$$\alpha = x + \langle x^2 + 1\rangle.$$

29.20 Table

$+$	0	1	α	$1+\alpha$
0	0	1	α	$1+\alpha$
1	1	0	$1+\alpha$	α
α	α	$1+\alpha$	0	1
$1+\alpha$	$1+\alpha$	α	1	0

29.21 Table

	0	1	α	$1+\alpha$
0	0	0	0	0
1	0	1	α	$1+\alpha$
α	0	α	$1+\alpha$	1
$1+\alpha$	0	$1+\alpha$	1	α

Then $\mathbb{R}(\alpha) = \mathbb{R}[x]/\langle x^2 + 1\rangle$ and consists of all elements of the form $a + b\alpha$ for $a, b \in \mathbb{R}$, by Theorem 29.18. But since $\alpha^2 + 1 = 0$, we see that α plays the role of $i \in \mathbb{C}$, and $a + b\alpha$ plays the role of $(a + bi) \in \mathbb{C}$. Thus $\mathbb{R}(\alpha) \simeq \mathbb{C}$. *This is the elegant algebraic way to construct* $\mathbb{C}$ *from* $\mathbb{R}$.

EXERCISES 29

Computations

In Exercises 1 through 5, show that the given number $\alpha \in \mathbb{C}$ is algebraic over $\mathbb{Q}$ by finding $f(x) \in \mathbb{Q}[x]$ such that $f(\alpha) = 0$.

1. $1 + \sqrt{2}$

2. $\sqrt{2} + \sqrt{3}$

3. $1 + i$

4. $\sqrt{1 + \sqrt[3]{2}}$

5. $\sqrt{\sqrt[3]{2} - i}$

In Exercises 6 through 8, find $\text{irr}(\alpha, \mathbb{Q})$ and $\deg(\alpha, \mathbb{Q})$ for the given algebraic number $\alpha \in \mathbb{C}$. Be prepared to prove that your polynomials are irreducible over $\mathbb{Q}$ if challenged to do so.

6. $\sqrt{3 - \sqrt{6}}$

7. $\sqrt{(\frac{1}{3}) + \sqrt{7}}$

8. $\sqrt{2} + i$

In Exercises 9 through 16, classify the given $\alpha \in \mathbb{C}$ as algebraic or transcendental over the given field F. If α is algebraic over F, find $\deg(\alpha, F)$.

9. $\alpha = i$, $F = \mathbb{Q}$

10. $\alpha = 1 + i$, $F = \mathbb{R}$

11. $\alpha = \sqrt{\pi}$, $F = \mathbb{Q}$

12. $\alpha = \sqrt{\pi}$, $F = \mathbb{R}$

13. $\alpha = \sqrt{\pi}$, $F = \mathbb{Q}(\pi)$

14. $\alpha = \pi^2$, $F = \mathbb{Q}$

15. $\alpha = \pi^2$, $F = \mathbb{Q}(\pi)$

16. $\alpha = \pi^2$, $F = \mathbb{Q}(\pi^3)$

17. Refer to Example 29.19 of the text. The polynomial $x^2 + x + 1$ has a zero α in $\mathbb{Z}_2(\alpha)$ and thus must factor into a product of linear factors in $(\mathbb{Z}_2(\alpha))[x]$. Find this factorization. [*Hint:* Divide $x^2 + x + 1$ by $x - \alpha$ by long division, using the fact that $\alpha^2 = \alpha + 1$.]

18. a. Show that the polynomial $x^2 + 1$ is irreducible in $\mathbb{Z}_3[x]$.

b. Let α be a zero of $x^2 + 1$ in an extension field of $\mathbb{Z}_3$. As in Example 29.19, give the multiplication and addition tables for the nine elements of $\mathbb{Z}_3(\alpha)$, written in the order $0, 1, 2, \alpha, 2\alpha, 1 + \alpha, 1 + 2\alpha, 2 + \alpha$, and $2 + 2\alpha$.

Concepts

In Exercises 19 through 22, correct the definition of the italicized term without reference to the text, if correction is needed, so that it is in a form acceptable for publication.

19. An element α of an extension field E of a field F if *algebraic over* F if and only if α is a zero of some polynomial.

20. An element β of an extension field E of a field F is *transcendental over* F if and only if β is not a zero of any polynomial in $F[x]$.

21. A *monic polynomial in* $F[x]$ is one having all coefficients equal to 1.

22. A field E is a *simple extension* of a subfield F if and only if there exists some $\alpha \in E$ such that no proper subfield of E contains α.

23. Mark each of the following true or false.

_______ **a.** The number π is transcendental over $\mathbb{Q}$.
_______ **b.** $\mathbb{C}$ is a simple extension of $\mathbb{R}$.
_______ **c.** Every element of a field F is algebraic over F.
_______ **d.** $\mathbb{R}$ is an extension field of $\mathbb{Q}$.
_______ **e.** $\mathbb{Q}$ is an extension field of $\mathbb{Z}_2$.
_______ **f.** Let $\alpha \in \mathbb{C}$ be algebraic over $\mathbb{Q}$ of degree n. If $f(\alpha) = 0$ for nonzero $f(x) \in \mathbb{Q}[x]$, then (degree $f(x)) \geq n$.
_______ **g.** Let $\alpha \in \mathbb{C}$ be algebraic over $\mathbb{Q}$ of degree n. If $f(\alpha) = 0$ for nonzero $f(x) \in \mathbb{R}[x]$, then (degree $f(x)) \geq n$.
_______ **h.** Every nonconstant polynomial in $F[x]$ has a zero in some extension field of F.
_______ **i.** Every nonconstant polynomial in $F[x]$ has a zero in every extension field of F.
_______ **j.** If x is an indeterminate, $\mathbb{Q}[\pi] \simeq \mathbb{Q}[x]$.

24. We have stated without proof that π and e are transcendental over $\mathbb{Q}$.

a. Find a subfield F of $\mathbb{R}$ such that π is algebraic of degree 3 over F.
b. Find a subfield E of $\mathbb{R}$ such that e^2 is algebraic of degree 5 over E.

25. **a.** Show that $x^3 + x^2 + 1$ is irreducible over $\mathbb{Z}_2$.
b. Let α be a zero of $x^3 + x^2 + 1$ in an extension field of $\mathbb{Z}_2$. Show that $x^3 + x^2 + 1$ factors into three linear factors in $(\mathbb{Z}_2(\alpha))[x]$ by actually finding this factorization. [*Hint:* Every element of $\mathbb{Z}_2(\alpha)$ is of the form

$$a_0 + a_1\alpha + a_2\alpha^2 \quad \text{for} \quad a_i = 0, 1.$$

Divide $x^3 + x^2 + 1$ by $x - \alpha$ by long division. Show that the quotient also has a zero in $\mathbb{Z}_2(\alpha)$ by simply trying the eight possible elements. Then complete the factorization.]

26. Let E be an extension field of $\mathbb{Z}_2$ and let $\alpha \in E$ be algebraic of degree 3 over $\mathbb{Z}_2$. Classify the groups $\langle \mathbb{Z}_2(\alpha), + \rangle$ and $\langle (\mathbb{Z}_2(\alpha))^*, \cdot \rangle$ according to the Fundamental Theorem of finitely generated abelian groups. As usual, $(\mathbb{Z}_2(\alpha))^*$ is the set of nonzero elements of $\mathbb{Z}_2(\alpha)$.

27. Let E be an extension field of a field F and let $\alpha \in E$ be algebraic over F. The polynomial $\operatorname{irr}(\alpha, F)$ is sometimes referred to as the **minimal polynomial for α over** F. Why is this designation appropriate?

Proof Synopsis

28. Give a two- or three-sentence synopsis of Theorem 29.3.

Theory

29. Let E be an extension field of F, and let $\alpha, \beta \in E$. Suppose α is transcendental over F but algebraic over $F(\beta)$. Show that β is algebraic over $F(\alpha)$.

30. Let E be an extension field of a finite field F, where F has q elements. Let $\alpha \in E$ be algebraic over F of degree n. Prove that $F(\alpha)$ has q^n elements.

31. **a.** Show that there exists an irreducible polynomial of degree 3 in $\mathbb{Z}_3[x]$.
b. Show from part (a) that there exists a finite field of 27 elements. [*Hint:* Use Exercise 30.]

32. Consider the prime field $\mathbb{Z}_p$ of characteristic $p \neq 0$.

a. Show that, for $p \neq 2$, not every element in $\mathbb{Z}_p$ is a square of an element of $\mathbb{Z}_p$. [*Hint:* $1^2 = (p-1)^2 = 1$ in $\mathbb{Z}_p$. Deduce the desired conclusion *by counting*.]

b. Using part (a), show that there exist finite fields of p^2 elements for every prime p in $\mathbb{Z}^+$.

33. Let E be an extension field of a field F and let $\alpha \in E$ be transcendental over F. Show that every element of $F(\alpha)$ that is not in F is also transcendental over F.

34. Show that $\{a + b(\sqrt[3]{2}) + c(\sqrt[3]{2})^2 \mid a, b, c \in \mathbb{Q}\}$ is a subfield of $\mathbb{R}$ by using the ideas of this section, rather than by a formal verification of the field axioms. [*Hint:* Use Theorem 29.18.]

35. Following the idea of Exercise 31, show that there exists a field of 8 elements; of 16 elements; of 25 elements.

36. Let F be a finite field of characteristic p. Show that every element of F is algebraic over the prime field $\mathbb{Z}_p \leq F$. [*Hint:* Let F^* be the set of nonzero elements of F. Apply group theory to the group $\langle F^*, \cdot \rangle$ to show that every $\alpha \in F^*$ is a zero of some polynomial in $\mathbb{Z}_p[x]$ of the form $x^n - 1$.]

37. Use Exercises 30 and 36 to show that every finite field is of prime-power order, that is, it has a prime-power number of elements.

SECTION 30 Vector Spaces

The notions of a vector space, scalars, independent vectors, and bases may be familiar. In this section, we present these ideas where the scalars may be elements of any field. We use Greek letters like α and β for vectors since, in our application, the vectors will be elements of an extension field E of a field F. The proofs are all identical with those often given in a first course in linear algebra. If these ideas are familiar, we suggest studying Examples 30.4, 30.8, 30.11, 30.14, and 30.22, and then reading Theorem 30.23 and its proof. If the examples and the theorem are understood, then do some exercises and proceed to the next section.

Definition and Elementary Properties

The topic of vector spaces is the cornerstone of linear algebra. Since linear algebra is not the subject for study in this text, our treatment of vector spaces will be brief, designed to develop only the concepts of linear independence and dimension that we need for our field theory.

The terms *vector* and *scalar* are probably familiar from calculus. Here we allow scalars to be elements of any field, not just the real numbers, and develop the theory by axioms just as for the other algebraic structures we have studied.

30.1 Definition Let F be a field. A **vector space over** F (or F-**vector space**) consists of an abelian group V under addition together with an operation of scalar multiplication of each element of V by each element of F on the left, such that for all $a, b \in F$ and $\alpha, \beta \in V$ the following conditions are satisfied:

$\mathscr{V}_1$. $a\alpha \in V$.

$\mathscr{V}_2$. $a(b\alpha) = (ab)\alpha$.

$\mathscr{V}_3$. $(a + b)\alpha = (a\alpha) + (b\alpha)$.

$\mathscr{V}_4$. $a(\alpha + \beta) = (a\alpha) + (a\beta)$.

$\mathscr{V}_5$. $1\alpha = \alpha$.

The elements of V are **vectors** and the elements of F are **scalars.** When only one field F is under discussion, we drop the reference to F and refer to a *vector space*. ■

Note that scalar multiplication for a vector space is not a binary operation on one set in the sense we defined it in Section 2. It associates an element $a\alpha$ of V with each ordered pair (a, α), consisting of an element a of F and an element α of V. Thus scalar multiplication is a *function* mapping $F \times V$ into V. Both the additive identity for V, the 0-vector, and the additive identity for F, the 0-scalar, will be denoted by 0.

30.2 Example Consider the abelian group $\langle \mathbb{R}_n, + \rangle = \mathbb{R} \times \mathbb{R} \times \cdots \times \mathbb{R}$ for n factors, which consists of ordered n-tuples under addition by components. Define scalar multiplication for scalars in $\mathbb{R}$ by

$$r\alpha = (ra_1, \cdots, ra_n)$$

Historical Note

The ideas behind the abstract notion of a vector space occurred in many concrete examples during the nineteenth century and earlier. For example, William Rowan Hamilton dealt with complex numbers explicitly as pairs of real numbers and, as noted in Section 24, also dealt with triples and eventually quadruples of real numbers in his invention of the quaternions. In these cases, the "vectors" turned out to be objects which could both be added and multiplied by scalars, using "reasonable" rules for both of these operations. Other examples of such objects included differential forms (expressions under integral signs) and algebraic integers.

Although Hermann Grassmann (1809–1877) succeeded in working out a detailed theory of n-dimensional spaces in his *Die Lineale Ausdehnungslehre* of 1844 and 1862, the first mathematician to give an abstract definition of a vector space equivalent to Definition 30.1 was Giuseppe Peano (1858–1932) in his *Calcolo Geometrico* of 1888. Peano's aim in the book, as the title indicates, was to develop a geometric calculus. According to Peano, such a calculus "consists of a system of operations analogous to those of algebraic calculus, but in which the objects with which the calculations are performed are, instead of numbers, geometrical objects." Curiously, Peano's work had no immediate effect on the mathematical scene. Although Hermann Weyl (1885–1955) essentially repeated Peano's definition in his *Space-Time-Matter* of 1918, the definition of a vector space did not enter the mathematical mainstream until it was announced for a third time by Stefan Banach (1892–1945) in the 1922 publication of his dissertation dealing with what we now call *Banach spaces,* complete normed vector spaces.

for $r \in \mathbb{R}$ and $\alpha = (a_1, \cdots, a_n) \in \mathbb{R}^n$. With these operations, $\mathbb{R}^n$ becomes a vector space over $\mathbb{R}$. The axioms for a vector space are readily checked. In particular, $\mathbb{R}^2 = \mathbb{R} \times \mathbb{R}$ as a vector space over $\mathbb{R}$ can be viewed as all "vectors whose starting points are the origin of the Euclidean plane" in the sense often studied in calculus courses. ▲

30.3 Example For any field F, $F[x]$ can be viewed as a vector space over F, where addition of vectors is ordinary addition of polynomials in $F[x]$ and scalar multiplication $a\alpha$ of an element of $F[x]$ by an element of F is ordinary multiplication in $F[x]$. The axioms $\mathscr{V}_1$ through $\mathscr{V}_5$ for a vector space then follow immediately from the fact that $F[x]$ is a ring with unity. ▲

30.4 Example Let E be an extension field of a field F. Then E can be regarded as a vector space over F, where addition of vectors is the usual addition in E and scalar multiplication $a\alpha$ is the usual field multiplication in E with $a \in F$ and $\alpha \in E$. The axioms follow at once from the field axioms for E. Here our field of scalars is actually a subset of our space of vectors. *It is this example that is the important one for us.* ▲

We are assuming nothing about vector spaces from previous work and shall prove everything we need from the definition, even though the results may be familiar from calculus.

30.5 Theorem If V is a vector space over F, then $0\alpha = 0$, $a0 = 0$ and $(-a)\alpha = a(-\alpha) = -(a\alpha)$ for all $a \in F$ and $\alpha \in V$.

Proof The equation $0\alpha = 0$ is to be read "(0-scalar)α = 0-vector." Likewise, $a0 = 0$ is to be read "a(0-vector) = 0-vector." The proofs here are very similar to those in Theorem 18.8 for a ring and again depend heavily on the distributive laws $\mathscr{V}_3$ and $\mathscr{V}_4$. Now

$$(0\alpha) = (0 + 0)\alpha = (0\alpha) + (0\alpha)$$

is an equation in the abelian group $\langle V, +\rangle$, so by the group cancellation law, $0 = 0\alpha$. Likewise, from

$$a0 = a(0 + 0) = a0 + a0,$$

we conclude that $a0 = 0$. Then

$$0 = 0\alpha = (a + (-a))\alpha = a\alpha + (-a)\alpha,$$

so $(-a)\alpha = -(a\alpha)$. Likewise, from

$$0 = a0 = a(\alpha + (-\alpha)) = a\alpha + a(-\alpha),$$

we conclude that $a(-\alpha) = -(a\alpha)$ also. ◆

Linear Independence and Bases

30.6 Definition Let V be a vector space over F. The vectors in a subset $S = \{\alpha_i \mid i \in I\}$ of V **span** (or **generate**) V if for every $\beta \in V$, we have

$$\beta = a_1\alpha_{i_1} + a_2\alpha_{i_2} + \cdots + a_n\alpha_{i_n}$$

for some $a_j \in F$ and $\alpha_{i_j} \in S$, $j = 1, \cdots, n$. A vector $\sum_{j=1}^{n} a_j\alpha_{i_j}$ is a **linear combination of the** α_{i_j}. ■

30.7 Example In the vector space $\mathbb{R}^n$ over $\mathbb{R}$ of Example 30.2, the vectors

$$(1, 0, \cdots, 0), (0, 1, \cdots, 0), \cdots, (0, 0, \cdots, 1)$$

clearly span $\mathbb{R}^n$, for

$$(a_1, a_2, \cdots, a_n) = a_1(1, 0, \cdots, 0) + a_2(0, 1, \cdots, 0) + \cdots + a_n(0, 0, \cdots, 1).$$

Also, the monomials x^m for $m \geq 0$ span $F[x]$ over F, the vector space of Example 30.3. ▲

30.8 Example Let F be a field and E an extension field of F. Let $\alpha \in E$ be algebraic over F. Then $F(\alpha)$ is a vector space over F and by Theorem 30.18, it is spanned by the vectors in $\{1, \alpha, \cdots, \alpha^{n-1}\}$, where $n = \deg(\alpha, F)$. *This is the important example for us.* ▲

30.9 Definition A vector space V over a field F is **finite dimensional** if there is a finite subset of V whose vectors span V. ■

30.10 Example Example 30.7 shows that $\mathbb{R}^n$ is finite dimensional. The vector space $F[x]$ over F is *not* finite dimensional, since polynomials of arbitrarily large degree could not be linear combinations of elements of any *finite* set of polynomials. ▲

30.11 Example If $F \leq E$ and $\alpha \in E$ is algebraic over the field F, Example 30.8 shows that $F(\alpha)$ is a finite-dimensional vector space over F. *This is the most important example for us.* ▲

The next definition contains the most important idea in this section.

30.12 Definition The vectors in a subset $S = \{\alpha_i \mid i \in I\}$ of a vector space V over a field F are **linearly independent over** F if, for any distinct vectors $\alpha_{i_j} \in S$, coefficients $a_j \in F$ and $n \in \mathbb{Z}^+$, we have $\sum_{j=1}^{n} a_j\alpha_{i_j} = 0$ in V only if $a_j = 0$ for $j = 1, \cdots, n$. If the vectors are not linearly independent over F, they are **linearly dependent over** F. ■

Thus the vectors in $\{\alpha_i \mid i \in I\}$ are linearly independent over F if the only way the 0-vector can be expressed as a linear combination of the vectors α_i is to have all scalar coefficients equal to 0. If the vectors are linearly dependent over F, then there exist $a_j \in F$ for $j = 1, \cdots, n$ such that $\sum_{j=1}^{n} a_j\alpha_{i_j} = 0$, where not all $a_j = 0$.

30.13 Example Observe that the vectors spanning the space $\mathbb{R}^n$ that are given in Example 30.7 are linearly independent over $\mathbb{R}$. Likewise, the vectors in $\{x^m \mid m \geq 0\}$ are linearly independent vectors of $F[x]$ over F. Note that $(1, -1)$, $(2, 1)$, and $(-3, 2)$ are linearly dependent in $\mathbb{R}^2$ over $\mathbb{R}$, since

$$7(1, -1) + (2, 1) + 3(-3, 2) = (0, 0) = 0.$$ ▲

30.14 Example Let E be an extension field of a field F, and let $\alpha \in E$ be algebraic over F. If $\deg(\alpha, F) = n$, then by Theorem 29.18, every element of $F(\alpha)$ can be *uniquely* expressed in the form

$$b_0 + b_1\alpha + \cdots + b_{n-1}\alpha^{n-1}$$

for $b_i \in F$. In particular, $0 = 0 + 0\alpha + \cdots + 0\alpha^{n-1}$ must be a *unique* such expression for 0. Thus the elements $1, \alpha, \cdots, \alpha^{n-1}$ are linearly independent vectors in $F(\alpha)$ over the field F. They also span $F(\alpha)$, so by the next definition, $1, \alpha, \cdots, \alpha^{n-1}$ form a *basis* for $F(\alpha)$ over F. *This is the important example for us*. In fact, this is the reason we are doing this material on vector spaces. ▲

30.15 Definition If V is a vector space over a field F, the vectors in a subset $B = \{\beta_i \mid i \in I\}$ of V form a **basis for V over F** if they span V and are linearly independent. ■

Dimension

The only other results we wish to prove about vector spaces are that every finite-dimensional vector space has a basis, and that any two bases of a finite-dimensional vector space have the same number of elements. Both these facts are true without the assumption that the vector space is finite dimensional, but the proofs require more knowledge of set theory than we are assuming, and the finite-dimensional case is all we need. First we give an easy lemma.

30.16 Lemma Let V be a vector space over a field F, and let $\alpha \in V$. If α is a linear combination of vectors β_i in V for $i = 1, \cdots, m$ and each β_i is a linear combination of vectors γ_j in V for $j = 1, \cdots, n$, then α is a linear combination of the γ_j.

Proof Let $\alpha = \sum_{i=1}^{m} a_i\beta_i$, and let $\beta_i = \sum_{j=1}^{n} b_{ij}\gamma_j$, where a_i and b_{ij} are in F. Then

$$\alpha = \sum_{i=1}^{m} a_i \left(\sum_{j=1}^{n} b_{ij}\gamma_j \right) = \sum_{j=1}^{n} \left(\sum_{i=1}^{m} a_i b_{ij} \right) \gamma_j,$$

and $(\sum_{i=1}^{m} a_i b_{ij}) \in F$. ◆

30.17 Theorem In a finite-dimensional vector space, every finite set of vectors spanning the space contains a subset that is a basis.

Proof Let V be finite dimensional over F, and let vectors $\alpha_1, \cdots, \alpha_n$ in V span V. Let us list the α_i in a row. Examine each α_i in succession, starting at the left with $i = 1$, and discard the first α_j that is some linear combination of the preceding α_i for $i < j$. Then continue, starting with the following α_{j+1}, and discard the next α_k that is some linear combination of its remaining predecessors, and so on. When we reach α_n after a finite number of steps, those α_i remaining in our list are such that none is a linear combination of the preceding α_i in this reduced list. Lemma 30.16 shows that any vector that is a linear combination of the original collection of α_i is still a linear combination of our reduced, and possibly smaller, set in which no α_i is a linear combination of its predecessors. Thus the vectors in the reduced set of α_i again span V.

For the reduced set, suppose that

$$a_1\alpha_{i_1} + \cdots + a_r\alpha_{i_r} = 0$$

for $i_1 < i_2 < \cdots < i_r$ and that some $a_j \neq 0$. We may assume from Theorem 30.5 that $a_r \neq 0$, or we could drop $a_r\alpha_{i_r}$ from the left side of the equation. Then, using

Theorem 30.5 again, we obtain

$$\alpha_{i_r} = \left(-\frac{a_1}{a_r}\right)\alpha_{i_1} + \cdots + \left(-\frac{a_{r-1}}{a_r}\right)\alpha_{i_{r-1}},$$

which shows that α_{i_r} is a linear combination of its predecessors, contradicting our construction. Thus the vectors α_i in the reduced set both span V and are linearly independent, so they form a basis for V over F. ◆

30.18 Corollary A finite-dimensional vector space has a finite basis.

Proof By definition, a finite-dimensional vector space has a finite set of vectors that span the space. Theorem 30.17 completes the proof. ◆

The next theorem is the culmination of our work on vector spaces.

30.19 Theorem Let $S = \{\alpha_1, \cdots, \alpha_r\}$ be a finite set of linearly independent vectors of a finite-dimensional vector space V over a field F. Then S can be enlarged to a basis for V over F. Furthermore, if $B = \{\beta_1, \cdots, \beta_n\}$ is any basis for V over F, then $r \leq n$.

Proof By Corollary 30.18, there is a basis $B = \{\beta_1, \cdots, \beta_n\}$ for V over F. Consider the finite sequence of vectors

$$\alpha_1, \cdots, \alpha_r, \beta_1, \cdots, \beta_n.$$

These vectors span V, since B is a basis. Following the technique, used in Theorem 30.17, of discarding in turn each vector that is a linear combination of its remaining predecessors, working from left to right, we arrive at a basis for V. Observe that no α_i is cast out, since the α_i are linearly independent. Thus S can be enlarged to a basis for V over F.

For the second part of the conclusion, consider the sequence

$$\alpha_1, \beta_1, \cdots, \beta_n.$$

These vectors are not linearly independent over F, because α_1 is a linear combination

$$\alpha_1 = b_1\beta_1 + \cdots + b_n\beta_n,$$

since the β_i form a basis. Thus

$$\alpha_1 + (-b_1)\beta_1 + \cdots + (-b_n)\beta_n = 0.$$

The vectors in the sequence do span V, and if we form a basis by the technique of working from left to right and casting out in turn each vector that is a linear combination of its remaining predecessors, at least one β_i must be cast out, giving a basis

$$\{\alpha_1, \beta_1^{(1)}, \cdots, \beta_m^{(1)}\},$$

where $m \leq n - 1$. Applying the same technique to the sequence of vectors

$$\alpha_1, \alpha_2, \beta_1^{(1)}, \cdots, \beta_m^{(1)},$$

we arrive at a new basis

$$\{\alpha_1, \alpha_2, \beta_1^{(2)}, \cdots, \beta_s^{(2)}\},$$

with $s \le n - 2$. Continuing, we arrive finally at a basis

$$\{\alpha_1, \cdots, \alpha_r, \beta_1^{(r)}, \cdots, \beta_t^{(r)}\},$$

where $0 \le t \le n - r$. Thus $r \le n$. ◆

30.20 Corollary Any two bases of a finite-dimensional vector space V over F have the same number of elements.

Proof Let $B = \{\beta_1, \cdots, \beta_n\}$ and $B' = \{\beta_1', \cdots, \beta_m'\}$ be two bases. Then by Theorem 30.19, regarding B as an independent set of vectors and B' as a basis, we see that $n \le m$. A symmetric argument gives $m \le n$, so $m = n$. ◆

30.21 Definition If V is a finite-dimensional vector space over a field F, the number of elements in a basis (independent of the choice of basis, as just shown) is the **dimension of** V **over** F. ■

30.22 Example Let E be an extension field of a field F, and let $\alpha \in E$. Example 30.14 shows that if α is algebraic over F and $\deg(\alpha, F) = n$, then the dimension of $F(\alpha)$ as a vector space over F is n. *This is the important example for us.* ▲

An Application to Field Theory

We collect the results of field theory contained in Examples 30.4, 30.8, 30.11, 30.14, and 30.22, and incorporate them into one theorem. The last sentence of this theorem gives an additional nice application of these vector space ideas to field theory.

30.23 Theorem Let E be an extension field of F, and let $\alpha \in E$ be algebraic over F. If $\deg(\alpha, F) = n$, then $F(\alpha)$ is an n-dimensional vector space over F with basis $\{1, \alpha, \cdots, \alpha^{n-1}\}$. Furthermore, every element β of $F(\alpha)$ is algebraic over F, and $\deg(\beta, F) \le \deg(\alpha, F)$.

Proof We have shown everything in the preceding examples except the very important result stated in the last sentence of the above theorem. Let $\beta \in F(\alpha)$, where α is algebraic over F of degree n. Consider the elements

$$1, \beta, \beta^2, \cdots, \beta^n.$$

These cannot be $n + 1$ distinct elements of $F(\alpha)$ that are linearly independent over F, for by Theorem 30.19, any basis of $F(\alpha)$ over F would have to contain at least as many elements as are in any set of linearly independent vectors over F. However, the basis $\{1, \alpha, \cdots, \alpha^{n-1}\}$ has just n elements. If $\beta^i = \beta^j$, then $\beta^i - \beta^j = 0$, so in any case there exist $b_i \subset F$ such that

$$b_0 + b_1\beta + b_2\beta^2 + \cdots + b_n\beta^n = 0,$$

where not all $b_i = 0$. Then $f(x) = b_nx^n + \cdots + b_1x + b_0$ is a nonzero element of $F[x]$ such that $f(\beta) = 0$. Therefore, β is algebraic over F and $\deg(\beta, F)$ is at most n. ◆

EXERCISES 30

Computations

1. Find three bases for $\mathbb{R}^2$ over $\mathbb{R}$, no two of which have a vector in common.

In Exercises 2 and 3, determine whether the given set of vectors is a basis for $\mathbb{R}^3$ over $\mathbb{R}$.

2. $\{(1, 1, 0), (1, 0, 1), (0, 1, 1)\}$

3. $\{(-1, 1, 2), (2, -3, 1), (10, -14, 0)\}$

In Exercises 4 through 9, give a basis for the indicated vector space over the field.

4. $\mathbb{Q}(\sqrt{2})$ over $\mathbb{Q}$

5. $\mathbb{R}(\sqrt{2})$ over $\mathbb{R}$

6. $\mathbb{Q}(\sqrt[3]{2})$ over $\mathbb{Q}$

7. $\mathbb{C}$ over $\mathbb{R}$

8. $\mathbb{Q}(i)$ over $\mathbb{Q}$

9. $\mathbb{Q}(\sqrt[4]{2})$ over $\mathbb{Q}$

10. According to Theorem 30.23, the element $1 + \alpha$ of $\mathbb{Z}_2(\alpha)$ of Example 29.19 is algebraic over $\mathbb{Z}_2$. Find the irreducible polynomial for $1 + \alpha$ in $\mathbb{Z}_2[x]$.

Concepts

In Exercises 11 through 14, correct the definition of the italicized term without reference to the text, if correction is needed, so that it is in a form acceptable for publication.

11. The vectors in a subset S of a vector space V over a field F *span* V if and only if each $\beta \in V$ can be expressed uniquely as a linear combination of the vectors in S.

12. The vectors in a subset S of a vector space V over a field F are *linearly independent over* F if and only if the zero vector cannot be expressed as a linear combination of vectors in S.

13. The *dimension over* F of a finite-dimensional vector space V over a field F is the minimum number of vectors required to span V.

14. A *basis* for a vector space V over a field F is a set of vectors in V that span V and are linearly dependent.

15. Mark each of the following true or false.

_______ **a.** The sum of two vectors is a vector.

_______ **b.** The sum of two scalars is a vector.

_______ **c.** The product of two scalars is a scalar.

_______ **d.** The product of a scalar and a vector is a vector.

_______ **e.** Every vector space has a finite basis.

_______ **f.** The vectors in a basis are linearly dependent.

_______ **g.** The 0-vector may be part of a basis.

_______ **h.** If $F \leq E$ and $\alpha \in E$ is algebraic over the field F, then α^2 is algebraic over F.

_______ **i.** If $F \leq E$ and $\alpha \in E$ is algebraic over the field F, then $\alpha + \alpha^2$ is algebraic over F.

_______ **j.** Every vector space has a basis.

The exercises that follow deal with the further study of vector spaces. In many cases, we are asked to define for vector spaces some concept that is analogous to one we have studied for other algebraic structures. These exercises should improve our ability to recognize parallel and related situations in algebra. Any of these exercises may assume knowledge of concepts defined in the preceding exercises.

16. Let V be a vector space over a field F.

a. Define a *subspace of the vector space* V *over* F.

b. Prove that an intersection of subspaces of V is again a subspace of V over F.

17. Let V be a vector space over a field F, and let $S = \{\alpha_i \mid i \in I\}$ be a nonempty collection of vectors in V.

a. Using Exercise 16(b), define the *subspace of* V *generated by* S.

b. Prove that the vectors in the subspace of V generated by S are precisely the (finite) linear combinations of vectors in S. (Compare with Theorem 7.6.)

18. Let $V_1, \cdots, V_n$ be vector spaces over the same field F. Define the *direct sum* $V_1 \oplus \cdots \oplus V_n$ *of the vectors spaces* V_i for $i = 1, \cdots, n$, and show that the direct sum is again a vector space over F.

19. Generalize Example 30.2 to obtain the vector space F^n of ordered n-tuples of elements of F over the field F, for any field F. What is a basis for F^n?

20. Define an *isomorphism* of a vector space V over a field F with a vector space V' over the same field F.

Theory

21. Prove that if V is a finite-dimensional vector space over a field F, then a subset $\{\beta_i, \beta_2, \cdots, \beta_n\}$ of V is a basis for V over F if and only if every vector in V can be expressed *uniquely* as a linear combination of the β_i.

22. Let F be any field. Consider the "system of m simultaneous linear equations in n unknowns"

$$\begin{aligned} a_{11}X_1 + a_{12}X_2 + \cdots + a_{1n}X_n &= b_1, \\ a_{21}X_1 + a_{22}X_2 + \cdots + a_{2n}X_n &= b_2, \\ &\vdots \\ a_{m1}X_1 + a_{m2}X_2 + \cdots + a_{mn}X_n &= b_m, \end{aligned}$$

where $a_{ij}, b_i \in F$.

a. Show that the "system has a solution," that is, there exist $X_1, \cdots, X_n \in F$ that satisfy all m equations, if and only if the vector $\beta = (b_1, \cdots, b_m)$ of F^m is a linear combination of the vectors $\alpha_j = (a_{1j}, \cdots, a_{mj})$. (This result is straightforward to prove, being practically the definition of a solution, but should really be regarded as the *fundamental existence theorem for a simultaneous solution of a system of linear equations*.)

b. From part (a), show that if $n = m$ and $\{\alpha_j \mid j = 1, \cdots, n\}$ is a basis for F^n, then the system always has a unique solution.

23. Prove that every finite-dimensional vector space V of dimension n over a field F is isomorphic to the vector space F^n of Exercise 19.

24. Let V and V' be vector spaces over the same field F. A function $\phi : V \to V'$ is a **linear transformation of** V **into** V' if the following conditions are satisfied for all $\alpha, \beta \in V$ and $a \in F$:

$$\begin{aligned} \phi(\alpha + \beta) &= \phi(\alpha) + \phi(\beta). \\ \phi(a\alpha) &= a(\phi(\alpha)). \end{aligned}$$

a. If $\{\beta_i \mid i \in I\}$ is a basis for V over F, show that a linear transformation $\phi : V \to V'$ is completely determined by the vectors $\phi(\beta_i) \in V'$.

b. Let $\{\beta_i \mid i \in I\}$ be a basis for V, and let $\{\beta_i' \mid i \in I\}$ be any set of vectors, not necessarily distinct, of V'. Show that there exists exactly one linear transformation $\phi : V \to V'$ such that $\phi(\beta_i) = \beta_i'$.

25. Let V and V' be vector spaces over the same field F, and let $\phi : V \to V'$ be a linear transformation.

a. To what concept that we have studied for the algebraic structures of groups and rings does the concept of a *linear transformation* correspond?

b. Define the *kernel* (or *nullspace*) of ϕ, and show that it is a subspace of V.

c. Describe when ϕ is an isomorphism of V with V'.

26. Let V be a vector space over a field F, and let S be a subspace of V. Define the *quotient space* V/S, and show that it is a vector space over F.

27. Let V and V' be vector spaces over the same field F, and let V be finite dimensional over F. Let $\dim(V)$ be the dimension of the vector space V over F. Let $\phi : V \to V'$ be a linear transformation.

a. Show that $\phi[V]$ is a subspace of V'.

b. Show that $\dim(\phi[V]) = \dim(V) - \dim(\text{Ker}(\phi))$. [*Hint:* Choose a convenient basis for V, using Theorem 30.19. For example, enlarge a basis for $\text{Ker}(\phi)$ to a basis for V.]

SECTION 31 ALGEBRAIC EXTENSIONS

Finite Extensions

In Theorem 30.23 we saw that if E is an extension field of a field F and $\alpha \in E$ is algebraic over F, then every element of $F(\alpha)$ is algebraic over F. In studying zeros of polynomials in $F[x]$, we shall be interested almost exclusively in extensions of F containing only elements algebraic over F.

31.1 Definition An extension field E of a field F is an **algebraic extension of** F if every element in E is algebraic over F. ■

31.2 Definition If an extension field E of a field F is of finite dimension n as a vector space over F, then E is a **finite extension of degree n over** F. We shall let $[E : F]$ be the degree n of E over F. ■

To say that a field E is a finite extension of a field F does *not* mean that E is finite field. It just asserts that E is a finite-dimensional vector space over F, that is, that $[E : F]$ is finite.

We shall often use the fact that if E is a finite extension of F, then, $[E : F] = 1$ if and only if $E = F$. We need only observe that by Theorem 30.19, $\{1\}$ can always be enlarged to a basis for E over F. Thus $[E : F] = 1$ if and only if $E = F(1) = F$.

Let us repeat the argument of Theorem 30.23 to show that a finite extension E of a field F must be an algebraic extension of F.

31.3 Theorem A finite extension field E of a field F is an algebraic extension of F.

Proof We must show that for $\alpha \in E$, α is algebraic over F. By Theorem 30.19 if $[E : F] = n$, then

$$1, \alpha, \cdots, \alpha^n$$

cannot be linearly independent elements, so there exist $a_i \in F$ such that

$$a_n\alpha^n + \cdots + a_1\alpha + a_0 = 0,$$

and not all $a_i = 0$. Then $f(x) = a_nx^n + \cdots + a_1x + a_0$ is a nonzero polynomial in $F[x]$, and $f(\alpha) = 0$. Therefore, α is algebraic over F. ◆

We cannot overemphasize the importance of our next theorem. It plays a role in field theory analogous to the role of the theorem of Lagrange in group theory. While its proof follows easily from our brief work with vector spaces, it is a tool of incredible power. An elegant application of it in the section that follows shows the impossibility of performing certain geometric constructions with a straightedge and a compass. *Never underestimate a theorem that counts something.*

31.4 Theorem If E is a finite extension field of a field F, and K is a finite extension field of E, then K is a finite extension of F, and

$$[K : F] = [K : E][E : F].$$

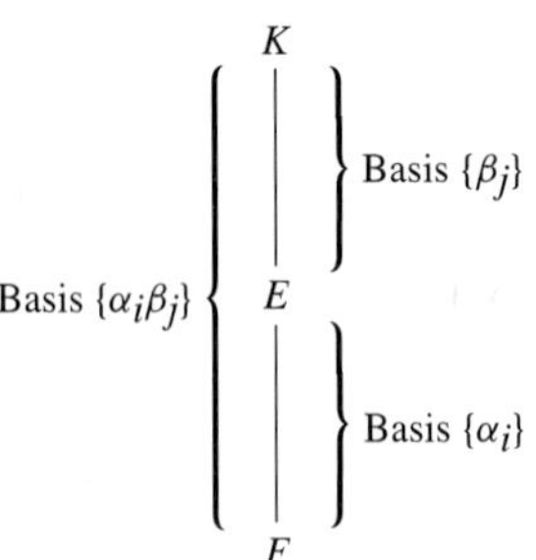

31.5 Figure

Proof Let $\{\alpha_i \mid i = 1, \cdots, n\}$ be a basis for E as a vector space over F, and let the set $\{\beta_j \mid j = 1, \cdots, m\}$ be a basis for K as a vector space over E. The theorem will be proved if we can show that the mn elements $\alpha_i\beta_j$ form a basis for K, viewed as a vector space over F. (See Fig. 31.5.)

Let γ be any element of K. Since the β_j form a basis for K over E, we have

$$\gamma = \sum_{j=1}^{m} b_j\beta_j$$

for $b_j \in E$. Since the α_i form a basis for E over F, we have

$$b_j = \sum_{i=1}^{n} a_{ij}\alpha_i$$

for $a_{ij} \in F$. Then

$$\gamma = \sum_{j=1}^{m}\left(\sum_{i=1}^{n} a_{ij}\alpha_i\right)\beta_j = \sum_{i,j} a_{ij}(\alpha_i\beta_j),$$

so the mn vectors $\alpha_i\beta_j$ span K over F.

It remains for us to show that the mn elements $\alpha_i\beta_j$ are independent over F. Suppose that $\Sigma_{i,j}c_{ij}(\alpha_i\beta_j) = 0$, with $c_{ij} \in F$. Then

$$\sum_{j=1}^{m}\left(\sum_{i=1}^{n} c_{ij}\alpha_i\right)\beta_j = 0,$$

and $(\Sigma_{i=1}^{n}c_{ij}\alpha_i) \in E$. Since the elements β_j are independent over E, we must have

$$\sum_{i=1}^{n} c_{ij}\alpha_i = 0$$

for all j. But now the α_i are independent over F, so $\sum_{i=1}^{n} c_{ij}\alpha_i = 0$ implies that $c_{ij} = 0$ for all i and j. Thus the $\alpha_i\beta_j$ not only span K over F but also are independent over F. Thus they form a basis for K over F. ◆

Note that we proved this theorem by actually exhibiting a basis. It is worth remembering that if $\{\alpha_i \mid i = 1, \cdots, n\}$ is a basis for E over F and $\{\beta_j \mid j = 1, \cdots, m\}$ is a basis for K over E, for fields $F \leq E \leq K$, then the set $\{\alpha_i\beta_j\}$ of mn products is a basis for K over F. Figure 31.5 gives a diagram for this situation. We shall illustrate this further in a moment.

31.6 Corollary If F_i is a field for $i = 1, \cdots, r$ and F_{i+1} is a finite extension of F_i, then F_r is a finite extension of F_1, and

$$[F_r : F_1] = [F_r : F_{r-1}][F_{r-1} : F_{r-2}] \cdots [F_2 : F_1].$$

Proof The proof is a straightforward extension of Theorem 31.4 by induction. ◆

31.7 Corollary If E is an extension field of F, $\alpha \in E$ is algebraic over F, and $\beta \in F(\alpha)$, then $\deg(\beta, F)$ divides $\deg(\alpha, F)$.

Proof By Theorem 30.23, $\deg(\alpha, F) = [F(\alpha) : F]$ and $\deg(\beta, F) = [F(\beta) : F]$. We have $F \leq F(\beta) \leq F(\alpha)$, so by Theorem 31.4 $[F(\beta) : F]$ divides $[F(\alpha) : F]$. ◆

The following example illustrates a type of argument one often makes using Theorem 31.4 or its corollaries.

31.8 Example By Corollary 31.7, there is no element of $\mathbb{Q}(\sqrt{2})$ that is a zero of $x^3 - 2$. Note that $\deg(\sqrt{2}, \mathbb{Q}) = 2$, while a zero of $x^3 - 2$ is of degree 3 over $\mathbb{Q}$, but 3 does not divide 2. ▲

Let E be an extension field of a field F, and let α_1, α_2 be elements of E, not necessarily algebraic over F. By definition, $F(\alpha_1)$ is the smallest extension field of F in E that contains α_1. Similarly, $(F(\alpha_1))(\alpha_2)$ can be characterized as the smallest extension field of F in E containing both α_1 and α_2. We could equally have started with α_2, so $(F(\alpha_1))(\alpha_2) = (F(\alpha_2))(\alpha_1)$. We denote this field by $F(\alpha_1, \alpha_2)$. Similarly, for $\alpha_i \in E$, $F(\alpha_1, \cdots \alpha_n)$ is the smallest extension field of F in E containing all the α_i for $i = 1, \cdots, n$. We obtain the field $F(\alpha_1, \cdots, \alpha_n)$ from the field F by **adjoining to** F **the elements** α_i in E. Exercise 49 of Section 18 shows that, analogous to an intersection of subgroups of a group, an intersection of subfields of a field E is again a subfield of E. Thus $F(\alpha_1, \cdots, \alpha_n)$ can be characterized as the intersection of all subfields of E containing F and all the α_i for $i = 1, \cdots, n$.

31.9 Example Consider $\mathbb{Q}(\sqrt{2})$. Theorem 30.23 shows that $\{1, \sqrt{2}\}$ is a basis for $\mathbb{Q}(\sqrt{2})$ over $\mathbb{Q}$. Using the technique demonstrated in Example 29.10, we can easily discover that $\sqrt{2} + \sqrt{3}$ is a zero of $x^4 - 10x^2 + 1$. By the method demonstrated in Example 23.14, we can show that this polynomial is irreducible in $\mathbb{Q}[x]$. Thus $\operatorname{irr}(\sqrt{2} + \sqrt{3}, \mathbb{Q}) = x^4 - 10x^2 + 1$, so $[\mathbb{Q}(\sqrt{2} + \sqrt{3}) : \mathbb{Q}] = 4$. Thus $(\sqrt{2} + \sqrt{3}) \notin \mathbb{Q}(\sqrt{2})$, so $\sqrt{3} \notin \mathbb{Q}(\sqrt{2})$. Consequently, $\{1, \sqrt{3}\}$ is a basis for $\mathbb{Q}(\sqrt{2}, \sqrt{3}) = (\mathbb{Q}(\sqrt{2}))(\sqrt{3})$ over $\mathbb{Q}(\sqrt{2})$. The proof of Theorem 31.4 (see the comment following the theorem) then shows that $\{1, \sqrt{2}, \sqrt{3}, \sqrt{6}\}$ is a basis for $\mathbb{Q}(\sqrt{2}, \sqrt{3})$ over $\mathbb{Q}$. ▲

31.10 Example Let $2^{1/3}$ be the real cube root of 2 and $2^{1/2}$ be the positive square root of 2. Then $2^{1/2} \notin \mathbb{Q}(2^{1/3})$ because $\deg(2^{1/2}, \mathbb{Q}) = 2$ and 2 is not a divisor of $3 = \deg(2^{1/3}, \mathbb{Q})$. Thus $[\mathbb{Q}(2^{1/3}, 2^{1/2}) : \mathbb{Q}(2^{1/3})] = 2$. Hence $\{1, 2^{1/3}, 2^{2/3}\}$ is a basis for $\mathbb{Q}(2^{1/3})$ over $\mathbb{Q}$ and $\{1, 2^{1/2}\}$ is a basis for $\mathbb{Q}(2^{1/3}, 2^{1/2})$ over $\mathbb{Q}(2^{1/3})$. Furthermore, by Theorem 31.4 (see the comment following the theorem),

$$\{1, 2^{1/2}, 2^{1/3}, 2^{5/6}, 2^{2/3}, 2^{7/6}\}$$

is a basis for $\mathbb{Q}(2^{1/2}, 2^{1/3})$ over $\mathbb{Q}$. Because $2^{7/6} = 2(2^{1/6})$, we have $2^{1/6} \in \mathbb{Q}(2^{1/2}, 2^{1/3})$. Now $2^{1/6}$ is a zero of $x^6 - 2$, which is irreducible over $\mathbb{Q}$, by Eisenstein's criterion, with

$p = 2$. Thus

$$\mathbb{Q} \leq \mathbb{Q}(2^{1/6}) \leq \mathbb{Q}(2^{1/2}, 2^{1/3})$$

and by Theorem 31.4

$$\begin{aligned} 6 = [\mathbb{Q}(2^{1/2}, 2^{1/3}) : \mathbb{Q}] &= [\mathbb{Q}(2^{1/2}, 2^{1/3}) : \mathbb{Q}(2^{1/6})][\mathbb{Q}(2^{1/6}) : \mathbb{Q}] \\ &= [\mathbb{Q}(2^{1/2}, 2^{1/3}) : \mathbb{Q}(2^{1/6})](6). \end{aligned}$$

Therefore, we must have

$$[\mathbb{Q}(2^{1/2}, 2^{1/3}) : \mathbb{Q}(2^{1/6})] = 1,$$

so $\mathbb{Q}(2^{1/2}, 2^{1/3}) = \mathbb{Q}(2^{1/6})$, by the comment preceding Theorem 31.3. ▲

Example 31.10 shows that it is possible for an extension $F(\alpha_1, \cdots, \alpha_n)$ of a field F to be actually a simple extension, even though $n > 1$.

Let us characterize extensions of F of the form $F(\alpha_1, \cdots, \alpha_n)$ in the case that all the α_i are algebraic over F.

31.11 Theorem Let E be an algebraic extension of a field F. Then there exist a finite number of elements $\alpha_1, \cdots, \alpha_n$ in E such that $E = F(\alpha_1, \cdots, \alpha_n)$ if and only if E is a finite-dimensional vector space over F, that is, if and only if E is a finite extension of F.

Proof Suppose that $E = F(\alpha_1, \cdots, \alpha_n)$. Since E is an algebraic extension of F, each α_i, is algebraic over F, so each α_i is algebraic over every extension field of F in E. Thus $F(\alpha_1)$ is algebraic over F, and in general, $F(\alpha_1, \cdots, \alpha_j)$ is algebraic over $F(\alpha_i, \cdots, \alpha_{j-1})$ for $j = 2, \cdots, n$. Corollary 31.6 applied to the sequence of finite extensions

$$F,\ F(\alpha_1),\ F(\alpha_1, \alpha_2),\ \cdots,\ F(\alpha_1, \cdots, \alpha_n) = E$$

then shows that E is a finite extension of F.

Conversely, suppose that E is a finite algebraic extension of F. If $[E : F] = 1$, then $E = F(1) = F$, and we are done. If $E \neq F$, let $\alpha_1 \in E$, where $\alpha_1 \notin F$. Then $[F(\alpha_1) : F] > 1$. If $F(\alpha_1) = E$, we are done; if not, let $\alpha_2 \in E$, where $\alpha_2 \notin F(\alpha_1)$. Continuing this process, we see from Theorem 31.4 that since $[E : F]$ is finite, we must arrive at α_n such that

$$F(\alpha_1, \cdots, \alpha_n) = E.$$ ◆

Algebraically Closed Fields and Algebraic Closures

We have not yet observed that if E is an extension of a field F and $\alpha, \beta \in E$ are algebraic over F, then so are $\alpha + \beta$, $\alpha\beta$, $\alpha - \beta$, and α/β, if $\beta \neq 0$. This follows from Theorem 31.3 and is also included in the following theorem.

31.12 Theorem Let E be an extension field of F. Then

$$\overline{F}_E = \{\alpha \in E \mid \alpha \text{ is algebraic over } F\}$$

is a subfield of E, the **algebraic closure of F in E**.

Proof Let $\alpha, \beta \in \overline{F}_E$. Then Theorem 31.11 shows that $F(\alpha, \beta)$ is a finite extension of F, and by Theorem 31.3 every element of $F(\alpha, \beta)$ is algebraic over F, that is, $F(\alpha, \beta) \subseteq \overline{F}_E$. Thus

$\bar{F}_E$ contains $\alpha + \beta, \alpha\beta, \alpha - \beta$, and also contains α/β for $\beta \neq 0$, so $\bar{F}_E$ is a subfield of E. ◆

31.13 Corollary The set of all algebraic numbers forms a field.

Proof Proof of this corollary is immediate from Theorem 31.12, because the set of all algebraic numbers is the algebraic closure of $\mathbb{Q}$ in $\mathbb{C}$. ◆

It is well known that the complex numbers have the property that every nonconstant polynomial in $\mathbb{C}[x]$ has a zero in $\mathbb{C}$. This is known as the *Fundamental Theorem of Algebra*. An analytic proof of this theorem is given in Theorem 31.18. We now give a definition generalizing this important concept to other fields.

31.14 Definition A field F is **algebraically closed** if every nonconstant polynomial in $F[x]$ has a zero in F. ■

Note that a field F can be the algebraic closure of F in an extension field E without F being algebraically closed. For example, $\mathbb{Q}$ is the algebraic closure of $\mathbb{Q}$ in $\mathbb{Q}(x)$, but $\mathbb{Q}$ is not algebraically closed because $x^2 + 1$ has no zero in $\mathbb{Q}$.

The next theorem shows that the concept of a field being algebraically closed can also be defined in terms of factorization of polynomials over the field.

31.15 Theorem A field F is algebraically closed if and only if every nonconstant polynomial in $F[x]$ factors in $F[x]$ into linear factors.

Proof Let F be algebraically closed, and let $f(x)$ be a nonconstant polynomial in $F[x]$ Then $f(x)$ has a zero $a \in F$. By Corollary 23.3, $x - a$ is a factor of $f(x)$, so $f(x) = (x - a)g(x)$. Then if $g(x)$ is nonconstant, it has a zero $b \in F$, and we have $f(x) = (x - a)(x - b)h(x)$. Continuing, we get a factorization of $f(x)$ in $F[x]$ into linear factors.

Conversely, suppose that every nonconstant polynomial of $F[x]$ has a factorization into linear factors. If $ax - b$ is a linear factor of $f(x)$, then b/a is a zero of $f(x)$. Thus F is algebraically closed. ◆

31.16 Corollary An algebraically closed field F has no proper algebraic extensions, that is, no algebraic extensions E with $F < E$.

Proof Let E be an algebraic extension of F, so $F \leq E$. Then if $\alpha \in E$, we have $\text{irr}(\alpha, F) = x - \alpha$, by Theorem 31.15, since F is algebraically closed. Thus $\alpha \in F$, and we must have $F = E$. ◆

In a moment we shall show that just as there exists an algebraically closed extension $\mathbb{C}$ of the real numbers $\mathbb{R}$, for any field F there exists similarly an algebraic extension $\overline{F}$ of F, with the property that $\overline{F}$ is algebraically closed. Naively, to find $\overline{F}$ we proceed as follows. If a polynomial $f(x)$ in $F[x]$ has a no zero in F, then adjoin a zero α of such an $f(x)$ to F, thus obtaining the field $F(\alpha)$. *Theorem 29.3, Kronecker's theorem, is strongly used here, of course*. If $F(\alpha)$ is still not algebraically closed, then continue the process further. The trouble is that, contrary to the situation for the algebraic closure $\mathbb{C}$ of $\mathbb{R}$, we may have to do this a (possibly large) infinite number of times. It can be shown (see Exercises 33 and 36) that $\overline{\mathbb{Q}}$ is isomorphic to the field of all algebraic numbers, and that we cannot obtain $\overline{\mathbb{Q}}$ from $\mathbb{Q}$ by adjoining a finite number of algebraic numbers. We shall have to first discuss some set-theoretic machinery, *Zorn's lemma*, in order to be able to handle such a situation. This machinery is a bit complex, so we are putting the proof under a separate heading. The existence theorem for $\overline{F}$ is very important, and we state it here so that we will know this fact, even if we do not study the proof.

31.17 Theorem Every field F has an **algebraic closure**, that is, an algebraic extension $\overline{F}$ that is algebraically closed.

It is well known that $\mathbb{C}$ is an algebraically closed field. We recall an analytic proof for the student who has had a course in functions of a complex variable. There are algebraic proofs, but they are much longer.

31.18 Theorem **(Fundamental Theorem of Algebra)** The field $\mathbb{C}$ of complex numbers is an algebraically closed field.

Proof Let the polynomial $f(z) \in \mathbb{C}[z]$ have no zero in $\mathbb{C}$. Then $1/f(z)$ gives an entire function; that is, $1/f$ is analytic everywhere. Also if $f \notin \mathbb{C}$, $\lim_{|z|\to\infty} |f(z)| = \infty$, so $\lim_{|z|\to\infty} |1/f(z)| = 0$. Thus $1/f$ must be bounded in the plane. Hence by Liouville's theorem of complex function theory, $1/f$ is constant, and thus f is constant. Therefore, a nonconstant polynomial in $\mathbb{C}[z]$ must have a zero in $\mathbb{C}$, so $\mathbb{C}$ is algebraically closed. ◆

Proof of the Existence of an Algebraic Closure

We shall prove that every field has an algebraic extension that is algebraically closed. Mathematics students should have the opportunity to see some proof involving the *Axiom of Choice* by the time they finish college. This is a natural place for such a proof. We shall use an equivalent form, *Zorn's lemma*, of the Axiom of Choice. To state Zorn's lemma, we have to give a set-theoretic definition.

31.19 Definition A **partial ordering of a set** S is given by a relation $\leq$ defined for certain ordered pairs of elements of S such that the following conditions are satisfied:

1. $a \leq a$ for all $a \in S$ **(reflexive law).**
2. If $a \leq b$ and $b \leq a$, then $a = b$ **(antisymmetric law).**
3. If $a \leq b$ and $b \leq c$, then $a \leq c$ **(transitive law).** ■

In a *partially* ordered set, not every two elements need be **comparable;** that is, for $a, b \in S$, we need not have either $a \leq b$ or $b \leq a$. As usual, $a < b$ denotes $a \leq b$ but $a \neq b$.

A subset T of a partially ordered set S is a **chain** if every two elements a and b in T are comparable, that is, either $a \leq b$ or $b \leq a$ (or both). An element $u \in S$ is an **upper bound for a subset** A of partially ordered set S if $a \leq u$ for all $a \in A$. Finally, an element m of a partially ordered set S is **maximal** if there is no $s \in S$ such that $m < s$.

31.20 Example The collection of all subsets of a set forms a partially ordered set under the relation $\leq$ given by $\subseteq$. For example, if the whole set is $\mathbb{R}$, we have $\mathbb{Z} \subseteq \mathbb{Q}$. Note, however, that for $\mathbb{Z}$ and $\mathbb{Q}^+$, neither $\mathbb{Z} \subseteq \mathbb{Q}^+$ nor $\mathbb{Q}^+ \subseteq \mathbb{Z}$. ▲

31.21 Zorn's Lemma If S is a partially ordered set such that every chain in S has an upper bound in S, then S has at least one maximal element.

There is no question of *proving* Zorn's lemma. The lemma is equivalent to the Axiom of Choice. Thus we are really taking Zorn's lemma here as an *axiom* for our set theory. Refer to the literature for a statement of the Axiom of Choice and a proof of its equivalence to Zorn's lemma. (See Edgerton [47].)

Zorn's lemma is often useful when we want to show the existence of a largest or maximal structure of some kind. If a field F has an algebraic extension $\overline{F}$ that is algebraically closed, then $\overline{F}$ will certainly be a maximal algebraic extension of F, for since $\overline{F}$ is algebraically closed, it can have no proper algebraic extensions.

The idea of our proof of Theorem 31.17 is very simple. Given a field F, we shall first describe a class of algebraic extensions of F that is so large that it must contain (up to isomorphism) any conceivable algebraic extension of F. We then define a partial ordering, the ordinary subfield ordering, on this class, and show that the hypotheses of Zorn's lemma are satisfied. By Zorn's lemma, there will exist a maximal algebraic extension $\overline{F}$ of F in this class. We shall then argue that, as a maximal element, this extension $\overline{F}$ can have no proper algebraic extensions, so it must be algebraically closed.

Our proof differs a bit from the one found in many texts. We like it because it uses no algebra other than that derived from Theorems 29.3 and 31.4. Thus it throws into sharp relief the tremendous strength of both Kronecker's theorem and Zorn's lemma. The proof looks long, but only because we are writing out every little step. To the professional mathematician, the construction of the proof from the information in the preceding paragraph is a routine matter. This proof was suggested to the author during his graduate student days by a fellow graduate student, Norman Shapiro, who also had a strong preference for it.

HISTORICAL NOTE

The Axiom of Choice, although used implicitly in the 1870s and 1880s, was first stated explicitly by Ernst Zermelo in 1904 in connection with his proof of the well-ordering theorem, the result that for any set A, there exists an order–relation $<$ such that every nonempty subset B of A contains a least element with respect to $<$. Zermelo's Axiom of Choice asserted that, given any set M and the set S of all subsets of M, there always exists a "choice" function, a function $f : S \to M$ such that $f(M') \in M'$ for every M' in S. Zermelo noted, in fact, that "this logical principal cannot ... be reduced to a still simpler one, but it is applied without hesitation everywhere in mathematical deduction." A few years later he included this axiom in his collection of axioms for set theory, a collection which was slightly modified in 1930 into what is now called Zermelo–Fraenkel set theory, the axiom system generally used today as a basis of that theory.

Zorn's lemma was introduced by Max Zorn (1906–1993) in 1935. Although he realized that it was equivalent to the well-ordering theorem (itself equivalent to the Axiom of Choice), he claimed that his lemma was more natural to use in algebra because the well-ordering theorem was somehow a "transcendental" principal. Other mathematicians soon agreed with his reasoning. The lemma appeared in 1939 in the first volume of Nicolas Bourbaki's *Eléments de Mathématique: Les Structures Fondamentales de l'Analyse*. It was used consistently in that work and quickly became an essential part of the mathematician's toolbox.

We are now ready to carry out our proof of Theorem 31.17, which we restate here.

31.22 Restated Theorem 31.17 Every field F has an algebraic closure $\overline{F}$.

Proof It can be shown in set theory that given any set, there exists a set with *strictly more* elements. Suppose we form a set

$$A = \{\omega_{f,i} \mid f \in F[x]; i = 0, \cdots, (\text{degree } f)\}$$

that has an element for every possible zero of any $f(x) \in F[x]$. Let Ω be a set with strictly more elements than A. Replacing Ω by $\Omega \cup F$ if necessary, we can assume $F \subset \Omega$. Consider all possible fields that are algebraic extension of F and that, as sets, consist of elements of Ω. One such algebraic extension is F itself. If E is any extension field of F, and if $\gamma \in E$ is a zero $f(x) \in F[x]$ for $\gamma \notin F$ and $\deg(\gamma, F) = n$, then renaming γ by ω for $\omega \in \Omega$ and $\omega \notin F$, and renaming elements $a_0 + a_1\gamma + \cdots + a_{n-1}\gamma^{n-1}$ of $F(\gamma)$ by distinct elements of Ω as the a_i range over F, we can consider our renamed $F(\gamma)$ to be an algebraic extension field $F(\omega)$ of F, with $F(\omega) \subset \Omega$ and $f(\omega) = 0$. The set Ω has enough elements to form $F(\omega)$, since Ω has more than enough elements to provide n different zeros for each element of each degree n in any subset of $F[x]$.

All algebraic extension fields E_j of F, with $E_j \subseteq \Omega$, form a set

$$S = \{E_j \mid j \in J\}$$

that is partially ordered under our usual subfield inclusion $\leq$. One element of S is F itself. The preceding paragraphs shows that if F is far away from being algebraically closed, there will be many fields E_j in S.

Let $T = \{E_{j_k}\}$ be a chain in S, and let $W = \cup_k E_{j_k}$. We now make W into a field. Let $\alpha, \beta \in W$. Then there exist $E_{j_1}, E_{j_2} \in S$, with $\alpha \in E_{j_1}$ and $\beta \in E_{j_2}$. Since T is a chain, one of the fields E_{j_1} and E_{j_2} is a subfield of the other, say $E_{j_1} \leq E_{j_2}$. Then $\alpha, \beta \in E_{j_2}$, and we use the field operations of E_{j_2} to *define* the sum of α and β in W as $(\alpha + \beta) \in E_{j_2}$ and, likewise, the product as $(\alpha\beta) \in E_{j_2}$. These operations are well defined in W; they are independent of our choice of E_{j_2}, since if $\alpha, \beta \in E_{j_3}$ also, for E_{j_3} in T, then one of the fields E_{j_2} and E_{j_3} is a subfield of the other, since T is a chain. Thus we have operations of addition and multiplication defined on W.

All the field axioms for W under these operations now follow from the fact that these operations were defined in terms of addition and multiplication in fields. Thus, for example, $1 \in F$ serves as multiplicative identity in W, since for $\alpha \in W$, if $1, \alpha \in E_{j_1}$, then we have $1\alpha = \alpha$ in E_{j_1}, so $1\alpha = \alpha$ in W, by definition of multiplication in W. As further illustration, to check the distributive laws, let $\alpha, \beta, \gamma \in W$. Since T is a chain, we can find one field in T containing all three elements α, β, and γ, and in this field the distributive laws for α, β, and γ hold. Thus they hold in W. Therefore, we can view W as a field, and by construction, $E_{j_k} \leq W$ for every $E_{j_k} \in T$.

If we can show that W is algebraic over F, then $W \in S$ will be an upper bound for T. But if $\alpha \in W$, then $\alpha \in E_{j_1}$ for some E_{j_1} in T, so α is algebraic over F. Hence W is an algebraic extension of F and is an upper bound for T.

The hypotheses of Zorn's lemma are thus fulfilled, so there is a maximal element $\overline{F}$ of S. We claim that $\overline{F}$ is algebraically closed. Let $f(x) \in \overline{F}[x]$, where $f(x) \notin \overline{F}$. Suppose that $f(x)$ has no zero in $\overline{F}$. Since Ω has many more elements than $\overline{F}$ has, we can take $\omega \in \Omega$, where $\omega \notin \overline{F}$, and form a field $\overline{F}(\omega) \subseteq \Omega$, with ω a zero of $f(x)$, as we saw in the first paragraph of this proof. Let β be in $\overline{F}(\omega)$. Then by Theorem 30.23, β is a zero of a polynomial

$$g(x) = \alpha_0 + \alpha_1 x + \cdots + \alpha_n x^n$$

in $\overline{F}[x]$, with $\alpha_i \in \overline{F}$, and hence α_i algebraic over F. Then by Theorem 31.11 the field $F(\alpha_0, \cdots, \alpha_n)$ is a finite extension of F, and since β is algebraic over $F(\alpha_0, \cdots, \alpha_n)$, we also see that $F(\alpha_0, \cdots, \alpha_n, \beta)$ is a finite extension over $F(\alpha_0, \cdots, \alpha_n)$. Theorem 31.4 then shows that $F(\alpha_0, \cdots, \alpha_n, \beta)$ is a finite extension of F, so by Theorem 31.3, β is algebraic over F. Hence $\overline{F}(\omega) \in S$ and $\overline{F} < \overline{F}(\omega)$, which contradicts the choice of $\overline{F}$ as maximal in S. Thus $f(x)$ must have had a zero in $\overline{F}$, so $\overline{F}$ is algebraically closed. ◆

The mechanics of the preceding proof are routine to the professional mathematician. Since it may be the first proof that we have ever seen using Zorn's lemma, we wrote the proof out in detail.

EXERCISES 31

Computations

In Exercises 1 through 13, find the degree and a basis for the given field extension. Be prepared to justify your answers.

1. $\mathbb{Q}(\sqrt{2})$ over $\mathbb{Q}$
2. $\mathbb{Q}(\sqrt{2}, \sqrt{3})$ over $\mathbb{Q}$
3. $\mathbb{Q}(\sqrt{2}, \sqrt{3}, \sqrt{18})$ over $\mathbb{Q}$
4. $\mathbb{Q}(\sqrt[3]{2}, \sqrt{3})$ over $\mathbb{Q}$
5. $\mathbb{Q}(\sqrt{2}, \sqrt[3]{2})$ over $\mathbb{Q}$
6. $\mathbb{Q}(\sqrt{2} + \sqrt{3})$ over $\mathbb{Q}$
7. $\mathbb{Q}(\sqrt{2}\sqrt{3})$ over $\mathbb{Q}$
8. $\mathbb{Q}(\sqrt{2}, \sqrt[3]{5})$ over $\mathbb{Q}$
9. $\mathbb{Q}(\sqrt[3]{2}, \sqrt[3]{6}, \sqrt[3]{24})$ over $\mathbb{Q}$
10. $\mathbb{Q}(\sqrt{2}, \sqrt{6})$ over $\mathbb{Q}(\sqrt{3})$
11. $\mathbb{Q}(\sqrt{2} + \sqrt{3})$ over $\mathbb{Q}(\sqrt{3})$
12. $\mathbb{Q}(\sqrt{2}, \sqrt{3})$ over $\mathbb{Q}(\sqrt{2} + \sqrt{3})$
13. $\mathbb{Q}(\sqrt{2}, \sqrt{6} + \sqrt{10})$ over $\mathbb{Q}(\sqrt{3} + \sqrt{5})$

Concepts

In Exercises 14 through 17, correct the definition of the italicized term without reference to the text, if correction is needed, so that it is in a form acceptable for publication.

14. An *algebraic extension* of a field F is a field $F(\alpha_1, \alpha_2, \cdots, \alpha_n)$ where each α_i is a zero of some polynomial in $F[x]$.
15. A *finite extension field* of a field F is one that can be obtained by adjoining a finite number of elements to F.
16. The *algebraic closure* $\overline{F}_E$ *of a field F in an extension field E of F* is the field consisting of all elements of E that are algebraic over F.
17. A field F is *algebraically closed* if and only if every polynomial has a zero in F.
18. Show by an example that for a proper extension field E of a field F, the algebraic closure of F in E need not be algebraically closed.

19. Mark each of the following true or false.

_____ **a.** If a field E is a finite extension of a field F, then E is a finite field.
_____ **b.** Every finite extension of a field is an algebraic extension.
_____ **c.** Every algebraic extension of a field is a finite extension.
_____ **d.** The top field of a finite tower of finite extensions of fields is a finite extension of the bottom field.
_____ **e.** $\mathbb{Q}$ is its own algebraic closure in $\mathbb{R}$, that is $\mathbb{Q}$ is **algebraically closed in** $\mathbb{R}$.
_____ **f.** $\mathbb{C}$ is algebraically closed in $\mathbb{C}(x)$, where x is an indeterminate.
_____ **g.** $\mathbb{C}(x)$ is algebraically closed, where x is an indeterminate.
_____ **h.** The field $\mathbb{C}(x)$ has no algebraic closure, since $\mathbb{C}$ already contains all algebraic numbers.
_____ **i.** An algebraically closed field must be of characteristic 0.
_____ **j.** If E is an algebraically closed extension field of F, then E is an algebraic extension of F.

Proof Synopsis

20. Give a one-sentence synopsis of the proof of Theorem 31.3.

21. Give a one- or two-sentence synopsis of the proof of Theorem 31.4.

Theory

22. Let $(a + bi) \in \mathbb{C}$ where $a, b \in \mathbb{R}$ and $b \neq 0$. Show that $\mathbb{C} = \mathbb{R}(a + bi)$.

23. Show that if E is a finite extension of a field F and $[E : F]$ is a prime number, then E is a simple extension of F and, indeed, $E = F(\alpha)$ for every $\alpha \in E$ not in F.

24. Prove that $x^2 - 3$ is irreducible over $\mathbb{Q}(\sqrt[3]{2})$.

25. What degree field extensions can we obtain by successively adjoining to a field F a square root of an element of F not a square in F, then square root of some nonsquare in this new field, and so on? Argue from this that a zero of $x^{14} - 3x^2 + 12$ over $\mathbb{Q}$ can never be expressed as a rational function of square roots of rational functions of square roots, and so on, of elements of $\mathbb{Q}$.

26. Let E be a finite extension field of F. Let D be an integral domain such that $F \subseteq D \subseteq E$. Show that D is a field.

27. Prove in detail that $\mathbb{Q}(\sqrt{3} + \sqrt{7}) = \mathbb{Q}(\sqrt{3}, \sqrt{7})$.

28. Generalizing Exercise 27, show that if $\sqrt{a} + \sqrt{b} \neq 0$, then $\mathbb{Q}(\sqrt{a} + \sqrt{b}) = \mathbb{Q}(\sqrt{a}, \sqrt{b})$ for all a and b in $\mathbb{Q}$. [*Hint:* Compute $(a - b)/(\sqrt{a} + \sqrt{b})$.]

29. Let E be a finite extension of a field F, and let $p(x) \in F[x]$ be irreducible over F and have degree that is not a divisor of $[E : F]$. Show that $p(x)$ has no zeros in E.

30. Let E be an extension field of F. Let $\alpha \in E$ be algebraic of odd degree over F. Show that α^2 is algebraic of odd degree over F, and $F(\alpha) = F(\alpha^2)$.

31. Show that if F, E, and K are fields with $F \leq E \leq K$, then K is algebraic over F if and only if E is algebraic over F, and K is algebraic over E. (You must *not* assume the extensions are finite.)

32. Let E be an extension field of a field F. Prove that every $\alpha \in E$ that is not in the algebraic closure $\overline{F}_E$ of F in E is transcendental over $\overline{F}_E$.

33. Let E be an algebraically closed extension field of a field F. Show that the algebraic closure $\overline{F}_E$ of F in E is algebraically closed. (Applying this exercise to $\mathbb{C}$ and $\mathbb{Q}$, we see that the field of all algebraic numbers is an algebraically closed field.)

34. Show that if E is an algebraic extension of a field F and contains all zeros in $\overline{F}$ of every $f(x) \in F[x]$, then E is an algebraically closed field.

35. Show that no finite field of odd characteristic is algebraically closed. (Actually, no finite field of characteristic 2 is algebraically closed either.) [*Hint*: By counting, show that for such a finite field F, some polynomial $x^2 - a$, for some $a \in F$, has no zero in F. See Exercise 32, Section 29.]

36. Prove that, as asserted in the text, the algebraic closure of $\mathbb{Q}$ in $\mathbb{C}$ is not a finite extension of $\mathbb{Q}$.

37. Argue that every finite extension field of $\mathbb{R}$ is either $\mathbb{R}$ itself or is isomorphic to $\mathbb{C}$.

38. Use Zorn's lemma to show that every proper ideal of a ring R with unity is contained in some maximal ideal.

SECTION 32 †GEOMETRIC CONSTRUCTIONS

In this section we digress briefly to give an application demonstrating the power of Theorem 31.4. For a more detailed study of geometric constructions, you are referred to Courant and Robbins [44, Chapter III].

We are interested in what types of figures can be constructed with a compass and a straightedge in the sense of classical Euclidean plane geometry. We shall discuss the impossibility of trisecting certain angles and other classical questions.

Constructible Numbers

Let us imagine that we are given only a single line segment that we shall define to be *one unit* in length. A real number α is **constructible** if we can construct a line segment of length $|\alpha|$ in a finite number of steps from this given segment of unit length by using a straightedge and a compass.

The rules of the game are pretty strict. We suppose that we are given just two points at the moment, the endpoints of our unit line segment, let us suppose that they correspond to the points $(0, 0)$ and $(1, 0)$ in the Euclidean plane. We are allowed to draw a line only with our straightedge through two points that we have already located. Thus we can start by using the straightedge and drawing the line through $(0, 0)$ and $(1, 0)$. We are allowed to open our compass only to a distance between points we have already found. Let us open our compass to the distance between $(0, 0)$ and $(1, 0)$. We can then place the point of the compass at $(1, 0)$ and draw a circle of radius 1, which passes through the point $(2, 0)$. Thus we now have located a third point, $(2, 0)$. Continuing in this way, we can locate points $(3, 0), (4, 0), (-1, 0), (-2, 0)$, and so on. Now open the compass the distance from $(0, 0)$ to $(0, 2)$, put the point at $(1, 0)$, and draw a circle of radius 2. Do the same with the point at $(-1, 0)$. We have now found two new points, where these circles intersect, and we can put our straightedge on them to draw what we think of as the y-axis. Then opening our compass to the distance from $(0, 0)$ to $(1, 0)$, we draw a circle with center at $(0, 0)$ and locate the point $(0, 1)$ where the circle intersects the y-axis. Continuing in this fashion, we can locate all points (x, y) with integer coordinates in any rectangle containing the point $(0, 0)$. Without going into more detail, it can be shown that it is possible, among other things, to erect a perpendicular to a given line at a known point

† This chapter is not used in the remainder of the text.

on the line, and find a line passing through a known point and parallel to a given line. Our first result is the following theorem.

32.1 Theorem If α and β are constructible real numbers, then so are $\alpha + \beta$, $\alpha - \beta$, $\alpha\beta$, and α/β, if $\beta \neq 0$.

Proof We are given that α and β are constructible, so there are line segments of lengths $|\alpha|$ and $|\beta|$ available to us. For $\alpha, \beta > 0$, extend a line segment of length α with the straightedge. Start at one end of the original segment of length α, and lay off on the extension the length β with the compass. This constructs a line segment of length $\alpha + \beta$; $\alpha - \beta$ is similiarly constructible (see Fig. 32.2). If α and β are not both positive, an obvious breakdown into cases according to their signs shows that $\alpha + \beta$ and $\alpha - \beta$ are still constructible.

The construction of $\alpha\beta$ is indicated in Fig. 32.3. We shall let $\overline{OA}$ be the line segment from the point O to the point A, and shall let $|\overline{OA}|$ be the length of this line segment. If $\overline{OA}$ is of length $|\alpha|$, construct a line l through O not containing $\overline{OA}$. (Perhaps, if O is at $(0, 0)$ and A is at $(a, 0)$, you use the line through $(0, 0)$ and $(4, 2)$.) Then find the points P and B on l such that $\overline{OP}$ is of length 1 and $\overline{OB}$ is of length $|\beta|$. Draw $\overline{PA}$ and construct l' through B, parallel to $\overline{PA}$ and intersecting $\overline{OA}$ extended at Q. By similar triangles, we have

$$\frac{1}{|\alpha|} = \frac{|\beta|}{|\overline{OQ}|},$$

so $\overline{OQ}$ is of length $|\alpha\beta|$.

Finally, Fig. 32.4 shows that α/β is constructible if $\beta \neq 0$. Let $\overline{OA}$ be of length $|\alpha|$, and construct l through O not containing OA. Then find B and P on l such that $\overline{OB}$ is of length $|\beta|$ and $\overline{OP}$ is of length 1. Draw BA and construct l' through P, parallel to $\overline{BA}$, and intersecting $\overline{OA}$ at Q. Again by similar triangles, we have

$$\frac{|\overline{OQ}|}{1} = \frac{|\alpha|}{|\beta|},$$

so $\overline{OQ}$ is of length $|\alpha/\beta|$. ◆

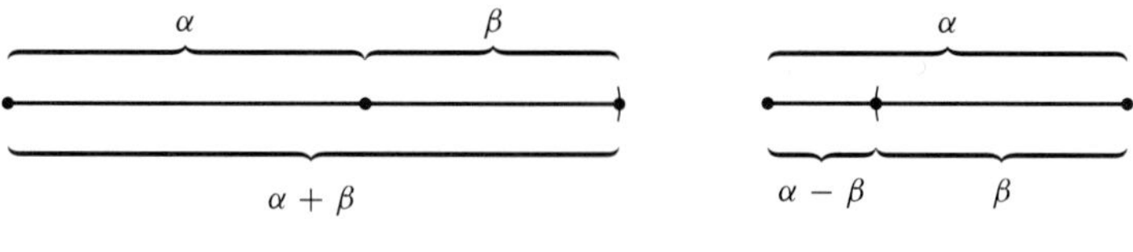

32.2 Figure

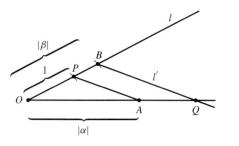

32.3 Figure

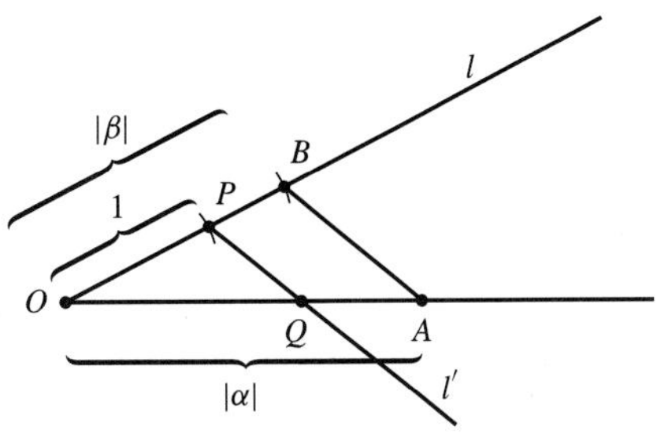

32.4 Figure

32.5 Corollary The set of all constructible real numbers forms a subfield F of the field of real numbers.

Proof Proof of this corollary is immediate from Theorem 32.1. ◆

Thus the field F of all constructible real numbers contains $\mathbb{Q}$, the field of rational numbers, since $\mathbb{Q}$ is the smallest subfield of $\mathbb{R}$.

From now on, we proceed analytically. We can construct any rational number. Regarding our given segment

0______1

of length 1 as the basic unit on an x-axis, we can locate any point (q_1, q_2) in the plane with both coordinates rational. Any further point in the plane that we can locate by using a compass and a straightedge can be found in one of the following three ways:

1. as an intersection of two lines, each of which passes through two known points having rational coordinates,
2. as an intersection of a line that passes through two points having rational coordinates and a circle whose center has rational coordinates and whose radius is rational.
3. as an intersection of two circles whose centers have rational coordinates and whose radii are rational.

Equations of lines and circles of the type discussed in 1, 2, and 3 are of the form

$$ax + by + c = 0$$

and

$$x^2 + y^2 + dx + ey + f = 0,$$

where a, b, c, d, e, and f are all in $\mathbb{Q}$. Since in Case 3 the intersection of two circles with equations

$$x^2 + y^2 + d_1x + e_1y + f_1 = 0$$

and

$$x^2 + y^2 + d_2x + e_2y + f_2 = 0$$

is the same as the intersection of the first circle having equation

$$x^2 + y^2 + d_1x + e_1y + f_1 = 0,$$

and the line (the common chord) having equation

$$(d_1 - d_2)x + (e_1 - e_2)y + f_1 - f_2 = 0,$$

we see that Case 3 can be reduced to Case 2. For Case 1, a simultaneous solution of two linear equations with rational coefficients can only lead to rational values of x and y, giving us no new points. However, finding a simultaneous solution of a linear equation with rational coefficients and a quadratic equation with rational coefficients, as in Case 2, leads, upon substitution, to a quadratic equation. Such an equation, when solved by the quadratic formula, may have solutions involving square roots of numbers that are not squares in $\mathbb{Q}$.

In the preceding argument, nothing was really used involving $\mathbb{Q}$ except field axioms. If H is the smallest field containing those real numbers constructed so far, the argument shows that the "next new number" constructed lies in a field $H(\sqrt{\alpha})$ for some $\alpha \in H$, where $\alpha > 0$. We have proved half of our next theorem.

32.6 Theorem The field F of constructible real numbers consists precisely of all real numbers that we can obtain from $\mathbb{Q}$ by taking square roots of positive numbers a finite number of times and applying a finite number of field operations.

Proof We have shown that F can contain no numbers except those we obtain from $\mathbb{Q}$ by taking a finite number of square roots of positive numbers and applying a finite number of field operations. However, if $\alpha > 0$ is constructible, then Fig. 32.7 shows that $\sqrt{\alpha}$ is constructible. Let $\overline{OA}$ have length α, and find P on $\overline{OA}$ extended so that $\overline{OP}$ has length 1. Find the midpoint of $\overline{PA}$ and draw a semicircle with $\overline{PA}$ as diameter. Erect a perpendicular to $\overline{PA}$ at O, intersecting the semicircle at Q. Then the triangles OPQ and OQA are similar, so

$$\frac{|\overline{OQ}|}{|\overline{OA}|} = \frac{|\overline{OP}|}{|\overline{OQ}|},$$

and $|\overline{OQ}|^2 = 1\alpha = \alpha$. Thus $\overline{OQ}$ is of length $\sqrt{\alpha}$. Therefore square roots of constructible numbers are constructible.

Theorem 32.1 showed that field operations are possible by construction. ◆

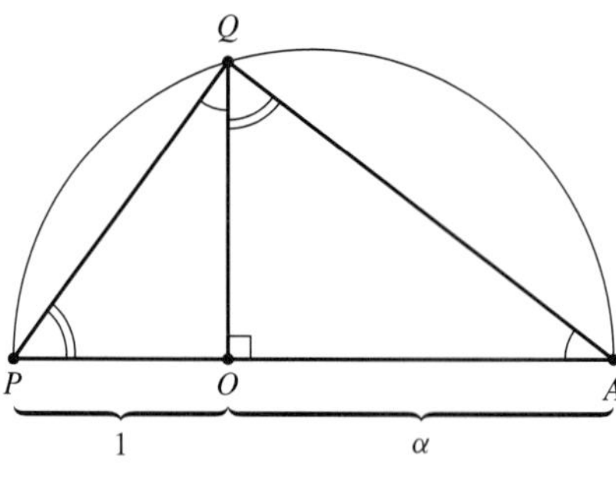

32.7 Figure

32.8 Corollary If γ is constructible and $\gamma \notin \mathbb{Q}$, then there is a finite sequence of real numbers $\alpha_1, \cdots, \alpha_n = \gamma$ such that $\mathbb{Q}(\alpha_1, \cdots, \alpha_i)$ is an extension of $\mathbb{Q}(\alpha_1, \cdots, \alpha_{i-1})$ of degree 2. In particular, $[\mathbb{Q}(\gamma) : \mathbb{Q}] = 2^r$ for some integer $r \geq 0$.

Proof The existence of the α_i is immediate from Theorem 32.6. Then

$$\begin{aligned} 2^n &= [\mathbb{Q}(\alpha_1, \cdots, \alpha_n) : \mathbb{Q}] \\ &= [\mathbb{Q}(\alpha_1, \cdots, \alpha_n) : \mathbb{Q}(\gamma)][\mathbb{Q}(\gamma) : \mathbb{Q}], \end{aligned}$$

by Theorem 32.4, which completes the proof. ◆

The Impossibility of Certain Constructions

We can now show the impossibility of certain geometric constructions.

32.9 Theorem *Doubling the cube is impossible,* that is, given a side of a cube, it is not always possible to construct with a straightedge and a compass the side of a cube that has double the volume of the original cube.

Proof Let the given cube have a side of length 1, and hence a volume of 1. The cube being sought would have to have a volume of 2, and hence a side of length $\sqrt[3]{2}$. But $\sqrt[3]{2}$ is a zero of irreducible $x^3 - 2$ over $\mathbb{Q}$, so

$$[\mathbb{Q}(\sqrt[3]{2}) : \mathbb{Q}] = 3.$$

Corollary 32.8 shows that to double this cube of volume 1, we would need to have $3 = 2^r$ for some integer r, but no such r exists. ◆

32.10 Theorem *Squaring the circle is impossible;* that is, given a circle, it is not always possible to construct with a straightedge and a compass a square having area equal to the area of the given circle.

Proof Let the given circle have a radius of 1, and hence an area of π. We would need to construct a square of side $\sqrt{\pi}$. But π is transcendental over $\mathbb{Q}$, so $\sqrt{\pi}$ is transcendental over $\mathbb{Q}$ also. ◆

32.11 Theorem *Trisecting the angle is impossible;* that is, there exists an angle that cannot be trisected with a straightedge and a compass.

Proof Figure 32.12 indicates that the angle θ can be constructed if and only if a segment of length $|\cos\theta|$ can be constructed. Now $60°$ is a constructible angle, and we shall show that it cannot be trisected. Note that

$$\begin{aligned} \cos 3\theta &= \cos(2\theta + \theta) \\ &= \cos 2\theta \cos\theta - \sin 2\theta \sin\theta \\ &= (2\cos^2\theta - 1)\cos\theta - 2\sin\theta\cos\theta\sin\theta \\ &= (2\cos^2\theta - 1)\cos\theta - 2\cos\theta(1 - \cos^2\theta) \\ &= 4\cos^3\theta - 3\cos\theta. \end{aligned}$$

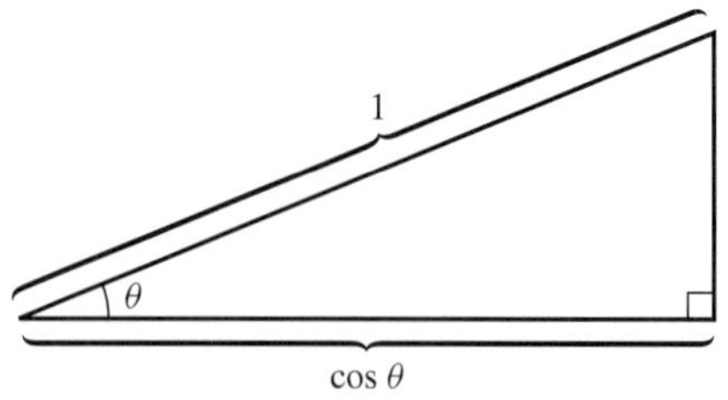

32.12 Figure

[We realize that many students today have not seen the trigonometric identities we just used. Exercise 1 repeats Exercise 40 of Section 1 and asks you to prove the identity $\cos 3\theta = 4\cos^3\theta - 3\cos\theta$ from Euler's formula.]

Let $\theta = 20°$, so that $\cos 3\theta = \frac{1}{2}$, and let $\alpha = \cos 20°$. From the identity $4\cos^3\theta - 3\cos\theta = \cos 3\theta$, we see that

$$4\alpha^3 - 3\alpha = \frac{1}{2}.$$

Thus α is a zero of $8x^3 - 6x - 1$. This polynomial is irreducible in $\mathbb{Q}[x]$, since, by Theorem 23.11, it is enough to show that it does not factor in $\mathbb{Z}[x]$. But a factorization in $\mathbb{Z}[x]$ would entail a linear factor of the form $(8x \pm 1)$, $(4x \pm 1)$, $(2x \pm 1)$, or $(x \pm 1)$. We can quickly check that none of the numbers $\pm\frac{1}{8}, \pm\frac{1}{4}, \pm\frac{1}{2}$, and ± 1 is a zero of $8x^3 - 6x - 1$. Thus

$$[\mathbb{Q}(\alpha) : \mathbb{Q}] = 3,$$

so by Corollary 32.8, α is not constructible. Hence $60°$ cannot be trisected. ◆

Historical Note

Greek mathematicians as far back as the fourth century B.C. had tried without success to find geometric constructions using straightedge and compass to trisect the angle, double the cube, and square the circle. Although they were never able to prove that such constructions were impossible, they did manage to construct the solutions to these problems using other tools, including the conic sections.

It was Carl Gauss in the early nineteenth century who made a detailed study of constructibility in connection with his solution of cyclotomic equations, the equations of the form $x^p - 1 = 0$ with p prime whose roots form the vertices of a regular p-gon. He showed that although all such equations are solvable using radicals, if $p - 1$ is not a power of 2, then the solutions must involve roots higher than the second. In fact, Gauss asserted that anyone who attempted to find a geometric construction for a p-gon where $p - 1$ is not a power of 2 would "spend his time uselessly." Interestingly, Gauss did not prove the assertion that such constructions were impossible. That was accomplished in 1837 by Pierre Wantzel (1814–1848), who in fact proved Corollary 32.8 and also demonstrated Theorems 32.9 and 32.11. The proof of Theorem 32.10, on the other hand, requires a proof that π is transcendental, a result finally achieved in 1882 by Ferdinand Lindemann (1852–1939).

Note that the regular n-gon is constructible for $n \geq 3$ if and only if the angle $2\pi/n$ is constructible, which is the case if and only if a line segment of length $\cos(2\pi/n)$ is constructible.

EXERCISES 32

Computations

1. Prove the trigonometric identity $\cos 3\theta = 4\cos^3\theta - 3\cos\theta$ from the Euler formula, $e^{i\theta} = \cos\theta + i\sin\theta$.

Concepts

2. Mark each of the following true or false.

_____ **a.** It is impossible to double any cube of constructible edge by compass and straightedge constructions.

_____ **b.** It is impossible to double every cube of constructible edge by compass and straightedge constructions.

_____ **c.** It is impossible to square any circle of constructible radius by straightedge and compass constructions.

_____ **d.** No constructible angle can be trisected by straightedge and compass constructions.

_____ **e.** Every constructible number is of degree 2^r over $\mathbb{Q}$ for some integer $r \geq 0$.

_____ **f.** We have shown that every real number of degree 2^r over $\mathbb{Q}$ for some integer $r \geq 0$ is constructible.

_____ **g.** The fact that factorization of a positive integer into a product of primes is unique (up to order) was used strongly at the conclusion of Theorems 32.9 and 32.11.

_____ **h.** Counting arguments are exceedingly powerful mathematical tools.

_____ **i.** We can find any constructible number in a finite number of steps by starting with a given segment of unit length and using a straightedge and a compass.

_____ **j.** We can find the totality of all constructible numbers in a finite number of steps by starting with a given segment of unit length and using a straightedge and a compass.

Theory

3. Using the proof of Theorem 32.11, show that the regular 9-gon is not constructible.

4. Show *algebraically* that it is possible to construct an angle of 30°.

5. Referring to Fig. 32.13, where $\overline{AQ}$ bisects angle OAP, show that the regular 10-gon is constructible (and therefore that the regular pentagon is also). [*Hint:* Triangle OAP is similar to triangle APQ. Show algebraically that r is constructible.]

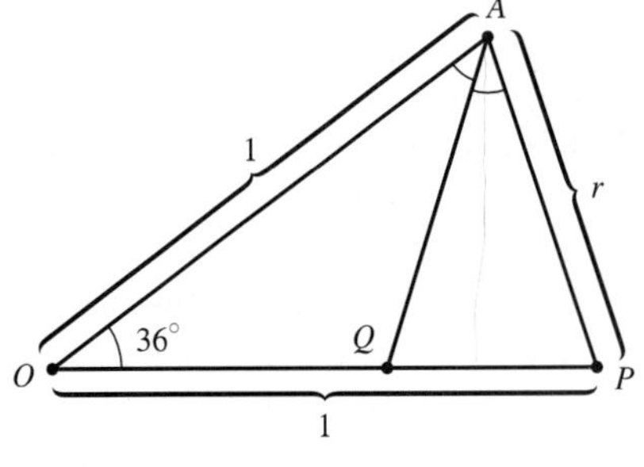

32.13 Figure

In Exercises 6 through 9 use the results of Exercise 5 where needed to show that the statement is true.

6. The regular 20-gon is constructible.

7. The regular 30-gon is constructible.

8. The angle 72° can be trisected.

9. The regular 15-gon can be constructed.

10. Suppose you wanted to explain roughly in just three or four sentences, for a high school plane geometry teacher who never had a course in abstract algebra, how it can be shown that it is impossible to trisect an angle of 60°. Write down what you would say.

SECTION 33 FINITE FIELDS

The purpose of this section is to determine the structure of all finite fields. We shall show that for every prime p and positive integer n, there is exactly one finite field (up to isomorphism) of order p^n. This field $\mathrm{GF}(p^n)$ is usually referred to as the **Galois field of order** p^n. We shall be using quite a bit of our material on cyclic groups. The proofs are simple and elegant.

The Structure of a Finite Field

We now show that all finite fields must have prime-power order.

33.1 Theorem Let E be a finite extension of degree n over a finite field F. If F has q elements, then E has q^n elements.

Proof Let $\{\alpha_1, \cdots, \alpha_n\}$ be a basis for E as a vector space over F. By Exercise 21 of Section 30, every $\beta \in E$ can be *uniquely* written in the form

$$\beta = b_1\alpha_1 + \cdots + b_n\alpha_n$$

for $b_i \in F$. Since each b_i may be any of the q elements of F, the total number of such distinct linear combinations of the α_i is q^n. ◆

33.2 Corollary If E is a finite field of characteristic p, then E contains exactly p^n elements for some positive integer n.

Proof Every finite field E is a finite extension of a prime field isomorphic to the field $\mathbb{Z}_p$, where p is the characteristic of E. The corollary follows at once from Theorem 33.1. ◆

We now turn to the study of the multiplicative structure of a finite field. The following theorem will show us how any finite field can be formed from the prime subfield.

33.3 Theorem Let E be a field of p^n elements contained in an algebraic closure $\overline{\mathbb{Z}}_p$ of $\mathbb{Z}_p$. The elements of E are precisely the zeros in $\overline{\mathbb{Z}}_p$ of the polynomial $x^{p^n} - x$ in $\mathbb{Z}_p[x]$.

Proof The set E^* of nonzero elements of E forms a multiplicative group of order $p^n - 1$ under the field multiplication. For $\alpha \in E^*$, the order of α in this group divides the order $p^n - 1$ of the group. Thus for $\alpha \in E^*$, we have $\alpha^{p^n - 1} = 1$, so $\alpha^{p^n} = \alpha$. Therefore, every element in E is a zero of $x^{p^n} - x$. Since $x^{p^n} - x$ can have at most p^n zeros, we see that E contains precisely the zeros of $x^{p^n} - x$ in $\overline{\mathbb{Z}}_p$. ◆

33.4 Definition An element α of a field is an ***n*th root of unity** if $\alpha^n = 1$. It is a **primitive *n*th root of unity** if $\alpha^n = 1$ and $\alpha^m \neq 1$ for $0 < m < n$. ■

Thus the nonzero elements of a finite field of p^n elements are all $(p^n - 1)$th roots of unity.

Recall that in Corollary 23.6, we showed that the multiplicative group of nonzero elements of a finite field is cyclic. This is a very important fact about finite fields; it has actually been applied to algebraic coding. For the sake of completeness in this section, we now state it here as a theorem, give a corollary, and illustrate with an example.

33.5 Theorem The multiplicative group $\langle F^*, \cdot\rangle$ of nonzero elements of a finite field F is cyclic.

Proof See Corollary 23.6. ◆

33.6 Corollary A finite extension E of a finite field F is a simple extension of F.

Proof Let α be a generator for the cyclic group E^* of nonzero elements of E. Then $E = F(\alpha)$ ◆

33.7 Example Consider the finite field $\mathbb{Z}_{11}$. By Theorem 33.5 $\langle \mathbb{Z}_{11}{}^*, \cdot\rangle$ is cyclic. Let us try to find a generator of $\mathbb{Z}_{11}{}^*$ by brute force and ignorance. We start by trying 2. Since $|\mathbb{Z}_{11}{}^*| = 10$, 2 must be an element of $\mathbb{Z}_{11}{}^*$ of order dividing 10, that is, either 2, 5, or 10. Now

$$2^2 = 4, \quad 2^4 = 4^2 = 5, \quad \text{and} \quad 2^5 = (2)(5) = 10 = -1.$$

Thus neither 2^2 nor 2^5 is 1, but, of course, $2^{10} = 1$, so 2 is a generator of $\mathbb{Z}_{11}{}^*$, that is, 2 is a primitive 10th root of unity in $\mathbb{Z}_{11}$. We were lucky.

By the theory of cyclic groups, all the generators of $Z_{11}{}^*$, that is, all the primitive 10th roots of unity in $\mathbb{Z}_{11}$, are of the form 2^n, where n is relatively prime to 10. These elements are

$$2^1 = 2, \quad 2^3 = 8, \quad 2^7 = 7, \quad 2^9 = 6.$$

Historical Note

Although Carl F. Gauss had shown that the set of residues modulo a prime p satisfied the field properties, it was Evariste Galois (1811–1832) who first dealt with what he called "incommensurable solutions" to the congruence $F(x) \equiv 0 \pmod{p}$, where $F(x)$ is an nth degree irreducible polynomial modulo p. He noted in a paper written in 1830 that one should consider the roots of this congruence as "a variety of imaginary symbols" that one can use in calculations just as one uses $\sqrt{-1}$. Galois then showed that if α is any solution of $F(x) \equiv 0 \pmod{p}$, the expression $a_0 + a_1\alpha + a_2\alpha^2 + \cdots + a_{n-1}\alpha^{n-1}$ takes on precisely p^n different values. Finally, he proved results equivalent to Theorems 33.3 and 33.5 of the text.

Galois' life was brief and tragic. He showed brilliance in mathematics early on, publishing several papers before he was 20 and essentially established the basic ideas of Galois theory. He was, however, active in French revolutionary politics following the July revolution of 1830. In May 1831, he was arrested for threatening the life of King Louis-Philippe. Though he was acquitted, he was rearrested for participating, heavily armed, in a republican demonstration on Bastille Day of that year. Two months after his release from prison the following March, he was killed in a duel, "the victim of an infamous coquette and her two dupes"; the previous night he had written a letter to a friend clarifying some of his work in the theory of equations and requesting that it be studied by other mathematicians. Not until 1846, however, were his major papers published; it is from that date that his work became influential.

The primitive 5th roots of unity in $\mathbb{Z}_{11}$ are of the form 2^m, where the gcd of m and 10 is 2, that is,

$$2^2 = 4, \quad 2^4 = 5, \quad 2^6 = 9, \quad 2^8 = 3.$$

The primitive square root of unity in $\mathbb{Z}_{11}$ is $2^5 = 10 = -1$. ▲

The Existence of GF(p^n)

We turn now to the question of the existence of a finite field of order p^r for every prime power p^r, $r > 0$. We need the following lemma.

33.8 Lemma If F is a field of prime characteristic p with algebraic closure $\overline{F}$, then $x^{p^n} - x$ has p^n distinct zeros in $\overline{F}$. ◆

Proof Because $\overline{F}$ is algebraically closed, $x^{p^n} - x$ factors over that field into a product of linear factors $x - \alpha$, so it suffices to show that none of these factors occurs more than once in the factorization.

Since we have not introduced an algebraic theory of derivatives, this elegant technique is not available to us, so we proceed by long division. Observe that 0 is a zero of $x^{p^n} - x$ of multiplicity 1. Suppose $\alpha \neq 0$ is a zero of $x^{p^n} - x$, and hence is a zero of $f(x) = x^{p^n-1} - 1$. Then $x - \alpha$ is a factor of $f(x)$ in $\overline{F}[x]$, and by long division, we find that

$$\frac{f(x)}{(x-\alpha)} = g(x)$$
$$= x^{p^n-2} + \alpha x^{p^n-3} + \alpha^2 x^{p^n-4} + \cdots + \alpha^{p^n-3}x + \alpha^{p^n-2}.$$

Now $g(x)$ has $p^n - 1$ summands, and in $g(\alpha)$, each summand is

$$\alpha^{p^n-2} = \frac{\alpha^{p^n-1}}{\alpha} = \frac{1}{\alpha}.$$

Thus

$$g(\alpha) = [(p^n - 1)\cdot 1]\frac{1}{\alpha} = -\frac{1}{\alpha}.$$

since we are in a field of characteristic p. Therefore, $g(\alpha) \neq 0$, so α is a zero of $f(x)$ of multiplicity 1. ◆

33.9 Lemma If F is a field of prime characteristic p, then $(\alpha+\beta)^{p^n} = \alpha^{p^n} + \beta^{p^n}$ for all $\alpha, \beta \in F$ and all positive integers n. ◆

Proof Let $\alpha, \beta \in F$. Applying the binomial theorem to $(\alpha+\beta)^p$, we have

$$\begin{aligned}(\alpha+\beta)^p &= \alpha^p + (p\cdot 1)\alpha^{p-1}\beta + \left(\frac{p(p-1)}{2}\cdot 1\right)\alpha^{p-2}\beta^2 \\ &\quad + \cdots + (p\cdot 1)\alpha\beta^{p-1} + \beta^p \\ &= \alpha^p + 0\alpha^{p-1}\beta + 0\alpha^{p-2}\beta^2 + \cdots + 0\alpha\beta^{p-1} + \beta^p \\ &= \alpha^p + \beta^p.\end{aligned}$$

Proceeding by induction on n, suppose that we have $(\alpha+\beta)^{p^{n-1}} = \alpha^{p^{n-1}} + \beta^{p^{n-1}}$. Then $(\alpha+\beta)^{p^n} = [(\alpha+\beta)^{p^{n-1}}]^p = (\alpha^{p^{n-1}} + \beta^{p^{n-1}})^p = \alpha^{p^n} + \beta^{p^n}$. ◆

33.10 Theorem A finite field GF(p^n) of p^n elements exists for every prime power p^n.

Proof Let $\overline{\mathbb{Z}}_p$ be an algebraic closure of $\mathbb{Z}_p$, and let K be the subset of $\overline{\mathbb{Z}}_p$ consisting of all zeros of $x^{p^n} - x$ in $\overline{\mathbb{Z}}_p$. Let $\alpha, \beta \in K$. Lemma 33.9 shows that $(\alpha+\beta) \in K$, and the equation $(\alpha\beta)^{p^n} = \alpha^{p^n}\beta^{p^n} = \alpha\beta$ shows that $\alpha\beta \in K$. From $\alpha^{p^n} = \alpha$ we obtain $(-\alpha)^{p^n} = (-1)^{p^n}\alpha^{p^n} = (-1)^{p^n}\alpha$. If p is an odd prime, then $(-1)^{p^n} = -1$ and if $p = 2$ then $-1 = 1$. Thus $(-\alpha)^{p^n} = -\alpha$ so $-\alpha \in K$. Now 0 and 1 are zeros of $x^{p^n} - x$. For $\alpha \neq 0$, $\alpha^{p^n} = \alpha$ implies that $(1/\alpha)^{p^n} = 1/\alpha$. Thus K is a subfield of $\overline{\mathbb{Z}}_p$ containing $\mathbb{Z}_p$. Therefore, K is the desired field of p^n elements, since Lemma 33.8 showed that $x^{p^n} - x$ has p^n distinct zeros in $\overline{\mathbb{Z}}_p$. ◆

33.11 Corollary If F is any finite field, then for every positive integer n, there is an irreducible polynomial in $F[x]$ of degree n.

Proof Let F have $q = p^r$ elements, where p is the characteristic of F. By Theorem 33.10, there is a field $K \leq \bar{F}$ containing $\mathbb{Z}_p$ (up to isomorphism) and consisting precisely of the zeros of $x^{p^{rn}} - x$. We want to show $F \leq K$. Every element of F is a zero of $x^{p^r} - x$, by Theorem 33.3. Now $p^{rs} = p^r p^{r(s-1)}$. Applying this equation repeatedly to the exponents and using the fact that for $\alpha \in F$ we have $\alpha^{p^r} = \alpha$, we see that for $\alpha \in F$,

$$\alpha^{p^{rn}} = \alpha^{p^{r(n-1)}} = \alpha^{p^{r(n-2)}} = \cdots = \alpha^{p^r} = \alpha.$$

Thus $F \leq K$. Then Theorem 33.1 shows that we must have $[K : F] = n$. We have seen that K is simple over F in Corollary 33.6 so $K = F(\beta)$ for some $\beta \in K$. Therefore, $\text{irr}(\beta, F)$ must be of degree n. ◆

33.12 Theorem Let p be a prime and let $n \in \mathbb{Z}^+$. If E and E' are fields of order p^n, then $E \simeq E'$.

Proof Both E and E' have $\mathbb{Z}_p$ as prime field, up to isomorphism. By Corollary 33.6, E is a simple extension of $\mathbb{Z}_p$ of degree n, so there exists an irreducible polynomial $f(x)$ of degree n in $\mathbb{Z}_p[x]$ such that $E \simeq \mathbb{Z}_p[x]/\langle f(x)\rangle$. Because the elements of E are zeros of $x^{p^n} - x$, we see that $f(x)$ is a factor of $x^{p^n} - x$ in $\mathbb{Z}_p[x]$. Because E' also consists of zeros of $x^{p^n} - x$, we see that E' also contains zeros of irreducible $f(x)$ in $\mathbb{Z}_p[x]$. Thus, because E' also contains exactly p^n elements, E' is also isomorphic to $\mathbb{Z}_p[x]/\langle f(x)\rangle$. ◆

Finite fields have been used in algebraic coding. In an article in the *American Mathematical Monthly* 77 (1970): 249–258, Norman Levinson constructs a linear code that can correct up to three errors using a finite field of order 16.

EXERCISES 33

Computations

In Exercises 1 through 3, determine whether there exists a finite field having the given number of elements. (A calculator may be useful.)

1. 4096 **2.** 3127 **3.** 68,921

4. Find the number of primitive 8th roots of unity in GF(9).

5. Find the number of primitive 18th roots of unity in GF(19).

6. Find the number of primitive 15th roots of unity in GF(31).

7. Find the number of primitive 10th roots of unity in GF(23).

Concepts

8. Mark each of the following true or false.

_____ **a.** The nonzero elements of every finite field form a cyclic group under multiplication.

_____ **b.** The elements of every finite field form a cyclic group under addition.

_____ **c.** The zeros in $\mathbb{C}$ of $(x^{28} - 1) \in \mathbb{Q}[x]$ form a cyclic group under multiplication.

_____ **d.** There exists a finite field of 60 elements.

_____ **e.** There exists a finite field of 125 elements.

_____ **f.** There exists a finite field of 36 elements.

______ **g.** The complex number i is a primitive 4th root of unity.

______ **h.** There exists an irreducible polynomial of degree 58 in $\mathbb{Z}_2[x]$.

______ **i.** The nonzero elements of $\mathbb{Q}$ form a cyclic group $\mathbb{Q}^*$ under field multiplication.

______ **j.** If F is a finite field, then every isomorphism mapping F onto a subfield of an algebraic closure $\overline{F}$ of F is an automorphism of F.

Theory

9. Let $\overline{\mathbb{Z}_2}$ be an algebraic closure of $\mathbb{Z}_2$, and let $\alpha, \beta \in \overline{\mathbb{Z}_2}$ be zeros of $x^3 + x^2 + 1$ and of $x^3 + x + 1$, respectively. Using the results of this section, show that $\mathbb{Z}_2(\alpha) = \mathbb{Z}_2(\beta)$.

10. Show that every irreducible polynomial in $\mathbb{Z}_p[x]$ is a divisor of $x^{p^n} - x$ for some n.

11. Let F be a finite field of p^n elements containing the prime subfield $\mathbb{Z}_p$. Show that if $\alpha \in F$ is a generator of the cyclic group $\langle F^*, \cdot \rangle$ of nonzero elements of F, then $\deg(\alpha, \mathbb{Z}_p) = n$.

12. Show that a finite field of p^n elements has exactly one subfield of p^m elements for each divisor m of n.

13. Show that $x^{p^n} - x$ is the product of all monic irreducible polynomials in $\mathbb{Z}_p[x]$ of a degree d dividing n.

14. Let p be an odd prime.

a. Show that for $a \in \mathbb{Z}$, where $a \not\equiv 0 \pmod{p}$, the congruence $x^2 \equiv a \pmod{p}$ has a solution in $\mathbb{Z}$ if and only if $a^{(p-1)/2} \equiv 1 \pmod{p}$. [*Hint:* Formulate an equivalent statement in the finite field $\mathbb{Z}_p$, and use the theory of cyclic groups.]

b. Using part (a), determine whether or not the polynomial $x^2 - 6$ is irreducible in $\mathbb{Z}_{17}[x]$.

PART VII

Advanced Group Theory

SECTION 34 Isomorphism Theorems

There are several theorems concerning isomorphic factor groups that are known as the *isomorphism theorems* of group theory. The first of these is Theorem 14.11, which we restate for easy reference. The theorem is diagrammed in Fig. 34.1.

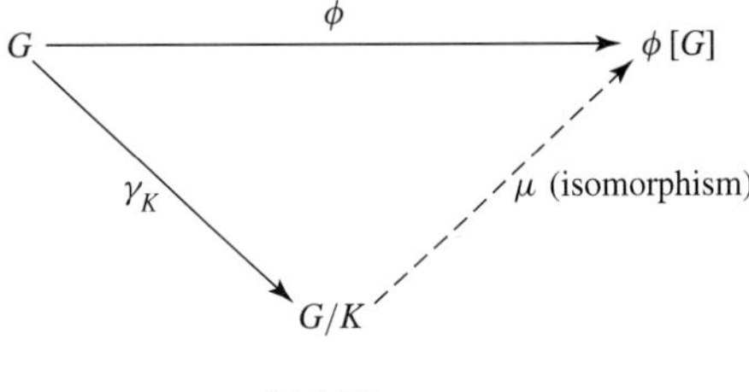

34.1 Figure

34.2 Theorem **(First Isomorphism Theorem)** Let $\phi : G \to G'$ be a homomorphism with kernel K, and let $\gamma_K : G \to G/K$ be the canonical homomorphism. There is a unique isomorphism $\mu : G/K \to \phi[G]$ such that $\phi(x) = \mu(\gamma_K(x))$ for each $x \in G$.

The lemma that follows will be of great aid in our proof and intuitive understanding of the other two isomorphism theorems.

34.3 Lemma Let N be a normal subgroup of a group G and let $\gamma : G \to G/N$ be the canonical homomorphism. Then the map ϕ from the set of normal subgroups of G containing N to the set of normal subgroups of G/N given by $\phi(L) = \gamma[L]$ is one to one and onto.

Proof Theorem 15.16 shows that if L is a normal subgroup of G containing N, then $\phi(L) = \gamma[L]$ is a normal subgroup of G/N. Because $N \leq L$, for each $x \in L$ the entire coset xN in G is contained in L. Thus by Theorem 13.15, $\gamma^{-1}[\phi(L)] = L$. Consequently, if L and M are normal subgroups of G, both containing N, and if $\phi(L) = \phi(M) = H$, then $L = \gamma^{-1}[H] = M$. Therefore ϕ is one to one.

If H is a normal subgroup of G/N, then $\gamma^{-1}[H]$ is a normal subgroup of G by Theorem 15.16. Because $N \in H$ and $\gamma^{-1}[\{N\}] = N$, we see that $N \subseteq \gamma^{-1}[H]$. Then $\phi(\gamma^{-1}[H]) = \gamma[\gamma^{-1}[H]] = H$. This shows that ϕ is onto the set of normal subgroups of G/N. ◆

If H and N are subgroups of a group G, then we let

$$HN = \{hn \mid h \in H, n \in N\}.$$

We define the **join** $H \vee N$ of H and N as the intersection of all subgroups of G that contain HN; thus $H \vee N$ is the smallest subgroup of G containing HN. Of course $H \vee N$ is also the smallest subgroup of G containing both H and N, since any such subgroup must contain HN. In general, HN need not be a subgroup of G. However, we have the following lemma.

34.4 Lemma If N is a normal subgroup of G, and if H is any subgroup of G, then $H \vee N = HN = NH$. Furthermore, if H is also normal in G, then HN is normal in G.

Proof We show that HN is a subgroup of G, from which $H \vee N = HN$ follows at once. Let $h_1, h_2 \in H$ and $n_1, n_2 \in N$. Since N is a normal subgroup, we have $n_1h_2 = h_2n_3$ for some $n_3 \in N$. Then $(h_1n_1)(h_2n_2) = h_1(n_1h_2)n_2 = h_1(h_2n_3)n_2 = (h_1h_2)(n_3n_2) \in HN$, so HN is closed under the induced operation in G. Clearly $e = ee$ is in HN. For $h \in H$ and $n \in N$, we have $(hn)^{-1} = n^{-1}h^{-1} = h^{-1}n_4$ for some $n_4 \in N$, since N is a normal subgroup. Thus $(hn)^{-1} \in HN$, so $HN \leq G$. A similar argument shows that NH is a subgroup, so $NH = H \vee N = HN$.

Now suppose that H is also normal in G, and let $h \in H, n \in N$, and $g \in G$. Then $ghng^{-1} = (ghg^{-1})(gng^{-1}) \in HN$, so HN is indeed normal in G. ◆

We are now ready for the second isomorphism theorem.

34.5 Theorem **(Second Isomorphism Theorem)** Let H be a subgroup of G and let N be a normal subgroup of G. Then $(HN)/N \simeq H/(H \cap N)$.

Proof Let $\gamma : G \to G/N$ be the canonical homomorphism and let $H \leq G$. Then $\gamma[H]$ is a subgroup of G/N by Theorem 13.12. Now the action of γ on just the elements of H (called γ **restricted to** H) provides us with a homomorphism mapping H onto $\gamma[H]$, and the kernel of this restriction is clearly the set of elements of N that are also in H, that is, the intersection $H \cap N$. Theorem 34.2 then shows that there is an isomorphism $\mu_1 : H/(H \cap N) \to \gamma[H]$.

On the other hand, γ restricted to HN also provides a homomorphism mapping HN onto $\gamma[H]$, because $\gamma(n)$ is the identity N of G/N for all $n \in N$. The kernel of γ restricted to HN is N. Theorem 34.2 then provides us with an isomorphism $\mu_2 : (HN)/N \to \gamma[H]$.

Because $(HN)/N$ and $H/(H \cap N)$ are both isomorphic to $\gamma[H]$, they are isomorphic to each other. Indeed, $\phi : (HN)/N \to H/(H \cap N)$ where $\phi = \mu_1^{-1}\mu_2$ will be an isomorphism. More explicitly,

$$\phi((hn)N) = \mu_1^{-1}(\mu_2((hn)N)) = \mu_1^{-1}(h) = h(H \cap N). \quad \blacklozenge$$

34.6 Example Let $G = \mathbb{Z} \times \mathbb{Z} \times \mathbb{Z}$, $H = \mathbb{Z} \times \mathbb{Z} \times \{0\}$, and $N = \{0\} \times \mathbb{Z} \times \mathbb{Z}$. Then clearly $HN = \mathbb{Z} \times \mathbb{Z} \times \mathbb{Z}$ and $H \cap N = \{0\} \times \mathbb{Z} \times \{0\}$. We have $(HN)/N \simeq \mathbb{Z}$ and we also have $H/(H \cap N) \simeq \mathbb{Z}$. ▲

If H and K are two normal subgroups of G and $K \leq H$, then H/K is a normal subgroup of G/K. The third isomorphism theorem concerns these groups.

34.7 Theorem **(Third Isomorphism Theorem)** Let H and K be normal subgroups of a group G with $K \leq H$. Then $G/H \simeq (G/K)/(H/K)$.

Proof Let $\phi : G \to (G/K)/(H/K)$ be given by $\phi(a) = (aK)(H/K)$ for $a \in G$. Clearly ϕ is onto $(G/K)/(H/K)$, and for $a, b \in G$,

$$\begin{aligned}\phi(ab) &= [(ab)K](H/K) = [(aK)(bK)](H/K) \\ &= [(aK)(H/K)][(bK)(H/K)] \\ &= \phi(a)\phi(b),\end{aligned}$$

so ϕ is a homomorphism. The kernel consists of those $x \in G$ such that $\phi(x) = H/K$. These x are just the elements of H. Then Theorem 34.2 shows that $G/H \simeq (G/K)/(H/K)$. ◆

A nice way of viewing Theorem 34.7 is to regard the canonical map $\gamma_H : G \to G/H$ as being factored via a normal subgroup K of G, $K \leq H \leq G$, to give

$$\gamma_H = \gamma_{H/K}\gamma_K,$$

up to a natural isomorphism, as illustrated in Fig. 34.8. Another way of visualizing this theorem is to use the subgroup diagram in Fig. 34.9, where each group is a normal subgroup of G and is contained in the one above it. *The larger the normal subgroup, the smaller the factor group*. Thus we can think of G collapsed by H, that is, G/H, as being smaller than G collapsed by K. Theorem 34.7 states that we can collapse G all the way down to G/H in two steps. First, collapse to G/K, and then, using H/K, collapse this to $(G/K)/(H/K)$. The overall result is the same (up to isomorphism) as collapsing G by H.

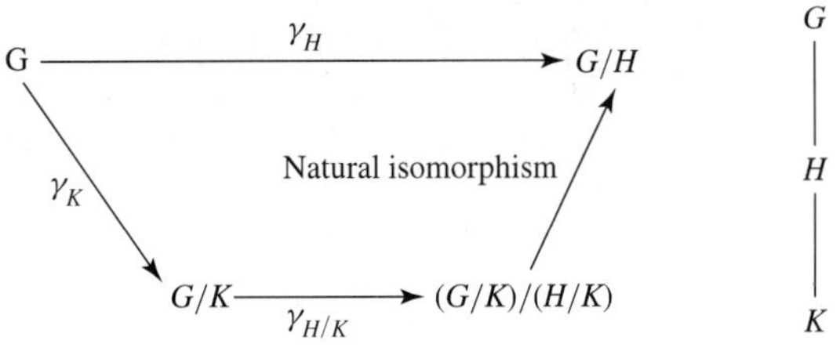

34.8 Figure **34.9 Figure**

34.10 Example Consider $K = 6\mathbb{Z} < H = 2\mathbb{Z} < G = \mathbb{Z}$. Then $G/H = \mathbb{Z}/2\mathbb{Z} \simeq \mathbb{Z}_2$. Now $G/K = \mathbb{Z}/6\mathbb{Z}$ has elements

$$6\mathbb{Z}, \quad 1+6\mathbb{Z}, \quad 2+6\mathbb{Z}, \quad 3+6\mathbb{Z}, \quad 4+6\mathbb{Z}, \quad \text{and} \quad 5+6\mathbb{Z}.$$

Of these six cosets, $6\mathbb{Z}$, $2+6\mathbb{Z}$, and $4+6\mathbb{Z}$ lie in $2\mathbb{Z}/6\mathbb{Z}$. Thus $(\mathbb{Z}/6\mathbb{Z})/(2\mathbb{Z}/6\mathbb{Z})$ has two elements and is isomorphic to $\mathbb{Z}_2$ also. Alternatively, we see that $\mathbb{Z}/6\mathbb{Z} \simeq \mathbb{Z}_6$, and $2\mathbb{Z}/6\mathbb{Z}$ corresponds *under this isomorphism* to the cyclic subgroup $\langle 2 \rangle$ of $\mathbb{Z}_6$. Thus $(\mathbb{Z}/6\mathbb{Z})/(2\mathbb{Z}/6\mathbb{Z}) \simeq \mathbb{Z}_6/\langle 2 \rangle \simeq \mathbb{Z}_2 \simeq \mathbb{Z}/2\mathbb{Z}$. ▲

EXERCISES 34

Computations

In using the three isomorphism theorems, it is often necessary to know the actual correspondence given by the isomorphism and not just the fact that the groups are isomorphic. The first six exercises give us training for this.

1. Let $\phi : \mathbb{Z}_{12} \to \mathbb{Z}_3$ be the homomorphism such that $\phi(1) = 2$.

- **a.** Find the kernel K of ϕ.
- **b.** List the cosets in $\mathbb{Z}_{12}/K$, showing the elements in each coset.
- **c.** Give the correspondence between $\mathbb{Z}_{12}/K$ and $\mathbb{Z}_3$ given by the map μ described in Theorem 34.2.

2. Let $\phi : \mathbb{Z}_{18} \to \mathbb{Z}_{12}$ be the homomorphism where $\phi(1) = 10$.

- **a.** Find the kernel K of ϕ.
- **b.** List the cosets in $\mathbb{Z}_{18}/K$, showing the elements in each coset.
- **c.** Find the group $\phi[\mathbb{Z}_{18}]$.
- **d.** Give the correspondence between $\mathbb{Z}_{18}/K$ and $\phi[\mathbb{Z}_{18}]$ given by the map μ described in Theorem 34.2.

3. In the group $\mathbb{Z}_{24}$, let $H = \langle 4 \rangle$ and $N = \langle 6 \rangle$.

- **a.** List the elements in HN (which we might write $H + N$ for these additive groups) and in $H \cap N$.
- **b.** List the cosets in HN/N, showing the elements in each coset.
- **c.** List the cosets in $H/(H \cap N)$, showing the elements in each coset.
- **d.** Give the correspondence between HN/N and $H/(H \cap N)$ described in the proof of Theorem 34.5.

4. Repeat Exercise 3 for the group $\mathbb{Z}_{36}$ with $H = \langle 6 \rangle$ and $N = \langle 9 \rangle$.

5. In the group $G = \mathbb{Z}_{24}$, let $H = \langle 4 \rangle$ and $K = \langle 8 \rangle$.

- **a.** List the cosets in G/H, showing the elements in each coset.
- **b.** List the cosets in G/K, showing the elements in each coset.
- **c.** List the cosets in H/K, showing the elements in each coset.
- **d.** List the cosets in $(G/K)/(H/K)$, showing the elements in each coset.
- **e.** Give the correspondence between G/H and $(G/K)/(H/K)$ described in the proof of Theorem 34.7.

6. Repeat Exercise 5 for the group $G = \mathbb{Z}_{36}$ with $H = \langle 9 \rangle$ and $K = \langle 18 \rangle$.

Theory

7. Show directly from the definition of a normal subgroup that if H and N are subgroups of a group G, and N is normal in G, then $H \cap N$ is normal in H.

8. Let H, K, and L be normal subgroups of G with $H < K < L$. Let $A = G/H$, $B = K/H$, and $C = L/H$.

a. Show that B and C are normal subgroups of A, and $B < C$.

b. To what factor group of G is $(A/B)/(C/B)$ isomorphic?

9. Let K and L be normal subgroups of G with $K \vee L = G$, and $K \cap L = \{e\}$. Show that $G/K \simeq L$ and $G/L \simeq K$.

SECTION 35 SERIES OF GROUPS

Subnormal and Normal Series

This section is concerned with the notion of a *series* of a group G, which gives insight into the structure of G. The results hold for both abelian and nonabelian groups. They are not too important for finitely generated abelian groups because of our strong Theorem 11.12. Many of our illustrations will be taken from abelian groups, however, for ease of computation.

35.1 Definition A **subnormal** (or **subinvariant**) **series of a group** G is a finite sequence $H_0, H_1, \cdots, H_n$ of subgroups of G such that $H_i < H_{i+1}$ and H_i is a normal subgroup of H_{i+1} with $H_0 = \{e\}$ and $H_n = G$. A **normal** (or **invariant**) **series of** G is a finite sequence $H_0, H_1, \cdots, H_n$ of normal subgroups of G such that $H_i < H_{i+1}$, $H_0 = \{e\}$, and $H_n = G$. ■

Note that for abelian groups the notions of subnormal and normal series coincide, since every subgroup is normal. A normal series is always subnormal, but the converse need not be true. We defined a subnormal series before a normal series, since the concept of a subnormal series is more important for our work.

35.2 Example Two examples of normal series of $\mathbb{Z}$ under addition are

$$\{0\} < 8\mathbb{Z} < 4\mathbb{Z} < \mathbb{Z}$$

and

$$\{0\} < 9\mathbb{Z} < \mathbb{Z}.$$

▲

35.3 Example Consider the group D_4 of symmetries of the square in Example 8.10. The series

$$\{\rho_0\} < \{\rho_0, \mu_1\} < \{\rho_0, \rho_2, \mu_1, \mu_2\} < D_4$$

is a subnormal series, as we could check using Table 8.12. It is not a normal series since $\{\rho_0, \mu_1\}$ is not normal in D_4. ▲

35.4 Definition A subnormal (normal) series $\{K_j\}$ is a **refinement of a subnormal (normal) series** $\{H_i\}$ of a group G if $\{H_i\} \subseteq \{K_j\}$, that is, if each H_i is one of the K_j. ■

35.5 Example The series

$$\{0\} < 72\mathbb{Z} < 24\mathbb{Z} < 8\mathbb{Z} < 4\mathbb{Z} < \mathbb{Z}$$

is a refinement of the series

$$\{0\} < 72\mathbb{Z} < 8\mathbb{Z} < \mathbb{Z}.$$

Two new terms, $4\mathbb{Z}$ and $24\mathbb{Z}$, have been inserted. ▲

Of interest in studying the structure of G are the factor groups H_{i+1}/H_i. These are defined for both normal and subnormal series, since H_i is normal in H_{i+1} in either case.

35.6 Definition Two subnormal (normal) series $\{H_i\}$ and $\{K_j\}$ of the same group G are **isomorphic** if there is a one-to-one correspondence between the collections of factor groups $\{H_{i+1}/H_i\}$ and $\{K_{j+1}/K_j\}$ such that corresponding factor groups are isomorphic. ■

Clearly, two isomorphic subnormal (normal) series must have the same number of groups.

35.7 Example The two series of $\mathbb{Z}_{15}$,

$$\{0\} < \langle 5\rangle < \mathbb{Z}_{15}$$

and

$$\{0\} < \langle 3\rangle < \mathbb{Z}_{15},$$

are isomorphic. Both $\mathbb{Z}_{15}/\langle 5\rangle$ and $\langle 3\rangle/\{0\}$ are isomorphic to $\mathbb{Z}_5$, and $\mathbb{Z}_{15}/\langle 3\rangle$ is isomorphic to $\langle 5\rangle/\{0\}$, or to $\mathbb{Z}_3$. ▲

The Schreier Theorem

We proceed to prove that two subnormal series of a group G have isomorphic refinements. This is a fundamental result in the theory of series. The proof is not too difficult. However, we know from experience that some students get lost in the proof, and then tend to feel that they cannot understand the theorem. We now give an illustration of the theorem before we proceed to its proof.

35.8 Example Let us try to find isomorphic refinements of the series

$$\{0\} < 8\mathbb{Z} < 4\mathbb{Z} < \mathbb{Z}$$

and

$$\{0\} < 9\mathbb{Z} < \mathbb{Z}$$

given in Example 35.2. Consider the refinement

$$\{0\} < 72\mathbb{Z} < 8\mathbb{Z} < 4\mathbb{Z} < \mathbb{Z}$$

of $\{0\} < 8\mathbb{Z} < 4\mathbb{Z} < \mathbb{Z}$ and the refinement

$$\{0\} < 72\mathbb{Z} < 18\mathbb{Z} < 9\mathbb{Z} < \mathbb{Z}$$

of $\{0\} < 9\mathbb{Z} < \mathbb{Z}$. In both cases the refinements have four factor groups isomorphic to $\mathbb{Z}_4$, $\mathbb{Z}_2$, $\mathbb{Z}_9$, and $72\mathbb{Z}$ or $\mathbb{Z}$. The *order* in which the factor groups occur is different to be sure. ▲

We start with a rather technical lemma developed by Zassenhaus. This lemma is sometimes called the *butterfly lemma*, since Fig. 35.9, which accompanies the lemma, has a butterfly shape.

Let H and K be subgroups of a group G, and let H^* be a normal subgroup of H and K^* be a normal subgroup of K. Applying the first statement in Lemma 34.4 to H^* and $H \cap K$ as subgroups of H, we see that $H^*(H \cap K)$ is a group. Similar arguments show that $H^*(H \cap K^*)$, $K^*(H \cap K)$, and $K^*(H^* \cap K)$ are also groups. It is not hard to show that $H^* \cap K$ is a normal subgroup of $H \cap K$ (see Exercise 22). The same argument using Lemma 34.4 applied to $H^* \cap K$ and $H \cap K^*$ as subgroups of $H \cap K$ shows that $L = (H^* \cap K)(H \cap K^*)$ is a group. Thus we have the diagram of subgroups shown in Fig. 35.9. It is not hard to verify the inclusion relations indicated by the diagram.

Since both $H \cap K^*$ and $H^* \cap K$ are normal subgroups of $H \cap K$, the second statement in Lemma 34.4 shows that $L = (H^* \cap K)(H \cap K^*)$ is a normal subgroup of $H \cap K$. We have denoted this particular normal subgroup relationship by the heavy middle line in Fig. 35.9. We claim the other two heavy lines also indicate normal subgroup relationships, and that the three factor groups given by the three normal subgroup relations are all isomorphic. To show this, we shall define a homomorphism $\phi : H^*(H \cap K) \to (H \cap K)/L$, and show that ϕ is onto $(H \cap K)/L$ with kernel $H^*(H \cap K^*)$. It will then follow at once from Theorem 34.2 that $H^*(H \cap K^*)$ is normal

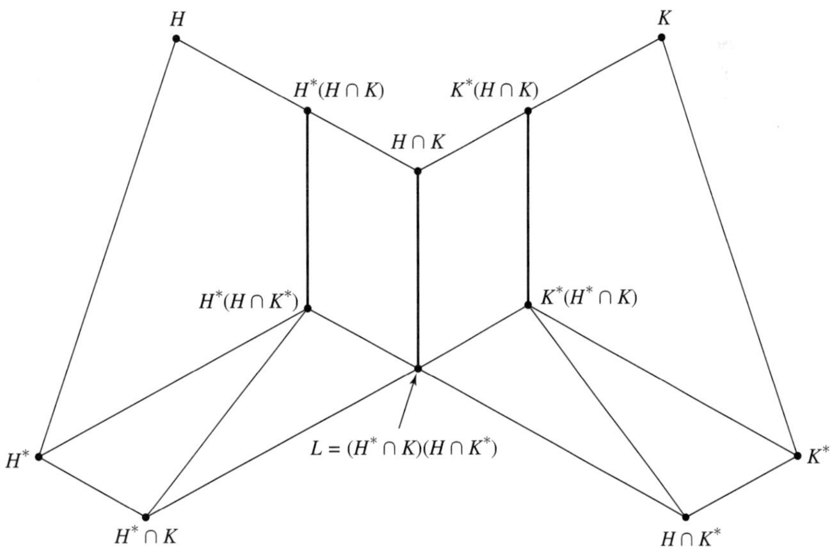

35.9 Figure

in $H^*(H \cap K)$, and that $H^*(H \cap K)/H^*(H \cap K^*) \simeq (H \cap K)/L$. A similar result for the groups on the right-hand heavy line in Fig. 35.9 then follows by symmetry.

Let $\phi : H^*(H \cap K) \to (H \cap K)/L$ be defined as follows. For $h \in H^*$ and $x \in H \cap K$, let $\phi(hx) = xL$. We show ϕ is well-defined and a homomorphism. Let $h_1, h_2 \in H^*$ and $x_1, x_2 \in H \cap K$. If $h_1x_1 = h_2x_2$, then $h_2^{-1}h_1 = x_2x_1^{-1} \in H^* \cap (H \cap K) = H^* \cap K \subseteq L$, so $x_1L = x_2L$. Thus ϕ is well defined. Since H^* is normal in H, there is h_3 in H^* such that $x_1h_2 = h_3x_1$. Then

$$\begin{aligned}\phi((h_1x_1)(h_2x_2)) &= \phi((h_1h_3)(x_1x_2)) = (x_1x_2)L\\ &= (x_1L)(x_2L) = \phi(h_1x_1)\cdot\phi(h_2x_2).\end{aligned}$$

Thus ϕ is a homomorphism.

Obviously ϕ is onto $(H \cap K)/L$. Finally if $h \in H^*$ and $x \in H \cap K$, then $\phi(hx) = xL = L$ if and only if $x \in L$, or if and only if $hx \in H^*L = H^*(H^* \cap K)(H \cap K^*) = H^*(H \cap K^*)$. Thus $\mathrm{Ker}(\phi) = H^*(H \cap K^*)$.

We have proved the following lemma.

35.10 Lemma **(Zassenhaus Lemma)** Let H and K be subgroups of a group G and let H^* and K^* be normal subgroups of H and K, respectively. Then

1. $H^*(H \cap K^*)$ is a normal subgroup of $H^*(H \cap K)$.
2. $K^*(H^* \cap K)$ is a normal subgroup of $K^*(H \cap K)$.
3. $$\begin{aligned}H^*(H \cap K)/H^*(H \cap K^*) &\simeq K^*(H \cap K)/K^*(H^* \cap K)\\ &\simeq (H \cap K)/[(H^* \cap K)(H \cap K^*)].\end{aligned}$$

35.11 Theorem **(Schreier Theorem)** Two subnormal (normal) series of a group G have isomorphic refinements.

Proof Let G be a group and let

$$\{e\} = H_0 < H_1 < H_2 < \cdots < H_n = G \qquad \textbf{(1)}$$

and

$$\{e\} = K_0 < K_1 < K_2 < \cdots < K_m = G \qquad \textbf{(2)}$$

be two subnormal series for G. For i where $0 \le i \le n-1$, form the chain of groups

$$H_i = H_i(H_{i+1} \cap K_0) \le H_i(H_{i+1} \cap K_1) \le \cdots \le H_i(H_{i+1} \cap K_m) = H_{i+1}.$$

This inserts $m-1$ not necessarily distinct groups between H_i and H_{i+1}. If we do this for each i where $0 \le i \le n-1$ and let $H_{ij} = H_i(H_{i+1} \cap K_j)$, then we obtain the chain of groups

$$\begin{aligned}\{e\} = H_{0,0} &\le H_{0,1} \le H_{0,2} \le \cdots \le H_{0,m-1} \le H_{1,0}\\ &\le H_{1,1} \le H_{1,2} \le \cdots \le H_{1,m-1} \le H_{2,0}\\ &\le H_{2,1} \le H_{2,2} \le \cdots \le H_{2,m-1} \le H_{3,0}\\ &\le \cdots\\ &\le H_{n-1,1} \le H_{n-1,2} \le \cdots \le H_{n-1,m-1} \le H_{n-1,m}\\ &= G.\end{aligned} \qquad \textbf{(3)}$$

This chain (3) contains $nm + 1$ not necessarily distinct groups, and $H_{i,0} = H_i$ for each i. By the Zassenhaus lemma, chain (3) is a subnormal chain, that is, each group is normal in the following group. This chain refines the series (1).

In a symmetric fashion, we set $K_{j,i} = K_j(K_{j+1} \cap H_i)$ for $0 \leq j \leq m - 1$ and $0 \leq i \leq n$. This gives a subnormal chain

$$\begin{aligned}\{e\} = K_{0,0} &\leq K_{0,1} \leq K_{0,2} \leq \cdots \leq K_{0,n-1} \leq K_{1,0} \\ &\leq K_{1,1} \leq K_{1,2} \leq \cdots \leq K_{1,n-1} \leq K_{2,0} \\ &\leq K_{2,1} \leq K_{2,2} \leq \cdots \leq K_{2,n-1} \leq K_{3,0} \\ &\leq \cdots \\ &\leq K_{m-1,1} \leq K_{m-1,2} \leq \cdots \leq K_{m-1,n-1} \leq K_{m-1,n} \\ &= G. \end{aligned} \tag{4}$$

This chain (4) contains $mn + 1$ not necessarily distinct groups, and $K_{j,0} = K_j$ for each j. This chain refines the series (2).

By the Zassenhaus lemma 35.10, we have

$$H_i(H_{i+1} \cap K_{j+1})/H_i(H_{i+1} \cap K_j) \simeq K_j(K_{j+1} \cap H_{i+1})/K_j(K_{j+1} \cap H_i),$$

or

$$H_{i,j+1}/H_{i,j} \simeq K_{j,i+1}/K_{j,i} \tag{5}$$

for $0 \leq i \leq n - 1$ and $0 \leq j \leq m - 1$. The isomorphisms of relation (5) give a one-to-one correspondence of isomorphic factor groups between the subnormal chains (3) and (4). To verify this correspondence, note that $H_{i,0} = H_i$ and $H_{i,m} = H_{i+1}$, while $K_{j,0} = K_j$ and $K_{j,n} = K_{j+1}$. Each chain in (3) and (4) contains a rectangular array of mn symbols $\leq$. Each $\leq$ gives rise to a factor group. The factor groups arising from the rth *row* of $\leq$'s in chain (3) correspond to the factor groups arising from the rth *column* of $\leq$'s in chain (4). Deleting repeated groups from the chains in (3) and (4), we obtain subnormal series of distinct groups that are isomorphic refinements of chains (1) and (2). This establishes the theorem for subnormal series.

For normal series, where all H_i and K_j are normal in G, we merely observe that all the groups $H_{i,j}$ and $K_{j,i}$ formed above are also normal in G, so the same proof applies. This normality of $H_{i,j}$ and $K_{j,i}$ follows at once from the second assertion in Lemma 34.4 and from the fact that intersections of normal subgroups of a group yield normal subgroups. ◆

The Jordan–Hölder Theorem

We now come to the real meat of the theory.

35.12 Definition A subnormal series $\{H_i\}$ of a group G is a **composition series** if all the factor groups H_{i+1}/H_i are simple. A normal series $\{H_i\}$ of G is a **principal** or **chief series** if all the factor groups H_{i+1}/H_i are simple. ■

Note that for abelian groups the concepts of composition and principal series coincide. Also, since every normal series is subnormal, every principal series is a composition series for any group, abelian or not.

35.13 Example We claim that $\mathbb{Z}$ has no composition (and also no principal) series. For if

$$\{0\} = H_0 < H_1 < \cdots < H_{n-1} < H_n = \mathbb{Z}$$

is a subnormal series, H_1 must be of the form $r\mathbb{Z}$ for some $r \in \mathbb{Z}^+$. But then H_1/H_0 is isomorphic to $r\mathbb{Z}$, which is infinite cyclic with many nontrivial proper normal subgroups, for example, $2r\mathbb{Z}$. Thus $\mathbb{Z}$ has no composition (and also no principal) series. ▲

35.14 Example The series

$$\{e\} < A_n < S_n$$

for $n \geq 5$ is a composition series (and also a principal series) of S_n, because $A_n/\{e\}$ is isomorphic to A_n, which is simple for $n \geq 5$, and S_n/A_n is isomorphic to $\mathbb{Z}_2$, which is simple. Likewise, the two series given in Example 35.7 are composition series (and also principal series) of $\mathbb{Z}_{15}$. They are isomorphic, as shown in that example. This illustrates our main theorem, which will be stated shortly. ▲

Observe that by Theorem 15.18, H_{i+1}/H_i is simple if and only if H_i is a maximal normal subgroup of H_{i+1}. Thus for a composition series, each H_i must be a maximal normal subgroup of H_{i+1}. *To form a composition series of a group G, we just hunt for a maximal normal subgroup H_{n-1} of G, then for a maximal normal subgroup H_{n-2} of H_{n-1}, and so on. If this process terminates in a finite number of steps, we have a composition series.* Note that by Theorem 15.18, a composition series cannot have any further refinement. *To form a principal series, we have to hunt for a maximal normal subgroup H_{n-1} of G, then for a maximal normal subgroup H_{n-2} of H_{n-1} that is also normal in G, and so on.* The main theorem is as follows.

35.15 Theorem **(Jordan–Hölder Theorem)** Any two composition (principal) series of a group G are isomorphic.

Proof Let $\{H_i\}$ and $\{K_i\}$ be two composition (principal) series of G. By Theorem 35.11, they have isomorphic refinements. But since all factor groups are already simple, Theorem 15.18 shows that neither series has any further refinement. Thus $\{H_i\}$ and $\{K_i\}$ must already be isomorphic. ◆

For a finite group, we should regard a composition series as a type of factorization of the group into simple factor groups, analogous to the factorization of a positive integer into primes. In both cases, the factorization is unique, up to the order of the factors.

HISTORICAL NOTE

This first appearance of what became the Jordan–Hölder theorem occurred in 1869 in a commentary on the work of Galois by the brilliant French algebraist Camille Jordan (1838–1922). The context of its appearance is the study of permutation groups associated with the roots of polynomial equations. Jordan asserted that even though the sequence of normal subgroups $G, I, J, \cdots$ of the group of the equation is not necessarily unique, nevertheless the sequence of indices of this composition series is unique. Jordan gave a proof in his monumental 1870 *Treatise on Substitutions and Algebraic Equations*. This latter work, though restricted to what we now call permutation groups, remained the standard treatise on group theory for many years.

The Hölder part of the theorem, that the sequence of factor groups in a composition series is unique up to order, was due to Otto Hölder (1859–1937), who played a very important role in the development of group theory once the completely abstract definition of a group had been given. Among his other contributions, he gave the first abstract definition of a "factor group" and determined the structure of all finite groups of square-free order.

35.16 Theorem If G has a composition (principal) series, and if N is a proper normal subgroup of G, then there exists a composition (principal) series containing N.

Proof The series

$$\{e\} < N < G$$

is both a subnormal and a normal series. Since G has a composition series $\{H_i\}$, then by Theorem 35.11 there is a refinement of $\{e\} < N < G$ to a subnormal series isomorphic to a refinement of $\{H_i\}$. But as a composition series, $\{H_i\}$ can have no further refinement. Thus $\{e\} < N < G$ can be refined to a subnormal series all of whose factor groups are simple, that is, to a composition series. A similar argument holds if we start with a principal series $\{K_j\}$ of G. ◆

35.17 Example A composition (and also a principal) series of $\mathbb{Z}_4 \times \mathbb{Z}_9$ containing $\langle(0, 1)\rangle$ is

$$\{(0, 0)\} < \langle(0, 3)\rangle < \langle(0, 1)\rangle < \langle 2\rangle \times \langle 1\rangle < \langle 1\rangle \times \langle 1\rangle = \mathbb{Z}_4 \times \mathbb{Z}_9.$$ ▲

The next definition is basic to the characterization of those polynomial equations whose solutions can be expressed in terms of radicals.

35.18 Definition A group G is **solvable** if it has a composition series $\{H_i\}$ such that all factor groups H_{i+1}/H_i are abelian. ■

By the Jordan–Hölder theorem, we see that for a solvable group, *every* composition series $\{H_i\}$ must have abelian factor groups H_{i+1}/H_i.

35.19 Example The group S_3 is solvable, because the composition series

$$\{e\} < A_3 < S_3$$

has factor groups isomorphic to $\mathbb{Z}_3$ and $\mathbb{Z}_2$, which are abelian. The group S_5 is not solvable, for since A_5 is simple, the series

$$\{e\} < A_5 < S_5$$

is a composition series, and $A_5/\{e\}$, which is isomorphic to A_5, is not abelian. *This group A_5 of order 60 can be shown to be the smallest group that is not solvable.* This fact is closely connected with the fact that a polynomial equation of degree 5 is not in general solvable by radicals, but a polynomial equation of degree ≤ 4 is. ▲

The Ascending Central Series

We mention one subnormal series for a group G that can be formed using centers of groups. Recall from Section 15 that the center $Z(G)$ of a group G is defined by

$$Z(G) = \{z \in G \mid zg = gz \text{ for all } g \in G\},$$

and that $Z(G)$ is a normal subgroup of G. If we have the table for a finite group G, it is easy to find the center. An element a will be in the center of G if and only if the elements in the row opposite a at the extreme left are given in the same order as the elements in the column under a at the very top of the table.

Now let G be a group, and let $Z(G)$ be the center of G. Since $Z(G)$ is normal in G, we can form the factor group $G/Z(G)$ and find the center $Z(G/Z(G))$ of this factor group. Since $Z(G/Z(G))$ is normal in $G/Z(G)$, if $\gamma : G \to G/Z(G)$ is the canonical map, then by Theorem 15.16, $\gamma^{-1}[Z(G/Z(G))]$ is a normal subgroup $Z_1(G)$ of G. We can then form the factor group $G/Z_1(G)$ and find its center, take $(\gamma_1)^{-1}$ of it to get $Z_2(G)$, and so on.

35.20 Definition The series

$$\{e\} \leq Z(G) \leq Z_1(G) \leq Z_2(G) \leq \cdots$$

described in the preceding discussion is the **ascending central series of the group** G. ■

35.21 Example The center of S_3 is just the identity $\{\rho_0\}$. Thus the ascending central series of S_3 is

$$\{\rho_0\} \leq \{\rho_0\} \leq \{\rho_0\} \leq \cdots.$$

The center of the group D_4 of symmetries of the square in Example 8.10 is $\{\rho_0, \rho_2\}$. (Do you remember that we said that this group would give us nice examples of many things we discussed?) Since $D_4/\{\rho_0, \rho_2\}$ is of order 4 and hence abelian, its center is all of $D_4/\{\rho_0, \rho_2\}$. Thus the ascending central series of D_4 is

$$\{\rho_0\} \leq \{\rho_0, \rho_2\} \leq D_4 \leq D_4 \leq D_4 \leq \cdots. \quad ▲$$

EXERCISES 35

Computations

In Exercises 1 through 5, give isomorphic refinements of the two series.

1. $\{0\} < 10\mathbb{Z} < \mathbb{Z}$ and $\{0\} < 25\mathbb{Z} < \mathbb{Z}$

2. $\{0\} < 60\mathbb{Z} < 20\mathbb{Z} < \mathbb{Z}$ and $\{0\} < 245\mathbb{Z} < 49\mathbb{Z} < \mathbb{Z}$

3. $\{0\} < \langle 3 \rangle < \mathbb{Z}_{24}$ and $\{0\} < \langle 8 \rangle < \mathbb{Z}_{24}$

4. $\{0\} < \langle 18 \rangle < \langle 3 \rangle < \mathbb{Z}_{72}$ and $\{0\} < \langle 24 \rangle < \langle 12 \rangle < \mathbb{Z}_{72}$

5. $\{(0, 0)\} < (60\mathbb{Z}) \times \mathbb{Z} < (10\mathbb{Z}) \times \mathbb{Z} < \mathbb{Z} \times \mathbb{Z}$ and $\{(0, 0)\} < \mathbb{Z} \times (80\mathbb{Z}) < \mathbb{Z} \times (20\mathbb{Z}) < \mathbb{Z} \times \mathbb{Z}$

6. Find all composition series of $\mathbb{Z}_{60}$ and show that they are isomorphic.

7. Find all composition series of $\mathbb{Z}_{48}$ and show that they are isomorphic.

8. Find all composition series of $\mathbb{Z}_5 \times \mathbb{Z}_5$.

9. Find all composition series of $S_3 \times \mathbb{Z}_2$.

10. Find all composition series of $\mathbb{Z}_2 \times \mathbb{Z}_5 \times \mathbb{Z}_7$.

11. Find the center of $S_3 \times \mathbb{Z}_4$.

12. Find the center of $S_3 \times D_4$.

13. Find the ascending central series of $S_3 \times \mathbb{Z}_4$.

14. Find the ascending central series of $S_3 \times D_4$.

Concepts

In Exercises 15 and 16, correct the definition of the italicized term without reference to the text, if correction is needed, so that it is in a form acceptable for publication.

15. A *composition series* of a group G is a finite sequence

$$\{e\} = H_0 < H_1 < H_2 < \cdots < H_{n-1} < H_n = G$$

of subgroups of G such that H_i is a maximal normal subgroup of H_{i+1} for $i = 0, 1, 2, \cdots, n-1$.

16. A *solvable group* is one that has a composition series of abelian groups.

17. Mark each of the following true or false.

_____ **a.** Every normal series is also subnormal.

_____ **b.** Every subnormal series is also normal.

_____ **c.** Every principal series is a composition series.

_____ **d.** Every composition series is a principal series.

_____ **e.** Every abelian group has exactly one composition series.

_____ **f.** Every finite group has a composition series.

_____ **g.** A group is solvable if and only if it has a composition series with simple factor groups.

_____ **h.** S_7 is a solvable group.

_____ **i.** The Jordan–Hölder theorem has some similarity with the Fundamental Theorem of Arithmetic, which states that every positive integer greater than 1 can be factored into a product of primes uniquely up to order.

_____ **j.** Every finite group of prime order is solvable.

18. Find a composition series of $S_3 \times S_3$. Is $S_3 \times S_3$ solvable?

19. Is the group D_4 of symmetries of the square in Example 8.10 solvable?

20. Let G be $\mathbb{Z}_{36}$. Refer to the proof of Theorem 35.11. Let the subnormal series (1) be

$$\{0\} < \langle 12 \rangle < \langle 3 \rangle < \mathbb{Z}_{36}$$

and let the subnormal series (2) be

$$\{0\} < \langle 18 \rangle < \mathbb{Z}_{36}.$$

Find chains (3) and (4) and exhibit the isomorphic factor groups as described in the proof. Write chains (3) and (4) in the rectangular array shown in the text.

21. Repeat Exercise 20 for the group $\mathbb{Z}_{24}$ with the subnormal series (1)

$$\{0\} < \langle 12 \rangle < \langle 4 \rangle < \mathbb{Z}_{24}$$

and (2)

$$\{0\} < \langle 6 \rangle < \langle 3 \rangle < \mathbb{Z}_{24}.$$

Theory

22. Let H^*, H, and K be subgroups of G with H^* normal in H. Show that $H^* \cap K$ is normal in $H \cap K$.

23. Show that if

$$H_0 = \{e\} < H_1 < H_2 < \cdots < H_n = G$$

is a subnormal (normal) series for a group G, and if H_{i+1}/H_i is of finite order s_{i+1}, then G is of finite order $s_1 s_2 \cdots s_n$.

24. Show that an infinite abelian group can have no composition series. [*Hint:* Use Exercise 23, together with the fact that an infinite abelian group always has a proper normal subgroup.]

25. Show that a finite direct product of solvable groups is solvable.

26. Show that a subgroup K of a solvable group G is solvable. [*Hint:* Let $H_0 = \{e\} < H_1 < \cdots < H_n = G$ be a composition series for G. Show that the distinct groups among $K \cap H_i$ for $i = 0, \cdots, n$ form a composition series for K. Observe that

$$(K \cap H_i)/(K \cap H_{i-1}) \simeq [H_{i-1}(K \cap H_i)]/[H_{i-1}],$$

by Theorem 34.5, with $H = K \cap H_i$ and $N = H_{i-1}$, and that $H_{i-1}(K \cap H_i) \leq H_i$.]

27. Let $H_0 = \{e\} < H_1 < \cdots < H_n = G$ be a composition series for a group G. Let N be a normal subgroup of G, and suppose that N is a simple group. Show that the distinct groups among H_0, $H_i N$ for $i = 0, \cdots, n$ also form a composition series for G. [*Hint:* $H_i N$ is a group by Lemma 34.4. Show that $H_{i-1}N$ is normal in $H_i N$. By Theorem 34.5

$$(H_i N)/(H_{i-1} N) \simeq H_i/[H_i \cap (H_{i-1} N)],$$

and the latter group is isomorphic to

$$[H_i/H_{i-1}]/[(H_i \cap (H_{i-1} N))/H_{i-1}],$$

by Theorem 34.7. But H_i/H_{i-1} is simple.]

28. Let G be a group, and let $H_0 = \{e\} < H_1 < \cdots < H_n = G$ be a composition series for G. Let N be a normal subgroup of G, and let $\gamma : G \to G/N$ be the canonical map. Show that the distinct groups among $\gamma[H_i]$ for $i = 0, \cdots, n$, form a composition series for G/N. [*Hint:* Observe that the map

$$\psi : H_i N \to \gamma[H_i]/\gamma[H_{i-1}]$$

defined by

$$\psi(h_i n) = \gamma(h_i n)\gamma[H_{i-1}]$$

is a homomorphism with kernel $H_{i-1}N$. By Theorem 34.2.

$$\gamma[H_i]/\gamma[H_{i-1}] \simeq (H_i N)/(H_{i-1}N).$$

Proceed via Theorem 34.5, as shown in the hint for Exercise 27.]

29. Prove that a homomorphic image of a solvable group is solvable. [*Hint:* Apply Exercise 28 to get a composition series for the homomorphic image. The hints for Exercises 27 and 28 then show how the factor groups of this composition series in the image look.]

SECTION 36 SYLOW THEOREMS

The fundamental theorem for finitely generated abelian groups (Theorem 11.12) gives us complete information about all finite abelian groups. The study of finite nonabelian groups is much more complicated. The Sylow theorems give us some important information about them.

We know the order of a subgroup of a finite group G must divide $|G|$. If G is abelian, then there exist subgroups of every order dividing $|G|$. We showed in Example 15.6 that A_4, which has order 12, has no subgroup of order 6. Thus a nonabelian group G may have no subgroup of some order d dividing $|G|$; the "converse of the theorem of Lagrange" does not hold. The Sylow theorems give a weak converse. Namely, they show that if d is a power of a prime and d divides $|G|$, then G does contain a subgroup of order d. (Note that 6 is not a power of a prime.) The Sylow theorems also give some information concerning the number of such subgroups and their relationship to each other. We will see that these theorems are very useful in studying finite nonabelian groups.

Proofs of the Sylow theorems give us another application of action of a group on a set described in Section 16. This time, the set itself is formed from the group; in some instances the set is the group itself, sometimes it is a collection of cosets of a subgroup, and sometimes it is a collection of subgroups.

p-Groups

Section 17 gave applications of Burnside's formula that counted the number of orbits in a finite G-set. Most of our results in this section flow from an equation that counts the number of elements in a finite G-set.

Let X be a finite G-set. Recall that for $x \in X$, the orbit of x in X under G is $Gx = \{gx \mid g \in G\}$. Suppose that there are r orbits in X under G, and let $\{x_1, x_2, \cdots, x_r\}$ contain one element from each orbit in X. Now every element of X is in precisely one orbit, so

$$|X| = \sum_{i=1}^{r} |Gx_i|. \quad \textbf{(1)}$$

There may be one-element orbits in X. Let $X_G = \{x \in X \mid gx = x \text{ for all } g \in G\}$. Thus X_G is precisely the union of the one-element orbits in X. Let us suppose there are s one-element orbits, where $0 \leq s \leq r$. Then $|X_G| = s$, and reordering the x_i if necessary, we may rewrite Eq. (1) as

$$|X| = |X_G| + \sum_{i=s+1}^{r} |Gx_i|. \quad \textbf{(2)}$$

Most of the results of this section will flow from Eq. (2). We shall develop Sylow theory as in Hungerford [10], where credit is given to R. J. Nunke for the line of proof. The proof of Theorem 36.3 (Cauchy's theorem) is credited there to J. H. McKay.

Theorem 36.1, which follows, is not quite a counting theorem, but it does have a numerical conclusion. It counts modulo p. The theorem seems to be amazingly powerful. In the rest of the chapter, if we choose the correct set, the correct group action on it, and apply Theorem 36.1, what we want seems to fall right into our lap! Compared with older proofs, the arguments are extremely pretty and elegant.

Throughout this section, p will always be a prime integer.

36.1 Theorem Let G be a group of order p^n and let X be a finite G-set. Then $|X| \equiv |X_G| \pmod{p}$.

Proof In the notation of Eq. (2), we know that $|Gx_i|$ divides $|G|$ by Theorem 16.16. Consequently p divides $|Gx_i|$ for $s + 1 \leq i \leq r$. Equation (2) then shows that $|X| - |X_G|$ is divisible by p, so $|X| \equiv |X_G| \pmod{p}$. ◆

36.2 Definition Let p be a prime. A group G is a p-**group** if every element in G has order a power of the prime p. A subgroup of a group G is a p-**subgroup of** G if the subgroup is itself a p-group. ■

Our goal in this section is to show that a finite group G has a subgroup of every prime-power order dividing $|G|$. As a first step, we prove Cauchy's theorem, which says that if p divides $|G|$, then G has a subgroup of order p.

36.3 Theorem **(Cauchy's Theorem)** Let p be a prime. Let G be a finite group and let p divide $|G|$. Then G has an element of order p and, consequently, a subgroup of order p.

Proof We form the set X of all p-tuples $(g_1, g_2, \cdots, g_p)$ of elements of G having the property that the product of the coordinates in G is e. That is,

$$X = \{(g_1, g_2, \cdots, g_p) \mid g_i \in G \text{ and } g_1 g_2 \cdots g_p = e\}.$$

We claim p divides $|X|$. In forming a p-tuple in X, we may let $g_1, g_2, \cdots, g_{p-1}$ be any elements of G, and g_p is then uniquely determined as $(g_1 g_2 \cdots g_{p-1})^{-1}$. Thus $|X| = |G|^{p-1}$ and since p divides $|G|$, we see that p divides $|X|$.

Let σ be the cycle $(1, 2, 3, \cdots, p)$ in S_p. We let σ act on X by

$$\sigma(g_1, g_2, \cdots, g_p) = (g_{\sigma(1)}, g_{\sigma(2)}, \cdots, g_{\sigma(p)}) = (g_2, g_3, \cdots, g_p, g_1).$$

Note that $(g_2, g_3, \cdots, g_p, g_1) \in X$, for $g_1(g_2 g_3 \cdots g_p) = e$ implies that $g_1 = (g_2 g_3 \cdots g_p)^{-1}$, so $(g_2 g_3 \cdots g_p)g_1 = e$ also. Thus σ acts on X, and we consider the subgroup $\langle\sigma\rangle$ of S_p to act on X by iteration in the natural way.

Now $|\langle\sigma\rangle| = p$, so we may apply Theorem 36.1, and we know that $|X| \equiv |X_{\langle\sigma\rangle}|$ (mod p). Since p divides $|X|$, it must be that p divides $|X_{\langle\sigma\rangle}|$ also. Let us examine $X_{\langle\sigma\rangle}$. Now $(g_1, g_2, \cdots, g_p)$ is left fixed by σ, and hence by $\langle\sigma\rangle$, if and only if $g_1 = g_2 = \cdots = g_p$. We know at least one element in $X_{\langle\sigma\rangle}$, namely $(e, e, \cdots, e)$. Since p divides $|X_{\langle\sigma\rangle}|$, there must be at least p elements in $X_{\langle\sigma\rangle}$. Hence there exists some element $a \in G$, $a \neq e$, such that $(a, a, \cdots, a) \in X_{\langle\sigma\rangle}$ and hence $a^p = e$, so a has order p. Of course, $\langle a\rangle$ is a subgroup of G of order p. ◆

36.4 Corollary Let G be a finite group. Then G is a p-group if and only if $|G|$ is a power of p.

Proof We leave the proof of this corollary to Exercise 14. ◆

The Sylow Theorems

Let G be a group, and let $\mathscr{S}$ be the collection of all subgroups of G. We make $\mathscr{S}$ into a G-set by letting G act on $\mathscr{S}$ by conjugation. That is, if $H \in \mathscr{S}$ so $H \leq G$ and $g \in G$, then g acting on H yields the conjugate subgroup gHg^{-1}. (To avoid confusion, we will never write this action as gH.) Now $G_H = \{g \in G \mid gHg^{-1} = H\}$ is easily seen to be a subgroup of G (Exercise 11), and H is a normal subgroup of G_H. Since G_H consists of *all* elements of G that leave H invariant under conjugation, G_H is the largest subgroup of G having H as a normal subgroup.

36.5 Definition The subgroup G_H just discussed is the **normalizer of** H **in** G and will be denoted $N[H]$ from now on. ■

In the proof of the lemma that follows, we will use the fact that if H is a *finite* subgroup of a group G, then $g \in N[H]$ if $ghg^{-1} \in H$ for all $h \in H$. To see this, note that if $gh_1g^{-1} = gh_2g^{-1}$, then $h_1 = h_2$ by cancellation in the group G. Thus the conjugation map $i_g : H \to H$ given by $i_g(h) = ghg^{-1}$ is one to one. Because $|H|$ is finite, i_g must then map H *onto* H, so $gHg^{-1} = H$ and $g \in N[H]$.

36.6 Lemma Let H be a p-subgroup of a finite group G. Then

$$(N[H] : H) \equiv (G : H) (\text{mod } p).$$

Historical Note

The Sylow theorems are due to the Norwegian mathematician Peter Ludvig Mejdell Sylow (1832–1918), who published them in a brief paper in 1872. Sylow stated the theorems in terms of permutation groups (since the abstract definition of a group had not yet been given). Georg Frobenius re-proved the theorems for abstract groups in 1887, even though he noted that in fact every group can be considered as a permutation group (Cayley's theorem [Theorem 8.16]). Sylow himself immediately applied the theorems to the question of solving algebraic equations and showed that any equation whose Galois group has order a power of a prime p is solvable by radicals.

Sylow spent most of his professional life as a high school teacher in Halden, Norway, and was only appointed to a position at Christiana University in 1898. He devoted eight years of his life to the project of editing the mathematical works of his countryman Niels Henrik Abel.

Proof Let $\mathscr{L}$ be the set of left cosets of H in G, and let H act on $\mathscr{L}$ by left translation, so that $h(xH) = (hx)H$. Then $\mathscr{L}$ becomes an H-set. Note that $|\mathscr{L}| = (G : H)$.

Let us determine $\mathscr{L}_H$, that is, those left cosets that are fixed under action by all elements of H. Now $xH = h(xH)$ if and only if $H = x^{-1}hxH$, or if and only if $x^{-1}hx \in H$. Thus $xH = h(xH)$ for all $h \in H$ if and only if $x^{-1}hx = x^{-1}h(x^{-1})^{-1} \in H$ for all $h \in H$, or if and only if $x^{-1} \in N[H]$ (see the comment before the lemma), or if and only if $x \in N[H]$. Thus the left cosets in $\mathscr{L}_H$ are those contained in $N[H]$. The number of such cosets is $(N[H] : H)$, so $|\mathscr{L}_H| = (N[H] : H)$.

Since H is a p-group, it has order a power of p by Corollary 36.4. Theorem 36.1 then tells us that $|\mathscr{L}| \equiv |\mathscr{L}_H| \pmod{p}$, that is, that $(G : H) \equiv (N[H] : H) \pmod{p}$. ◆

36.7 Corollary Let H be a p-subgroup of a finite group G. If p divides $(G : H)$, then $N[H] \neq H$.

Proof It follows from Lemma 36.6 that p divides $(N[H] : H)$, which must then be different from 1. Thus $H \neq N[H]$. ◆

We are now ready for the first of the Sylow theorems, which asserts the existence of prime-power subgroups of G for any prime power dividing $|G|$.

36.8 Theorem **(First Sylow Theorem)** Let G be a finite group and let $|G| = p^n m$ where $n \geq 1$ and where p does not divide m. Then

1. G contains a subgroup of order p^i for each i where $1 \leq i \leq n$,
2. Every subgroup H of G of order p^i is a normal subgroup of a subgroup of order p^{i+1} for $1 \leq i < n$.

Proof

1. We know G contains a subgroup of order p by Cauchy's theorem (Theorem 36.3). We use an induction argument and show that the existence of a subgroup of order p^i for $i < n$ implies the existence of a subgroup of order p^{i+1}. Let H be a subgroup of order p^i. Since $i < n$, we see p divides $(G : H)$. By Lemma 36.6, we then know p divides $(N[H] : H)$. Since H is a normal

subgroup of $N[H]$, we can form $N[H]/H$, and we see that p divides $|N[H]/H|$. By Cauchy's theorem, the factor group $N[H]/H$ has a subgroup K which is of order p. If $\gamma : N[H] \to N[H]/H$ is the canonical homomorphism, then $\gamma^{-1}[K] = \{x \in N[H] \mid \gamma(x) \in K\}$ is a subgroup of $N[H]$ and hence of G. This subgroup contains H and is of order p^{i+1}.

2. We repeat the construction in part 1 and note that $H < \gamma^{-1}[K] \leq N[H]$ where $|\gamma^{-1}[K]| = p^{i+1}$. Since H is normal in $N[H]$, it is of course normal in the possibly smaller group $\gamma^{-1}[K]$. ◆

36.9 Definition A **Sylow p-subgroup** P of a group G is a maximal p-subgroup of G, that is, a p-subgroup contained in no larger p-subgroup. ■

Let G be a finite group, where $|G| = p^n m$ as in Theorem 36.8. The theorem shows that the Sylow p-subgroups of G are precisely those subgroups of order p^n. If P is a Sylow p-subgroup, every conjugate gPg^{-1} of P is also a Sylow p-subgroup. The second Sylow theorem states that every Sylow p-subgroup can be obtained from P in this fashion; that is, any two Sylow p-subgroups are conjugate.

36.10 Theorem **(Second Sylow Theorem)** Let P_1 and P_2 be Sylow p-subgroups of a finite group G. Then P_1 and P_2 are conjugate subgroups of G.

Proof Here we will let one of the subgroups act on left cosets of the other, and use Theorem 36.1. Let $\mathscr{L}$ be the collection of left cosets of P_1, and let P_2 act on $\mathscr{L}$ by $y(xP_1) = (yx)P_1$ for $y \in P_2$. Then $\mathscr{L}$ is a P_2-set. By Theorem 36.1, $|\mathscr{L}_{P_2}| \equiv |\mathscr{L}| \pmod{p}$, and $|\mathscr{L}| = (G : P_1)$ is not divisible by p, so $|\mathscr{L}_{P_2}| \neq 0$. Let $xP_1 \in \mathscr{L}_{P_2}$. Then $yxP_1 = xP_1$ for all $y \in P_2$, so $x^{-1}yxP_1 = P_1$ for all $y \in P_2$. Thus $x^{-1}yx \in P_1$ for all $y \in P_2$, so $x^{-1}P_2x \leq P_1$. Since $|P_1| = |P_2|$, we must have $P_1 = x^{-1}P_2x$, so P_1 and P_2 are indeed conjugate subgroups. ◆

The final Sylow theorem gives information on the number of Sylow p-subgroups. A few illustrations are given after the theorem, and many more are given in the next section.

36.11 Theorem **(Third Sylow Theorem)** If G is a finite group and p divides $|G|$, then the number of Sylow p-subgroups is congruent to 1 modulo p and divides $|G|$.

Proof Let P be one Sylow p-subgroup of G. Let $\mathscr{S}$ be the set of all Sylow p-subgroups and let P act on $\mathscr{S}$ by conjugation, so that $x \in P$ carries $T \in \mathscr{S}$ into xTx^{-1}. By Theorem 36.1, $|\mathscr{S}| \equiv |\mathscr{S}_P| \pmod{p}$. Let us find $\mathscr{S}_P$. If $T \in \mathscr{S}_P$, then $xTx^{-1} = T$ for all $x \in P$. Thus $P \leq N[T]$. Of course $T \leq N[T]$ also. Since P and T are both Sylow p-subgroups of G, they are also Sylow p-subgroups of $N[T]$. But then they are conjugate in $N[T]$ by Theorem 36.10. Since T is a normal subgroup of $N[T]$, it is its only conjugate in $N[T]$. Thus $T = P$. Then $\mathscr{S}_P = \{P\}$. Since $|\mathscr{S}| \equiv |\mathscr{S}_P| = 1 \pmod{p}$, we see the number of Sylow p-subgroups is congruent to 1 modulo p.

Now let G act on $\mathscr{S}$ by conjugation. Since all Sylow p-subgroups are conjugate, there is only one orbit in $\mathscr{S}$ under G. If $P \in \mathscr{S}$, then $|\mathscr{S}| = |\text{orbit of } P| = (G : G_P)$ by Theorem 16.16. (G_P is, in fact, the normalizer of P.) But $(G : G_P)$ is a divisor of $|G|$, so the number of Sylow p-subgroups divides $|G|$. ◆

36.12 Example The Sylow 2-subgroups of S_3 have order 2. The subgroups of order 2 in S_3 in Example 8.7 are

$$\{\rho_0, \mu_1\}, \qquad \{\rho_0, \mu_2\}, \quad \{\rho_0, \mu_3\}.$$

Note that there are three subgroups and that $3 \equiv 1 \pmod{2}$. Also, 3 divides 6, the order of S_3. We can readily check that

$$i_{\rho_2}[\{\rho_0, \mu_1\}] = \{\rho_0, \mu_3\} \qquad \text{and} \qquad i_{\rho_1}[\{\rho_0, \mu_1\}] = \{\rho_0, \mu_2\}$$

where $i_{\rho_j}(x) = \rho_j x \rho_j^{-1}$, illustrating that they are all conjugate. ▲

36.13 Example Let us use the Sylow theorems to show that no group of order 15 is simple. Let G have order 15. We claim that G has a normal subgroup of order 5. By Theorem 36.8 G has at least one subgroup of order 5, and by Theorem 36.11 the number of such subgroups is congruent to 1 modulo 5 and divides 15. Since 1, 6, and 11 are the only positive numbers less than 15 that are congruent to 1 modulo 5, and since among these only the number 1 divides 15, we see that G has exactly one subgroup P of order 5. But for each $g \in G$, the inner automorphism i_g of G with $i_g(x) = gxg^{-1}$ maps P onto a subgroup gPg^{-1}, again of order 5. Hence we must have $gPg^{-1} = P$ for all $g \in G$, so P is a normal subgroup of G. Therefore, G is not simple. (Example 37.10 will show that G must actually be abelian and therefore must be cyclic.) ▲

We trust that Example 36.13 gives some inkling of the power of Theorem 36.11. *Never underestimate a theorem that counts something, even modulo p.*

EXERCISES 36

Computations

In Exercises 1 through 4, fill in the blanks.

1. A Sylow 3-subgroup of a group of order 12 has order ______.
2. A Sylow 3-subgroup of a group of order 54 has order ______.
3. A group of order 24 must have either ______ or ______ Sylow 2-subgroups. (Use only the information given in Theorem 36.11.)
4. A group of order $255 = (3)(5)(17)$ must have either ______ or ______ Sylow 3-subgroups and ______ or ______ Sylow 5-subgroups. (Use only the information given in Theorem 36.11.)
5. Find all Sylow 3-subgroups of S_4 and demonstrate that they are all conjugate.
6. Find two Sylow 2-subgroups of S_4 and show that they are conjugate.

Concepts

In Exercises 7 through 9, correct the definition of the italicized term without reference to the text, if correction is needed, so that it is in a form acceptable for publication.

7. Let p be a prime. A *p-group* is a group with the property that every element has order p.
8. The *normalizer* $N[H]$ of a subgroup H of a group G is the set of all inner automorphisms that carry H onto itself.

9. Let G be a group whose order is divisible by a prime p. The *Sylow p-subgroup* of a group is the largest subgroup P of G with the property that P has some power of p as its order.

10. Mark each of the following true or false.

_______**a.** Any two Sylow p-subgroups of a finite group are conjugate.

_______**b.** Theorem 36.11 shows that a group of order 15 has only one Sylow 5-subgroup.

_______**c.** Every Sylow p-subgroup of a finite group has order a power of p.

_______**d.** Every p-subgroup of every finite group is a Sylow p-subgroup.

_______**e.** Every finite abelian group has exactly one Sylow p-subgroup for each prime p dividing the order of G.

_______**f.** The normalizer in G of a subgroup H of G is always a normal subgroup of G.

_______**g.** If H is a subgroup of G, then H is always a normal subgroup of $N[H]$.

_______**h.** A Sylow p-subgroup of a finite group G is normal in G if and only if it is the only Sylow p-subgroup of G.

_______**i.** If G is an abelian group and H is a subgroup of G, then $N[H] = H$.

_______**j.** A group of prime-power order p^n has no Sylow p-subgroup.

Theory

11. Let H be a subgroup of a group G. Show that $G_H = \{g \in G \mid gHg^{-1} = H\}$ is a subgroup of G.

12. Let G be a finite group and let primes p and $q \neq p$ divide $|G|$. Prove that if G has precisely one proper Sylow p-subgroup, it is a normal subgroup, so G is not simple.

13. Show that every group of order 45 has a normal subgroup of order 9.

14. Prove Corollary 36.4.

15. Let G be a finite group and let p be a prime dividing $|G|$. Let P be a Sylow p-subgroup of G. Show that $N[N[P]] = N[P]$. [*Hint:* Argue that P is the only Sylow p-subgroup of $N[N[P]]$, and use Theorem 36.10.]

16. Let G be a finite group and let a prime p divide $|G|$. Let P be a Sylow p-subgroup of G and let H be any p-subgroup of G. Show there exists $g \in G$ such that $gHg^{-1} \leq P$.

17. Show that every group of order $(35)^3$ has a normal subgroup of order 125.

18. Show that there are no simple groups of order $255 = (3)(5)(17)$.

19. Show that there are no simple groups of order $p^r m$, where p is a prime, r is a positive integer, and $m < p$.

20. Let G be a finite group. Regard G as a G-set where G acts on itself by conjugation.

a. Show that G_G is the center $Z(G)$ of G. (See Section 15.)

b. Use Theorem 36.1 to show that the center of a finite nontrivial p-group is nontrivial.

21. Let p be a prime. Show that a finite group of order p^n contains *normal* subgroups H_i for $0 \leq i \leq n$ such that $|H_i| = p^i$ and $H_i < H_{i+1}$ for $0 \leq i < n$. [*Hint:* See Exercise 20 and get an idea from Section 35.]

22. Let G be a finite group and let P be a normal p-subgroup of G. Show that P is contained in every Sylow p-subgroup of G.

SECTION 37 Applications of the Sylow Theory

In this section we give several applications of the Sylow theorems. It is intriguing to see how easily certain facts about groups of particular orders can be deduced. However, we should realize that we are working only with groups of finite order and really making only a small dent in the general problem of determining the structure of all finite groups. If the order of a group has only a few factors, then the techniques illustrated in this section may be of some use in determining the structure of the group. This will be demonstrated further in Section 40, where we shall show how it is sometimes possible to describe all groups (up to isomorphism) of certain orders, even when some of the groups are not abelian. However, if the order of a finite group is highly composite, that is, has a large number of factors, the problem is in general much harder.

Applications to p-Groups and the Class Equation

37.1 Theorem Every group of prime-power order (that is, every finite p-group) is solvable.

Proof If G has order p^r, it is immediate from Theorem 36.8 that G has a subgroup H_i of order p^i normal in a subgroup H_{i+1} of order p^{i+1} for $1 \leq i < r$. Then

$$\{e\} = H_0 < H_1 < H_2 < \cdots < H_r = G$$

is a composition series, where the factor groups are of order p, and hence abelian and actually cyclic. Thus, G is solvable. ◆

The older proofs of the Sylow theorems used the *class equation*. The line of proof in Section 36 avoided explicit mention of the class equation, although Eq. (2) there is a general form of it. We now develop the classic class equation so you will be familiar with it.

Let X be a finite G-set where G is a finite group. Then Eq. (2) of Section 36 tells us that

$$|X| = |X_G| + \sum_{i=s+1}^{r} |Gx_i| \tag{1}$$

where x_i is an element in the ith orbit in X. Consider now the special case of Eq. (1), where $X = G$ and the action of G on G is by conjugation, so $g \in G$ carries $x \in X = G$ into gxg^{-1}. Then

$$\begin{aligned} X_G &= \{x \in G \mid gxg^{-1} = x \text{ for all } g \in G\} \\ &= \{x \in G \mid xg = gx \text{ for all } g \in G\} = Z(G), \end{aligned}$$

the center of G. If we let $c = |Z(G)|$ and $n_i = |Gx_i|$ in Eq. (1), then we obtain

$$|G| = c + n_{c+1} + \cdots + n_r \tag{2}$$

where n_i is the number of elements in the ith orbit of G under conjugation by itself. Note that n_i divides $|G|$ for $c + 1 \leq i \leq r$ since in Eq. (1) we know $|Gx_i| = (G : G_{x_i})$, which is a divisor of $|G|$.

37.2 Definition Equation (2) is the **class equation of** G. Each orbit in G under conjugation by G is a **conjugate class in** G. ■

37.3 Example It is readily checked that for S_3 of Example 8.7, the conjugate classes are

$$\{\rho_0\}, \qquad \{\rho_1, \rho_2\}, \qquad \{\mu_1, \mu_2, \mu_3\}.$$

The class equation of S_3 is

$$6 = 1 + 2 + 3. \qquad \blacktriangle$$

For illustration of the use of the class equation, we prove a theorem that Exercise 20(b) in Section 36 asked us to prove.

37.4 Theorem The center of a finite nontrivial p-group G is nontrivial.

Proof In Eq. (2) for G, each n_i divides $|G|$ for $c + 1 \leq i \leq r$, so p divides each n_i, and p divides $|G|$. Therefore p divides c. Now $e \in Z(G)$, so $c \geq 1$. Therefore $c \geq p$, and there exists some $a \in Z(G)$ where $a \neq e$. ◆

We turn now to a lemma on direct products that will be used in some of the theorems that follow.

37.5 Lemma Let G be a group containing normal subgroups H and K such that $H \cap K = \{e\}$ and $H \vee K = G$. Then G is isomorphic to $H \times K$.

Proof We start by showing that $hk = kh$ for $k \in K$ and $h \in H$. Consider the commutator $hkh^{-1}k^{-1} = (hkh^{-1})k^{-1} = h(kh^{-1}k^{-1})$. Since H and K are normal subgroups of G, the two groupings with parentheses show that $hkh^{-1}k^{-1}$ is in both K and H. Since $K \cap H = \{e\}$, we see that $hkh^{-1}k^{-1} = e$, so $hk = kh$.

Let $\phi : H \times K \to G$ be defined by $\phi(h, k) = hk$. Then

$$\begin{aligned}\phi((h, k)(h', k')) &= \phi(hh', kk') = hh'kk' \\ &= hkh'k' = \phi(h, k)\phi(h', k'),\end{aligned}$$

so ϕ is a homomorphism.

If $\phi(h, k) = e$, then $hk = e$, so $h = k^{-1}$, and both h and k are in $H \cap K$. Thus $h = k = e$, so $\text{Ker}(\phi) = \{(e, e)\}$ and ϕ is one to one.

By Lemma 34.4, we know that $HK = H \vee K$, and $H \vee K = G$ by hypothesis. Thus ϕ is onto G, and $H \times K \simeq G$. ◆

37.6 Theorem For a prime number p, every group G of order p^2 is abelian.

Proof If G is not cyclic, then every element except e must be of order p. Let a be such an element. Then the cyclic subgroup $\langle a \rangle$ of order p does not exhaust G. Also let $b \in G$ with $b \notin \langle a \rangle$. Then $\langle a \rangle \cap \langle b \rangle = \{e\}$, since an element c in $\langle a \rangle \cap \langle b \rangle$ with $c \neq e$ would generate both $\langle a \rangle$ and $\langle b \rangle$, giving $\langle a \rangle = \langle b \rangle$, contrary to construction. From Theorem 36.8, $\langle a \rangle$ is normal in some subgroup of order p^2 of G, that is, normal in all of G. Likewise $\langle b \rangle$ is normal in G. Now $\langle a \rangle \vee \langle b \rangle$ is a subgroup of G properly containing $\langle a \rangle$ and of order dividing p^2. Hence $\langle a \rangle \vee \langle b \rangle$ must be all of G. Thus the hypotheses of Lemma 37.5 are satisfied, and G is isomorphic to $\langle a \rangle \times \langle b \rangle$ and therefore abelian. ◆

Further Applications

We turn now to a discussion of whether there exist simple groups of certain orders. We have seen that every group of prime order is simple. We also asserted that A_n is simple for $n \geq 5$ and that A_5 is the smallest simple group that is not of prime order. There was a famous conjecture of Burnside that every finite simple group of nonprime order must be of even order. It was a triumph when this was proved by Thompson and Feit [21].

37.7 Theorem If p and q are distinct primes with $p < q$, then every group G of order pq has a single subgroup of order q and this subgroup is normal in G. Hence G is not simple. If q is not congruent to 1 modulo p, then G is abelian and cyclic.

Proof Theorems 36.8 and 36.11 tell us that G has a Sylow q-subgroup and that the number of such subgroups is congruent to 1 modulo q and divides pq, and therefore must divide p. Since $p < q$, the only possibility is the number 1. Thus there is only one Sylow q-subgroup Q of G. This group Q must be normal in G, for under an inner automorphism it would be carried into a group of the same order, hence itself. Thus G is not simple.

Likewise, there is a Sylow p-subgroup P of G, and the number of these divides pq and is congruent to 1 modulo p. This number must be either 1 or q. If q is not congruent to 1 modulo p, then the number must be 1 and P is normal in G. Let us assume that $q \not\equiv 1 \pmod{p}$. Since every element in Q other than e is of order q and every element in P other than e is of order p, we have $Q \cap P = \{e\}$. Also $Q \vee P$ must be a subgroup of G properly containing Q and of order dividing pq. Hence $Q \vee P = G$ and by Lemma 37.5 is isomorphic to $Q \times P$ or $\mathbb{Z}_q \times \mathbb{Z}_p$. Thus G is abelian and cyclic. ◆

We need another lemma for some of the counting arguments that follow.

37.8 Lemma If H and K are finite subgroups of a group G, then

$$|HK| = \frac{(|H|)(|K|)}{|H \cap K|}.$$

Proof Recall that $HK = \{hk \mid h \in H, k \in K\}$. Let $|H| = r$, $|K| = s$, and $|H \cap K| = t$. Now HK has at most rs elements. However, it is possible for h_1k_1 to equal h_2k_2, for $h_1, h_2 \in H$ and $k_1, k_2 \in K$; that is, there may be some collapsing. If $h_1k_1 = h_2k_2$, then let

$$x = (h_2)^{-1}h_1 = k_2(k_1)^{-1}$$

Now $x = (h_2)^{-1}h_1$ shows that $x \in H$, and $x = k_2(k_1)^{-1}$ shows that $x \in K$. Hence $x \in (H \cap K)$, and

$$h_2 = h_1x^{-1} \quad \text{and} \quad k_2 = xk_1.$$

On the other hand, if for $y \in (H \cap K)$ we let $h_3 = h_1y^{-1}$ and $k_3 = yk_1$, then clearly $h_3k_3 = h_1k_1$, with $h_3 \in H$ and $k_3 \in K$. Thus each element $hk \in HK$ can be represented in the form h_ik_i, for $h_i \in H$ and $k_i \in K$, as many times as there are elements of $H \cap K$, that is, t times. Therefore, the number of elements in HK is rs/t. ◆

Lemma 37.8 is another result that counts something, so do not underestimate it. The lemma will be used in the following way: A finite group G cannot have subgroups H and K that are too large with intersections that are too small, or the order of HK would have to exceed the order of G, which is impossible. For example, a group of order 24 cannot have two subgroups of orders 12 and 8 with an intersection of order 2.

The remainder of this section consists of several examples illustrating techniques of proving that all groups of certain orders are abelian or that they have nontrivial proper normal subgroups, that is, that they are not simple. We will use one fact we mentioned before only in the exercises. *A subgroup H of index 2 in a finite group G is always normal*, for by counting, we see that there are only the left cosets H itself and the coset consisting of all elements in G not in H. The right cosets are the same. Thus every right coset is a left coset, and H is normal in G.

37.9 Example No group of order p^r for $r > 1$ is simple, where p is a prime. For by Theorem 36.8 such a group G contains a subgroup of order p^{r-1} normal in a subgroup of order p^r, which must be all of G. Thus a group of order 16 is not simple; it has a normal subgroup of order 8. ▲

37.10 Example Every group of order 15 is cyclic (hence abelian and not simple, since 15 is not a prime). This is because $15 = (5)(3)$, and 5 is not congruent to 1 modulo 3. By Theorem 37.7 we are done. ▲

37.11 Example No group of order 20 is simple, for such a group G contains Sylow 5-subgroups in number congruent to 1 modulo 5 and a divisor of 20, hence only 1. This Sylow 5-subgroup is then normal, since all conjugates of it must be itself. ▲

37.12 Example No group of order 30 is simple. We have seen that if there is only one Sylow p-subgroup for some prime p dividing 30, we are done. By Theorem 36.11 the possibilities for the number of Sylow 5-subgroups are 1 or 6, and those for Sylow 3-subgroups are 1 or 10. But if G has six Sylow 5-subgroups, then the intersection of any two is a subgroup of each of order dividing 5, and hence just $\{e\}$. Thus each contains 4 elements of order 5 that are in none of the others. Hence G must contain 24 elements of order 5. Similarly, if G has 10 Sylow 3-subgroups, it has at least 20 elements of order 3. The two types of Sylow subgroups together would require at least 44 elements in G. Thus there is a normal subgroup either of order 5 or of order 3. ▲

37.13 Example No group of order 48 is simple. Indeed, we shall show that a group G of order 48 has a normal subgroup of either order 16 or order 8. By Theorem 36.11 G has either one or three Sylow 2-subgroups of order 16. If there is only one subgroup of order 16, it is normal in G, by a now familiar argument.

Suppose that there are three subgroups of order 16, and let H and K be two of them. Then $H \cap K$ must be of order 8, for if $H \cap K$ were of order ≤ 4, then by Lemma 37.8 HK would have at least $(16)(16)/4 = 64$ elements, contradicting the fact that G has only 48 elements. Therefore, $H \cap K$ is normal in both H and K (being of index 2, or by Theorem 36.8). Hence the normalizer of $H \cap K$ contains both H and K and must have order a multiple >1 of 16 and a divisor of 48, therefore 48. Thus $H \cap K$ must be normal in G. ▲

37.14 Example No group of order 36 is simple. Such a group G has either one or four subgroups of order 9. If there is only one such subgroup, it is normal in G. If there are four such subgroups, let H and K be two of them. As in Example 37.13, $H \cap K$ must have at least 3 elements, or HK would have to have 81 elements, which is impossible. Thus the normalizer of $H \cap K$ has as order a multiple of >1 of 9 and a divisor of 36; hence the order must be either 18 or 36. If the order is 18, the normalizer is then of index 2 and therefore is normal in G. If the order is 36, then $H \cap K$ is normal in G. ▲

37.15 Example Every group of order $255 = (3)(5)(17)$ is abelian (hence cyclic by the Fundamental Theorem 11.12 and not simple, since 255 is not a prime). By Theorem 36.11 such a group G has only one subgroup H of order 17. Then G/H has order 15 and is abelian by Example 37.10. By Theorem 15.20, we see that the commutator subgroup C of G is contained in H. Thus as a subgroup of H, C has either order 1 or 17. Theorem 36.11 also shows that G has either 1 or 85 subgroups of order 3 and either 1 or 51 subgroups of order 5. However, 85 subgroups of order 3 would require 170 elements of order 3, and 51 subgroups of order 5 would require 204 elements of order 5 in G; both together would then require 375 elements in G, which is impossible. Hence there is a subgroup K having either order 3 or order 5 and normal in G. Then G/K has either order (5)(17) or order (3)(17), and in either case Theorem 37.7 shows that G/K is abelian. Thus $C \leq K$ and has order either 3, 5, or 1. Since $C \leq H$ showed that C has order 17 or 1, we conclude that C has order 1. Hence $C = \{e\}$, and $G/C \simeq G$ is abelian. The Fundamental Theorem 11.12 then shows that G is cyclic. ▲

EXERCISES 37

Computations

1. Let D_4 be the group of symmetries of the square in Example 8.10.

   a. Find the decomposition of D_4 into conjugate classes.

   b. Write the class equation for D_4.

2. By arguments similar to those used in the examples of this section, convince yourself that every nontrivial group of order not a prime and less than 60 contains a nontrivial proper normal subgroup and hence is not simple. You need not write out the details. (The hardest cases were discussed in the examples.)

Concepts

3. Mark each of the following true or false.

   _____ **a.** Every group of order 159 is cyclic.

   _____ **b.** Every group of order 102 has a nontrivial proper normal subgroup.

   _____ **c.** Every solvable group is of prime-power order.

   _____ **d.** Every group of prime-power order is solvable.

   _____ **e.** It would become quite tedious to show that no group of nonprime order between 60 and 168 is simple by the methods illustrated in the text.

   _____ **f.** No group of order 21 is simple.

   _____ **g.** Every group of 125 elements has at least 5 elements that commute with every element in the group.

   _____ **h.** Every group of order 42 has a normal subgroup of order 7.

   _____ **i.** Every group of order 42 has a normal subgroup of order 8.

   _____ **j.** The only simple groups are the groups $\mathbb{Z}_p$ and A_n where p is a prime and $n \neq 4$.

Theory

4. Prove that every group of order (5)(7)(47) is abelian and cyclic.

5. Prove that no group of order 96 is simple.

6. Prove that no group of order 160 is simple.

7. Show that every group of order 30 contains a subgroup of order 15. [*Hint:* Use the last sentence in Example 37.12, and go to the factor group.]

8. This exercise determines the conjugate classes of S_n for every integer $n \geq 1$.

   a. Show that if $\sigma = (a_1, a_2, \cdots, a_m)$ is a cycle in S_n and τ is any element of S_n then $\tau\sigma\tau^{-1} = (\tau a_1, \tau a_2, \cdots, \tau a_m)$.

   b. Argue from (a) that any two cycles in S_n of the same length are conjugate.

   c. Argue from (a) and (b) that a product of s disjoint cycles in S_n of lengths r_i for $i = 1, 2, \cdots, s$ is conjugate to every other product of s disjoint cycles of lengths r_i in S_n.

   d. Show that the number of conjugate classes in S_n is $p(n)$, where $p(n)$ is the number of ways, neglecting the order of the summands, that n can be expressed as a sum of positive integers. The number $p(n)$ is the **number of partitions of** n.

   e. Compute $p(n)$ for $n = 1, 2, 3, 4, 5, 6, 7$.

9. Find the conjugate classes and the class equation for S_4. [*Hint:* Use Exercise 8.]

10. Find the class equation for S_5 and S_6. [*Hint:* Use Exercise 8.]

11. Show that the number of conjugate classes in S_n is also the number of different abelian groups (up to isomorphism) of order p^n, where p is a prime number. [*Hint:* Use Exercise 8.]

12. Show that if $n > 2$, the center of S_n is the subgroup consisting of the identity permutation only. [*Hint:* Use Exercise 8.]

SECTION 38 Free Abelian Groups

In this section we introduce the concept of free abelian groups and prove some results concerning them. The section concludes with a demonstration of the Fundamental Theorem of finitely generated abelian groups (Theorem 11.12).

Free Abelian Groups

We should review the notions of a generating set for a group G and a finitely generated group, as given in Section 7. In this section we shall deal exclusively with abelian groups and use additive notations as follows:

$$0 \text{ for the identity, } + \text{ for the operation,}$$

$$\left.\begin{aligned} na &= \underbrace{a + a + \cdots + a}_{n \text{ summands}} \\ -na &= \underbrace{(-a) + (-a) + \cdots + (-a)}_{n \text{ summands}} \end{aligned}\right\} \text{ for } n \in \mathbb{Z}^+ \text{ and } a \in G.$$

$$0a = 0 \text{ for the first 0 in } \mathbb{Z} \text{ and the second in } G.$$

We shall continue to use the symbol $\times$ for direct product of groups rather than change to direct sum notation.

Notice that $\{(1, 0), (0, 1)\}$ is a generating set for the group $\mathbb{Z} \times \mathbb{Z}$ since $(n, m) = n(1, 0) + m(0, 1)$ for any (n, m) in $\mathbb{Z} \times \mathbb{Z}$. This generating set has the property that each element of $\mathbb{Z} \times \mathbb{Z}$ can be *uniquely* expressed in the form $n(1, 0) + m(0, 1)$. That is, the coefficients n and m in $\mathbb{Z}$ are unique.

38.1 Theorem Let X be a subset of a nonzero abelian group G. The following conditions on X are equivalent.

1. Each nonzero element a in G can be expressed *uniquely* (up to order of summands) in the form $a = n_1x_1 + n_2x_2 + \cdots + n_rx_r$ for $n_i \neq 0$ in $\mathbb{Z}$ and distinct x_i in X.
2. X generates G, and $n_1x_1 + n_2x_2 + \cdots + n_rx_r = 0$ for $n_i \in \mathbb{Z}$ and distinct $x_i \in X$ if and only if $n_1 = n_2 = \cdots = n_r = 0$.

Proof Suppose Condition 1 is true. Since $G \neq \{0\}$, we have $X \neq \{0\}$. It follows from 1 that $0 \notin X$, for if $x_i = 0$ and $x_j \neq 0$, then $x_j = x_i + x_j$, which would contradict the uniqueness of the expression for x_j. From Condition 1, X generates G, and $n_1x_1 + n_2x_2 + \cdots + n_rx_r = 0$ if $n_1 = n_2 = \cdots = n_r = 0$. Suppose that $n_1x_1 + n_2x_2 + \cdots + n_rx_r = 0$ with some $n_i \neq 0$; by dropping terms with zero coefficients and renumbering, we can assume all $n_i \neq 0$. Then

$$\begin{aligned} x_1 &= x_1 + (n_1x_1 + n_2x_2 + \cdots + n_rx_r) \\ &= (n_1 + 1)x_1 + n_2x_2 + \cdots + n_rx_r, \end{aligned}$$

which gives two ways of writing $x_1 \neq 0$, contradicting the uniqueness assumption in Condition 1. Thus Condition 1 implies Condition 2.

We now show that Condition 2 implies Condition 1. Let $a \in G$. Since X generates G, we see a can be written in the form $a = n_1x_1 + n_2x_2 + \cdots + n_rx_r$. Suppose a has another such expression in terms of elements of X. By using some zero coefficients in the two expressions, we can assume they involve the same elements in X and are of the form

$$a = n_1x_1 + n_2x_2 + \cdots n_rx_r$$

$$a = m_1x_1 + m_2x_2 + \cdots m_rx_r.$$

Subtracting, we obtain

$$0 = (n_1 - m_1)x_1 + (n_2 - m_2)x_2 + \cdots + (n_r - m_r)x_r,$$

so $n_i - m_i = 0$ by Condition 2, and $n_i = m_i$ for $i = 1, 2, \cdots, r$. Thus the coefficients are unique. ◆

38.2 Definition An abelian group having a generating set X satisfying the conditions described in Theorem 38.1 is a **free abelian group**, and X is a **basis** for the group. ■

38.3 Example The group $\mathbb{Z} \times \mathbb{Z}$ is free abelian and $\{(1, 0), (0, 1)\}$ is a basis. Similarly, a basis for the free abelian group $\mathbb{Z} \times \mathbb{Z} \times \mathbb{Z}$ is $\{(1, 0, 0), (0, 1, 0), (0, 0, 1)\}$, and so on. Thus finite direct products of the group $\mathbb{Z}$ with itself are free abelian groups. ▲

38.4 Example The group $\mathbb{Z}_n$ is not free abelian, for $nx = 0$ for every $x \in \mathbb{Z}_n$, and $n \neq 0$, which would contradict Condition 2. ▲

Suppose a free abelian group G has a finite basis $X = \{x_1, x_2, \cdots, x_r\}$. If $a \in G$ and $a \neq 0$, then a has a *unique* expression of the form

$$a = n_1x_1 + n_2x_2 + \cdots + n_rx_r \quad \text{for} \quad n_i \in \mathbb{Z}.$$

(Note that in the preceding expression for a, we included all elements x_i of our finite basis X, as opposed to the expression for a in Condition 1 of Theorem 38.1 where the basis may be infinite. Thus in the preceding expression for a we must allow the possibility that some of the coefficients n_i are zero, whereas in Condition 1 of Theorem 38.1, we specified that each $n_i \neq 0$.)
We define

$$\phi : G \to \underbrace{\mathbb{Z} \times \mathbb{Z} \times \cdots \times \mathbb{Z}}_{r \text{ factors}}$$

by $\phi(a) = (n_1, n_2, \cdots, n_r)$ and $\phi(0) = (0, 0, \cdots, 0)$. It is straightforward to check that ϕ is an isomorphism. We leave the details to the exercises (see Exercise 9) and state the result as a theorem.

38.5 Theorem If G is a nonzero free abelian group with a basis of r elements, then G is isomorphic to $\mathbb{Z} \times \mathbb{Z} \times \cdots \times \mathbb{Z}$ for r factors.

It is a fact that any two bases of a free abelian group G contain the same number of elements. We shall prove this only if G has a finite basis, although it is also true if every basis of G is infinite. The proof is really lovely; it gives an easy characterization of the number of elements in a basis in terms of the size of a factor group.

38.6 Theorem Let $G \neq \{0\}$ be a free abelian group with a finite basis. Then every basis of G is finite, and all bases of G have the same number of elements.

Proof Let G have a basis $\{x_1, x_2, \cdots, x_r\}$. Then G is isomorphic to $\mathbb{Z} \times \mathbb{Z} \times \cdots \times \mathbb{Z}$ for r factors. Let $2G = \{2g \mid g \in G\}$. It is readily checked that $2G$ is a subgroup of G. Since $G \simeq \mathbb{Z} \times \mathbb{Z} \times \cdots \times \mathbb{Z}$ for r factors, we have

$$\begin{aligned} G/2G &\simeq (\mathbb{Z} \times \mathbb{Z} \times \cdots \times \mathbb{Z})/(2\mathbb{Z} \times 2\mathbb{Z} \times \cdots \times 2\mathbb{Z}) \\ &\simeq \mathbb{Z}_2 \times \mathbb{Z}_2 \times \cdots \times \mathbb{Z}_2 \end{aligned}$$

for r factors. Thus $|G/2G| = 2^r$, so the number of elements in any finite basis X is $\log_2 |G/2G|$. Thus any two finite bases have the same number of elements.

It remains to show that G cannot also have an infinite basis. Let Y be any basis for G, and let $\{y_1, y_2, \cdots, y_s\}$ be distinct elements in Y. Let H be the subgroup of G generated by $\{y_1, y_2, \cdots, y_s\}$, and let K be the subgroup of G generated by the remaining elements of Y. It is readily checked that $G \simeq H \times K$, so

$$G/2G \simeq (H \times K)/(2H \times 2K) \simeq (H/2H) \times (K/2K).$$

Since $|H/2H| = 2^s$, we see $|G/2G| \geq 2^s$. Since we have $|G/2G| = 2^r$, we see that $s \leq r$. Then Y cannot be an infinite set, for we could take $s > r$. ◆

38.7 Definition If G is a free abelian group, the **rank** of G is the number of elements in a basis for G. (All bases have the same number of elements.) ■

Proof of the Fundamental Theorem

We shall prove the Fundamental Theorem (Theorem 11.12) by showing that any finitely generated abelian group is isomorphic to a factor group of the form

$$(\mathbb{Z} \times \mathbb{Z} \times \cdots \times \mathbb{Z})/(d_1\mathbb{Z} \times d_2\mathbb{Z} \times \cdots \times d_s\mathbb{Z} \times \{0\} \times \cdots \times \{0\}),$$

where both "numerator" and "denominator" have n factors, and d_1 divides d_2, which divides $d_3 \cdots$, which divides d_s. The prime-power decomposition of Theorem 11.12 will then follow.

To show that G is isomorphic to such a factor group, we will show that there is a homomorphism of $\mathbb{Z} \times \mathbb{Z} \times \cdots \times \mathbb{Z}$ onto G with kernel of the form $d_1\mathbb{Z} \times d_2\mathbb{Z} \times \cdots \times d_s\mathbb{Z} \times \{0\} \times \cdots \times \{0\}$. The result will then follow by Theorem 14.11. The theorems that follow give the details of the argument. Our purpose in these introductory paragraphs is to let us see where we are going as we read what follows.

38.8 Theorem Let G be a finitely generated abelian group with generating set $\{a_1, a_2, \cdots, a_n\}$. Let

$$\phi : \underbrace{\mathbb{Z} \times \mathbb{Z} \times \cdots \times \mathbb{Z}}_{n \text{ factors}} \to G$$

be defined by $\phi(h_1, h_2, \cdots, h_n) = h_1a_1 + h_2a_2 + \cdots + h_na_n$. Then ϕ is a homomorphism onto G.

Proof From the meaning of h_ia_i for $h_i \in \mathbb{Z}$ and $a_i \in G$, we see at once that

$$\begin{aligned}
\phi[(h_1, \cdots, h_n) + (k_1, \cdots, k_n)] &= \phi(h_1 + k_1, \cdots, h_n + k_n) \\
&= (h_1 + k_1)a_1 + \cdots + (h_n + k_n)a_n \\
&= (h_1a_1 + k_1a_1) + \cdots + (h_na_n + k_na_n) \\
&= (h_1a_1 + \cdots + h_na_n) + (k_1a_1 + \cdots + k_na_n) \\
&= \phi(k_1, \cdots, k_n) + \phi(h_1, \cdots, h_n).
\end{aligned}$$

Since $\{a_1, \cdots, a_n\}$ generates G, clearly the homomorphism ϕ is onto G. ◆

We now prove a "replacement property" that makes it possible for us to adjust a basis.

38.9 Theorem If $X = \{x_1, \cdots, x_r\}$ is a basis for a free abelian group G and $t \in \mathbb{Z}$, then for $i \neq j$, the set

$$Y = \{x_1, \cdots, x_{j-1}, x_j + tx_i, x_{j+1}, \cdots, x_r\}$$

is also a basis for G.

Proof Since $x_j = (-t)x_i + (1)(x_j + tx_i)$, we see that x_j can be recovered from Y, which thus also generates G. Suppose

$$n_1x_1 + \cdots + n_{j-1}x_{j-1} + n_j(x_j + tx_i) + n_{j+1}x_{j+1} + \cdots + n_rx_r = 0.$$

Then

$$n_1x_1 + \cdots + (n_i + n_jt)x_i + \cdots + n_jx_j + \cdots + n_rx_r = 0.$$

and since X is a basis, $n_1 = \cdots = n_i + n_jt = \cdots = n_j = \cdots = n_r = 0$. From $n_j = 0$ and $n_i + n_jt = 0$, it follows that $n_i = 0$ also, so $n_1 = \cdots = n_i = \cdots = n_j = \cdots = n_r = 0$, and Condition 2 of Theorem 38.1 is satisfied. Thus Y is a basis. ◆

38.10 Example A basis for $\mathbb{Z} \times \mathbb{Z}$ is $\{(1, 0), (0, 1)\}$. Another basis is $\{(1, 0), (4, 1)\}$ for $(4, 1) = 4(1, 0) + (0, 1)$. However, $\{(3, 0), (0, 1)\}$ is not a basis. For example, we cannot express $(2, 0)$ in the form $n_1(3, 0) + n_2(0, 1)$, for $n, n_2 \in \mathbb{Z}$. Here $(3, 0) = (1, 0) + 2(1, 0)$, and a multiple of a basis element was added to *itself*, rather than to a *different* basis element. ▲

A free abelian group G of finite rank may have many bases. We show that if $K \leq G$, then K is also free abelian with rank not exceeding that of G. Equally important, there exist bases of G and K nicely related to each other.

38.11 Theorem Let G be a nonzero free abelian group of finite rank n, and let K be a nonzero subgroup of G. Then K is free abelian of rank $s \leq n$. Furthermore, there exists a basis $\{x_1, x_2, \cdots, x_n\}$ for G and positive integers, $d_1, d_2, \cdots, d_s$ where d_i divides d_{i+1} for $i = 1, \cdots, s - 1$, such that $\{d_1x_1, d_2x_2, \cdots, d_sx_s\}$ is a basis for K.

Proof We show that K has a basis of the described form, which will show that K is free abelian of rank at most n. Suppose $Y = \{y_1, \cdots, y_n\}$ is a basis for G. All nonzero elements in K can be expressed in the form

$$k_1y_1 + \cdots + k_ny_n,$$

where some $|k_i|$ is nonzero. Among *all* bases Y for G, select one Y_1 that yields the minimal such nonzero value $|k_i|$ as all nonzero elements of K are written in terms of the basis elements in Y_1. By renumbering the elements of Y_1 if necessary, we can assume there is $w_1 \in K$ such that

$$w_1 = d_1y_1 + k_2y_2 + \cdots + k_ny_n$$

where $d_1 > 0$ and d_1 is the minimal attainable coefficient as just described. Using the division algorithm, we write $k_j = d_1q_j + r_j$ where $0 \leq r_j < d_1$ for $j = 2, \cdots, n$. Then

$$w_1 = d_1(y_1 + q_2y_2 + \cdots + q_ny_n) + r_2y_2 + \cdots + r_ny_n. \qquad \textbf{(1)}$$

Now let $x_1 = y_1 + q_2y_2 + \cdots + q_ny_n$. By Theorem 38.9 $\{x_1, y_2, \cdots, y_n\}$ is also a basis for G. From Eq. (1) and our choice of Y_1 for minimal coefficient d_1, we see that $r_2 = \cdots = r_n = 0$. Thus $d_1x_1 \in K$.

We now consider bases for G of the form $\{x_1, y_2, \cdots, y_n\}$. Each element of K can be expressed in the form

$$h_1x_1 + k_2y_2 + \cdots + k_ny_n.$$

Since $d_1x_1 \in K$, we can subtract a suitable multiple of d_1x_1 and then using the minimality of d_1 to see that h_1 is a multiple of d_1, we see we actually have $k_2y_2 + \cdots + k_ny_n$ in K. Among all such bases $\{x_1, y_2, \cdots, y_n\}$, we choose one Y_2 that leads to some $k_i \neq 0$ of minimal magnitude. (It is possible all k_i are always zero. In this case, K is generated by d_1x_1 and we are done.) By renumbering the elements of Y_2 we can assume that there is $w_2 \in K$ such that

$$w_2 = d_2y_2 + \cdots + k_ny_n$$

where $d_2 > 0$ and d_2 is minimal as just described. Exactly as in the preceding paragraph, we can modify our basis from $Y_2 = \{x_1, y_2, \cdots, y_n\}$ to a basis $\{x_1, x_2, y_3, \cdots, y_n\}$ for G where $d_1x_1 \in K$ and $d_2x_2 \in K$. Writing $d_2 = d_1q + r$ for $0 \leq r < d_1$, we see that $\{x_1 + qx_2, x_2, y_3, \cdots, y_n\}$ is a basis for G, and $d_1x_1 + d_2x_2 = d_1(x_1 + qx_2) + rx_2$ is in K. By our minimal choice of d_1, we see $r = 0$, so d_1 divides d_2.

We now consider all bases of the form $\{x_1, x_2, y_3, \cdots, y_n\}$ for G and examine elements of K of the form $k_3y_3 + \cdots + k_ny_n$. The pattern is clear. The process continues until we obtain a basis $\{x_1, x_2, \cdots, x_s, y_{s+1}, \cdots, y_n\}$ where the only element of K of the form $k_{s+1}y_{s+1} + \cdots + k_ny_n$ is zero, that is, all k_i are zero. We then let $x_{s+1} = y_{s+1}, \cdots, x_n = y_n$ and obtain a basis for G of the form described in the statement of Theorem 38.11. ◆

38.12 Theorem Every finitely generated abelian group is isomorphic to a group of the form

$$\mathbb{Z}_{m_1} \times \mathbb{Z}_{m_2} \times \cdots \times \mathbb{Z}_{m_r} \times \mathbb{Z} \times \mathbb{Z} \times \cdots \times \mathbb{Z},$$

where m_i divides m_{i+1} for $i = 1, \cdots, r-1$.

Proof For the purposes of this proof, it will be convenient to use as notations $\mathbb{Z}/1\mathbb{Z} = \mathbb{Z}/\mathbb{Z} \simeq \mathbb{Z}_1 = \{0\}$. Let G be finitely generated by n elements. Let $F = \mathbb{Z} \times \mathbb{Z} \times \cdots \times \mathbb{Z}$ for n factors. Consider the homomorphism $\phi : F \to G$ of Theorem 38.8, and let K be the kernel of this homomorphism. Then there is a basis for F of the form $\{x_1, \cdots, x_n\}$, where $\{d_1x_1, \cdots, d_sx_s\}$ is a basis for K and d_i divides d_{i+1} for $i = 1, \cdots, s-1$. By Theorem 14.11, G is isomorphic to F/K. But

$$\begin{aligned} F/K &\simeq (\mathbb{Z} \times \mathbb{Z} \times \cdots \times \mathbb{Z})/(d_1\mathbb{Z} \times d_2\mathbb{Z} \times \cdots \times d_s\mathbb{Z} \times \{0\} \times \cdots \times \{0\}) \\ &\simeq \mathbb{Z}_{d_1} \times \mathbb{Z}_{d_2} \times \cdots \times \mathbb{Z}_{d_s} \times \mathbb{Z} \times \cdots \times \mathbb{Z}. \end{aligned}$$

It is possible that $d_1 = 1$, in which case $\mathbb{Z}_{d_1} = \{0\}$ and can be dropped (up to isomorphism) from this product. Similarly, d_2 may be 1, and so on. We let m_1 be the first $d_i > 1$, m_2 be the next d_i, and so on, and our theorem follows at once. ◆

We have demonstrated the toughest part of the Fundamental Theorem (Theorem 11.12). Of course, a prime-power decomposition exists since we can break the groups $\mathbb{Z}_{m_i}$ into prime-power factors. The only remaining part of Theorem 11.12 concerns the uniqueness of the Betti number, of the torsion coefficients, and of the prime powers. The Betti number appears as the rank of the free abelian group G/T, where T is the torsion subgroup of G. This rank is invariant by Theorem 38.6 which shows the uniqueness of the Betti number. The uniqueness of the torsion coefficients and of prime powers is a bit more difficult to show. We give some exercises that indicate their uniqueness (see Exercises 14 through 22).

EXERCISES 38

Computations

1. Find a basis $\{(a_1, a_2, a_3), (b_1, b_2, b_3), (c_1, c_2, c_3)\}$ for $\mathbb{Z} \times \mathbb{Z} \times \mathbb{Z}$ with all $a_i \neq 0$, all $b_i \neq 0$, and all $c_i \neq 0$. (Many answers are possible.)

2. Is $\{(2, 1), (3, 1)\}$ a basis for $\mathbb{Z} \times \mathbb{Z}$? Prove your assertion.

3. Is $\{(2, 1), (4, 1)\}$ a basis for $\mathbb{Z} \times \mathbb{Z}$? Prove your assertion.

4. Find conditions on $a, b, c, d \in \mathbb{Z}$ for $\{$(a, b), (c, d)$\}$ to be a basis for $\mathbb{Z} \times \mathbb{Z}$. [*Hint:* Solve $x(a, b) + y(c, d) = (e, f)$ in $\mathbb{R}$, and see when the x and y lie in $\mathbb{Z}$.]

Concepts

In Exercises 5 and 6, correct the definition of the italicized term without reference to the text, if correction is needed, so that it is in a form acceptable for publication.

5. The *rank* of a free abelian group G is the number of elements in a generating set for G.

6. A *basis* for a nonzero abelian group G is a generating set $X \subseteq G$ such that $n_1x_1 + n_2x_2 + \cdots + n_mx_m = 0$ for distinct $x_i \in X$ and $n_i \in \mathbb{Z}$ only if $n_1 = n_2 = \cdots = n_m = 0$.

7. Show by example that it is possible for a proper subgroup of a free abelian group of finite rank r also to have rank r.

8. Mark each of the following true or false.

_____ **a.** Every free abelian group is torsion free.

_____ **b.** Every finitely generated torsion-free abelian group is a free abelian group.

_____ **c.** There exists a free abelian group of every positive integer rank.

_____ **d.** A finitely generated abelian group is free abelian if its Betti number equals the number of elements in some generating set.

_____ **e.** If X generates a free abelian group G and $X \subseteq Y \subseteq G$, then Y generates G.

_____ **f.** If X is a basis for a free abelian group G and $X \subseteq Y \subseteq G$, then Y is a basis for G.

_____ **g.** Every nonzero free abelian group has an infinite number of bases.

_____ **h.** Every free abelian group of rank at least 2 has an infinite number of bases.

_____ **i.** If K is a nonzero subgroup of a finitely generated free abelian group, then K is free abelian.

_____ **j.** If K is a nonzero subgroup of a finitely generated free abelian group, then G/K is free abelian.

Theory

9. Complete the proof of Theorem 38.5 (See the two sentences preceding the theorem).

10. Show that a free abelian group contains no nonzero elements of finite order.

11. Show that if G and G' are free abelian groups, then $G \times G'$ is free abelian.

12. Show that free abelian groups of finite rank are precisely the finitely generated abelian groups containing no nonzero elements of finite order.

13. Show that $\mathbb{Q}$ under addition is not a free abelian group. [*Hint:* Show that no two distinct rational numbers n/m and r/s could be contained in a set satisfying Condition 2 of Thorem 38.1.]

Exercises 14 through 19 deal with showing the uniqueness of the prime powers appearing in the prime-power decomposition of the torsion subgroup T of a finitely generated abelian group.

14. Let p be a fixed prime. Show that the elements of T having as order some power of p, together with zero, form a subgroup T_p of T.

15. Show that in any prime-power decomposition of T, the subgroup T_p in the preceding exercise is isomorphic to the direct product of those cyclic factors of order some power of the prime p. [This reduces our problem to showing that the group T_p cannot have essentially different decompositions into products of cyclic groups.]

16. Let G be any abelian group and let n be any positive integer. Show that $G[n] = \{x \in G \mid nx = 0\}$ is a subgroup of G. (In multiplicative notation, $G[n] = \{x \in G \mid x^n = e\}$.)

17. Referring to Exercise 16, show that $\mathbb{Z}_{p^r}[p] \simeq \mathbb{Z}_p$ for any $r \geq 1$ and prime p.

18. Using Exercise 17, show that

$$(\mathbb{Z}_{p^{r_1}} \times \mathbb{Z}_{p^{r_2}} \times \cdots \times \mathbb{Z}_{p^{r_m}})[p] \simeq \underbrace{\mathbb{Z}_p \times \mathbb{Z}_p \times \cdots \times \mathbb{Z}_p}_{m \text{ factors}}$$

provided each $r_i \geq 1$.

19. Let G be a finitely generated abelian group and T_p the subgroup defined in Exercise 14. Suppose $T_p \simeq \mathbb{Z}_{p^{r_1}} \times \mathbb{Z}_{p^{r_2}} \times \cdots \times \mathbb{Z}_{p^{r_m}} \simeq \mathbb{Z}_{p^{s_1}} \mathbb{Z}_{p^{s_2}} \times \cdots \times \mathbb{Z}_{p^{s_n}}$, where $1 \leq r_1 \leq r_2 \leq \cdots \leq r_m$ and $1 \leq s_1 \leq s_2 \leq \cdots \leq s_n$. We need to show that $m = n$ and $r_i = s_i$ for $i = 1, \cdots, n$ to complete the demonstration of uniqueness of the prime-power decomposition.

a. Use Exercise 18 to show that $n = m$.

b. Show $r_1 = s_1$. Suppose $r_i = s_i$ for all $i < j$. Show $r_j = s_j$, which will complete the proof. [*Hint:* Suppose $r_j < s_j$. Consider the subgroup $p^{r_j} T_p = \{p^{r_j} x \mid x \in T_p\}$, and show that this subgroup would then have two prime-power decompositions involving different numbers of nonzero factors. Then argue that this is impossible by part (a) of this exercise.]

Let T be the torsion subgroup of a finitely generated abelian group. Suppose $T \simeq \mathbb{Z}_{m_1} \times \mathbb{Z}_{m_2} \times \cdots \times \mathbb{Z}_{m_r} \simeq \mathbb{Z}_{n_1} \times \mathbb{Z}_{n_2} \times \cdots \times \mathbb{Z}_{n_s}$, where m_i divides m_{i+1} for $i = 1, \cdots, r-1$, and n_j divides n_{j+1} for $n = 1, \cdots, s-1$, and $m_1 > 1$ and $n_1 > 1$. We wish to show that $r = s$ and $m_k = n_k$ for $k = 1, \cdots, r$, demonstrating the uniqueness of the torsion coefficients. This is done in Exercises 20 through 22.

20. Indicate how a prime-power decomposition can be obtained from a torsion-coefficient decomposition. (Observe that the preceding exercises show the prime powers obtained are unique.)

21. Argue from Exercise 20 that m_r and n_s can both be characterized as follows. Let $p_1, \cdots, p_t$ be the distinct primes dividing $|T|$, and let $p_1^{h_1}, \cdots, p_t^{h_t}$ be the highest powers of these primes appearing in the (unique) prime-power decomposition. Then $m_r = n_s = p_1^{h_1} p_2^{h_2} \cdots p_t^{h_t}$.

22. Characterize m_{r-1} and n_{s-1}, showing that they are equal, and continue to show $m_{r-i} = n_{s-1}$ for $i = 1, \cdots, r-1$, and then $r = s$.

SECTION 39 FREE GROUPS

In this section and Section 40 we discuss a portion of group theory that is of great interest not only in algebra but in topology as well. In fact, an excellent and readable discussion of free groups and presentations of groups is found in Crowell and Fox [46, Chapters 3 and 4].

Words and Reduced Words

Let A be any (not necessarily finite) set of elements a_i for $i \in I$. We think of A as an **alphabet** and of the a_i as **letters** in the alphabet. Any symbol of the form $a_i{}^n$ with $n \in \mathbb{Z}$ is a **syllable** and a finite string w of syllables written in juxtaposition is a **word.** We also introduce the **empty word** 1, which has no syllables.

39.1 Example Let $A = \{a_1, a_2, a_3\}$. Then

$$a_1 a_3{}^{-4} a_2{}^2 a_3, \qquad a_2{}^3 a_2{}^{-1} a_3 a_1{}^2 a_1{}^{-7}, \qquad \text{and} \qquad a_3{}^2$$

are all words, if we follow the convention of understanding that $a_i{}^1$ is the same as a_i. ▲

There are two natural types of modifications of certain words, the **elementary contractions.** The first type consists of replacing an occurrence of $a_i{}^m a_i{}^n$ in a word by $a_i{}^{m+n}$. The second type consists of replacing an occurrence of $a_i{}^0$ in a word by 1, that is, dropping it out of the word. By means of a finite number of elementary contractions, every word can be changed to a **reduced word,** one for which no more elementary contractions are possible. Note that these elementary contractions formally amount to the usual manipulations of integer exponents.

39.2 Example The reduced form of the word $a_2{}^3 a_2{}^{-1} a_3 a_1{}^2 a_1{}^{-7}$ of Example 39.1 is $a_2{}^2 a_3 a_1{}^{-5}$. ▲

It should be said here once and for all that we are going to gloss over several points that some books spend pages proving, usually by complicated induction arguments broken down into many cases. For example, suppose we are given a word and wish to find its reduced form. There may be a variety of elementary contractions that could be performed first. How do we know that the reduced word we end up with is the same no matter in what order we perform the elementary contractions? The student will probably say this is obvious. Some authors spend considerable effort proving this. The author tends to agree here with the student. Proofs of this sort he regards as tedious, and they have never made him more comfortable about the situation. However, the author is the first to acknowledge that he is not a great mathematician. In deference to the fact that many mathematicians feel that these things do need considerable discussion, we shall mark an occasion when we just state such facts by the phrase, "It would seem obvious that," keeping the quotation marks.

Free Groups

Let the set of all reduced words formed from our alphabet A be $F[A]$. We now make $F[A]$ into a group in a natural way. For w_1 and w_2 in $F[A]$, define $w_1 \cdot w_2$ to be the reduced form of the word obtained by the juxtaposition $w_1 w_2$ of the two words.

39.3 Example If

$$w_1 = a_2{}^3 a_1{}^{-5} a_3{}^2$$

and

$$w_2 = a_3{}^{-2} a_1{}^2 a_3 a_2{}^{-2},$$

then $w_1 \cdot w_2 = a_2{}^3 a_1{}^{-3} a_3 a_2{}^{-2}$. ▲

"It would seem obvious that" this operation of multiplication on $F[A]$ is well defined and associative. The empty word 1 acts as an identity element. "It would seem obvious that" given a reduced word $w \in F[A]$, if we form the word obtained by first writing the syllables of w in the opposite order and second by replacing each $a_i{}^n$ by $a_i{}^{-n}$, then the resulting word w^{-1} is a reduced word also, and

$$w \cdot w^{-1} = w^{-1} \cdot w = 1.$$

39.4 Definition The group $F[A]$ just described is the **free group generated** by A. ■

Look back at Theorem 7.6 and the definition preceding it to see that the present use of the term *generated* is consistent with the earlier use.

Starting with a group G and a generating set $\{a_i \mid i \in I\}$ which we will abbreviate by $\{a_i\}$, we might ask if G is *free* on $\{a_i\}$, that is, if G is essentially the free group generated by $\{a_i\}$. We define precisely what this is to mean.

39.5 Definition If G is a group with a set $A = \{a_i\}$ of generators, and if G is isomorphic to $F[A]$ under a map $\phi : G \to F[A]$ such that $\phi(a_i) = a_i$, then G is **free on** A, and the a_i are **free generators of** G. A group is **free** if it is free on some nonempty set A. ■

39.6 Example The only example of a free group that has occurred before is $\mathbb{Z}$, which is free on one generator. Note that every free group is infinite. ▲

Refer to the literature for proofs of the next three theorems. We will not be using these results. They are stated simply to inform us of these interesting facts.

39.7 Theorem If a group G is free on A and also on B, then the sets A and B have the same number of elements; that is, any two sets of free generators of a free group have the same cardinality.

39.8 Definition If G is free on A, the number of elements in A is the **rank of the free group** G. ■

Actually, the next theorem is quite evident from Theorem 39.7.

39.9 Theorem Two free groups are isomorphic if and only if they have the same rank.

39.10 Theorem A nontrivial proper subgroup of a free group is free.

39.11 Example Let $F[\{x, y\}]$ be the free group on $\{x, y\}$. Let

$$y_k = x^k y x^{-k}$$

for $k \geq 0$. The y_k for $k \geq 0$ are free generators for the subgroup of $F[\{x, y\}]$ that they generate. This illustrates that although a subgroup of a free group is free, the rank of the subgroup may be much greater than the rank of the whole group! ▲

Homomorphisms of Free Groups

Our work in this section will be concerned primarily with homomorphisms defined on a free group. The results here are simple and elegant.

39.12 Theorem Let G be generated by $A = \{a_i \mid i \in I\}$ and let G' be any group. If $a_i{}'$ for $i \in I$ are any elements in G', not necessarily distinct, then there is at most one homomorphism $\phi : G \to G'$ such that $\phi(a_i) = a_i{}'$. If G is free on A, then there is exactly one such homomorphism.

Proof Let ϕ be a homomorphism from G into G' such that $\phi(a_i) = a_i{}'$. Now by Theorem 7.6, for any $x \in G$ we have

$$x = \prod_j a_{i_j}{}^{n_j}$$

for some finite product of the generators a_i, where the a_{i_j} appearing in the product need not be distinct. Then since ϕ is a homomorphism, we must have

$$\phi(x) = \prod_j \phi\left(a_{i_j}{}^{n_j}\right) = \prod_j \left(a_{i_j}{}'\right)^{n_j}.$$

Thus a homomorphism is completely determined by its values on elements of a generating set. This shows that there is at most one homomorphism such that $\phi(a_i) = a_i{}'$.

Now suppose G is free on A; that is, $G = F[A]$. For

$$x = \prod_j a_{i_j}{}^{n_j}$$

in G, define $\psi : G \to G'$ by

$$\psi(x) = \prod_j \left(a_{i_j}{}'\right)^{n_j}.$$

The map is well defined, since $F[A]$ consists precisely of reduced words; no two different formal products in $F[A]$ are equal. Since the rules for computation involving exponents in G' are formally the same as those involving exponents in G, it is clear that $\psi(xy) = \psi(x)\psi(y)$ for any elements x and y in G, so ψ is indeed a homomorphism. ◆

Perhaps we should have proved the first part of this theorem earlier, rather than having relegated it to the exercises. Note that the theorem states that *a homomorphism of a group is completely determined if we know its value on each element of a generating set*. This was Exercise 46 of Section 13. In particular, a homomorphism of a cyclic group is completely determined by its value on any single generator of the group.

39.13 Theorem Every group G' is a homomorphic image of a free group G.

Proof Let $G' = \{a_i' \mid i \in I\}$, and let $A = \{a_i \mid i \in I\}$ be a set with the same number of elements as G'. Let $G = F[A]$. Then by Theorem 39.12 there exists a homomorphism ψ mapping G into G' such that $\psi(a_i) = a_i'$. Clearly the image of G under ψ is all of G'. ◆

Another Look at Free Abelian Groups

It is important that we do not confuse the notion of a free group with the notion of a free abelian group. A free group on more than one generator is not abelian. In the preceding section, we defined a free abelian group as an abelian group that has a basis, that is, a generating set satisfying properties described in Theorem 38.1. There is another approach, via free groups, to free abelian groups. We now describe this approach.

Let $F[A]$ be the free group on the generating set A. We shall write F in place of $F[A]$ for the moment. Note that F is not abelian if A contains more than one element. Let C be the commutator subgroup of F. Then F/C is an abelian group, and it is not hard to show that F/C is free abelian with basis $\{aC \mid a \in A\}$. If aC is renamed a, we can view F/C as a free abelian group with basis A. This indicates how a free abelian group having a given set as basis can be constructed. Every free abelian group can be constructed in this fashion, up to isomorphism. That is, if G is free abelian with basis X, form the free group $F[X]$, form the factor group of $F[X]$ modulo its commutator subgroup, and we have a group isomorphic to G.

Theorems 39.7, 39.9, and 39.10 hold for free abelian groups as well as for free groups. In fact, the abelian version of Theorem 39.10 was proved for the finite rank case in Theorem 38.11. In contrast to Example 39.11 for free groups, it is true that for a free abelian group the rank of a subgroup is at most the rank of the entire group. Theorem 38.11 also showed this for the finite rank case.

EXERCISES 39

Computations

1. Find the reduced form and the inverse of the reduced form of each of the following words.

   a. $a^2b^{-1}b^3a^3c^{-1}c^4b^{-2}$ **b.** $a^2a^{-3}b^3a^4c^4c^2a^{-1}$

2. Compute the products given in parts (a) and (b) of Exercise 1 in the case that $\{a, b, c\}$ is a set of generators forming a basis for a free abelian group. Find the inverse of these products.

3. How many different homomorphisms are there of a free group of rank 2 into

   a. $\mathbb{Z}_4$? **b.** $\mathbb{Z}_6$? **c.** S_3?

4. How many different homomorphisms are there of a free group of rank 2 onto

   a. $\mathbb{Z}_4$? **b.** $\mathbb{Z}_6$? **c.** S_3?

5. How many different homomorphisms are there of a free abelian group of rank 2 into

   a. $\mathbb{Z}_4$? **b.** $\mathbb{Z}_6$? **c.** S_3?

6. How many different homomorphisms are there of a free abelian group of rank 2 onto

   a. $\mathbb{Z}_4$? **b.** $\mathbb{Z}_6$? **c.** S_3?

Concepts

In Exercises 7 and 8, correct the definition of the italicized term without reference to the text, if correction is needed, so that it is in a form acceptable for publication.

7. A *reduced word* is one in which there are no appearances in juxtaposition of two syllables having the same letter and also no appearances of a syllable with exponent 0.

8. The *rank of a free group* is the number of elements in a set of generators for the group.

9. Take one of the instances in this section in which the phrase "It would seem obvious that" was used and discuss your reaction in that instance.

10. Mark each of the following true or false.

______ **a.** Every proper subgroup of a free group is a free group.

______ **b.** Every proper subgroup of every free abelian group is a free group.

______ **c.** A homomorphic image of a free group is a free group.

______ **d.** Every free abelian group has a basis.

______ **e.** The free abelian groups of finite rank are precisely the finitely generated abelian groups.

______ **f.** No free group is free.

______ **g.** No free abelian group is free.

______ **h.** No free abelian group of rank >1 is free.

______ **i.** Any two free groups are isomorphic.

______ **j.** Any two free abelian groups of the same rank are isomorphic.

Theory

11. Let G be a finitely generated abelian group with identity 0. A finite set $\{b_1, \cdots, b_n\}$, where $b_i \in G$, is a **basis for** G if $\{b_1, \cdots, b_n\}$ generates G and $\sum_{i=1}^{n} m_i b_i = 0$ if and only if each $m_i b_i = 0$, where $m_i \in \mathbb{Z}$.

a. Show that $\{2, 3\}$ is not a basis for $\mathbb{Z}_4$. Find a basis for $\mathbb{Z}_4$.

b. Show that both $\{1\}$ and $\{2, 3\}$ are bases for $\mathbb{Z}_6$. (This shows that for a finitely generated abelian group G with torsion, the number of elements in a basis may vary; that is, it need not be an *invariant* of the group G.)

c. Is a basis for a free abelian group as we defined it in Section 38 a basis in the sense in which it is used in this exercise?

d. Show that every finite abelian group has a basis $\{b_1, \cdots, b_n\}$, where the order of b_i divides the order of b_{i+1}.

In present-day expositions of algebra, a frequently used technique (particularly by the disciples of N. Bourbaki) for introducing a new algebraic entity is the following:

1. Describe algebraic properties that this algebraic entity is to possess.
2. Prove that any two algebraic entities with these properties are isomorphic, that is, that these properties characterize the entity.
3. Show that at least one such entity exists.

The next three exercises illustrate this technique for three algebraic entities, each of which we have met before. So that we do not give away their identities, we use fictitious names for them in the first two exercises. The last part of these first two exercises asks us to give the usual name for the entity.

12. Let G be any group. An abelian group G^* is a **blip group of** G if there exists a fixed homomorphism ϕ of G onto G^* such that each homomorphism ψ of G into an abelian group G' can be factored as $\psi = \theta\phi$, where θ is a homomorphism of G^* into G' (see Fig. 39.14).

- **a.** Show that any two blip groups of G are isomorphic. [*Hint:* Let ${G_1}^*$ and ${G_2}^*$ be two blip groups of G. Then each of the fixed homomorphisms $\phi_1 : G \to {G_1}^*$ and $\phi_2 : G \to {G_2}^*$ can be factored via the other blip group according to the definition of a blip group; that is, $\phi_1 = \theta_1\phi_2$ and $\phi_2 = \theta_2\phi_1$. Show that θ_1 is an isomorphism of ${G_2}^*$ onto ${G_1}^*$ by showing that both $\theta_1\theta_2$ and $\theta_2\theta_1$ are identity maps.]
- **b.** Show for every group G that a blip group G^* of G exists.
- **c.** What concept that we have introduced before corresponds to this idea of a blip group of G?

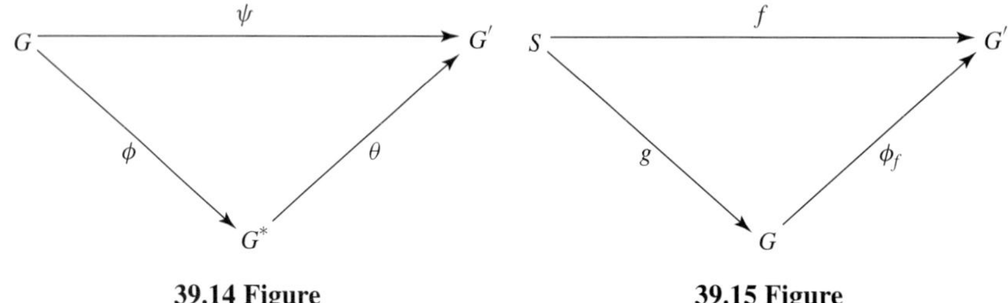

39.14 Figure **39.15 Figure**

13. Let S be any set. A group G together with a fixed function $g : S \to G$ constitutes a **blop group on** S if for each group G' and map $f : S \to G'$ there exists a *unique* homomorphism ϕ_f of G into G' such that $f = \phi_f g$ (see Fig. 39.15).

- **a.** Let S be a fixed set. Show that if both G_1, together with $g_1 : S \to G_1$, and G_2, together with $g_2 : S \to G_2$, are blop groups on S, then G_1 and G_2 are isomorphic. [*Hint:* Show that g_1 and g_2 are one-to-one maps and that g_1S and g_2S generate G_1 and G_2, respectively. Then proceed in a way analogous to that given by the hint for Exercise 12.]
- **b.** Let S be a set. Show that a blop group on S exists. You may use any theorems of the text.
- **c.** What concept that we have introduced before corresponds to this idea of a blop group on S?

14. Characterize a free abelian group by properties in a fashion similar to that used in Exercise 13.

SECTION 40 GROUP PRESENTATIONS

Definition

Following most of the literature on group presentations, in this section we let 1 be the identity of a group. The idea of a *group presentation* is to form a group by giving a set of generators for the group and certain equations or relations that we want the generators to satisfy. We want the group to be as free as it possibly can be on the generators, subject to these relations.

40.1 Example Suppose G has generators x and y and is *free except for the relation* $xy = yx$, which we may express as $xyx^{-1}y^{-1} = 1$. Note that the condition $xy = yx$ is exactly what is needed to make G abelian, even though $xyx^{-1}y^{-1}$ is just one of the many possible commutators of $F[\{x, y\}]$. Thus G is free abelian on two generators and is isomorphic to $F[\{x, y\}]$ modulo its commutator subgroup. This commutator subgroup of $F[\{x, y\}]$ is the smallest normal subgroup containing $xyx^{-1}y^{-1}$, since any normal subgroup containing $xyx^{-1}y^{-1}$ gives rise to a factor group that is abelian and thus contains the commutator subgroup by Theorem 15.20. ▲

The preceding example illustrates the general situation. Let $F[A]$ be a free group and suppose that we want to form a new group as much like $F[A]$ as it can be, subject to certain equations that we want satisfied. Any equation can be written in a form in which the right-hand side is 1. Thus we can consider the equations to be $r_i = 1$ for $i \in I$, where $r_i \in F[A]$. If we require that $r_i = 1$, then we will have to have

$$x(r_i{}^n)x^{-1} = 1$$

for any $x \in F[A]$ and $n \in \mathbb{Z}$. Also any product of elements equal to 1 will again have to equal 1. Thus any finite product of the form

$$\prod_j x_j(r_{i_j}{}^{n_j})x_j^{-1},$$

where the r_{i_j} need not be distinct, will have to equal 1 in the new group. It is readily checked that the set of all these finite products is a normal subgroup R of $F[A]$. Thus any group looking as much as possible like $F[A]$, subject to the requirements $r_i = 1$, also has $r = 1$ for every $r \in R$. But $F[A]/R$ looks like $F[A]$ (remember that we multiply cosets by choosing representatives), except that R has been collapsed to form the identity 1. Hence the group we are after is (at least isomorphic to) $F[A]/R$. We can view this group as described by the generating set A and the set $\{r_i \mid i \in I\}$, which we will abbreviate $\{r_i\}$.

Historical Note

The idea of a group presentation already appears in Arthur Cayley's 1859 paper, "On the Theory of Groups as Depending on the Symbolic Equation $\theta^n = 1$. Third Part." In this article, Cayley gives a complete enumeration of the five groups of order 8, both by listing all the elements of each and by giving for each a presentation. For example, his third example is what is here called the *octic group;* Cayley notes that this group is generated by the two elements α, β with the relations $\alpha^4 = 1, \beta^2 = 1, \alpha\beta = \beta\alpha^3$. He also shows more generally that a group of order mn is generated by α, β with the relations $\alpha^m = 1, \beta^n = 1, \alpha\beta = \beta\alpha^s$ if and only if $s^n \equiv 1 \pmod{m}$ (see Exercise 13).

In 1878, Cayley returned to the theory of groups and noted that a central problem in that theory is the determination of all groups of a given order n. In the early 1890s, Otto Hölder published several papers attempting to solve Cayley's problem. Using techniques similar to those discussed in Sections 36, 37, and 40, Hölder determined all simple groups of order up to 200 and characterized all the groups of orders p^3, pq^2, pqr, and p^4, where p, q, r are distinct prime numbers. Furthermore, he developed techniques for determining the possible structures of a group G, if one is given the structure of a normal subgroup H and the structure of the factor group G/H. Interestingly, since the notion of an abstract group was still fairly new at this time, Hölder typically began his papers with the definition of a group and also emphasized that isomorphic groups are essentially one and the same object.

40.2 Definition Let A be a set and let $\{r_i\} \subseteq F[A]$. Let R be the least normal subgroup of $F[A]$ containing the r_i. An isomorphism ϕ of $F[A]/R$ onto a group G is a **presentation of** G. The sets

A and $\{r_i\}$ give a **group presentation.** The set A is the set of **generators for the presentation** and each r_i is a **relator.** Each $r \in R$ is a **consequence of** $\{r_i\}$. An equation $r_i = 1$ is a **relation.** A **finite presentation** is one in which both A and $\{r_i\}$ are finite sets. ■

This definition may seem complicated, but it really is not. In Example 40.1, $\{x, y\}$ is our set of generators and $xyx^{-1}y^{-1}$ is the only relator. The equation $xyx^{-1}y^{-1} = 1$, or $xy = yx$, is a relation. This was an example of a finite presentation.

If a group presentation has generators x_j and relators r_i, we shall use the notations

$$(x_j : r_i) \quad \text{or} \quad (x_j : r_i = 1)$$

to denote the group presentation. We may refer to $F[\{x_j\}]/R$ as *the group with presentation* $(x_j : r_i)$.

Isomorphic Presentations

40.3 Example Consider the group presentation with

$$A = \{a\} \quad \text{and} \quad \{r_i\} = \{a^6\},$$

that is, the presentation

$$(a : a^6 = 1).$$

This group defined by one generator a, with the relation $a^6 = 1$, is isomorphic to $\mathbb{Z}_6$.

Now consider the group defined by two generators a and b, with $a^2 = 1$, $b^3 = 1$, and $ab = ba$, that is, the group with presentation

$$(a, b : a^2, b^3, aba^{-1}b^{-1}).$$

The condition $a^2 = 1$ gives $a^{-1} = a$. Also $b^3 = 1$ gives $b^{-1} = b^2$. Thus every element in this group can be written as a product of nonnegative powers of a and b. The relation $aba^{-1}b^{-1} = 1$, that is, $ab = ba$, allows us to write first all the factors involving a and then the factors involving b. Hence every element of the group is equal to some $a^m b^n$. But then $a^2 = 1$ and $b^3 = 1$ show that there are just six distinct elements,

$$1, b, b^2, a, ab, ab^2.$$

Therefore this presentation also gives a group of order 6 that is abelian, and by the Fundamental Theorem 11.12, it must again be cyclic and isomorphic to $\mathbb{Z}_6$. ▲

The preceding example illustrates that different presentations may give isomorphic groups. When this happens, we have **isomorphic presentations.** To determine whether two presentations are isomorphic may be very hard. It has been shown (see Rabin [22]) that a number of such problems connected with this theory are not generally solvable; that is, there is no *routine* and well-defined way of discovering a solution in all cases. These unsolvable problems include the problem of deciding whether two presentations are isomorphic, whether a group given by a presentation is finite, free, abelian, or trivial, and the famous *word problem* of determining whether a given word w is a consequence of a given set of relations $\{r_i\}$.

The importance of this material is indicated by our Theorem 39.13, which guarantees that *every group has a presentation.*

40.4 Example Let us show that

$$(x, y : y^2x = y, yx^2y = x)$$

is a presentation of the trivial group of one element. We need only show that x and y are consequences of the relators y^2xy^{-1} and yx^2yx^{-1}, or that $x = 1$ and $y = 1$ can be deduced from $y^2x = y$ and $yx^2y = x$. We illustrate both techniques.

As a consequence of y^2xy^{-1}, we get yx upon conjugation by y^{-1}. From yx we deduce $x^{-1}y^{-1}$, and then $(x^{-1}y^{-1})(yx^2yx^{-1})$ gives xyx^{-1}. Conjugating xyx^{-1} by x^{-1}, we get y. From y we get y^{-1}, and $y^{-1}(yx)$ is x.

Working with relations instead of relators, from $y^2x = y$ we deduce $yx = 1$ upon multiplication by y^{-1} on the left. Then substituting $yx = 1$ into $yx^2y = x$, that is, $(yx)(xy) = x$, we get $xy = x$. Then multiplying by x^{-1} on the left, we have $y = 1$. Substituting this in $yx = 1$, we get $x = 1$.

Both techniques amount to the same work, but it somehow seems more natural to most of us to work with relations. ▲

Applications

We conclude this chapter with two applications.

40.5 Example Let us determine all groups of order 10 up to isomorphism. We know from the Fundamental Theorem 11.12 that every abelian group of order 10 is isomorphic to $\mathbb{Z}_{10}$. Suppose that G is nonabelian of order 10. By Sylow theory, G contains a normal subgroup H of order 5, and H must be cyclic. Let a be a generator of H. Then G/H is of order 2 and thus isomorphic to Z_2. If $b \in G$ and $b \notin H$, we must then have $b^2 \in H$. Since every element of H except 1 has order 5, if b^2 were not equal to 1, then b^2 would have order 5, so b would have order 10. This would mean that G would be cyclic, contradicting our assumption that G is not abelian. Thus $b^2 = 1$. Finally, since H is a normal subgroup of G, $bHb^{-1} = H$, so in particular, $bab^{-1} \in H$. Since conjugation by b is an automorphism of H, bab^{-1} must be another element of H of order 5, hence bab^{-1} equals a, a^2, a^3, or a^4. But $bab^{-1} = a$ would give $ba = ab$, and then G would be abelian, since a and b generate G. Thus the possibilities for presentations of G are:

1. $(a, b : a^5 = 1, b^2 = 1, ba = a^2b)$,

2. $(a, b : a^5 = 1, b^2 = 1, ba = a^3b)$,

3. $(a, b : a^5 = 1, b^2 = 1, ba = a^4b)$.

Note that all three of these presentations can give groups of order at most 10, since the last relation $ba = a^ib$ enables us to express every product of a's and b's in G in the form a^sb^t. Then $a^5 = 1$ and $b^2 = 1$ show that the set

$$S = \{a^0b^0, a^1b^0, a^2b^0, a^3b^0, a^4b^0, a^0b^1, a^1b^1, a^2b^1, a^3b^1, a^4b^1\}$$

includes all elements of G.

It is not yet clear that all these elements in S are distinct, so that we have in all three cases a group of order 10. For example, the group presentation

$$(a, b : a^5 = 1, b^2 = 1, ba = a^2b)$$

gives a group in which, using the associative law, we have

$$\begin{aligned} a &= b^2a = (bb)a = b(ba) = b(a^2b) = (ba)(ab) \\ &= (a^2b)(ab) = a^2(ba)b = a^2(a^2b)b = a^4b^2 = a^4 \end{aligned}$$

Thus in this group, $a = a^4$, so $a^3 = 1$, which, together with $a^5 = 1$, yields $a^2 = 1$. But $a^2 = 1$, together with $a^3 = 1$, means that $a = 1$. Hence every element in the group with presentation

$$(a, b : a^5 = 1, b^2 = 1, ba = a^2b)$$

is equal to either 1 or b; that is, this group is isomorphic to $\mathbb{Z}_2$. A similar study of

$$(bb)a = b(ba)$$

for

$$(a, b : a^5 = 1, b^2 = 1, ba = a^3b)$$

shows that $a = a^4$ again, so this also yields a group isomorphic to $\mathbb{Z}_2$.

This leaves just

$$(a, b : a^5 = 1, b^2 = 1, ba = a^4b)$$

as a candidate for a nonabelian group of order 10. In this case, it can be shown that all elements of S are distinct, so this presentation does give a nonabelian group G of order 10. How can we show that all elements in S represent distinct elements of G? The easy way is to observe that we know that there is at least one nonabelian group of order 10, the dihedral group D_5. Since G is the only remaining candidate, we must have $G \simeq D_5$. Another attack is as follows. Let us try to make S into a group by defining $(a^sb^t)(a^ub^v)$ to be a^xb^y, where x is the remainder of $s + u(4^t)$ when divided by 5, and y is the remainder of $t + v$ when divided by 2, in the sense of the division algorithm (Theorem 6.3). In other words, we use the relation $ba = a^4b$ as a guide in defining the product $(a^sb^t)(a^ub^v)$ of two elements of S. We see that a^0b^0 acts as identity, and that given a^ub^v, we can determine t and s successively by letting

$$t \equiv -v \pmod 2$$

and then

$$s \equiv -u(4^t) \pmod 5,$$

giving a^sb^t, which is a left inverse for a^ub^v. We will then have a group structure on S if and only if the associative law holds. Exercise 13 asks us to carry out the straight-forward computation for the associative law and to discover a condition for S to be a group under such a definition of multiplication. The criterion of the exercise in this case amounts to the valid congruence

$$4^2 \equiv 1 \pmod 5.$$

Thus we do get a group of order 10. Note that

$$2^2 \not\equiv 1 \pmod{5}$$

and

$$3^2 \not\equiv 1 \pmod{5},$$

so Exercise 13 also shows that

$$(a, b : a^5 = 1, b^2 = 1, ba = a^2b)$$

and

$$(a, b : a^5 = 1, b^2 = 1, ba = a^3b)$$

do not give groups of order 10. ▲

40.6 Example Let us determine all groups of order 8 up to isomorphism. We know the three abelian ones:

$$\mathbb{Z}_8, \qquad \mathbb{Z}_2 \times \mathbb{Z}_4, \qquad \mathbb{Z}_2 \times \mathbb{Z}_2 \times \mathbb{Z}_2.$$

Using generators and relations, we shall give presentations of the nonabelian groups.

Let G be nonabelian of order 8. Since G is nonabelian, it has no elements of order 8, so each element but the identity is of order either 2 or 4. If every element were of order 2, then for $a, b \in G$, we would have $(ab)^2 = 1$, that is, $abab = 1$. Then since $a^2 = 1$ and $b^2 = 1$ also, we would have

$$ba = a^2bab^2 = a(ab)^2b = ab,$$

contrary to our assumption that G is not abelian. Thus G must have an element of order 4.

Let $\langle a \rangle$ be a subgroup of G of order 4. If $b \notin \langle a \rangle$, the cosets $\langle a \rangle$ and $b\langle a \rangle$ exhaust all of G. Hence a and b are generators for G and $a^4 = 1$. Since $\langle a \rangle$ is normal in G (by Sylow theory, or because it is of index 2), $G/\langle a \rangle$ is isomorphic to $\mathbb{Z}_2$ and we have $b^2 \in \langle a \rangle$. If $b^2 = a$ or $b^2 = a^3$, then b would be of order 8. Hence $b^2 = 1$ or $b^2 = a^2$. Finally, since $\langle a \rangle$ is normal, we have $bab^{-1} \in \langle a \rangle$, and since $b\langle a \rangle b^{-1}$ is a subgroup conjugate to $\langle a \rangle$ and hence isomorphic to $\langle a \rangle$, we see that bab^{-1} must be an element of order 4. Thus $bab^{-1} = a$ or $bab^{-1} = a^3$. If bab^{-1} were equal to a, then ba would equal ab, which would make G abelian. Hence $bab^{-1} = a^3$, so $ba = a^3b$. Thus we have two possibilities for G, namely,

$$G_1 : (a, b : a^4 = 1, b^2 = 1, ba = a^3b)$$

and

$$G_2 : (a, b : a^4 = 1, b^2 = a^2, ba = a^3b).$$

Note that $a^{-1} = a^3$, and that b^{-1} is b in G_1 and b^3 in G_2. These facts, along with the relation $ba = a^3b$, enable us to express every element in G_i in the form $a^m b^n$, as in Examples 40.3 and 40.5. Since $a^4 = 1$ and either $b^2 = 1$ or $b^2 = a^2$, the possible elements in each group are

$$1, \quad a, \quad a^2, \quad a^3, \quad b, \quad ab, \quad a^2b, \quad a^3b.$$

Thus G_1 and G_2 each have order at most 8. That G_1 is a group of order 8 can be seen from Exercise 13. An argument similar to that used in Exercise 13 shows that G_2 has order 8 also.

Since $ba = a^3b \neq ab$, we see that both G_1 and G_2 are nonabelian. That the two groups are not isomorphic follows from the fact that a computation shows that G_1 has only two elements of order 4, namely, a and a^3. On the other hand, in G_2 all elements but 1 and a^2 are of order 4. We leave the computations of the tables for these groups to Exercise 3. To illustrate suppose we wish to compute $(a^2b)(a^3b)$. Using $ba = a^3b$ repeatedly, we get

$$(a^2b)(a^3b) = a^2(ba)a^2b = a^5(ba)ab = a^8(ba)b = a^{11}b^2.$$

Then for G_1, we have

$$a^{11}b^2 = a^{11} = a^3,$$

but if we are in G_2, we get

$$a^{11}b^2 = a^{13} = a.$$

The group G_1 is the **octic group** and is isomorphic to our old friend, the group D_4 of symmetries of the square. The group G_2 is the **quaternion group;** it is isomorphic to the multiplicative group $\{1, -1, i, -i, j, -j, k, -k\}$ of quaternions. Quaternions were discussed in Section 24. ▲

EXERCISES 40

Computations

1. Give a presentation of $\mathbb{Z}_4$ involving one generator; involving two generators; involving three generators.
2. Give a presentation of S_3 involving three generators.
3. Give the tables for both the octic group

$$(a, b : a^4 = 1, b^2 = 1, ba = a^3b)$$

and the quaternion group

$$(a, b : a^4 = 1, b^2 = a^2, ba = a^3b).$$

In both cases, write the elements in the order $1, a, a^2, a^3, b, ab, a^2b, a^3b$. (Note that we do not have to compute *every* product. We know that these presentations give groups of order 8, and once we have computed enough products the rest are forced so that each row and each column of the table has each element exactly once.)

4. Determine all groups of order 14 up to isomorphism. [*Hint:* Follow the outline of Example 40.5 and use Exercise 13, part (b).]
5. Determine all groups of order 21 up to isomorphism. [*Hint:* Follow the outline of Example 40.5 and use Exercise 13, part (b). It may seem that there are two presentations giving nonabelian groups. Show that they are isomorphic.]

Concepts

In Exercises 6 and 7, correct the definition of the italicized term without reference to the text, if correction is needed, so that it is in a form acceptable for publication.

6. A *consequence* of the set of relators is any finite product of relators raised to powers.

7. Two group presentations are *isomorphic* if and only if there is a one-to-one correspondence of the generators of the first presentation with the generators of the second that yields, by renaming generators, a one-to-one correspondence of the relators of the first presentation with those of the second.

8. Mark each of the following true or false.

_______ **a.** Every group has a presentation.

_______ **b.** Every group has many different presentations.

_______ **c.** Every group has two presentations that are not isomorphic.

_______ **d.** Every group has a finite presentation.

_______ **e.** Every group with a finite presentation is of finite order.

_______ **f.** Every cyclic group has a presentation with just one generator.

_______ **g.** Every conjugate of a relator is a consequence of the relator.

_______ **h.** Two presentations with the same number of generators are always isomorphic.

_______ **i.** In a presentation of an abelian group, the set of consequences of the relators contains the commutator subgroup of the free group on the generators.

_______ **j.** Every presentation of a free group has 1 as the only relator.

Theory

9. Use the methods of this section and Exercise 13, part (b), to show that there are no nonabelian groups of order 15. (See also Example 37.10).

10. Show, using Exercise 13, that

$$(a, b : a^3 = 1, b^2 = 1, ba = a^2b)$$

gives a group of order 6. Show that it is nonabelian.

11. Show that the presentation

$$(a, b : a^3 = 1, b^2 = 1, ba = a^2b)$$

of Exercise 10 gives (up to isomorphism) the only nonabelian group of order 6, and hence gives a group isomorphic to S_3.

12. We showed in Example 15.6 that A_4 has no subgroup of order 6. The preceding exercise shows that such a subgroup of A_4 would have to be isomorphic to either $\mathbb{Z}_6$ or S_3. Show again that this is impossible by considering orders of elements.

13. Let

$$S = \{a^i b^j \mid 0 \leq i < m, 0 \leq j < n\},$$

that is, S consists of all formal products $a^i b^j$ starting with $a^0 b^0$ and ending with $a^{m-1} b^{n-1}$. Let r be a positive integer, and define multiplication on S by

$$(a^s b^t)(a^u b^v) = a^x b^y,$$

where x is the remainder of $s + u(r^t)$ when divided by m, and y is the remainder of $t + v$ when divided by n, in the sense of the division algorithm (Theorem 6.3).

a. Show that a necessary and sufficient condition for the associative law to hold and for S to be a group under this multiplication is that $r^n \equiv 1 \pmod{m}$.

b. Deduce from part (a) that the group presentation

$$(a, b : a^m = 1, b^n = 1, ba = a^r b)$$

gives a group of order mn if and only if $r^n \equiv 1 \pmod{m}$. (See the Historical Note on page xxx.)

14. Show that if $n = pq$, with p and q primes and $q > p$ and $q \equiv 1 \pmod{p}$, then there is exactly one nonabelian group (up to isomorphism) of order n. Recall that the $q - 1$ nonzero elements of $\mathbb{Z}_q$ form a cyclic group $\mathbb{Z}_q^*$ under multiplication modulo q. [*Hint:* The solutions of $x^p \equiv 1 \pmod{q}$ form a cyclic subgroup of $\mathbb{Z}_q^*$ with elements $1, r, r^2, \cdots, r^{p-1}$. In the group with presentation $(a, b : a^q = 1, b^p = 1, ba = a^r b)$, we have $bab^{-1} = a^r$, so $b^j ab^{-j} = a^{(r^j)}$. Thus, since b^j generates $\langle b \rangle$ for $j = 1, \cdots, p - 1$, this presentation is isomorphic to

$$(a, b^j : a^q = 1, (b^j)^p = 1, (b^j)a = a^{(r^j)}(b^j)),$$

so all the presentations $(a, b : a^q = 1, b^p = 1, ba = a^{(r^j)}b)$ are isomorphic.]

Groups in Topology[†]

SECTION 41 SIMPLICIAL COMPLEXES AND HOMOLOGY GROUPS

Motivation

Topology concerns sets for which we have enough of an idea of when two points are close together to be able to define a continuous function. Two such sets, or *topological spaces,* are structurally the same if there is a one-to-one function mapping one onto the other such that both this function and its inverse are continuous. Naively, this means that one space can be stretched, twisted, and otherwise deformed, without being torn or cut, to look just like the other. Thus a big sphere is topologically the same structure as a small sphere, the boundary of a circle the same structure as the boundary of a square, and so on. Two spaces that are structurally the same in this sense are **homeomorphic.** Hopefully the student recognizes that *the concept of homeomorphism is to topology as the concept of isomorphism* (*where sets have the same algebraic structure*) *is to algebra.*

The *main problem of topology* is to find useful, necessary and sufficient conditions, other than just the definition, for two spaces to be homeomorphic. Sufficient conditions are hard to come by in general. Necessary conditions are a dime a dozen, but some are very important and useful. A "nice" space has associated with it various kinds of groups, namely *homology groups, cohomology groups, homotopy groups*, and *cohomotopy groups.* If two spaces are homeomorphic, it can be shown that the groups of one are isomorphic to the corresponding groups associated with the other. Thus a necessary condition for spaces to be homeomorphic is that their groups be isomorphic. Some of these groups may reflect very interesting properties of the spaces. Moreover, a continuous mapping of one space into another gives rise to homomorphisms from the groups

[†] Part VIII is not required for the remainder of the text.

of one into the groups of the other. These group homomorphisms may reflect interesting properties of the mapping.

If the student could make neither head nor tail out of the preceding paragraphs, he need not worry. The above paragraphs were just intended as motivation for what follows. It is the purpose of this section to describe some groups, *homology groups*, that are associated with certain simple spaces, in our work, usually some subset of the familiar Euclidean 3-space $\mathbb{R}^3$.

Preliminary Notions

First we introduce the idea of an *oriented n-simplex* in Euclidean 3-space $\mathbb{R}^3$ for $n = 0, 1, 2,$ and 3. An **oriented 0-simplex** is just a point P. An **oriented 1-simplex** is a *directed* line segment P_1P_2 joining the points P_1 and P_2 and viewed as traveled in the direction from P_1 to P_2. Thus $P_1P_2 \neq P_2P_1$. We will agree, however, that $P_1P_2 = -P_2P_1$. An **oriented 2-simplex** is a triangular region $P_1P_2P_3$, as in Fig. 41.1, together with a prescribed order of movement around the triangle, e.g., indicated by the arrow in Fig. 41.1 as the order $P_1P_2P_3$. The order $P_1P_2P_3$ is clearly the same order as $P_2P_3P_1$ and $P_3P_1P_2$, but the opposite order from $P_1P_3P_2$, $P_3P_2P_1$ and $P_2P_1P_3$. We will agree that

$$P_1P_2P_3 = P_2P_3P_1 = P_3P_1P_2 = -P_1P_3P_2 = -P_3P_2P_1 = -P_2P_1P_3.$$

Note that $P_iP_jP_k$ is equal to $P_1P_2P_3$ if

$$\begin{pmatrix} 1 & 2 & 3 \\ i & j & k \end{pmatrix}$$

is an even permutation, and is equal to $-P_1P_2P_3$ if the permutation is odd. The same could be said for an oriented 1-simplex P_1P_2. Note also that for $n = 0, 1, 2$, an oriented n-simplex is an n-dimensional object.

The definition of an oriented 3-simplex should now be clear: An **oriented 3-simplex** is given by an ordered sequence $P_1P_2P_3P_4$ of four vertices of a solid tetrahedron, as in Fig. 41.2. We agree that $P_1P_2P_3P_4 = \pm P_iP_jP_rP_s$, depending on whether the permutation

$$\begin{pmatrix} 1 & 2 & 3 & 4 \\ i & j & r & s \end{pmatrix}$$

is even or odd. Similar definitions hold for $n > 3$, but we shall stop here with dimensions that we can visualize. These simplexes are **oriented,** or have an **orientation,** meaning that we are concerned with the *order* of the vertices as well as with the actual points where the vertices are located. All our simplexes will be oriented, and we shall drop the adjective from now on.

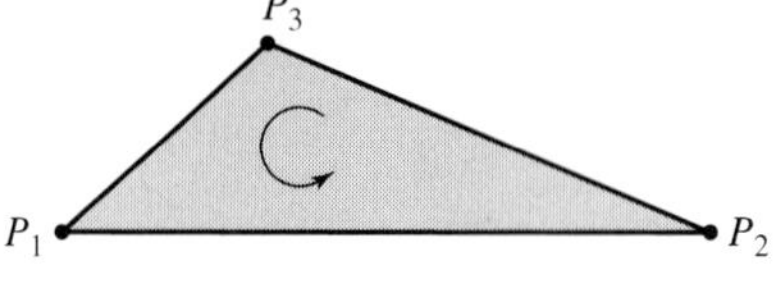

41.1 Figure

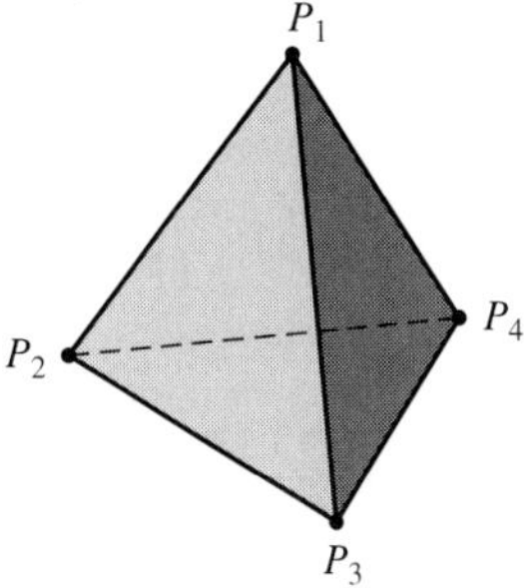

41.2 Figure

We are now going to define the *boundary of an n-simplex* for $n = 0, 1, 2, 3$. The term *boundary* is intuitive. We define the **boundary of a 0-simplex** P to be the empty simplex, which we denote this time by "0." The notation is

$$\text{“}\partial_0(P) = 0.\text{”}$$

The **boundary of a 1-simplex** P_1P_2 is defined by

$$\partial_1(P_1P_2) = P_2 - P_1,$$

that is, the formal difference of the end point and the beginning point. Likewise, the **boundary of a 2-simplex** is defined by

$$\partial_2(P_1P_2P_3) = P_2P_3 - P_1P_3 + P_1P_2,$$

which we can remember by saying that it is the formal sum of terms that we obtain by dropping each P_i in succession from the 2-simplex $P_1P_2P_3$ and taking the sign to be $+$ if the first term is omitted, $-$ if the second is omitted, and $+$ if the third is omitted. Referring to Fig. 41.1, we see that this corresponds to going around what we naturally would call the boundary in the direction indicated by the orientation arrow. Note also that the equation $\partial_1(P_1P_2) = P_2 - P_1$ can be remembered in the same way. Thus we are led to the following definition of the **boundary of a 3-simplex:**

$$\partial_3(P_1P_2P_3P_4) = P_2P_3P_4 - P_1P_3P_4 + P_1P_2P_4 - P_1P_2P_3.$$

Similar definitions hold for the definition of ∂_n for $n > 3$. Each individual *summand* of the boundary of a simplex is a **face of the simplex.** Thus, $P_2P_3P_4$ is a face of $P_1P_2P_3P_4$, but $P_1P_3P_4$ is not a face. However, $P_1P_4P_3 = -P_1P_3P_4$ is a face of $P_1P_2P_3P_4$.

Suppose that you have a subset of $\mathbb{R}^3$ that is divided up "nicely" into simplexes, as, for example, the *surface S* of the tetrahedron in Fig. 41.2, which is split up into four 2-simplexes nicely fitted together. Thus on the surface of the tetrahedron, we have some 0-simplexes, or the vertices, of the tetrahedron; some 1-simplexes, or the edges of the tetrahedron; and some 2-simplexes, or the triangles of the tetrahedron. In general, for a space to be divided up "nicely" into simplexes, we require that the following be true:

1. Each point of the space belongs to at least one simplex.
2. Each point of the space belongs to only a finite number of simplexes.

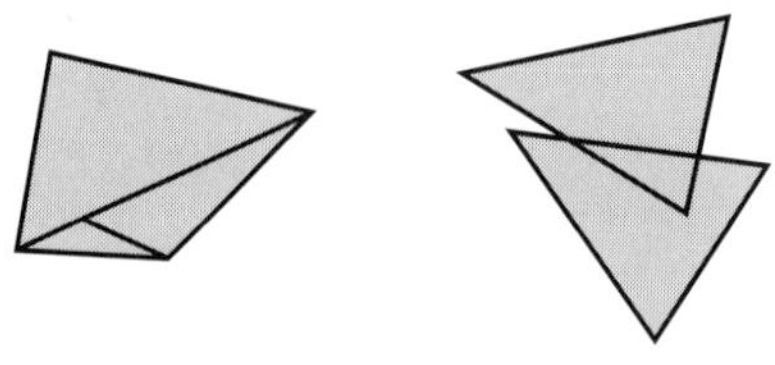

41.3 Figure

3. Two different (up to orientation) simplexes either have no points in common or one is (except possibly for orientation) a face of the other or a face of a face of the other, etc., or the set of points in common is (except possibly for orientation) a face, or a face of a face, etc., of each simplex.

Condition (3) excludes configurations like those shown in Fig. 41.3. A space divided up into simplexes according to these requirements is a **simplicial complex.**

Chains, Cycles, and Boundaries

Let us now describe some groups associated with a simplicial complex X. We shall illustrate each definition with the case of the *surface* S of our tetrahedron in Fig. 41.2. The **group** $C_n(X)$ **of (oriented)** $\boldsymbol{n}$**-chains of** $\boldsymbol{X}$ is the free abelian group generated by the (oriented) n-simplexes of X. Thus every element of $C_n(X)$ is a finite sum of the form $\sum_i m_i\sigma_i$, where the σ_i are n-simplexes of X and $m_i \in \mathbf{Z}$. We accomplish addition of chains by taking the algebraic sum of the coefficients of each occurrence in the chains of a fixed simplex.

41.4 Example For the surface S of our tetrahedron, every element of $C_2(S)$ is of the form

$$m_1P_2P_3P_4 + m_2P_1P_3P_4 + m_3P_1P_2P_4 + m_4P_1P_2P_3$$

for $m_i \in \mathbf{Z}$. As an illustration of addition, note that

$$(3P_2P_3P_4 - 5P_1P_2P_3) + (6P_2P_3P_4 - 4P_1P_3P_4) \\ = 9P_2P_3P_4 - 4P_1P_3P_4 - 5P_1P_2P_3.$$

An element of $C_1(S)$ is of the form

$$m_1P_1P_2 + m_2P_1P_3 + m_3P_1P_4 + m_4P_2P_3 + m_5P_2P_4 + m_6P_3P_4,$$

and an element of $C_0(S)$ is of the form

$$m_1P_1 + m_2P_2 + m_3P_3 + m_4P_4.$$

▲

Now if σ is an n-simplex, $\partial_n(\sigma) \in C_{n-1}(X)$ for $n = 1, 2, 3$. Let us define $C_{-1}(X) = \{0\}$, the trivial group of one element, and then we will also have $\partial_0(\sigma) \in C_{-1}(X)$. Since $C_n(X)$ is free abelian, and since we can specify a homomorphism of such a group by giving its values on generators, we see that ∂_n gives a unique **boundary homomorphism,** which we denote again by "∂_n," mapping $C_n(X)$ into $C_{n-1}(X)$ for $n = 0, 1, 2, 3$.

41.5 Example We have

$$\partial_n\left(\sum_i m_i\sigma_i\right) = \sum_i m_i\partial_n(\sigma_i).$$

For example,

$$\begin{aligned}\partial_1(3P_1P_2 - 4P_1P_3 + 5P_2P_4) &= 3\partial_1(P_1P_2) - 4\partial_1(P_1P_3) + 5\partial_1(P_2P_4)\\ &= 3(P_2 - P_1) - 4(P_3 - P_1) + 5(P_4 - P_2)\\ &= P_1 - 2P_2 - 4P_3 + 5P_4.\end{aligned}$$

▲

The student is reminded again that *any time you have a homomorphism, two things are of great interest, the kernel and the image.* The kernel of ∂_n consists of those n-chains with boundary 0. The elements of the kernel are ***n*-cycles.** The usual notation for the kernel of ∂_n, that is, the **group of *n*-cycles,** is "$Z_n(X)$."

41.6 Example If $z = P_1P_2 + P_2P_3 + P_3P_1$, then

$$\partial_1(z) = (P_2 - P_1) + (P_3 - P_2) + (P_1 - P_3) = 0.$$

Thus z is a 1-cycle. However, if we let $c = P_1P_2 + 2P_2P_3 + P_3P_1$, then

$$\partial_1(c) = (P_2 - P_1) + 2(P_3 - P_2) + (P_1 - P_3) = -P_2 + P_3 \neq 0.$$

Thus $c \notin Z_1(X)$. ▲

Note that $z = P_1P_2 + P_2P_3 + P_3P_1$ of Example 41.6 corresponds to one circuit, or *cycle*, around a triangle with vertices P_1, P_2, and P_3.

The image under ∂_n, the **group of $(n-1)$-boundaries,** consists exactly of those $(n-1)$-chains that are boundaries of n-chains. This group is denoted by "$B_{n-1}(X)$."

41.7 Example Referring to Example 41.6, we see that if

$$P_1P_2 + 2P_2P_3 + P_3P_1$$

is a 1-chain in $C_1(X)$, then $P_3 - P_2$ is a 0-boundary. Note that $P_3 - P_2$ *bounds* P_2P_3. ▲

Let us now compute $Z_n(X)$ and $B_n(X)$ for a more complicated example. In topology, if a group is the trivial group consisting just of the identity 0, one usually denotes it by "0" rather than "$\{0\}$." We shall follow this convention.

41.8 Example Let us compute for $n = 0, 1, 2$ the groups $Z_n(S)$ and $B_n(S)$ for the *surface* S of the tetrahedron of Fig. 41.2.

First, for the easier cases, since the highest dimensional simplex for the surface is a 2-simplex, we have $C_3(S) = 0$, so

$$B_2(S) = \partial_3[C_3(S)] = 0.$$

Also, since $C_{-1}(S) = 0$ by our definition, we see that

$$Z_0(S) = C_0(S).$$

Thus $Z_0(S)$ is free abelian on four generators, P_1, P_2, P_3, and P_4. It is easily seen that the image of a group under a homomorphism is generated by the images of generators of the original group. Thus, since $C_1(S)$ is generated by P_1P_2, P_1P_3, P_1P_4, P_2P_3, P_2P_4, and P_3P_4, we see that $B_0(S)$ is generated by

$$P_2 - P_1,\ P_3 - P_1,\ P_4 - P_1,\ P_3 - P_2,\ P_4 - P_2,\ P_4 - P_3.$$

However, $B_0(S)$ is not free abelian on these generators. For example, $P_3 - P_2 = (P_3 - P_1) - (P_2 - P_1)$. It is easy to see that $B_0(S)$ is free abelian on $P_2 - P_1$, $P_3 - P_1$, and $P_4 - P_1$.

Now let us go after the tougher group $Z_1(S)$. An element c of $C_1(S)$ is a formal sum of integral multiples of edges P_iP_j. It is clear that $\partial_1(c) = 0$ if and only if each vertex that is the beginning point of a total (counting multiplicity) of r edges of c is also the end point of exactly r edges. Thus

$$\begin{aligned} z_1 &= P_2P_3 + P_3P_4 + P_4P_2, \\ z_2 &= P_1P_4 + P_4P_3 + P_3P_1, \\ z_3 &= P_1P_2 + P_2P_4 + P_4P_1, \\ z_4 &= P_1P_3 + P_3P_2 + P_2P_1 \end{aligned}$$

are all 1-cycles. These are exactly the boundaries of the individual 2-simplexes. We claim that the z_i generate $Z_1(S)$. Let $z \in Z_1(S)$, and choose a particular vertex, say P_1; let us work on edges having P_1 as an end point. These edges are P_1P_2, P_1P_3, and P_1P_4. Let the coefficient of P_1P_j in z be m_j. Then

$$z + m_2z_4 - m_4z_2$$

is again a cycle, but does not contain the edges P_1P_2 or P_1P_4. Thus the only edge having P_1 as a vertex in the cycle $z + m_2z_4 - m_4z_2$ is possibly P_1P_3, but this edge could not appear with a nonzero coefficient as it would contribute a nonzero multiple of the vertex P_1 to the boundary, contradicting the fact that a cycle has boundary 0. Thus $z + m_2z_4 - m_4z_2$ consists of the edges of the 2-simplex $P_2P_3P_4$. Since in a 1-cycle each of P_2, P_3, and P_4 must serve the same number of times as a beginning and an end point of edges in the cycle, counting multiplicity, we see that

$$z + m_2z_4 - m_4z_2 = rz_1$$

for some integer r. Thus $Z_1(S)$ is generated by the z_i, actually by any three of the z_i. Since the z_i are the individual boundaries of the 2-simplexes, as we observed, we see that

$$Z_1(S) = B_1(S).$$

The student should see geometrically what this computation means in terms of Fig. 41.2.

Finally, we describe $Z_2(S)$. Now $C_2(S)$ is generated by the simplexes $P_2P_3P_4$, $P_3P_1P_4$, $P_1P_2P_4$, and $P_2P_1P_3$. If $P_2P_3P_4$ has coefficient r_1 and $P_3P_1P_4$ has coefficient r_2 in a 2-cycle, then the common edge P_3P_4 has coefficient $r_1 - r_2$ in its boundary.

Thus we must have $r_1 = r_2$, and by a similar argument, in a cycle each one of the four 2-simplexes appears with the same coefficient. Thus $Z_2(S)$ is generated by

$$P_2P_3P_4 + P_3P_1P_4 + P_1P_2P_4 + P_2P_1P_3,$$

that is, $Z_2(S)$ is infinite cyclic. Again, the student should interpret this computation geometrically in terms of Fig. 41.2. Note that the orientation of each summand corresponds to going around that triangle clockwise, when viewed from the outside of the tetrahedron. ▲

$\partial^2 = 0$ and Homology Groups

We now come to one of the most important equations in all of mathematics. We shall state it only for $n = 1, 2$, and 3, but it holds for all $n > 0$.

41.9 Theorem Let X be a simplicial complex, and let $C_n(X)$ be the n-chains of X for $n = 0, 1, 2, 3$. Then the composite homomorphism $\partial_{n-1}\partial_n$ mapping $C_n(X)$ into $C_{n-2}(X)$ maps everything into 0 for $n = 1, 2, 3$. That is, for each $c \in C_n(X)$ we have $\partial_{n-1}(\partial_n(c)) = 0$. We use the notation "$\partial_{n-1}\partial_n = 0$," or, more briefly, "$\partial^2 = 0$."

Proof Since a homomorphism is completely determined by its values on generators, it is enough to check that for an n-simplex σ, we have $\partial_{n-1}(\partial_n(\sigma)) = 0$. For $n = 1$ this is obvious, since ∂_0 maps everything into 0. For $n = 2$,

$$\begin{aligned}\partial_1(\partial_2(P_1P_2P_3)) &= \partial_1(P_2P_3 - P_1P_3 + P_1P_2)\\ &= (P_3 - P_2) - (P_3 - P_1) + (P_2 - P_1)\\ &= 0.\end{aligned}$$

The case $n = 3$ will make an excellent exercise for the student in the definition of the boundary operator (see Exercise 2). ◆

41.10 Corollary For $n = 0, 1, 2$, and 3, $B_n(X)$ is a subgroup of $Z_n(X)$.

Proof For $n = 0, 1$, and 2, we have $B_n(X) = \partial_{n+1}[C_{n+1}(X)]$. Then if $b \in B_n(X)$, we must have $b = \partial_{n+1}(c)$ for some $c \in C_{n+1}(X)$. Thus

$$\partial_n(b) = \partial_n(\partial_{n+1}(c)) = 0,$$

so $b \in Z_n(X)$.

For $n = 3$, since we are not concerned with simplexes of dimension greater than 3, $B_3(X) = 0$. ◆

41.11 Definition The factor group $H_n(X) = Z_n(X)/B_n(X)$ is the ***n*-dimensional homology group of *X***. ■

41.12 Example Let us calculate $H_n(S)$ for $n = 0, 1, 2$, and 3 and where S is the *surface* of the tetrahedron in Fig. 41.2.

We found $Z_n(S)$ and $B_n(S)$ in Example 41.8. Now $C_3(S) = 0$, so $Z_3(S)$ and $B_3(S)$ are both 0, and hence

$$H_3(S) = 0.$$

Also, $Z_2(S)$ is infinite cyclic and we saw that $B_2(S) = 0$. Thus $H_2(S)$ is infinite cyclic, that is,

$$H_2(S) \simeq \mathbb{Z}.$$

We saw that $Z_1(S) = B_1(S)$, so the factor group $Z_1(S)/B_1(S)$ is the trivial group of one element, that is,

$$H_1(S) = 0.$$

Finally, $Z_0(S)$ was free abelian on P_1, P_2, P_3, and P_4, while $B_0(S)$ was generated by $P_2 - P_1, P_3 - P_1, P_4 - P_1, P_3 - P_2, P_4 - P_2$, and $P_4 - P_3$. We claim that every coset of $Z_0(S)/B_0(S)$ contains exactly one element of the form rP_1. Let $z \in Z_0(S)$, and suppose that the coefficient of P_2 in z is s_2, of P_3 is s_3, and of P_4 is s_4. Then

$$z - [s_2(P_2 - P_1) + s_3(P_3 - P_1) + s_4(P_4 - P_1)] = rP_1$$

for some r, so $z \in [rP_1 + B_0(S)]$, that is, any coset does contain an element of the form rP_1. If the coset also contains $r'P_1$, then $r'P_1 \in [rP_1 + B_0(S)]$, so $(r' - r)P_1$ is in $B_0(S)$. Clearly, the only multiple of P_1 that is a boundary is zero, so $r = r'$ and the coset contains exactly one element of the form rP_1. We may then choose the rP_1 as representatives of the cosets in computing $H_0(S)$. Thus $H_0(S)$ is infinite cyclic, that is,

$$H_0(S) \simeq \mathbb{Z}.$$ ▲

These definitions and computations probably seem very complicated to the student. The ideas are very natural, but we admit that they are a bit messy to write down. However, the arguments used in these calculations are typical for homology theory, i.e., if you can understand them, you will understand all our others. Furthermore, we can make them *geometrically*, looking at the picture of the space. The next section will be devoted to further computations of homology groups of certain simple but important spaces.

EXERCISES 41

Suggested Exercises

1. Assume that $c = 2P_1P_3P_4 - 4P_3P_4P_6 + 3P_3P_2P_4 + P_1P_6P_4$ is a 2-chain of a certain simplicial complex X.

a. Compute $\partial_2(c)$. **b.** Is c a 2-cycle? **c.** Is $\partial_2(c)$ a 1-cycle?

2. Compute $\partial_2(\partial_3(P_1P_2P_3P_4))$ and show that it is 0, completing the proof of Theorem 41.9.

3. Describe $C_i(P)$, $Z_i(P)$, $B_i(P)$, and $H_i(P)$ for the space consisting of just the 0-simplex P. (This is really a trivial problem.)

4. Describe $C_i(X)$, $Z_i(X)$, $B_i(X)$, and $H_i(X)$ for the space X consisting of two distinct 0-simplexes, P and P'. (*Note:* The line segment joining the two points is *not* part of the space.)

5. Describe $C_i(X)$, $Z_i(X)$, $B_i(X)$, and $H_i(X)$ for the space X consisting of the 1-simplex P_1P_2.

6. Mark each of the following true or false.

_____ **a.** Every boundary is a cycle.

_____ **b.** Every cycle is a boundary.

_____ **c.** $C_n(X)$ is always a free abelian group.

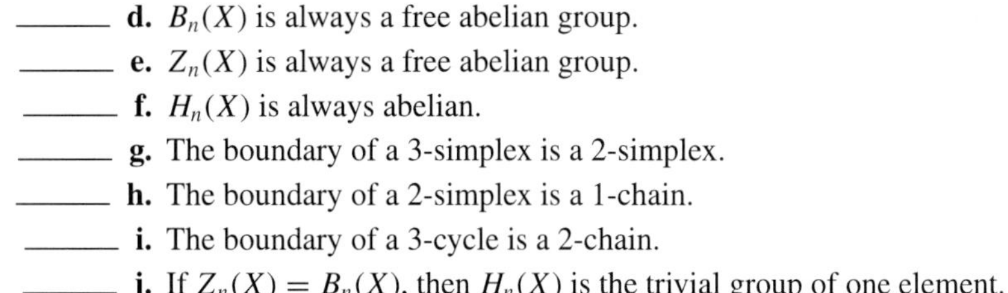

_____ **d.** $B_n(X)$ is always a free abelian group.

_____ **e.** $Z_n(X)$ is always a free abelian group.

_____ **f.** $H_n(X)$ is always abelian.

_____ **g.** The boundary of a 3-simplex is a 2-simplex.

_____ **h.** The boundary of a 2-simplex is a 1-chain.

_____ **i.** The boundary of a 3-cycle is a 2-chain.

_____ **j.** If $Z_n(X) = B_n(X)$, then $H_n(X)$ is the trivial group of one element.

More Exercises

7. Define the following concepts so as to generalize naturally the definitions in the text given for dimensions 0, 1, 2, and 3.

a. An oriented n-simplex

b. The boundary of an oriented n-simplex

c. A face of an oriented n-simplex

8. Continuing the idea of Exercise 7, what would be an easy way to answer a question asking you to define $C_n(X)$, $\partial_n : C_n(X) \to C_{n-1}(X)$, $Z_n(X)$, and $B_n(X)$ for a simplicial complex X perhaps containing some simplexes of dimension greater than 3?

9. Following the ideas of Exercises 7 and 8, prove that $\partial^2 = 0$ in general, i.e., that $\partial_{n-1}(\partial_n(c)) = 0$ for every $c \in C_n(X)$, where n may be greater than 3.

10. Let X be a simplicial complex. For an (oriented) n-simplex σ of X, the **coboundary** $\delta^{(n)}(\sigma)$ **of** σ is the $(n+1)$-chain $\sum \tau$, where the sum is taken over all $(n+1)$-simplexes τ that have σ as a face. That is, the simplexes τ appearing in the sum are precisely those that have σ as a *summand* of $\partial_{n+1}(\tau)$. *Orientation is important here.* Thus P_2 is a face of P_1P_2, but P_1 is not. However, P_1 is a face of P_2P_1. Let X be the simplicial complex consisting of the *solid tetrahedron* of Fig. 41.2.

a. Compute $\delta^{(0)}(P_1)$ and $\delta^{(0)}(P_4)$.

b. Compute $\delta^{(1)}(P_3P_2)$.

c. Compute $\delta^{(2)}(P_3P_2P_4)$.

11. Following the idea of Exercise 10, let X be a simplicial complex, and let the group $C^{(n)}(X)$ of ***n*-cochains** be the same as the group $C_n(X)$.

a. Define $\delta^{(n)} : C^{(n)}(X) \to C^{(n+1)}(X)$ in a way analogous to the way we defined $\partial_n : C_n(X) \to C_{n-1}(X)$.

b. Show that $\delta^2 = 0$, that is, that $\delta^{(n+1)}(\delta^{(n)}(c)) = 0$ for each $c \in C^{(n)}(X)$.

12. Following the ideas of Exercises 10 and 11, define the *group* $Z^{(n)}(X)$ *of n-cocycles* of X, the *group* $B^{(n)}(X)$ *of n-coboundaries* of X, and show that $B^{(n)}(X) \leq Z^{(n)}(X)$.

13. Following the ideas of Exercises 10, 11, and 12, define the *n-dimensional cohomology group* $H^{(n)}(X)$ of X. Compute $H^{(n)}(S)$ for the *surface S* of the tetrahedron of Fig. 41.2.

SECTION 42 Computations of Homology Groups

Triangulations

Suppose you wish to calculate homology groups for the surface of a sphere. The first thing you probably will say, if you are alert, is that the surface of a sphere is not a simplicial complex, since this surface is curved and a triangle is a plane surface. Remember that two spaces are topologically the same if one can be obtained from the other by bending, twisting, and so on. Imagine our 3-simplex, the tetrahedron, to have a rubber surface and to be filled with air. If the rubber surface is flexible, like the rubber of a balloon, it will promptly deform itself into a sphere and the four faces of the tetrahedron will then appear as "triangles" drawn on the surface of the sphere. This illustrates what is meant by a *triangulation* of a space. The term *triangulation* need not refer to a division into 2-simplexes only, but is also used for a division into n-simplexes for any $n \geq 0$. If a space is divided up into pieces in such a way that near each point the space can be deformed to look like a part of some Euclidean space $\mathbb{R}^n$ and the pieces into which the space was divided appear after this deformation as part of a simplicial complex, then the original division of the space is a **triangulation of the space.** The homology groups of the space are then defined formally just as in the last section.

Invariance Properties

There are two very important *invariance properties* of homology groups, the proofs of which require quite a lot of machinery, but that are easy for us to explain roughly. First, the homology groups of a space are defined in terms of a triangulation, but actually they are the same (i.e., isomorphic) groups no matter how the space is triangulated. For example, a square region can be triangulated in many ways, two of which are shown in Fig. 42.1. The homology groups are the same no matter which triangulation is used to compute them. *This is not obvious!*

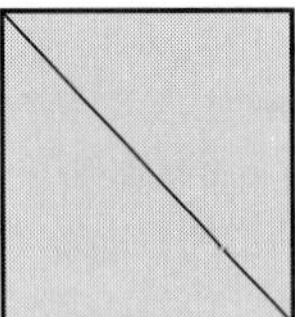
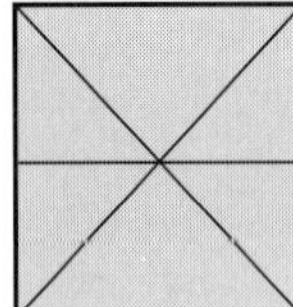

42.1 Figure

For the second invariance property, if one triangulated space is homeomorphic to another (e.g., can be deformed into the other without being torn or cut), the homology groups of the two spaces are the same (i.e., isomorphic) in each dimension n. *This is, again, not obvious.* We shall use both of these facts without proof.

42.2 Example The homology groups of the surface of a sphere are the same as those for the surface of our tetrahedron in Example 41.12, since the two spaces are homeomorphic. ▲

Two important types of spaces in topology are the spheres and the cells. Let us introduce them and the usual notations. The ***n*-sphere** S^n is the set of all points a distance of 1 unit from the origin in $(n+1)$-dimensional Euclidean space $\mathbb{R}^{n+1}$. Thus the 2-sphere S^2 is what is usually called the *surface* of a sphere in $\mathbb{R}^3$, S^1 is the rim of a circle, and S^0 is two points. Of course, the choice of 1 for the distance from the origin is not important. A 2-sphere of radius 10 is homeomorphic to one of radius 1 and homeomorphic to the surface of an ellipsoid for that matter. The ***n*-cell** or ***n*-ball** E^n is the set of all points in $\mathbb{R}^n$ a distance ≤ 1 from the origin. Thus E^3 is what you usually think of as a solid sphere, E^2 is a circular region, and E^1 is a line segment.

42.3 Example The above remarks and the computations of Example 41.12 show that $H_2(S^2)$ and $H_0(S^2)$ are both isomorphic to $\mathbb{Z}$, and $H_1(S^2) = 0$. ▲

Connected and Contractible Spaces

There is a very nice interpretation of $H_0(X)$ for a space X with a triangulation. A space is **connected** if any two points in it can be joined by a path (a concept that we will not define) lying totally in the space. If a space is not connected, then it is split up into a number of pieces, each of which is connected but no two of which can be joined by a path in the space. These pieces are the **connected components of the space.**

42.4 Theorem If a space X is triangulated into a finite number of simplexes, then $H_0(X)$ is isomorphic to $\mathbb{Z} \times \mathbb{Z} \times \cdots \times \mathbb{Z}$, and the Betti number m of factors $\mathbb{Z}$ is the number of connected components of X.

Proof Now $C_0(X)$ is the free abelian group generated by the finite number of vertices P_i in the triangulation of X. Also, $B_0(X)$ is generated by expressions of the form

$$P_{i_2} - P_{i_1},$$

where $P_{i_1}P_{i_2}$ is an edge in the triangulation. Fix P_{i_1}. Any vertex P_{i_r} in the same connected component of X as P_{i_1} can be joined to P_{i_1} by a finite sequence

$$P_{i_1}P_{i_2}, P_{i_2}P_{i_3}, \ldots, P_{i_{r-1}}P_{i_r}$$

of edges. Then

$$P_{i_r} = P_{i_1} + (P_{i_2} - P_{i_1}) + (P_{i_3} - P_{i_2}) + \cdots + (P_{i_r} - P_{i_{r-1}}),$$

showing that $P_{i_r} \in [P_{i_1} + B_0(X)]$. Clearly, if P_{i_s} is not in the same connected component with P_{i_1}, then $P_{i_s} \notin [P_{i_1} + B_0(X)]$, since no edge joins the two components. Thus, if we select one vertex from each connected component, each coset of $H_0(X)$ contains exactly one representative that is an integral multiple of one of the selected vertices. The theorem follows at once. ◆

42.5 Example We have at once that

$$H_0(S^n) \simeq \mathbb{Z}$$

for $n > 0$, since S^n is connected for $n > 0$. However,

$$H_0(S^0) \simeq \mathbb{Z} \times \mathbb{Z}$$

(see Chapter 41, Exercise 4). Also,

$$H_0(E^n) \simeq \mathbb{Z}$$

for $n \geq 1$. ▲

A space is **contractible** if it can be compressed to a point without being torn or cut, *but always kept within the space it originally occupied.* We just state the next theorem.

42.6 Theorem If X is a contractible space triangulated into a finite number of simplexes, then $H_n(X) = 0$ for $n \geq 1$.

42.7 Example It is a fact that S^2 is not contractible. It is not too easy to prove this fact. The student will, however, probably be willing to take it as self-evident that you cannot compress the "surface of a sphere" to a point without tearing it, *keeping it always within the original space S^2 that it occupied.* It is not fair to compress it all to the "center of the sphere." We saw that $H_2(S^2) \neq 0$ but is isomorphic to $\mathbb{Z}$.

Suppose, however, we consider $H_2(E^3)$, where we can regard E^3 as our solid tetrahedron of Fig. 41.2, for it is homomorphic to E^3. The surface S of this tetrahedron is homomorphic to S^2. The simplexes here for E^3 are the same as they are for S (or S^2), which we examined in Examples 41.8 and 41.12, except for the whole 3-simplex σ that now appears. Remember that a generator of $Z_2(S)$, and hence of $Z_2(E^3)$, was exactly the entire boundary of σ. Viewed in E^3, this is $\partial_3(\sigma)$, an element of $B_2(E^3)$, so $Z_2(E^3) = B_2(E^3)$ and $H_2(E^3) = 0$. Since E^3 is obviously contractible, this is consistent with Theorem 42.6. ▲

In general, E^n is contractible for $n \geq 1$, so we have by Theorem 42.6,

$$H_i(E^n) = 0$$

for $i > 0$.

Further Computations

We have seen a nice interpretation for $H_0(X)$ in Theorem 42.4. As the preceding examples illustrate, the 1-cycles in a triangulated space are generated by closed curves of the space formed by edges of the triangulation. The 2-cycles can be thought of as generated by 2-spheres or other closed 2-dimensional surfaces in the space. Forming the factor group

$$H_1(X) = Z_1(X)/B_1(X)$$

amounts roughly to counting the closed curves that appear in the space that are not there simply because they appear as the boundary of a 2-dimensional piece (i.e., a collection of 2-simplexes) of the space. Similarly, forming $H_2(X) = Z_2(X)/B_2(X)$ amounts roughly to counting the closed 2-dimensional surfaces in the space that cannot be "filled in solid" within the space, i.e., are not boundaries of some collection of 3-simplexes. Thus for $H_1(S^2)$, every closed curve drawn on the surface of the 2-sphere bounds a 2-dimensional piece of the sphere, so $H_1(S^2) = 0$. However, the only possible closed 2-dimensional surface, S^2 itself, cannot be "filled in solid" *within the whole space S^2 itself*, so $H_2(S^2)$ is free abelian on one generator.

42.8 Example According to the reasoning above, one would expect $H_1(S^1)$ to be free abelian on one generator, i.e., isomorphic to $\mathbb{Z}$, since the circle itself is not the boundary of a 2-dimensional part of S^1. You see, there *is* no 2-dimensional part of S^1. We compute and see whether this is indeed so.

A triangulation of S^1 is given in Fig. 42.9. Now $C_1(S^1)$ is generated by P_1P_2, P_2P_3, and P_3P_1. If a 1-chain is a cycle so that its boundary is zero, then it must contain P_1P_2

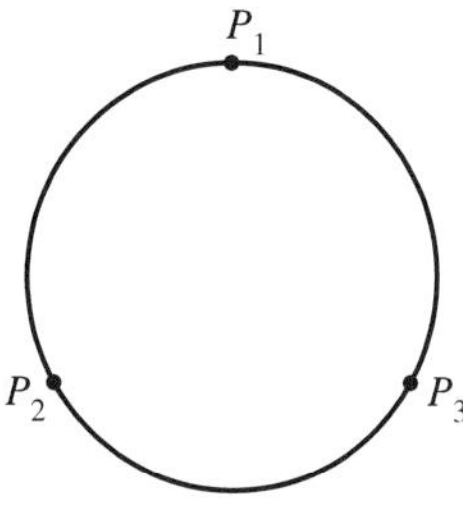

42.9 Figure

and P_2P_3 the same number of times, otherwise its boundary would contain a nonzero multiple of P_2. A similar argument holds for any two edges. Thus $Z_1(S^1)$ is generated by $P_1P_2 + P_2P_3 + P_3P_1$. Since $B_1(S^1) = \partial_2[C_2(S^1)] = 0$, there being no 2-simplexes, we see that $H_1(S^1)$ is free abelian on one generator, that is,

$$H_1(S^1) \simeq \mathbb{Z}. \qquad \blacktriangle$$

It can be proved that for $n > 0$, $H_n(S^n)$ and $H_0(S^n)$ are isomorphic to $\mathbb{Z}$, while $H_i(S^n) = 0$ for $0 < i < n$.

To conform to topological terminology, we shall call an element of $H_n(X)$, that is, a coset of $B_n(X)$ in $Z_n(X)$, a **"homology class."** Cycles in the same homology class are **homologous.**

42.10 Example Let us compute the homology groups of a plane annular region X between two concentric circles. A triangulation is indicated in Fig. 42.11. Of course, since X is connected, it follows that

$$H_0(X) \simeq \mathbb{Z}.$$

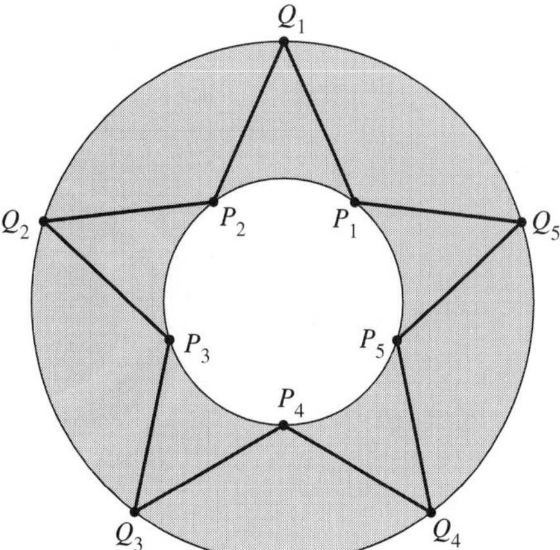

42.11 Figure

If z is any 1-cycle, and if P_1P_2 has coefficient r in z, then $z - r\partial_2(P_1P_2Q_1)$ is a cycle without P_1P_2 homologous to z. By continuing this argument, we find that there is a 1-cycle homologous to z containing no edge on the inner circle of the annulus. Using the "outside" triangles, we can adjust further by multiples of $\partial_2(Q_iP_iQ_j)$, and we arrive at z' containing no edge Q_iP_i either. But then if Q_5P_1 appears in z' with nonzero coefficient, P_1 appears with nonzero coefficient in $\partial_1(z')$, contradicting the fact that z' is a cycle. Similarly, no edge Q_iP_{i+1} can occur for $i = 1, 2, 3, 4$. Thus z is homologous to a cycle made up of edges only on the outer circle. By a familiar argument, such a cycle must be of the form

$$n(Q_1Q_2 + Q_2Q_3 + Q_3Q_4 + Q_4Q_5 + Q_5Q_1).$$

It is then clear that

$$H_1(X) \simeq \mathbb{Z}.$$

We showed that we could "push" any 1-cycle to the outer circle. Of course, we could have pushed it to the inner circle equally well.

For $H_2(X)$, note that $Z_2(X) = 0$, since every 2-simplex has in its boundary an edge on either the inner or the outer circle of the annulus that appears in no other 2-simplex. The boundary of any nonzero 2-chain must then contain some nonzero multiples of these edges. Hence

$$H_2(X) = 0.$$

▲

42.12 Example We shall compute the homology groups of the torus surface X which looks like the surface of a doughnut, as in Fig. 42.13. To visualize a triangulation of the torus, imagine that you cut it on the circle marked a, then cut it all around the circle marked b, and flatten it out as in Fig. 42.14. Then draw the triangles. To recover the torus from Fig. 42.14, join the left edge b to the right edge b in such a way that the arrows are going in the same direction. This gives a cylinder with circle a at each end. Then bend the cylinder around and join the circles a, again keeping the arrows going the same way around the circles.

Since the torus is connected, $H_0(X) \simeq \mathbb{Z}$.

For $H_1(X)$, let z be a 1-cycle. By changing z by a multiple of the boundary of the triangle numbered 1 in Fig. 42.14, you can get a homologous cycle not containing the side / of triangle 1. Then by changing this new 1-cycle by a suitable multiple of the boundary of triangle 2, you can further eliminate the side | of 2. Continuing, we can then

42.13 Figure

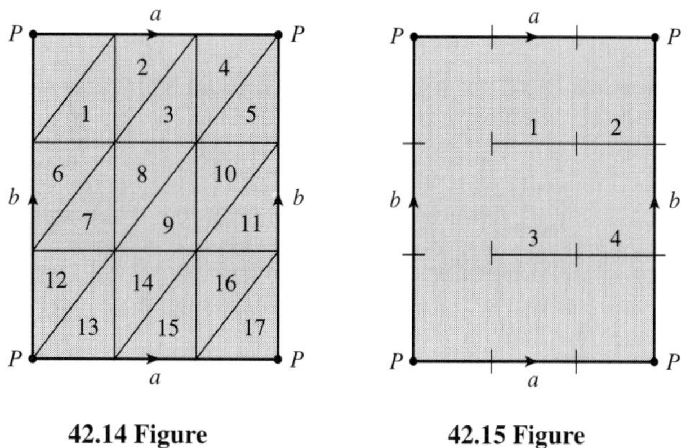

42.14 Figure **42.15 Figure**

eliminate / of 3, | of 4, / of 5, — of 6, / of 7, | of 8, / of 9, | of 10, / of 11, — of 12, / of 13, | of 14, / of 15, | of 16, and / of 17. The resulting cycle, homologous to z, can then only contain the edges shown in Fig. 42.15. But such a cycle could not contain, with nonzero coefficient, any of the edges we have numbered in Fig. 42.15, or it would not have boundary 0. Thus z is homologous to a 1-cycle having edges only on the circle a or the circle b (refer to Fig. 42.13). By a now hopefully familiar argument, every edge on circle a must appear the same number of times, and the same is also true for edges on circle b; however, an edge on circle b need not appear the same number of times as an edge appears on a. Furthermore, if a 2-chain is to have a boundary just containing a and b, all the triangles oriented counterclockwise must appear with the same coefficient so that the inner edges will cancel out. The boundary of such a 2-chain is 0. Thus every homology class (coset) contains one and only one element

$$ra + sb,$$

where r and s are integers. Hence $H_1(X)$ is free abelian on two generators, represented by the two circles a and b. Therefore,

$$H_1(X) \simeq \mathbb{Z} \times \mathbb{Z}.$$

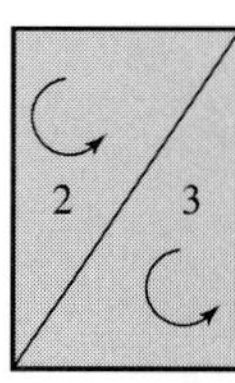

42.16 Figure

Finally, for $H_2(X)$, a 2-cycle must contain the triangle numbered 2 of Fig. 42.14 with counterclockwise orientation the same number of times as it contains the triangle numbered 3, also with counterclockwise orientation, in order for the common edge / of these triangles not to be in the boundary. These orientations are illustrated in Fig. 42.16. The same holds true for any two adjacent triangles, and thus every triangle with the counterclockwise orientation must appear the same number of times in a 2-cycle. Clearly, any multiple of the formal sum of all the 2-simplexes, all with counterclockwise orientation, is a 2-cycle. Thus $Z_2(X)$ is infinite cyclic, isomorphic to $\mathbb{Z}$. Also, $B_2(X) = 0$, there being no 3-simplexes, so

$$H_2(X) \simeq \mathbb{Z}. \qquad \blacktriangle$$

EXERCISES 42

In these exercises, you need not write out in detail your computations or arguments.

Computations

1. Compute the homology groups of the space consisting of two tangent 1-spheres, i.e., a figure eight.
2. Compute the homology groups of the space consisting of two tangent 2-spheres.
3. Compute the homology groups of the space consisting of a 2-sphere with an annular ring (as in Fig. 42.11) that does not touch the 2-sphere.
4. Compute the homology groups of the space consisting of a 2-sphere with an annular ring whose inner circle is a great circle of the 2-sphere.
5. Compute the homology groups of the space consisting of a circle touching a 2-sphere at one point.
6. Compute the homology groups of the surface consisting of a 2-sphere with a handle (see Fig. 42.17).
7. Mark each of the following true or false.

_____ **a.** Homeomorphic simplicial complexes have isomorphic homology groups.
_____ **b.** If two simplicial complexes have isomorphic homology groups, then the spaces are homeomorphic.
_____ **c.** S^n is homeomorphic to E^n.
_____ **d.** $H_n(X)$ is trivial for $n > 0$ if X is a connected space with a finite triangulation.
_____ **e.** $H_n(X)$ is trivial for $n > 0$ if X is a contractible space with a finite triangulation.
_____ **f.** $H_n(S^n) = 0$ for $n > 0$.
_____ **g.** $H_n(E^n) = 0$ for $n > 0$.
_____ **h.** A torus is homeomorphic to S^2.
_____ **i.** A torus is homeomorphic to E^2.
_____ **j.** A torus is homeomorphic to a sphere with a handle on it (see Fig. 42.17).

8. Compute the homology groups of the space consisting of two torus surfaces having no points in common.

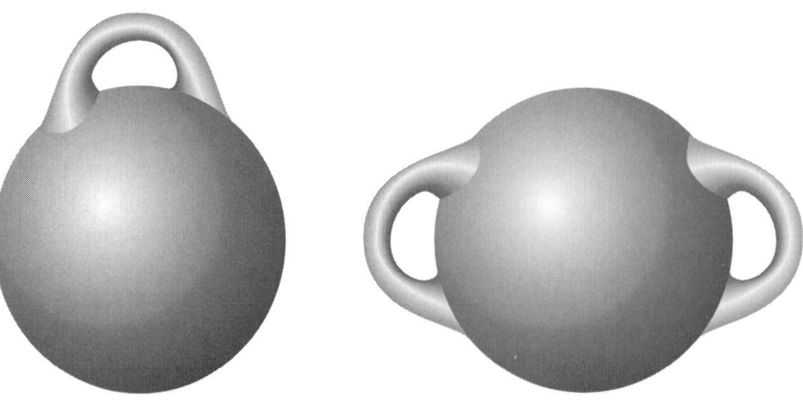

42.17 Figure **42.18 Figure**

9. Compute the homology groups of the space consisting of two stacked torus surfaces, stacked as one would stack two inner tubes.

10. Compute the homology groups of the space consisting of a torus tangent to a 2-sphere at all points of a great circle of the 2-sphere, i.e., a balloon wearing an inner tube.

11. Compute the homology groups of the surface consisting of a 2-sphere with two handles (see Fig. 42.18).

12. Compute the homology groups of the surface consisting of a 2-sphere with n handles (generalizing Exercises 6 and 11.

SECTION 43 More Homology Computations and Applications

One-Sided Surfaces

Thus far all the homology groups we have found have been free abelian, so that there were no nonzero elements of finite order. This can be shown always to be the case for the homology groups of a *closed surface* (i.e., a surface like S^2, which has no boundary) that has two sides. Our next example is of a *one-sided closed surface,* the *Klein bottle*. Here the 1-dimensional homology group will have a nontrivial torsion subgroup reflecting the twist in the surface.

43.1 Example Let us calculate the homology groups of the Klein bottle X. Figure 43.2 represents the Klein bottle cut apart, just as Fig. 42.14 represents the torus cut apart. The only difference is that the arrows on the top and bottom edge a of the rectangle are in *opposite* directions this time. To recover a Klein bottle from Fig. 43.2, again bend the rectangle joining the edges labeled b so that the directions of the arrows match up. This gives a cylinder that is shown somewhat deformed, with the bottom end pushed a little way up inside the cylinder, in Fig. 43.3. Such deformations are legitimate in topology. Now the circles a must be joined so that the arrows go around the same way. *This cannot be done in* $\mathbb{R}^3$. You must imagine that you are in $\mathbb{R}^4$, so that you can bend the neck of the bottle around and "through" the side without intersecting the side, as shown in Fig. 43.4. With a little thought, you can see that this resulting surface really has only one side. That is, if you

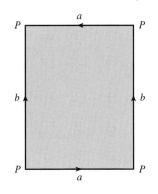

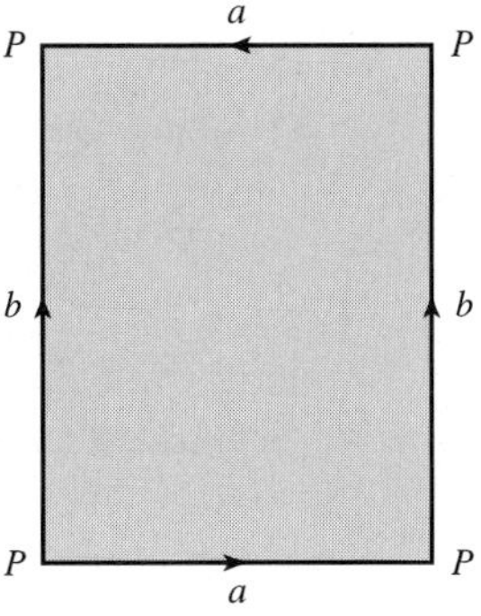

43.2 Figure

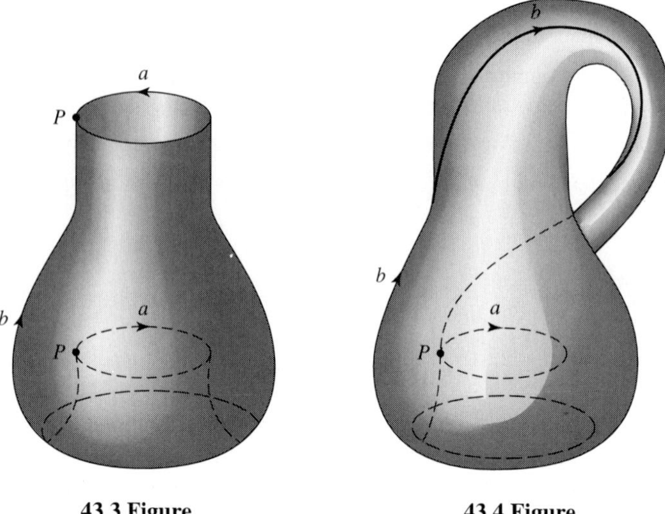

43.3 Figure

43.4 Figure

start at any place and begin to paint "one side," you will wind up painting the whole thing. There is no concept of an *inside* of a Klein bottle.

We can calculate the homology groups of the Klein bottle much as we calculated the homology groups of the torus in Example 42.12, by splitting Fig. 43.2 into triangles exactly as we did for the torus. Of course,

$$H_0(X) \simeq \mathbb{Z},$$

since X is connected. As we found for the torus, if we triangulate the Klein bottle by dividing Fig. 43.2 into triangles, every 1-cycle is homologous to a cycle of the form

$$ra + sb$$

for r and s integers. If a 2-chain is to have a boundary containing just a and b, again, all the triangles oriented counterclockwise must appear with the same coefficient so that the inner edges will cancel each other. In the case of the torus, the boundary of such a 2-chain was 0. Here, however, it is $k(2a)$, where k is the number of times each triangle appears. Thus $H_1(X)$ is an abelian group with generators the homology classes of a and b and the relations $a + b = b + a$ and $2a = 0$. Therefore,

$$H_1(X) \simeq \mathbb{Z}_2 \times \mathbb{Z},$$

a group with torsion coefficient 2 and Betti number 1. Our argument above regarding 2-chains shows that there are no 2-cycles this time, so

$$H_2(X) = 0. \qquad \blacktriangle$$

A torsion coefficient does not have to be present in some homology group of a one-sided surface *with boundary*. Mostly for the sake of completeness, we give this standard example of the *Möbius strip*.

43.5 Example Let X be the Möbius strip, which we can form by taking a rectangle of paper and joining the two ends marked a with a half twist so that the arrows match up, as indicated in Fig. 43.6. Note that the Möbius strip is a surface with a boundary, and the boundary is just one closed curve (homomorphic to a circle) made up of l and l'. It is clear that the Möbius strip, like the Klein bottle, has just one side, in the sense that if you were asked to color only one side of it, you would wind up coloring the whole thing.

Of course, since X is connected,

$$H_0(X) \simeq \mathbb{Z}.$$

Let z be any 1-cycle. By subtracting in succession suitable multiples of the triangles numbered 2, 3, and 4 in Fig. 43.6, we can eliminate edges/of triangle 2, | of triangle 3, and \ of triangle 4. Thus z is homologous to a cycle z' having edges on only l, l', and a, and as before, both edges on l' must appear the same number of times. But if c is a 2-chain consisting of the formal sum of the triangles oriented as shown in Fig. 43.6, we see that $\partial_2(c)$ consists of the edges on l and l' plus $2a$. Since both edges on l' must appear in z' the same number of times, by subtracting a suitable multiple of $\partial_2(c)$, we see that z is homologous to a cycle with edges just lying on l and a. By a familiar argument, all these edges properly oriented must appear the same number of times in this new cycle, and thus the homology class containing their formal sum is a generator for $H_1(X)$. Therefore,

$$H_1(X) \simeq \mathbb{Z}.$$

This generating cycle starts at Q and goes around the strip, then cuts across it at P via a, and arrives at its starting point.

If z'' were a 2-cycle, it would have to contain the triangles 1, 2, 3, and 4 of Fig. 43.6 an equal number r of times with the indicated orientation. But then $\partial_2(z'')$ would be $r(2a + l + l') \neq 0$. Thus $Z_2(X) = 0$, so

$$H_2(X) = 0. \qquad \blacktriangle$$

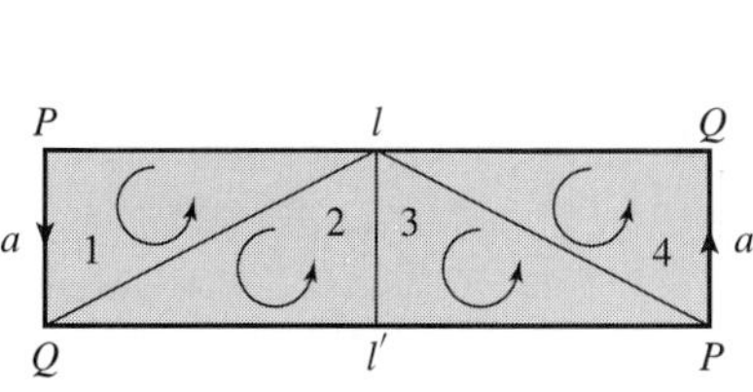

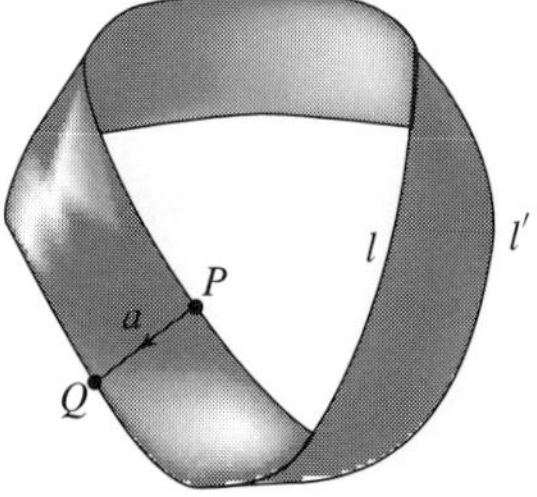

43.6 Figure

The Euler Characteristic

Let us turn from the computation of homology groups to a few interesting facts and applications. Let X be a finite simplicial complex (or triangulated space) consisting of simplexes of dimension 3 and less. Let n_0 be the total number of vertices in the triangulation, n_1 the number of edges, n_2 the number of 2-simplexes, and n_3 the number of 3-simplexes. The number

$$n_0 - n_1 + n_2 - n_3 = \sum_{i=0}^{3} (-1)^i n_i$$

can be shown to be the same no matter how the space X is triangulated. This number is the **Euler characteristic** $\chi(X)$ of the space. We just state the following fascinating theorem.

43.7 Theorem Let X be a finite simplicial complex (or triangulated space) of dimension ≤ 3. Let $\chi(X)$ be the Euler characteristic of the space X, and let β_j be the Betti number of $H_j(X)$. Then

$$\chi(X) = \sum_{j=0}^{3} (-1)^j \beta_j.$$

This theorem holds also for X of dimension greater than 3, with the obvious extension of the definition of the Euler characteristic to dimension greater than 3.

43.8 Example Consider the solid tetrahedron E^3 of Fig. 41.2. Here $n_0 = 4$, $n_1 = 6$, $n_2 = 4$, and $n_3 = 1$, so

$$\chi(E^3) = 4 - 6 + 4 - 1 = 1.$$

Remember that we saw that $H_3(E^3) = H_2(E^3) = H_1(E^3) = 0$ and $H_0(E^3) \simeq \mathbb{Z}$. Thus $\beta_3 = \beta_2 = \beta_1 = 0$ and $\beta_0 = 1$, so

$$\sum_{j=0}^{3} (-1)^j \beta_j = 1 = \chi(E^3).$$

For the surface S^2 of the tetrahedron in Fig. 41.2, we have $n_0 = 4$, $n_1 = 6$, $n_2 = 4$, and $n_3 = 0$, so

$$\chi(S^2) = 4 - 6 + 4 = 2.$$

Also, $H_3(S^2) = H_1(S^2) = 0$, and $H_2(S^2)$ and $H_0(S^2)$ are both isomorphic to $\mathbb{Z}$. Thus $\beta_3 = \beta_1 = 0$ and $\beta_2 = \beta_0 = 1$, so

$$\sum_{j=0}^{3} (-1)^j \beta_j = 2 = \chi(S^2).$$

Finally, for S^1 in Fig. 42.9, $n_0 = 3$, $n_1 = 3$, and $n_2 = n_3 = 0$, so

$$\chi(S^1) = 3 - 3 = 0.$$

Here $H_1(S^1)$ and $H_0(S^1)$ are both isomorphic to $\mathbb{Z}$, and $H_3(S^1) = H_2(S^1) = 0$. Thus $\beta_0 = \beta_1 = 1$ and $\beta_2 = \beta_3 = 0$, giving

$$\sum_{j=0}^{3} (-1)^j \beta_j = 0 = \chi(S^1).$$

▲

Mappings of Spaces

A continuous function f mapping a space X into a space Y gives rise to a homomorphism f_{*n} mapping $H_n(X)$ into $H_n(Y)$ for $n \geq 0$. The demonstration of the existence of this homomorphism takes more machinery than we wish to develop here, but let us attempt to describe how these homomorphisms can be computed in certain cases. The following is true:

> If $z \in \mathbb{Z}_n(X)$, and if $f(z)$, regarded as the result of picking up z and setting it down in Y in the naively obvious way, should be an n-cycle in Y, then
>
> $$f_{*n}(z + B_n(X)) = f(z) + B_n(Y).$$
>
> That is, if z represents a homology class in $H_n(X)$ and $f(z)$ is an n-cycle in Y, then $f(z)$ represents the image homology class under f_{*n} of the homology class containing z.

Let us illustrate this and attempt to show just what we mean here by $f(z)$.

43.9 Example Consider the unit circle

$$S^1 = \{(x, y) \mid x^2 + y^2 = 1\}$$

in $\mathbb{R}^2$. Any point in S^1 has coordinates $(\cos\theta, \sin\theta)$, as indicated in Fig. 43.10. Let $f : S^1 \to S^1$ be given by

$$f((\cos\theta, \sin\theta)) = (\cos 3\theta, \sin 3\theta).$$

Obviously, this function f is continuous. Now f should induce

$$f_{*1} : H_1(S^1) \to H_1(S^1).$$

Here $H_1(S^1)$ is isomorphic to $\mathbb{Z}$ and has as generator the homology class of $z = P_1P_2 + P_2P_3 + P_3P_1$, as seen in Example 42.8. Now if P_1, P_2, and P_3 are evenly spaced about the circle, then f maps each of the arcs P_1P_2, P_2P_3, and P_3P_1 onto the whole perimeter of the circle, that is,

$$f(P_1P_2) = f(P_2P_3) = f(P_3P_1) = P_1P_2 + P_2P_3 + P_3P_1.$$

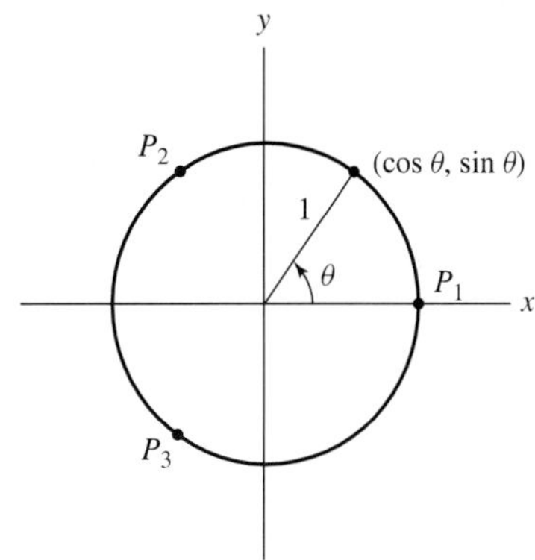

43.10 Figure

Thus

$$f_{*1}(z + B_1(S^1)) = 3(P_1P_2 + P_2P_3 + P_3P_1) + B_1(S^1)$$
$$= 3z + B_1(S^1),$$

that is, f_{*1} maps a generator of $H_1(S^1)$ onto three times itself. This obviously reflects the fact that f winds S^1 around itself three times. ▲

Example 43.9 illustrates our previous assertion that the homomorphisms of homology groups associated with a continuous mapping f may mirror important properties of the mapping.

Finally, we use these concepts to indicate a proof of the famous *Brouwer Fixed-Point Theorem*. This theorem states that a continuous map f of E^n into itself has a *fixed point*, i.e., there is some $x \in E^n$ such that $f(x) = x$. Let us see what this means for E^2, a circular region. Imagine that you have a thin sheet of rubber stretched out on a table to form a circular disk. Mark with a pencil the outside boundary of the rubber circle on the table. Then stretch, compress, bend, twist, and fold the rubber in any fashion without tearing it, but keep it always within the penciled circle on the table. When you finish, some point on the rubber will be over exactly the same point on the table at which it first started.

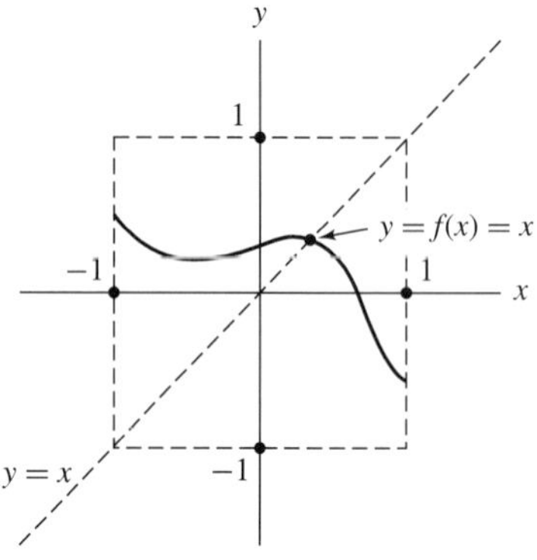

43.11 Figure

The proof we outline is good for any $n > 1$. For $n = 1$, looking at the graph of a function $f : E^1 \to E^1$, we find that the theorem simply states that any continuous path joining the left and right sides of a square must cross the diagonal somewhere, as indicated in Fig. 43.11. The student should visualize the construction of our proof with E^3 having boundary S^2 and E^2 having boundary S^1. The proof contains a figure illustrating the construction for the case of E^2.

43.12 Theorem (Brouwer Fixed-Point Theorem). A continuous map f of E^n into itself has a fixed point for $n \geq 1$.

Proof The case $n = 1$ was considered above. Let f be a map of E^n into E^n for $n > 1$. We shall assume that f has no fixed point and shall derive a contradiction.

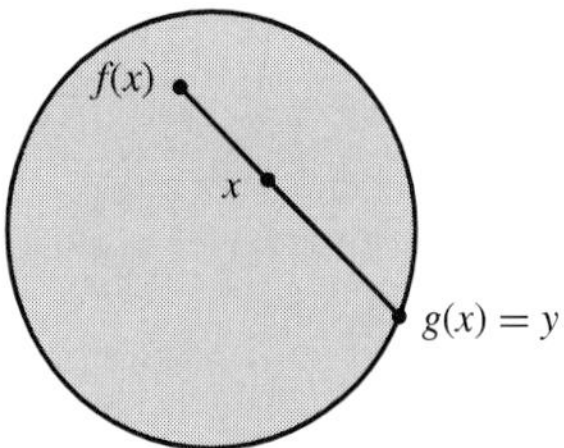

43.13 Figure

If $f(x) \neq x$ for all $x \in E^n$, we can consider the line segment from $f(x)$ to x. Let us extend this line segment *in the direction from* $f(x)$ *to* x until it goes through the boundary S^{n-1} of E^n at some point y. This defines for us a function $g : E^n \to S^{n-1}$ with $g(x) = y$, as illustrated in Fig. 43.13. Note that for y on the boundary, we have $g(y) = y$. Now since f is continuous, it is pretty obvious that g is also continuous. (A **continuous function** is roughly one that maps points that are sufficiently close together into points that are close together. If x_1 and x_2 are sufficiently close together, then $f(x_1)$ and $f(x_2)$ are sufficiently close together so that the line segment joining $f(x_1)$ and x_1 is so close to the line segment joining $f(x_2)$ and x_2 that $y_1 = g(x_1)$ is close to $y_2 = g(x_2)$.) Then g is a continuous mapping of E^n into S^{n-1}, and thus induces a homomorphism

$$g_{*(n-1)} : H_{n-1}(E^n) \to H_{n-1}(S^{n-1}).$$

Now we said that $H_{n-1}(E^n) = 0$, for $n > 1$, since E^n is contractible, and we checked it for $n = 2$ and $n = 3$. Since $g_{*(n-1)}$ is a homomorphism, we must have $g_{*(n-1)}(0) = 0$. But an $(n-1)$-cycle representing the homology class 0 of $H_{n-1}(E^n)$ is the whole complex S^{n-1} with proper orientation of simplexes, and $g(S^{n-1}) = S^{n-1}$, since $g(y) = y$ for all $y \in S^{n-1}$. Thus

$$g_{*(n-1)}(0) = S^{n-1} + B_{n-1}(S^{n-1}).$$

which is a generator $\neq 0$ of $H_{n-1}(S^{n-1})$, a contradiction. ◆

We find the preceding proof very satisfying aesthetically, and hope you agree.

EXERCISES 43

Suggested Exercises

1. Verify by direct calculation that both triangulations of the square region X in Fig. 42.1 give the same value for the Euler characteristic $\chi(X)$.
2. Illustrate Theorem 43.7, as we did in Example 43.8, for each of the following spaces.

   a. The annular region of Example 42.10

   b. The torus of Example 42.12

   c. The Klein bottle of Example 43.1

3. Will every continuous map of a square region into itself have a fixed point? Why or why not? Will every continuous map of a space consisting of two disjoint 2-cells into itself have a fixed point? Why or why not?
4. Compute the homology groups of the space consisting of a 2-sphere touching a Klein bottle at one point.
5. Compute the homology groups of the space consisting of two Klein bottles with no points in common.
6. Mark each of the following true or false.
   _____ **a.** Every homology group of a contractible space is the trivial group of one element.
   _____ **b.** A continuous map from a simplicial complex X into a simplicial complex Y induces a homomorphism of $H_n(X)$ into $H_n(Y)$.
   _____ **c.** All homology groups are abelian.
   _____ **d.** All homology groups are free abelian.
   _____ **e.** All 0-dimensional homology groups are free abelian.
   _____ **f.** If a space X has n-simplexes but none of dimension greater than n and $H_n(X) \neq 0$, then $H_n(X)$ is free abelian.
   _____ **g.** The boundary of an n-chain is an $(n - 1)$-chain.
   _____ **h.** The boundary of an n-chain is an $(n - 1)$-cycle.
   _____ **i.** The n-boundaries form a subgroup of the n-cycles.
   _____ **j.** The n-dimensional homology group of a simplicial complex is always a subgroup of the group of n-chains.

More Exercises

7. Find the Euler characteristic of a 2-sphere with n handles (see Section 42, Exercise 12).
8. We can form the topological *real projective plane* X, using Fig. 43.14, by joining the semicircles a so that diametrically opposite points come together and the directions of the arrows match up. *This cannot be done in Euclidean 3-space* $\mathbb{R}^3$. One must go to $\mathbb{R}^4$. Triangulate this space X, starting with the form exhibited in Fig. 43.14, and compute its homology groups.
9. The circular disk shown in Fig. 43.14 can be deformed topologically to appear as a 2-sphere with a hole in it, as shown in Fig. 43.15. We form the real projective plane from this configuration by sewing up the hole in such a way that only diametrically opposite points on the rim of the hole are sewn together. This cannot be done in $\mathbb{R}^3$.

   Extending this idea, a 2-sphere with q holes in it, which are then sewn up by bringing together diametrically opposite points on the rims of the holes, gives a **2-sphere with q cross caps.** Find the homology groups of a 2-sphere with q cross caps. (To see a triangulation, view the space as the disk in Fig. 43.14 but with $q - 1$ holes in it to be sewn up as described above. Then triangulate this disk with these holes.)

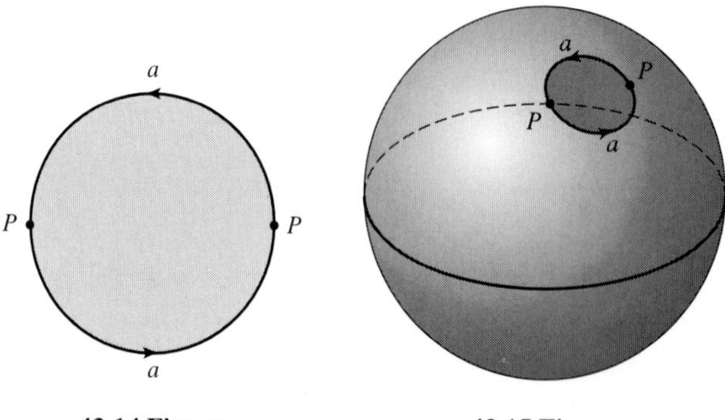

43.14 Figure **43.15 Figure**

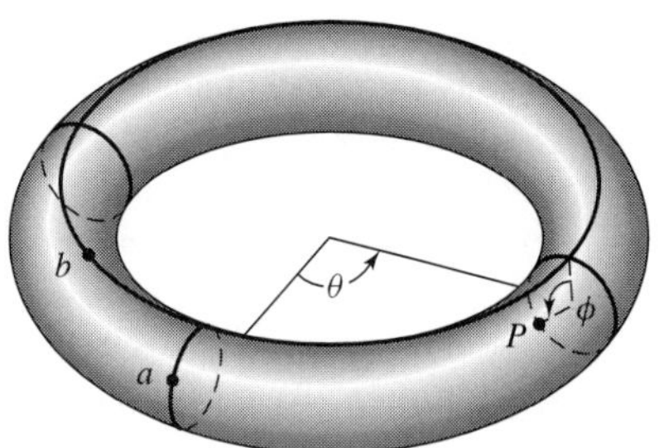

43.16 Figure

Comment: It can be shown that every sufficiently nice closed surface, namely a *closed 2-manifold,* is homeomorphic to a 2-sphere with some number $h \geq 0$ of handles if the surface is two sided, and is homeomorphic to a 2-sphere with $q > 0$ cross caps if it is one sided. The number h or q, as the case may be, is the **genus of the surface.**

10. Every point P on a regular torus X can be described by means of two angles θ and ϕ, as shown in Fig. 43.16. That is, we can associate *coordinates* (θ, ϕ) with P. For each of the mappings f of the torus X onto itself given below, describe the induced map f_{*n} of $H_n(X)$ into $H_n(X)$ for $n = 0$, 1, and 2, by finding the images of the generators for $H_n(X)$ described in Example 42.12. Interpret these group homomorphisms geometrically as we did in Example 43.9.

a. $f : X \to X$ given by $f((\theta, \phi)) = (2\theta, \phi)$
b. $f : X \to X$ given by $f((\theta, \phi)) = (\theta, 2\phi)$
c. $f : X \to X$ given by $f((\theta, \phi)) = (2\theta, 2\phi)$

11. With reference to Exercise 10, the torus X can be mapped onto its circle b (which is homeomorphic to S^1) by a variety of maps. For each such map $f : X \to b$ given below, describe the homomorphism $f_{*n} : H_n(X) \to H_n(b)$ for $n = 0$, 1, and 2, by describing the image of generators of $H_n(X)$ as in Exercise 10.

a. $f : X \to b$ given by $f((\theta, \phi)) = (\theta, 0)$
b. $f : X \to b$ given by $f((\theta, \phi)) = (2\theta, 0)$

12. Repeat Exercise 11, but view the map f as a map of the torus X into itself, inducing maps $f_{*n} : H_n(X) \to H_n(X)$.

13. Consider the map f of the Klein bottle in Fig. 43.2 given by mapping a point Q of the rectangle in Fig. 43.2 onto the point of b directly opposite (closest to) it. Note that b is topologically a 1-sphere. Compute the induced maps $f_{*n} : H_n(X) \to H_n(b)$ for $n = 0$, 1, and 2, by describing images of generators of $H_n(X)$.

SECTION 44 HOMOLOGICAL ALGEBRA

Chain Complexes and Mappings

The subject of algebraic topology was responsible for a surge in a new direction in algebra. You see, if you have a simplicial complex X, then you naturally get chain groups $C_k(X)$ and maps ∂_k, as indicated in the diagram

$$C_n(X) \xrightarrow{\partial_n} C_{n-1}(X) \xrightarrow{\partial_{n-1}} \cdots \xrightarrow{\partial_2} C_1(X) \xrightarrow{\partial_1} C_0(X) \xrightarrow{\partial_0} 0,$$

with $\partial_{k-1}\partial_k = 0$. You then abstract the purely algebraic portion of this situation and consider any sequence of abelian groups A_k and homomorphisms $\partial_k : A_k \to A_{k-1}$ such that $\partial_{k-1}\partial_k = 0$ for $k \geq 1$. So that you do not always have to require $k \geq 1$ in $\partial_{k-1}\partial_k = 0$, it is convenient to consider "doubly infinite" sequences of groups A_k for all $k \in \mathbb{Z}$. Often, $A_k = 0$ for $k < 0$ and $k > n$ in applications. The study of such sequences and maps of such sequences is a topic of *homological algebra.*

44.1 Definition **A chain complex** $\langle A, \partial \rangle$ is a doubly infinite sequence

$$A = \{\cdots, A_2, A_1, A_0, A_{-1}, A_{-2}, \cdots\}$$

of abelian groups A_k, together with a collection $\partial = \{\partial_k \mid k \in \mathbb{Z}\}$ of homomorphisms such that $\partial_k : A_k \to A_{k-1}$ and $\partial_{k-1}\partial_k = 0$. ■

As a convenience similar to our notation in group theory, we shall be sloppy and let "A" denote the chain complex $\langle A, \partial \rangle$. We can now imitate in a completely algebraic setting our constructions and definitions of Section 41.

44.2 Theorem If A is a chain complex, then the image under ∂_k is a subgroup of the kernel of ∂_{k-1}.

Proof Consider

$$A_k \xrightarrow{\partial_k} A_{k-1} \xrightarrow{\partial_{k-1}} A_{k-2}.$$

Now $\partial_{k-1}\partial_k = 0$, since A is a chain complex. That is, $\partial_{k-1}[\partial_k[A_k]] = 0$. This tells us at once that $\partial_k[A_k]$ is contained in the kernel of ∂_{k-1}, which is what we wished to prove. ◆

44.3 Definition If A is a chain complex, then the kernel $Z_k(A)$ of ∂_k is the **group of k-cycles,** and the image $B_k(A) = \partial_{k+1}[A_{k+1}]$ is the **group of k-boundaries.** The factor group $H_k(A) = Z_k(A)/B_k(A)$ is the **kth homology group of** A. ■

We stated in the last section that for simplicial complexes X and Y, a continuous mapping f from X into Y induces a homomorphism of $H_k(X)$ into $H_k(Y)$. This mapping of the homology groups arises in the following way. For suitable triangulations of X and Y, the mapping f gives rise to a homomorphism f_k of $C_k(X)$ into $C_k(Y)$, which has the important property that *it commutes with* ∂_k, that is,

$$\partial_k f_k = f_{k-1}\partial_k.$$

Let us turn to the purely algebraic situation and see how this induces a map of the homology groups.

44.4 Theorem **(Fundamental Lemma)** Let A and A' with collections ∂ and ∂' of homomorphisms be chain complexes, and suppose that there is a collection f of homomorphisms $f_k : A_k \to A'_k$ as indicated in the diagram

$$\begin{array}{ccccccccc}
\cdots & \xrightarrow{\partial_{k+2}} & A_{k+1} & \xrightarrow{\partial_{k+1}} & A_k & \xrightarrow{\partial_k} & A_{k-1} & \xrightarrow{\partial_{k-1}} & \cdots \\
 & & \downarrow f_{k+1} & & \downarrow f_k & & \downarrow f_{k-1} & & \\
\cdots & \xrightarrow{\partial'_{k+2}} & A'_{k+1} & \xrightarrow{\partial'_{k+1}} & A'_k & \xrightarrow{\partial'_k} & A'_{k-1} & \xrightarrow{\partial'_{k-1}} & \cdots.
\end{array}$$

Suppose, furthermore, that every square is commutative, that is,

$$f_{k-1}\partial_k = \partial'_k f_k$$

for all k. Then f_k induces a natural homomorphism $f_{*k} : H_k(A) \to H_k(A')$.

Proof Let $z \in Z_k(A)$. Now

$$\partial'_k(f_k(z)) = f_{k-1}(\partial_k(z)) = f_{k-1}(0) = 0,$$

so $f_k(z) \in Z_k(A')$. Let us attempt to define $f_{*k} : H_k(A) \to H_k(A')$ by

$$f_{*k}(z + B_k(A)) = f_k(z) + B_k(A'). \quad \textbf{(1)}$$

We must first show that f_{*k} is well defined, i.e., independent of our choice of a representative of $z + B_k(A)$. Suppose that $z_1 \in (z + B_k(A))$. Then $(z_1 - z) \in B_k(A)$, so there exists $c \in A_{k+1}$ such that $z_1 - z = \partial_{k+1}(c)$. But then

$$f_k(z_1) - f_k(z) = f_k(z_1 - z) = f_k(\partial_{k+1}(c)) = \partial'_{k+1}(f_{k+1}(c)),$$

and this last term is an element of $\partial'_{k+1}[A'_{k+1}] = B_k(A')$. Hence

$$f_k(z_1) \in (f_k(z) + B_k(A')).$$

Thus two representatives of the same coset in $H_k(A) = Z_k(A)/B_k(A)$ are mapped into representatives of just one coset in $H_k(A') = Z_k(A')/B_k(A')$. This shows that $f_{*k} : H_k(A) \to H_k(A')$ is well defined by equation (1).

Now we compute f_{*k} by taking f_k of representatives of cosets, and we define the group operation of a factor group by applying the group operation of the original group to representatives of cosets. It follows at once from the fact that the action of f_k on $Z_k(A)$ is a homomorphism of $Z_k(A)$ into $Z_k(A')$ that f_{*k} is a homomorphism of $H_k(A)$ into $H_k(A')$. ◆

If the collections of maps f, ∂, and ∂' have the property, given in Theorem 44.4, that the squares are commutative, then f **commutes with** ∂.

After another definition, we shall give a seemingly trivial but very important illustration of Theorem 44.4.

44.5 Definition A chain complex $\langle A', \partial' \rangle$ is a **subcomplex of a chain complex** $\langle A, \partial \rangle$, if, for all k, A'_k is a subgroup of A_k and $\partial'_k(c) = \partial_k(c)$ for every $c \in A'_k$, that is, ∂'_k and ∂_k have the same effect on elements of the subgroup A'_k of A_k. ■

44.6 Example Let A be a chain complex, and let A' be a subcomplex of A. Let i be the collection of injection mappings $i_k : A'_k \to A_k$ given by $i_k(c) = c$ for $c \in A'_k$. It is obvious that i commutes with ∂. Thus we have induced homomorphisms $i_{*k} : H_k(A') \to H_k(A)$. One might naturally suspect that i_{*k} must be an isomorphic mapping of $H_k(A')$ into $H_k(A)$. *This need not be so!* For example, let us consider the 2-sphere S^2 as a subcomplex of the 3-cell E^3. This gives rise to $i_2 : C_2(S^2) \to C_2(E^3)$ and induces

$$i_{*2} : H_2(S^2) \to H_2(E^3).$$

But we have seen that $H_2(S^2) \simeq \mathbb{Z}$, while $H_2(E^3) = 0$. Thus i_{*2} cannot possibly be an isomorphic mapping. ▲

Relative Homology

Suppose that A' is a subcomplex of the chain complex A. The topological situation from which this arises is the consideration of a *simplicial subcomplex* Y (in the obvious sense) of a simplicial complex X. We can then naturally consider $C_k(Y)$ a subgroup of $C_k(X)$, just as in the algebraic situation where we have A'_k a subgroup of A_k. Clearly, we would have

$$\partial_k[C_k(Y)] \leq C_{k-1}(Y).$$

Let us deal now with the algebraic situation and remember that it can be applied to our topological situation at any time.

If A' is a subcomplex of the chain complex A, we can form the collection A/A' of factor groups A_k/A'_k. We claim that A/A' again gives rise to a chain complex in a natural way, and we must exhibit a collection $\bar{\partial}$ of homomorphisms

$$\bar{\partial}_k : (A_k/A'_k) \to (A_{k-1}/A'_{k-1})$$

such that $\bar{\partial}_{k-1}\bar{\partial}_k = 0$. The definition of $\bar{\partial}_k$ to attempt is obvious, namely, define

$$\bar{\partial}_k(c + A'_k) = \partial_k(c) + A'_{k-1}$$

for $c \in A_k$. We have to show three things: that $\bar{\partial}_k$ is well defined, that it is a homomorphism, and that $\bar{\partial}_{k-1}\bar{\partial}_k = 0$.

First, to show that $\bar{\partial}_k$ is well defined, let c_1 also be in $c + A'_k$. Then $(c_1 - c) \in A'_k$, so $\partial_k(c_1 - c) \in A'_{k-1}$. Thus

$$\partial_k(c_1) \in (\partial_k(c) + A'_{k-1})$$

also. This shows that $\bar{\partial}_k$ is well defined.

The equation

$$\begin{aligned}
\bar{\partial}_k((c_1 + A'_k) + (c_2 + A'_k)) &= \bar{\partial}_k((c_1 + c_2) + A'_k) \\
&= \partial_k(c_1 + c_2) + A'_{k-1} \\
&= (\partial_k(c_1) + \partial_k(c_2)) + A'_{k-1} \\
&= \bar{\partial}_k(c_1 + A'_k) + \bar{\partial}_k(c_2 + A'_k)
\end{aligned}$$

shows that $\bar{\partial}_k$ is a homomorphism.

Finally,

$$\begin{aligned}
\bar{\partial}_{k-1}(\bar{\partial}_k(c + A'_k)) &= \bar{\partial}_{k-1}(\partial_k(c) + A'_{k-1}) \\
&= \partial_{k-1}(\partial_k(c)) + A'_{k-2} = 0 + A'_{k-2},
\end{aligned}$$

so $\bar{\partial}_{k-1}\bar{\partial}_k = 0$.

The preceding arguments are typical routine computations to the homological algebraist, just as addition and multiplication of integers are routine to you. We gave them in great detail. One has to be a little careful to keep track of *dimension,* i.e., to keep track of subscripts. Actually, the expert in homological algebra usually does not write most of these indices, but he always knows precisely with which group he is working. We gave all the indices so that you could keep track of exactly which groups were under consideration. Let us summarize the above work in a theorem.

44.7 Theorem If A' is a subcomplex of the chain complex A, then the collection A/A' of factor groups A_k/A'_k, together with the collection $\bar{\partial}$ of homomorphisms $\bar{\partial}_k$ defined by

$$\bar{\partial}_k(c + A'_k) = \partial_k(c) + A'_{k-1}$$

for $c \in A_k$, is a chain complex.

Since A/A' is a chain complex, we can then form the homology groups $H_k(A/A')$.

44.8 Definition The homology group $H_k(A/A')$ is the ***k*th relative homology group of *A* modulo *A'*.** ■

In our topological situation where Y is a subcomplex of a simplicial complex X, we shall conform to the usual notation of topologists and denote the kth relative homology group arising from the subcomplex $C(Y)$ of the chain complex $C(X)$ by "$H_k(X, Y)$." All the chains of Y are thus "set equal to 0." Geometrically, this corresponds to shrinking Y to a point.

44.9 Example Let X be the simplicial complex consisting of the edges (excluding the inside) of the triangle in Fig. 44.10, and let Y be the subcomplex consisting of the edge P_2P_3. We have seen that $H_1(X) \simeq H_1(S^1) \simeq \mathbb{Z}$. Shrinking P_2P_3 to a point collapses the rim of the triangle, as shown in Fig. 44.11. The result is still topologically the same as S^1. Thus, we would expect again to have $H_1(X, Y) \simeq \mathbb{Z}$.

Generators for $C_1(X)$ are P_1P_2, P_2P_3, and P_3P_1. Since $P_2P_3 \in C_1(Y)$, we see that generators of $C_1(X)/C_1(Y)$ are

$$P_1P_2 + C_1(Y) \qquad \text{and} \qquad P_3P_1 + C_1(Y).$$

To find $Z_1(X, Y)$ we compute

$$\begin{aligned}\bar{\partial}_1(nP_1P_2 + mP_3P_1 + C_1(Y)) &= \partial_1(nP_1P_2) + \partial_1(mP_3P_1) + C_0(Y)\\ &= n(P_2 - P_1) + m(P_1 - P_3) + C_0(Y)\\ &= (m - n)P_1 + C_0(Y),\end{aligned}$$

since $P_2, P_3 \in C_0(Y)$. Thus for a cycle, we must have $m = n$, so a generator of $Z_1(X, Y)$ is $(P_1P_2 + P_3P_1) + C_1(Y)$. Since $B_1(X, Y) = 0$, we see that indeed

$$H_1(X, Y) \simeq \mathbb{Z}.$$

Since $P_1 + C_0(Y)$ generates $Z_0(X, Y)$ and

$$\bar{\partial}_1(P_2P_1 + C_1(Y)) = (P_1 - P_2) + C_0(Y) = P_1 + C_0(Y),$$

we see that $H_0(X, Y) = 0$. This is characteristic of relative homology groups of dimension 0 for connected simplicial complexes. ▲

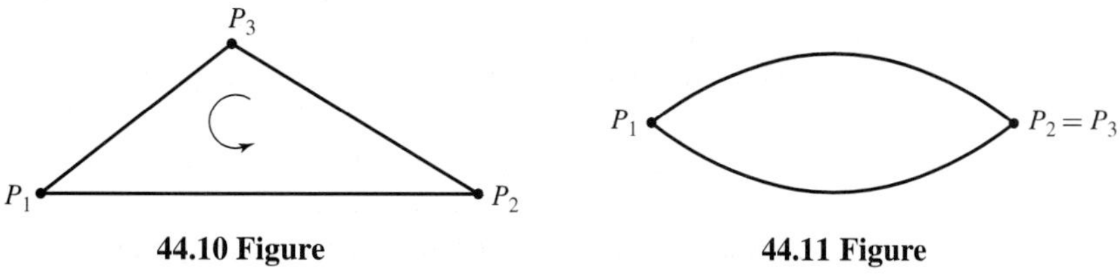

44.10 Figure **44.11 Figure**

44.12 Example Let us consider S^1 as a subcomplex (the boundary) of E^2 and compute $H_2(E^2, S^1)$. Remember that E^2 is a circular disk, so S^1 can be indeed thought of as its boundary (see Fig. 44.13). You can demonstrate the shrinking of S^1 to a point by putting a drawstring around the edge of a circular piece of cloth and then drawing the string so that the rim of the circle comes in to one point. The resulting space is then a closed bag or S^2. Thus, while $H_2(E^2) = 0$, since E^2 is a contractible space, we would expect

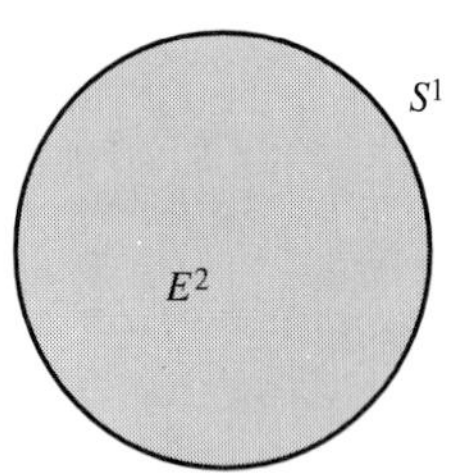

44.13 Figure

$$H_2(E^2, S^1) \simeq \mathbb{Z}.$$

For purposes of computation, we can regard E^2 topologically as the triangular region of Fig. 44.10 and S^1 as the rim of the triangle. Then $C_2(E^2, S^1)$ is generated by $P_1P_2P_3 + C_2(S^1)$, and

$$\begin{aligned}\bar{\partial}_2\big(P_1P_2P_3 + C_2(S^1)\big) &= \partial_2(P_1P_2P_3) + C_1(S^1)\\ &= (P_2P_3 - P_1P_3 + P_1P_2) + C_1(S^1).\end{aligned}$$

But $(P_2P_3 - P_1P_3 + P_1P_2) \in C_1(S^1)$, so we have

$$\bar{\partial}_2\big(P_1P_2P_3 + C_2(S^1)\big) = 0.$$

Hence $P_1P_2P_3 + C_2(S^1)$ is an element of $Z_2(E^2, S^1)$. Since

$$B_2(E^2, S^1) = 0,$$

we see that

$$H_2(E^2, S^1) \simeq \mathbb{Z},$$

as we expected. ▲

The Exact Homology Sequence of a Pair

We now describe the *exact homology sequence of a pair* and give an application. We shall not carry out all the details of the computations. The computations are routine and straightforward. We shall give all the necessary definitions, and shall let the student supply the details in the exercises.

44.14 Lemma Let A' be a subcomplex of a chain complex A. Let j be the collection of natural homomorphisms $j_k : A_k \to (A_k/A'_k)$. Then

$$j_{k-1}\partial_k = \bar{\partial}_k j_k,$$

that is, j commutes with ∂.

Proof We leave this easy computation to the exercises (see Exercise 12). ◆

44.15 Theorem The map j_k of Lemma 44.14 induces a natural homomorphism

$$j_{*k} : H_k(A) \to H_k(A/A').$$

Proof This is immediate from Lemma 44.14 and Theorem 44.4 ◆

Let A' be a subcomplex of the chain complex A. Let $h \in H_k(A/A')$. Then $h = z + B_k(A/A')$ for $z \in Z_k(A/A')$, and in turn $z = c + A'_k$ for some $c \in A_k$. (Note that we arrive at c from h by two successive choices of representatives.) Now $\bar{\partial}_k(z) = 0$, which implies that $\partial_k(c) \in A'_{k-1}$. This, together with $\partial_{k-1}\partial_k = 0$, gives us $\partial_k(c) \in Z_{k-1}(A')$. Define

$$\partial_{*k} : H_k(A/A') \to H_{k-1}(A')$$

by

$$\partial_{*k}(h) = \partial_k(c) + B_{k-1}(A').$$

This definition of ∂_{*k} looks very complicated. Think of it as follows. Start with an element of $H_k(A/A')$. Now such an element is represented by a relative k-cycle modulo A'. To say it is a relative k-cycle modulo A' is to say that its boundary is in A'_{k-1}. Since its boundary is in A'_{k-1} and is a boundary of something in A_k, this boundary must be a $(k-1)$-cycle in A'_{k-1}. Thus starting with $h \in H_k(A/A')$, we have arrived at a $(k-1)$-cycle representing a homology class in $H_{k-1}(A')$.

44.16 Lemma The map $\partial_{*k} : H_k(A/A') \to H_{k-1}(A')$, which we have just defined, is well defined, and is a homomorphism of $H_k(A/A')$ into $H_{k-1}(A')$.

Proof We leave this proof to the exercises (see Exercise 13). ◆

Let i_{*k} be the map of Example 44.6. We now can construct the following diagram.

$$\begin{aligned} \cdots \xrightarrow{\partial_{*k+1}} H_k(A') &\xrightarrow{i_{*k}} H_k(A) \xrightarrow{j_{*k}} H_k(A/A') \\ \xrightarrow{\partial_{*k}} H_{k-1}(A') &\xrightarrow{i_{*k-1}} H_{k-1}(A) \xrightarrow{j_{*k-1}} H_{k-1}(A/A') \xrightarrow{\partial_{*k-1}} \cdots \end{aligned} \quad \textbf{(1)}$$

44.17 Lemma The groups in diagram (1), together with the given maps, form a chain complex.

Proof You need only check that a sequence of two consecutive maps always gives 0. We leave this for the exercises (see Exercise 14). ◆

Since diagram (1) gives a chain complex, we could (horrors!) ask for the homology groups of this chain complex. We have been aiming at this question, the answer to which is actually quite easy. *All the homology groups of this chain complex are* 0. You may think that such a chain complex is uninteresting. Far from it. Such a chain complex even has a special name.

44.18 Definition A sequence of groups A_k and homomorphisms ∂_k forming a chain complex is an **exact sequence** if all the homology groups of the chain complex are 0, that is, if for all k we have that the image under ∂_k is equal to the kernel of ∂_{k-1}.

Exact sequences are of great importance in topology. We shall give some elementary properties of them in the exercises. ■

44.19 Theorem The groups and maps of the chain complex in diagram (1) form an exact sequence.

Proof We leave this proof to the exercises (see Exercise 15). ◆

44.20 Definition The exact sequence in diagram (1) is the **exact homology sequence of the pair** (A, A'). ■

44.21 Example Let us now give an application of Theorem 44.19 to topology. We have stated without proof that $H_n(S^n) \simeq \mathbb{Z}$ and $H_0(S^n) \simeq \mathbb{Z}$, but that $H_k(S^n) = 0$ for $k \neq 0, n$. We have also stated without proof that $H_k(E^n) = 0$ for $k \neq 0$, since E^n is contractible. Let us assume the result for E^n and now *derive* from this the result for S^n.

We can view S^n as a subcomplex of the simplicial complex E^{n+1}. For example, E^{n+1} is topologically equivalent to an $(n+1)$-simplex, and S^n is topologically equivalent to its boundary. Let us form the exact homology sequence of the pair (E^{n+1}, S^n). We have

$$\underbrace{H_{n+1}(S^n)}_{=0} \xrightarrow{i_{*n+1}} \underbrace{H_{n+1}(E^{n+1})}_{=0} \xrightarrow{j_{*n+1}} \underbrace{H_{n+1}(E^{n+1}, S^n)}_{\simeq \mathbb{Z}} \xrightarrow{\partial_{*n+1}}$$

$$\underbrace{H_n(S^n)}_{=?} \xrightarrow{i_{*n}} \underbrace{H_n(E^{n+1})}_{=0} \xrightarrow{j_{*n}} \underbrace{H_n(E^{n+1}, S^n)}_{=0} \xrightarrow{\partial_{*n}} \cdots \xrightarrow{j_{*k+1}}$$

$$\underbrace{H_{k+1}(E^{n+1}, S^n)}_{=0} \xrightarrow{\partial_{*k+1}} \underbrace{H_k(S^n)}_{=?} \xrightarrow{i_{*k}} \underbrace{H_k(E^{n+1})}_{=0} \xrightarrow{j_{*k}} \cdots \qquad \textbf{(2)}$$

for $1 \leq k < n$. The fact that E^{n+1} is contractible gives $H_k(E^{n+1}) = 0$ for $k \geq 1$. We have indicated this on diagram (2). Viewing E^{n+1} as an $(n+1)$-simplex and S^n as its boundary, we see that $C_k(E^{n+1}) \leq C_k(S^n)$ for $k \leq n$. Therefore $H_k(E^{n+1}, S^n) = 0$ for $k \leq n$. We also indicated this on diagram (2). Just as in Example 44.12, one sees that $H_{n+1}(E^{n+1}, S^n) \simeq \mathbb{Z}$, with a generating homology class containing as representative

$$P_1 P_2 \cdots P_{n+2} + C_{n+1}(S^n).$$

For $1 \leq k < n$, the exact sequence in the last row of diagram (2) tells us that $H_k(S^n) = 0$, for from $H_k(E^{n+1}) = 0$, we see that

$$(\text{kernel } i_{*k}) = H_k(S^n).$$

But from $H_{k+1}(E^{n+1}, S^n) = 0$, we see that $(\text{image } \partial_{*k+1}) = 0$. From exactness, $(\text{kernel } i_{*k}) = (\text{image } \partial_{*k+1})$, so $H_k(S^n) = 0$ for $1 \leq k < n$.

The following chain of reasoning leads to $H_n(S^n) \simeq \mathbb{Z}$. Refer to diagram (2) above.

1. Since $H_{n+1}(E^{n+1}) = 0$, we have $(\text{image } j_{*n+1}) = 0$.
2. Hence $(\text{kernel } \partial_{*n+1}) = (\text{image } j_{*n+1}) = 0$ by exactness, that is, ∂_{*n+1} is an isomorphic mapping.
3. Therefore $(\text{image } \partial_{*n+1}) \simeq \mathbb{Z}$.
4. Since $H_n(E^{n+1}) = 0$, we have $(\text{kernel } i_{*n}) = H_n(S^n)$.
5. By exactness, $(\text{image } \partial_{*n+1}) = (\text{kernel } i_{*n})$, so $H_n(S^n) \simeq \mathbb{Z}$.

Thus we see that $H_n(S^n) \simeq \mathbb{Z}$ and $H_k(S^n) = 0$ for $1 \leq k < n$.

Since S^n is connected, $H_0(S^n) \simeq \mathbb{Z}$. This fact could also be deduced from the exact sequence

$$\underbrace{H_1(E^{n+1}, S^n)}_{=0} \xrightarrow{\partial_{*1}} H_0(S^n) \xrightarrow{i_{*0}} \underbrace{H_0(E^{n+1})}_{\simeq \mathbb{Z}} \xrightarrow{j_{*0}} \underbrace{H_0(E^{n+1}, S^n)}_{=0}.$$ ▲

EXERCISES 44

Suggested Exercises

1. Let A and B be additive groups, and suppose that the sequence

$$0 \to A \xrightarrow{f} B \to 0$$

is exact. Show that $A \simeq B$.

2. Let A, B, and C be additive groups and suppose that the sequence

$$0 \to A \xrightarrow{i} B \xrightarrow{j} C \to 0$$

is exact. Show that

a. j maps B onto C

b. i is an isomorphism of A into B

c. C is isomorphic to $B/i[A]$

3. Let A, B, C, and D be additive groups and let

$$A \xrightarrow{i} B \xrightarrow{j} C \xrightarrow{k} D$$

be an exact sequence. Show that the following three conditions are equivalent:

a. i is onto B

b. j maps all of B onto 0

c. k is a one-to-one map

4. Show that if

$$A \xrightarrow{g} B \xrightarrow{h} C \xrightarrow{i} D \xrightarrow{j} E \xrightarrow{k} F$$

is an exact sequence of additive groups, then the following are equivalent:

a. h and j both map everything onto 0

b. i is an isomorphism of C onto D

c. g is onto B and k is one to one

More Exercises

5. Theorem 44.4 and Theorem 44.7 are closely connected with Exercise 39 of Section 14. Show the connection.

6. In a computation analogous to Examples 44.9 and 44.12 of the text, find the relative homology groups $H_n(X, a)$ for the torus X with subcomplex a, as shown in Figs. 42.13 and Fig. 42.14. (Since we can regard these relative homology groups as the homology groups of the space obtained from X by shrinking a to a point, these should be the homology groups of the *pinched torus*.)

7. For the simplicial complex X and subcomplex a of Exercise 6, form the exact homology sequence of the pair (X, a) and verify by direct computation that this sequence is exact.

8. Repeat Exercise 6 with X the Klein bottle of Fig. 43.2 and Fig. 43.4. (This should give the homology groups of the *pinched Klein bottle*.)

9. For the simplicial complex X and subcomplex a of Exercise 8, form the exact homology sequence of the pair (X, a) and verify by direct computation that this sequence is exact.

10. Find the relative homology groups $H_n(X, Y)$, where X is the annular region of Fig. 42.11 and Y is the subcomplex consisting of the two boundary circles.

11. For the simplicial complex X and subcomplex Y of Exercise 10, form the exact homology sequence of the pair (X, Y) and verify by direct computation that this sequence is exact.

12. Prove Lemma 44.14

13. Prove Lemma 44.16

14. Prove Lemma 44.17

15. Prove Theorem 44.19 by means of the following steps.

a. Show (image i_{*k}) $\subseteq$ (kernel j_{*k}).
b. Show (kernel j_{*k}) $\subseteq$ (image i_{*k}).
c. Show (image j_{*k}) $\subseteq$ (kernel ∂_{*k}).
d. Show (kernel ∂_{*k}) $\subseteq$ (image j_{*k}).
e. Show (image ∂_{*k}) $\subseteq$ (kernel i_{*k-1}).
f. Show (kernel i_{*k-1}) $\subseteq$ (image ∂_{*k}).

16. Let $\langle A, \partial\rangle$ and $\langle A', \partial'\rangle$ be chain complexes, and let f and g be collections of homomorphisms $f_k : A_k \to A'_k$ and $g_k : A_k \to A'_k$ such that both f and g commute with ∂. An **algebraic homotopy** between f and g is a collection D of homomorphisms $D_k : A_k \to A'_{k+1}$ such that for all $c \in A_k$, we have

$$f_k(c) - g_k(c) = \partial'_{k+1}(D_k(c)) + D_{k-1}(\partial_k(c)).$$

(One abbreviates this condition by $f - g = \partial D + D\partial$.) Show that if there exists an algebraic homotopy between f and g, that is, if f and g are **homotopic,** then f_{*k} and g_{*k} are the same homomorphism of $H_k(A)$ into $H_k(A')$.

Factorization

SECTION 45 UNIQUE FACTORIZATION DOMAINS

The integral domain $\mathbb{Z}$ is our standard example of an integral domain in which there is unique factorization into primes (irreducibles). Section 23 showed that for a field F, $F[x]$ is also such an integral domain with unique factorization. In order to discuss analogous ideas in an arbitrary integral domain, we shall give several definitions, some of which are repetitions of earlier ones. It is nice to have them all in one place for reference.

45.1 Definition Let R be a commutative ring with unity and let $a, b \in R$. If there exists $c \in R$ such that $b = ac$, then a **divides** b (or a **is a factor of** b), denoted by $a \mid b$. We read $a \nmid b$ as "a does not divide b." ■

45.2 Definition An element u of a commutative ring with unity R is a **unit of** R if u divides 1, that is, if u has a multiplicative inverse in R. Two elements $a, b \in R$ are **associates in** R if $a = bu$, where u is a unit in R.

Exercise 27 asks us to show that this criterion for a and b to be associates is an equivalence relation on R. ■

45.3 Example The only units in $\mathbb{Z}$ are 1 and -1. Thus the only associates of 26 in $\mathbb{Z}$ are 26 and -26. ▲

45.4 Definition A nonzero element p that is not a unit of an integral domain D is an **irreducible of** D if in every factorization $p = ab$ in D has the property that either a or b is a unit. ■

Note that an associate of an irreducible p is again an irreducible, for if $p = uc$ for a unit u, then any factorization of c provides a factorization of p.

■ Historical Note

The question of unique factorization in an integral domain other than the integers was first raised in public in connection with the attempted proof by Gabriel Lamé (1795–1870) of Fermat's Last Theorem, the conjecture that $x^n + y^n = z^n$ has no nontrivial integral solutions for $n > 2$. It is not hard to show that the conjecture is true if it can be proved for all odd primes p. At a meeting of the Paris Academy on March 1, 1847, Lamé announced that he had proved the theorem and presented a sketch of the proof. Lamé's idea was first to factor $x^p + y^p$ over the complex numbers as

$$x^p + y^p = (x + y)(x + \alpha y)(x + \alpha^2 y) \cdots (x + \alpha^{p-1} y)$$

where α is a primitive pth root of unity. He next proposed to show that if the factors in this expression are relatively prime and if $x^p + y^p = z^p$, then each of the p factors must be a pth power. He could then demonstrate that this Fermat equation would be true for a triple x', y', z', each number smaller than the corresponding number in the original triple. This would lead to an infinite descending sequence of positive integers, an impossibility that would prove the theorem.

After Lamé finished his announcement, however, Joseph Liouville (1809–1882) cast serious doubts on the purported proof, noting that the conclusion that each of the relatively prime factors was a pth power because their product was a pth power depended on the result that any integer can be uniquely factored into a product of primes. It was by no means clear that "integers" of the form $x + \alpha^k y$ had this unique factorization property. Although Lamé attempted to overcome Liouville's objections, the matter was settled on May 24, when Liouville produced a letter from Ernst Kummer noting that in 1844 he had already proved that unique factorization failed in the domain $\mathbb{Z}[\alpha]$, where α is a 23rd root of unity.

It was not until 1994 that Fermat's Last Theorem was proved, and by techniques of algebraic geometry unknown to Lamé and Kummer. In the late 1950s, Yutaka Taniyama and Goro Shimura noticed a curious relationship between two seemingly disparate fields of mathematics, elliptic curves and modular forms. A few years after Taniyama's tragic death at age 31, Shimura clarified this idea and eventually formulated what became known as the Taniyama–Shimura Conjecture. In 1984, Gerhard Frey asserted and in 1986 Ken Ribet proved that the Taniyama–Shimura Conjecture would imply the truth of Fermat's Last Theorem. But it was finally Andrew Wiles of Princeton University who, after secretly working on this problem for seven years, gave a series of lectures at Cambridge University in June 1993 in which he announced a proof of enough of the Taniyama–Shimura Conjecture to derive Fermat's Last Theorem. Unfortunately, a gap in the proof was soon discovered, and Wiles went back to work. It took him more than a year, but with the assistance of his student Richard Taylor, he finally was able to fill the gap. The result was published in the *Annals of Mathematics* in May 1995, and this 350-year-old problem was now solved.

45.5 Definition An integral domain D is a **unique factorization domain** (abbreviated UFD) if the following conditions are satisfied:

1. Every element of D that is neither 0 nor a unit can be factored into a product of a finite number of irreducibles.
2. If $p_1 \cdots p_r$ and $q_1 \cdots q_s$ are two factorizations of the same element of D into irreducibles, then $r = s$ and the q_j can be renumbered so that p_i and q_i are associates.

■

45.6 Example Theorem 23.20 shows that for a field F, $F[x]$ is a UFD. Also we know that $\mathbb{Z}$ is a UFD; we have made frequent use of this fact, although we have never proved it. For example, in $\mathbb{Z}$ we have

$$24 = (2)(2)(3)(2) = (-2)(-3)(2)(2).$$

Here 2 and -2 are associates, as are 3 and -3. Thus except for order and associates, the irreducible factors in these two factorizations of 24 are the same. ▲

Recall that the *principal ideal* $\langle a \rangle$ of D consists of all multiples of the element a. After just one more definition we can describe what we wish to achieve in this section.

45.7 Definition An integral domain D is a **principal ideal domain** (abbreviated PID) if every ideal in D is a principal ideal. ■

We know that $\mathbb{Z}$ is a PID because every ideal is of the form $n\mathbb{Z}$, generated by some integer n. Theorem 27.24 shows that if F is a field, then $F[x]$ is a PID. Our purpose in this section is to prove two exceedingly important theorems:

1. Every PID is a UFD. (Theorem 45.17)

2. If D is a UFD, then $D[x]$ is a UFD. (Theorem 45.29)

The fact that $F[x]$ is a UFD, where F is a field (by Theorem 23.20), illustrates both theorems. For by Theorem 27.24, $F[x]$ is a PID. Also, since F has no nonzero elements that are not units, F satisfies our definition for a UFD. Thus Theorem 45.29 would give another proof that $F[x]$ is a UFD, except for the fact that we shall actually use Theorem 23.20 in proving Theorem 45.29. In the following section we shall study properties of a certain special class of UFDs, the *Euclidean domains*.

Let us proceed to prove the two theorems.

Every PID Is a UFD

The steps leading up to Theorem 23.20 and its proof indicate the way for our proof of Theorem 45.17. Much of the material will be repetitive. We inefficiently handled the special case of $F[x]$ separately in Theorem 23.20, since it was easy and was the only case we needed for our field theory in general.

To prove that an integral domain D is a UFD, it is necessary to show that both Conditions 1 and 2 of the definition of a UFD are satisfied. For our special case of $F[x]$ in Theorem 23.20, Condition 1 was very easy and resulted from an argument that in a factorization of a polynomial of degree > 0 into a product of two nonconstant polynomials, the degree of each factor was less than the degree of the original polynomial. Thus we couldn't keep on factoring indefinitely without running into unit factors, that is, polynomials of degree 0. For the general case of a PID, it is harder to show that this is so. We now turn to this problem. We shall need one more set-theoretic concept.

45.8 Definition If $\{A_i \mid i \in I\}$ is a collection of sets, then the **union** $\cup_{i \in I} A_i$ **of the sets** A_i is the set of all x such that $x \in A_i$ for at least one $i \in I$. ■

45.9 Lemma Let R be a commutative ring and let $N_1 \subseteq N_2 \subseteq \cdots$ be an ascending chain of ideals N_i in R. Then $N = \cup_i N_i$ is an ideal of R.

Proof Let $a, b \in N$. Then there are ideals N_i and N_j in the chain, with $a \in N_i$ and $b \in N_j$. Now either $N_i \subseteq N_j$ or $N_j \subseteq N_i$; let us assume that $N_i \subseteq N_j$, so both a and b are in N_j. This implies that $a \pm b$ and ab are in N_j, so $a \pm b$ and ab are in N. Taking $a = 0$, we see that $b \in N$ implies $-b \in N$, and $0 \in N$ since $0 \in N_i$. Thus N is a subring of D. For $a \in N$ and $d \in D$, we must have $a \in N_i$ for some N_i. Then since N_i is an ideal, $da = ad$ is in N_i. Therefore, $da \in \cup_i N_i$, that is, $da \in N$. Hence N is an ideal. ◆

45.10 Lemma **(Ascending Chain Condition for a PID)** Let D be a PID. If $N_1 \subseteq N_2 \subseteq \cdots$ is an ascending chain of ideals N_i, then there exists a positive integer r such that $N_r = N_s$ for all $s \geq r$. Equivalently, every strictly ascending chain of ideals (all inclusions proper) in a PID is of finite length. We express this by saying that the **ascending chain condition** (ACC) holds for ideals in a PID.

Proof By Lemma 45.9, we know that $N = \cup_i N_i$ is an ideal of D. Now as an ideal in D, which is a PID, $N = \langle c \rangle$ for some $c \in D$. Since $N = \cup_i N_i$, we must have $c \in N_r$, for some $r \in \mathbb{Z}^+$. For $s \geq r$, we have

$$\langle c \rangle \subseteq N_r \subseteq N_s \subseteq N = \langle c \rangle.$$

Thus $N_r = N_s$ for $s \geq r$.

The equivalence with the ACC is immediate. ◆

In what follows, it will be useful to remember that for elements a and b of a domain D,

$\langle a \rangle \subseteq \langle b \rangle$ if and only if b divides a, and

$\langle a \rangle = \langle b \rangle$ if and only if a and b are associates.

For the first property, note that $\langle a \rangle \subseteq \langle b \rangle$ if and only if $a \in \langle b \rangle$, which is true if and only if $a = bd$ for some $d \in D$, so that b divides a. Using this first property, we see that $\langle a \rangle = \langle b \rangle$ if and only if $a = bc$ and $b = ad$ for some $c, d \in D$. But then $a = adc$ and by canceling, we obtain $1 = dc$. Thus d and c are units so a and b are associates.

We can now prove Condition 1 of the definition of a UFD for an integral domain that is a PID.

45.11 Theorem Let D be a PID. Every element that is neither 0 nor a unit in D is a product of irreducibles.

Proof Let $a \in D$, where a is neither 0 nor a unit. We first show that a has at least one irreducible factor. If a is an irreducible, we are done. If a is not an irreducible, then $a = a_1 b_1$, where *neither* a_1 nor b_1 is a unit. Now

$$\langle a \rangle \subset \langle a_1 \rangle,$$

for $\langle a \rangle \subseteq \langle a_1 \rangle$ follows from $a = a_1 b_1$, and if $\langle a \rangle = \langle a_1 \rangle$, then a and a_1 would be associates and b_1 would be a unit, contrary to construction. Continuing this procedure then, starting now with a_1, we arrive at a strictly ascending chain of ideals

$$\langle a \rangle \subset \langle a_1 \rangle \subset \langle a_2 \rangle \subset \cdots.$$

By the ACC in Lemma 45.10, this chain terminates with some $\langle a_r \rangle$, and a_r must then be irreducible. Thus a has an irreducible factor a_r.

By what we have just proved, for an element a that is neither 0 nor a unit in D, either a is irreducible or $a = p_1c_1$ for p_1 an irreducible and c_1 not a unit. By an argument similar to the one just made, in the latter case we can conclude that $\langle a \rangle \subset \langle c_1 \rangle$. If c_1 is not irreducible, then $c_1 = p_2c_2$ for an irreducible p_2 with c_2 not a unit. Continuing, we get a strictly ascending chain of ideals

$$\langle a \rangle \subset \langle c_1 \rangle \subset \langle c_2 \rangle \subset \cdots.$$

This chain must terminate, by the ACC in Lemma 45.10, with some $c_r = q_r$ that is an irreducible. Then $a = p_1p_2 \cdots p_rq_r$. ◆

This completes our demonstration of Condition 1 of the definition of a UFD. Let us turn to Condition 2. Our arguments here are parallel to those leading to Theorem 23.20. The results we encounter along the way are of some interest in themselves.

45.12 Lemma **(Generalization of Theorem 27.25)** An ideal $\langle p \rangle$ in a PID is maximal if and only if p is an irreducible.

Proof Let $\langle p \rangle$ be a maximal ideal of D, a PID. Suppose that $p = ab$ in D. Then $\langle p \rangle \subseteq \langle a \rangle$. Suppose that $\langle a \rangle = \langle p \rangle$. Then a and p would be associates, so b must be a unit. If $\langle a \rangle \neq \langle p \rangle$, then we must have $\langle a \rangle = \langle 1 \rangle = D$, since $\langle p \rangle$ is maximal. But then a and 1 are associates, so a is a unit. Thus, if $p = ab$, either a or b must be a unit. Hence p is an irreducible of D.

Conversely, suppose that p is an irreducible in D. Then if $\langle p \rangle \subseteq \langle a \rangle$, we must have $p = ab$. Now if a is a unit, then $\langle a \rangle = \langle 1 \rangle = D$. If a is not a unit, then b must be a unit, so there exists $u \in D$ such that $bu = 1$. Then $pu = abu = a$, so $\langle a \rangle \subseteq \langle p \rangle$, and we have $\langle a \rangle = \langle p \rangle$. Thus $\langle p \rangle \subseteq \langle a \rangle$ implies that either $\langle a \rangle = D$ or $\langle a \rangle = \langle p \rangle$, and $\langle p \rangle \neq D$ or p would be a unit. Hence $\langle p \rangle$ is a maximal ideal. ◆

45.13 Lemma **(Generalization of Theorem 27.27)** In a PID, if an irreducible p divides ab, then either $p \mid a$ or $p \mid b$.

Proof Let D be a PID and suppose that for an irreducible p in D we have $p \mid ab$. Then $(ab) \in \langle p \rangle$. Since every maximal ideal in D is a prime ideal by Corollary 27.16, $(ab) \in \langle p \rangle$ implies that either $a \in \langle p \rangle$ or $b \in \langle p \rangle$, giving either $p \mid a$ or $p \mid b$. ◆

45.14 Corollary If p is an irreducible in a PID and p divides the product $a_1a_2 \cdots a_n$ for $a_i \in D$, then $p \mid a_i$ for at least one i.

Proof Proof of this corollary is immediate from Lemma 45.13 if we use mathematical induction. ◆

45.15 Definition A nonzero nonunit element p of an integral domain D is a **prime** if, for all $a, b \in D$, $p \mid ab$ implies either $p \mid a$ or $p \mid b$. ■

Lemma 45.13 focused our attention on the defining property of a prime. In Exercises 25 and 26, we ask you to show that a prime in an integral domain is always an irreducible and that in a UFD an irreducible is also a prime. Thus the concepts of prime and irreducible coincide in a UFD. Example 45.16 will exhibit an integral domain containing some irreducibles that are not primes, so the concepts do not coincide in every domain.

45.16 Example Let F be a field and let D be the subdomain $F[x^3, xy, y^3]$ of $F[x, y]$. Then x^3, xy, and y^3 are irreducibles in D, but

$$(x^3)(y^3) = (xy)(xy)(xy).$$

Since xy divides x^3y^3 but not x^3 or y^3, we see that xy is not a prime. Similar arguments show that neither x^3 nor y^3 is a prime. ▲

The defining property of a prime is precisely what is needed to establish uniqueness of factorization, Condition 2 in the definition of a UFD. We now complete the proof of Theorem 45.17 by demonstrating the uniqueness of factorization in a PID.

45.17 Theorem **(Generalization of Theorem 23.20)** Every PID is a UFD.

Proof Theorem 45.11 shows that if D is a PID, then each $a \in D$, where a is neither 0 nor a unit, has a factorization

$$a = p_1 p_2 \cdots p_r$$

into irreducibles. It remains for us to show uniqueness. Let

$$a = q_1 q_2 \cdots q_s$$

be another such factorization into irreducibles. Then we have $p_1 \mid (q_1 q_2 \cdots q_s)$, which implies that $p_1 \mid q_j$ for some j by Corollary 45.14. By changing the order of the q_j if necessary, we can assume that $j = 1$ so $p_1 \mid q_1$. Then $q_1 = p_1 u_1$, and since p_1 is an irreducible, u_1 is a unit, so p_1 and q_1 are associates. We have then

$$p_1 p_2 \cdots p_r = p_1 u_1 q_2 \cdots q_s,$$

so by the cancellation law in D,

$$p_2 \cdots p_r = u_1 q_2 \cdots q_s.$$

Continuing this process, starting with p_2 and so on, we finally arrive at

$$1 = u_1 u_2 \cdots u_r q_{r+1} \cdots q_s.$$

Since the q_j are irreducibles, we must have $r = s$. ◆

Example 45.31 at the end of this section will show that the converse to Theorem 45.17 is false. That is, a UFD need not be a PID.

Many algebra texts start by proving the following corollary of Theorem 45.17. We have assumed that you were familiar with this corollary and used it freely in our other work.

45.18 Corollary **(Fundamental Theorem of Arithmetic)** The integral domain $\mathbb{Z}$ is a UFD.

Proof We have seen that all ideals in $\mathbb{Z}$ are of the form $n\mathbb{Z} = \langle n \rangle$ for $n \in \mathbb{Z}$. Thus $\mathbb{Z}$ is a PID, and Theorem 45.17 applies. ◆

It is worth noting that the proof that $\mathbb{Z}$ is a PID was really way back in Corollary 6.7. We proved Theorem 6.6 by using the division algorithm for $\mathbb{Z}$ exactly as we proved, in Theorem 27.24, that $F[x]$ is a PID by using the division algorithm for $F[x]$. In Section 46, we shall examine this parallel more closely.

If D Is a UFD, then $D[x]$ Is a UFD

We now start the proof of Theorem 45.29, our second main result for this section. The idea of the argument is as follows. Let D be a UFD. We can form a field of quotients F of D. Then $F[x]$ is a UFD by Theorem 23.20, and we shall show that we can recover a factorization for $f(x) \in D[x]$ from its factorization in $F[x]$. It will be necessary to compare the irreducibles in $F[x]$ with those in $D[x]$, of course. This approach, which we prefer as more intuitive than some more efficient modern ones, is essentially due to Gauss.

45.19 Definition Let D be a UFD and let $a_1, a_2, \cdots, a_n$ be nonzero elements of D. An element d of D is a **greatest common divisor** (abbreviated gcd) **of all of the** a_i if $d \mid a_i$ for $i = 1, \cdots, n$ and any other $d' \in D$ that divides all the a_i also divides d. ■

In this definition, we called d "a" gcd rather than "the" gcd because gcd's are only defined up to units. Suppose that d and d' are two gcd's of a_i for $i = 1, \cdots, n$. Then $d \mid d'$ and $d' \mid d$ by our definition. Thus $d = q'd'$ and $d' = qd$ for some $q, q' \in D$, so $1d = q'qd$. By cancellation in D, we see that $q'q = 1$ so q and q' are indeed units.

The technique in the example that follows shows that gcd's exist in a UFD.

45.20 Example Let us find a gcd of 420, -168, and 252 in the UFD $\mathbb{Z}$. Factoring, we obtain $420 = 2^2 \cdot 3 \cdot 5 \cdot 7$, $-168 = 2^3 \cdot (-3) \cdot 7$, and $252 = 2^2 \cdot 3^2 \cdot 7$. We choose one of these numbers, say 420, and find the highest power of each of its irreducible factors (up to associates) that divides all the numbers, 420, -168 and 252 in our case. We take as gcd the product of these highest powers of irreducibles. For our example, these powers of irreducible factors of 420 are $2^2, 3^1, 5^0$, and 7^1 so we take as gcd $d = 4 \cdot 3 \cdot 1 \cdot 7 = 84$. The only other gcd of these numbers in $\mathbb{Z}$ is -84, because 1 and -1 are the only units. ▲

Execution of the technique in Example 45.20 depends on being able to factor an element of a UFD into a product of irreducibles. This can be a tough job, even in $\mathbb{Z}$. Section 46 will exhibit a technique, the Euclidean Algorithm, that will allow us to find gcd's without factoring in a class of UFD's that includes $\mathbb{Z}$ and $F[x]$ for a field F.

45.21 Definition Let D be a UFD. A nonconstant polynomial

$$f(x) = a_0 + a_1x + \cdots + a_nx^n$$

in $D[x]$ is **primitive** if 1 is a gcd of the a_i for $i = 0, 1, \cdots, n$. ■

45.22 Example In $\mathbb{Z}[x]$, $4x^2 + 3x + 2$ is primitive, but $4x^2 + 6x + 2$ is not, since 2, a nonunit in $\mathbb{Z}$, is a common divisor of 4, 6, and 2. ▲

Observe that every nonconstant irreducible in $D[x]$ must be a primitive polynomial.

45.23 Lemma If D is a UFD, then for every nonconstant $f(x) \in D[x]$ we have $f(x) = (c)g(x)$, where $c \in D$, $g(x) \in D[x]$, and $g(x)$ is primitive. The element c is unique up to a unit factor in D and is the **content of** $f(x)$. Also $g(x)$ is unique up to a unit factor in D.

Proof Let $f(x) \in D[x]$ be given where $f(x)$ is a nonconstant polynomial with coefficients $a_0, a_1, \cdots, a_n$. Let c be a gcd of the a_i for $i = 0, 1, \cdots, n$. Then for each i, we have $a_i = cq_i$ for some $q_i \in D$. By the distributive law, we have $f(x) = (c)g(x)$, where no irreducible in D divides all of the coefficients $q_0, q_1, \cdots, q_n$ of $g(x)$. Thus $g(x)$ is a primitive polynomial.

For uniqueness, if also $f(x) = (d)h(x)$ for $d \in D$, $h(x) \in D[x]$, and $h(x)$ primitive, then each irreducible factor of c must divide d and conversely. By setting $(c)g(x) = (d)h(x)$ and canceling irreducible factors of c into d, we arrive at $(u)g(x) = (v)h(x)$ for a unit $u \in D$. But then v must be a unit of D or we would be able to cancel irreducible factors of v into u. Thus u and v are both units, so c is unique up to a unit factor. From $f(x) = (c)g(x)$, we see that the primitive polynomial $g(x)$ is also unique up to a unit factor. ◆

45.24 Example In $\mathbb{Z}[x]$,

$$4x^2 + 6x - 8 = (2)(2x^2 + 3x - 4),$$

where $2x^3 + 3x - 4$ is primitive. ▲

45.25 Lemma **(Gauss's Lemma)** If D is a UFD, then a product of two primitive polynomials in $D[x]$ is again primitive.

Proof Let

$$f(x) = a_0 + a_1x + \cdots + a_nx^n$$

and

$$g(x) = b_0 + b_1x + \cdots + b_mx^m$$

be primitive in $D[x]$, and let $h(x) = f(x)g(x)$. Let p be an irreducible in D. Then p does not divide all a_i and p does not divide all b_j, since $f(x)$ and $g(x)$ are primitive. Let a_r be the first coefficient of $f(x)$ not divisible by p; that is, $p \mid a_i$ for $i < r$, but $p \nmid a_r$ (that is, p does not divide a_r). Similarly, let $p \mid b_j$ for $j < s$, but $p \nmid b_s$. The coefficient of x^{r+s} in $h(x) = f(x)g(x)$ is

$$c_{r+s} = (a_0b_{r+s} + \cdots + a_{r-1}b_{s+1}) + a_rb_s + (a_{r+1}b_{s-1} + \cdots + a_{r+s}b_0).$$

Now $p \mid a_i$ for $i < r$ implies that

$$p \mid (a_0 b_{r+s} + \cdots + a_{r-1} b_{s+1}),$$

and also $p \mid b_j$ for $j < s$ implies that

$$p \mid (a_{r+1} b_{s-1} + \cdots + a_{r+s} b_0).$$

But p does not divide a_r or b_s, so p does not divide $a_r b_s$, and consequently p does not divide c_{r+s}. This shows that given an irreducible $p \in D$, there is some coefficient of $f(x)g(x)$ not divisible by p. Thus $f(x)g(x)$ is primitive. ◆

45.26 Corollary If D is a UFD, then a finite product of primitive polynomials in $D[x]$ is again primitive.

Proof This corollary follows from Lemma 45.25 by induction. ◆

Now let D be a UFD and let F be a field of quotients of D. By Theorem 23.20, $F[x]$ is a UFD. As we said earlier, we shall show that $D[x]$ is a UFD by carrying a factorization in $F[x]$ of $f(x) \in D[x]$ back into one in $D[x]$. The next lemma relates the nonconstant irreducibles of $D[x]$ to those of $F[x]$. This is the last important step.

45.27 Lemma Let D be a UFD and let F be a field of quotients of D. Let $f(x) \in D[x]$, where (degree $f(x)) > 0$. If $f(x)$ is an irreducible in $D[x]$, then $f(x)$ is also an irreducible in $F[x]$. Also, if $f(x)$ is primitive in $D[x]$ and irreducible in $F[x]$, then $f(x)$ is irreducible in $D[x]$.

Proof Suppose that a nonconstant $f(x) \in D[x]$ factors into polynomials of lower degree in $F[x]$, that is,

$$f(x) = r(x)s(x)$$

for $r(x), s(x) \in F[x]$. Then since F is a field of quotients of D, each coefficient in $r(x)$ and $s(x)$ is of the form a/b for some $a, b \in D$. By clearing denominators, we can get

$$(d)f(x) = r_1(x)s_1(x)$$

for $d \in D$, and $r_1(x), s_1(x) \in D[x]$, where the degrees of $r_1(x)$ and $s_1(x)$ are the degrees of $r(x)$ and $s(x)$, respectively. By Lemma 45.23, $f(x) = (c)g(x)$, $r_1(x) = (c_1)r_2(x)$, and $s_1(x) = (c_2)s_2(x)$ for primitive polynomials $g(x)$, $r_2(x)$, and $s_2(x)$, and $c, c_1, c_2 \in D$. Then

$$(dc)g(x) = (c_1c_2)r_2(x)s_2(x),$$

and by Lemma 45.25, $r_2(x)s_2(x)$ is primitive. By the uniqueness part of Lemma 45.23, $c_1c_2 = dcu$ for some unit u in D. But then

$$(dc)g(x) = (dcu)r_2(x)s_2(x),$$

so

$$f(x) = (c)g(x) = (cu)r_2(x)s_2(x).$$

We have shown that if $f(x)$ factors nontrivially in $F[x]$, then $f(x)$ factors nontrivially into polynomials of the same degrees in $D[x]$. Thus if $f(x) \in D[x]$ is irreducible in $D[x]$, it must be irreducible in $F[x]$.

A nonconstant $f(x) \in D[x]$ that is primitive in $D[x]$ and irreducible in $F[x]$ is also irreducible in $D[x]$, since $D[x] \subseteq F[x]$. ◆

Lemma 45.27 shows that if D is a UFD, the irreducibles in $D[x]$ are precisely the irreducibles in D, together with the nonconstant primitive polynomials that are irreducible in $F[x]$, where F is a field of quotients of $D[x]$.

The preceding lemma is very important in its own right. This is indicated by the following corollary, a special case of which was our Theorem 23.11. (We admit that it does not seem very sensible to call a special case of a corollary of a lemma a theorem. The label assigned to a result depends somewhat on the context in which it appears.)

45.28 Corollary If D is a UFD and F is a field of quotients of D, then a nonconstant $f(x) \in D[x]$ factors into a product of two polynomials of lower degrees r and s in $F[x]$ if and only if it has a factorization into polynomials of the same degrees r and s in $D[x]$.

Proof It was shown in the proof of Lemma 45.27 that if $f(x)$ factors into a product of two polynomials of lower degree in $F[x]$, then it has a factorization into polynomials of the same degrees in $D[x]$ (see the next to last sentence of the first paragraph of the proof).

The converse holds since $D[x] \subseteq F[x]$. ◆

We are now prepared to prove our main theorem.

45.29 Theorem If D is a UFD, then $D[x]$ is a UFD.

Proof Let $f(x) \in D[x]$, where $f(x)$ is neither 0 nor a unit. If $f(x)$ is of degree 0, we are done, since D is a UFD. Suppose that (degree $f(x)) > 0$. Let

$$f(x) = g_1(x)g_2(x)\cdots g_r(x)$$

be a factorization of $f(x)$ in $D[x]$ having the greatest number r of factors of positive degree. (There is such a greatest number of such factors because r cannot exceed the degree of $f(x)$.) Now factor each $g_i(x)$ in the form $g_i(x) = c_i h_i(x)$ where c_i is the content of $g_i(x)$ and $h_i(x)$ is a primitive polynomial. Each of the $h_i(x)$ is irreducible, because if it could be factored, none of the factors could lie in D, hence all would have positive degree leading to a corresponding factorzation of $g_i(x)$, and then to a factorization of $f(x)$ with more than r factors of positive degree, contradicting our choice of r. Thus we now have

$$f(x) = c_1h_1(x)c_2h_2(x)\cdots c_rh_r(x)$$

where the $h_i(x)$ are irreducible in $D[x]$. If we now factor the c_i into irreducibles in D, we obtain a factorization of $f(x)$ into a product of irreducibles in $D[x]$.

The factorization of $f(x) \in D[x]$, where $f(x)$ has degree 0, is unique since D is a UFD; see the comment following Lemma 45.27. If $f(x)$ has degree greater than 0, we can view any factorization of $f(x)$ into irreducibles in $D[x]$ as a factorization in $F[x]$ into units (that is, the factors in D) and irreducible polynomials in $F[x]$ by Lemma 45.27. By Theorem 23.20, these polynomials are unique, except for possible constant factors in F. But as an irreducible in $D[x]$, each polynomial of degree >0 appearing in the factorization of $f(x)$ in $D[x]$ is primitive. By the uniqueness part of Lemma 45.23, this shows that these polynomials are unique in $D[x]$ up to unit factors, that is, associates. The product of the irreducibles in D in the factorization of $f(x)$ is the content of $f(x)$, which is again unique up to a unit factor by Lemma 45.23. Thus all irreducibles in $D[x]$ appearing in the factorization are unique up to order and associates. ◆

45.30 Corollary If F is a field and $x_1, \cdots, x_n$ are indeterminates, then $F[x_1, \cdots, x_n]$ is a UFD.

Proof By Theorem 23.20, $F[x_1]$ is a UFD. By Theorem 45.29, so is $(F[x_1])[x_2] = F[x_1, x_2]$. Continuing in this procedure, we see (by induction) that $F[x_1, \cdots, x_n]$ is a UFD. ◆

We have seen that a PID is a UFD. Corollary 45.30 makes it easy for us to give an example that shows that *not every* UFD *is a* PID.

45.31 Example Let F be a field and let x and y be indeterminates. Then $F[x, y]$ is a UFD by Corollary 45.30. Consider the set N of all polynomials in x and y in $F[x, y]$ having constant term 0. Then N is an ideal, but not a principal ideal. Thus $F[x, y]$ is not a PID. ▲

Another example of a UFD that is not a PID is $\mathbb{Z}[x]$, as shown in Exercise 12, Section 46.

EXERCISES 45

Computations

In Exercises 1 through 8, determine whether the element is an irreducible of the indicated domain.

1. 5 in $\mathbb{Z}$

2. -17 in $\mathbb{Z}$

3. 14 in $\mathbb{Z}$

4. $2x - 3$ in $\mathbb{Z}[x]$

5. $2x - 10$ in $\mathbb{Z}[x]$

6. $2x - 3$ in $\mathbb{Q}[x]$

7. $2x - 10$ in $\mathbb{Q}[x]$

8. $2x - 10$ in $\mathbb{Z}_{11}[x]$

9. If possible, give four different associates of $2x - 7$ viewed as an element of $\mathbb{Z}[x]$; of $\mathbb{Q}[x]$; of $\mathbb{Z}_{11}[x]$.

10. Factor the polynomial $4x^2 - 4x + 8$ into a product of irreducibles viewing it as an element of the integral domain $\mathbb{Z}[x]$; of the integral domain $\mathbb{Q}[x]$; of the integral domain $\mathbb{Z}_{11}[x]$.

In Exercises 11 through 13, find all gcd's of the given elements of $\mathbb{Z}$.

11. 234, 3250, 1690

12. 784, -1960, 448

13. 2178, 396, 792, 594

In Exercises 14 through 17, express the given polynomial as the product of its content with a primitive polynomial in the indicated UFD.

14. $18x^2 - 12x + 48$ in $\mathbb{Z}[x]$

15. $18x^2 - 12x + 48$ in $\mathbb{Q}[x]$

16. $2x^2 - 3x + 6$ in $\mathbb{Z}[x]$

17. $2x^2 - 3x + 6$ in $\mathbb{Z}_7[x]$

Concepts

In Exercises 18 through 20, correct the definition of the italicized term without reference to the text, if correction is needed, so that it is in a form acceptable for publication.

18. Two elements a and b in an integral domain D are *associates* in D if and only if their quotient a/b in D is a unit.

19. An element of an integral domain D is an *irreducible* of D if and only if it cannot be factored into a product of two elements of D.

20. An element of an integral domain D is a *prime* of D if and only if it cannot be factored into a product of two smaller elements of D.

21. Mark each of the following true or false.

_______ **a.** Every field is a UFD.
_______ **b.** Every field is a PID.
_______ **c.** Every PID is a UFD.
_______ **d.** Every UFD is a PID.
_______ **e.** $\mathbb{Z}[x]$ is a UFD.
_______ **f.** Any two irreducibles in any UFD are associates.
_______ **g.** If D is a PID, then $D[x]$ is a PID.
_______ **h.** If D is a UFD, then $D[x]$ is a UFD.
_______ **i.** In any UFD, if $p \mid a$ for an irreducible p, then p itself appears in every factorization of a.
_______ **j.** A UFD has no divisors of 0.

22. Let D be a UFD. Describe the irreducibles in $D[x]$ in terms of the irreducibles in D and the irreducibles in $F[x]$, where F is a field of quotients of D.

23. Lemma 45.26 states that if D is a UFD with a field of quotients F, then a nonconstant irreducible $f(x)$ of $D[x]$ is also an irreducible of $F[x]$. Show by an example that a $g(x) \in D[x]$ that is an irreducible of $F[x]$ need not be an irreducible of $D[x]$.

24. All our work in this section was restricted to integral domains. Taking the same definition in this section but for a commutative ring with unity, consider factorizations into irreducibles in $\mathbb{Z} \times \mathbb{Z}$. What can happen? Consider in particular $(1, 0)$.

Theory

25. Prove that if p is a prime in an integral domain D, then p is an irreducible.

26. Prove that if p is an irreducible in a UFD, then p is a prime.

27. For a commutative ring R with unity show that the relation $a \sim b$ if a is an associate of b (that is, if $a = bu$ for u a unit in R is an equivalence relation on R.

28. Let D be an integral domain. Exercise 37, Section 18 showed that $\langle U, \cdot \rangle$ is a group where U is the set of units of D. Show that the set $D^* - U$ of nonunits of D excluding 0 is closed under multiplication. Is this set a group under the multiplication of D?

29. Let D be a UFD. Show that a nonconstant divisor of a primitive polynomial in $D[x]$ is again a primitive polynomial.

30. Show that in a PID, every proper ideal is contained in a maximal ideal. [*Hint:* Use Lemma 45.10.]

31. Factor $x^3 - y^3$ into irreducibles in $\mathbb{Q}[x, y]$ and prove that each of the factors is irreducible.

There are several other concepts often considered that are similar in character to the ascending chain condition on ideals in a ring. The following three exercises concern some of these concepts.

32. Let R be any ring. The **ascending chain condition** (ACC) **for ideals** holds in R if every strictly increasing sequence $N_1 \subset N_2 \subset N_3 \subset \cdots$ of ideals in R is of finite length. The **maximum condition** (MC) **for ideals** holds in R if every nonempty set S of ideals in R contains an ideal not properly contained in any other ideal of the set S. The **finite basis condition** (FBC) **for ideals** holds in R if for each ideal N in R, there is a finite set $B_N = \{b_1, \cdots, b_n\} \subseteq N$ such that N is the intersection of all ideals of R containing B_N. The set B_N is a **finite generating set for** N.

Show that for every ring R, the conditions ACC, MC, and FBC are equivalent.

33. Let R be any ring. The **descending chain condition** (DCC) **for ideals** holds in R if every strictly decreasing sequence $N_1 \supset N_2 \supset N_3 \supset \cdots$ of ideals in R is of finite length. The **minimum condition** (mC) **for ideals** holds in R if given any set S of ideals of R, there is an ideal of S that does not properly contain any other ideal in the set S.

Show that for every ring, the conditions DCC and mC are equivalent.

34. Give an example of a ring in which ACC holds but DCC does not hold. (See Exercises 32 and 33.)

SECTION 46 EUCLIDEAN DOMAINS

We have remarked several times on the importance of division algorithms. Our first contact with them was the *division algorithm for* $\mathbb{Z}$ in Section 6. This algorithm was immediately used to prove the important theorem that a subgroup of a cyclic group is cyclic, that is, has a single generator. Of course, this shows at once that $\mathbb{Z}$ is a PID. The *division algorithm for* $F[x]$ appeared in Theorem 23.1 and was used in a completely analogous way to show that $F[x]$ is a PID. Now a modern technique of mathematics is to take some clearly related situations and to try to bring them under one roof by abstracting the important ideas common to them. The following definition is an illustration of this technique, as is this whole text! Let us see what we can develop by starting with the existence of a fairly general division algorithm in an integral domain.

46.1 Definition A **Euclidean norm** on an integral domain D is a function ν mapping the nonzero elements of D into the nonnegative integers such that the following conditions are satisfied:

1. For all $a, b \in D$ with $b \neq 0$, there exist q and r in D such that $a = bq + r$, where either $r = 0$ or $\nu(r) < \nu(b)$.
2. For all $a, b \in D$, where neither a nor b is 0, $\nu(a) \leq \nu(ab)$.

An integral domain D is a **Euclidean domain** if there exists a Euclidean norm on D. ■

The importance of Condition 1 is clear from our discussion. The importance of Condition 2 is that it will enable us to characterize the units of a Euclidean domain D.

46.2 Example The integral domain $\mathbb{Z}$ is a Euclidean domain, for the function ν defined by $\nu(n) = |n|$ for $n \neq 0$ in $\mathbb{Z}$ is a Euclidean norm on $\mathbb{Z}$. Condition 1 holds by the division algorithm for $\mathbb{Z}$. Condition 2 follows from $|ab| = |a||b|$ and $|a| \geq 1$ for $a \neq 0$ in $\mathbb{Z}$. ▲

46.3 Example If F is a field, then $F[x]$ is a Euclidean domain, for the function ν defined by $\nu(f(x)) =$ (degree $f(x)$) for $f(x) \in F[x]$, and $f(x) \neq 0$ is a Euclidean norm. Condition 1 holds by Theorem 23.1, and Condition 2 holds since the degree of the product of two polynomials is the sum of their degrees. ▲

Of course, we should give some examples of Euclidean domains other than these familiar ones that motivated the definition. We shall do this in Section 47. In view of the opening remarks, we anticipate the following theorem.

46.4 Theorem Every Euclidean domain is a PID.

Proof Let D be a Euclidean domain with a Euclidean norm ν, and let N be an ideal in D. If $N = \{0\}$, then $N = \langle 0 \rangle$ and N is principal. Suppose that $N \neq \{0\}$. Then there exists $b \neq 0$ in N. Let us choose b such that $\nu(b)$ is minimal among all $\nu(n)$ for $n \in N$. We claim that $N = \langle b \rangle$. Let $a \in N$. Then by Condition 1 for a Euclidean domain, there exist q and r in D such that

$$a = bq + r,$$

where either $r = 0$ or $\nu(r) < \nu(b)$. Now $r = a - bq$ and $a, b \in N$, so that $r \in N$ since N is an ideal. Thus $\nu(r) < \nu(b)$ is impossible by our choice of b. Hence $r = 0$, so $a = bq$. Since a was any element of N, we see that $N = \langle b \rangle$. ◆

46.5 Corollary A Euclidean domain is a UFD.

Proof By Theorem 46.4, a Euclidean domain is a PID and by Theorem 45.17, a PID is a UFD. ◆

Finally, we should mention that while a Euclidean domain is a PID by Theorem 46.4, not every PID is a Euclidean domain. Examples of PIDs that are not Euclidean are not easily found, however.

Arithmetic in Euclidean Domains

We shall now investigate some properties of Euclidean domains related to their multiplicative structure. We emphasize that the arithmetic structure of a Euclidean domain is not affected in any way by a Euclidean norm ν on the domain. A Euclidean norm is merely a useful tool for possibly throwing some light on this arithmetic structure of the domain. The arithmetic structure of a domain D is completely determined by the set D and the two binary operations $+$ and $\cdot$ on D.

Let D be a Euclidean domain with a Euclidean norm ν. We can use Condition 2 of a Euclidean norm to characterize the units of D.

46.6 Theorem For a Euclidean domain with a Euclidean norm ν, $\nu(1)$ is minimal among all $\nu(a)$ for nonzero $a \in D$, and $u \in D$ is a unit if and only if $\nu(u) = \nu(1)$.

Proof Condition 2 for ν tells us at once that for $a \neq 0$,

$$\nu(1) \leq \nu(1a) = \nu(a).$$

On the other hand, if u is a unit in D, then

$$\nu(u) \leq \nu(uu^{-1}) = \nu(1).$$

Thus

$$\nu(u) = \nu(1)$$

for a unit u in D.

Conversely, suppose that a nonzero $u \in D$ is such that $\nu(u) = \nu(1)$. Then by the division algorithm, there exist q and r in D such that

$$1 = uq + r,$$

where either $r = 0$ or $\nu(r) < \nu(u)$. But since $\nu(u) = \nu(1)$ is minimal over all $\nu(d)$ for nonzero $d \in D$, $\nu(r) < \nu(u)$ is impossible. Hence $r = 0$ and $1 = uq$, so u is a unit. ◆

46.7 Example For $\mathbb{Z}$ with $\nu(n) = |n|$, the minimum of $\nu(n)$ for nonzero $n \in \mathbb{Z}$ is 1, and 1 and -1 are the only elements of $\mathbb{Z}$ with $\nu(n) = 1$. Of course, 1 and -1 are exactly the units of $\mathbb{Z}$. ▲

46.8 Example For $F[x]$ with $\nu(f(x)) = (\text{degree } f(x))$ for $f(x) \neq 0$, the minimum value of $\nu(f(x))$ for all nonzero $f(x) \in F[x]$ is 0. The nonzero polynomials of degree 0 are exactly the nonzero elements of F, and these are precisely the units of $F[x]$. ▲

We emphasize that everything we prove here holds in *every* Euclidean domain, in particular in $\mathbb{Z}$ and $F[x]$. As indicated in Example 45.20, we can show that any a and b in a UFD have a gcd and actually compute one by factoring a and b into irreducibles, but such factorizations can be very tough to find. However, if a UFD is actually Euclidean, and we know an easily computed Euclidean norm, there is an easy constructive way to find gcd's, as the next theorem shows.

■ Historical Note

The Euclidean algorithm appears in Euclid's *Elements* as propositions 1 and 2 of Book VII, where it is used as here to find the greatest common divisor of two integers. Euclid uses it again in Book X (propositions 2 and 3) to find the greatest common measure of two magnitudes (if it exists) and to determine whether two magnitudes are incommensurable.

The algorithm appears again in the *Brahmesphutasiddhanta* (Correct Astronomical System of Brahma) (628) of the seventh-century Indian mathematician and astronomer Brahmagupta. To solve the indeterminate equation $rx + c = sy$ in integers, Brahmagupta uses Euclid's procedure to "reciprocally divide" r by s until he reaches the final nonzero remainder. By then using, in effect, a substitution procedure based on the various quotients and remainders, he produces a straightforward algorithm for finding the smallest positive solution to his equation.

The thirteenth-century Chinese algebraist Qin Jiushao also used the Euclidean algorithm in his solution of the so-called Chinese Remainder problem published in the *Shushu jiuzhang* (Mathematical Treatise in Nine Sections) (1247). Qin's goal was to display a method for solving the system of congruences $N \equiv r_i \pmod{m_i}$. As part of that method he needed to solve congruences of the form $Nx \equiv 1 \pmod{m}$, where N and m are relatively prime. The solution to a congruence of this form is again found by a substitution procedure, different from the Indian one, using the quotients and remainders from the Euclidean algorithm applied to N and m. It is not known whether the common element in the Indian and Chinese algorithms, the Euclidean algorithm itself, was discovered independently in these cultures or was learned from Greek sources.

46.9 Theorem **(Euclidean Algorithm)** Let D be a Euclidean domain with a Euclidean norm ν, and let a and b be nonzero elements of D. Let r_1 be as in Condition 1 for a Euclidean norm, that is,

$$a = bq_1 + r_1,$$

where either $r_1 = 0$ or $\nu(r_1) < \nu(b)$. If $r_1 \neq 0$, let r_2 be such that

$$b = r_1 q_2 + r_2,$$

where either $r_2 = 0$ or $\nu(r_2) < \nu(r_1)$. In general, let r_{i+1} be such that

$$r_{i-1} = r_i q_{i+1} + r_{i+1},$$

where either $r_{i+1} = 0$ or $\nu(r_{i+1}) < \nu(r_i)$. Then the sequence $r_i, r_2, \cdots$ must terminate with some $r_s = 0$. If $r_1 = 0$, then b is a gcd of a and b. If $r_1 \neq 0$ and r_s is the first $r_i = 0$, then a gcd of a and b is r_{s-1}.

Furthermore, if d is a gcd of a and b, then there exist λ and μ in D such that $d = \lambda a + \mu b$.

Proof Since $\nu(r_i) < \nu(r_{i-1})$ and $\nu(r_i)$ is a nonnegative integer, it follows that after some finite number of steps we must arrive at some $r_s = 0$.

If $r_1 = 0$, then $a = bq_1$, and b is a gcd of a and b. Suppose $r_1 \neq 0$. Then if $d \mid a$ and $d \mid b$, we have

$$d \mid (a - bq_1),$$

so $d \mid r_1$. However, if $d_1 \mid r_1$ and $d_1 \mid b$, then

$$d_1 \mid (bq_1 + r_1),$$

so $d_1 \mid a$. Thus the set of common divisors of a and b is the same set as the set of common divisors of b and r_1. By a similar argument, if $r_2 \neq 0$, the set of common divisors of b and r_1 is the same set as the set of common divisors of r_1 and r_2. Continuing this process, we see finally that the set of common divisors of a and b is the same set as the set of common divisors of r_{s-2} and r_{s-1}, where r_s is the first r_i equal to 0. Thus a gcd of r_{s-2} and r_{s-1} is also a gcd of a and b. But the equation

$$r_{s-2} = q_s r_{s-1} + r_s = q_s r_{s-1}$$

shows that a gcd of r_{s-2} and r_{s-1} is r_{s-1}.

It remains to show that we can express a gcd d of a and b as $d = \lambda a + \mu b$. In terms of the construction just given, if $d = b$, then $d = 0a + 1b$ and we are done. If $d = r_{s-1}$, then, working backward through our equations, we can express each r_i in the form $\lambda_i r_{i-1} + \mu_i r_{i-2}$ for some $\lambda_i, \mu_i \in D$. To illustrate using the first step, from the equation

$$r_{s-3} = q_{s-1} r_{s-2} + r_{s-1}$$

we obtain

$$d = r_{s-1} = r_{s-3} - q_{s-1} r_{s-2}. \tag{1}$$

We then express r_{s-2} in terms of r_{s-3} and r_{s-4} and substitute in Eq. (1) to express d in terms of r_{s-3} and r_{s-4}. Eventually, we will have

$$\begin{aligned} d &= \lambda_3 r_2 + \mu_3 r_1 = \lambda_3(b - r_1 q_2) + \mu_3 r_1 = \lambda_3 b + (\mu_3 - \lambda_3 q_2) r_1 \\ &= \lambda_3 b + (\mu_3 - \lambda_3 q_2)(a - b q_1) \end{aligned}$$

which can be expressed in the form $d = \lambda a + \mu b$. If d' is any other gcd of a and b, then $d' = ud$ for some unit u, so $d' = (\lambda u)a + (\mu u)b$. ◆

The nice thing about Theorem 46.9 is that it can be implemented on a computer. Of course, we anticipate that of anything that is labeled an "algorithm."

46.10 Example Let us illustrate the Euclidean algorithm for the Euclidean norm $|\ |$ on $\mathbb{Z}$ by computing a gcd of 22,471 and 3,266. We just apply the division algorithm over and over again, and the last nonzero remainder is a gcd. We label the numbers obtained as in Theorem 46.9 to further illustrate the statement and proof of the theorem. The computations are easily checked.

$$\begin{aligned}
& & a &= 22{,}471 \\
& & b &= 3{,}266 \\
22{,}471 &= (3{,}266)6 + 2{,}875 & r_1 &= 2{,}875 \\
3{,}266 &= (2{,}875)1 + 391 & r_2 &= 391 \\
2{,}875 &= (391)7 + 138 & r_3 &= 138 \\
391 &= (138)2 + 115 & r_4 &= 115 \\
138 &= (115)1 + 23 & r_5 &= 23 \\
115 &= (23)5 + 0 & r_6 &= 0
\end{aligned}$$

Thus $r_5 = 23$ is a gcd of 22,471 and 3,266. We found a gcd without factoring! This is important, for sometimes it is very difficult to find a factorization of an integer into primes. ▲

46.11 Example Note that the division algorithm Condition 1 in the definition of a Euclidean norm says nothing about r being "positive." In computing a gcd in $\mathbb{Z}$ by the Euclidean algorithm for $|\ |$, as in Example 46.10, it is surely to our interest to make $|r_i|$ as small as possible in each division. Thus, repeating Example 46.10, it would be more efficient to write

$$\begin{aligned}
& & a &= 22{,}471 \\
& & b &= 3{,}266 \\
22{,}471 &= (3{,}266)7 - 391 & r_1 &= -391 \\
3{,}266 &= (391)8 + 138 & r_2 &= 138 \\
391 &= (138)3 - 23 & r_3 &= -23 \\
138 &= (23)6 + 0 & r_4 &= 0
\end{aligned}$$

We can change the sign of r_i from negative to positive when we wish since the divisors of r_i and $-r_i$ are the same. ▲

EXERCISES 46

Computations

In Exercises 1 through 5, state whether the given function ν is a Euclidean norm for the given integral domain.

1. The function ν for $\mathbb{Z}$ given by $\nu(n) = n^2$ for nonzero $n \in \mathbb{Z}$

2. The function ν for $\mathbb{Z}[x]$ given by $\nu(f(x)) =$ (degree of $f(x)$) for $f(x) \in \mathbb{Z}[x]$, $f(x) \neq 0$

3. The function ν for $\mathbb{Z}[x]$ given by $\nu(f(x)) =$ (the absolute value of the coefficient of the highest degree nonzero term of $f(x)$) for nonzero $f(x) \in \mathbb{Z}[x]$

4. The function ν for $\mathbb{Q}$ given by $\nu(a) = a^2$ for nonzero $a \in \mathbb{Q}$

5. The function ν for $\mathbb{Q}$ given by $\nu(a) = 50$ for nonzero $a \in \mathbb{Q}$

6. By referring to Example 46.11, actually express the gcd 23 in the form $\lambda(22{,}471) + \mu(3{,}266)$ for $\lambda, \mu \in \mathbb{Z}$. [*Hint:* From the next to the last line of the computation in Example 46.11, $23 = (138)3 - 391$. From the line before that, $138 = 3{,}266 - (391)8$, so substituting, you get $23 = [3{,}266 - (391)8]3 - 391$, and so on. That is, work your way back up to actually find values for λ and μ.]

7. Find a gcd of 49,349 and 15,555 in $\mathbb{Z}$.

8. Following the idea of Exercise 6 and referring to Exercise 7, express the positive gcd of 49,349 and 15,555 in $\mathbb{Z}$ in the form $\lambda(49{,}349) + \mu(15{,}555)$ for $\lambda, \mu \in \mathbb{Z}$.

9. Find a gcd of

$$x^{10} - 3x^9 + 3x^8 - 11x^7 + 11x^6 - 11x^5 + 19x^4 - 13x^3 + 8x^2 - 9x + 3$$

and

$$x^6 - 3x^5 + 3x^4 - 9x^3 + 5x^2 - 5x + 2$$

in $\mathbb{Q}[x]$.

10. Describe how the Euclidean Algorithm can be used to find the gcd of n members $a_1, a_2, \cdots, a_n$ of a Euclidean domain.

11. Using your method devised in Exercise 10, find the gcd of 2178, 396, 792, and 726.

Concepts

12. Let us consider $\mathbb{Z}[x]$.

- **a.** Is $\mathbb{Z}[x]$ a UFD? Why?
- **b.** Show that $\{a + xf(x) \mid a \in 2\mathbb{Z}, f(x) \in \mathbb{Z}[x]\}$ is an ideal in $\mathbb{Z}[x]$.
- **c.** Is $\mathbb{Z}[x]$ a PID? (Consider part (b).)
- **d.** Is $\mathbb{Z}[x]$ a Euclidean domain? Why?

13. Mark each of the following true or false.

_____ **a.** Every Euclidean domain is a PID.

_____ **b.** Every PID is a Euclidean domain.

_____ **c.** Every Euclidean domain is a UFD.

_____ **d.** Every UFD is a Euclidean domain.

_____ **e.** A gcd of 2 and 3 in $\mathbb{Q}$ is $\frac{1}{2}$.

_____ **f.** The Euclidean algorithm gives a constructive method for finding a gcd of two integers.

_____ **g.** If ν is a Euclidean norm on a Euclidean domain D, then $\nu(1) \leq \nu(a)$ for all nonzero $a \in D$.

_______ **h.** If ν is a Euclidean norm on a Euclidean domain D, then $\nu(1) < \nu(a)$ for all nonzero $a \in D$, $a \neq 1$.

_______ **i.** If ν is a Euclidean norm on a Euclidean domain D, then $\nu(1) < \nu(a)$ for all nonzero nonunits $a \in D$.

_______ **j.** For any field F, $F[x]$ is a Euclidean domain.

14. Does the choice of a particular Euclidean norm ν on a Euclidean domain D influence the arithmetic structure of D in any way? Explain.

Theory

15. Let D be a Euclidean domain and let ν be a Euclidean norm on D. Show that if a and b are associates in D, then $\nu(a) = \nu(b)$.

16. Let D be a Euclidean domain and let ν be a Euclidean norm on D. Show that for nonzero $a, b \in D$, one has $\nu(a) < \nu(ab)$ if and only if b is not a unit of D. [*Hint:* Argue from Exercise 15 that $\nu(a) < \nu(ab)$ implies that b is not a unit of D. Using the Euclidean algorithm, show that $\nu(a) = \nu(ab)$ implies $\langle a \rangle = \langle ab \rangle$. Conclude that if b is not a unit, then $\nu(a) < \nu(ab)$.]

17. Prove or disprove the following statement: If ν is a Euclidean norm on Euclidean domain D, then $\{a \in D \mid \nu(a) > \nu(1)\} \cup \{0\}$ is an ideal of D.

18. Show that every field is a Euclidean domain.

19. Let ν be a Euclidean norm on a Euclidean domain D.

a. Show that if $s \in \mathbb{Z}$ such that $s + \nu(1) > 0$, then $\eta : D^* \to \mathbb{Z}$ defined by $\eta(a) = \nu(a) + s$ for nonzero $a \in D$ is a Euclidean norm on D. As usual, D^* is the set of nonzero elements of D.

b. Show that for $t \in \mathbb{Z}^+$, $\lambda : D^* \to \mathbb{Z}$ given by $\lambda(a) = t \cdot \nu(a)$ for nonzero $a \in D$ is a Euclidean norm on D.

c. Show that there exists a Euclidean norm μ on D such that $\mu(1) = 1$ and $\mu(a) > 100$ for all nonzero nonunits $a \in D$.

20. Let D be a UFD. An element c in D is a **least common multiple** (abbreviated lcm) of two elements a and b in D if $a \mid c$, $b \mid c$ and if c divides every element of D that is divisible by both a and b. Show that every two nonzero elements a and b of a Euclidean domain D have an lcm in D. [*Hint:* Show that all common multiples, in the obvious sense, of both a and b form an ideal of D.]

21. Use the last statement in Theorem 46.9 to show that two nonzero elements $r, s \in \mathbb{Z}$ generate the group $\langle \mathbb{Z}, + \rangle$ if and only if r and s, viewed as integers in the domain $\mathbb{Z}$, are **relatively prime,** that is, have a gcd of 1.

22. Using the last statement in Theorem 46.9, show that for nonzero $a, b, n \in \mathbb{Z}$, the congruence $ax \equiv b \pmod{n}$ has a solution in $\mathbb{Z}$ if a and n are relatively prime.

23. Generalize Exercise 22 by showing that for nonzero $a, b, n \in \mathbb{Z}$, the congruence $ax \equiv b \pmod{n}$ has a solution in $\mathbb{Z}$ if and only if the positive gcd of a and n in $\mathbb{Z}$ divides b. Interpret this result in the ring $\mathbb{Z}_n$.

24. Following the idea of Exercises 6 and 23, outline a constructive method for finding a solution in $\mathbb{Z}$ of the congruence $ax \equiv b \pmod{n}$ for nonzero $a, b, n \in \mathbb{Z}$, if the congruence does have a solution. Use this method to find a solution of the congruence $22x \equiv 18 \pmod{42}$.

SECTION 47 Gaussian Integers and Multiplicative Norms

Gaussian Integers

We should give an example of a Euclidean domain different from $\mathbb{Z}$ and $F[x]$.

47.1 Definition A **Gaussian integer** is a complex number $a + bi$, where $a, b \in \mathbb{Z}$. For a Gaussian integer $\alpha = a + bi$, the **norm** $N(\alpha)$ **of** α is $a^2 + b^2$. ■

We shall let $\mathbb{Z}[i]$ be the set of all Gaussian integers. The following lemma gives some basic properties of the norm function N on $\mathbb{Z}[i]$ and leads to a demonstration that the function ν defined by $\nu(\alpha) = N(\alpha)$ for nonzero $\alpha \in \mathbb{Z}[i]$ is a Euclidean norm on $\mathbb{Z}[i]$. Note that the Gaussian integers include all the **rational integers,** that is, all the elements of $\mathbb{Z}$.

Historical Note

In his *Disquisitiones Arithmeticae*, Gauss studied in detail the theory of quadratic residues, that is, the theory of solutions to the congruence $x^2 \equiv p \pmod{q}$ and proved the famous quadratic reciprocity theorem showing the relationship between the solutions of the congruences $x^2 \equiv p \pmod{q}$ and $x^2 \equiv q \pmod{p}$ where p and q are primes. In attempting to generalize his results to theories of quartic residues, however, Gauss realized that it was much more natural to consider the Gaussian integers rather than the ordinary integers.

Gauss's investigations of the Gaussian integers are contained in a long paper published in 1832 in which he proved various analogies between them and the ordinary integers. For example, after noting that there are four units (invertible elements) among the Gaussian integers, namely $1, -1, i$, and $-i$, and defining the norm as in Definition 47.1, he generalized the notion of a prime integer by defining a prime Gaussian integer to be one that cannot be expressed as the product of two other integers, neither of them units. He was then able to determine which Gaussian integers are prime: A Gaussian integer that is not real is prime if and only if its norm is a real prime, which can only be 2 or of the form $4n + 1$. The real prime $2 = (1 + i)(1 - i)$ and real primes congruent to 1 modulo 4 like $13 = (2 + 3i)(2 - 3i)$ factor as the product of two Gaussian primes. Real primes of the form $4n + 3$ like 7 and 11 are still prime in the domain of Gaussian integers. See Exercise 10.

47.2 Lemma In $\mathbb{Z}[i]$, the following properties of the norm function N hold for all $\alpha, \beta \in \mathbb{Z}[i]$:

1. $N(\alpha) \geq 0$.
2. $N(\alpha) = 0$ if and only if $\alpha = 0$.
3. $N(\alpha\beta) = N(\alpha)N(\beta)$.

Proof If we let $\alpha = a_1 + a_2 i$ and $\beta = b_1 + b_2 i$, these results are all straightforward computations. We leave the proof of these properties as an exercise (see Exercise 11). ◆

47.3 Lemma $\mathbb{Z}[i]$ is an integral domain.

Proof It is obvious that $\mathbb{Z}[i]$ is a commutative ring with unity. We show that there are no divisors of 0. Let $\alpha, \beta \in \mathbb{Z}[i]$. Using Lemma 47.2, if $\alpha\beta = 0$ then

$$N(\alpha)N(\beta) = N(\alpha\beta) = N(0) = 0.$$

Thus $\alpha\beta = 0$ implies that $N(\alpha) = 0$ or $N(\beta) = 0$. By Lemma 47.2 again, this implies that either $\alpha = 0$ or $\beta = 0$. Thus $\mathbb{Z}[i]$ has no divisors of 0, so $\mathbb{Z}[i]$ is an integral domain. ◆

Of course, since $\mathbb{Z}[i]$ is a subring of $\mathbb{C}$, where $\mathbb{C}$ is the field of complex numbers, it is really obvious that $\mathbb{Z}[i]$ has no 0 divisors. We gave the argument of Lemma 47.3 to illustrate the use of the multiplicative property 3 of the norm function N and to avoid going outside of $\mathbb{Z}[i]$ in our argument.

47.4 Theorem The function ν given by $\nu(\alpha) = N(\alpha)$ for nonzero $\alpha \in \mathbb{Z}[i]$ is a Euclidean norm on $\mathbb{Z}[i]$. Thus $\mathbb{Z}[i]$ is a Euclidean domain.

Proof Note that for $\beta = b_1 + b_2 i \neq 0$, $N(b_1 + b_2 i) = b_1{}^2 + b_2{}^2$, so $N(\beta) \geq 1$. Then for all $\alpha, \beta \neq 0$ in $\mathbb{Z}[i]$, $N(\alpha) \leq N(\alpha)N(\beta) = N(\alpha\beta)$. This proves Condition 2 for a Euclidean norm in Definition 46.1.

It remains to prove the division algorithm, Condition 1, for N. Let $\alpha, \beta \in \mathbb{Z}[i]$, with $\alpha = a_1 + a_2 i$ and $\beta = b_1 + b_2 i$, where $\beta \neq 0$. We must find σ and ρ in $\mathbb{Z}[i]$ such that $\alpha = \beta\sigma + \rho$, where either $\rho = 0$ or $N(\rho) < N(\beta) = b_1{}^2 + b_2{}^2$. Let $\alpha/\beta = r + si$ for $r, s \in \mathbb{Q}$. Let q_1 and q_2 be integers in $\mathbb{Z}$ as close as possible to the rational numbers r and s, respectively. Let $\sigma = q_1 + q_2 i$ and $\rho = \alpha - \beta\sigma$. If $\rho = 0$, we are done. Otherwise, by construction of σ, we see that $|r - q_1| \leq \frac{1}{2}$ and $|s - q_2| \leq \frac{1}{2}$. Therefore

$$\begin{aligned} N\left(\frac{\alpha}{\beta} - \sigma\right) &= N((r + si) - (q_1 + q_2 i)) \\ &= N((r - q_1) + (s - q_2)i) \leq \left(\frac{1}{2}\right)^2 + \left(\frac{1}{2}\right)^2 = \frac{1}{2}. \end{aligned}$$

Thus we obtain

$$N(\rho) = N(\alpha - \beta\sigma) = N\left(\beta\left(\frac{\alpha}{\beta} - \sigma\right)\right) = N(\beta)N\left(\frac{\alpha}{\beta} - \sigma\right) \leq N(\beta)\frac{1}{2},$$

so we do indeed have $N(\rho) < N(\beta)$ as desired. ◆

47.5 Example We can now apply all our results of Section 46 to $\mathbb{Z}[i]$. In particular, since $N(1) = 1$, the units of $\mathbb{Z}[i]$ are exactly the $\alpha = a_1 + a_2 i$ with $N(\alpha) = a_1{}^2 + a_2{}^2 = 1$. From the fact that a_1 and a_2 are integers, it follows that the only possibilities are $a_1 = \pm 1$ with $a_2 = 0$, or $a_1 = 0$ with $a_2 = \pm 1$. Thus the units of $\mathbb{Z}[i]$ are ± 1 and $\pm i$. One can also use the Euclidean Algorithm to compute a gcd of two nonzero elements. We leave such computations to the exercises. Finally, note that while 5 is an irreducible in $\mathbb{Z}$, 5 is no longer an irreducible in $\mathbb{Z}[i]$, for $5 = (1 + 2i)(1 - 2i)$, and neither $1 + 2i$ nor $1 - 2i$ is a unit. ▲

Multiplicative Norms

Let us point out again that for an integral domain D, *the arithmetic concepts of irreducibles and units are not affected in any way by a norm that may be defined on the domain.* However, as the preceding section and our work thus far in this section show, a suitably defined norm may be of help in determining the arithmetic structure of D. This is strikingly illustrated in *algebraic number theory*, where for a domain of *algebraic integers* we consider many different norms of the domain, each doing its part in helping to determine the arithmetic structure of the domain. In a domain of algebraic integers, we have essentially one norm for each irreducible (up to associates), and each such norm gives information concerning the behavior in the integral domain of the irreducible to which it corresponds. This is an example of the importance of studying properties of elements in an algebraic structure by means of mappings associated with them.

Let us study integral domains that have a multiplicative norm satisfying Properties 2 and 3 of N on $\mathbb{Z}[i]$ given in Lemma 47.2.

47.6 Definition Let D be an integral domain. A **multiplicative norm** N **on** D is a function mapping D into the integers $\mathbb{Z}$ such that the following conditions are satisfied:

1. $N(\alpha) = 0$ if and only if $\alpha = 0$.

2. $N(\alpha\beta) = N(\alpha)N(\beta)$ for all $\alpha, \beta \in D$. ■

47.7 Theorem If D is an integral domain with a multiplicative norm N, then $N(1) = 1$ and $|N(u)| = 1$ for every unit u in D. If, furthermore, every α such that $|N(\alpha)| = 1$ is a unit in D, then an element π in D, with $|N(\pi)| = p$ for a prime $p \in \mathbb{Z}$, is an irreducible of D.

Proof Let D be an integral domain with a multiplicative norm N. Then

$$N(1) = N((1)(1)) = N(1)N(1)$$

shows that $N(1) = 1$. Also, if u is a unit in D, then

$$1 = N(1) = N(uu^{-1}) = N(u)N(u^{-1}).$$

Since $N(u)$ is an integer, this implies that $|N(u)| = 1$.

Now suppose that the units of D are *exactly* the elements of norm ± 1. Let $\pi \in D$ be such that $|N(\pi)| = p$, where p is a prime in $\mathbb{Z}$. Then if $\pi = \alpha\beta$, we have

$$p = |N(\pi)| = |N(\alpha)N(\beta)|,$$

so either $|N(\alpha)| = 1$ or $|N(\beta)| = 1$. By assumption, this means that either α or β is a unit of D. Thus π is an irreducible of D. ◆

47.8 Example On $\mathbb{Z}[i]$, the function N defined by $N(a + bi) = a^2 + b^2$ gives a multiplicative norm in the sense of our definition. We saw that the function ν given by $\nu(\alpha) = N(\alpha)$ for nonzero $\alpha \in \mathbb{Z}[i]$ is a Euclidean norm on $\mathbb{Z}[i]$, so the units are precisely the elements α of $\mathbb{Z}[i]$ with $N(\alpha) = N(1) = 1$. Thus the second part of Theorem 47.7 applies in $\mathbb{Z}[i]$. We saw in Example 47.5 that 5 is not an irreducible in $\mathbb{Z}[i]$, for $5 = (1 + 2i)(1 - 2i)$. Since $N(1 + 2i) = N(1 - 2i) = 1^2 + 2^2 = 5$ and 5 is a prime in $\mathbb{Z}$, we see from Theorem 47.7 that $1 + 2i$ and $1 - 2i$ are both irreducibles in $\mathbb{Z}[i]$.

As an application of mutiplicative norms, we shall now give another example of an integral domain that is *not* a UFD. We saw one example in Example 45.16. The following is the standard illustration.

47.9 Example Let $\mathbb{Z}[\sqrt{-5}] = \{a + ib\sqrt{5} \mid a, b \in \mathbb{Z}\}$. As a subset of the complex numbers closed under addition, subtraction, and multiplication, and containing 0 and 1, $\mathbb{Z}[\sqrt{-5}]$ is an integral domain. Define N on $\mathbb{Z}[\sqrt{-5}]$ by

$$N(a + b\sqrt{-5}) = a^2 + 5b^2.$$

(Here $\sqrt{-5} = i\sqrt{5}$.) Clearly, $N(\alpha) = 0$ if and only if $\alpha = a + b\sqrt{-5} = 0$. That $N(\alpha\beta) = N(\alpha)N(\beta)$ is a straightforward computation that we leave to the exercises (see Exercise 12). Let us find all candidates for units in $\mathbb{Z}[\sqrt{-5}]$ by finding all elements α in $\mathbb{Z}[\sqrt{-5}]$ with $N(\alpha) = 1$. If $\alpha = a + b\sqrt{-5}$, and $N(\alpha) = 1$, we must have $a^2 + 5b^2 = 1$ for *integers* a and b. This is only possible if $b = 0$ and $a = \pm 1$. Hence ± 1 are the only candidates for units. Since ± 1 are units, they are then precisely the units in $\mathbb{Z}[\sqrt{-5}]$.

Now in $\mathbb{Z}[\sqrt{-5}]$, we have $21 = (3)(7)$ and also

$$21 = (1 + 2\sqrt{-5})(1 - 2\sqrt{-5}).$$

If we can show that $3, 7, 1 + 2\sqrt{-5}$, and $1 - 2\sqrt{-5}$ are all irreducibles in $\mathbb{Z}[\sqrt{-5}]$, we will then know that $\mathbb{Z}[\sqrt{-5}]$ cannot be a UFD, since neither 3 nor 7 is $\pm(1 + 2\sqrt{-5})$.

Suppose that $3 = \alpha\beta$. Then

$$9 = N(3) = N(\alpha)N(\beta)$$

shows that we must have $N(\alpha) = 1$, 3, or 9. If $N(\alpha) = 1$, then α is a unit. If $\alpha = a + b\sqrt{-5}$, then $N(\alpha) = a^2 + 5b^2$, and for no choice of integers a and b is $N(\alpha) = 3$. If $N(\alpha) = 9$, then $N(\beta) = 1$, so β is a unit. Thus from $3 = \alpha\beta$, we can conclude that either α or β is a unit. Therefore, 3 is an irreducible in $\mathbb{Z}[\sqrt{-5}]$. A similar argument shows that 7 is also an irreducible in $\mathbb{Z}[\sqrt{-5}]$.

If $1 + 2\sqrt{-5} = \gamma\delta$, we have

$$21 = N(1 + 2\sqrt{-5}) = N(\gamma)N(\delta).$$

so $N(\gamma) = 1$, 3, 7, or 21. We have seen that there is no element of $\mathbb{Z}[\sqrt{-5}]$ of norm 3 or 7. This either $N(\gamma) = 1$, and γ is a unit, or $N(\gamma) = 21$, so $N(\delta) = 1$, and δ is a unit. Therefore, $1 + 2\sqrt{-5}$ is an irreducible in $\mathbb{Z}[\sqrt{-5}]$. A parallel argument shows that $1 - 2\sqrt{-5}$ is also an irreducible in $\mathbb{Z}[\sqrt{-5}]$.

In summary, we have shown that

$$\mathbb{Z}[\sqrt{-5}] = \{a + ib\sqrt{5} \mid a, b \in \mathbb{Z}\}$$

is an integral domain but not a UFD. In particular, there are two *different* factorizations

$$21 = 3 \cdot 7 = (1 + 2\sqrt{-5})(1 - 2\sqrt{-5})$$

of 21 into irreducibles. These irreducibles cannot be primes, for the property of a prime enables us to prove uniqueness of factorization (see the proof of Theorem 45.17). ▲

We conclude with a classical application, determining which primes p in $\mathbb{Z}$ are equal to a sum of squares of two integers in $\mathbb{Z}$. For example, $2 = 1^2 + 1^2$, $5 = 1^2 + 2^2$, and $13 = 2^2 + 3^2$ are sums of squares. Since we have now answered this question for the only even prime number, 2, we can restrict ourselves to odd primes.

47.10 Theorem **(Fermat's $p = a^2 + b^2$ Theorem)** Let p be an odd prime in $\mathbb{Z}$. Then $p = a^2 + b^2$ for integers a and b in $\mathbb{Z}$ if and only if $p \equiv 1 \pmod 4$.

Proof First, suppose that $p = a^2 + b^2$. Now a and b cannot both be even or both be odd since p is an odd number. If $a = 2r$ and $b = 2s + 1$, then $a^2 + b^2 = 4r^2 + 4(s^2 + s) + 1$, so $p \equiv 1 \pmod 4$. This takes care of one direction for this "if and only if" theorem.

For the other direction, we assume that $p \equiv 1 \pmod 4$. Now the multiplicative group of nonzero elements of the finite field $\mathbb{Z}_p$ is cyclic, and has order $p - 1$. Since 4 is a divisor of $p - 1$, we see that $\mathbb{Z}_p$ contains an element n of multiplicative order 4. It follows that n^2 has multiplicative order 2, so $n^2 = -1$ in $\mathbb{Z}_p$. Thus in $\mathbb{Z}$, we have $n^2 \equiv -1$ (mod p), so p divides $n^2 + 1$ in $\mathbb{Z}$.

Viewing p and $n^2 + 1$ in $\mathbb{Z}[i]$, we see that p divides $n^2 + 1 = (n + i)(n - i)$. Suppose that p is irreducible in $\mathbb{Z}[i]$; then p would have to divide $n + i$ or $n - i$. If p divides $n + i$, then $n + i = p(a + bi)$ for some $a, b \in \mathbb{Z}$. Equating coefficients of i, we obtain $1 = pb$, which is impossible. Similarly, p divides $n - i$ would lead to an impossible equation $-1 = pb$. Thus our assumption that p is irreducible in $\mathbb{Z}[i]$ must be false.

Since p is not irreducible in $\mathbb{Z}[i]$, we have $p = (a + bi)(c + di)$ where neither $a + bi$ nor $c + di$ is a unit. Taking norms, we have $p^2 = (a^2 + b^2)(c^2 + d^2)$ where neither $a^2 + b^2 = 1$ nor $c^2 + d^2 = 1$. Consequently, we have $p = a^2 + b^2$, which completes our proof. [Since $a^2 + b^2 = (a + bi)(a - bi)$, we see that this is the factorization of p, that is, $c + di = a - bi$.] ◆

Exercise 10 asks you to determine which primes p in $\mathbb{Z}$ remain irreducible in $\mathbb{Z}[i]$.

EXERCISES 47

Computations

In Exercises 1 through 4, factor the Gaussian integer into a product of irreducibles in $\mathbb{Z}[i]$. [*Hint:* Since an irreducible factor of $\alpha \in \mathbb{Z}[i]$ must have norm >1 and dividing $N(\alpha)$, there are only a finite number of Gaussian integers $a + bi$ to consider as possible irreducible factors of a given α. Divide α by each of them in $\mathbb{C}$, and see for which ones the quotient is again in $\mathbb{Z}[i]$.]

1. 5 **2.** 7 **3.** $4 + 3i$ **4.** $6 - 7i$

5. Show that 6 does not factor uniquely (up to associates) into irreducibles in $\mathbb{Z}[\sqrt{-5}]$. Exhibit two different factorizations.

6. Consider $\alpha = 7 + 2i$ and $\beta = 3 - 4i$ in $\mathbb{Z}[i]$. Find σ and ρ in $\mathbb{Z}[i]$ such that

$$\alpha = \beta\sigma + \rho \qquad \text{with} \qquad N(\rho) < N(\beta).$$

[*Hint:* Use the construction in the proof of Theorem 47.4.

7. Use a Euclidean algorithm in $\mathbb{Z}[i]$ to find a gcd of $8 + 6i$ and $5 - 15i$ in $\mathbb{Z}[i]$. [*Hint:* Use the construction in the proof of Theorem 47.4.]

Concepts

8. Mark each of the following true or false.

_____ **a.** $\mathbb{Z}[i]$ is a PID.

_____ **b.** $\mathbb{Z}[i]$ is a Euclidean domain.

_____ **c.** Every integer in $\mathbb{Z}$ is a Gaussian integer.

_____ **d.** Every complex number is a Gaussian integer.

_____ **e.** A Euclidean algorithm holds in $\mathbb{Z}[i]$.

_____ **f.** A multiplicative norm on an integral domain is sometimes an aid in finding irreducibles of the domain.

_______ **g.** If N is a multiplicative norm on an integral domain D, then $|N(u)| = 1$ for every unit u of D.

_______ **h.** If F is a field, then the function N defined by $N(f(x)) =$ (degree of $f(x)$) is a multiplicative norm on $F[x]$.

_______ **i.** If F is a field, then the function defined by $N(f(x)) = 2^{(\text{degree of } f(x))}$ for $f(x) \neq 0$ and $N(0) = 0$ is a multiplicative norm on $F[x]$ according to our definition.

_______ **j.** $\mathbb{Z}[\sqrt{-5}]$ is an integral domain but not a UFD.

9. Let D be an integral domain with a multiplicative norm N such that $|N(\alpha)| = 1$ for $\alpha \in D$ if and only if α is a unit of D. Let π be such that $|N(\pi)|$ is minimal among all $|N(\beta)| > 1$ for $\beta \in D$. Show that π is an irreducible of D.

10. **a.** Show that 2 is equal to the product of a unit and the square of an irreducible in $\mathbb{Z}[i]$.

b. Show that an odd prime p in $\mathbb{Z}$ is irreducible in $\mathbb{Z}[i]$ if and only if $p \equiv 3 \pmod 4$. (Use Theorem 47.10.)

11. Prove Lemma 47.2.

12. Prove that N of Example 47.9 is multiplicative, that is, that $N(\alpha\beta) = N(\alpha)N(\beta)$ for $\alpha, \beta \in \mathbb{Z}[\sqrt{-5}]$.

13. Let D be an integral domain with a multiplicative norm N such that $|N(\alpha)| = 1$ for $\alpha \in D$ if and only if α is a unit of D. Show that every nonzero nonunit of D has a factorization into irreducibles in D.

14. Use a Euclidean algorithm in $\mathbb{Z}[i]$ to find a gcd of $16 + 7i$ and $10 - 5i$ in $\mathbb{Z}[i]$. [*Hint:* Use the construction in the proof of Theorem 47.4.]

15. Let $\langle\alpha\rangle$ be a nonzero principal ideal in $\mathbb{Z}[i]$.

a. Show that $\mathbb{Z}[i]/\langle\alpha\rangle$ is a finite ring. [*Hint:* Use the division algorithm.]

b. Show that if π is an irreducible of $\mathbb{Z}[i]$, then $\mathbb{Z}[i]/\langle\pi\rangle$ is a field.

c. Referring to part (b), find the order and characteristic of each of the following fields.

i. $\mathbb{Z}[i]/\langle 3\rangle$ **ii.** $\mathbb{Z}[i]/\langle 1 + i\rangle$ **iii.** $\mathbb{Z}[i]/\langle 1 + 2i\rangle$

16. Let $n \in \mathbb{Z}^+$ be square free, that is, not divisible by the square of any prime integer. Let $\mathbb{Z}[\sqrt{-n}] = \{a + ib\sqrt{n} \mid a, b \in \mathbb{Z}\}$.

a. Show that the norm N, defined by $N(\alpha) = a^2 + nb^2$ for $\alpha = a + ib\sqrt{n}$, is a multiplicative norm on $\mathbb{Z}[\sqrt{-n}]$.

b. Show that $N(\alpha) = 1$ for $\alpha \in \mathbb{Z}[\sqrt{-n}]$ if and only if α is a unit of $\mathbb{Z}[\sqrt{-n}]$.

c. Show that every nonzero $\alpha \in \mathbb{Z}[\sqrt{-n}]$ that is not a unit has a factorization into irreducibles in $\mathbb{Z}[\sqrt{-n}]$. [*Hint:* Use part (b).]

17. Repeat Exercise 16 for $\mathbb{Z}[\sqrt{n}] = \{a + b\sqrt{n} \mid a, b \in \mathbb{Z}\}$ for square free $n > 1$, with N defined by $N(\alpha) = a^2 - nb^2$ for $\alpha = a + b\sqrt{n}$ in $\mathbb{Z}[\sqrt{n}]$. For part b show $|\,N(\alpha)\,| = 1$.

18. Show by a construction analogous to that given in the proof of Theorem 47.4 that the division algorithm holds in the integral domain $\mathbb{Z}[\sqrt{-2}]$ for $\nu(\alpha) = N(\alpha)$ for nonzero α in this domain (see Exercise 16). (Thus this domain is Euclidean. See Hardy and Wright [29] for a discussion of which domains $\mathbb{Z}[\sqrt{n}]$ and $\mathbb{Z}[\sqrt{-n}]$ are Euclidean.)

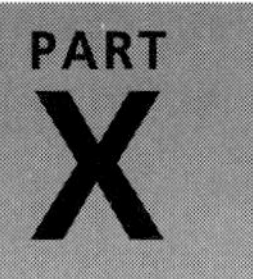

Automorphisms and Galois Theory

PART X

SECTION 48 AUTOMORPHISMS OF FIELDS

The Conjugation Isomorphisms of Algebraic Field Theory

Let F be a field, and let $\overline{F}$ be an algebraic closure of F, that is, an algebraic extension of F that is algebraically closed. Such a field $\overline{F}$ exists, by Theorem 31.17. Our selection of a particular $\overline{F}$ is not critical, since, as we shall show in Section 49, any two algebraic closures of F are isomorphic under a map leaving F fixed. *From now on in our work, we shall assume that all algebraic extensions and all elements algebraic over a field F under consideration are contained in one fixed algebraic closure $\overline{F}$ of F.*

Remember that we are engaged in the study of zeros of polynomials. In the terminology of Section 31, studying zeros of polynomials in $F[x]$ amounts to studying the structure of algebraic extensions of F and of elements algebraic over F. We shall show that if E is an algebraic extension of F with $\alpha, \beta \in E$, then α and β have the same algebraic properties if and only if $\operatorname{irr}(\alpha, F) = \operatorname{irr}(\beta, F)$. We shall phrase this fact in terms of mappings, as we have been doing all along in field theory. We achieve this by showing that if $\operatorname{irr}(\alpha, F) = \operatorname{irr}(\beta, F)$, then there exists an isomorphism $\psi_{\alpha,\beta}$ of $F(\alpha)$ onto $F(\beta)$ that maps each element of F onto itself and maps α onto β. The next theorem exhibits this isomorphism $\psi_{\alpha,\beta}$. These isomorphisms will become our fundamental tools for the study of algebraic extensions; they supplant the *evaluation homomorphisms* ϕ_α of Theorem 22.4, which make their last contribution in defining these isomorphisms. Before stating and proving this theorem, let us introduce some more terminology.

† Section 52 is not required for the remainder of the text.

48.1 Definition Let E be an algebraic extension of a field F. Two elements $\alpha, \beta \in E$ are **conjugate over** F if $\text{irr}(\alpha, F) = \text{irr}(\beta, F)$, that is, if α and β are zeros of the same irreducible polynomial over F. ■

48.2 Example The concept of conjugate elements just defined conforms with the classic idea of *conjugate complex numbers* if we understand that by conjugate complex numbers we mean numbers that are *conjugate over* $\mathbb{R}$. If $a, b \in \mathbb{R}$ and $b \neq 0$, the conjugate complex numbers $a + bi$ and $a - bi$ are both zeros of $x^2 - 2ax + a^2 + b^2$, which is irreducible in $\mathbb{R}[x]$. ▲

48.3 Theorem **(The Conjugation Isomorphisms)** Let F be a field, and let α and β be algebraic over F with $\deg(\alpha, F) = n$. The map $\psi_{\alpha,\beta} : F(\alpha) \to F(\beta)$ defined by

$$\psi_{\alpha,\beta}(c_0 + c_i\alpha + \cdots + c_{n-1}\alpha^{n-1}) = c_0 + c_1\beta + \cdots + c_{n-1}\beta^{n-1}$$

for $c_i \in F$ is an isomorphism of $F(\alpha)$ onto $F(\beta)$ if and only if α and β are conjugate over F.

Proof Suppose that $\psi_{\alpha,\beta} : F(\alpha) \to F(\beta)$ as defined in the statement of the theorem is an isomorphism. Let $\text{irr}(\alpha, F) = a_0 + a_1x + \cdots + a_nx^n$. Then $a_0 + a_1\alpha + \cdots + a_n\alpha^n = 0$, so

$$\psi_{\alpha,\beta}(a_0 + a_1\alpha + \cdots + a_n\alpha^n) = a_0 + a_1\beta + \cdots + a_n\beta^n = 0.$$

By the last assertion in the statement of Theorem 29.13 this implies that $\text{irr}(\beta, F)$ divides $\text{irr}(\alpha, F)$. A similar argument using the isomorphism $(\psi_{\alpha,\beta})^{-1} = \psi_{\beta,\alpha}$ shows that $\text{irr}(\alpha, F)$ divides $\text{irr}(\beta, F)$. Therefore, since both polynomials are monic, $\text{irr}(\alpha, F) = \text{irr}(\beta, F)$, so α and β are conjugate over F.

Conversely, suppose $\text{irr}(\alpha, F) = \text{irr}(\beta, F) = p(x)$. Then the evaluation homomorphisms $\phi_\alpha : F[x] \to F(\alpha)$ and $\phi_\beta : F[x] \to F(\beta)$ both have the same kernel $\langle p(x) \rangle$. By Theorem 26.17, corresponding to $\phi_\alpha : F[x] \to F(\alpha)$, there is a natural isomorphism ψ_α mapping $F[x]/\langle p(x) \rangle$ onto $\phi_\alpha[F[x]] = F(\alpha)$. Similarly, ϕ_β gives rise to an isomorphism ψ_β mapping $F[x]/\langle p(x) \rangle$ onto $F(\beta)$. Let $\psi_{\alpha,\beta} = \psi_\beta(\psi_\alpha)^{-1}$. These mappings are diagrammed in Fig. 48.4 where the dashed lines indicate corresponding elements under the mappings. As the composition of two isomorphisms, $\psi_{\alpha,\beta}$ is again an isomorphism and maps $F(\alpha)$ onto $F(\beta)$. For $(c_0 + c_1\alpha + \cdots + c_{n-1}\alpha^{n-1}) \in F(\alpha)$, we have

$$\begin{aligned}\psi_{\alpha,\beta}&(c_0 + c_1\alpha + \cdots + c_{n-1}\alpha^{n-1}) \\ &= (\psi_\beta{\psi_\alpha}^{-1})(c_0 + c_1\alpha + \cdots + c_{n-1}\alpha^{n-1})\end{aligned}$$

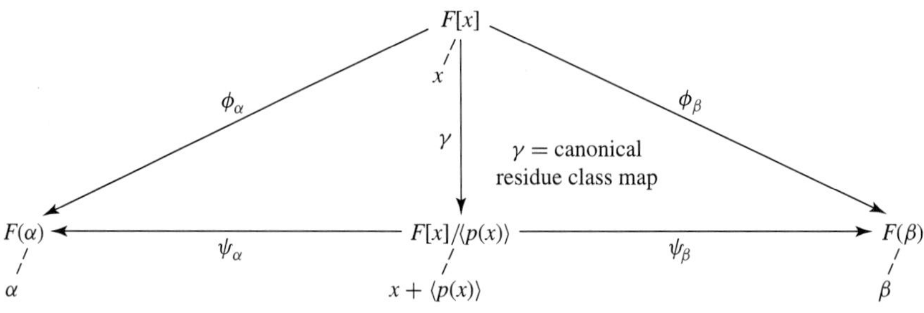

48.4 Figure

$$= \psi_\beta((c_0 + c_1x + \cdots + c_{n-1}x^{n-1}) + \langle p(x)\rangle)$$
$$= c_0 + c_1\beta + \cdots + c_{n-1}\beta^{n-1}.$$

Thus $\psi_{\alpha,\beta}$ is the map defined in the statement of the theorem. ◆

The following corollary of Theorem 48.3 is the cornerstone of our proof of the important Isomorphism Extension Theorem of Section 49 and of most of the rest of our work.

48.5 Corollary Let α be algebraic over a field F. Every isomorphism ψ mapping $F(\alpha)$ onto a subfield of $\bar{F}$ such that $\psi(a) = a$ for $a \in F$ maps α onto a conjugate β of α over F. Conversely, for each conjugate β of α over F, there exists exactly one isomorphism $\psi_{\alpha,\beta}$ of $F(\alpha)$ onto a subfield of $\bar{F}$ mapping α onto β and mapping each $a \in F$ onto itself.

Proof Let ψ be an isomorphism of $F(\alpha)$ onto a subfield of $\bar{F}$ such that $\psi(a) = a$ for $a \in F$. Let $\mathrm{irr}(\alpha, F) = a_0 + a_1x + \cdots + a_nx^n$. Then

$$a_0 + a_1\alpha + \cdots + a_n\alpha^n = 0,$$

so

$$0 = \psi(a_0 + a_1\alpha + \cdots + a_n\alpha^n) = a_0 + a_1\psi(\alpha) + \cdots + a_n\psi(\alpha)^n,$$

and $\beta = \psi(\alpha)$ is a conjugate of α.

Conversely, for each conjugate β of α over F, the conjugation isomorphism $\psi_{\alpha,\beta}$ of Theorem 48.3 is an isomorphism with the desired properties. That $\psi_{\alpha,\beta}$ is the only such isomorphism follows from the fact that an isomorphism of $F(\alpha)$ is completely determined by its values on elements of F and its value on α. ◆

As a second corollary of Theorem 48.3, we can prove a familiar result.

48.6 Corollary Let $f(x) \in \mathbb{R}[x]$. If $f(a + bi) = 0$ for $(a + bi) \in \mathbb{C}$, where $a, b \in \mathbb{R}$, then $f(a - bi) = 0$ also. Loosely, complex zeros of polynomials with real coefficients occur in conjugate pairs.

Proof We have seen that $\mathbb{C} = \mathbb{R}(i)$. Now

$$\mathrm{irr}(i, \mathbb{R}) = \mathrm{irr}(-i, \mathbb{R}) = x^2 + 1,$$

so i and $-i$ are conjugate over $\mathbb{R}$. By Theorem 48.3, the conjugation map $\psi_{i,-i} : \mathbb{C} \to \mathbb{C}$ where $\psi_{i,-i}(a + bi) = a - bi$ is an isomorphism. Thus, if for $a_i \in \mathbb{R}$,

$$f(a + bi) = a_0 + a_1(a + bi) + \cdots + a_n(a + bi)^n = 0,$$

then

$$0 = \psi_{i,-i}(f(a + bi)) = a_0 + a_1(a - bi) + \cdots + a_n(a - bi)^n$$
$$= f(a - bi),$$

that is, $f(a - bi) = 0$ also. ◆

48.7 Example Consider $\mathbb{Q}(\sqrt{2})$ over $\mathbb{Q}$. The zeros of $\text{irr}(\sqrt{2}, \mathbb{Q}) = x^2 - 2$ are $\sqrt{2}$, and $-\sqrt{2}$, so $\sqrt{2}$ and $-\sqrt{2}$ are conjugate over $\mathbb{Q}$. According to Theorem 48.3 the map $\psi_{\sqrt{2},-\sqrt{2}} : \mathbb{Q}(\sqrt{2}) \to \mathbb{Q}(\sqrt{2})$ defined by

$$\psi_{\sqrt{2},-\sqrt{2}}(a + b\sqrt{2}) = a - b\sqrt{2}$$

is an isomorphism of $\mathbb{Q}(\sqrt{2})$ onto itself. ▲

Automorphisms and Fixed Fields

As illustrated in the preceding corollary and example, a field may have a nontrivial isomorphism onto itself. *Such maps will be of utmost importance in the work that follows.*

48.8 Definition An isomorphism of a field onto itself is an **automorphism of the field.** ■

48.9 Definition If σ is an isomorphism of a field E onto some field, then an element a of E is **left fixed by** σ if $\sigma(a) = a$. A collection S of isomorphisms of E **leaves a subfield** F **of** E **fixed** if each $a \in F$ is left fixed by every $\sigma \in S$. If $\{\sigma\}$ leaves F fixed, then σ **leaves** F **fixed**. ■

48.10 Example Let $E = \mathbb{Q}(\sqrt{2}, \sqrt{3})$. The map $\sigma : E \to E$ defined by

$$\sigma(a + b\sqrt{2} + c\sqrt{3} + d\sqrt{6}) = a + b\sqrt{2} - c\sqrt{3} - d\sqrt{6}$$

for $a, b, c, d \in \mathbb{Q}$ is an automorphism of E; it is the conjugation isomorphism $\psi_{\sqrt{3},-\sqrt{3}}$ of E onto itself if we view E as $(\mathbb{Q}(\sqrt{2}))(\sqrt{3})$. We see that σ leaves $\mathbb{Q}(\sqrt{2})$ fixed. ▲

It is our purpose to study the structure of an algebraic extension E of a field F by studying the automorphisms of E that leave fixed each element of F. We shall presently show that these automorphisms form a group in a natural way. We can then apply the results concerning group structure to get information about the structure of our field extension. Thus much of our preceding work is now being brought together. The next three theorems are readily proved, but the ideas contained in them form the foundation for everything that follows. These theorems are therefore of great importance to us. They really amount to observations, rather than theorems; it is the *ideas* contained in them that are important. A big step in mathematics does not always consist of proving a *hard* theorem, but may consist of noticing how certain known mathematics may relate to new situations. Here we are bringing group theory into our study of zeros of polynomials. Be sure to understand the concepts involved. Unlikely as it may seem, they are the key to the solution of our *final goal* in this text.

Final Goal (to be more precisely stated later): To show that not all zeros of every quintic (degree 5) polynomial $f(x)$ can be expressed in terms of radicals starting with elements in the field containing the coefficients of $f(x)$.

Historical Note

It was Richard Dedekind who first developed the idea of an automorphism of a field, what he called a "permutation of the field," in 1894. The earlier application of group theory to the theory of equations had been through groups of permutations of the roots of certain polynomials. Dedekind extended this idea to mappings of the entire field and proved several of the theorems of this section.

Though Heinrich Weber continued Dedekind's approach to groups acting on fields in his algebra text of 1895, this method was not pursued in other texts near the turn of the century. It was not until the 1920s, after Emmy Noether's abstract approach to algebra became influential at Gottingen, that Emil Artin (1898–1962) developed this relationship of groups and fields in great detail. Artin emphasized that the goal of what is now called Galois theory should not be to determine solvability conditions for algebraic equations, but to explore the relationship between field extensions and groups of automorphisms. Artin detailed his approach in a lecture given in 1926; his method was first published in B. L. Van der Waerden's *Modern Algebra* text of 1930 and later by Artin himself in lecture notes in 1938 and 1942. In fact, the remainder of this text is based on Artin's development of Galois theory.

If $\{\sigma_i \mid i \in I\}$ is a collection of automorphisms of a field E, the elements of E about which $\{\sigma_i \mid i \in I\}$ gives the least information are those $a \in E$ left fixed by every σ_i for $i \in I$. This first of our three theorems contains almost all that can be said about these fixed elements of E.

48.11 Theorem Let $\{\sigma_i \mid i \in I\}$ be a collection of automorphisms of a field E. Then the set $E_{\{\sigma_i\}}$ of all $a \in E$ left fixed by every σ_i for $i \in I$ forms a subfield of E.

Proof If $\sigma_i(a) = a$ and $\sigma_i(b) = b$ for all $i \in I$, then

$$\sigma_i(a \pm b) = \sigma_i(a) \pm \sigma_i(b) = a \pm b$$

and

$$\sigma_i(ab) = \sigma_i(a)\sigma_i(b) = ab$$

for all $i \in I$. Also, if $b \neq 0$, then

$$\sigma_i(a/b) = \sigma_i(a)/\sigma_i(b) = a/b$$

for all $i \in I$. Since the σ_i are automorphisms, we have

$$\sigma_i(0) = 0 \qquad \text{and} \qquad \sigma_i(1) = 1$$

for all $i \in I$. Hence $0, 1 \in E_{\{\sigma_i\}}$ Thus $E_{\{\sigma_i\}}$ is a subfield of E. ◆

48.12 Definition The field $E_{\{\sigma_i\}}$ of Theorem 48.11 is the **fixed field of** $\{\sigma_i \mid i \in I\}$. For a single automorphism σ, we shall refer to $E_{\{\sigma\}}$ as the **fixed field of** σ. ■

48.13 Example Consider the automorphism $\psi_{\sqrt{2},-\sqrt{2}}$ of $\mathbb{Q}(\sqrt{2})$ given in Example 48.7. For $a, b \in \mathbb{Q}$, we have

$$\psi_{\sqrt{2},-\sqrt{2}}(a + b\sqrt{2}) = a - b\sqrt{2},$$

and $a - b\sqrt{2} = a + b\sqrt{2}$ if and only if $b = 0$. Thus the fixed field of $\psi_{\sqrt{2},-\sqrt{2}}$ is $\mathbb{Q}$. ▲

Note that an automorphism of a field E is in particular a one-to-one mapping of E onto E, that is, a *permutation of* E. If σ and τ are automorphisms of E, then the permutation $\sigma\tau$ is again an automorphism of E, since, in general, composition of homomorphisms again yields a homomorphism. This is how group theory makes its entrance.

48.14 Theorem The set of all automorphisms of a field E is a group under function composition.

Proof Multiplication of automorphisms of E is defined by function composition, and is thus associative (it is *permutation multiplication*). The identity permutation $\iota : E \to E$ given by $\iota(\alpha) = \alpha$ for $\alpha \in E$ is an automorphism of E. If σ is an automorphism, then the permutation σ^{-1} is also an automorphism. Thus all automorphisms of E form a subgroup of S_E, the group of all permutations of E given by Theorem 8.5. ◆

48.15 Theorem Let E be a field, and let F be a subfield of E. Then the set $G(E/F)$ of all automorphisms of E leaving F fixed forms a subgroup of the group of all automorphisms of E. Furthermore, $F \leq E_{G(E/F)}$.

Proof For $\sigma, \tau \in G(E/F)$ and $a \in F$, we have

$$(\sigma\tau)(a) = \sigma(\tau(a)) = \sigma(a) = a,$$

so $\sigma\tau \in G(E/F)$. Of course, the identity automorphism ι is in $G(E/F)$. Also, if $\sigma(a) = a$ for $a \in F$, then $a = \sigma^{-1}(a)$ so $\sigma \in G(E/F)$ implies that $\sigma^{-1} \in G(E/F)$. Thus $G(E/F)$ is a subgroup of the group of all automorphisms of E.

Since every element of F is left fixed by every element of $G(E/F)$, it follows immediately that the field $E_{G(E/F)}$ of *all* elements of E left fixed by $G(E/F)$ contains F. ◆

48.16 Definition The group $G(E/F)$ of the preceding theorem is the **group of automorphisms of E leaving F fixed**, or, more briefly, the **group of E over F**.

Do not think of E/F in the notation $G(E/F)$ as denoting a quotient space of some sort, but rather as meaning that E is an extension field of the field F.

The ideas contained in the preceding three theorems are illustrated in the following example. We urge you to study this example carefully. ■

48.17 Example Consider the field $\mathbb{Q}(\sqrt{2}, \sqrt{3})$. Example 31.9 shows that $[\mathbb{Q}(\sqrt{2}, \sqrt{3}) : \mathbb{Q}] = 4$. If we view $\mathbb{Q}(\sqrt{2}, \sqrt{3})$ as $(\mathbb{Q}(\sqrt{3}))(\sqrt{2})$, the conjugation isomorphism $\psi_{\sqrt{2},-\sqrt{2}}$ of Theorem 48.3 defined by

$$\psi_{\sqrt{2},-\sqrt{2}}(a + b\sqrt{2}) = a - b\sqrt{2}$$

for $a, b \in \mathbb{Q}(\sqrt{3})$ is an automorphism of $\mathbb{Q}(\sqrt{2}, \sqrt{3})$ having $\mathbb{Q}(\sqrt{3})$ as fixed field. Similarly, we have the automorphism $\psi_{\sqrt{3},-\sqrt{3}}$ of $\mathbb{Q}(\sqrt{2}, \sqrt{3})$ having $\mathbb{Q}(\sqrt{2})$ as fixed field. Since the product of two automorphisms is an automorphism, we can consider $\psi_{\sqrt{2},-\sqrt{2}}\psi_{\sqrt{3},-\sqrt{3}}$ which **moves** both $\sqrt{2}$ and $\sqrt{3}$, that is, leaves neither number fixed. Let

$$\begin{aligned}
\iota &= \text{the identity automorphism,} \\
\sigma_1 &= \psi_{\sqrt{2},-\sqrt{2}}, \\
\sigma_2 &= \psi_{\sqrt{3},-\sqrt{3}}, \quad \text{and} \\
\sigma_3 &= \psi_{\sqrt{2},-\sqrt{2}}\psi_{\sqrt{3},-\sqrt{3}}.
\end{aligned}$$

48.18 Table

	ι	σ_1	σ_2	σ_3
ι	ι	σ_1	σ_2	σ_3
σ_1	σ_1	ι	σ_3	σ_2
σ_2	σ_2	σ_3	ι	σ_1
σ_3	σ_3	σ_2	σ_1	ι

The group of all automorphisms of $\mathbb{Q}(\sqrt{2}, \sqrt{3})$ has a fixed field, by Theorem 48.11. This fixed field must contain $\mathbb{Q}$, since every automorphism of a field leaves 1 and hence the prime subfield fixed. A basis for $\mathbb{Q}(\sqrt{2}, \sqrt{3})$ over $\mathbb{Q}$ is $\{1, \sqrt{2}, \sqrt{3}, \sqrt{6}\}$. Since $\sigma_1(\sqrt{2}) = -\sqrt{2}$, $\sigma_1(\sqrt{6}) = -\sqrt{6}$ and $\sigma_2(\sqrt{3}) = -\sqrt{3}$, we see that $\mathbb{Q}$ is exactly the fixed field of $\{\iota, \sigma_1, \sigma_2, \sigma_3\}$. It is readily checked that $G = \{\iota, \sigma_1, \sigma_2, \sigma_3\}$ is a group under automorphism multiplication (function composition). The group table for G is given in Table 48.18. For example,

$$\sigma_1\sigma_3 = \psi_{\sqrt{2},-\sqrt{2}}(\psi_{\sqrt{2},-\sqrt{2}}\psi_{\sqrt{3},-\sqrt{3}}) = \psi_{\sqrt{3},-\sqrt{3}} = \sigma_2.$$

The group G is isomorphic to the Klein 4-group. We can show that G is the full group $\mathbb{G}(\mathbb{Q}(\sqrt{2}, \sqrt{3})/\mathbb{Q})$, because every automorphism τ of $\mathbb{Q}(\sqrt{2}, \sqrt{3})$ maps $\sqrt{2}$ onto either $\pm\sqrt{2}$, by Corollary 48.5. Similarly, τ maps $\sqrt{3}$ onto either $\pm\sqrt{3}$. But since $\{1, \sqrt{2}, \sqrt{3}, \sqrt{2}\sqrt{3}\}$ is a basis for $\mathbb{Q}(\sqrt{2}, \sqrt{3})$ over $\mathbb{Q}$, an automorphism of $\mathbb{Q}(\sqrt{2}, \sqrt{3})$ leaving $\mathbb{Q}$ fixed is determined by its values on $\sqrt{2}$ and $\sqrt{3}$. Now, ι, σ_1, σ_2, and σ_3 give all possible combinations of values on $\sqrt{2}$ and $\sqrt{3}$, and hence are all possible automorphisms of $\mathbb{Q}(\sqrt{2}, \sqrt{3})$.

Note that $G(\mathbb{Q}(\sqrt{2}, \sqrt{3})/\mathbb{Q})$ has order 4, and $[\mathbb{Q}(\sqrt{2}, \sqrt{3}) : \mathbb{Q}] = 4$. *This is no accident*, but rather an instance of a general situation, as we shall see later. ▲

The Frobenius Automorphism

Let F be a finite field. We shall show later that the group of all automorphisms of F is cyclic. Now a cyclic group has by definition a generating element, and it may have several generating elements. For an abstract cyclic group there is no way of distinguishing any one generator as being more important than any other. However, for the cyclic group of all automorphisms of a finite field there is a canonical (natural) generator, the *Frobenius automorphism* (classically, the *Frobenius substitution*). This fact is of considerable importance in some advanced work in algebra. The next theorem exhibits this Frobenius automorphism.

48.19 Theorem Let F be a finite field of characteristic p. Then the map $\sigma_p : F \to F$ defined by $\sigma_p(a) = a^p$ for $a \in F$ is an automorphism, the **Frobenius automorphism,** of F. Also, $F_{\{\sigma_p\}} \simeq \mathbb{Z}_p$.

Proof Let $a, b \in F$. Taking $n = 1$ in Lemma 33.9, we see that $(a + b)^p = a^p + b^p$. Thus we have

$$\sigma_p(a + b) = (a + b)^p = a^p + b^p = \sigma_p(a) + \sigma_p(b).$$

Of course,

$$\sigma_p(ab) = (ab)^p = a^p b^p = \sigma_p(a)\sigma_p(b),$$

so σ_p is at least a homomorphism. If $\sigma_p(a) = 0$, then $a^p = 0$, and $a = 0$, so the kernel of σ_p is $\{0\}$, and σ_p is a one-to-one map. Finally, since F is finite, σ_p is onto, by counting. Thus σ_p is an automorphism of F.

The prime field $\mathbb{Z}_p$ must be contained (up to isomorphism) in F, since F is of characteristic p. For $c \in \mathbb{Z}_p$, we have $\sigma_p(c) = c^p = c$, by Fermat's theorem (see Corollary 20.2). Thus the polynomial $x^p - x$ has p zeros in F, namely the elements of $\mathbb{Z}_p$. By Corollary 23.5, a polynomial of degree n over a field can have at most n zeros in the field. Since the elements fixed under σ_p are precisely the zeros in F of $x^p - x$, we see that

$$\mathbb{Z}_p = F_{\{\sigma_p\}}.$$ ◆

Freshmen in college still sometimes make the error of saying that $(a+b)^n = a^n + b^n$. Here we see that this *freshman exponentiation*, $(a+b)^p = a^p + b^p$ with exponent p, is actually valid in a field F of characteristic p.

EXERCISES 48

Computations

In Exercises 1 through 8, find all conjugates in $\mathbb{C}$ of the given number over the given field.

1. $\sqrt{2}$ over $\mathbb{Q}$

2. $\sqrt{2}$ over $\mathbb{R}$

3. $3 + \sqrt{2}$ over $\mathbb{Q}$

4. $\sqrt{2} - \sqrt{3}$ over $\mathbb{Q}$

5. $\sqrt{2} + i$ over $\mathbb{Q}$

6. $\sqrt{2} + i$ over $\mathbb{R}$

7. $\sqrt{1+\sqrt{2}}$ over $\mathbb{Q}$

8. $\sqrt{1+\sqrt{2}}$ over $\mathbb{Q}(\sqrt{2})$

In Exercises 9 through 14, we consider the field $E = \mathbb{Q}(\sqrt{2}, \sqrt{3}, \sqrt{5})$. It can be shown that $[E : \mathbb{Q}] = 8$. In the notation of Theorem 48.3, we have the following conjugation isomorphisms (which are here automorphisms of E):

$$\psi_{\sqrt{2},-\sqrt{2}} : (\mathbb{Q}(\sqrt{3}, \sqrt{5}))(\sqrt{2}) \to (\mathbb{Q}(\sqrt{3}, \sqrt{5}))(-\sqrt{2}),$$
$$\psi_{\sqrt{3},-\sqrt{3}} : (\mathbb{Q}(\sqrt{2}, \sqrt{5}))(\sqrt{3}) \to (\mathbb{Q}(\sqrt{2}, \sqrt{5}))(-\sqrt{3}),$$
$$\psi_{\sqrt{5},-\sqrt{5}} : (\mathbb{Q}(\sqrt{2}, \sqrt{3}))(\sqrt{5}) \to (\mathbb{Q}(\sqrt{2}, \sqrt{3}))(-\sqrt{5}).$$

For shorter notation, let $\tau_2 = \psi_{\sqrt{2},-\sqrt{2}}$, $\tau_3 = \psi_{\sqrt{3},-\sqrt{3}}$, and $\tau_5 = \psi_{\sqrt{5},-\sqrt{5}}$. Compute the indicated element of E.

9. $\tau_2(\sqrt{3})$

10. $\tau_2(\sqrt{2} + \sqrt{5})$

11. $(\tau_3\tau_2)(\sqrt{2} + 3\sqrt{5})$

12. $(\tau_5\tau_3)\left(\dfrac{\sqrt{2} - 3\sqrt{5}}{2\sqrt{3} - \sqrt{2}}\right)$

13. $({\tau_5}^2\tau_3\tau_2)(\sqrt{2} + \sqrt{45})$

14. $\tau_3[\tau_5(\sqrt{2} - \sqrt{3} + (\tau_2\tau_5)(\sqrt{30}))]$

15. Referring to Example 48.17, find the following fixed fields in $E = \mathbb{Q}(\sqrt{2}, \sqrt{3})$.

a. $E_{\{\sigma_1,\sigma_3\}}$ **b.** $E_{\{\sigma_3\}}$ **c.** $E_{\{\sigma_2,\sigma_3\}}$

In Exercises 16 through 21, refer to the directions for Exercises 9 through 14 and find the fixed field of the automorphism or set of automorphisms of E.

16. τ_3

17. τ_3^2

18. $\{\tau_2, \tau_3\}$

19. $\tau_5\tau_2$

20. $\tau_5\tau_3\tau_2$

21. $\{\tau_2, \tau_3, \tau_5\}$

22. Refer to the directions for Exercises 9 through 14 for this exercise.

a. Show that each of the automorphisms τ_2, τ_3 and τ_5 is of order 2 in $G(E/\mathbb{Q})$. (Remember what is meant by the *order* of an element of a group.)

b. Find the subgroup H of $G(E/\mathbb{Q})$ generated by the elements τ_2, τ_3, and τ_5, and give the group table. [*Hint:* There are eight elements.]

c. Just as was done in Example 48.17, argue that the group H of part (b) is the full group $G(E/\mathbb{Q})$.

Concepts

In Exercises 23 and 24, correct the definition of the italicized term without reference to the text, if correction is needed, so that it is in a form acceptable for publication.

23. Two elements, α and β, of an algebraic extension E of a field F are *conjugate over* F if and only if they are both zeros of the same polynomial $f(x)$ in $F[x]$.

24. Two elements, α and β, of an algebraic extension E of a field F are *conjugate over* F if and only if the evaluation homomorphisms $\phi_\alpha : F[x] \to E$ and $\phi_\beta : F[x] \to E$ have the same kernel.

25. The fields $\mathbb{Q}(\sqrt{2})$ and $\mathbb{Q}(3 + \sqrt{2})$ are the same, of course. Let $\alpha = 3 + \sqrt{2}$.

a. Find a conjugate $\beta \neq \alpha$ of α over $\mathbb{Q}$.

b. Referring to part (a), compare the conjugation automorphism $\psi_{\sqrt{2},-\sqrt{2}}$ of $\mathbb{Q}(\sqrt{2})$ with the conjugation automorphism $\psi_{\alpha,\beta}$.

26. Describe the value of the Frobenius automorphism σ_2 on each element of the finite field of four elements given in Example 29.19. Find the fixed field of σ_2.

27. Describe the value of the Frobenius automorphism σ_3 on each element of the finite field of nine elements given in Exercise 18 of Section 29. Find the fixed field of σ_3.

28. Let F be a field of characteristic $p \neq 0$. Give an example to show that the map $\sigma_p : F \to F$ given by $\sigma_p(a) = a^p$ for $a \in F$ need not be an automorphism in the case that F is infinite. What may go wrong?

29. Mark each of the following true or false.

______ **a.** For all $\alpha, \beta \in E$, there is always an automorphism of E mapping α onto β.

______ **b.** For α, β algebraic over a field F, there is always an isomorphism of $F(\alpha)$ onto $F(\beta)$.

______ **c.** For α, β algebraic and conjugate over a field F, there is always an isomorphism of $F(\alpha)$ onto $F(\beta)$.

______ **d.** Every automorphism of every field E leaves fixed every element of the prime subfield of E.

______ **e.** Every automorphism of every field E leaves fixed an infinite number of elements of E.

______ **f.** Every automorphism of every field E leaves fixed at least two elements of E.

______ **g.** Every automorphism of every field E of characteristic 0 leaves fixed an infinite number of elements of E.

______ **h.** All automorphisms of a field E form a group under function composition.

______ **i.** The set of all elements of a field E left fixed by a single automorphism of E forms a subfield of E.

______ **j.** For fields $F \leq E \leq K$, $G(K/E) \leq G(K/F)$.

Proof Synopsis

30. Give a one-sentence synopsis of the "if" part of Theorem 48.3.

31. Give a one-sentence synopsis of the "only if" part of Theorem 48.3.

Theory

32. Let α be algebraic of degree n over F. Show from Corollary 48.5 that there are at most n different isomorphisms of $F(\alpha)$ onto a subfield of $\overline{F}$ and leaving F fixed.

33. Let $F(\alpha_1, \cdots, \alpha_n)$ be an extension field of F. Show that any automorphism σ of $F(\alpha_1, \cdots, \alpha_n)$ leaving F fixed is completely determined by the n values $\sigma(\alpha_i)$.

34. Let E be an algebraic extension of a field F, and let σ be an automorphism of E leaving F fixed. Let $\alpha \in E$. Show that σ induces a permutation of the set of all zeros of $\operatorname{irr}(\alpha, F)$ that are in E.

35. Let E be an algebraic extension of a field F. Let $S = \{\sigma_i \mid i \in I\}$ be a collection of automorphisms of E such that every σ_i leaves each element of F fixed. Show that if S generates the subgroup H of $G(E/F)$, then $E_S = E_H$.

36. We saw in Corollary 23.17 that the cyclotomic polynomial

$$\Phi_p(x) = \frac{x^p - 1}{x - 1} = x^{p-1} + x^{p-2} + \cdots + x + 1$$

is irreducible over $\mathbb{Q}$ for every prime p. Let ζ be a zero of $\Phi_p(x)$, and consider the field $\mathbb{Q}(\zeta)$.

a. Show that $\zeta, \zeta^2, \cdots, \zeta^{p-1}$ are distinct zeros of $\Phi_p(x)$, and conclude that they are all the zeros of $\Phi_p(x)$.

b. Deduce from Corollary 48.5 and part (a) of this exercise that $G(\mathbb{Q}(\zeta)/\mathbb{Q})$ is abelian of order $p - 1$.

c. Show that the fixed field of $G(\mathbb{Q}(\zeta)/\mathbb{Q})$ is $\mathbb{Q}$. [*Hint:* Show that

$$\{\zeta, \zeta^2, \cdots, \zeta^{p-1}\}$$

is a basis for $\mathbb{Q}(\zeta)$ over $\mathbb{Q}$, and consider which linear combinations of $\zeta, \zeta^2, \cdots, \zeta^{p-1}$ are left fixed by all elements of $G(\mathbb{Q}(\zeta)/\mathbb{Q})$.

37. Theorem 48.3 described conjugation isomorphisms for the case where α and β were conjugate algebraic elements over F. Is there a similar isomorphism of $F(\alpha)$ with $F(\beta)$ in the case that α and β are both transcendental over F?

38. Let F be a field, and let x be an indeterminate over F. Determine all automorphisms of $F(x)$ leaving F fixed, by describing their values on x.

39. Prove the following sequence of theorems.

a. An automorphism of a field E carries elements that are squares of elements in E onto elements that are squares of elements of E.

b. An automorphism of the field $\mathbb{R}$ of real numbers carries positive numbers onto positive numbers.

c. If σ is an automorphism of $\mathbb{R}$ and $a < b$, where $a, b \in \mathbb{R}$, then $\sigma(a) < \sigma(b)$.

d. The only automorphism of $\mathbb{R}$ is the identity automorphism.

SECTION 49 The Isomorphism Extension Theorem

The Extension Theorem

Let us continue studying automorphisms of fields. In this section and the next, we shall be concerned with both the existence and the number of automorphisms of a field E.

Suppose that E is an algebraic extension of F and that we want to find some automorphisms of E. We know from Theorem 48.3 that if $\alpha, \beta \in E$ are conjugate over F, then there is an isomorphism $\psi_{\alpha,\beta}$ of $F(\alpha)$ onto $F(\beta)$. Of course, $\alpha, \beta \in E$ implies both $F(\alpha) \leq E$ and $F(\beta) \leq E$. It is natural to wonder whether the domain of definition of $\psi_{\alpha,\beta}$ can be enlarged from $F(\alpha)$ to a larger field, perhaps all of E, and whether this might perhaps lead to an automorphism of E. A mapping diagram of this situation is shown in Fig. 49.1. Rather than speak of "enlarging the domain of definition of $\psi_{\alpha,\beta}$," it

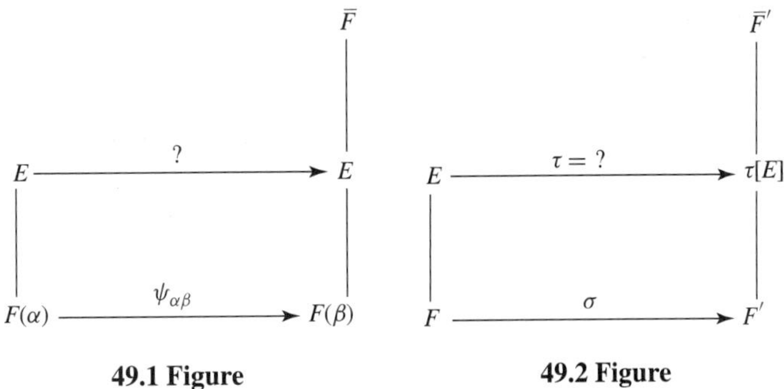

49.1 Figure **49.2 Figure**

is customary to speak of "**extending the map $\psi_{\alpha,\beta}$ to a map τ**," which is a mapping of all of E.

Remember that we are always assuming that all algebraic extension of F under consideration are contained in a fixed algebraic closure $\overline{F}$ of F. The Isomorphism Extension Theorem shows that the mapping $\psi_{\alpha,\beta}$ can indeed always be extended to an *isomorphism* of E onto a subfield of $\overline{F}$. Whether this extension gives an *automorphism* of E, that is, maps E into itself, is a question we shall study in Section 50. Thus this extension theorem, used in conjunction with our conjugation isomorphisms $\psi_{\alpha,\beta}$ will guarantee the existence of lots of *isomorphism mappings*, at least, for many fields. Extension theorems are very important in mathematics, particularly in algebraic and topological situations.

Let us take a more general look at this situation. Suppose that E is an algebraic extension of a field F and that we have an isomorphism σ of F onto a field F'. Let $\overline{F'}$ be an algebraic closure of F'. We would like to extend σ to an isomorphism τ of E onto a subfield of $\overline{F'}$. This situation is shown in Fig. 49.2. Naively, we pick $\alpha \in E$ but not in F and try to extend σ to $F(\alpha)$. If

$$p(x) = \text{irr}(\alpha, F) = a_0 + a_1x + \cdots + a_nx^n,$$

let β be a zero in $\overline{F'}$ of

$$q(x) = \sigma(a_0) + \sigma(a_1)x + \cdots + \sigma(a_n)x^n.$$

Here $q(x) \in F'[x]$. Since σ is an isomorphism, we know that $q(x)$ is irreducible in $F'[x]$. It seems reasonable that $F(\alpha)$ can be mapped isomorphically onto $F'(\beta)$ by a map extending σ and mapping α onto β. (This is not quite Theorem 48.3, but it is close to it; a few elements have been renamed by the isomorphism σ.) If $F(\alpha) = E$, we are done. If $F(\alpha) \neq E$, we have to find another element in E not in $F(\alpha)$ and continue the process. It is a situation very much like that in the construction of an algebraic closure $\overline{F}$ of a field F. Again the trouble is that, in general, where E is not a finite extension, the process may have to be repeated a (possibly large) infinite number of times, so we need Zorn's lemma to handle it. For this reason, we postpone the general proof of Theorem 49.3 to the end of this section.

49.3 Theorem **(Isomorphism Extension Theorem)** Let E be an algebraic extension of a field F. Let σ be an isomorphism of F onto a field F'. Let $\overline{F'}$ be an algebraic closure of F'. Then σ

can be extended to an isomorphism τ of E onto a subfield of $\overline{F'}$ such that $\tau(a) = \sigma(a)$ for all $a \in F$.

We give as a corollary the existence of an extension of one of our conjugation isomorphisms $\psi_{\alpha,\beta}$, as discussed at the start of this section.

49.4 Corollary If $E \leq \overline{F}$ is an algebraic extension of F and $\alpha, \beta \in E$ are conjugate over F, then the conjugation isomorphism $\psi_{\alpha,\beta} : F(\alpha) \to F(\beta)$, given by Theorem 48.3, can be extended to an isomorphism of E onto a subfield of $\overline{F}$.

Proof Proof of this corollary is immediate from Theorem 49.3 if in the statement of the theorem we replace F by $F(\alpha)$, F' by $F(\beta)$, and $\overline{F'}$ by $\overline{F}$. ◆

As another corollary, we can show, as we promised earlier, that an algebraic closure of F is unique, up to an isomorphism leaving F fixed.

49.5 Corollary Let $\overline{F}$ and $\overline{F'}$ be two algebraic closures of F. Then $\overline{F}$ is isomorphic to $\overline{F'}$ under an isomorphism leaving each element of F fixed.

Proof By Theorem 49.3, the identity isomorphism of F onto F can be extended to an isomorphism τ mapping $\overline{F}$ onto a subfield of $\overline{F'}$ that leaves F fixed (see Fig. 49.6). We need only show that τ is onto $\overline{F'}$. But by Theorem 49.3, the map $\tau^{-1} : \tau[\overline{F}] \to \overline{F}$ can be extended to an isomorphism of $\overline{F'}$ onto a subfield of $\overline{F}$. Since τ^{-1} is already onto $\overline{F}$, we must have $\tau[\overline{F}] = \overline{F'}$. ◆

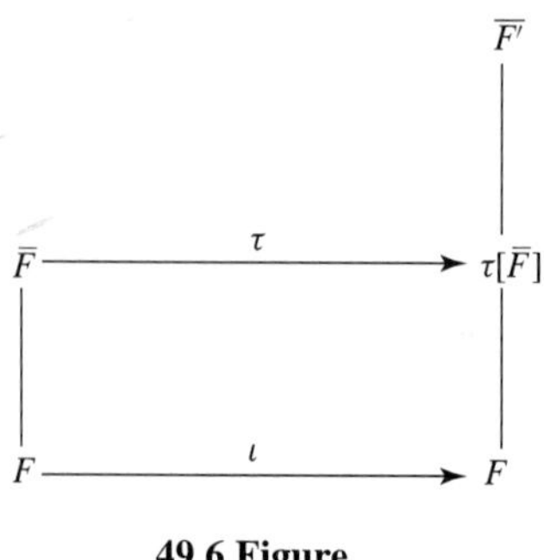

49.6 Figure

The Index of a Field Extension

Having discussed the question of *existence*, we turn now to the question of *how many*. For a *finite* extension E of a field F, we would like to count how many isomorphisms there are of E onto a subfield of $\overline{F}$ that leave F fixed. We shall show that there are only a finite number of isomorphisms. Since every automorphism in $G(E/F)$ is such an isomorphism, a count of these isomorphisms will include all these automorphisms. Example 48.17 showed that $G(\mathbb{Q}(\sqrt{2}, \sqrt{3})/\mathbb{Q})$ has four elements, and that $4 = [\mathbb{Q}(\sqrt{2}, \sqrt{3}) : \mathbb{Q}]$. While such an equality is not always true, it is true in a very important case. The next theorem takes the first big step in proving this. We state the theorem in more general terms than we shall need, but it does not make the proof any harder.

49.7 Theorem Let E be a finite extension of a field F. Let σ be an isomorphism of F onto a field F', and let $\overline{F'}$ be an algebraic closure of F'. Then the number of extensions of σ to an isomorphism τ of E onto a subfield of $\overline{F'}$ is finite, and independent of F', $\overline{F'}$, and σ. That is, the number of extensions is completely determined by the two fields E and F; it is intrinsic to them.

Proof The diagram in Fig. 49.8 may help us to follow the construction that we are about to make. This diagram is constructed in the following way. Consider two isomorphisms

$$\sigma_1 : F \xrightarrow{\text{onto}} F_1', \qquad \sigma_2 : F \xrightarrow{\text{onto}} F_2',$$

where $\overline{F_1'}$ and $\overline{F_2'}$ are algebraic closures of F_1' and F_2', respectively. Now $\sigma_2\sigma_1^{-1}$ is an isomorphism of F_1' onto F_2'. Then by Theorem 49.3 and Corollary 49.5 there is an isomorphism

$$\lambda : \overline{F_1'} \xrightarrow{\text{onto}} \overline{F_2'}$$

extending this isomorphism $\sigma_2\sigma_1^{-1} : F_1' \xrightarrow{\text{onto}} F_2'$. Referring to Fig. 49.8, corresponding to each $\tau_1 : E \to \overline{F_1'}$ that extends σ_1 we obtain an isomorphism $\tau_2 : E \to \overline{F_2'}$, by starting at E and going first to the left, then up, and then to the right. Written algebraically,

$$\tau_2(\alpha) = (\lambda\tau_1)(\alpha)$$

for $\alpha \in E$. Clearly τ_2 extends σ_2. The fact that we could have *started* with τ_2 and recovered τ_1 by defining

$$\tau_1(\alpha) = (\lambda^{-1}\tau_2)(\alpha),$$

that is, by chasing the other way around the diagram, shows that the correspondence between $\tau_1 : E \to \overline{F_1'}$ and $\tau_2 : E \to \overline{F_2'}$ is one to one. In view of this one-to-one correspondence, the number of τ extending σ is independent of F', $\overline{F'}$ and σ.

That the number of mappings extending σ is finite follows from the fact that since E is a finite extension of F, $E = F(\alpha_1, \cdots, \alpha_n)$ for some $\alpha_1, \cdots, \alpha_n$ in E, by Theorem 31.11.

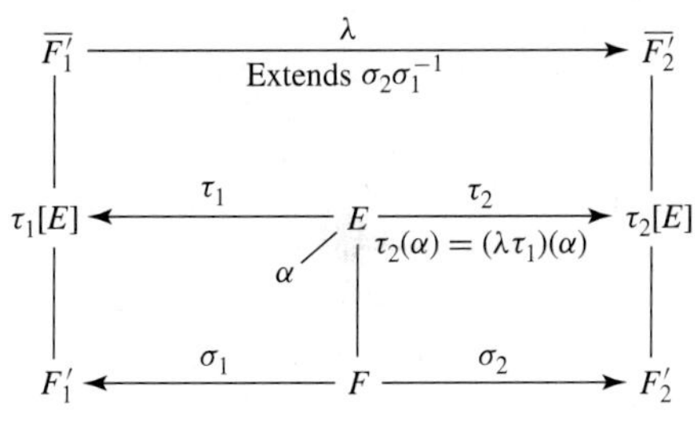

49.8 Figure

There are only a finite number of possible candiates for the images $\tau(\alpha_i)$ in $\overline{F'}$, for if

$$\operatorname{irr}(\alpha_i, F) = a_{i0} + a_{i1}x + \cdots + a_{im_i}x^{m_i},$$

where $a_{ik} \in F$, then $\tau(\alpha_i)$ must be one of the zeros in $\overline{F'}$ of

$$[\sigma(a_{i0}) + \sigma(a_{i1})x + \cdots + \sigma(a_{im_i})x^{m_i}] \in F'[x].$$

◆

49.9 Definition Let E be a finite extension of a field F. The number of isomorphisms of E onto a subfield of $\overline{F}$ leaving F fixed is the **index** $\{E : F\}$ **of** E **over** F. ■

49.10 Corollary If $F \leq E \leq K$, where K is a finite extension field of the field F, then $\{K : F\} = \{K : E\}\{E : F\}$.

Proof It follows from Theorem 49.7 that each of the $\{E : F\}$ isomorphisms τ_i of E onto a subfield of $\overline{F}$ leaving F fixed has $\{K : E\}$ extensions to an isomorphism of K onto a subfield of $\overline{F}$. ◆

The preceding corollary was really the main thing we were after. Note that it counts something. *Never underestimate a result that counts something*, even if it is only called a "corollary."

We shall show in Section 51 that unless F is an infinite field of characteristics $p \neq 0$, we always have $[E : F] = \{E : F\}$ for every finite extension field E of F. For the case $E = F(\alpha)$, the $\{F(\alpha) : F\}$ extensions of the identity map $\iota : F \to F$ to maps of $F(\alpha)$ onto a subfield of $\overline{F}$ are given by the conjugation isomorphisms $\psi_{\alpha,\beta}$ for each conjugate β in $\overline{F}$ of α over F. Thus if $\operatorname{irr}(\alpha, F)$ has n *distinct* zeros in $\overline{F}$, we have $\{E : F\} = n$. We shall show later that unless F is infinite and of characteristic $p \neq 0$, the number of distinct zeros of $\operatorname{irr}(\alpha, F)$ is $\deg(\alpha, F) = [F(\alpha) : F]$.

49.11 Example Consider $E = \mathbb{Q}(\sqrt{2}, \sqrt{3})$ over $\mathbb{Q}$, as in Example 48.17. Our work in that example shows that $\{E : \mathbb{Q}\} = [E : \mathbb{Q}] = 4$. Also, $\{E : \mathbb{Q}(\sqrt{2})\} = 2$, and $\{\mathbb{Q}(\sqrt{2}) : \mathbb{Q}\} = 2$, so

$$4 = \{E : \mathbb{Q}\} = \{E : \mathbb{Q}(\sqrt{2})\}\{\mathbb{Q}(\sqrt{2}) : \mathbb{Q}\} = (2)(2).$$

This illustrates Corollary 49.10 ▲

Proof of the Extension Theorem

We restate the Isomorphism Extension Theorem 49.3.

Isomorphism Extension Theorem Let E be an algebraic extension of a field F. Let σ be an isomorphism of F onto a field F'. Let $\overline{F'}$ be an algebraic closure of F'. Then σ can be extended to an isomorphism τ of E onto a subfield of $\overline{F'}$ such that $\tau(a) = \sigma(a)$ for $a \in F$.

Proof Consider all pairs (L, λ), where L is a field such that $F \leq L \leq E$ and λ is an isomorphism of L onto a subfield of $\overline{F'}$ such that $\lambda(a) = \sigma(a)$ for $a \in F$. The set S of such pairs (L, λ) is nonempty, since (F, σ) is such a pair. Define a partial ordering on S by $(L_1, \lambda_1) \leq (L_2, \lambda_2)$, if $L_1 \leq L_2$ and $\lambda_1(a) = \lambda_2(a)$ for $a \in L_1$. It is readily checked that this relation $\leq$ does give a partial ordering of S.

Let $T = \{(H_i, \lambda_i) \mid i \in I\}$ be a chain of S. We claim that $H = \bigcup_{i \in I} H_i$ is a subfield of E. Let $a, b \in H$, where $a \in H_1$ and $b \in H_2$; then either $H_1 \leq H_2$ or $H_2 \leq H_1$, since T is a chain. If, say, $H_1 \leq H_2$, then $a, b, \in H_2$, so $a \pm b$, ab, and a/b for $b \neq 0$ are all in H_2 and hence in H. Since for each $i \in I$, $F \subseteq H_i \subseteq E$, we have $F \subseteq H \subseteq E$. Thus H is a subfield of E.

Define $\lambda : H \to \overline{F'}$ as follows. Let $c \in H$. Then $c \in H_i$ for some $i \in I$, and let

$$\lambda(c) = \lambda_i(c).$$

The map λ is well defined because if $c \in H_1$ and $c \in H_2$, then either $(H_1, \lambda_1) \leq (H_2, \lambda_2)$ or $(H_2, \lambda_2) \leq (H_1, \lambda_1)$, since T is a chain. In either case, $\lambda_1(c) = \lambda_2(c)$. We claim that λ is an isomorphism of H onto a subfield of $\overline{F'}$. If $a, b \in H$ then there is an H_i such that $a, b \in H_i$, and

$$\lambda(a + b) = \lambda_i(a + b) = \lambda_i(a) + \lambda_i(b) = \lambda(a) + \lambda(b).$$

Similarly,

$$\lambda(ab) = \lambda_i(ab) = \lambda_i(a)\lambda_i(b) = \lambda(a)\lambda(b).$$

If $\lambda(a) = 0$, then $a \in H_i$ for some i implies that $\lambda_i(a) = 0$, so $a = 0$. Therefore, λ is an isomorphism. Thus $(H, \lambda) \in S$, and it is clear from our definitions of H and λ that (H, λ) is an upper bound for T.

We have shown that every chain of S has an upper bound in S, so the hypotheses of Zorn's lemma are satisfied. Hence there exists a maximal element (K, τ) of S. Let $\tau(K) = K'$, where $K' \leq \overline{F'}$. Now if $K \neq E$, let $\alpha \in E$ but $\alpha \notin K$. Now α is algebraic over F, so α is algebraic over K. Also, let $p(x) = \text{irr}(\alpha, K)$. Let ψ_α be the canonical isomorphism

$$\psi_\alpha : K[x]/\langle p(x) \rangle \to K(\alpha),$$

corresponding to the evaluation homomorphism $\phi_\alpha : K[x] \to K(\alpha)$. If

$$p(x) = a_0 + a_1 x + \cdots + a_n x^n,$$

consider

$$q(x) = \tau(a_0) + \tau(a_1)x + \cdots + \tau(a_n)x^n$$

in $K'[x]$. Since τ is an isomorphism, $q(x)$ is irreducible in $K'[x]$. Since $K' \leq \overline{F'}$, there is a zero α' of $q(x)$ in $\overline{F'}$. Let

$$\psi_{\alpha'} : K'[x]/\langle q(x) \rangle \to K'(\alpha')$$

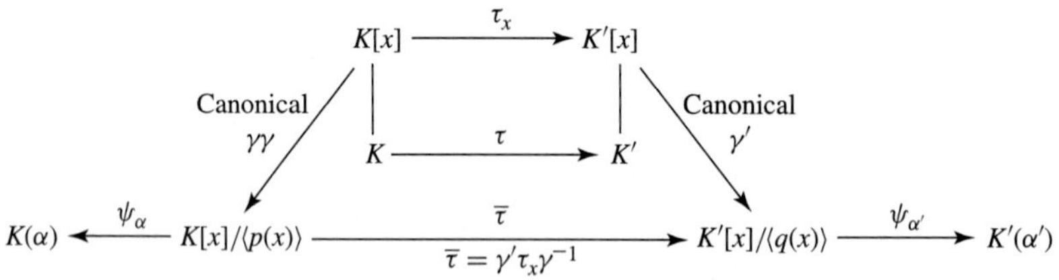

$$K[x] \xrightarrow{\tau_x} K'[x]$$

$$K \xrightarrow{\tau} K'$$

Canonical γ Canonical γ'

$$K(\alpha) \xleftarrow{\psi_\alpha} K[x]/\langle p(x)\rangle \xrightarrow[\overline{\tau} = \gamma'\tau_x\gamma^{-1}]{\overline{\tau}} K'[x]/\langle q(x)\rangle \xrightarrow{\psi_{\alpha'}} K'(\alpha')$$

49.12 Figure

be the isomorphism analogous to ψ_α. Finally, let

$$\overline{\tau} : K[x]/\langle p(x)\rangle \to K'[x]/\langle q(x)\rangle$$

be the isomorphism extending τ on K and mapping $x + \langle p(x)\rangle$ onto $x + \langle q(x)\rangle$. (See Fig. 49.12.) Then the composition of maps

$$\psi_{\alpha'}\overline{\tau}\psi_\alpha^{-1} : K(\alpha) \to K'(\alpha')$$

is an isomorphism of $K(\alpha)$ onto a subfield of $\overline{F'}$. Clearly, $(K, \tau) < (K(\alpha), \psi_{\alpha'}\overline{\tau}\psi_\alpha^{-1})$, which contradicts that (K, τ) is maximal. Therefore we must have had $K = E$. ◆

EXERCISES 49

Computations

Let $E = \mathbb{Q}(\sqrt{2}, \sqrt{3}, \sqrt{5})$. It can be shown that $[E : \mathbb{Q}] = 8$. In Exercises 1 through 3, for the given isomorphic mapping of a subfield of E, give all extensions of the mapping to an isomorphic mapping of E onto a subfield of $\overline{\mathbb{Q}}$. Describe the extensions by giving values on the generating set $\{\sqrt{2}, \sqrt{3}, \sqrt{5}\}$ for E over $\mathbb{Q}$.

1. $\iota : \mathbb{Q}(\sqrt{2}, \sqrt{15}) \to \mathbb{Q}(\sqrt{2}, \sqrt{15})$, where ι is the identity map

2. $\sigma : \mathbb{Q}(\sqrt{2}, \sqrt{15}) \to \mathbb{Q}(\sqrt{2}, \sqrt{15})$ where $\sigma(\sqrt{2}) = \sqrt{2}$ and $\sigma(\sqrt{15}) = -\sqrt{15}$

3. $\psi_{\sqrt{30}, -\sqrt{30}} : \mathbb{Q}(\sqrt{30}) \to \mathbb{Q}(\sqrt{30})$

It is a fact, which we can verify by cubing, that the zeros of $x^3 - 2$ in $\overline{\mathbb{Q}}$ are

$$\alpha_1 = \sqrt[3]{2}, \qquad \alpha_2 = \sqrt[3]{2}\frac{-1 + i\sqrt{3}}{2}, \qquad \text{and} \qquad \alpha_3 = \sqrt[3]{2}\frac{-1 - i\sqrt{3}}{2},$$

where $\sqrt[3]{2}$, as usual, is the real cube root of 2. Use this information in Exercises 4 through 6.

4. Describe all extensions of the identity map of $\mathbb{Q}$ to an isomorphism mapping $\mathbb{Q}(\sqrt[3]{2})$ onto a subfield of $\overline{\mathbb{Q}}$.

5. Describe all extensions of the identity map of $\mathbb{Q}$ to an isomorphism mapping $\mathbb{Q}(\sqrt[3]{2}, \sqrt{3})$ onto a subfield of $\overline{\mathbb{Q}}$.

6. Describe all extensions of the automorphism $\psi_{\sqrt{3}, -\sqrt{3}}$ of $\mathbb{Q}(\sqrt{3})$ to an isomorphism mapping $\mathbb{Q}(i, \sqrt{3}, \sqrt[3]{2})$ onto a subfield of $\overline{\mathbb{Q}}$.

7. Let σ be the automorphism of $\mathbb{Q}(\pi)$ that maps π onto $-\pi$.

a. Describe the fixed field of σ.

b. Describe all extensions of σ to an isomorphism mapping the field $\mathbb{Q}(\sqrt{\pi})$ onto a subfield of $\overline{\mathbb{Q}(\pi)}$.

Concepts

8. Mark each of the following true or false.

_____ **a.** Let $F(\alpha)$ be any simple extension of a field F. Then every isomorphism of F onto a subfield of $\overline{F}$ has an extension to an isomorphism of $F(\alpha)$ onto a subfield of $\overline{F}$.

_____ **b.** Let $F(\alpha)$ be any simple algebraic extension of a field F. Then every isomorphism of F onto a subfield of $\overline{F}$ has an extension to an isomorphism of $F(\alpha)$ onto a subfield of $\overline{F}$.

_____ **c.** An isomorphism of F onto a subfield of $\overline{F}$ has the same number of extensions to each simple algebraic extension of F.

_____ **d.** Algebraic closures of isomorphic fields are always isomorphic.

_____ **e.** Algebraic closures of fields that are not isomorphic are never isomorphic.

_____ **f.** Any algebraic closure of $\mathbb{Q}(\sqrt{2})$ is isomorphic to any algebraic closure of $\mathbb{Q}(\sqrt{17})$.

_____ **g.** The index of a finite extension E over a field F is finite.

_____ **h.** The index behaves multiplicatively with respect to finite towers of finite extensions of fields.

_____ **i.** Our remarks prior to the first statement of Theorem 49.3 essentially constitute a proof of this theorem for a finite extension E over F.

_____ **j.** Corollary 49.5 shows that $\mathbb{C}$ is isomorphic to $\overline{\mathbb{Q}}$.

Theory

9. Let K be an algebraically closed field. Show that every isomorphism σ of K onto a subfield of itself such that K is algebraic over $\sigma[K]$ is an automorphism of K, that is, is an onto map. [*Hint:* Apply Theorem 49.3 to σ^{-1}.]

10. Let E be an algebraic extension of a field F. Show that every isomorphism of E onto a subfield of $\overline{F}$ leaving F fixed can be extended to an automorphism of $\overline{F}$.

11. Prove that if E is an algebraic extension of a field F, then two algebraic closures $\overline{F}$ and $\overline{E}$ of F and E, respectively, are isomorphic.

12. Prove that the algebraic closure of $\mathbb{Q}(\sqrt{\pi})$ in $\mathbb{C}$ is isomorphic to any algebraic closure of $\overline{\mathbb{Q}}(x)$, where $\overline{\mathbb{Q}}$ is the field of algebraic numbers and x is an indeterminate.

13. Prove that if E is a finite extension of a field F, then $\{E : F\} \leq [E : F]$. [*Hint:* The remarks preceding Example 49.11 essentially showed this for a simple algebraic extension $F(\alpha)$ of F. Use the fact that a finite extension is a tower of simple extensions, together with the multiplicative properties of the index and degree.]

SECTION 50 SPLITTING FIELDS

We are going to be interested chiefly in *automorphisms* of a field E, rather than mere isomorphic mappings of E onto a subfield of $\overline{E}$. It is the *automorphisms* of a field that form a group. We wonder whether for some extension field E of a field F, *every* isomorphic mapping of E onto a subfield of $\overline{F}$ leaving F fixed is actually an automorphism of E.

Suppose E is an algebraic extension of a field E. If $\alpha \in E$ and $\beta \in \overline{F}$ is a conjugate of α over F, then there is a conjugation isomorphism

$$\psi_{\alpha,\beta} : F(\alpha) \to F(\beta).$$

By Corollary 49.4, $\psi_{\alpha,\beta}$ can be extended to an isomorphic mapping of E onto a subfield of $\overline{F}$. Now if $\beta \notin E$, such an isomorphic mapping of E can't be an automorphism of E. *Thus, if an algebraic extension E of a field F is such that all its isomorphic mappings onto* *a subfield of $\overline{F}$ leaving F fixed are actually automorphisms of E, then for every $\alpha \in E$, all conjugates of α over F must be in E also.* This observation seemed to come very easily. We point out that we used a lot of power, namely the existence of the conjugation isomorphisms and the Isomorphism Extension Theorem 49.3.

These ideas suggest the formulation of the following definition.

50.1 Definition Let F be a field with algebraic closure $\overline{F}$. Let $\{f_i(x) \mid i \in I\}$ be a collection of polynomials in $F[x]$. A field $E \leq \overline{F}$ is the **splitting field of** $\{f_i(x) \mid i \in I\}$ **over** F if E is the smallest subfield of $\overline{F}$ containing F and all the zeros in $\overline{F}$ of each of the $f_i(x)$ for $i \in I$. A field $K \leq \overline{F}$ is a **splitting field over** F if it is the splitting field of some set of polynomials in $F[x]$. ■

50.2 Example We see that $\mathbb{Q}[\sqrt{2}, \sqrt{3}]$ is a splitting field of $\{x^2 - 2, x^2 - 3\}$ and also of $\{x^4 - 5x^2 + 6\}$. ▲

For one polynomial $f(x) \in F[x]$, we shall often refer to the splitting field of $\{f(x)\}$ over F as the **splitting field of** $f(x)$ **over** F. Note that the splitting field of $\{f_i(x) \mid i \in I\}$ over F in $\overline{F}$ is the intersection of all subfields of $\overline{F}$ containing F and all zeros in $\overline{F}$ of each $f_i(x)$ for $i \in I$. Thus such a splitting field surely does exist.

We now show that splitting fields over F are precisely those fields $E \leq \overline{F}$ with the property that all isomorphic mappings of E onto a subfield of $\overline{F}$ leaving F fixed are automorphisms of E. This will be a corollary of the next theorem. *Once more, we are characterizing a concept in terms of mappings.* Remember, we are always assuming that all algebraic extensions of a field F under consideration are in one fixed algebraic closure $\overline{F}$ of F.

50.3 Theorem A field E, where $F \leq E \leq \overline{F}$, is a splitting field over F if and only if every automorphism of $\overline{F}$ leaving F fixed maps E onto itself and thus induces an automorphism of E leaving F fixed.

Proof Let E be a splitting field over F in $\overline{F}$ of $\{f_i(x) \mid i \in I\}$, and let σ be an automorphism of $\overline{F}$ leaving F fixed. Let $\{\alpha_j \mid j \in J\}$ be the collection of all zeros in $\overline{F}$ of all the $f_i(x)$ for $i \in I$. Now our previous work shows that for a fixed α_j, the field $F(\alpha_j)$ has as elements all expressions of the form

$$g(\alpha_j) = a_0 + a_1\alpha_j + \cdots + a_{n_j-1}\alpha_j^{n_j-1},$$

where n_j is the degree of $\mathrm{irr}(\alpha_j, F)$ and $a_k \in F$. Consider the set S of all *finite* sums of *finite* products of elements of the form $g(\alpha_j)$ for all $j \in J$. The set S is a subset of E closed under addition and multiplication and containing 0, 1, and the additive inverse of each element. Since each element of S is in some $F(\alpha_{j_1}, \cdots, \alpha_{j_r}) \subseteq S$, we see that S also contains the multiplicative inverse of each nonzero element. Thus S is a subfield of E containing all α_j for $j \in J$. By definition of the splitting field E of $\{f_i(x) \mid i \in I\}$, we see that we must have $S = E$. All this work was just to show that $\{\alpha_j \mid j \in J\}$ *generates* E over F, in the sense of taking *finite* sums and *finite* products. Knowing this, we see immediately that the value of σ on any element of E is completely determined by the values $\sigma(\alpha_j)$. But by Corollary 48.5, $\sigma(\alpha_j)$ must also be a zero of $\mathrm{irr}(\alpha_j, F)$. By

Theorem 29.13, $\operatorname{irr}(\alpha_j, F)$ divides the $f_i(x)$ for which $f_i(\alpha_j) = 0$, so $\sigma(\alpha_j) \in E$ also. Thus σ maps E onto a subfield of E isomorphically. However, the same is true of the automorphism σ^{-1} of $\overline{F}$. Since for $\beta \in E$,

$$\beta = \sigma(\sigma^{-1}(\beta)),$$

we see that σ maps E onto E, and thus induces an automorphism of E.

Suppose, conversely, that every automorphism of $\overline{F}$ leaving F fixed induces an automorphism of E. Let $g(x)$ be an *irreducible* polynomial in $F[x]$ having a zero α in E. If β is any zero of $g(x)$ in $\overline{F}$, then by Theorem 48.3, there is a conjugation isomorphism $\psi_{\alpha,\beta}$ of $F(\alpha)$ onto $F(\beta)$ leaving F fixed. By Theorem 49.3, $\psi_{\alpha,\beta}$ can be extended to an isomorphism τ of $\overline{F}$ onto a subfield of $\overline{F}$. But then

$$\tau^{-1} : \tau[\overline{F}] \to \overline{F}$$

can be extended to an isomorphism mapping $\overline{F}$ onto a subfield of $\overline{F}$. Since the image of τ^{-1} is already all of $\overline{F}$, we see that τ must have been onto $\overline{F}$, so τ is an automorphism of $\overline{F}$ leaving F fixed. Then by assumption, τ induces an automorphism of E, so $\tau(\alpha) = \beta$ is in E. We have shown that if $g(x)$ is an irreducible polynomial in $F[x]$ having one zero in E, then all zeros of $g(x)$ in $\overline{F}$ are in E. Hence if $\{g_k(x)\}$ is the set of *all* irreducible polynomials in $F[x]$ having a zero in E, then E is the splitting field of $\{g_k(x)\}$. ◆

50.4 Definition Let E be an extension field of a field F. A polynomial $f(x) \in F[x]$ **splits in** E if it factors into a product of linear factors in $E[x]$. ■

50.5 Example The polynomial $x^4 - 5x^2 + 6$ in $\mathbb{Q}[x]$ splits in the field $\mathbb{Q}[\sqrt{2}, \sqrt{3}]$ into $(x - \sqrt{2})(x + \sqrt{2})(x - \sqrt{3})(x + \sqrt{3})$. ▲

50.6 Corollary If $E \leq \overline{F}$ is a splitting field over F, then every irreducible polynomial in $F[x]$ having a zero in E splits in E.

Proof If E is a splitting field over F in $\overline{F}$, then every automorphism of $\overline{F}$ induces an automorphism of E. The second half of the proof of Theorem 50.3 showed precisely that E is also the splitting field over F of the set $\{g_k(x)\}$ of *all* irreducible polynomials in $F[x]$ having a zero in E. Thus an irreducible polynomial $f(x)$ of $F[x]$ having a zero in E has all its zeros in $\overline{F}$ in E. Therefore, its factorization into linear factors in $\overline{F}[x]$, given by Theorem 31.15, actually takes place in $E[x]$, so $f(x)$ splits in E. ◆

50.7 Corollary If $E \leq \overline{F}$ is a splitting field over F, then every isomorphic mapping of E onto a subfield of $\overline{F}$ and leaving F fixed is actually an automorphism of E. In particular, if E is a splitting field of finite degree over F, then

$$\{E : F\} = |G(E/F)|.$$

Proof Every isomorphism σ mapping E onto a subfield of $\overline{F}$ leaving F fixed can be extended to an automorphism τ of $\overline{F}$, by Theorem 49.3, together with the *onto* argument of the second half of the proof of Theorem 50.3. If E is a splitting field over F, then by Theorem 50.3, τ restricted to E, that is σ, is an automorphism of E. Thus for a splitting field E over F, every isomorphic mapping of E onto a subfield of $\overline{F}$ leaving F fixed is an automorphism of E.

The equation $\{E : F\} = |G(E/F)|$ then follows immediately for a splitting field E of finite degree over F, since $\{E : F\}$ was defined as the number of different isomorphic mappings of E onto a subfield of $\bar{F}$ leaving F fixed. ◆

50.8 Example Observe that $\mathbb{Q}(\sqrt{2}, \sqrt{3})$ is the splitting field of

$$\{x^2 - 2, x^2 - 3\}$$

over $\mathbb{Q}$. Example 48.17 showed that the mappings $\iota, \sigma_1, \sigma_2$, and σ_3 are all the automorphisms of $\mathbb{Q}(\sqrt{2}, \sqrt{3})$ leaving $\mathbb{Q}$ fixed. (Actually, since every automorphism of a field must leave the prime subfield fixed, we see that these are the only automorphisms of $\mathbb{Q}(\sqrt{2}, \sqrt{3})$.) Then

$$\{\mathbb{Q}(\sqrt{2}, \sqrt{3}) : \mathbb{Q}\} = |G(\mathbb{Q}(\sqrt{2}, \sqrt{3})/\mathbb{Q})| = 4.$$

illustrating Corollary 50.7. ▲

We wish to determine conditions under which

$$|G(E/F)| = \{E : F\} = [E : F]$$

for finite extensions E of F. This is our next topic. We shall show in the following section that this equation always holds when E is a splitting field over a field F of characteristic 0 or when F is a finite field. This equation need not be true when F is an infinite field of characteristic $p \neq 0$.

50.9 Example Let $\sqrt[3]{2}$ be the real cube root of 2, as usual. Now $x^3 - 2$ does not split in $\mathbb{Q}(\sqrt[3]{2})$, for $\mathbb{Q}(\sqrt[3]{2}) < \mathbb{R}$ and only one zero of $x^3 - 2$ is real. Thus $x^3 - 2$ factors in $(\mathbb{Q}(\sqrt[3]{2}))[x]$ into a linear factor $x - \sqrt[3]{2}$ and an irreducible quadratic factor. The splitting field E of $x^3 - 2$ over $\mathbb{Q}$ is therefore of degree 2 over $\mathbb{Q}(\sqrt[3]{2})$. Then

$$[E : \mathbb{Q}] = [E : \mathbb{Q}(\sqrt[3]{2})][\mathbb{Q}(\sqrt[3]{2}) : \mathbb{Q}] = (2)(3) = 6.$$

We have shown that the splitting field over $\mathbb{Q}$ of $x^3 - 2$ is of degree 6 over $\mathbb{Q}$.

We can verify by cubing that

$$\sqrt[3]{2}\frac{-1 + i\sqrt{3}}{2} \quad \text{and} \quad \sqrt[3]{2}\frac{-1 - i\sqrt{3}}{2}$$

are the other zeros of $x^3 - 2$ in $\mathbb{C}$. Thus the splitting field E of $x^3 - 2$ over $\mathbb{Q}$ is $\mathbb{Q}(\sqrt[3]{2}, i\sqrt{3})$. (This is *not* the same field as $\mathbb{Q}(\sqrt[3]{2}, i, \sqrt{3})$, which is of degree 12 over $\mathbb{Q}$.) Further study of this interesting example is left to the exercises (see Exercises 7, 8, 9, 16, 21, and 23). ▲

EXERCISES 50

Computations

In Exercises 1 through 6, find the degree over $\mathbb{Q}$ of the splitting field over $\mathbb{Q}$ of the given polynomial in $\mathbb{Q}[x]$.

1. $x^2 + 3$ **2.** $x^4 - 1$ **3.** $(x^2 - 2)(x^2 - 3)$

4. $x^3 - 3$ **5.** $x^3 - 1$ **6.** $(x^2 - 2)(x^3 - 2)$

Refer to Example 50.9 for Exercises 7 through 9.

7. What is the order of $G(\mathbb{Q}(\sqrt[3]{2})/\mathbb{Q})$?

8. What is the order of $G(\mathbb{Q}(\sqrt[3]{2}, i\sqrt{3})/\mathbb{Q})$?

9. What is the order of $G(\mathbb{Q}(\sqrt[3]{2}, i\sqrt{3})/\mathbb{Q}(\sqrt[3]{2}))$?

10. Let α be a zero of $x^3 + x^2 + 1$ over $\mathbb{Z}_2$. Show that $x^3 + x^2 + 1$ splits in $\mathbb{Z}_2(\alpha)$. [*Hint:* There are eight elements in $\mathbb{Z}_2(\alpha)$. Exhibit two more zeros of $x^3 + x^2 + 1$, in addition to α, among these eight elements. Alternatively, use the results of Section 33.]

Concepts

In Exercises 11 and 12, correct the definition of the italicized term without reference to the text, if correction is needed, so that it is in a form acceptable for publication.

11. Let $F \leq E \leq \overline{F}$ where $\overline{F}$ is an algebraic closure of a field F. The field E is a *splitting field* over F if and only if E contains all the zeros in $\overline{F}$ of every polynomial in $F[x]$ that has a zero in E.

12. A polynomial $f(x)$ in $F[x]$ *splits in an extension field* E of F if and only if it factors in $E[x]$ into a product of polynomials of lower degree.

13. Let $f(x)$ be a polynomial in $F[x]$ of degree n. Let $E \leq \overline{F}$ be the splitting field of $f(x)$ over F in $\overline{F}$. What bounds can be put on $[E : F]$?

14. Mark each of the following true or false.

_____ **a.** Let $\alpha, \beta \in E$, where $E \leq \overline{F}$ is a splitting field over F. Then there exists an automorphism of E leaving F fixed and mapping α onto β if and only if $\text{irr}(\alpha, F) = \text{irr}(\beta, F)$.

_____ **b.** $\mathbb{R}$ is a splitting field over $\mathbb{Q}$.

_____ **c.** $\mathbb{R}$ is a splitting field over $\mathbb{R}$.

_____ **d.** $\mathbb{C}$ is a splitting field over $\mathbb{R}$.

_____ **e.** $\mathbb{Q}(i)$ is a splitting field over $\mathbb{Q}$.

_____ **f.** $\mathbb{Q}(\pi)$ is a splitting field over $\mathbb{Q}(\pi^2)$.

_____ **g.** For every splitting field E over F, where $E \leq \overline{F}$, every isomorphic mapping of E is an automorphism of E.

_____ **h.** For every splitting field E over F, where $E \leq \overline{F}$, every isomorphism mapping E onto a subfield of $\overline{F}$ is an automorphism of E.

_____ **i.** For every splitting field E over F, where $E \leq \overline{F}$, every isomorphism mapping E onto a subfield of $\overline{F}$ and leaving F fixed is an automorphism of E.

_____ **j.** Every algebraic closure $\overline{F}$ of a field F is a splitting field over F.

15. Show by an example that Corollary 50.6 is no longer true if the word *irreducible* is deleted.

16. **a.** Is $|G(E/F)|$ multiplicative for finite towers of finite extensions, that is, is

$$|G(K/F)| = |G(K/E)||G(E/F)| \quad \text{for} \quad F \leq E \leq K \leq \overline{F}?$$

Why or why not? [*Hint:* Use Exercises 7 through 9.]

b. Is $|G(E/F)|$ multiplicative for finite towers of finite extensions, each of which is a splitting field over the bottom field? Why or why not?

Theory

17. Show that if a finite extension E of a field F is a splitting field over F, then E is a splitting field of one polynomial in $F[x]$.

18. Show that if $[E : F] = 2$, then E is a splitting field over F.

19. Show that for $F \leq E \leq \overline{F}$, E is a splitting field over F if and only if E contains all conjugates over F in $\overline{F}$ for each of its elements.

20. Show that $\mathbb{Q}(\sqrt[3]{2})$ has only the identity automorphism.

21. Referring to Example 50.9, show that

$$G(\mathbb{Q}(\sqrt[3]{2}, i\sqrt{3})/\mathbb{Q}(i\sqrt{3})) \simeq \langle \mathbb{Z}_3, +\rangle.$$

22. **a.** Show that an automorphism leaving F fixed of a splitting field E over F of a polynomial $f(x) \in F[x]$ permutes the zeros of $f(x)$ in E.

b. Show that an automorphism leaving F fixed of a splitting field E over F of a polynomial $f(x) \in F[x]$ is completely determined by the permutation of the zeros of $f(x)$ in E given in part (a).

c. Show that if E is a splitting field over F of a polynomial $f(x) \in F[x]$, then $G(E/F)$ can be viewed in a natural way as a certain group of permutations.

23. Let E be the splitting field of $x^3 - 2$ over $\mathbb{Q}$, as in Example 50.9.

a. What is the order of $G(E/\mathbb{Q})$? [*Hint:* Use Corollary 50.7 and Corollary 49.4 applied to the tower $\mathbb{Q} \leq \mathbb{Q}(i\sqrt{3}) \leq E$.]

b. Show that $G(E/\mathbb{Q}) = S_3$, the symmetric group on three letters. [*Hint:* Use Exercise 22, together with part (a).]

24. Show that for a prime p, the splitting field over $\mathbb{Q}$ of $x^p - 1$ is of degree $p - 1$ over $\mathbb{Q}$. [*Hint:* Refer to Corollary 23.17.]

25. Let $\overline{F}$ and $\overline{F'}$ be two algebraic closures of a field F, and let $f(x) \in F[x]$. Show that the splitting field E over F of $f(x)$ in $\overline{F}$ is isomorphic to the splitting field E' over F of $f(x)$ in $\overline{F'}$. [*Hint:* Use Corollary 49.5.]

SECTION 51 Separable Extensions

Multiplicity of Zeros of a Polynomial

Remember that we are now always assuming that all algebraic extensions of a field F under consideration are contained in one fixed algebraic closure $\overline{F}$ of F.

Our next aim is to determine, for a finite extension E of F, under what conditions $\{E : F\} = [E : F]$. The key to answering this question is to consider the multiplicity of zeros of polynomials.

51.1 Definition Let $f(x) \in F[x]$. An element α of $\overline{F}$ such that $f(\alpha) = 0$ is a **zero of $f(x)$ of multiplicity ν** if ν is the greatest integer such that $(x - \alpha)^\nu$ is a factor of $f(x)$ in $\overline{F}[x]$. ■

The next theorem shows that the multiplicities of the zeros of one given *irreducible* polynomial over a field are all the same. The ease with which we can prove this theorem is a further indication of the power of our conjugation isomorphisms and of our whole approach to the study of zeros of polynomials by means of mappings.

51.2 Theorem Let $f(x)$ be irreducible in $F[x]$. Then all zeros of $f(x)$ in $\overline{F}$ have the same multiplicity.

Proof Let α and β be zeros of $f(x)$ in $\overline{F}$. Then by Theorem 48.3, there is a conjugation isomorphism $\psi_{\alpha,\beta} : F(\alpha) \xrightarrow{\text{onto}} F(\beta)$. By Corollary 49.4, $\psi_{\alpha,\beta}$ can be extended to an isomorphism $\tau : \overline{F} \to \overline{F}$. Then τ induces a natural isomorphism $\tau_x : \overline{F}[x] \to \overline{F}[x]$, with $\tau_x(x) = x$. Now τ_x leaves $f(x)$ fixed, since $f(x) \in F[x]$ and $\psi_{\alpha,\beta}$ leaves F fixed. However,

$$\tau_x((x-\alpha)^\nu) = (x-\beta)^\nu,$$

which shows that the multiplicity of β in $f(x)$ is greater than or equal to the multiplicity of α. A symmetric argument gives the reverse inequality, so the multiplicity of α equals that of β. ◆

51.3 Corollary If $f(x)$ is irreducible in $F[x]$, then $f(x)$ has a factorization in $\overline{F}[x]$ of the form

$$a \prod_i (x - \alpha_i)^\nu,$$

where the α_i are the distinct zeros of $f(x)$ in $\overline{F}$ and $a \in F$.

Proof The corollary is immediate from Theorem 51.2. ◆

At this point, we should probably show by an example that the phenomenon of a zero of multiplicity greater than 1 of an irreducible polynomial can occur. We shall show later in this section that it can only occur for a polynomial over an infinite field of characteristic $p \neq 0$.

51.4 Example Let $E = \mathbb{Z}_p(y)$, where y is an indeterminate. Let $t = y^p$, and let F be the subfield $\mathbb{Z}_p(t)$ of E. (See Fig. 51.5.) Now $E = F(y)$ is algebraic over F, for y is a zero of $(x^p - t) \in F[x]$. By Theorem 29.13, $\text{irr}(y, F)$ must divide $x^p - t$ in $F[x]$. [Actually, $\text{irr}(y, F) = x^p - t$. We leave a proof of this to the exercises (see Exercise 10).] Since $F(y)$ is not equal to F, we must have the degree of $\text{irr}(y, F) \geq 2$. But note that

$$x^p - t = x^p - y^p = (x - y)^p,$$

since E has characteristic p (see Theorem 48.19 and the following comment). Thus y is a zero of $\text{irr}(y, F)$ of multiplicity > 1. Actually, $x^p - t = \text{irr}(y, F)$, so the multiplicity of y is p. ▲

$E = \mathbb{Z}_p(y) = F(y)$

$F = \mathbb{Z}_p(t) = \mathbb{Z}_p(y^p)$

$\mathbb{Z}_p$

51.5 Figure

From here on we rely heavily on Theorem 49.7 and its corollary. Theorem 48.3 and its corollary show that for a simple algebraic extension $F(\alpha)$ of F there is one extension of the identity isomorphism ι mapping F into F for every distinct zero of $\text{irr}(\alpha, F)$ and that these are the only extensions of ι. *Thus* $\{F(\alpha) : F\}$ *is the number of distinct zeros of* $\text{irr}(\alpha, F)$.

In view of our work with the theorem of Lagrange and Theorem 31.4, we should recognize the potential of a theorem like this next one.

51.6 Theorem If E is a finite extension of F, then $\{E : F\}$ divides $[E : F]$.

Proof By Theorem 31.11, if E is finite over F, then $E = F(\alpha_1, \cdots, \alpha_n)$, where $\alpha_i \in \overline{F}$. Let $\text{irr}(\alpha_i, F(\alpha_1, \cdots, \alpha_{i-1}))$ have α_i as one of n_i distinct zeros that are all of a common multiplicity ν_i, by Theorem 51.2. Then

$$[F(\alpha_1, \cdots, \alpha_i) : F(\alpha_1, \cdots, \alpha_{i-1})] = n_i\nu_i = \{F(\alpha_1, \cdots, \alpha_i) : F(\alpha_1, \cdots, \alpha_{i-1})\}\nu_i.$$

By Theorem 31.4 and Corollary 49.10,

$$[E : F] = \prod_i n_i\nu_i,$$

and

$$\{E : F\} = \prod_i n_i.$$

Therefore, $\{E : F\}$ divides $[E : F]$. ◆

Separable Extensions

51.7 Definition A finite extension E of F is a **separable extension of** F if $\{E : F\} = [E : F]$. An element α of $\overline{F}$ is **separable over** F if $F(\alpha)$ is a separable extension of F. An irreducible polynomial $f(x) \in F[x]$ is **separable over** F if every zero of $f(x)$ in $\overline{F}$ is separable over F. ■

51.8 Example The field $E = \mathbb{Q}[\sqrt{2}, \sqrt{3}]$ is separable over $\mathbb{Q}$ since we saw in Example 50.8 that $\{E : \mathbb{Q}\} = 4 = [E : \mathbb{Q}]$. ▲

To make things a little easier, we have restricted our definition of a separable extension of a field F to *finite* extensions E of F. For the corresponding definition for infinite extensions, see Exercise 12.

We know that $\{F(\alpha) : F\}$ is the number of distinct zeros of $\text{irr}(\alpha, F)$. Also, the multiplicity of α in $\text{irr}(\alpha, F)$ is the same as the multiplicity of each conjugate of α over F, by Theorem 51.2. *Thus α is separable over F if and only if* $\text{irr}(\alpha, F)$ *has all zeros of multiplicity 1. This tells us at once that an irreducible polynomial $f(x) \in F[x]$ is separable over F if and only if $f(x)$ has all zeros of multiplicity 1.*

51.9 Theorem If K is a finite extension of E and E is a finite extension of F, that is, $F \leq E \leq K$, then K is separable over F if and only if K is separable over E and E is separable over F.

Proof Now

$$[K : F] = [K : E][E : F],$$

and

$$\{K : F\} = \{K : E\}\{E : F\}.$$

Then if K is separable over F, so that $[K : F] = \{K : F\}$, we must have $[K : E] = \{K : E\}$ and $[E : F] = \{E : F\}$, since in each case the index divides the degree, by Theorem 51.6. Thus, if K is separable over F, then K is separable over E and E is separable over F.

For the converse, note that $[K : E] = \{K : E\}$ and $[E : F] = \{E : F\}$ imply that

$$[K : F] = [K : E][E : F] = \{K : E\}\{E : F\} = \{K : F\}. \quad \blacklozenge$$

Theorem 51.9 can be extended in the obvious way, by induction, to any finite tower of finite extensions. The top field is a separable extension of the bottom one if and only if each field is a separable extension of the one immediately under it.

51.10 Corollary If E is a finite extension of F, then E is separable over F if and only if each α in E is separable over F.

Proof Suppose that E is separable over F, and let $\alpha \in E$. Then

$$F \leq F(\alpha) \leq E,$$

and Theorem 51.9 shows that $F(\alpha)$ is separable over F.

Suppose, conversely, that every $\alpha \in E$ is separable over F. Since E is a finite extension of F, there exist $\alpha_1, \cdots, \alpha_n$ such that

$$F < F(\alpha_1) < F(\alpha_1, \alpha_2) < \cdots < E = F(\alpha_1, \cdots, \alpha_n).$$

Now since α_i is separable over F, α_i is separable over $F(\alpha_1, \cdots, \alpha_{i-1})$, because

$$q(x) = \operatorname{irr}(\alpha_i, F(\alpha_1, \cdots, \alpha_{i-1}))$$

divides $\operatorname{irr}(\alpha_i, F)$, so that α_i is a zero of $q(x)$ of multiplicity 1. Thus $F(\alpha_1, \cdots, \alpha_i)$ is separable over $F(\alpha_1, \cdots, \alpha_{i-1})$, so E is separable over F by Theorem 51.9, extended by induction. ◆

Perfect Fields

We now turn to the task of proving that α can fail to be separable over F only if F is an infinite field of characteristic $p \neq 0$. One method is to introduce formal derivatives of polynomials. While this is an elegant technique, and also a useful one, we shall, for the sake of brevity, use the following lemma instead. Formal derivatives are developed in Exercises 15 through 22.

51.11 Lemma Let $\overline{F}$ be an algebraic closure of F, and let

$$f(x) = x^n + a_{n-1}x^{n-1} + \cdots + a_1 x + a_0$$

be any monic polynomial in $\overline{F}[x]$. If $(f(x))^m \in F[x]$ and $m \cdot 1 \neq 0$ in F, then $f(x) \in F[x]$, that is, all $a_i \in F$.

Proof We must show that $a_i \in F$, and we proceed, by induction on r, to show that $a_{n-r} \in F$. For $r = 1$,

$$(f(x))^m = x^{mn} + (m \cdot 1)a_{n-1}x^{mn-1} + \cdots + a_0{}^m.$$

Since $(f(x))^m \in F[x]$, we have, in particular,

$$(m \cdot 1)a_{n-1} \in F.$$

Thus $a_{n-1} \in F$, since $m \cdot 1 \neq 0$ in F.

As induction hypothesis, suppose that $a_{n-r} \in F$ for $r = 1, 2, \cdots, k$. Then the coefficient of $x^{mn-(k+1)}$ in $(f(x))^m$ is of the form

$$(m \cdot 1)a_{n-(k+1)} + g_{k+1}(a_{n-1}, a_{n-2}, \cdots, a_{n-k}),$$

where $g_{k+1}(a_{n-1}, a_{n-2}, \cdots, a_{n-k})$ is a formal polynomial expression in $a_{n-1}, a_{n-2}, \cdots, a_{n-k}$. By the induction hypothesis that we just stated, $g_{k+1}(a_{n-1}, a_{n-2}, \cdots, a_{n-k}) \in F$, so $a_{n-(k+1)} \in F$, since $m \cdot 1 \neq 0$ in F. ◆

We are now in a position to handle fields F of characteristic zero and to show that for a finite extension E of F, we have $\{E : F\} = [E : F]$. By definition, this amounts to proving that every finite extension of a field of characteristic zero is a separable extension. First, we give a definition.

51.12 Definition A field is **perfect** if every finite extension is a separable extension. ■

51.13 Theorem Every field of characteristic zero is perfect.

Proof Let E be a finite extension of a field F of characteristic zero, and let $\alpha \in E$. Then $f(x) = \text{irr}(\alpha, F)$ factors in $\overline{F}[x]$ into $\prod_i (x - \alpha_i)^\nu$, where the α_i are the distinct zeros of $\text{irr}(\alpha, F)$, and, say, $\alpha = \alpha_1$. Thus

$$f(x) = \left(\prod_i (x - \alpha_i)\right)^\nu,$$

and since $\nu \cdot 1 \neq 0$ for a field F of characteristic 0, we must have

$$\left(\prod_i (x - \alpha_i)\right) \in F[x]$$

by Lemma 51.11. Since $f(x)$ is irreducible and of minimal degree in $F[x]$ having α as a zero, we then see that $\nu = 1$. Therefore, α is separable over F for all $\alpha \in E$. By Corollary 51.10, this means that E is a separable extension of F. ◆

Lemma 51.11 will also get us through for the case of a finite field, although the proof is a bit harder.

51.14 Theorem Every finite field is perfect.

Proof Let F be a finite field of characteristic p, and let E be a finite extension of F. Let $\alpha \in E$. We need to show that α is separable over F. Now $f(x) = \text{irr}(\alpha, F)$ factors in $\overline{F}$ into $\prod_i (x - \alpha_i)^\nu$, where the α_i are the distinct zeros of $f(x)$, and, say, $\alpha = \alpha_1$. Let $\nu = p^t e$, where p does not divide e. Then

$$f(x) = \prod_i (x - \alpha_i)^\nu = \left(\prod_i (x - \alpha_i)^{p^t}\right)^e$$

is in $F[x]$, and by Lemma 51.11, $\prod_i (x - \alpha_i)^{p^t}$ is in $F[x]$ since $e \cdot 1 \neq 0$ in F. Since $f(x) = \text{irr}(\alpha, F)$ is of minimal degree over F having α as a zero, we must have $e = 1$.

Theorem 48.19 and the remark following it show then that

$$f(x) = \prod_i (x - \alpha_i)^{p^t} = \prod_i \left(x^{p^t} - \alpha_i^{p^t}\right).$$

Thus, if we regard $f(x)$ as $g(x^{p^t})$, we must have $g(x) \in F[x]$. Now $g(x)$ is separable over F with distinct zeros $\alpha_i^{p^t}$. Consider $F(\alpha_1^{p^t}) = F(\alpha^{p^t})$. Then $F(\alpha^{p^t})$ is separable over F. Since $x^{p^t} - \alpha^{p^t} = (x - \alpha)^{p^t}$, we see that α is the only zero of $x^{p^t} - \alpha^{p^t}$ in $\overline{F}$. As a finite-dimensional vector space over a finite field F, $F(\alpha^{p^t})$ must be again a finite field. Hence the map

$$\sigma_p : F(\alpha^{p^t}) \to F(\alpha^{p^t})$$

given by $\sigma_p(a) = a^p$ for $a \in F(\alpha^{p^t})$ is an automorphism of $F(\alpha^{p^t})$ by Theorem 48.19. Consequently, $(\sigma_p)^t$ is also an automorphism of $F(\alpha^{p^t})$, and

$$(\sigma_p)^t(a) = a^{p^t}.$$

Since an automorphism of $F(\alpha^{p^t})$ is an onto map, there is $\beta \in F(\alpha^{p^t})$ such that $(\sigma_p)^t(\beta) = \alpha^{p^t}$. But then $\beta^{p^t} = \alpha^{p^t}$, and we saw that α was the only zero of $x^{p^t} - \alpha^{p^t}$, so we must have $\beta = \alpha$. Since $\beta \in F(\alpha^{p^t})$, we have $F(\alpha) = F(\alpha^{p^t})$. Since $F(\alpha^{p^t})$ was separable over F, we now see that $F(\alpha)$ is separable over F. Therefore, α is separable over F and $t = 0$.

We have shown that for $\alpha \in E$, α is separable over F. Then by Corollary 51.10, E is a separable extension of F. ◆

We have completed our aim, which was to show that fields of characteristic 0 and finite fields have only separable finite extensions, that is, these fields are perfect. *For finite extensions E of such perfect fields F, we then have $[E : F] = \{E : F\}$.*

The Primitive Element Theorem

The following theorem is a classic of field theory.

51.15 Theorem **(Primitive Element Theorem)** Let E be a finite separable extension of a field F. Then there exists $\alpha \in E$ such that $E = F(\alpha)$. (Such an element α is a **primitive element.**) That is, a finite separable extension of a field is a simple extension.

Proof If F is a finite field, then E is also finite. Let α be a generator for the cylic group E^* of nonzero elements of E under multiplication. (See Theorem 33.5.) Clearly, $E = F(\alpha)$, so α is a primitive element in this case.

We now assume that F is infinite, and prove our theorem in the case that $E = F(\beta, \gamma)$. The induction argument from this to the general case is straightforward. Let $\text{irr}(\beta, F)$ have distinct zeros $\beta = \beta_1, \cdots, \beta_n$, and let $\text{irr}(\gamma, F)$ have distinct zeros $\gamma = \gamma_1, \cdots, \gamma_m$ in $\overline{F}$, where all zeros are of multiplicity 1, since E is a separable extension of F. Since F is infinite, we can find $a \in F$ such that

$$a \neq (\beta_i - \beta)/(\gamma - \gamma_j)$$

for all i and j, with $j \neq 1$. That is, $a(\gamma - \gamma_j) \neq \beta_i - \beta$. Letting $\alpha = \beta + a\gamma$, we have $\alpha = \beta + a\gamma \neq \beta_i + a\gamma_j$, so

$$\alpha - a\gamma_j \neq \beta_i$$

for all i and all $j \neq 1$. Let $f(x) = \text{irr}(\beta, F)$, and consider

$$h(x) = f(\alpha - ax) \in (F(\alpha))[x].$$

Now $h(\gamma) = f(\beta) = 0$. However, $h(\gamma_j) \neq 0$ for $j \neq 1$ by construction, since the β_i were the only zeros of $f(x)$. Hence $h(x)$ and $g(x) = \text{irr}(\gamma, F)$ have a common factor in $(F(\alpha))[x]$, namely $\text{irr}(\gamma, F(\alpha))$, which must be linear, since γ is the only common zero of $g(x)$ and $h(x)$. Thus $\gamma \in F(\alpha)$, and therefore $\beta = \alpha - a\gamma$ is in $F(\alpha)$. Hence $F(\beta, \gamma) = F(\alpha)$. ◆

51.16 Corollary A finite extension of a field of characteristic zero is a simple extension.

Proof This corollary follows at once from Theorems 51.13 and 51.15. ◆

We see that the only possible "bad case" where a finite extension may not be simple is a finite extension of an infinite field of characteristic $p \neq 0$.

EXERCISES 51

Computations

In Exercises 1 through 4, find α such that the given field is $\mathbb{Q}(\alpha)$. Show that your α is indeed in the given field. Verify by direct computation that the given generators for the extension of $\mathbb{Q}$ can indeed be expressed as formal polynomials in your α with coefficients in $\mathbb{Q}$.

1. $\mathbb{Q}(\sqrt{2}, \sqrt[3]{2})$

2. $\mathbb{Q}(\sqrt[4]{2}, \sqrt[6]{2})$

3. $\mathbb{Q}(\sqrt{2}, \sqrt{3})$

4. $\mathbb{Q}(i, \sqrt[3]{2})$

Concepts

In Exercises 5 and 6, correct the definition of the italicized term without reference to the text, if correction is needed, so that it is in a form acceptable for publication.

5. Let $\overline{F}$ be an algebraic closure of a field F. The *multiplicity of a zero* $\alpha \in \overline{F}$ of a polynomial $f(x) \in F[x]$ is $\nu \in \mathbb{Z}^+$ if and only if $(x - \alpha)^\nu$ is the highest power of $x - \alpha$ that is a factor of $f(x)$ in $F[x]$.

6. Let $\overline{F}$ be an algebraic closure of a field F. An element α in $\overline{F}$ is *separable over* F if and only if α is a zero of multiplicity 1 of $\text{irr}(\alpha, F)$.

7. Give an example of an $f(x) \in \mathbb{Q}[x]$ that has no zeros in $\mathbb{Q}$ but whose zeros in $\mathbb{C}$ are all of multiplicity 2. Explain how this is consistent with Theorem 51.13, which shows that $\mathbb{Q}$ is perfect.

8. Mark each of the following true or false.

_______ **a.** Every finite extension of every field F is separable over F.

_______ **b.** Every finite extension of every finite field F is separable over F.

_______ **c.** Every field of characteristic 0 is perfect.

_______ **d.** Every polynomial of degree n over every field F always has n distinct zeros in $\overline{F}$.

_______ **e.** Every polynomial of degree n over every perfect field F always has n distinct zeros in $\overline{F}$.

_______ **f.** Every irreducible polynomial of degree n over every perfect field F always has n distinct zeros in $\overline{F}$.

_______ **g.** Every algebraically closed field is perfect.

_______ **h.** Every field F has an algebraic extension E that is perfect.

_______ **i.** If E is a finite separable splitting field extension of F, then $|G(E/F)| = [E : F]$.

_______ **j.** If E is a finite splitting field extension of F, then $|G(E/F)|$ divides $[E : F]$.

Theory

9. Show that if $\alpha, \beta \in \overline{F}$ are both separable over F, then $\alpha \pm \beta$, $\alpha\beta$, and α/β, if $\beta \neq 0$, are all separable over F. [*Hint:* Use Theorem 51.9 and its corollary.]

10. Show that $\{1, y, \cdots, y^{p-1}\}$ is a basis for $\mathbb{Z}_p(y)$ over $\mathbb{Z}_p(y^p)$, where y is an indeterminate. Referring to Example 51.4, conclude by a degree argument that $x^p - t$ is irreducible over $\mathbb{Z}_p(t)$, where $t = y^p$.

11. Prove that if E is an algebraic extension of a perfect field F, then E is perfect.

12. A (possibly infinite) algebraic extension E of a field F is a **separable extension of** F if for every $\alpha \in E$, $F(\alpha)$ is a separable extension of F, in the sense defined in the text. Show that if E is a (possibly infinite) separable extension of F and K is a (possibly infinite) separable extension of E, then K is a separable extension of F.

13. Let E be an algebraic extension of a field F. Show that the set of all elements in E that are separable over F forms a subfield of E, the **separable closure of** F in E. [*Hint:* Use Exercise 9.]

14. Let E be a finite field of order p^n.

a. Show that the Frobenius automorphism σ_p has order n.

b. Deduce from part (a) that $G(E/\mathbb{Z}_p)$ is cyclic of order n with generator σ_p. [*Hint:* Remember that

$$|G(E/F)| = \{E : F\} = [E : F]$$

for a finite separable splitting field extension E over F.]

Exercises 15 through 22 introduce formal derivatives in $F[x]$.

15. Let F be any field and let $f(x) = a_0 + a_1x + \cdots + a_ix^i + \cdots + a_nx^n$ be in $F[x]$. The **derivative** $f'(x)$ **of** $f(x)$ is the polynomial

$$f'(x) = a_1 + \cdots + (i \cdot 1)a_ix^{i-1} + \cdots + (n \cdot 1)a_nx^{n-1},$$

where $i \cdot 1$ has its usual meaning for $i \in \mathbb{Z}^+$ and $1 \in F$. *These are formal derivatives; no "limits" are involved here.*

a. Prove that the map $D : F[x] \to F[x]$ given by $D(f(x)) = f'(x)$ is a homomorphism of $\langle F[x], +\rangle$.

b. Find the kernel of D in the case that F is of characteristic 0.

c. Find the kernel of D in the case that F is of characteristic $p \neq 0$.

16. Continuing the ideas of Exercise 15, shows that:

a. $D(af(x)) = aD(f(x))$ for all $f(x) \in F[x]$ and $a \in F$.

b. $D(f(x)g(x)) = f(x)g'(x) + f'(x)g(x)$ for all $f(x), g(x) \in F[x]$. [*Hint:* Use part (a) of this exercise and the preceding exercise and proceed by induction on the degree of $f(x)g(x)$.]

c. $D((f(x))^m) = (m \cdot 1)f(x)^{m-1}f'(x)$ for all $f(x) \in F[x]$. [*Hint:* Use part (*b*).]

17. Let $f(x) \in F[x]$, and let $\alpha \in \overline{F}$ be a zero of $f(x)$ of multiplicity ν. Show that $\nu > 1$ if and only if α is also a zero of $f'(x)$. [*Hint:* Apply parts (*b*) and (*c*) of Exercise 16 to the factorization $f(x) = (x - \alpha)^\nu g(x)$ of $f(x)$ in the ring $\overline{F}[x]$.]

18. Show from Exercise 17 that every irreducible polynomial over a field F of characteristic 0 is separable. [*Hint:* Use the fact that irr(α, F) is the *minimal* polynomial for α over F.]

19. Show from Exercise 17 that an irreducible polynomial $q(x)$ over a field F of characteristic $p \neq 0$ is not separable if and only if each exponent of each term of $q(x)$ is divisible by p.

20. Generalize Exercise 17, showing that $f(x) \in F[x]$ has no zero of multiplicity >1 if and only if $f(x)$ and $f'(x)$ have no common factor in $\overline{F}[x]$ of degree >0.

21. Working a bit harder than in Exercise 20, show that $f(x) \in F[x]$ has no zero of multiplicity >1 if and only if $f(x)$ and $f'(x)$ have no common nonconstant factor in $F[x]$. [*Hint:* Use Theorem 46.9 to show that if 1 is a gcd of $f(x)$ and $f'(x)$ in $F[x]$, it is a gcd of these polynomials in $\overline{F}[x]$ also.]

22. Describe a feasible computational procedure for determining whether $f(x) \in F[x]$ has a zero of multiplicity >1, without actually finding the zeros of $f(x)$. [*Hint:* Use Exercise 21.]

SECTION 52 †Totally Inseparable Extensions

This section shows that a finite extension E of a field F can be split into two stages: a separable extension K of F, followed by a further extension of K to E that is as far from being separable as one can imagine.

We develop our theory of totally inseparable extensions in a fashion parallel to our development of separable extensions.

52.1 Definition A finite extension E of a field F is a **totally inseparable extension of** F if $\{E : F\} = 1 < [E : F]$. An element α of $\overline{F}$ is **totally inseparable over** F if $F(\alpha)$ is totally inseparable over F. ■

We know that $\{F(\alpha) : F\}$ is the number of distinct zeros of irr(α, F). Thus α is totally inseparable over F if and only if irr(α, F) has only one zero that is of multiplicity >1.

52.2 Example Referring to Example 51.4, we see that $\mathbb{Z}_p(y)$ is totally inseparable over $\mathbb{Z}_p(y^p)$, where y is an indeterminate. ▲

52.3 Theorem **(Counterpart of Theorem 51.9)** If K is a finite extension of E, E is a finite extension of F, and $F < E < K$, then K is totally inseparable over F if and only if K is totally inseparable over E and E is totally inseparable over F.

† This section is not used in the remainder of the text.

Proof Since $F < E < K$, we have $[K : E] > 1$ and $[E : F] > 1$. Suppose K is totally inseparable over F. Then $\{K : F\} = 1$, and

$$\{K : F\} = \{K : E\}\{E : F\},$$

so we must have

$$\{K : E\} = 1 < [K : E] \quad \text{and} \quad \{E : F\} = 1 < [E : F].$$

Thus K is totally inseparable over E, and E is totally inseparable over F.

Conversely, if K is totally inseparable over E and E is totally inseparable over F, then

$$\{K : F\} = \{K : E\}\{E : F\} = (1)(1) = 1,$$

and $[K : F] > 1$. Thus K is totally inseparable over F. ◆

Theorem 52.3 can be extended by induction, to any finite proper tower of finite extensions. The top field is a totally inseparable extension of the bottom one if and only if each field is a totally inseparable extension of the one immediately under it.

52.4 Corollary **(Counterpart of the Corollary of Theorem 51.10)** If E is a finite extension of F, then E is totally inseparable over F if and only if each α in E, $\alpha \notin F$, is totally inseparable over F.

Proof Suppose that E is totally inseparable over F, and let $\alpha \in E$, with $\alpha \notin F$. Then

$$F < F(\alpha) \leq E.$$

If $F(\alpha) = E$, we are done, by the definition of α totally inseparable over F. If $F < F(\alpha) < E$, then Theorem 52.3 shows that since E is totally inseparable over F, $F(\alpha)$ is totally inseparable over F.

Conversely, suppose that for every $\alpha \in E$, with $\alpha \notin F, \alpha$ is totally inseparable over F. Since E is finite over F, there exist $\alpha_1, \cdots, \alpha_n$ such that

$$F < F(\alpha_1) < F(\alpha_1, \alpha_2) < \cdots < E = F(\alpha_1, \cdots, \alpha_n).$$

Now since α_i is totally inseparable over F, α_i is totally inseparable over $F(\alpha_1, \cdots, \alpha_{i-1})$, because $q(x) = \text{irr}(\alpha_i, F(\alpha_1, \cdots, \alpha_{i-1}))$ divides $\text{irr}(\alpha_i, F)$ so that α_i is the only zero of $q(x)$ and is of multiplicity >1. Thus $F(\alpha_1, \cdots, \alpha_i)$ is totally inseparable over $F(\alpha_1, \cdots, \alpha_{i-1})$, and E is totally inseparable over F, by Theorem 52.3, extended by induction. ◆

Thus far we have so closely paralleled our work in Section 51 that we could have handled these ideas together.

Separable Closures

We now come to our main reason for including this material.

52.5 Theorem Let F have characteristic $p \neq 0$, and let E be a finite extension of F. Then $\alpha \in E, \alpha \notin F$, is totally inseparable over F if and only if there is some integer $t \geq 1$ such that $\alpha^{p^t} \in F$.

Furthermore, there is a unique extension K of F, with $F \leq K \leq E$, such that K is separable over F, and either $E = K$ or E is totally inseparable over K.

Proof Let $\alpha \in E$, $\alpha \notin F$, be totally inseparable over F. Then $\operatorname{irr}(\alpha, F)$ has just one zero α of multiplicity >1, and, as shown in the proof of Theorem 51.14, $\operatorname{irr}(\alpha, F)$ must be of the form

$$x^{p^t} - \alpha^{p^t}.$$

Hence $\alpha^{p^t} \in F$ for some $t \geq 1$.

Conversely, if $\alpha^{p^t} \in F$ for some $t \geq 1$, where $\alpha \in E$ and $\alpha \notin F$, then

$$x^{p^t} - \alpha^{p^t} = (x - \alpha)^{p^t},$$

and $(x^{p^t} - \alpha^{p^t}) \in F[x]$, showing that $\operatorname{irr}(\alpha, F)$ divides $(x - \alpha)^{p^t}$. Thus $\operatorname{irr}(\alpha, F)$ has α as its only zero and this zero is of multiplicity >1, so α is totally inseparable over F.

For the second part of the theorem, let $E = F(\alpha_1, \cdots, \alpha_n)$. Then if

$$\operatorname{irr}(\alpha_i, F) = \prod_j \left(x^{p^{t_i}} - \alpha_{ij}{}^{p^{t_i}}\right),$$

with $\alpha_{i1} = \alpha_i$, let $\beta_{ij} = \alpha_{ij}{}^{p^{t_i}}$. We have $F(\beta_{11}, \beta_{21}, \cdots, \beta_{n1}) \leq E$, and β_{i1} is a zero of

$$f_i(x) = \prod_j (x - \beta_{ij}),$$

where $f_i(x) \in F[x]$. Now since raising to the power p is an isomorphism σ_p of E onto a subfield of E, raising to the power of p^t is the isomorphic mapping $(\sigma_p)^t$ of E onto a subfield of E. Thus since the α_{ij} are all distinct for a fixed i, so are the β_{ij} for a fixed i. Therefore, β_{ij} is separable over F, because it is a zero of a polynomial $f_i(x)$ in $F[x]$ with zeros of multiplicity 1. Then

$$K = F(\beta_{11}, \beta_{21}, \cdots, \beta_{n1})$$

is separable over F, by the proof of Corollary 51.10. If all $p^{t_i} = 1$, then $K = E$. If some $p^{t_i} \neq 1$, then $K \neq E$, and $\alpha_i^{p^{t_i}} = \beta_{i1}$ is in K, showing that each $\alpha_i \notin K$ is totally inseparable over K, by the first part of this theorem. Hence $E = K(\alpha_1, \cdots, \alpha_n)$ is totally inseparable over K, by the proof of Corollary 52.4.

It follows from Corollaries 51.10 and 52.4 that the field K consists of all elements α in E that are separable over F. Thus K is unique. ◆

52.6 Definition The unique field K of Theorem 52.5 is the **separable closure of** F **in** E. ■

The preceding theorem shows the precise structure of totally inseparable extensions of a field of characteristic p. Such an extension can be obtained by repeatedly adjoining pth roots of elements that are not already pth powers.

We remark that Theorem 52.5 is true for infinite algebraic extensions E of F. The proof of the first assertion of the theorem is valid for the case of infinite extensions also. For the second part, since $\alpha \pm \beta$, $\alpha\beta$, and α/β, for $\beta \neq 0$, are all contained in the field $F(\alpha, \beta)$, all elements of E separable over F form a subfield K of E, the **separable closure of** F **in** E. It follows that an $\alpha \in E$, $\alpha \notin K$, is totally inseparable over K, since α and all coefficients of $\operatorname{irr}(\alpha, K)$ are in a finite extension of F, and then Theorem 52.5 can be applied.

■ EXERCISES 52

Concepts

1. Let y and z be indeterminates, and let $u = y^{12}$ and $v = z^{18}$. Describe the separable closure of $\mathbb{Z}_3(u, v)$ in $\mathbb{Z}_3(y, z)$.
2. Let y and z be indeterminates, and let $u = y^{12}$ and $v = y^2z^{18}$. Describe the separable closure of $\mathbb{Z}_3(u, v)$ in $\mathbb{Z}_3(y, z)$.
3. Referring to Exercise 1, describe the totally inseparable closure (see Exercise 6) of $\mathbb{Z}_3(u, v)$ in $\mathbb{Z}_3(y, z)$.
4. Referring to Exercise 2, describe the totally inseparable closure of $\mathbb{Z}_3(u, v)$ in $\mathbb{Z}_3(y, z)$. (See Exercise 6.)
5. Mark each of the following true or false.

_______ **a.** No proper algebraic extension of an infinite field of characteristic $p \neq 0$ is ever a separable extension.
_______ **b.** If $F(\alpha)$ is totally inseparable over F of characteristic $p \neq 0$, then $\alpha^{p^t} \in F$ for some $t > 0$.
_______ **c.** For an indeterminate y, $\mathbb{Z}_5(y)$ is separable over $\mathbb{Z}_5(y^5)$.
_______ **d.** For an indeterminate y, $\mathbb{Z}_5(y)$ is separable over $\mathbb{Z}_5(y^{10})$.
_______ **e.** For an indeterminate y, $\mathbb{Z}_5(y)$ is totally inseparable over $\mathbb{Z}_5(y^{10})$.
_______ **f.** If F is a field and α is algebraic over F, then α is either separable or totally inseparable over F.
_______ **g.** If E is an algebraic extension of a field F, then F has a separable closure in E.
_______ **h.** If E is an algebraic extension of a field F, then E is totally inseparable over the separable closure of F in E.
_______ **i.** If E is an algebraic extension of a field F and E is not a separable extension of F, then E is totally inseparable over the separable closure of F in E.
_______ **j.** If α is totally inseparable over F, then α is the only zero of $\mathrm{irr}(\alpha, F)$.

Theory

6. Show that if E is an algebraic extension of a field F, then the union of F with the set of all elements of E totally inseparable over F forms a subfield of E, the **totally inseparable closure of F in E**.
7. Show that a field F of characteristic $p \neq 0$ is perfect if and only if $F^p = F$, that is, every element of F is a pth power of some element of F.
8. Let E be a finite extension of a field F of characteristic p. In the notation of Exercise 7, show that $E^p = E$ if and only if $F^p = F$. [*Hint:* The map $\sigma_p : E \to E$ defined by $\sigma_p(\alpha) = \alpha^p$ for $\alpha \in E$ is an isomorphism onto a subfield of E. Consider the diagram in Fig. 52.7, and make degree arguments.]

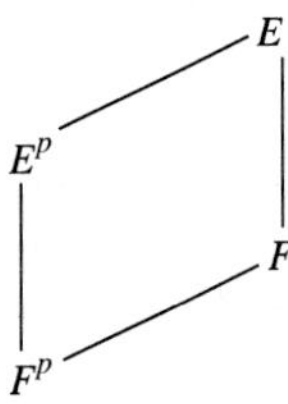

52.7 Figure

SECTION 53 Galois Theory

Résumé

This section is perhaps the climax in elegance of the subject matter of the entire text. The Galois theory gives a beautiful interplay of group and field theory. Starting with Section 48, our work has been aimed at this goal. We shall start by recalling the main results we have developed and should have well in mind.

1. Let $F \le E \le \overline{F}$, $\alpha \in E$, and let β be a conjugate of α over F, that is, $\text{irr}(\alpha, F)$ has β as a zero also. Then there is an isomorphism $\psi_{\alpha,\beta}$ mapping $F(\alpha)$ onto $F(\beta)$ that leaves F fixed and maps α onto β.
2. If $F \le E \le \overline{F}$ and $\alpha \in E$, then an automorphism σ of $\overline{F}$ that leaves F fixed *must* map α onto some conjugate of α over F.
3. If $F \le E$, the collection of all automorphisms of E leaving F fixed forms a group $G(E/F)$. For any subset S of $G(E/F)$, the set of all elements of E left fixed by all elements of S is a field E_S. Also, $F \le E_{G(E/F)}$.
4. A field E, $F \le E \le \overline{F}$, is a splitting field over F if and only if every isomorphism of E onto a subfield of $\overline{F}$ leaving F fixed is an automorphism of E. If E is a finite extension and a splitting field over F, then $|G(E/F)| = \{E : F\}$.
5. If E is a finite extension of F, then $\{E : F\}$ divides $[E : F]$. If E is also separable over F, then $\{E : F\} = [E : F]$. Also, E is separable over F if and only if $\text{irr}(\alpha, F)$ has all zeros of multiplicity 1 for every $\alpha \in E$.
6. If E is a finite extension of F and is a separable splitting field over F, then $|G(E/F)| = \{E : F\} = [E : F]$.

Normal Extensions

We are going to be interested in finite extensions K of F such that every isomorphism of K onto a subfield of $\overline{F}$ leaving F fixed is an automorphism of K and such that

$$[K : F] = \{K : F\}.$$

In view of results 4 and 5, these are the finite extensions of F that are separable splitting fields over F.

53.1 Definition A finite extension K of F is a **finite normal extension of** F if K is a separable splitting field over F. ■

Suppose that K is a finite normal extension of F, where $K \le \overline{F}$, as usual. Then by result 4, every automorphism of $\overline{F}$ leaving F fixed induces an automorphism of K. As before, we let $G(K/F)$ be the group of all automorphisms of K leaving F fixed. After one more result, we shall be ready to illustrate the main theorem.

53.2 Theorem Let K be a finite normal extension of F, and let E be an extension of F, where $F \le E \le K \le \overline{F}$. Then K is a finite normal extension of E, and $G(K/E)$ is precisely the subgroup of $G(K/F)$ consisting of all those automorphisms that leave E fixed. Moreover, two automorphisms σ and τ in $G(K/F)$ induce the same isomorphism of E onto a subfield of $\overline{F}$ if and only if they are in the same left coset of $G(K/E)$ in $G(K/F)$.

Proof If K is the splitting field of a set $\{f_i(x) \mid i \in I\}$ of polynomials in $F[x]$, then K is the splitting field over E of this same set of polynomials viewed as elements of $E[x]$. Theorem 51.9 shows that K is separable over E, since K is separable over F. Thus K is a normal extension of E. This establishes our first contention.

Now every element of $G(K/E)$ is an automorphism of K leaving F fixed, since it even leaves the possibly larger field E fixed. Thus $G(K/E)$ can be viewed as a subset of $G(K/F)$. Since $G(K/E)$ is a group under function composition also, we see that $G(K/E) \leq G(K/F)$.

Finally, for σ and τ in $G(K/F)$, σ and τ are in the same left coset of $G(K/E)$ if and only if $\tau^{-1}\sigma \in G(K/E)$ or if and only if $\sigma = \tau\mu$ for $\mu \in G(K/E)$. But if $\sigma = \tau\mu$ for $\mu \in G(K/E)$, then for $\alpha \in E$, we have

$$\sigma(\alpha) = (\tau\mu)(\alpha) = \tau(\mu(\alpha)) = \tau(\alpha),$$

since $\mu(\alpha) = \alpha$ for $\alpha \in E$. Conversely, if $\sigma(\alpha) = \tau(\alpha)$ for all $\alpha \in E$, then

$$(\tau^{-1}\sigma)(\alpha) = \alpha$$

for all $\alpha \in E$, so $\tau^{-1}\sigma$ leaves E fixed, and $\mu = \tau^{-1}\sigma$ is thus in $G(K/E)$. ◆

The preceding theorem shows that there is a one-to-one correspondence between left cosets of $G(K/E)$ in $G(K/F)$ and isomorphisms of E onto a subfield of K leaving F fixed. Note that we cannot say that these left cosets correspond to *automorphisms* of E over F, since E may not be a splitting field over F. Of course, if E is a *normal* extension of F, then these isomorphisms would be automorphisms of E over F. We might guess that this will happen if and only if $G(K/E)$ is a *normal* subgroup of $G(K/F)$, and this is indeed the case. That is, the two different uses of the word *normal* are really closely related. Thus if E is a normal extension of F, then the left cosets of $G(K/E)$ in $G(K/F)$ can be viewed as elements of the *factor group* $G(K/F)/G(K/E)$, which is then a group of automorphisms acting on E and leaving F fixed. We shall show that this factor group is isomorphic to $G(E/F)$.

The Main Theorem

The Main Theorem of Galois Theory states that for a finite normal extension K of a field F, there is a one-to-one correspondence between the subgroups of $G(K/F)$ and the intermediate fields E, where $F \leq E \leq K$. *This correspondence associates with each intermediate field E the subgroup $G(K/E)$. We can also go the other way and start with a subgroup H of $G(K/F)$ and associate with H its fixed field K_H.* We shall illustrate this with an example, then state the theorem and discuss its proof.

53.3 Example Let $K = \mathbb{Q}(\sqrt{2}, \sqrt{3})$. Now K is a normal extension of $\mathbb{Q}$, and Example 48.17 showed that there are four automorphisms of K leaving $\mathbb{Q}$ fixed. We recall them by giving their values on the basis $\{1, \sqrt{2}, \sqrt{3}, \sqrt{6}\}$ for K over $\mathbb{Q}$.

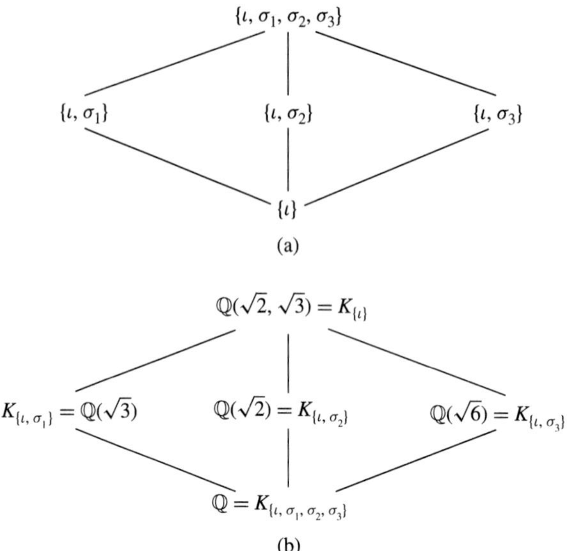

53.4 Figure (a) Group diagram. (b) Field diagram.

ι: The identity map
σ_1: Maps $\sqrt{2}$ onto $-\sqrt{2}$, $\sqrt{6}$ onto $-\sqrt{6}$, and leaves the others fixed
σ_2: Maps $\sqrt{3}$ onto $-\sqrt{3}$, $\sqrt{6}$ onto $-\sqrt{6}$, and leaves the others fixed
σ_3: Maps $\sqrt{2}$ onto $-\sqrt{2}$, $\sqrt{3}$ onto $-\sqrt{3}$, and leaves the others fixed

We saw that $\{\iota, \sigma_1, \sigma_2, \sigma_3\}$ is isomorphic to the Klein 4-group. The complete list of subgroups, with each subgroup paired off with the corresponding intermediate field that it leaves fixed, is as follows:

$$\begin{aligned} \{\iota, \sigma_1, \sigma_2, \sigma_3\} &\leftrightarrow \mathbb{Q}, \\ \{\iota, \sigma_1\} &\leftrightarrow \mathbb{Q}(\sqrt{3}), \\ \{\iota, \sigma_2\} &\leftrightarrow \mathbb{Q}(\sqrt{2}), \\ \{\iota, \sigma_3\} &\leftrightarrow \mathbb{Q}(\sqrt{6}), \\ \{\iota\} &\leftrightarrow \mathbb{Q}(\sqrt{2}, \sqrt{3}). \end{aligned}$$

All subgroups of the abelian group $\{\iota, \sigma_1, \sigma_2, \sigma_3\}$ are normal subgroups, and all the intermediate fields are normal extensions of $\mathbb{Q}$. Isn't that elegant?

Note that if one subgroup is contained in another, then the larger of the two subgroups corresponds to the smaller of the two corresponding fixed fields. The larger the subgroup, that is, the more automorphisms, the smaller the fixed field, that is, the fewer elements left fixed. In Fig. 53.4 we give the corresponding diagrams for the subgroups and intermediate fields. *Note again that the groups near the top correspond to the fields near the bottom.* That is, one diagram looks like the other *inverted* or turned upside down. Since here each diagram actually looks like itself turned upside down, this is not a good example for us to use to illustrate this *inversion principle*. Turn ahead to Fig. 54.6 to see diagrams that do not look like their own inversions. ▲

53.5 Definition If K is a finite normal extension of a field F, then $G(K/F)$ is the **Galois group of K over F.** ■

We shall now state the main theorem, then give another example, and finally, complete the proof of the main theorem.

53.6 Theorem **(Main Theorem of Galois Theory)** Let K be a finite normal extension of a field F, with Galois group $G(K/F)$. For a field E, where $F \leq E \leq K$, let $\lambda(E)$ be the subgroup of $G(K/F)$ leaving E fixed. Then λ is a one-to-one map of the set of all such intermediate fields E onto the set of all subgroups of $G(K/F)$. The following properties hold for λ:

1. $\lambda(E) = G(K/E)$.
2. $E = K_{G(K/E)} = K_{\lambda(E)}$.
3. For $H \leq G(K/F)$, $\lambda(K_H) = H$.
4. $[K : E] = |\lambda(E)|$ and $[E : F] = (G(K/F) : \lambda(E))$, the number of left cosets of $\lambda(E)$ in $G(K/F)$.
5. E is a normal extension of F if and only if $\lambda(E)$ is a normal subgroup of $G(K/F)$. When $\lambda(E)$ is a normal subgroup of $G(K/F)$, then
$$G(E/F) \simeq G(K/F)/G(K/E).$$
6. The diagram of subgroups of $G(K/F)$ is the inverted diagram of intermediate fields of K over F.

Observations on the Proof We have really already proved a substantial part of this theorem. Let us see just how much we have left to prove.

Property 1 is just the definition of λ found in the statement of the theorem. For Property 2, Theorem 48.15 shows that
$$E \leq K_{G(K/E)}.$$
Let $\alpha \in K$, where $\alpha \notin E$. Since K is a normal extension of E, by using a conjugation isomorphism and the Isomorphism Extension Theorem, we can find an automorphism of K leaving E fixed and mapping α onto a different zero of $\text{irr}(\alpha, F)$. This implies that
$$K_{G(K/E)} \leq E,$$
so $E = K_{G(K/E)}$. This disposes of Property 2 and also tells us that λ is one to one, for if $\lambda(E_1) = \lambda(E_2)$, then by Property 2, we have
$$E_1 = K_{\lambda(E_1)} = K_{\lambda(E_2)} = E_2.$$

Now Property 3 is going to be our main job. This amounts exactly to showing that λ is an onto map. Of course, for $H \leq G(K/F)$, we have $H \leq \lambda(K_H)$, for H surely is included in the set of all automorphisms leaving K_H fixed. Here we will be using strongly our property $[K : E] = \{K : E\}$.

Property 4 follows from $[K : E] = \{K : E\}$, $[E : F] = \{E : F\}$, and the last statement in Theorem 53.2.

We shall have to show that the two senses of the word *normal* correspond for Property 5.

We have already disposed of Property 6 in Example 53.3. *Thus only Properties 3 and 5 remain to be proved.*

The Main Theorem of Galois Theory is a strong tool in the study of zeros of polynomials. If $f(x) \in F[x]$ is such that every irreducible factor of $f(x)$ is separable over F, then the splitting field K of $f(x)$ over F is a normal extension of F. The Galois group $G(K/F)$ is the **group of the polynomial** $f(x)$ **over** F. The structure of this group may give considerable information regarding the zeros of $f(x)$. This will be strikingly illustrated in Section 56 when we achieve our *final goal.*

Galois Groups over Finite Fields

Let K be a finite extension of a *finite field* F. We have seen that K is a separable extension of F (a finite field is perfect). Suppose that the order of F is p^r and $[K : F] = n$, so the order of K is p^{rn}. Then we have seen that K is the splitting field of $x^{p^{rn}} - x$ over F. Hence K is a normal extension of F.

Now one automorphism of K that leaves F fixed is σ_{p^r}, where for $\alpha \in K$, $\sigma_{p^r}(\alpha) = \alpha^{p^r}$. Note that $(\sigma_{p^r})^i(\alpha) = \alpha^{p^{ri}}$. Since a polynomial of degree p^{ri} can have at most p^{ri} zeros in a field, we see that the smallest power of σ_{p^r} that could possibly leave all p^{rn} elements of K fixed is the nth power. That is, the order of the element σ_{p^r} in $G(K/F)$ is at least n. Therefore, since $|G(K/F)| = [K : F] = n$, it must be that $G(K/F)$ is cyclic and generated by σ_{p^r}. We summarize these arguments in a theorem.

53.7 Theorem Let K be a finite extension of degree n of a finite field F of p^r elements. Then $G(K/F)$ is cyclic of order n, and is generated by σ_{p^r}, where for $\alpha \in K$, $\sigma_{p^r}(\alpha) = \alpha^{p^r}$.

We use this theorem to give another illustration of the Main Theorem of Galois Theory.

53.8 Example Let $F = \mathbb{Z}_p$, and let $K = \mathrm{GF}(p^{12})$, so $[K : F] = 12$. Then $G(K/F)$ is isomorphic to the cyclic group $\langle \mathbb{Z}_{12}, +\rangle$. The diagrams for the subgroups and for the intermediate fields are given in Fig. 53.9. Again, each diagram is not only the inversion of the other, but unfortunately, also looks like the inversion of itself. Examples where the diagrams do not look like their own inversion are given in next Section 54. We describe the cyclic

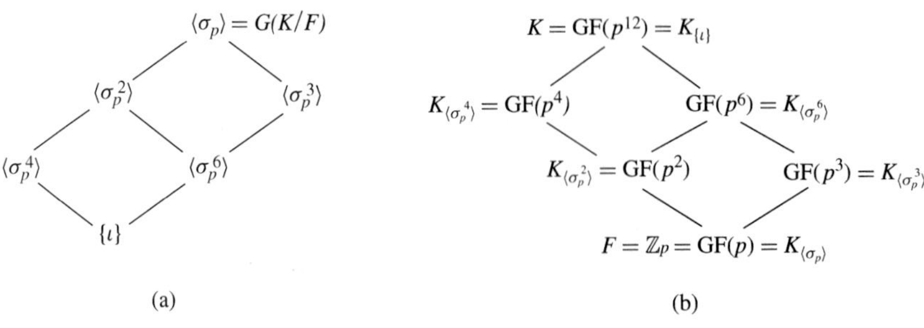

53.9 Figure (a) Group diagram. (b) Field diagram.

subgroups of $G(K/F) = \langle \sigma_p \rangle$ by giving generators, for example,

$$\langle {\sigma_p}^4 \rangle = \{\iota, {\sigma_p}^4, {\sigma_p}^8\}. \qquad \blacktriangle$$

Proof of the Main Theorem Completed

We saw that Properties 3 and 5 are all that remain to be proved in the Main Theorem of Galois Theory.

Proof Turning to Property 3, we must show that for $H \leq G(K/F)$, $\lambda(K_H) = H$. We know that $H \leq \lambda(K_H) \leq G(K/F)$. Thus what we really must show is that it is impossible to have H a *proper* subgroup of $\lambda(K_H)$. We shall suppose that

$$H < \lambda(K_H)$$

and shall derive a contradiction. As a finite separable extension, $K = K_H(\alpha)$ for some $\alpha \in K$, by Theorem 51.15. Let

$$n = [K : K_H] = \{K : K_H\} = |G(K/K_H)|.$$

Then $H < G(K/K_H)$ implies that $|H| < |G(K/K_H)| = n$. Thus we would have to have $|H| < [K : K_H] = n$. Let the elements of H be $\sigma_1, \cdots, \sigma_{|H|}$, and consider the polynomial

$$f(x) = \prod_{i=1}^{|H|}(x - \sigma_i(\alpha)).$$

Then $f(x)$ is of degree $|H| < n$. Now the coefficients of each power of x in $f(x)$ are *symmetric* expressions in the $\sigma_i(\alpha)$. For example, the coefficient of $x^{|H|-1}$ is $-\sigma_1(\alpha) - \sigma_2(\alpha) - \cdots - \sigma_{|H|}(\alpha)$. Thus these coefficients are invariant under each isomorphism $\sigma_i \in H$, since if $\sigma \in H$, then

$$\sigma\sigma_1, \cdots, \sigma\sigma_{|H|}$$

is again the sequence $\sigma_1, \cdots, \sigma_{|H|}$, except for order, H being a group. Hence $f(x)$ has coefficients in K_H, and since some σ_i is ι, we see that some $\sigma_i(\alpha)$ is α, so $f(\alpha) = 0$. Therefore, we would have

$$\deg(\alpha, K_H) \leq |H| < n = [K : K_H] = [K_H(\alpha) : K_H].$$

This is impossible. Thus we have proved Property 3.

We turn to Property 5. Every extension E of F, $F \leq E \leq K$, is separable over F, by Theorem 51.9. Thus E is normal over F if and only if E is a splitting field over F. By the Isomorphism Extension Theorem, every isomorphism of E onto a subfield of $\overline{F}$ leaving F fixed can be extended to an *automorphism* of K, since K is *normal* over F. Thus the automorphisms of $G(K/F)$ induce all possible isomorphisms of E onto a subfield of $\overline{F}$ leaving F fixed. By Theorem 50.3, this shows that E is a splitting field over F, and hence is normal over F, if and only if for all $\sigma \in G(K/F)$ and $\alpha \in E$,

$$\sigma(\alpha) \in E.$$

By Property 2, E is the fixed field of $G(K/E)$, so $\sigma(\alpha) \in E$ if and only if for all $\tau \in G(K/E)$

$$\tau(\sigma(\alpha)) = \sigma(\alpha).$$

This in turn holds if and only if

$$(\sigma^{-1}\tau\sigma)(\alpha) = \alpha$$

for all $\alpha \in E, \sigma \in G(K/F)$, and $\tau \in G(K/E)$. But this means that for all $\sigma \in G(K/F)$ and $\tau \in G(K/E)$, $\sigma^{-1}\tau\sigma$ leaves every element of E fixed, that is,

$$(\sigma^{-1}\tau\sigma) \in G(K/E).$$

This is precisely the condition that $G(K/E)$ be a normal subgroup of $G(K/F)$.

It remains for us to show that when E is a normal extension of F, $G(E/F) \simeq G(K/F)/G(K/E)$. For $\sigma \in G(K/F)$, let σ_E be the *automorphism* of E induced by σ (we are assuming that E is a *normal* extension of F). Thus $\sigma_E \in G(E/F)$. The map $\phi : G(K/F) \to G(E/F)$ given by

$$\phi(\sigma) = \sigma_E$$

for $\sigma \in G(K/F)$ is a homomorphism. By the Isomorphism Extension Theorem, every automorphism of E leaving F fixed can be extended to some automorphism of K; that is, it is τ_E for some $\tau \in G(K/F)$. Thus ϕ is onto $G(E/F)$. The kernel of ϕ is $G(K/E)$. Therefore, by the Fundamental Isomorphism Theorem, $G(E/F) \simeq G(K/F)/G(K/E)$. Furthermore, this isomorphism is a natural one. ◆

EXERCISES 53

Computations

The field $K = \mathbb{Q}(\sqrt{2}, \sqrt{3}, \sqrt{5})$ is a finite normal extension of $\mathbb{Q}$. It can be shown that $[K : \mathbb{Q}] = 8$. In Exercises 1 through 8, compute the indicated numerical quantity. The notation is that of Theorem 53.6.

1. $\{K : \mathbb{Q}\}$

2. $|G(K/\mathbb{Q})|$

3. $|\lambda(\mathbb{Q})|$

4. $|\lambda(\mathbb{Q}(\sqrt{2}, \sqrt{3}))|$

5. $|\lambda(\mathbb{Q}(\sqrt{6}))|$

6. $|\lambda(\mathbb{Q}(\sqrt{30}))|$

7. $|\lambda(\mathbb{Q}(\sqrt{2} + \sqrt{6}))|$

8. $|\lambda(K)|$

9. Describe the group of the polynomial $(x^4 - 1) \in \mathbb{Q}[x]$ over $\mathbb{Q}$.

10. Give the order and describe a generator of the group G(GF(729)/GF(9)).

11. Let K be the splitting field of $x^3 - 2$ over $\mathbb{Q}$. (Refer to Example 50.9.)

- **a.** Describe the six elements of $G(K/\mathbb{Q})$ by giving their values on $\sqrt[3]{2}$ and $i\sqrt{3}$. (By Example 50.9, $K = \mathbb{Q}(\sqrt[3]{2}, i\sqrt{3})$.)
- **b.** To what group we have seen before is $G(K/\mathbb{Q})$ isomorphic?
- **c.** Using the notation given in the answer to part (a) in the back of the text, give the diagrams for the subfields of K and for the subgroups of $G(K/\mathbb{Q})$, indicating corresponding intermediate fields and subgroups, as we did in Fig. 53.4.

12. Describe the group of the polynomial $(x^4 - 5x^2 + 6) \in \mathbb{Q}[x]$ over $\mathbb{Q}$.

13. Describe the group of the polynomial $(x^3 - 1) \in \mathbb{Q}[x]$ over $\mathbb{Q}$.

Concepts

14. Give an example of two finite normal extensions K_1 and K_2 of the same field F such that K_1 and K_2 are not isomorphic fields but $G(K_1/F) \simeq G(K_2/F)$.

15. Mark each of the following true or false.

______ **a.** Two different subgroups of a Galois group may have the same fixed field.

______ **b.** In the notation of Theorem 53.6, if $F \leq E < L \leq K$, then $\lambda(E) < \lambda(L)$.

______ **c.** If K is a finite normal extension of F, then K is a normal extension of E, where $F \leq E \leq K$.

______ **d.** If two finite normal extensions E and L of a field F have isomorphic Galois groups, then $[E : F] = [L : F]$.

______ **e.** If E is a finite normal extension of F and H is a normal subgroup of $G(E/F)$, then E_H is a normal extension of F.

______ **f.** If E is any finite normal simple extension of a field F, then the Galois group $G(E/F)$ is a simple group.

______ **g.** No Galois group is simple.

______ **h.** The Galois group of a finite extension of a finite field is abelian.

______ **i.** An extension E of degree 2 over a field F is always a normal extension of F.

______ **j.** An extension E of degree 2 over a field F is always a normal extension of F if the characteristic of F is not 2.

Theory

16. A finite normal extension K of a field F is **abelian over** F if $G(K/F)$ is an abelian group. Show that if K is abelian over F and E is a normal extension of F, where $F \leq E \leq K$, then K is abelian over E and E is abelian over F.

17. Let K be a finite normal extension of a field F. Prove that for every $\alpha \in K$, the **norm of** α **over** F, given by

$$N_{K/F}(\alpha) = \prod_{\sigma \in G(K/F)} \sigma(\alpha),$$

and the **trace of** α **over** F, given by

$$Tr_{K/F}(\alpha) = \sum_{\sigma \in G(K/F)} \sigma(\alpha),$$

are elements of F.

18. Consider $K = \mathbb{Q}(\sqrt{2}, \sqrt{3})$. Referring to Exercise 17, compute each of the following (see Example 53.3).

a. $N_{K/\mathbb{Q}}(\sqrt{2})$
b. $N_{K/\mathbb{Q}}(\sqrt{2} + \sqrt{3})$
c. $N_{K/\mathbb{Q}}(\sqrt{6})$
d. $N_{K/\mathbb{Q}}(2)$
e. $Tr_{K/\mathbb{Q}}(\sqrt{2})$
f. $Tr_{K/\mathbb{Q}}(\sqrt{2} + \sqrt{3})$
g. $Tr_{K/\mathbb{Q}}(\sqrt{6})$
h. $Tr_{K/\mathbb{Q}}(2)$

19. Let K be a normal extension of F, and let $K = F(\alpha)$. Let

$$\operatorname{irr}(\alpha, F) = x^n + a_{n-1}x^{n-1} + \cdots + a_1 x + a_0.$$

Referring to Exercise 17, show that

a. $N_{K/F}(\alpha) = (-1)^n a_0$,
b. $Tr_{K/F}(\alpha) = -a_{n-1}$.

20. Let $f(x) \in F[x]$ be a polynomial of degree n such that each irreducible factor is separable over F. Show that the order of the group of $f(x)$ over F divides $n!$.

21. Let $f(x) \in F[x]$ be a polynomial such that every irreducible factor of $f(x)$ is a separable polynomial over F. Show that the group of $f(x)$ over F can be viewed in a natural way as a group of permutations of the zeros of $f(x)$ in $\bar{F}$.

22. Let F be a field and let ζ be a primitive nth root of unity in $\bar{F}$, where the characteristic of F is either 0 or does not divide n.

a. Show that $F(\zeta)$ is a normal extension of F.

b. Show that $G(F(\zeta)/F)$ is abelian. [*Hint:* Every $\sigma \in G(F(\zeta)/F)$ maps ζ onto some ζ^r and is completely determined by this value r.]

23. A finite normal extension K of a field F is **cyclic over** F if $G(K/F)$ is a cyclic group.

a. Show that if K is cyclic over F and E is a normal extension of F, where $F \leq E \leq K$, then E is cyclic over F and K is cyclic over E.

b. Show that if K is cyclic over F, then there exists exactly one field E, $F \leq E \leq K$, of degree d over F for each divisor d of $[K : F]$.

24. Let K be a finite normal extension of F.

a. For $\alpha \in K$, show that

$$f(x) = \prod_{\sigma \in G(K/F)} (x - \sigma(\alpha))$$

is in $F[x]$.

b. Referring to part (a), show that $f(x)$ is a power of $\text{irr}(\alpha, F)$, and $f(x) = \text{irr}(\alpha, F)$ if and only if $K = F(\alpha)$.

25. The **join** $E \vee L$ of two extension fields E and L of F in $\bar{F}$ is the smallest subfield of $\bar{F}$ containing both E and L. That is, $E \vee L$ is the intersection of all subfields of $\bar{F}$ *containing both* E and L. Let K be a finite normal extension of a field F, and let E and L be extensions of F contained in K, as shown in Fig. 53.10. Describe $G(K/(E \vee L))$ in terms of $G(K/E)$ and $G(K/L)$.

26. With reference to the situation in Exercise 25, describe $G(K/(E \cap L))$ in terms of $G(K/E)$ and $G(K/L)$.

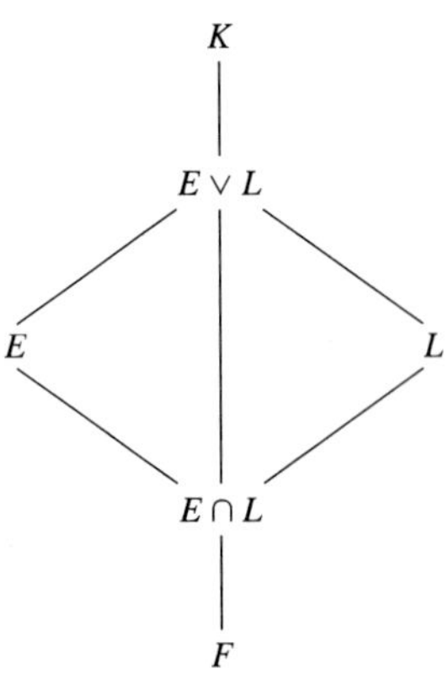

53.9 Figure

SECTION 54 ILLUSTRATIONS OF GALOIS THEORY

Symmetric Functions

Let F be a field, and let $y_1, \cdots, y_n$ be indeterminates. There are some natural automorphisms of $F(y_1, \cdots, y_n)$ leaving F fixed, namely, those defined by permutations of $\{y_1, \cdots, y_n\}$. To be more explicit, let σ be a permutation of $\{1, \cdots, n\}$, that is, $\sigma \in S_n$. Then σ gives rise to a natural map $\overline{\sigma} : F(y_1, \cdots, y_n) \to F(y_1, \cdots, y_n)$ given by

$$\overline{\sigma}\left(\frac{f(y_1, \cdots, y_n)}{g(y_1, \cdots, y_n)}\right) = \frac{f(y_{\sigma(1)}, \cdots, y_{\sigma(n)})}{g(y_{\sigma(1)}, \cdots, y_{\sigma(n)})}$$

for $f(y_1, \cdots, y_n), g(y_1, \cdots, y_n) \in F[y_1, \cdots, y_n]$, with $g(y_1, \cdots, y_n) \neq 0$. It is immediate that $\overline{\sigma}$ is an automorphism of $F(y_1, \cdots, y_n)$ leaving F fixed. The elements of $F(y_1, \cdots, y_n)$ left fixed by *all* $\overline{\sigma}$, for all $\sigma \in S_n$, are those rational functions that are *symmetric* in the indeterminates $y_1, \cdots, y_n$.

54.1 Definition An element of the field $F(y_1, \cdots, y_n)$ is a **symmetric function in** $y_1, \cdots, y_n$ **over** F, if it is left fixed by all permutations of $y_1, \cdots, y_n$, in the sense just explained. ■

Let $\overline{S_n}$ be the group of all the automorphisms $\overline{\sigma}$ for $\sigma \in S_n$. Observe that $\overline{S_n}$ is naturally isomorphic to S_n. Let K be the subfield of $F(y_1, \cdots, y_n)$ which is the fixed field of $\overline{S_n}$. Consider the polynomial

$$f(x) = \prod_{i=1}^{n}(x - y_i);$$

this polynomial $f(x) \in (F(y_1, \cdots, y_n))[x]$ is a **general polynomial of degree** n. Let $\overline{\sigma_x}$ be the extension of $\overline{\sigma}$, in the natural way, to $(F(y_1, \cdots, y_n))[x]$, where $\overline{\sigma_x}(x) = x$. Now $f(x)$ is left fixed by each map $\overline{\sigma_x}$ for $\sigma \in S_n$; that is,

$$\prod_{i=1}^{n}(x - y_i) = \prod_{i=1}^{n}(x - y_{\sigma(i)}).$$

Thus the coefficients of $f(x)$ are in K; they are *elementary symmetric functions* in the $y_1, \cdots, y_n$. As illustration, note that the constant term of $f(x)$ is

$$(-1)^n y_1 y_2 \cdots y_n,$$

the coefficient of x^{n-1} is $-(y_1 + y_2 + \cdots + y_n)$, and so on. These are symmetric functions in $y_1, \cdots, y_n$.

The first elementary symmetric function in $y_1, \cdots, y_n$ is

$$s_1 = y_1 + y_2 + \cdots + y_n,$$

the second is $s_2 = y_1y_2 + y_1y_3 + \cdots + y_{n-1}y_n$, and so on, and the nth is $s_n = y_1y_2 \cdots y_n$.

Consider the field $E = F(s_1, \cdots, s_n)$. Of course, $E \leq K$, where K is the field of all symmetric functions in $y_1, \cdots, y_n$ over F. But $F(y_1, \cdots, y_n)$ is a finite normal extension of E, namely, the splitting field of

$$f(x) = \prod_{i=1}^{n}(x - y_i)$$

over E. Since the degree of $f(x)$ is n, we have at once

$$[F(y_1, \cdots, y_n) : E] \leq n!$$

(see Exercise 13, Section 50). However, since K is the fixed field of $\overline{S_n}$ and

$$|\overline{S_n}| = |S_n| = n!,$$

we have also

$$n! \leq \{F(y_1, \cdots, y_n) : K\} \leq [F(y_1, \cdots, y_n) : K].$$

Therefore,

$$n! \leq [F(y_1, \cdots, y_n) : K] \leq [F(y_1, \cdots, y_n) : E] \leq n!,$$

so

$$K = E.$$

The full Galois group of $F(y_1, \cdots, y_n)$ over E is therefore $\overline{S_n}$. The fact that $K = E$ shows that every symmetric function can be expressed as a rational function of the elementary symmetric functions $s_1, \cdots, s_n$. We summarize these results in a theorem.

54.2 Theorem Let $s_1, \cdots, s_n$ be the elementary symmetric functions in the indeterminates $y_1, \cdots, y_n$. Then every symmetric function of $y_1, \cdots, y_n$ over F is a rational function of the elementary symmetric functions. Also, $F(y_1, \cdots, y_n)$ is a finite normal extension of degree $n!$ of $F(s_1, \cdots, s_n)$, and the Galois group of this extension is naturally isomorphic to S_n.

In view of Cayley's Theorem 8.16, it can be deduced from Theorem 54.2 that any finite group can occur as a Galois group (up to isomorphism). (See Exercise 11.)

Examples

Let us give our promised example of a finite normal extension having a Galois group whose subgroup diagram does not look like its own inversion.

54.3 Example Consider the splitting field in $\mathbb{C}$ of $x^4 - 2$ over $\mathbb{Q}$. Now $x^4 - 2$ is irreducible over $\mathbb{Q}$, by Eisenstein's criterion, with $p = 2$. Let $\alpha = \sqrt[4]{2}$ be the real positive zero of $x^4 - 2$. Then the four zeros of $x^4 - 2$ in $\mathbb{C}$ are α, $-\alpha$, $i\alpha$, and $-i\alpha$, where i is the usual zero of $x^2 + 1$ in $\mathbb{C}$. The splitting field K of $x^4 - 2$ over $\mathbb{Q}$ thus contains $(i\alpha)/\alpha = i$. Since α is a real number, $\mathbb{Q}(\alpha) < \mathbb{R}$, so $\mathbb{Q}(\alpha) \neq K$. However, since $\mathbb{Q}(\alpha, i)$ contains all zeros of $x^4 - 2$, we see that $\mathbb{Q}(\alpha, i) = K$. Letting $E = \mathbb{Q}(\alpha)$, we have the diagram in Fig. 54.4.

Now $\{1, \alpha, \alpha^2, \alpha^3\}$ is a basis for E over $\mathbb{Q}$, and $\{1, i\}$ is a basis for K over E. Thus

$$\{1, \alpha, \alpha^2, \alpha^3, i, i\alpha, i\alpha^2, i\alpha^3\}$$

$K = \mathbb{Q}(\alpha, i)$

$E = \mathbb{Q}(\alpha)$

$\mathbb{Q}$

54.4 Figure

is a basis for K over $\mathbb{Q}$. Since $[K : \mathbb{Q}] = 8$, we must have $|G(K/\mathbb{Q})| = 8$, so we need to find eight automorphisms of K leaving $\mathbb{Q}$ fixed. We know that any such automorphism σ is completely determined by its values on elements of the basis $\{1, \alpha, \alpha^2, \alpha^3, i, i\alpha, i\alpha^2, i\alpha^3\}$, and these values are in turn determined by $\sigma(\alpha)$ and $\sigma(i)$. But $\sigma(\alpha)$ must always be a conjugate of α over $\mathbb{Q}$, that is, one of the four zeros of $\text{irr}(\alpha, \mathbb{Q}) = x^4 - 2$. Likewise, $\sigma(i)$ must be a zero of $\text{irr}(i, \mathbb{Q}) = x^2 + 1$. Thus the four possibilities for $\sigma(\alpha)$, combined with the two possibilities for $\sigma(i)$, must give all eight automorphisms. We describe these in Table 54.5. For example, $\rho_3(\alpha) = -i\alpha$ and $\rho_3(i) = i$, while ρ_0 is the identity automorphism. Now

$$(\mu_1\rho_1)(\alpha) = \mu_1(\rho_1(\alpha)) = \mu_1(i\alpha) = \mu_1(i)\mu_1(\alpha) = -i\alpha,$$

and, similarly,

$$(\mu_1\rho_1)(i) = -i,$$

so $\mu_1\rho_1 = \delta_2$. A similar computation shows that

$$(\rho_1\mu_1)(\alpha) = i\alpha \qquad \text{and} \qquad (\rho_1\mu_1)(i) = -i.$$

Thus $\rho_1\mu_1 = \delta_1$, so $\rho_1\mu_1 \neq \mu_1\rho_1$ and $G(K/\mathbb{Q})$ is not abelian. Therefore, $G(K/\mathbb{Q})$ must be isomorphic to one of the two nonabelian groups of order 8 described in Example 40.6. Computing from Table 54.5, we see that ρ_1 is of order 4, μ_1 is of order 2, $\{\rho_1, \mu_1\}$ generates $G(K/\mathbb{Q})$, and $\rho_1\mu_1 = \mu_1{\rho_1}^3 = \delta_1$. Thus $G(K/\mathbb{Q})$ is isomorphic to the group G_1 of Example 40.6, the *octic group*. We chose our notation for the elements of $G(K/\mathbb{Q})$ so that its group table would coincide with the table for the octic group in Table 8.12. The diagram of subgroups H_i of $G(K/\mathbb{Q})$ is that given in Fig. 8.13. We repeat it here in Fig. 54.6 and also give the corresponding diagram of intermediate fields between $\mathbb{Q}$ and K. This finally illustrates nicely that one diagram is the inversion of the other.

The determination of the fixed fields K_{H_i} sometimes requires a bit of ingenuity. Let's illustrate. To find K_{H_2}, we merely have to find an extension of $\mathbb{Q}$ of degree 2 left fixed by $\{\rho_0, \rho_1, \rho_2, \rho_3\}$. Since all ρ_j leave i fixed, $\mathbb{Q}(i)$ is the field we are after. To find K_{H_4}, we have to find an extension of $\mathbb{Q}$ of degree 4 left fixed by ρ_0 and μ_1. Since μ_1 leaves α fixed and α is a zero of $\text{irr}(\alpha, \mathbb{Q}) = x^4 - 2$, we see that $\mathbb{Q}(\alpha)$ is of degree 4 over $\mathbb{Q}$ and is left fixed by $\{\rho_0, \mu_1\}$. By *Galois theory, it is the only such field*. Here we are using strongly the one-to-one correspondence given by the Galois theory. If we find one field that fits the bill, it is the one we are after. Finding K_{H_7} requires more ingenuity. Since $H_7 = \{\rho_0, \delta_1\}$ is a group, for any $\beta \in K$ we see that $\rho_0(\beta) + \delta_1(\beta)$ is left fixed by ρ_0 and δ_1. Taking $\beta = \alpha$, we see that $\rho_0(\alpha) + \delta_1(\alpha) = \alpha + i\alpha$ is left fixed by H_7. We can check and see that ρ_0 and δ_1 are the only automorphisms leaving $\alpha + i\alpha$ fixed. Thus

54.5 Table

	ρ_0	ρ_1	ρ_2	ρ_3	μ_1	δ_1	μ_2	δ_2
$\alpha \to$	α	$i\alpha$	$-\alpha$	$-i\alpha$	α	$i\alpha$	$-\alpha$	$-i\alpha$
$i \to$	i	i	i	i	$-i$	$-i$	$-i$	$-i$

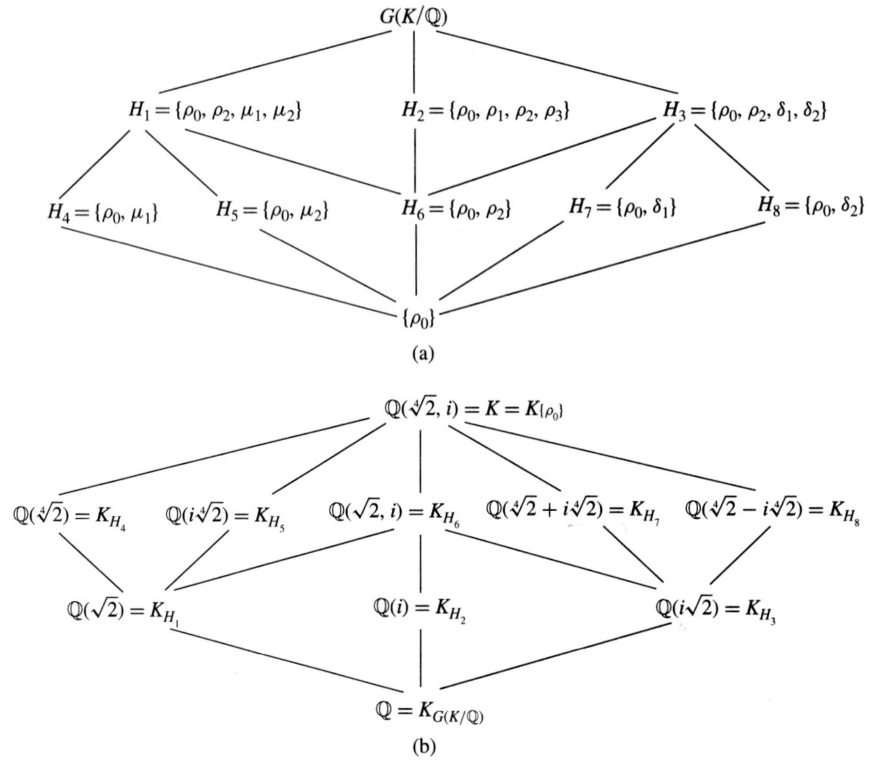

54.6 Figure (a) Group diagram. (b) Field diagram.

by the one-to-one correspondence, we must have

$$\mathbb{Q}(\alpha + i\alpha) = \mathbb{Q}(\sqrt[4]{2} + i\sqrt[4]{2}) = K_{H_7}.$$

Suppose we wish to find $\operatorname{irr}(\alpha + i\alpha, \mathbb{Q})$. If $\gamma = \alpha + i\alpha$, then for every conjugate of γ over $\mathbb{Q}$, there exists an automorphism of K mapping γ into that conjugate. Thus we need only compute the various different values $\sigma(\gamma)$ for $\sigma \in G(K/\mathbb{Q})$ to find the other zeros of $\operatorname{irr}(\gamma, \mathbb{Q})$. By Theorem 53.2, elements σ of $G(K/\mathbb{Q})$ giving these different values can be found by taking a set of representatives of the left cosets of $G(K/\mathbb{Q}(\gamma)) = \{\rho_0, \delta_1\}$ in $G(K/\mathbb{Q})$. A set of representatives for these left cosets is

$$\{\rho_0, \rho_1, \rho_2, \rho_3\}.$$

The conjugates of $\gamma = \alpha + i\alpha$ are thus $\alpha + i\alpha$, $i\alpha - \alpha$, $-\alpha - i\alpha$, and $-i\alpha + \alpha$. Hence

$$\begin{aligned}
\operatorname{irr}(\gamma, \mathbb{Q}) &= [(x - (\alpha + i\alpha))(x - (i\alpha - \alpha))] \\
&\quad \cdot [(x - (-\alpha - i\alpha))(x - (-i\alpha + \alpha))] \\
&= (x^2 - 2i\alpha x - 2\alpha^2)(x^2 + 2i\alpha x - 2\alpha^2) \\
&= x^4 + 4\alpha^4 = x^4 + 8.
\end{aligned}$$

▲

We have seen examples in which the splitting field of a quartic (4th degree) polynomial over a field F is an extension of F of degree 8 (Example 54.3) and of degree 24 (Theorem 54.2, with $n = 4$). The degree of an extension of a field F that is a splitting field of a quartic over F must always divide $4! = 24$. The splitting field of $(x - 2)^4$ over $\mathbb{Q}$ is $\mathbb{Q}$, an extension of degree 1, and the splitting field of $(x^2 - 2)^2$ over $\mathbb{Q}$ is $\mathbb{Q}(\sqrt{2})$, an extension of degree 2. Our last example will give an extension of degree 4 for the splitting field of a quartic.

54.7 Example Consider the splitting field of $x^4 + 1$ over $\mathbb{Q}$. By Theorem 23.11, we can show that $x^4 + 1$ is irreducible over $\mathbb{Q}$, by arguing that it does not factor in $\mathbb{Z}[x]$. (See Exercise 1.) The work on complex number in Section 1 shows that the zeros of $x^4 + 1$ are $(1 \pm i)/\sqrt{2}$ and $(-1 \pm i)/\sqrt{2}$. A computation shows that if

$$\alpha = \frac{1+i}{\sqrt{2}},$$

then

$$\alpha^3 = \frac{-1+i}{\sqrt{2}}, \qquad \alpha^5 = \frac{-1-i}{\sqrt{2}}, \qquad \text{and} \qquad \alpha^7 = \frac{1-i}{\sqrt{2}}.$$

Thus the splitting field K of $x^4 + 1$ over $\mathbb{Q}$ is $\mathbb{Q}(\alpha)$, and $[K : \mathbb{Q}] = 4$. Let us compute $G(K/\mathbb{Q})$ and give the group and field diagrams. Since there exist automorphisms of K mapping α onto each conjugate of α, and since an automorphism σ of $\mathbb{Q}(\alpha)$ is completely determined by $\sigma(\alpha)$, we see that the four elements of $G(K/\mathbb{Q})$ are defined by Table 54.8. Since

$$(\sigma_j \sigma_k)(\alpha) = \sigma_j(\alpha^k) = (\alpha^j)^k = \alpha^{jk}$$

and $\alpha^8 = 1$, we see that $G(K/\mathbb{Q})$ is isomorphic to the group $\{1, 3, 5, 7\}$ under multiplication modulo 8. This is the group G_8 of Theorem. 20.6. Since $\sigma_j^2 = \sigma_1$, the identity, for all j, $G(K/\mathbb{Q})$ must be isomorphic to the Klein 4-group. The diagrams are given in Fig. 54.9.

To find $K_{\{\sigma_1, \sigma_3\}}$, it is only necessary to find an element of K not in $\mathbb{Q}$ left fixed by $\{\sigma_1, \sigma_3\}$, since $[K_{\{\sigma_1, \sigma_3\}} : \mathbb{Q}] = 2$. Clearly $\sigma_1(\alpha) + \sigma_3(\alpha)$ is left fixed by both σ_1 and σ_3, since $\{\sigma_1, \sigma_3\}$ is a group. We have

$$\sigma_1(\alpha) + \sigma_3(\alpha) = \alpha + \alpha^3 = i\sqrt{2}.$$

Similarly,

$$\sigma_1(\alpha) + \sigma_7(\alpha) = \alpha + \alpha^7 = \sqrt{2}$$

54.8 Table

	σ_1	σ_3	σ_5	σ_7
$\alpha \to$	α	α^3	α^5	α^7

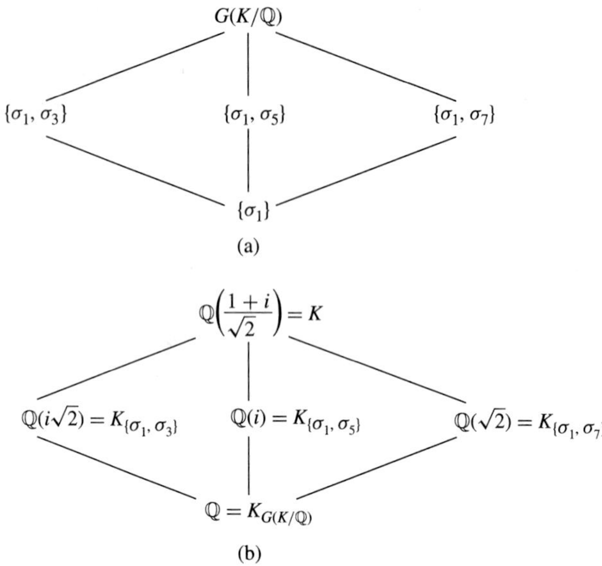

54.9 Figure (a) Group diagram. (b) Field diagram.

is left fixed by $\{\sigma_1, \sigma_7\}$. This technique is of no use in finding $E_{\{\sigma_1,\sigma_5\}}$, for

$$\sigma_1(\alpha) + \sigma_5(\alpha) = \alpha + \alpha^5 = 0,$$

and $0 \in \mathbb{Q}$. But by a similar argument, $\sigma_1(\alpha)\sigma_5(\alpha)$ is left fixed by both σ_1 and σ_5, and

$$\sigma_1(\alpha)\sigma_5(\alpha) = \alpha\alpha^5 = -i.$$

Thus $\mathbb{Q}(-i) = \mathbb{Q}(i)$ is the field we are after. ▲

EXERCISES 54

Computations (requiring more than the usual amount of theory)

1. Show that $x^4 + 1$ is irreducible in $\mathbb{Q}[x]$, as we asserted in Example 54.7.
2. Verify that the intermediate fields given in the field diagram in Fig. 54.6 are correct (Some are verified in the text. Verify the rest.)
3. For each field in the field diagram in Fig. 54.6, find a primitive element generating the field over $\mathbb{Q}$ (see Theorem 51.15) and give its irreducible polynomial over $\mathbb{Q}$.
4. Let ζ be a primitive 5th root of unity in $\mathbb{C}$.
   a. Show that $\mathbb{Q}(\zeta)$ is the splitting field of $x^5 - 1$ over $\mathbb{Q}$.
   b. Show that every automorphism of $K = \mathbb{Q}(\zeta)$ maps ζ onto some power ζ^r of ζ.
   c. Using part (b), describe the elements of $G(K/\mathbb{Q})$.
   d. Give the group and field diagrams for $\mathbb{Q}(\zeta)$ over $\mathbb{Q}$, computing the intermediate fields as we did in Examples 54.3 and 54.7.

5. Describe the group of the polynomial $(x^5 - 2) \in (\mathbb{Q}(\zeta))[x]$ over $\mathbb{Q}(\zeta)$, where ζ is a primitive 5th root of unity.

6. Repcat Exercise 4 for ζ a primitive 7th root of unity in $\mathbb{C}$.

7. In the easiest way possible, describe the group of the polynomial

$$(x^8 - 1) \in \mathbb{Q}[x]$$

over $\mathbb{Q}$.

8. Find the splitting field K in $\mathbb{C}$ of the polynomial $(x^4 - 4x^2 - 1) \in \mathbb{Q}[x]$. Compute the group of the polynomial over $\mathbb{Q}$ and exhibit the correspondence between the subgroups of $G(K/\mathbb{Q})$ and the intermediate fields. In other words, do the complete job.

9. Express each of the following symmetric functions in y_1, y_2, y_3 over $\mathbb{Q}$ as a rational function of the elementary symmetric functions s_1, s_2, s_3.

a. ${y_1}^2 + {y_2}^2 + {y_3}^2$

b. $\dfrac{y_1}{y_2} + \dfrac{y_2}{y_1} + \dfrac{y_1}{y_3} + \dfrac{y_3}{y_1} + \dfrac{y_2}{y_3} + \dfrac{y_3}{y_2}$

10. Let $\alpha_1, \alpha_2, \alpha_3$ be the zeros in $\mathbb{C}$ of the polynomial

$$(x^3 - 4x^2 + 6x - 2) \in \mathbb{Q}[x].$$

Find the polynomial having as zeros precisely the following:

a. $\alpha_1 + \alpha_2 + \alpha_3$

b. ${\alpha_1}^2, {\alpha_2}^2, {\alpha_3}^2$

Theory

11. Show that every finite group is isomorphic to some Galois group $G(K/F)$ for some finite normal extension K of some field F.

12. Let $f(x) \in F[x]$ be a monic polynomial of degree n having all its irreducible factors separable over F. Let $K \leq \bar{F}$ be the splitting field of $f(x)$ over F, and suppose that $f(x)$ factors in $K[x]$ into

$$\prod_{i=1}^{n}(x - \alpha_i).$$

Let

$$\Delta(f) = \prod_{i<j}(\alpha_i - \alpha_j);$$

the product $(\Delta(f))^2$ is the **discriminant of** $f(x)$.

a. Show that $\Delta(f) = 0$ if and only if $f(x)$ has as a factor the square of some irreducible polynomial in $F[x]$.

b. Show that $(\Delta(f))^2 \in F$.

c. $G(K/F)$ may be viewed as a subgroup of $\overline{S_n}$, where $\overline{S_n}$ is the group of all permutations of $\{\alpha_i \mid i = 1, \cdots, n\}$. Show that $G(K/F)$, when viewed in this fashion, is a subgroup of $\overline{A_n}$, the group formed by all even permutations of $\{\alpha_i \mid i = 1, \cdots, n\}$, if and only if $\Delta(f) \in F$.

13. An element of $\mathbb{C}$ is an **algebraic integer** if it is a zero of some *monic* polynomial in $\mathbb{Z}[x]$. Show that the set of all algebraic integers forms a subring of $\mathbb{C}$.

SECTION 55 Cyclotomic Extensions

The Galois Group of a Cyclotomic Extension

This section deals with extension fields of a field F obtained by adjoining to F some roots of unity. The case of a finite field F was covered in Section 33, so we shall be primarily concerned with the case where F is infinite.

55.1 Definition The splitting field of $x^n - 1$ over F is the ***n*th cyclotomic extension of *F***. ■

Suppose that F is any field, and consider $(x^n - 1) \in F[x]$. By long division, as in the proof of Lemma 33.8, we see that if α is a zero of $x^n - 1$ and $g(x) = (x^n - 1)/(x - \alpha)$, then $g(\alpha) = (n \cdot 1)(1/\alpha) \neq 0$, provided that the characteristic of F does not divide n. Therefore, under this condition, the splitting field of $x^n - 1$ is a separable and thus a normal extension of F.

Historical Note

Carl Gauss considered cyclotomic polynomials in the final chapter of his *Disquisitiones Arithmeticae* of 1801. In that chapter, he gave a constructive procedure for actually determining the roots of $\Phi_p(x)$ in the case where p is prime. Gauss's method, which became an important example for Galois in the development of the general theory, was to solve a series of auxiliary equations, each of degree a prime factor of $p - 1$, with the coefficients of each in turn being determined by the roots of the previous equation. Gauss, of course, knew that the roots of $\Phi_p(x)$ were all powers of one of them, say ζ. He determined the auxiliary equations by taking certain sets of sums of the roots ζ^j, which were the desired roots of these equations. For example, in the case where $p = 19$ (and $p - 1 = 18 = 3 \times 3 \times 2$), Gauss needed to find two equations of degree 3 and one of degree 2 as his auxiliaries. It turned out that the first one had the three roots, $\alpha_1 = \zeta + \zeta^8 + \zeta^7 + \zeta^{18} + \zeta^{11} + \zeta^{12}$, $\alpha_2 = \zeta^2 + \zeta^{16} + \zeta^{14} + \zeta^{17} + \zeta^3 + \zeta^5$, and $\alpha_3 = \zeta^4, + \zeta^{13} + \zeta^9 + \zeta^{15} + \zeta^6 + \zeta^{10}$. In fact, these three values are the roots of the cubic equation $x^3 + x^2 - 6x - 7$. Gauss then found a second cubic equation, with coefficients involving the α's, whose roots were sums of two of the powers of ζ, and finally a quadratic equation, whose coefficients involved the roots of the previous equation, which had ζ as one of its roots. Gauss then asserted (without a complete proof) that each auxiliary equation can in turn be reduced to an equation of the form $x^m - A$, which clearly can be solved by radicals. That is, he showed that the solvability of the Galois group in this case, the cyclic group of order $p - 1$, implied that the cyclotomic equation was solvable in terms of radicals. (See Section 56.)

Assume from now on that this is the case, and let K be the splitting field of $x^n - 1$ over F. Then $x^n - 1$ has n distinct zeros in K, and by Corollary 23.6, these form a cyclic group of order n under the field multiplication. We saw in Corollary 6.16 that a cyclic group of order n has $\varphi(n)$ generators, where φ is the Euler phi-function introduced prior to Theorem 20.8. For our situation here, these $\varphi(n)$ generators are exactly the primitive nth roots of unity.

55.2 Definition The polynomial

$$\Phi_n(x) = \prod_{i=1}^{\varphi(n)} (x - \alpha_i)$$

where the α_i are the primitive nth roots of unity in $\overline{F}$, is the ***n*th cyclotomic polynomial over F**. ■

Since an automorphism of the Galois group $G(K/F)$ must permute the primitive nth roots of unity, we see that $\Phi_n(x)$ is left fixed under every element of $G(K/F)$ regarded as extended in the natural way to $K[x]$. Thus $\Phi_n(x) \in F[x]$. In particular, for $F = \mathbb{Q}$, $\Phi_n(x) \in \mathbb{Q}[x]$, and $\Phi_n(x)$ is a divisor of $x^n - 1$. Thus over $\mathbb{Q}$, we must actually have $\Phi_n(x) \in \mathbb{Z}[x]$, by Theorem 23.11. We have seen that $\Phi_p(x)$ is irreducible over $\mathbb{Q}$, in Corollary 23.17. While $\Phi_n(x)$ need not be irreducible in the case of the fields $\mathbb{Z}_p$, it can be shown that over $\mathbb{Q}$, $\Phi_n(x)$ is irreducible.

Let us now limit our discussion to characteristic 0, in particular to subfields of the complex numbers. Let i be the usual complex zero of $x^2 + 1$. Our work with complex numbers in Section 1 shows that

$$\left(\cos\frac{2\pi}{n} + i\sin\frac{2\pi}{n}\right)^n = \cos 2\pi + i\sin 2\pi = 1,$$

so $\cos(2\pi/n) + i\sin(2\pi/n)$ is an nth root of unity. The least integer m such that $(\cos(2\pi/n) + i\sin(2\pi/n))^m = 1$ is n. *Thus* $\cos(2\pi/n) + i\sin(2\pi/n)$ *is a primitive* nth *root of unity, a zero of*

$$\Phi_n(x) \in \mathbb{Q}[x].$$

55.3 Example A primitive 8th root of unity in $\mathbb{C}$ is

$$\begin{aligned}\zeta &= \cos\frac{2\pi}{8} + i\sin\frac{2\pi}{8}\\ &= \cos\frac{\pi}{4} + i\sin\frac{\pi}{4}\\ &= \frac{1}{\sqrt{2}} + i\frac{1}{\sqrt{2}} = \frac{1+i}{\sqrt{2}}.\end{aligned}$$

By the theory of cyclic groups, in particular by Corollary 6.16 all the primitive 8th roots of unity in $\mathbb{Q}$ are ζ, ζ^3, ζ^5, and ζ^7, so

$$\Phi_8(x) = (x - \zeta)(x - \zeta^3)(x - \zeta^5)(x - \zeta^7).$$

We can compute, directly from this expression, $\Phi_8(x) = x^4 + 1$ (see Exercise 1). Compare this with Example 54.7. ▲

Let us still restrict our work to $F = \mathbb{Q}$, and let us assume, without proof, that $\Phi_n(x)$ is irreducible over $\mathbb{Q}$. Let

$$\zeta = \cos\frac{2\pi}{n} + i\sin\frac{2\pi}{n},$$

so that ζ is a primitive nth root of unity. Note that ζ is a generator of the cyclic multiplicative group of order n consisting of *all* nth roots of unity. All the primitive nth roots of unity, that is, all the generators of this group, are of the form ζ^m for $1 \leq m < n$ and m relatively prime to n. The field $\mathbb{Q}(\zeta)$ is the whole splitting field of $x^n - 1$ over $\mathbb{Q}$. Let $K = \mathbb{Q}(\zeta)$. If ζ^m is another primitive nth root of unity, then since ζ and ζ^m are conjugate over $\mathbb{Q}$, there is an automorphism τ_m in $G(K/\mathbb{Q})$ mapping ζ onto ζ^m. Let τ_r be the similar automorphism in $G(K/\mathbb{Q})$ corresponding to a primitive nth root of unity ζ^r. Then

$$(\tau_m\tau_r)(\zeta) = \tau_m(\zeta^r) = (\tau_m(\zeta))^r = (\zeta^m)^r = \zeta^{rm}.$$

This shows that the Galois group $G(K/\mathbb{Q})$ is isomorphic to the group G_n of Theorem 20.6 consisting of elements of $\mathbb{Z}_n$ relatively prime to n under multiplication modulo n. This group has $\varphi(n)$ elements and is abelian.

Special cases of this material have appeared several times in the text and exercises. For example, α of Example 54.7 is a primitive 8th root of unity, and we made arguments in that example identical to those given here. We summarize these results in a theorem.

55.4 Theorem The Galois group of the nth cyclotomic extension of $\mathbb{Q}$ has $\varphi(n)$ elements and is isomorphic to the group consisting of the positive integers less than n and relatively prime to n under multiplication modulo n.

55.5 Example Example 54.7 illustrates this theorem, for it is easy to see that the splitting field of $x^4 + 1$ is the same as the splitting field of $x^8 - 1$ over $\mathbb{Q}$. This follows from the fact that $\Phi_8(x) = x^4 + 1$ (see Example 55.3 and Exercise 1). ▲

55.6 Corollary The Galois group of the pth cyclotomic extension of $\mathbb{Q}$ for a prime p is cyclic of order $p - 1$.

Proof By Theorem 55.4, the Galois group of the pth cyclotomic extension of $\mathbb{Q}$ has $\varphi(p) = p - 1$ elements, and is isomorphic to the group of positive integers less than p and relatively prime to p under multiplication modulo p. This is exactly the multiplicative group $\langle \mathbb{Z}_p{}^*, \cdot \rangle$ of nonzero elements of the field $\mathbb{Z}_p$ under field multiplication. By Corollary 23.6, this group is cyclic. ◆

Constructible Polygons

We conclude with an application determining which regular n-gons are constructible with a compass and a straightedge. We saw in Section 32 that the regular n-gon is constructible if and only if $\cos(2\pi/n)$ is a constructible real number. Now let

$$\zeta = \cos\frac{2\pi}{n} + i\sin\frac{2\pi}{n}.$$

Then

$$\frac{1}{\zeta} = \cos\frac{2\pi}{n} - i\sin\frac{2\pi}{n},$$

for

$$\left(\cos\frac{2\pi}{n}+i\sin\frac{2\pi}{n}\right)\left(\cos\frac{2\pi}{n}-i\sin\frac{2\pi}{n}\right)=\cos^2\frac{2\pi}{n}+\sin^2\frac{2\pi}{n}=1.$$

But then

$$\zeta+\frac{1}{\zeta}=2\cos\frac{2\pi}{n}.$$

Thus Corollary 32.8 shows that the regular n-gon is constructible only if $\zeta + 1/\zeta$ generates an extension of $\mathbb{Q}$ of degree a power of 2.

If K is the splitting field of $x^n - 1$ over $\mathbb{Q}$, then $[K : \mathbb{Q}] = \varphi(n)$, by Theorem 55.4. If $\sigma \in G(K/\mathbb{Q})$ and $\sigma(\zeta) = \zeta^r$, then

$$\begin{aligned}\sigma\left(\zeta+\frac{1}{\zeta}\right)&=\zeta^r+\frac{1}{\zeta^r}\\&=\left(\cos\frac{2\pi r}{n}+i\sin\frac{2\pi r}{n}\right)+\left(\cos\frac{2\pi r}{n}-i\sin\frac{2\pi r}{n}\right)\\&=2\cos\frac{2\pi r}{n}.\end{aligned}$$

But for $1 < r < n$, we have $2\cos(2\pi r/n) = 2\cos(2\pi/n)$ only in the case that $r = n - 1$. Thus the only elements of $G(K/\mathbb{Q})$ carrying $\zeta + 1/\zeta$ onto itself are the identity automorphism and the automorphism τ, with $\tau(\zeta) = \zeta^{n-1} = 1/\zeta$. This shows that the subgroup of $G(K/\mathbb{Q})$ leaving $\mathbb{Q}(\zeta + 1/\zeta)$ fixed is of order 2, so by Galois theory,

$$\left[\mathbb{Q}\left(\zeta+\frac{1}{\zeta}\right):\mathbb{Q}\right]=\frac{\varphi(n)}{2}.$$

Hence the regular n-gon is constructible only if $\varphi(n)/2$, and therefore also $\varphi(n)$, is a power of 2.

It can be shown by elementary arguments in number theory that if

$$n = 2^\nu {p_1}^{s_1}\cdots {p_t}^{s_t},$$

where the p_i are the distinct odd primes dividing n, then

$$\varphi(n) = 2^{\nu-1}{p_1}^{s_1-1}\cdots {p_t}^{s_t-1}(p_1-1)\cdots(p_t-1). \qquad \textbf{(1)}$$

If $\varphi(n)$ is to be a power of 2, then every odd prime dividing n must appear only to the first power and must be one more than a power of 2. Thus we must have each

$$p_i = 2^m + 1$$

for some m. Since -1 is a zero of $x^q + 1$ for q an odd prime, $x + 1$ divides $x^q + 1$ for q an odd prime. Thus, if $m = qu$, where q is an odd prime, then $2^m + 1 = (2^u)^q + 1$ is divisible by $2^u + 1$. Therefore, for $p_i = 2^m + 1$ to be prime, it must be that m is divisible by 2 only, so p_i has to have the form

$$p_i = 2^{(2^k)} + 1,$$

a **Fermat prime.** Fermat conjectured that these numbers $2^{(2^k)}+1$ were prime for all nonnegative integers k. Euler showed that while $k=0, 1, 2, 3$, and 4 give the primes 3, 5, 17, 257, and 65537, for $k=5$, the integer $2^{(2^5)}+1$ is divisible by 641. It has been shown that for $5 \leq k \leq 19$, all the numbers $2^{(2^k)}+1$ are composite. The case $k=20$ is still unsolved as far as we know. For at least 60 values of k greater than 20, including $k=9448$, it has been shown that $2^{2^k}+1$ is composite. It is unknown whether the number of Fermat primes is finite or infinite.

We have thus shown that the only regular n-gons that might be constructible are those where the odd primes dividing n are Fermat primes whose squares do not divide n. In particular, the only regular p-gons that might be constructible for p a prime greater than 2 are those where p is a Fermat prime.

55.7 Example The regular 7-gon is not constructible, since 7 is not a Fermat prime. Similarly, the regular 18-gon is not constructible, for while 3 is a Fermat prime, its square divides 18. ▲

It is a fact that we now demonstrate that all these regular n-gons that are candidates for being constructible are indeed actually constructible. Let ζ again be the primitive nth root of unity $\cos(2\pi/n)+i\sin(2\pi/n)$. We saw above that

$$2\cos\frac{2\pi}{n}=\zeta+\frac{1}{\zeta},$$

and that

$$\left[\mathbb{Q}\left(\zeta+\frac{1}{\zeta}\right):\mathbb{Q}\right]=\frac{\varphi(n)}{2}.$$

Suppose now that $\varphi(n)$ is a power 2^s of 2. Let E be $\mathbb{Q}(\zeta+1/\zeta)$. We saw above that $\mathbb{Q}(\zeta+1/\zeta)$ is the subfield of $K=\mathbb{Q}(\zeta)$ left fixed by $H_1=\{\iota,\tau\}$, where ι is the identity element of $G(K/\mathbb{Q})$ and $\tau(\zeta)=1/\zeta$. By Sylow theory, there exist additional subgroups H_j of order 2^j of $G(\mathbb{Q}(\zeta)/\mathbb{Q})$ for $j=0, 2, 3, \cdots, s$ such that

$$\{\iota\}=H_0<H_1<\cdots<H_s=G(\mathbb{Q}(\zeta)/\mathbb{Q}).$$

By Galois theory,

$$\mathbb{Q}=K_{H_s}<K_{H_{s-1}}<\cdots<K_{H_1}=\mathbb{Q}\left(\zeta+\frac{1}{\zeta}\right),$$

and $[K_{H_{j-1}}:K_H]=2$. Note that $(\zeta+1/\zeta)\in\mathbb{R}$, so $\mathbb{Q}(\zeta+1/\zeta)<\mathbb{R}$. If $K_{H_{j-1}}=K_{H_j}(\alpha_j)$, then α_j is a zero of some $(a_jx^2+b_jx+c_j)\in K_{H_j}[x]$. By the familiar "quadratic formula," we have

$$K_{H_{j-1}}=K_{H_j}\left(\sqrt{b_j^2-4a_jc_j}\right).$$

Since we saw in Section 33 that construction of square roots of positive constructible numbers can be achieved by a straightedge and a compass, we see that every element in

$\mathbb{Q}(\zeta + 1/\zeta)$, in particular $\cos(2\pi/n)$, is constructible. Hence the regular n-gons where $\varphi(n)$ is a power of 2 are constructible.

We summarize our work under this heading in a theorem.

55.8 Theorem The regular n-gon is constructible with a compass and a straightedge if and only if all the odd primes dividing n are Fermat primes whose squares do not divide n.

55.9 Example The regular 60-gon is constructible, since $60 = (2^2)(3)(5)$ and 3 and 5 are both Fermat primes. ▲

EXERCISES 55

Computations

1. Referring to Example 55.3, complete the indicated computation, showing that $\Phi_8(x) = x^4 + 1$. [*Suggestion:* Compute the product in terms of ζ, and then use the fact that $\zeta^8 = 1$ and $\zeta^4 = -1$ to simplify the coefficients.]
2. Classify the group of the polynomial $(x^{20} - 1) \in \mathbb{Q}[x]$ over $\mathbb{Q}$ according to the Fundamental Theorem of finitely generated abelian groups. [*Hint:* Use Theorem 55.4.]
3. Using the formula for $\varphi(n)$ in terms of the factorization of n, as given in Eq. (1), compute the indicated value:

   a. $\varphi(60)$ **b.** $\varphi(1000)$ **c.** $\varphi(8100)$
4. Give the first 30 values of $n \geq 3$ for which the regular n-gon is constructible with a straightedge and a compass.
5. Find the smallest angle of integral degree, that is, $1°$, $2°$, $3°$, and so on, constructible with a straightedge and a compass. [*Hint:* Constructing a $1°$ angle amounts to constructing the regular 360-gon, and so on.]
6. Let K be the splitting field of $x^{12} - 1$ over $\mathbb{Q}$.

   a. Find $[K : \mathbb{Q}]$.

   b. Show that for $\sigma \in G(K/\mathbb{Q})$, σ^2 is the identity automorphism. Classify $G(K/\mathbb{Q})$ according to the Fundamental Theorem 11.12 of finitely generated abelian groups.
7. Find $\Phi_3(x)$ over $\mathbb{Z}_2$. Find $\Phi_8(x)$ over $\mathbb{Z}_3$.
8. How many elements are there in the splitting field of $x^6 - 1$ over $\mathbb{Z}_3$?

Concepts

9. Mark each of the following true or false.

   _____ **a.** $\Phi_n(x)$ is irreducible over every field of characteristic 0.

   _____ **b.** Every zero in $\mathbb{C}$ of $\Phi_n(x)$ is a primitive nth root of unity.

   _____ **c.** The group of $\Phi_n(x) \in \mathbb{Q}[x]$ over $\mathbb{Q}$ has order n.

   _____ **d.** The group of $\Phi_n(x) \in \mathbb{Q}[x]$ over $\mathbb{Q}$ is abelian.

   _____ **e.** The Galois group of the splitting field of $\Phi_n(x)$ over $\mathbb{Q}$ has order $\varphi(n)$.

   _____ **f.** The regular 25-gon is constructible with a straightedge and a compass.

   _____ **g.** The regular 17-gon is constructible with a straightedge and a compass.

   _____ **h.** For a prime p, the regular p-gon is constructible if and only if p is a Fermat prime.

   _____ **i.** All integers of the form $2^{(2^k)} + 1$ for nonnegative integers k are Fermat primes.

   _____ **j.** All Fermat primes are numbers of the form $2^{(2^k)} + 1$ for nonnegative integers k.

Theory

10. Show that if F is a field of characteristic not dividing n, then

$$x^n - 1 = \prod_{d \mid n} \Phi_d(x)$$

in $F[x]$, where the product is over all divisors d of n.

11. Find the cyclotomic polynomial $\Phi_n(x)$ over $\mathbb{Q}$ for $n = 1, 2, 3, 4, 5$, and 6. [*Hint:* Use Exercise 10.]

12. Find $\Phi_{12}(x)$ in $\mathbb{Q}[x]$. [*Hint:* Use Exercises 10 and 11.]

13. Show that in $\mathbb{Q}[x]$, $\Phi_{2n}(x) = \Phi_n(-x)$ for odd integers $n > 1$. [*Hint:* If ζ is a primitive nth root of unity for n odd, what is the order of $-\zeta$?]

14. Let $n, m \in \mathbb{Z}^+$ be relatively prime. Show that the splitting field in $\mathbb{C}$ of $x^{nm} - 1$ over $\mathbb{Q}$ is the same as the splitting field in $\mathbb{C}$ of $(x^n - 1)(x^m - 1)$ over $\mathbb{Q}$.

15. Let $n, m \in \mathbb{Z}^+$ be relatively prime. Show that the group of $(x^{nm} - 1) \in \mathbb{Q}[x]$ over $\mathbb{Q}$ is isomorphic to the direct product of the groups of $(x^n - 1) \in \mathbb{Q}[x]$ and of $(x^m - 1) \in \mathbb{Q}[x]$ over $\mathbb{Q}$. [*Hint:* Using Galois theory, show that the groups of $x^m - 1$ and $x^n - 1$ can both be regarded as subgroups of the group of $x^{nm} - 1$. Then use Exercises 50 and 51 of Section 11.]

SECTION 56 INSOLVABILITY OF THE QUINTIC

The Problem

We are familiar with the fact that a quadratic polynomial $f(x) = ax^2 + bx + c, a \neq 0$, with real coefficients has $(-b \pm \sqrt{b^2 - 4ac})/2a$ as zeros in $\mathbb{C}$. Actually, this is true for $f(x) \in F[x]$, where F is any field of characteristic $\neq 2$ and the zeros are in $\overline{F}$. Exercise 4 asks us to show this. Thus, for example, $(x^2 + 2x + 3) \in \mathbb{Q}[x]$ has its zeros in $\mathbb{Q}(\sqrt{-2})$. You may wonder whether the zeros of a cubic polynomial over $\mathbb{Q}$ can also always be expressed in terms of radicals. The answer is yes, and indeed, even the zeros of a polynomial of degree 4 over $\mathbb{Q}$ can be expressed in terms of radicals. After mathematicians had tried for years to find the "radical formula" for zeros of a 5th degree polynomial, it was a triumph when Abel proved that a quintic need not be solvable by radicals. Our first job will be to describe precisely what this means. A large amount of the algebra we have developed is used in the forthcoming discussion.

Extensions by Radicals

56.1 Definition An extension K of a field F is an **extension of F by radicals** if there are elements $\alpha_1, \cdots, \alpha_r \in K$ and positive integers $n_1, \cdots, n_r$ such that $K = F(\alpha_1, \cdots, \alpha_r)$, $\alpha_1^{n_1} \in F$ and $\alpha_i^{n_i} \in F(\alpha_1, \cdots, \alpha_{i-1})$ for $1 < i \leq r$. A polynomial $f(x) \in F[x]$ is **solvable by radicals over F** if the splitting field E of $f(x)$ over F is contained in an extension of F by radicals. ■

A polynomial $f(x) \in F(x)$ is thus solvable by radicals over F if we can obtain every zero of $f(x)$ by using a finite sequence of the operations of addition, subtraction, multiplication, division, and taking n_ith roots, starting with elements of F. Now to say that the quintic is not solvable in the classic case, that is, characteristic 0, is not to say that no quintic is solvable, as the following example shows.

Historical Note

The first publication of a formula for solving cubic equations in terms of radicals was in 1545 in the *Ars Magna* of Girolamo Cardano, although the initial discovery of the method is in part also due to Scipione del Ferro and Niccolo Tartaglia. Cardano's student, Lodovico Ferrari, discovered a method for solving quartic equations by radicals, which also appeared in Cardano's work.

After many mathematicians had attempted to solve quintics by similar methods, it was Joseph-Louis Lagrange who in 1770 first attempted a detailed analysis of the general principles underlying the solutions for polynomials of degree 3 and 4, and showed why these methods fail for those of higher degree. His basic insight was that in the former cases there were rational functions of the roots that took on two and three values, respectively, under all possible permutations of the roots, hence these rational functions could be written as roots of equations of degree less than that of the original. No such functions were evident in equations of higher degree.

The first mathematician to claim to have a proof of the insolvability of the quintic equation was Paolo Ruffini (1765–1822) in his algebra text of 1799. His proof was along the lines suggested by Lagrange, in that he in effect determined all of the subgroups of S_5 and showed how these subgroups acted on rational functions of the roots of the equation. Unfortunately, there were several gaps in his various published versions of the proof. It was Niels Henrik Abel who, in 1824 and 1826, published a complete proof, closing all of Ruffini's gaps and finally settling this centuries-old question.

56.2 Example The polynomial $x^5 - 1$ is solvable by radicals over $\mathbb{Q}$. The splitting field K of $x^5 - 1$ is generated over $\mathbb{Q}$ by a primitive 5th root ζ of unity. Then $\zeta^5 = 1$, and $K = \mathbb{Q}(\zeta)$. Similarly, $x^5 - 2$ is solvable by radicals over $\mathbb{Q}$, for its splitting field over $\mathbb{Q}$ is generated by $\sqrt[5]{2}$ and ζ, where $\sqrt[5]{2}$ is the real zero of $x^5 - 2$. ▲

To say that the quintic is insolvable in the classic case means that there exists *some* polynomial of degree 5 with real coefficients that is not solvable by radicals. We shall show this. *We assume throughout this section that all fields mentioned have characteristic 0.*

The outline of the argument is as follows, and it is worthwhile to try to remember it.

1. *We shall show that a polynomial $f(x) \in F[x]$ is solvable by radicals over F (if and) only if its splitting field E over F has a solvable Galois group.* Recall that a solvable group is one having a composition series with *abelian* quotients. While this theorem goes both ways, we shall not prove the "if" part.
2. *We shall show that there is a subfield F of the real numbers and a polynomial $f(x) \in F[x]$ of degree 5 with a splitting field E over F such that $G(E/F) \simeq S_5$, the symmetric group on 5 letters.* Recall that a composition series for S_5 is $\{\iota\} < A_5 < S_5$. Since A_5 is not abelian, we will be done.

The following lemma does most of our work for Step 1.

56.3 Lemma Let F be a field of characteristic 0, and let $a \in F$. If K is the splitting field of $x^n - a$ over F, then $G(K/F)$ is a solvable group.

Proof Suppose first that F contains all the nth roots of unity. By Corollary 23.6 the nth roots of unity form a cyclic subgroup of $\langle F^*, \cdot\rangle$. Let ζ be a generator of the subgroup. (Actually, the generators are exactly the *primitive* nth roots of unity.) Then the nth roots of unity are

$$1, \zeta, \zeta^2, \cdots, \zeta^{n-1}.$$

If $\beta \in \overline{F}$ is a zero of $(x^n - a) \in F[x]$, then all zeros of $x^n - a$ are

$$\beta, \zeta\beta, \zeta^2\beta, \cdots, \zeta^{n-1}\beta.$$

Since $K = F(\beta)$, an automorphism σ in $G(K/F)$ is determined by the value $\sigma(\beta)$ of the automorphism σ on β. Now if $\sigma(\beta) = \zeta^i\beta$ and $\tau(\beta) = \zeta^j\beta$, where $\tau \in G(K/F)$, then

$$(\tau\sigma)(\beta) = \tau(\sigma(\beta)) = \tau(\zeta^i\beta) = \zeta^i\tau(\beta) = \zeta^i\zeta^j\beta,$$

since $\zeta^i \in F$. Similarly,

$$(\sigma\tau)(\beta) = \zeta^j\zeta^i\beta.$$

Thus $\sigma\tau = \tau\sigma$, and $G(K/F)$ is abelian and therefore solvable.

Now suppose that F does not contain a primitive nth root of unity. Let ζ be a generator of the cyclic group of nth roots of unity under multiplication in $\overline{F}$. Let β again be a zero of $x^n - a$. Since β and $\zeta\beta$ are both in the splitting field K of $x^n - a$, $\zeta = (\zeta\beta)/\beta$ is in K. Let $F' = F(\zeta)$, so we have $F < F' \leq K$. Now F' is a normal extension of F, since F' is the splitting field of $x^n - 1$. Since $F' = F(\zeta)$, an automorphism η in $G(F'/F)$ is determined by $\eta(\zeta)$, and we must have $\eta(\zeta) = \zeta^i$ for some i, since all zeros of $x^n - 1$ are powers of ζ. If $\mu(\zeta) = \zeta^j$ for $\mu \in G(F'/F)$, then

$$(\mu\eta)(\zeta) = \mu(\eta(\zeta)) = \mu(\zeta^i) = \mu(\zeta)^i = (\zeta^j)^i = \zeta^{ij},$$

and, similarly,

$$(\eta\mu)(\zeta) = \zeta^{ij}.$$

Thus $G(F'/F)$ is abelian. By the Main Theorem of Galois Theory,

$$\{\iota\} \leq G(K/F') \leq G(K/F)$$

is a normal series and hence a subnormal series of groups. The first part of the proof shows that $G(K/F')$ is abelian, and Galois theory tells us that $G(K/F)/G(K/F')$ is isomorphic to $G(F'/F)$, which is abelian. Exercise 6 shows that if a group has a subnormal series of subgroups with abelian quotient groups, then any refinement of this series also has abelian quotient groups. Thus a composition series of $G(K/F)$ must have abelian quotient groups, so $G(K/F)$ is solvable. ◆

The following theorem will complete Part 1 of our program.

56.4 Theorem Let F be a field of characteristic zero, and let $F \leq E \leq K \leq \overline{F}$, where E is a normal extension of F and K is an extension of F by radicals. Then $G(E/F)$ is a solvable group.

Proof We first show that K is contained in a finite normal extension L of F by radicals and that the group $G(L/F)$ is solvable. Since K is an extension by radicals, $K = F(\alpha_1, \cdots, \alpha_r)$

where $\alpha_i^{n_i} \in F(\alpha_1, \cdots, \alpha_{i-1})$ for $1 < i \leq r$ and $\alpha_1^{n_1} \in F$. To form L, we first form the splitting field L_1 of $f_1(x) = x^{n_1} - \alpha_1^{n_1}$ over F. Then L_1 is a normal extension of F, and Lemma 56.3 shows that $G(L_1/F)$ is a solvable group. Now $\alpha_2^{n_2} \in L_1$ and we form the polynomial

$$f_2(x) = \prod_{\sigma \in G(L_1/F)} [(x^{n_2} - \sigma(\alpha_2)^{n_2}].$$

Since this polynomial is invariant under action by any σ in $G(L_1/F)$, we see that $f_2(x) \in F[x]$. We let L_2 be the splitting field of $f_2(x)$ over L_1. Then L_2 is a splitting field over F also and is a normal extension of F by radicals. We can form L_2 from L_1 via repeated steps as in Lemma 56.3, passing to a splitting field of $x^{n_2} - \sigma(\alpha_2)^{n_2}$ at each step. By Lemma 56.3 and Exercise 7, we see that the Galois group over F of each new extension thus formed continues to be solvable. We continue this process of forming splitting fields over F in this manner: At stage i, we form the splitting field of the polynomial

$$f_i(x) = \prod_{\alpha \in G(L_{i-1}/F)} [(x^{n_i} - \sigma(\alpha_i)^{n_i}]$$

over L_{i-1}. We finally obtain a field $L = L_r$ that is a normal extension of F by radicals, and we see that $G(L/F)$ is a solvable group. We see from construction that $K \leq L$.

To conclude, we need only note that by Theorem 53.6, we have $G(E/F) \simeq G(L/F)/G(L/E)$. Thus $G(E/F)$ is a factor group, and hence a homomorphic image, of $G(L/F)$. Since $G(L/F)$ is solvable, Exercise 29 of Section 35 shows that $G(E/F)$ is solvable. ◆

The Insolvability of the Quintic

It remains for us to show that there is a subfield F of the real numbers and a polynomial $f(x) \in F[x]$ of degree 5 such that the splitting field E of $f(x)$ over F has a Galois group isomorphic to S_5.

Let $y_1 \in \mathbb{R}$ be transcendental over $\mathbb{Q}$, $y_2 \in \mathbb{R}$ be transcendental over $\mathbb{Q}(y_1)$, and so on, until we get $y_5 \in \mathbb{R}$ transcendental over $\mathbb{Q}(y_1, \cdots, y_4)$. It can be shown by a counting argument that such transcendental real numbers exist. Transcendentals found in this fashion are **independent transcendental elements over** $\mathbb{Q}$. Let $E = \mathbb{Q}(y_1, \cdots, y_5)$, and let

$$f(x) = \prod_{i=1}^{5}(x - y_i).$$

Thus $f(x) \in E[x]$. Now the coefficients of $f(x)$ are, except possibly for sign, among the *elementary symmetry functions* in the y_i, namely

$$\begin{aligned}
s_1 &= y_1 + y_2 + \cdots + y_5, \\
s_2 &= y_1y_2 + y_1y_3 + y_1y_4 + y_1y_5 + y_2y_3 \\
&\quad + y_2y_4 + y_2y_5 + y_3y_4 + y_3y_5 + y_4y_5, \\
&\vdots \\
s_5 &= y_1y_2y_3y_4y_5.
\end{aligned}$$

$E = \mathbb{Q}(y_1, \dots, y_5)$

$F = \mathbb{Q}(s_1, \dots, s_5)$

$\mathbb{Q}$

56.5 Figure

The coefficient of x^i in $f(x)$ is $\pm s_{5-i}$. Let $F = \mathbb{Q}(s_1, s_2, \cdots, s_5)$; then $f(x) \in F[x]$ (see Fig. 56.5). Then E is the splitting field over F of $f(x)$. Since the y_i behave as indeterminates over $\mathbb{Q}$, for each $\sigma \in S_5$, the symmetric group on five letters, σ induces an automorphism $\overline{\sigma}$ of E defined by $\overline{\sigma}(a) = a$ for $a \in \mathbb{Q}$ and $\overline{\sigma}(y_i) = y_{\sigma(i)}$. Since $\Pi_{i=1}^5(x - y_i)$ is the same polynomial as $\Pi_{i=1}^5(x - y_{\sigma(i)})$, we have

$$\overline{\sigma}(s_i) = s_i$$

for each i, so $\overline{\sigma}$ leaves F fixed, and hence $\overline{\sigma} \in G(E/F)$. Now S_5 has order $5!$, so

$$|G(E/F)| \geq 5!.$$

Since the splitting field of a polynomial of degree 5 over F has degree at most 5! over F, we see that

$$|G(E/F)| \leq 5!.$$

Thus $|G(E/F)| = 5!$, and the automorphisms $\bar{\sigma}$ make up the full Galois group $G(E/F)$. Therefore, $G(E/F) \simeq S_5$, so $G(E/F)$ is not solvable. This completes our outline, and we summarize in a theorem.

56.6 Theorem Let $y_1, \cdots, y_5$ be independent transcendental real numbers over $\mathbb{Q}$. The polynomial

$$f(x) = \prod_{i=1}^{5}(x - y_i)$$

is not solvable by radicals over $F = \mathbb{Q}(s_1, \cdots, s_5)$, where s_i is the ith elementary symmetric function in $y_1, \cdots, y_5$.

It is evident that a generalization of these arguments shows that (*final goal*) a polynomial of degree n need not be solvable by radicals for $n \geq 5$.

In conclusion, we comment that there exist polynomials of degree 5 in $\mathbb{Q}[x]$ that are not solvable by radicals over $\mathbb{Q}$. A demonstration of this is left to the exercises (see Exercise 8).

EXERCISES 56

Concepts

1. Can the splitting field K of $x^2 + x + 1$ over $\mathbb{Z}_2$ be obtained by adjoining a square root to $\mathbb{Z}_2$ of an element in $\mathbb{Z}_2$? Is K an extension of $\mathbb{Z}_2$ by radicals?
2. Is every polynomial in $F[x]$ of the form $ax^8 + bx^6 + cx^4 + dx^2 + e$, where $a \neq 0$, solvable by radicals over F, if F is of characteristic 0? Why or why not?
3. Mark each of the following true of false.

_____ **a.** Let F be a field of characteristic 0. A polynomial in $F[x]$ is solvable by radicals if and only if its splitting field in $\overline{F}$ is contained in an extension of F by radicals.

_____ **b.** Let F be a field of characteristic 0. A polynomial in $F[x]$ is solvable by radicals if and only if its splitting field in $\overline{F}$ has a solvable Galois group over F.

_______ **c.** The splitting field of $x^{17} - 5$ over $\mathbb{Q}$ has a solvable Galois group.

_______ **d.** The numbers π and $\sqrt{\pi}$ are independent transcendental numbers over $\mathbb{Q}$.

_______ **e.** The Galois group of a finite extension of a finite field is solvable.

_______ **f.** No quintic polynomial is solvable by radicals over any field.

_______ **g.** Every 4th degree polynomial over a field of characteristic 0 is solvable by radicals.

_______ **h.** The zeros of a cubic polynomial over a field F of characteristic 0 can always be attained by means of a finite sequence of operations of addition, subtraction, multiplication, division, and taking square roots starting with elements in F.

_______ **i.** The zeros of a cubic polynomial over a field F of characteristic 0 can never be attained by means of a finite sequence of operations of addition, subtraction, multiplication, division, and taking square roots, starting with elements in F.

_______ **j.** The theory of subnormal series of groups play an important role in applications of Galois theory.

Theory

4. Let F be a field, and let $f(x) = ax^2 + bx + c$ be in $F[x]$, where $a \neq 0$. Show that if the characteristic of F is not 2, the splitting field of $f(x)$ over F is $F(\sqrt{b^2 - 4ac})$. [*Hint:* Complete the square, just as in your high school work, to derive the "quadratic formula."]

5. Show that if F is a field of characteristic different from 2 and

$$f(x) = ax^4 + bx^2 + c,$$

where $a \neq 0$, then $f(x)$ is solvable by radicals over F.

6. Show that for a finite group, every refinement of a subnormal series with abelian quotients also has abelian quotients, thus completing the proof of Lemma 56.3. [*Hint:* Use Theorem 34.7.]

7. Show that for a finite group, a subnormal series with solvable quotient groups can be refined to a composition series with abelian quotients, thus completing the proof of Theorem 56.4. [*Hint:* Use Theorem 34.7.]

8. This exercise exhibits a polynomial of degree 5 in $\mathbb{Q}[x]$ that is not solvable by radicals over $\mathbb{Q}$.

a. Show that if a subgroup H of S_5 contains a cycle of length 5 and a transposition τ, then $H = S_5$. [*Hint:* Show that H contains every transposition of S_5 and apply Corollary 9.12. See Exercise 39, Section 9.]

b. Show that if $f(x)$ is an irreducible polynomial in $\mathbb{Q}[x]$ of degree 5 having exactly two complex and three real zeros in $\mathbb{C}$, then the group of $f(x)$ over $\mathbb{Q}$ is isomorphic to S_5. [*Hint:* Use Sylow theory to show that the group has an element of order 5. Use the fact that $f(x)$ has exactly two complex zeros to show that the group has an element of order 2. Then apply part (a).]

c. The polynomial $f(x) = 2x^5 - 5x^4 + 5$ is irreducible in $\mathbb{Q}[x]$, by the Eisenstein criterion, with $p = 5$. Use the techniques of calculus to find relative maxima and minima and to "graph the polynomial function f" well enough to see that $f(x)$ must have exactly three real zeros in $\mathbb{C}$. Conclude from part (b) and Theorem 56.4 that $f(x)$ is not solvable by radicals over $\mathbb{Q}$.

Appendix: Matrix Algebra

We give a brief summary of matrix algebra here. Matrices appear in examples in some chapters of the text and also are involved in several exercises.

A **matrix** is a rectangular array of numbers. For example, the array

$$\begin{bmatrix} 2 & -1 & 4 \\ 3 & 1 & 2 \end{bmatrix} \tag{1}$$

is a matrix having two rows and three columns. A matrix having m rows and n columns is an $m \times n$ matrix, so Matrix (1) is a 2×3 matrix. If $m = n$, the matrix is **square**. Entries in a matrix may be any type of number—integer, rational, real, or complex. We let $M_{m\times n}(\mathbb{R})$ be the set of all $m \times n$ matrices with real number entries. If $m = n$, the notation is abbreviated to $M_n(\mathbb{R})$. We can similarly consider $M_n(\mathbb{Z})$, $M_{2\times 3}(\mathbb{C})$, etc.

Two matrices having the same number m of rows and the same number n of columns can be added in the obvious way: we add entries in corresponding positions.

A1 Example In $M_{2\times 3}(\mathbb{Z})$, we have

$$\begin{bmatrix} 2 & -1 & 4 \\ 3 & 1 & 2 \end{bmatrix} + \begin{bmatrix} 1 & 0 & -3 \\ 2 & -7 & 1 \end{bmatrix} = \begin{bmatrix} 3 & -1 & 1 \\ 5 & -6 & 3 \end{bmatrix}.$$ ▲

We will use uppercase letters to denote matrices. If A, B, and C are $m \times n$ matrices, it is easily seen that $A + B = B + A$ and that $A + (B + C) = (A + B) + C$.

Matrix multiplication, AB, is defined only if the number of columns of A is equal to the number of rows of B. That is, if A is an $m \times n$ matrix, then B must be an $n \times s$ matrix for some integer s. We start by defining as follows the product AB where A is a

$1 \times n$ matrix and B is an $n \times 1$ matrix:

$$AB = [a_1 \quad a_2 \quad \cdots \quad a_n] \begin{bmatrix} b_1 \\ b_2 \\ \vdots \\ b_n \end{bmatrix} = a_1b_1 + a_2b_2 + \cdots + a_nb_n. \qquad \textbf{(2)}$$

Note that the result is a number. (We shall not distinguish between a number and the 1×1 matrix having that number as its sole entry.) You may recognize this product as the *dot product* of vectors. Matrices having only one *row* or only one *column* are **row vectors** or **column vectors**, respectively.

A2 Example We find that

$$[3 \quad -7 \quad 2] \begin{bmatrix} 1 \\ 4 \\ 5 \end{bmatrix} = (3)(1) + (-7)(4) + (2)(5) = -15. \qquad ▲$$

Let A be an $m \times n$ matrix and let B be an $n \times s$ matrix. Note that the number n of entries in each row of A is the same as the number n of entries in each column of B. The product $C = AB$ is an $m \times s$ matrix. The entry in the ith row and jth column of AB is the product of the ith row of A times the jth column of B as defined by Eq. (2) and illustrated in Example A2.

A3 Example Compute

$$AB = \begin{bmatrix} 2 & -1 & 3 \\ 1 & 4 & 6 \end{bmatrix} \begin{bmatrix} 3 & 1 & 2 & 1 \\ 1 & 4 & 1 & -1 \\ -1 & 0 & 2 & 1 \end{bmatrix}.$$

Solution Note that A is 2×3 and B is 3×4. Thus AB will be 2×4. The entry in its second row and third column is

$$(\text{2nd row } A)(\text{3rd column } B) = [1 \quad 4 \quad 6] \begin{bmatrix} 2 \\ 1 \\ 2 \end{bmatrix} = 2 + 4 + 12 = 18.$$

Computing all eight entries of AB in this fashion, we obtain

$$AB = \begin{bmatrix} 2 & -2 & 9 & 6 \\ 1 & 17 & 18 & 3 \end{bmatrix}. \qquad ▲$$

A4 Example The product

$$\begin{bmatrix} 2 & -1 & 3 \\ 1 & 4 & 6 \end{bmatrix} \begin{bmatrix} 2 & 1 \\ 5 & 4 \end{bmatrix}$$

is not defined, since the number of entries in a row of the first matrix is not equal to the number of entries in a column of the second matrix. ▲

For square matrices of the same size, both addition and multiplication are always defined. Exercise 10 asks us to illustrate the following fact.

Matrix multiplication is not commutative.

That is, AB need not equal BA even when both products are defined, as for $A, B \in M_2(\mathbb{Z})$. It can be shown that $A(BC) = (AB)C$ and $A(B + C) = AB + AC$ whenever all these expressions are defined.

We let I_n be the $n \times n$ matrix with entries 1 along the diagonal from the upper-left corner to the lower-right corner, and entries 0 elsewhere. For example,

$$I_3 = \begin{bmatrix} 1 & 0 & 0 \\ 0 & 1 & 0 \\ 0 & 0 & 1 \end{bmatrix}.$$

It is easy to see that if A is any $n \times s$ matrix and B is any $r \times n$ matrix, then $I_n A = A$ and $BI_n = B$. That is, the matrix I_n acts much as the number 1 does for multiplication when multiplication by I_n is defined.

Let A be an $n \times n$ matrix and consider a matrix equation of the form $AX = B$, where A and B are known but X is unknown. If we can find an $n \times n$ matrix A^{-1} such that $A^{-1}A = AA^{-1} = I_n$, then we can conclude that

$$A^{-1}(AX) = A^{-1}B, \qquad (A^{-1}A)X = A^{-1}B, \qquad I_n X = A^{-1}B, \qquad X = A^{-1}B,$$

and we have found the desired matrix X. Such a matrix A^{-1} acts like the reciprocal of a number: $A^{-1}A = I_n$ and $(1/r)r = 1$. This is the reason for the notation A^{-1}.

If A^{-1} exists, the square matrix A is **invertible** and A^{-1} is the **inverse** of A. If A^{-1} does not exist, then A is said to be **singular**. It can be shown that if there exists a matrix A^{-1} such that $A^{-1}A = I_n$, then $AA^{-1} = I_n$ also, and furthermore, there is only one matrix A^{-1} having this property.

A5 Example Let

$$A = \begin{bmatrix} 2 & 9 \\ 1 & 4 \end{bmatrix}.$$

We can check that

$$\begin{bmatrix} -4 & 9 \\ 1 & -2 \end{bmatrix}\begin{bmatrix} 2 & 9 \\ 1 & 4 \end{bmatrix} = \begin{bmatrix} 2 & 9 \\ 1 & 4 \end{bmatrix}\begin{bmatrix} -4 & 9 \\ 1 & -2 \end{bmatrix} = \begin{bmatrix} 1 & 0 \\ 0 & 1 \end{bmatrix}.$$

Thus,

$$A^{-1} = \begin{bmatrix} -4 & 9 \\ 1 & -2 \end{bmatrix}.$$

▲

We leave the problems of determining the existence of A^{-1} and its computation to a course in linear algebra.

Associated with each square $n \times n$ matrix A is a number called the *determinant* of A and denoted by $\det(A)$. This number can be computed as sums and differences of certain products of the numbers that appear in the matrix A. For example, the determinant of the 2×2 matrix $\begin{bmatrix} a & b \\ c & d \end{bmatrix}$ is $ad - bc$. Note that an $n \times 1$ matrix with real number entries can be viewed as giving coordinates of a point in n-dimensional Euclidean space $\mathbb{R}^n$. Multiplication of such a single column matrix on the left by a real $n \times n$ matrix A produces another such single column matrix corresponding to another point in $\mathbb{R}^n$. This multiplication on the left by A thus gives a map of $\mathbb{R}^n$ into itself. It can be shown that a piece of $\mathbb{R}^n$ of volume V is mapped by this multiplication by A into a piece of volume $|\det(A)| \cdot V$. This is one of the reasons that determinants are important.

The following properties of determinants for $n \times n$ matrices A and B are of interest in this text:

1. $\det(I_n) = 1$
2. $\det(AB) = \det(A)\det(B))$
3. $\det(A) \neq 0$ if and only if A is an invertible matrix
4. If B is obtained from A by interchanging two rows (or two columns) of A, then $\det(B) = -\det(A)$
5. If every entry of A is zero above the *main diagonal* from the upper left corner to the lower right corner, then $\det(A)$ is the product of the entries on this diagonal. The same is true if all entries below the main diagonal are zero.

EXERCISES A

In Exercises 1 through 9, compute the given arithmetic matrix expression, if it is defined.

1. $\begin{bmatrix} -2 & 4 \\ 1 & 5 \end{bmatrix} + \begin{bmatrix} 4 & -3 \\ 1 & 2 \end{bmatrix}$

2. $\begin{bmatrix} 1+i & -2 & 3-i \\ 4 & i & 2-i \end{bmatrix} + \begin{bmatrix} 3 & i-1 & -2+i \\ 3-i & 1+i & 0 \end{bmatrix}$

3. $\begin{bmatrix} i & -1 \\ 4 & 1 \\ 3 & -2i \end{bmatrix} - \begin{bmatrix} 3-i & 4i \\ 2 & 1+i \\ 3 & -i \end{bmatrix}$

4. $\begin{bmatrix} 1 & -1 \\ 3 & 1 \end{bmatrix}\begin{bmatrix} 2 & 4 \\ -1 & 3 \end{bmatrix}$

5. $\begin{bmatrix} 3 & 1 \\ -4 & 2 \end{bmatrix}\begin{bmatrix} 1 & 5 & -3 \\ 2 & 1 & 6 \end{bmatrix}$

6. $\begin{bmatrix} 4 & -1 \\ 3 & 2 \end{bmatrix}\begin{bmatrix} 1 & 0 \\ -1 & 7 \\ 3 & 1 \end{bmatrix}$

7. $\begin{bmatrix} i & 1 \\ -2 & 1 \end{bmatrix}\begin{bmatrix} 3i & 1 \\ 4 & -2i \end{bmatrix}$

8. $\begin{bmatrix} 1 & -1 \\ 1 & 0 \end{bmatrix}^4$

9. $\begin{bmatrix} 1 & -i \\ i & 1 \end{bmatrix}^4$

10. Give an example in $M_2(\mathbb{Z})$ showing that matrix multiplication is not commutative.

11. Find $\begin{bmatrix} 0 & 1 \\ -1 & 0 \end{bmatrix}^{-1}$, by experimentation if necessary.

12. Find $\begin{bmatrix} 2 & 0 & 0 \\ 0 & 4 & 0 \\ 0 & 0 & -1 \end{bmatrix}^{-1}$, by experimentation if necessary.

13. If $A = \begin{bmatrix} 3 & 0 & 0 \\ 10 & -2 & 0 \\ 4 & 17 & 8 \end{bmatrix}$, find det (A).

14. Prove that if $A, B \in M_n(\mathbb{C})$ are invertible, then AB and BA are invertible also.

Bibliography

Classic Works

1. N. Bourbaki, *Eléments de Mathématique,* Book II of Part I, *Algèbre*. Paris: Hermann, 1942–58.
2. N. Jacobson, *Lectures in Abstract Algebra*. Princeton, N.J.: Van Nostrand, vols. I, 1951, II, 1953, and III, 1964.
3. O. Schreier and E. Sperner, *Introduction to Modern Algebra and Matrix Theory* (English translation), 2nd Ed. New York: Chelsea, 1959.
4. B. L. van der Waerden, *Modern Algebra* (English translation). New York: Ungar, vols. I, 1949, and II, 1950.

General Algebra Texts

5. M. Artin, *Algebra*. Englewood Cliffs, N.J.: Prentice-Hall, 1991.
6. A. A. Albert, *Fundamental Concepts of Higher Algebra*. Chicago: University of Chicago Press, 1956.
7. G. Birkhoff and S. MacLane, *A Survey of Modern Algebra,* 3rd Ed. New York: Macmillan, 1965.
8. J. A. Gallian, *Contemporary Abstract Algebra,* 2nd Ed. Lexington, Mass.: D. C. Heath, 1990.
9. I. N. Herstein, *Topics in Algebra*. New York: Blaisdell, 1964.
10. T. W. Hungerford, *Algebra*. New York: Springer, 1974.
11. S. Lang, *Algebra*. Reading, Mass.: Addison-Wesley, 1965.
12. S. MacLane and G. Birkhoff, *Algebra*. New York: Macmillan, 1967.
13. N. H. McCoy, *Introduction to Modern Algebra*. Boston: Allyn and Bacon, 1960.
14. G. D. Mostow, J. H. Sampson, and J. Meyer, *Fundamental Structures of Algebra*. New York: McGraw-Hill, 1963.
15. W. W. Sawyer, *A Concrete Approach to Abstract Algebra*. San Francisco: Freeman, 1959.

Group Theory

16. W. Burnside, *Theory of Groups of Finite Order,* 2nd Ed. New York: Dover, 1955.
17. H. S. M. Coxeter and W. O. Moser, *Generators and Relations for Discrete Groups,* 2nd Ed. Berlin: Springer, 1965.
18. M. Hall, Jr., *The Theory of Groups*. New York: Macmillan, 1959.
19. A. G. Kurosh, *The Theory of Groups* (English translation). New York: Chelsea, vols. I, 1955, and II, 1956.
20. W. Ledermann, *Introduction to the Theory of Finite Groups,* 4th rev. Ed. New York: Interscience, 1961.
21. J. G. Thompson and W. Feit, "Solvability of Groups of Odd Order." *Pac. J. Math.*, **13** (1963), 775–1029.
22. M. A. Rabin, "Recursive Unsolvability of Group Theoretic Problems." *Ann. Math.*, **67** (1958), 172–194.

Ring Theory

23. W. W. Adams and P. Loustaunau, *An Introduction to Gröbner Bases* (Graduate Studies in Mathematics, vol. 3). Providence, R.I.: American Mathematical Society, 1994.
24. E. Artin, C. J. Nesbitt, and R. M. Thrall, *Rings with Minimum Condition*. Ann Arbor: University of Michigan Press, 1944.
25. N. H. McCoy, *Rings and Ideals* (Carus Monograph No. 8). Buffalo: The Mathematical Association of America; LaSalle, III.: Open Court, 1948.
26. N. H. McCoy, *The Theory of Rings*. New York: Macmillan, 1964.

Field Theory

27. E. Artin, *Galois Theory* (Notre Dame Mathematical Lecture No. 2), 2nd Ed. Notre Dame, Ind.: University of Notre Dame Press, 1944.
28. O. Zariski and P. Samuel, *Commutative Algebra*. Princeton, N.J.: Van Nostrand, vol. I, 1958.

Number Theory

29. G. H. Hardy and E. M. Wright, *An Introduction to the Theory of Numbers,* 4th Ed. Oxford: Clarendon Press, 1960.
30. S. Lang, *Algebraic Numbers*. Reading, Mass.: Addison-Wesley, 1964.
31. W. J. LeVeque, *Elementary Theory of Numbers,* Reading, Mass.: Addison-Wesley, 1962.
32. W. J. LeVeque, *Topics in Number Theory*. Reading, Mass.: Addison-Wesley, 2 vols., 1956.
33. T. Nagell, *Introduction to Number Theory*. New York: Wiley, 1951.
34. I. Nivin and H. S. Zuckerman, *An Introduction to the Theory of Numbers*. New York: Wiley, 1960.
35. H. Pollard, *The Theory of Algebraic Numbers* (Carus Monograph No. 9). Buffalo: The Mathematical Association of America; New York: Wiley, 1950.
36. D. Shanks, *Solved and Unsolved Problems in Number Theory*. Washington, D.C.: Spartan Books, vol. I, 1962.
37. B. M. Stewart, *Theory of Numbers,* 2nd Ed. New York: Macmillan, 1964.
38. J. V. Uspensky and M. H. Heaslet, *Elementary Number Theory*. New York: McGraw-Hill, 1939.
39. E. Weiss, *Algebraic Number Theory*. New York: McGraw-Hill, 1963.

Homological Algebra

40. J. P. Jans, *Rings and Homology*. New York: Holt, 1964.
41. S. MacLane, *Homology*. Berlin: Springer, 1963.

Other References

42. A. A. Albert (ed.), *Studies in Modern Algebra* (MAA Studies in Mathematics, vol. 2). Buffalo: The Mathematical Association of America; Englewood Cliffs, N.J.: Prentice-Hall, 1963.
43. E. Artin, *Geometric Algebra*. New York: Interscience, 1957.
44. R. Courant and R. Robbins, *What Is Mathematics*? Oxford University Press, 1941.
45. H. S. M. Coxeter, *Introduction to Geometry*, 2nd Ed. New York: Wiley, 1969.
46. R. H. Crowell and R. H. Fox, *Introduction to Knot Theory*. New York: Ginn, 1963.
47. H. B. Edgerton, *Elements of Set Theory*. San Diego: Academic Press, 1977.
48. C. Schumacher, *Chapter Zero*. Reading, Mass.: Addison-Wesley, 1996.

Notations

$\in$, $a \in S$	membership, 1
$\varnothing$	empty set, 1
$\notin$, $a \notin S$	nonmembership, 1
$\{x \mid P(x)\}$	set of all x such that $P(x)$, 1
$B \subseteq A$	set inclusion, 2
$B \subset A$	subset $B \neq A$, 2
$A \times B$	Cartesian product of sets, 3
$\mathbb{Z}$	integers, 3
$\mathbb{Q}$	rational numbers, 3
$\mathbb{R}$	real numbers, 3
$\mathbb{C}$	complex numbers, 3
$\mathbb{Z}^+, \mathbb{Q}^+, \mathbb{R}^+$	positive elements of $\mathbb{Z}$, $\mathbb{Q}$, $\mathbb{R}$, 3
$\mathbb{Z}^*, \mathbb{Q}^*, \mathbb{R}^*, \mathbb{C}^*$	nonzero elements of $\mathbb{Z}$, $\mathbb{Q}$, $\mathbb{R}$, $\mathbb{C}$, 3
$\mathscr{R}$	relation, 3
$\|A\|$	number of elements in A, 4; as order of group, 50
$\phi : A \to B$	mapping of A into B by ϕ, 4
$\phi(a)$	image of element a under ϕ, 4
$\phi[A]$	image of set A under ϕ, 4
$\leftrightarrow$	one-to-one correspondence, 4
ϕ^{-1}	the inverse function of ϕ, 5
$\aleph_0$	cardinality of $\mathbb{Z}^+$, 5
$\bar{x}$	cell containing $x \in S$ in a partition of S, 6
$\equiv_n$, $a \equiv b (\text{mod } n)$	congruence modulo n, 7
$\mathscr{P}(A)$	power set of A, 9
U	set of all $z \in \mathbb{C}$ such that $\|z\| = 1$, 15
$\mathbb{R}_c$	set of all $x \in \mathbb{R}$ such that $0 \leq x < c$, 16
$+_c$	addition modulo c, 16
U_n	group of nth roots of unity, 18

$\mathbb{Z}_n$	$\{0, 1, 2, \cdots, n-1\}$, 18
	cyclic group $\{0, 1, \cdots, n-1\}$ under addition modulo n, 54
	group of residue classes modulo n, 137
	ring $\{0, 1, \cdots, n-1\}$ under addition and multiplication modulo n, 169
$*, a * b$	binary operation, 20
$\circ, f \circ g, \sigma\tau$	function composition, 22, 76
$\langle S, * \rangle$	binary structure, 29
$\simeq, S \simeq S'$	isomorphic structures, 30
e	identity element, 32
$M_{m\times n}(S)$	$m \times n$ matrices with entries from S, 40, 477
$M_n(S)$	$n \times n$ matrices with entries from S, 40, 477
$\mathrm{GL}(n, \mathbb{R})$	general linear group of degree n, 40
$\det(A)$	determinant of square matrix A, 46, 479
$a^{-1}, -a$	inverse of a, 49
$H \leq G; K \leq L$	subgroup inclusion, 50; substructure inclusion, 173
$H < G; K < L$	subgroup $H \neq G$, 50; substructure $K \neq L$, 173
$\langle a \rangle$	cyclic subgroup generated by a, 54
	principal ideal generated by a, 250
$n\mathbb{Z}$	subgroup of $\mathbb{Z}$ generated by n, 54
	subring (ideal) of $\mathbb{Z}$ generated by n, 169, 250
gcd	greatest common divisor, 62, 258, 395
$\bigcap_{i\in I} S_i$, $S_1 \cap S_2 \cap \cdots \cap S_n$	intersection of sets, 69
S_A	group of permutations of A, 77
ι	identity map, 77
S_n	symmetric group on n letters, 78
$n!$	n factorial, 78
D_n	nth dihedral group, 79
A_n	alternating group on n letters, 93
$aH, a + H$	left coset of H containing a, 97
$Ha, H + a$	right coset of H containing a, 97
$(G : H)$	index of H in G, 101
φ	Euler phi-function, 104, 187
$\prod_{i=1}^n S_i$, $S_1 \times S_2 \times \cdots \times S_n$	Cartesian product of sets, 104
$\prod_{i=1}^n G_i$	direct product of groups, 104, 105
$\bigoplus_{i=1}^n G_i$	direct sum of groups, 105
lcm	least common multiple, 107
$\overline{G}_i$	natural subgroup of $\prod_{i=1}^n G_i$, 107
ϕ_c	evaluation homomorphism, 126
π_i	projection onto ith component, 127
$\phi^{-1}[B]$	inverse image of the set B under ϕ, 128
$\mathrm{Ker}(\phi)$	kernel of homomorphism ϕ, 129
$G/N; R/N$	factor group, 137; factor ring, 242
γ	canonical residue class map, 139, 140
i_g	inner automorphism, 141
$Z(G)$	center of the group G, 150
C	commutator subgroup, 150
X_g	subset of elements of X left fixed by g, 157

PID	principal ideal domain, 391
$\bigcup_{i \in I} S_i$, $S_1 \cup S_2 \cup \cdots \cup S_n$	union of sets, 391
ν	Euclidean norm, 401
$N(\alpha)$	norm of α, 408, 410, 455
$\psi_{\alpha,\beta}$	conjugation isomorphism of $F(\alpha)$ with $F(\beta)$, 416
$E_{\{\sigma_i\}}$, E_H	subfield of E left fixed by all σ_i or all $\sigma \in H$, 419
$G(E/F)$	automorphism group of E over F, 420
$\{E : F\}$	index of E over F, 428

Answers to Odd-Numbered Exercises Not Asking for Definitions or Proofs

SECTION 0

1. $\{-\sqrt{3}, \sqrt{3}\}$

3. $\{1, -1, 2, -2, 3, -3, 4, -4, 5, -5, 6, -6, 10, -10, 12, -12, 15, -15, 20, -20, 30, -30, 60, -60\}$

5. Not a set (not well defined). A case can also be made for the empty set $\varnothing$.

7. The set $\varnothing$

9. The set $\mathbb{Q}$

11. $(a, 1), (a, 2), (a, c), (b, 1), (b, 2), (b, c), (c, 1), (c, 2), (c, c)$

13. Draw the line through P and x, and let y be the point where it intersects the line segment CD.

17. Conjecture: $n(\mathscr{P}(A)) = 2^s$. (Proofs are usually omitted from answers.)

21. $10^2, 10^5, 10^{\aleph_0} = 12^{\aleph_0} = 2^{\aleph_0} = |\mathbb{R}|$. (The numbers x where $0 \leq x \leq 1$ can be written to base 12 and to base 2 as well as to base 10.)

23. 1 **25.** 5 **27.** 52

29. Not an equivalence relation

31. An equivalence relation; $\overline{0} = \{0\}, \overline{a} = \{a, -a\}$ for each nonzero $a \in \mathbb{R}$

33. An equivalence relation;
$\overline{1} = \{1, 2, \cdots, 9\}$,
$\overline{10} = \{10, 11, \cdots, 99\}$,
$\overline{100} = \{100, 101, \cdots, 999\}$, and in general
$\overline{10^n} = \{10^n, 10^n + 1, \cdots, 10^{n+1} - 1\}$

35. **i.** $\{1, 3, 5, \cdots\}, \{2, 4, 6, \cdots\}$
ii. $\{1, 4, 7, \cdots\}, \{2, 5, 8, \cdots\}, \{3, 6, 9, \cdots\}$
iii. $\{1, 6, 11, \cdots\}, \{2, 7, 12, \cdots\}, \{3, 8, 13, \cdots\}, \{4, 9, 14, \cdots\}, \{5, 10, 15, \cdots\}$

37. The name *two-to-two function* suggests that such a function f should carry every pair of distinct points into two distinct points. Such a function is one to one in the conventional sense. (If the domain has only one element, a function cannot fail to be two to two, since the only way it can fail to be two to two is to carry two points into one point, and the set does not have two points.) Conversely, every function that is one to one in the conventional sense carries any pair of points into two distinct points. Thus the functions conventionally called one to one are precisely those that carry two points into two points, which is a much more intuitive unidirectional way of regarding them. Also, the standard way of trying to show a function is one to one is precisely to show that it does not carry two points into just one point. Thus, proving a function is one to one becomes more natural in the two-to-two terminology.

SECTION 1

1. $-i$ **3.** $-i$ **5.** $23 + 7i$
7. $17 - 15i$ **9.** $-4 + 4i$
11. $2\sqrt{13}$ **13.** $\sqrt{2}\left(-\frac{1}{\sqrt{2}} + \frac{1}{\sqrt{2}}i\right)$
15. $\sqrt{34}\left(\frac{-3}{\sqrt{34}} + \frac{5}{\sqrt{34}}i\right)$
17. $\frac{1}{\sqrt{2}} \pm \frac{1}{\sqrt{2}}i, -\frac{1}{\sqrt{2}} \pm \frac{1}{\sqrt{2}}i$ **19.** $3i, \pm\frac{3\sqrt{3}}{2} - \frac{3}{2}i$
21. $\sqrt{3} \pm i, \pm 2i, -\sqrt{3} \pm i$ **23.** 4 **25.** $\frac{3}{8}$ **27.** $\sqrt{2}$
29. 11 **31.** 5 **33.** 1, 7
35. $\zeta^0 \leftrightarrow 0, \zeta^3 \leftrightarrow 7, \zeta^4 \leftrightarrow 4, \zeta^5 \leftrightarrow 1, \zeta^6 \leftrightarrow 6, \zeta^7 \leftrightarrow 3$
37. With $\zeta \leftrightarrow 4$, we must have $\zeta^2 \leftrightarrow 2$, $\zeta^3 \leftrightarrow 0$, and $\zeta^4 \leftrightarrow 4$ again, which is impossible for a one-to-one correspondence.
39. Multiplying, we obtain

$$z_1 z_2 = |z_1||z_2|[(\cos\theta_1 \cos\theta_2 - \sin\theta_1 \sin\theta_2) + (\cos\theta_1 \sin\theta_2 + \sin\theta_1 \cos\theta_2)i]$$

and the desired result follows at once from Exercise 38 and the equation $|z_1||z_2| = |z_1 z_2|$.

SECTION 2

1. e, b, a **3.** a, c. $*$ is not associative.
5. Top row: d; second row: a; fourth row: c, b.
7. Not commutative, not associative
9. Commutative, associative
11. Not commutative, not associative
13. $8, 729, n^{[n(n+1)/2]}$
17. No. Condition 2 is violated. **19.** Yes
21. No. Condition 1 is violated.
23. **a.** Yes. **b.** Yes
25. Let $S = \{?, \Delta\}$. Define $*$ and $*'$ on S by $a * b = ?$ and $a *' b = \Delta$ for all $a, b \in S$. (Other answers are possible.)
27. True **29.** True
31. False. Let $f(x) = x^2$, $g(x) = x$, and $h(x) = 2x + 1$. Then
$(f(x) - g(x)) - h(x) = x^2 - 3x - 1$ but
$f(x) - (g(x) - h(x)) = x^2 - (-x - 1) = x^2 + x + 1$.
33. True **35.** False. Let $*$ be $+$ and let $*'$ be on $\mathbb{Z}$.

SECTION 3

1. **i.** ϕ must be one to one. **ii.** $\phi[S]$ must be all of S'.
iii. $\phi(a * b) = \phi(a) *' \phi(b)$ for all $a, b \in S$.
3. No, because ϕ does not map $\mathbb{Z}$ onto $\mathbb{Z}'$. $\phi(n) \neq 1$ for all $n \in \mathbb{Z}$.
5. Yes. **7.** Yes **9.** Yes
11. No, because $\phi(x^2) = \phi(x^2 + 1)$.
13. No, because $\phi(f) = x + 1$ has no solution $f \in F$.

15. No, because $\phi(f) = 1$ has no solution $f \in F$.

17. **a.** $m * n = mn - m - n + 2$; identity element 2
b. $m * n = mn + m + n$; identity element 0

19. **a.** $a * b = \frac{1}{3}(ab + a + b - 2)$; identity element 2
b. $a * b = 3ab - a - b + \frac{2}{3}$; identity element $\frac{2}{3}$

25. No. If $\langle S, *\rangle$ has a left identity element e_L and a right identity element e_R, then $e_L = e_R$. (It is our practice to omit proofs from answers.)

SECTION 4

1. No. $\mathscr{G}_3$ fails. **3.** No. $\mathscr{G}_1$ fails. **5.** No. $\mathscr{G}_1$ fails.

7. The group $\langle U_{1000}, \cdot\rangle$ of solutions of $z^{1000} = 1$ in $\mathbb{C}$ under multiplication has 1000 elements.

9. An equation of the form $x * x * x * x = e$ has four solutions in $\langle U, \cdot\rangle$, one solution in $\langle \mathbb{R}, +\rangle$, and two solutions in $\langle \mathbb{R}^*, \cdot\rangle$.

11. Yes **13.** Yes

15. No. The matrix with all entries 0 is upper triangular, but has no inverse.

17. Yes.

19. (Proofs are omitted.) **c.** $-1/3$

21. 2, 3. (It gets harder for 4 elements, where the answer is *not* 4.)

25. **a.** F **c.** T **e.** F **g.** T **i.** F

SECTION 5

1. Yes **3.** Yes **5.** Yes **7.** $\mathbb{Q}^+$ and $\{\pi^n \mid n \in \mathbb{Z}\}$ **9.** Yes

11. No. Not closed under multiplication.

13. Yes

15. **a.** Yes **b.** No. It is not even a subset of $\tilde{F}$.

17. **a.** No. Not closed under addition. **b.** Yes

19. **a.** Yes **b.** No. The zero constant function is not in $\tilde{F}$.

21. **a.** $-50, -25, 0, 25, 50$ **b.** 4, 2, 1, 1/2, 1/4 **c.** $1, \pi, \pi^2, 1/\pi, 1/\pi^2$

23. All matrices $\begin{bmatrix} 1 & n \\ 0 & 1 \end{bmatrix}$ for $n \in \mathbb{Z}$

25. All matrices of the form $\begin{bmatrix} 4^n & 0 \\ 0 & 4^n \end{bmatrix}$ or $\begin{bmatrix} 0 & -2^{2n+1} \\ -2^{2n+1} & 0 \end{bmatrix}$ for $n \in \mathbb{Z}$

27. 4 **29.** 3 **31.** 4 **33.** 2 **35.** 3

39. **a.** T **c.** T **e.** F **g.** F **i.** T

SECTION 6

1. $q = 4, r = 6$ **3.** $q = -7, r = 6$ **5.** 8 **7.** 60

9. 4 **11.** 16 **13.** 2 **15.** 2 **17.** 6 **19.** 4

21. An infinite cyclic group

23.

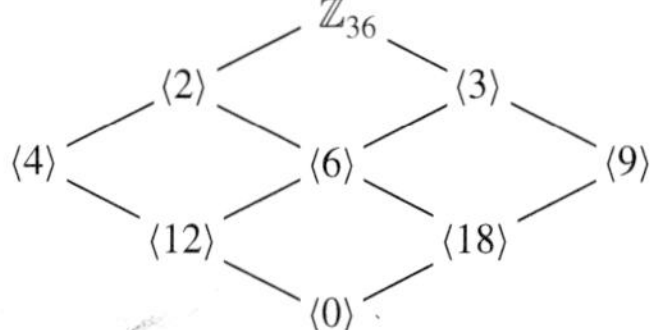

25. 1, 2, 3, 6 **27.** 1, 2, 3, 4, 6, 12 **29.** 1, 17

33. The Klein 4-group **35.** $\mathbb{Z}_2$ **37.** $\mathbb{Z}_8$

39. $\frac{1}{2}(1+i\sqrt{3})$ and $\frac{1}{2}(1-i\sqrt{3})$

41. $\frac{1}{2}(\sqrt{3}+i), \frac{1}{2}(\sqrt{3}-i), \frac{1}{2}(-\sqrt{3}+i), \frac{1}{2}(-\sqrt{3}-i)$

51. $(p-1)(q-1)$

SECTION 7

1. 0, 1, 2, 3, 4, 5, 6, 7, 8, 9, 10, 11 **3.** 0, 2, 4, 6, 8, 10, 12, 14, 16

5. $\cdots, -24, -18, -12, -6, 0, 6, 12, 18, 24, \cdots$

7. **a.** a^3b **b.** a^2 **c.** a^2

9.

	e	a	b	c	d	f
e	e	a	b	c	d	f
a	a	e	c	b	f	d
b	b	d	e	f	a	c
c	e	f	a	d	e	b
d	d	b	f	e	c	a
f	f	c	d	a	b	e

11. Choose a pair of generating directed arcs, call them *arc1* and *arc2*, start at any vertex of the digraph, and see if the sequences *arc1, arc2* and *arc2, arc1* lead to the same vertex. (This corresponds to asking if the two corresponding group generators commute.) The group is commutative if and only if these two sequences lead to the same vertex for *every pair* of generating directed arcs.

13. It is not obvious, since a digraph of a cyclic group might be formed using a generating set of two or more elements, no one of which generates the group.

15.

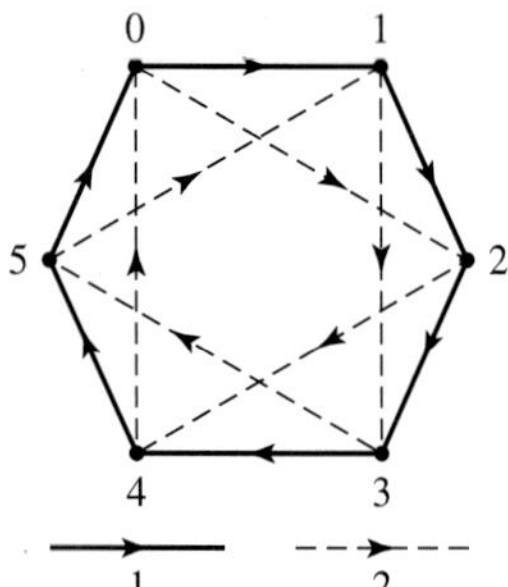

17. **a.** Starting from any vertex a, every path through the graph that terminates at that same vertex a represents a product of generators or their inverses that is equal to the identity and thus gives a relation.

b. $a^4 = e, b^2 = e, (ab)^2 = e$

SECTION 8

1. $\begin{pmatrix} 1 & 2 & 3 & 4 & 5 & 6 \\ 1 & 2 & 3 & 6 & 5 & 4 \end{pmatrix}$ **3.** $\begin{pmatrix} 1 & 2 & 3 & 4 & 5 & 6 \\ 3 & 4 & 1 & 6 & 2 & 5 \end{pmatrix}$

5. $\begin{pmatrix} 1 & 2 & 3 & 4 & 5 & 6 \\ 2 & 6 & 1 & 5 & 4 & 3 \end{pmatrix}$

7. 2 9. ι

11. $\{1, 2, 3, 4, 5, 6\}$ 13. $\{1, 5\}$

15. $\epsilon, \rho, \rho^2, \rho^3, \phi, \rho\phi, \rho^2\phi, \rho^3\phi$ where their ϕ is our μ_1. This gives our elements in the order $\rho_0, \rho_1, \rho_2, \rho_3, \mu_1, \delta_1, \mu_2, \delta_2$.

17. 24

19. Referring to Table 8.12, we find that $\langle\rho_0\rangle = \{\rho_0\}$, $\langle\rho_1\rangle = \langle\rho_3\rangle = \{\rho_0, \rho_1, \rho_2, \rho_3\}$, $\langle\rho_2\rangle = \{\rho_0, \rho_2\}$, $\langle\mu_1\rangle = \{\rho_0, \mu_1\}$, $\langle\mu_2\rangle = \{\rho_0, \mu_2\}$, $\langle\delta_1\rangle = \{\rho_0, \delta_1\}$, and $\langle\delta_2\rangle = \{\rho_0, \delta_2\}$. These are all the cyclic subgroups. A subgroup containing one of the "turn the square over" permutations μ_1, μ_2, δ_1, or δ_2 and also containing ρ_1 or ρ_3 will describe all positions of the square so it must be the entire group D_4. Checking the line of the table opposite μ_1, we see that the only other elements that can be in a proper subgroup with μ_1 are ρ_2, μ_2, and, of course, ρ_0. We check that $\{\rho_0, \rho_2, \mu_1, \mu_2\}$ is closed under multiplication and is a subgroup. Checking the row of the table opposite μ_2 gives the same subgroup. Checking the rows opposite δ_1 and opposite δ_2 gives the subgroup $\{\rho_0, \rho_2, \delta_1, \delta_2\}$ as the only remaining possibility, using the same reasoning.

21. **a.** These are "elementary permutation matrices," resulting from permuting the rows of the identity matrix. When another matrix A is multiplied on the left by one of these matrices P, the rows of A are permuted in the same fashion that the rows of the 3×3 identity matrix were permuted to obtain P. Because all 6 possible permutations of the three rows are present, we see they will act just like the elements of S_3 in permuting the entries 1, 2, 3 of the given column vector. Thus they form a group because S_3 is a group.
b. The symmetric group S_3.

23. $\mathbb{Z}_2$ 25. D_4

27. For $\mathbb{Z}_4$, $\lambda_0 = \begin{pmatrix} 0 & 1 & 2 & 3 \\ 0 & 1 & 2 & 3 \end{pmatrix}$, $\lambda_1 = \begin{pmatrix} 0 & 1 & 2 & 3 \\ 1 & 2 & 3 & 0 \end{pmatrix}$, $\lambda_2 = \begin{pmatrix} 0 & 1 & 2 & 3 \\ 2 & 3 & 0 & 1 \end{pmatrix}$, $\lambda_3 = \begin{pmatrix} 0 & 1 & 2 & 3 \\ 3 & 0 & 1 & 2 \end{pmatrix}$.
The table for the left regular representation is the same as the table for $\mathbb{Z}_4$ with n replaced by λ_n. For S_3, $\rho_0 = \begin{pmatrix} r_0 & r_1 & r_2 & m_1 & m_2 & m_3 \\ r_0 & r_1 & r_2 & m_1 & m_2 & m_3 \end{pmatrix}$, $\rho_1 = \begin{pmatrix} r_0 & r_1 & r_2 & m_1 & m_2 & m_3 \\ r_1 & r_2 & r_0 & m_2 & m_3 & m_1 \end{pmatrix}$, etc., where the bottom row in the permutation ρ_σ consists of the elements of S_3 in the order they appear down the column under σ in Table 8.8. The table for this right regular representation is the same as the table for S_3 with σ replaced by ρ_σ.

31. Not a permutation 33. Not a permutation

35. **a.** T **c.** T **e.** T **g.** F **i.** F

37. A monoid 41. No 43. Yes

SECTION 9

1. $\{1, 2, 5\}, \{3\}, \{4, 6\}$
3. $\{1, 2, 3, 4, 5\}, \{6\}, \{7, 8\}$
5. $\{2n \mid n \in \mathbb{Z}\}, \{2n + 1 \mid n \in \mathbb{Z}\}$
7. $\begin{pmatrix} 1 & 2 & 3 & 4 & 5 & 6 & 7 & 8 \\ 4 & 1 & 3 & 5 & 8 & 6 & 2 & 7 \end{pmatrix}$
9. $\begin{pmatrix} 1 & 2 & 3 & 4 & 5 & 6 & 7 & 8 \\ 5 & 4 & 3 & 7 & 8 & 6 & 2 & 1 \end{pmatrix}$
11. $(1, 3, 4)(2, 6)(5, 8, 7) = (1, 4)(1, 3)(2, 6)(5, 7)(5, 8)$
13. **a.** 4
b. A cycle of length n has order n.
c. σ has order 6; τ has order 4.
d. 6 in Exercises 10 and 11, 8 in Exercise 12.
e. The order of a permutation expressed as a product of disjoint cycles is the least common multiple of the lengths of the cycles.

15. 6 17. 30

19. 

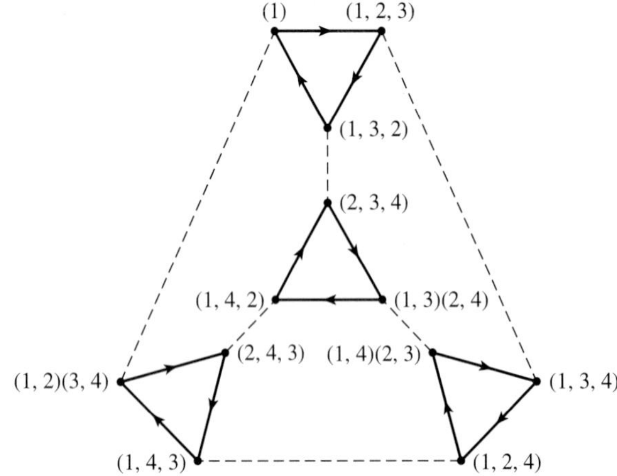

23. **a.** F **c.** F **e.** F **g.** T **i.** T

SECTION 10

1.
$$\begin{aligned} 4\mathbb{Z} &= \{\cdots, -8, -4, 0, 4, 8, \cdots\} \\ 1 + 4\mathbb{Z} &= \{\cdots, -7, -3, 1, 5, 9, \cdots\} \\ 2 + 4\mathbb{Z} &= \{\cdots, -6, -2, 2, 6, 10, \cdots\} \\ 3 + 4\mathbb{Z} &= \{\cdots, -5, -1, 3, 7, 11, \cdots\} \end{aligned}$$

3. $\langle 2 \rangle = \{0, 2, 4, 6, 8, 10\}$, $1 + \langle 2 \rangle = \{1, 3, 5, 7, 9, 11\}$

5. $\langle 18 \rangle = \{0, 18\}$, $1 + \langle 18 \rangle = \{1, 19\}$, $2 + \langle 18 \rangle = \{2, 20\}, \cdots, 17 + \langle 18 \rangle = \{17, 35\}$

7. $\{\rho_0, \mu_2\}, \{\rho_1, \delta_1\}, \{\rho_2, \mu_1\}, \{\rho_3, \delta_2\}$. Not the same.

9. $\{\rho_0, \rho_2\}, \{\rho_1, \rho_3\}, \{\mu_1, \mu_2\}, \{\delta_1, \delta_2\}$

11. Yes, we get a coset group isomorphic to the Klein 4-group V.

	ρ_0	ρ_2	ρ_1	ρ_3	μ_1	μ_2	δ_1	δ_2
ρ_0	ρ_0	ρ_2	ρ_1	ρ_3	μ_1	μ_2	δ_1	δ_2
ρ_2	ρ_2	ρ_0	ρ_3	ρ_1	μ_2	μ_1	δ_2	δ_1
ρ_1	ρ_1	ρ_3	ρ_2	ρ_0	δ_1	δ_2	μ_2	μ_1
ρ_3	ρ_3	ρ_1	ρ_0	ρ_2	δ_2	δ_1	μ_1	μ_2
μ_1	μ_1	μ_2	δ_2	δ_1	ρ_0	ρ_2	ρ_3	ρ_1
μ_2	μ_2	μ_1	δ_1	δ_2	ρ_2	ρ_0	ρ_1	ρ_3
δ_1	δ_1	δ_2	μ_1	μ_2	ρ_1	ρ_3	ρ_0	ρ_2
δ_2	δ_2	δ_1	μ_2	μ_1	ρ_3	ρ_1	ρ_2	ρ_0

13. 3 **15.** 24

19. **a.** T **c.** T **e.** T **g.** T **i.** F

21. $G = \mathbb{Z}_2$, subgroup $H = \mathbb{Z}_2$.

23. Impossible. The number of cells must divide the order of the group, and 12 does not divide 6.

SECTION 11

1.

Element	*Order*	*Element*	*Order*
(0, 0)	1	(0, 2)	2
(1, 0)	2	(1, 2)	2
(0, 1)	4	(0, 3)	4
(1, 1)	4	(1, 3)	4

The group is not cyclic

3. 2 **5.** 9 **7.** 60

9. $\{(0, 0), (0, 1)\}, \{(0, 0), (1, 0)\}, \{(0, 0), (1, 1)\}$

11. $\{(0, 0), (0, 1), (0, 2), (0, 3)\}$
$\{(0, 0), (0, 2), (1, 0), (1, 2)\}$
$\{(0, 0), (1, 1), (0, 2), (1, 3)\}$

13. $\mathbb{Z}_{20} \times \mathbb{Z}_3, \mathbb{Z}_{15} \times \mathbb{Z}_4, \mathbb{Z}_{12} \times \mathbb{Z}_5, \mathbb{Z}_5 \times \mathbb{Z}_3 \times \mathbb{Z}_4$

15. 12

17. 120

19. 180

21. $\mathbb{Z}_8, \mathbb{Z}_2 \times \mathbb{Z}_4, \mathbb{Z}_2 \times \mathbb{Z}_2 \times \mathbb{Z}_2$

23. $\mathbb{Z}_{32}, \mathbb{Z}_2 \times \mathbb{Z}_{16}, \mathbb{Z}_4 \times \mathbb{Z}_8, \mathbb{Z}_2 \times \mathbb{Z}_2 \times \mathbb{Z}_8, \mathbb{Z}_2 \times \mathbb{Z}_4 \times \mathbb{Z}_4,$
$\mathbb{Z}_2 \times \mathbb{Z}_2 \times \mathbb{Z}_2 \times \mathbb{Z}_4, \mathbb{Z}_2 \times \mathbb{Z}_2 \times \mathbb{Z}_2 \times \mathbb{Z}_2 \times \mathbb{Z}_2$

25. $\mathbb{Z}_9 \times \mathbb{Z}_{121}, \mathbb{Z}_3 \times \mathbb{Z}_3 \times \mathbb{Z}_{121}, \mathbb{Z}_9 \times \mathbb{Z}_{11} \times \mathbb{Z}_{11}, \mathbb{Z}_3 \times \mathbb{Z}_3 \times \mathbb{Z}_{11} \times \mathbb{Z}_{11}$

29. a.

n	2	3	4	5	6	7	8
number of groups	2	3	5	7	11	15	22

b. i) 225 **ii)** 225 **iii)** 110

31. a. It is abelian when the arrows on both n-gons have the same (clockwise or counterclockwise) direction.
b. $\mathbb{Z}_2 \times \mathbb{Z}_n$
c. When n is odd.
d. The dihedral group D_n.

33. $\mathbb{Z}_2$ is an example.

35. S_3 is an example.

37. The numbers are the same. **41.** $\{-1, 1\}$

SECTION 12

1. a. The only isometries of $\mathbb{R}$ leaving a number c fixed are the reflection through c that carries $c + x$ to $c - x$ for all $x \in \mathbb{R}$, and the identity map.
b. The isometries of $\mathbb{R}^2$ that leave a point P fixed are the rotations about P through any angle θ where $0 \leq \theta < 360^\circ$ and the reflections across any axis that passes through P.
c. The only isometries of $\mathbb{R}$ that carry a line segment into itself are the reflection through the midpoint of the line segment (see the answer to part (a)) and the identity map.
d. The isometries of $\mathbb{R}^2$ that carry a line segment into itself are a rotation of 180° about the midpoint of the line segment, a reflection in the axis containing the line segment, a reflection in the axis perpendicular to the line segment at its midpoint, and the identity map.
e. The isometries of $\mathbb{R}^3$ that carry a line segment into itself include rotations through any angle about an axis that contains the line segment, reflections across any plane that contains the line segment, and reflection across the plane perpendicular to the line segment at its midpoint.

3.

	τ	ρ	μ	γ
τ	τ	ρ	$\mu\gamma$	$\mu\gamma$
ρ	ρ	$\rho\tau$	$\mu\gamma$	$\mu\gamma$
μ	$\mu\gamma$	$\mu\gamma$	$\tau\rho$	$\tau\rho$
γ	$\mu\gamma$	$\mu\gamma$	$\tau\rho$	$\tau\rho$

5. **7.**

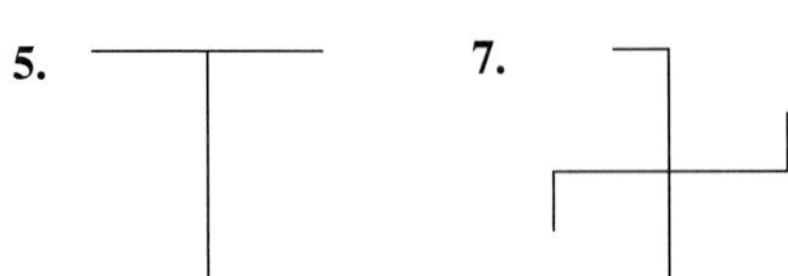

9. *Translation:* order ∞
Rotation: order any $n \geq 2$ or ∞
Reflection: order 2
Glide reflection: order ∞

11. Rotations **13.** Only the identity and reflections.

17. Yes. The product of two translations is a translation and the inverse of a translation is a translation.

19. Yes. There is only one reflection μ across one particular line L, and μ^2 is the identity, so we have a group isomorphic to $\mathbb{Z}_2$.

21. Only reflections and rotations (and the identity) because translations and glide reflections do not have finite order in the group of all plane isometries.

25. **a.** No **b.** No **c.** Yes **d.** No **e.** D_∞

27. **a.** Yes **b.** No **c.** No **d.** No **e.** D_∞

29. **a.** No **b.** No **c.** No **d.** Yes **e.** $\mathbb{Z}$

31. **a.** Yes. 90°, 180° **b.** Yes **c.** No

33. **a.** No **b.** No **c.** No

35. **a.** Yes. 180° **b.** Yes **c.** No

37. **a.** Yes. 120° **b.** Yes **c.** No

39. **a.** Yes. 90°, 180° **b.** Yes **c.** No **d.** $(-1, 1)$ and $(1, 1)$

41. **a.** Yes. 120° **b.** Yes **c.** No **d.** $(0, 1)$ and $(1, \sqrt{3})$

SECTION 13

1. Yes **3.** Yes **5.** No

7. Yes **9.** Yes

11. Yes **13.** Yes **15.** No

17. $\text{Ker}(\phi) = 7\mathbb{Z}$; $\phi(25) = 2$

19. $\text{Ker}(\phi) = 6\mathbb{Z}$; $\phi(20) = (1, 2, 7)(4, 5, 6)$

21. $\text{Ker}(\phi) = \{0, 4, 8, 12, 16, 20\}$; $\phi(14) = (1, 6)(4, 7)$

23. $\text{Ker}(\phi) = \{(0, 0)\}$; $\phi(4, 6) = (2, 18)$

25. 2 **27.** 2 **29.** For all $g \in G$

33. No nontrivial homomorphism. By Theorem 13.12, the image of ϕ would have to be a subgroup of $\mathbb{Z}_5$, and hence all of $\mathbb{Z}_5$ for a nontrivial ϕ. But the number of cosets of a subgroup of a finite group is a divisor of the order of the group, and 5 does not divide 12.

35. Let $\phi(m, n) = (m, 0)$ for $(m, n) \in \mathbb{Z}_2 \times \mathbb{Z}_4$.
37. Let $\phi(n) = \rho_n$ for $n \in \mathbb{Z}_3$, using our notation in the text for elements of S_3.
39. Let $\phi(m, n) = 2m$.
41. Viewing D_4 as a group of permutations, let $\phi(\sigma) = (1, 2)$ for odd $\sigma \in D_4$ and $\phi(\sigma)$ be the identity for even $\sigma \in D_4$.
43. Let $\phi(\sigma) = (1, 2)$ for odd $\sigma \in S_4$ and $\phi(\sigma)$ be the identity element for even $\sigma \in S_4$.
51. The image of ϕ is $\langle a \rangle$, and Ker(ϕ) must be some subgroup $n\mathbb{Z}$ of $\mathbb{Z}$.
53. $hk = kh$ **55.** h^n must be the identity e of G.

SECTION 14

1. 3 **3.** 4 **5.** 2 **7.** 2
9. 4 **11.** 3 **13.** 4 **15.** 1
21. **a.** When working with a factor group G/H, you would let a and b be elements of G, not elements of G/H. The student probably does not understand what elements of G/H look like and can write nothing sensible concerning them.
b. We must show that G/H is abelian. Let aH and bH be two elements of G/H.
23. **a.** T **c.** T **e.** T **g.** T **i.** T
29. $\{\rho_0, \mu_1\}$, $\{\rho_0, \mu_2\}$, and $\{\rho_0, \mu_3\}$
35. Example: Let $G = N = S_3$, and let $H = \{\rho_0, \mu_1\}$. Then N is normal in G, but $H \cap N = H$ is not normal in G.

SECTION 15

1. $\mathbb{Z}_2$ **3.** $\mathbb{Z}_4$ **5.** $\mathbb{Z}_4 \times \mathbb{Z}_8$ **7.** $\mathbb{Z}$ **9.** $\mathbb{Z}_3 \times \mathbb{Z} \times \mathbb{Z}_4$
11. $\mathbb{Z}_2 \times \mathbb{Z}$ **13.** $\mathbb{Z}(D_4) = C = \{\rho_0, \rho_2\}$
15. $Z(S_3 \times D_4) = \{(\rho_0, \rho_0), (\rho_0, \rho_2)\}$, using the notations for these groups in Section 8, $C = A_3 \times \{\rho_0, \rho_2\}$.
19. **a.** T **c.** F **e.** F **g.** F **i.** T
21. $\{f \in F^* \mid f(0) = 1\}$
23. Yes. Let $f(x) = 1$ for $x \geq 0$ and $f(x) = -1$ for $x < 0$. Then $f(x) \cdot f(x) = 1$ for all x, so $f^2 \in K^*$ but f is not in K^*. Thus fK^* has order 2 in F^*/K^*.
25. U
27. The multiplicative group U of complex numbers of absolute value 1
29. Let $G = \mathbb{Z}_2 \times \mathbb{Z}_4$. Then $H = \langle(1, 0)\rangle$ is isomorphic to $K = \langle(0, 2)\rangle$, but G/H is isomorphic to $\mathbb{Z}_4$ while G/K is isomorphic to $\mathbb{Z}_2 \times \mathbb{Z}_2$.
31. **a.** $\{e\}$ **b.** The whole group

SECTION 16

1. $X_{\rho_0} = X$, $X_{\rho_1} = \{C\}$, $X_{\rho_2} = \{m_1, m_2, d_1, d_2, C\}$, $X_{\rho_3} = \{C\}$,
$X_{\mu_1} = \{s_1, s_3, m_1, m_2, C, P_1, P_3\}$, $X_{\mu_2} = \{s_2, s_4, m_1, m_2, C, P_2, P_4\}$,
$X_{\delta_1} = \{2, 4, d_1, d_2, C\}$, $X_{\delta_2} = \{1, 3, d_1, d_2, C\}$.
3. $\{1, 2, 3, 4\}$, $\{s_1, s_2, s_3, s_4\}$, $\{m_1, m_2\}$, $\{d_1, d_2\}$, $\{C\}$, $\{P_1, P_2, P_3, P_4\}$
7. A transitive G-set has just one orbit.
9. **a.** $\{s_1, s_2, s_3, s_4\}$ and $\{P_1, P_2, P_3, P_4\}$
13. **b.** The set of points on the circle with center at the origin and passing through P
c. The cyclic subgroup $\langle 2\pi \rangle$ of $G = \mathbb{R}$
17. **a.** $K = g_0 H g_0^{-1}$.
b. *Conjecture:* H and K should be conjugate subgroups of G.

19.

	X	Y		Z		
	a	*a*	*b*	*a*	*b*	*c*
0	*a*	*a*	*b*	*a*	*b*	*c*
1	*a*	*b*	*a*	*b*	*c*	*a*
2	*a*	*a*	*b*	*c*	*a*	*b*
3	*a*	*b*	*a*	*a*	*b*	*c*
4	*a*	*a*	*b*	*b*	*c*	*a*
5	*a*	*b*	*a*	*c*	*a*	*b*

There are four of them: X, Y, Z, and $\mathbb{Z}_6$.

SECTION 17

1. 5 **3.** 2 **5.** 11,712
7. **a.** 45 **b.** 231
9. **a.** 90 **b.** 6,246

SECTION 18

1. 0 **3.** 1 **5.** $(1, 6)$
7. Commutative ring, no unity, not a field
9. Commutative ring with unity, not a field
11. Commutative ring with unity, not a field
13. No. $\{ri \mid r \in \mathbb{R}\}$ is not closed under multiplication.
15. $(1, 1)$,$(1, -1)$, $(-1, 1)$, $(-1, -1)$
17. All nonzero $q \in \mathbb{Q}$ **19.** 1, 3
21. Let $\mathbb{R} = \mathbb{Z}$ with unity 1 and $\mathrm{R}' = \mathbb{Z} \times \mathbb{Z}$ with unity $1' = (1, 1)$. Let $\phi : R \to R'$ be defined by $\phi(n) = (n, 0)$. Then $\phi(1) = (1, 0) \neq 1'$.
23. $\phi_1 : \mathbb{Z} \to \mathbb{Z}$ where $\phi_1(n) = 0$, $\phi_2 : \mathbb{Z} \to \mathbb{Z}$ where $\phi_2(n) = n$
25. $\phi_1 : \mathbb{Z} \times \mathbb{Z} \to \mathbb{Z}$ where $\phi_1(n, m) = 0$, $\phi_2 : \mathbb{Z} \times \mathbb{Z} \to \mathbb{Z}$ where $\phi_2(n, m) = n$
$\phi_3 : \mathbb{Z} \times \mathbb{Z} \to \mathbb{Z}$ where $\phi_3(n, m) = m$
27. The reasoning is not correct since a product $(X - I_3)(X + I_3)$ of two matrices may be the zero matrix 0 without having either matrix be 0. Counterexample:

$$\begin{bmatrix} 0 & 0 & 1 \\ 0 & 1 & 0 \\ 1 & 0 & 0 \end{bmatrix}^2 = I_3.$$

31. $a = 2, b = 3$ in $\mathbb{Z}_6$
33. **a.** T **c.** F **e.** T **g.** T **i.** T

SECTION 19

1. 0, 3, 5, 8, 9, 11 **3.** No solutions **5.** 0 **7.** 0 **9.** 12
11. $a^4 + 2a^2b^2 + b^4$ **13.** $a^6 + 2a^3b^3 + b^6$
17. **a.** F **c.** F **e.** T **g.** F **i.** F

19. **1.** $\text{Det}(A) = 0$. **2.** The column vectors of A are dependent.
3. The row vectors of A are dependent. **4.** Zero is an eigenvalue of A.
5. A is not invertible.

SECTION 20

1. 3 or 5 **3.** Any of 3, 5, 6, 7, 10, 11, 12, or 14. **5.** 2

7.

$\varphi(1) = 1$	$\varphi(7) = 6$	$\varphi(13) = 12$	$\varphi(19) = 18$	$\varphi(25) = 20$
$\varphi(2) = 1$	$\varphi(8) = 4$	$\varphi(14) = 6$	$\varphi(20) = 8$	$\varphi(26) = 12$
$\varphi(3) = 2$	$\varphi(9) = 6$	$\varphi(15) = 8$	$\varphi(21) = 12$	$\varphi(27) = 18$
$\varphi(4) = 2$	$\varphi(10) = 4$	$\varphi(16) = 8$	$\varphi(22) = 10$	$\varphi(28) = 12$
$\varphi(5) = 4$	$\varphi(11) = 10$	$\varphi(17) = 16$	$\varphi(23) = 22$	$\varphi(29) = 28$
$\varphi(6) = 2$	$\varphi(12) = 4$	$\varphi(18) = 6$	$\varphi(24) = 8$	$\varphi(30) = 8$

9. $(p-1)(q-1)$ **11.** $1 + 4\mathbb{Z}, 3 + 4\mathbb{Z}$ **13.** No solutions
15. No solutions
17. $3 + 65\mathbb{Z}, 16 + 65\mathbb{Z}, 29 + 65\mathbb{Z}, 42 + 65\mathbb{Z}, 55 + 65\mathbb{Z}$
19. 1 **21.** 9
23. **a.** *F* **c.** *T* **e.** *T* **g.** *F* **i.** *F*

SECTION 21

1. $\{q_1 + q_2 i \mid q_1, q_2 \in \mathbb{Q}\}$
15. It is isomorphic to the ring D of all rational numbers that can be expressed as a quotient of integers with denominator some power of 2.
17. It runs into trouble when we try to prove the transitive property in the proof of Lemma 5.4.2, for multiplicative cancellation may not hold. For $R = \mathbb{Z}_6$ and $T = \{1, 2, 4\}$ we have $(1, 2) \sim (2, 4)$ since $(1)(4) = (2)(2) = 4$ and $(2, 4) \sim (2, 1)$ since $(2)(1) = (4)(2)$ in $\mathbb{Z}_6$. However, $(1, 2)$ is not equivalent to $(2, 1)$ because $(1)(1) \neq (2)(2)$ in $\mathbb{Z}_6$.

SECTION 22

1. $f(x) + g(x) = 2x^2 + 5$, $f(x)g(x) = 6x^2 + 4x + 6$
3. $f(x) + g(x) = 5x^2 + 5x + 1$, $f(x)g(x) = x^3 + 5x$
5. 16 **7.** 7 **9.** 2 **11.** 0 **13.** 2, 3 **15.** 0, 2, 4
17. 0, 1, 2, 3
21. $0, x - 5, 2x - 10, x^2 - 25, x^2 - 5x, x^4 - 5x^3$. (Other answers are possible.)
23. **a.** *T* **c.** *T* **e.** *F* **g.** *T* **i.** *T*
25. **a.** They are the units of D. **b.** $1, -1$ **c.** 1, 2, 3, 4, 5, 6
27. **b.** *F* **c.** $F[x]$ **31.** **a.** 4, 27 **b.** $\mathbb{Z}_2 \times \mathbb{Z}_2, \mathbb{Z}_3 \times \mathbb{Z}_3 \times \mathbb{Z}_3$

SECTION 23

1. $q(x) = x^4 + x^3 + x^2 + x - 2$, $r(x) = 4x + 3$
3. $q(x) = 6x^4 + 7x^3 + 2x^2 - x + 2$, $r(x) = 4$
5. 2, 3 **7.** 3, 10, 5, 11, 14, 7, 12, 6
9. $(x-1)(x+1)(x-2)(x+2)$
11. $(x-3)(x+3)(2x+3)$

13. Yes. It is of degree 3 with no zeros in $\mathbb{Z}_5$.
$2x^3 + x^2 + 2x + 2$

15. *Partial answer:* $g(x)$ is irreducible over $\mathbb{R}$, but it is not irreducible over $\mathbb{C}$.

19. Yes. $p = 3$ **21.** Yes. $p = 5$

25. **a.** T **c.** T **e.** T **g.** T **i.** T

27. $x^2 + x + 1$

29. $x^2 + 1, x^2 + x + 2, x^2 + 2x + 2, 2x^2 + 2, 2x^2 + x + 1, 2x^2 + 2x + 1$

31. $p(p - 1)^2/2$

SECTION 24

1. $1e + 0a + 3b$ **3.** $2e + 2a + 2b$ **5.** j **7.** $(1/50)j - (3/50)k$

9. $\mathbb{R}^*$, that is, $\{a_1 + 0i + 0j + 0k \mid a_1 \in \mathbb{R}, a_1 \neq 0\}$

11. **a.** F **c.** F **e.** F **g.** T **i.** T
c. If $|A| = 1$, then $\text{End}(A) = \{0\}$. **e.** $0 \in \text{End}(A)$ is not in $\text{Iso}(A)$.

19. **a.** $K = \begin{bmatrix} i & 0 \\ 0 & -i \end{bmatrix}$.

b. Denoting by B the matrix with coefficient b and by C the matrix with coefficient c and the 2×2 identity matrix by I, we must check that

$$B^2 = -I, C^2 = -I, K^2 = -I,$$

$$CK = B, KB = C, CB = -K, KC = -B, \text{ and } BK = -C.$$

c. We should check that ϕ is one to one.

SECTION 25

1. $a < x < x^2 < x^3 < \cdots < x^n \cdots$ for any $a \in R$.

3. $m + n\sqrt{2}$ is positive if $m > 0$ and $n < 0$, or if $m > 0$ and $m^2 > 2n^2$, or if $n < 0$ and $2n^2 > m^2$.

5. **i.** a c e d b **ii.** e c b a d

7. **i.** d a b c e **ii.** d c e a b

9. **i.** c a e d b **ii.** e c b a d

11. d b a e c **13.** d e b a c

15. **a.** T **c.** F **e.** T **g.** T **i.** F

SECTION 26

1. There are just nine possibilities:
$\phi(1, 0) = (1, 0)$ while $\phi(0, 1) = (0, 0)$ or $(0, 1)$,
$\phi(1, 0) = (0, 1)$ while $\phi(0, 1) = (0, 0)$ or $(1, 0)$,
$\phi(1, 0) = (1, 1)$ while $\phi(0, 1) = (0, 0)$, and
$\phi(1, 0) = (0, 0)$ while $\phi(0, 1) = (0, 0), (1, 0), (0, 1)$, or $(1, 1)$.

3. $\langle 0 \rangle = \{0\}, \mathbb{Z}_{12}/\{0\} \simeq \mathbb{Z}_{12}$
$\langle 1 \rangle = \{0, 1, 2, 3, 4, 5, 6, 7, 8, 9, 10, 11\}, \mathbb{Z}_{12}/\langle 1 \rangle \simeq \{0\}$
$\langle 2 \rangle = \{0, 2, 4, 6, 8, 10\}, \mathbb{Z}_{12}/\langle 2 \rangle \simeq \mathbb{Z}_2$
$\langle 3 \rangle = \{0, 3, 6, 9\}, \mathbb{Z}_{12}/\langle 3 \rangle \simeq \mathbb{Z}_3$
$\langle 4 \rangle = \{0, 4, 8\}, \mathbb{Z}_{12}/\langle 4 \rangle \simeq \mathbb{Z}_4$
$\langle 6 \rangle = \{0, 6\}, \mathbb{Z}_{12}/\langle 6 \rangle \simeq \mathbb{Z}_6$

9. Let $\phi : \mathbb{Z} \to \mathbb{Z} \times \mathbb{Z}$ be given by $\phi(n) = (n, 0)$ for $n \in \mathbb{Z}$.

11. R/R and $R/\{0\}$ are not of real interest because R/R is the ring containing only the zero element, and $R/\{0\}$ is isomorphic to R.
13. $\mathbb{Z}$ is an integral domain. $\mathbb{Z}/4\mathbb{Z}$ is isomorphic to $\mathbb{Z}_4$, which has a divisor 2 of 0.
15. $\{(n, n) \mid n \in \mathbb{Z}\}$. (Other answers are possible.)
31. The nilradical of $\mathbb{Z}_{12}$ is $\{0, 6\}$. The nilradical of $\mathbb{Z}$ is $\{0\}$ and the nilradical of $\mathbb{Z}_{32}$ is $\{0, 2, 4, 6, 8, \cdots, 30\}$.
35. **a.** Let $R = \mathbb{Z}$ and let $N = 4\mathbb{Z}$. Then $\sqrt{N} = 2\mathbb{Z} \neq 4\mathbb{Z}$
b. Let $R = \mathbb{Z}$ and let $N = 2\mathbb{Z}$. Then $\sqrt{N} = N$.

SECTION 27

1. $\{0, 2, 4\}$ and $\{0, 3\}$ are both prime and maximal.
3. $\{(0, 0), (1, 0)\}$ and $\{(0, 0), (0, 1)\}$ are both prime and maximal.
5. 1 **7.** 2 **9.** 1, 4 **15.** $2\mathbb{Z} \times \mathbb{Z}$ **17.** $4\mathbb{Z} \times \{0\}$
19. Yes. $x^2 - 6x + 6$ is irreducible over $\mathbb{Q}$ by Eisenstein with $p = 2$.
21. Yes. $\mathbb{Z}_2 \times \mathbb{Z}_3$
23. No. Enlarging the domain to a field of quotients, you would have to have a field containing two different prime fields $\mathbb{Z}_p$ and $\mathbb{Z}_q$, which is impossible.

SECTION 28

1. $-3x^3 + 7x^2y^2z - 5x^2yz^3 + 2xy^3z^5$
3. $2x^2yz^2 - 2xy^2z^2 - 7x + 3y + 10z^3$
5. $2z^5y^3x - 5z^3yx^2 + 7zy^2x^2 - 3x^3$
7. $10z^3 - 2z^2y^2x + 2z^2yx^2 + 3y - 7x$
9. $1 < z < y < x < z^2 < yz < y^2 < xz < xy < x^2 < z^3 < yz^2 < y^2z < y^3 < xz^2 < xyz < xy^2 < x^2z < x^2y < x^3 < \cdots$
11. $3y^2z^5 - 8z^7 + 5y^3z^3 - 4x$ **13.** $3yz^3 - 8xy - 4xz + 2yz + 38$
15. $\langle y^5 + y^3, y^3 + z, x - y^4\rangle$ **17.** $\langle y^2z^3 + 3, -3y - 2z, y^2z^2 + 3\rangle$
19. $\{1\}$ **21.** $\{x - 1\}$
23. $\{2x + y - 5, y^2 - 9y + 18\}$
The algebraic variety is $\{(1, 3), (-\frac{1}{2}, 6)\}$.
25. $\{x + y, y^3 - y + 1\}$
The algebraic variety consists of one point $(a, -a)$ where $a \approx 1.3247$.
27. **a.** T **c.** T **e.** T **g.** T **i.** F

SECTION 29

1. $x^2 - 2x - 1$ **3.** $x^2 - 2x + 2$
5. $x^{12} + 3x^8 - 4x^6 + 3x^4 + 12x^2 + 5$
7. $\text{Irr}(\alpha, \mathbb{Q}) = x^4 - \frac{2}{3}x^2 - \frac{62}{9}$; $\deg(\alpha, \mathbb{Q}) = 4$
9. Algebraic, $\deg(\alpha, F) = 2$
11. Transcendental
13. Algebraic, $\deg(\alpha, F) = 2$
15. Algebraic, $\deg(\alpha, F) = 1$
17. $x^2 + x + 1 = (x - \alpha)(x + 1 + \alpha)$
23. **a.** T **c.** T **e.** F **g.** F **i.** F
25. **b.** $x^3 + x^2 + 1 = (x - \alpha)(x - \alpha^2)[x - (1 + \alpha + \alpha^2)]$
27. It is the monic polynomial in $F[x]$ of *minimal* degree having α as a zero.

SECTION 30

1. $\{(0, 1), (1, 0)\}$, $\{(1, 1), (-1, 1)\}$, $\{(2, 1), (1, 2)\}$. (Other answers are possible.)
3. No. $2(-1, 1, 2) - 4(2, -3, 1) + (10, -14, 0) = (0, 0, 0)$
5. $\{1\}$ **7.** $\{1, i\}$ **9.** $\{1, \sqrt[4]{2}, \sqrt{2}, (\sqrt[4]{2})^3\}$
15. **a.** T **c.** T **e.** F **g.** F **i.** T
17. **a.** The **subspace of** V **generated by** S is the intersection of all subspaces of V containing S.
19. *Partial answer:* A basis for F^n is

$$\{(1, 0, \cdots, 0), (0, 1, \cdots, 0), \cdots, (0, 0, \cdots, 1)\}$$

where 1 is the multiplicative identity of F.
25. **a.** A homomorphism
b. *Partial answer:* The **kernel** (or **nullspace**) of ϕ is $\{\alpha \in V \mid \phi(\alpha) = 0\}$.
c. ϕ is an isomorphism of V with V' if $\text{Ker}(\phi) = \{0\}$ and ϕ maps V onto V'.

SECTION 31

1. $2, \{1, \sqrt{2}\}$ **3.** $4, \{1, \sqrt{3}, \sqrt{2}, \sqrt{6}\}$
5. $6, \{1, \sqrt{2}, \sqrt[3]{2}, \sqrt{2}(\sqrt[3]{2}), (\sqrt[3]{2})^2, \sqrt{2}(\sqrt[3]{2})^2\}$ **7.** $2, \{1, \sqrt{6}\}$
9. $9, \{1, \sqrt[3]{2}, \sqrt[3]{4}, \sqrt[3]{3}, \sqrt[3]{6}, \sqrt[3]{12}, \sqrt[3]{9}, \sqrt[3]{18}, \sqrt[3]{36}\}$
11. $2, \{1, \sqrt{2}\}$ **13.** $2, \{1, \sqrt{2}\}$
19. **a.** F **c.** F **e.** F **g.** F **i.** F
23. *Partial answer:* Extensions of degree 2^n for $n \in \mathbb{Z}^+$ are obtained.

SECTION 32

All odd-numbered answers require proofs and are not listed here.

SECTION 33

1. Yes **3.** Yes **5.** 6 **7.** 0

SECTION 34

1. **a.** $K = \{0, 3, 6, 9\}$.
b. $0 + K = \{0, 3, 6, 9\}$, $1 + K = \{1, 4, 7, 10\}$, $2 + K = \{2, 5, 8, 11\}$.
c. $\mu(0 + K) = 0$, $\mu(1 + K) = 2$, $\mu(2 + K) = 1$.
3. **a.** $HN = \{0, 2, 4, 6, 8, 10, 12, 14, 16, 18, 20, 22\}$, $H \cap N = \{0, 12\}$.
b. $0 + N = \{0, 6, 12, 18\}$, $2 + N = \{2, 8, 14, 20\}$, $4 + N = \{4, 10, 16, 22\}$.
c. $0 + (H \cap N) = \{0, 12\}$, $4 + (H \cap N) = \{4, 16\}$, $8 + (H \cap N) = \{8, 20\}$.
d. $\phi(0 + N) = 0 + (H \cap N)$, $\phi(2 + N) = 8 + (H \cap N)$, $\phi(4 + N) = 4 + (H \cap N)$.
5. **a.** $0 + H = \{0, 4, 8, 12, 16, 20\}$, $1 + H = \{1, 5, 9, 13, 17, 21\}$,
$2 + H = \{2, 6, 10, 14, 18, 22\}$, $3 + H = \{3, 7, 11, 15, 19, 23\}$.
b. $0 + K = \{0, 8, 16\}$, $1 + K = \{1, 9, 17\}$, $2 + K = \{2, 10, 18\}$,
$3 + K = \{3, 11, 19\}$,
$4 + K = \{4, 12, 20\}$, $5 + K = \{5, 13, 21\}$, $6 + K = \{6, 14, 22\}$,
$7 + K = \{7, 15, 23\}$.

c. $0 + K = \{0, 8, 16\}, 4 + K = \{4, 12, 20\}$.
d. $(0 + K) + (H/K) = H/K = \{0 + K, 4 + K\} = \{\{0, 8, 16\}, 4, 12, 20\}\}$
$(1 + K) + (H/K) = \{1 + K, 5 + K\} = \{\{1, 9, 17\}, \{5, 13, 21\}\}$
$(2 + K) + (H/K) = \{2 + K, 6 + K\} = \{\{2, 10, 18\}, \{6, 14, 22\}\}$
$(3 + K) + (H/K) = \{3 + K, 7 + K\} = \{\{3, 11, 19\}, \{7, 15, 23\}\}$.
e. $\phi(0 + H) = (0 + K) + (H/K), \phi(1 + H) = (1 + K) + (H/K)$,
$\phi(2 + H) = (2 + K) + (H/K), \phi(3 + H) = (3 + K) + (H/K)$.

SECTION 35

1. The refinements $\{0\} < 250\mathbb{Z} < 10\mathbb{Z} < \mathbb{Z}$ of $\{0\} < 10\mathbb{Z} < \mathbb{Z}$ and $\{0\} < 250\mathbb{Z} < 25\mathbb{Z} < \mathbb{Z}$ of $0 < 25\mathbb{Z} < \mathbb{Z}$ are isomorphic.
3. The given series are isomorphic.
5. The refinements
$\{(0, 0)\} < (4800\mathbb{Z}) \times \mathbb{Z} < (240\mathbb{Z}) \times \mathbb{Z} < (60\mathbb{Z}) \times \mathbb{Z} < (10\mathbb{Z}) \times \mathbb{Z} < \mathbb{Z} \times \mathbb{Z}$ of the first series and
$\{(0, 0)\} < \mathbb{Z} \times (4800\mathbb{Z}) < \mathbb{Z} \times (480\mathbb{Z}) < \mathbb{Z} \times (80\mathbb{Z}) < \mathbb{Z} \times (20\mathbb{Z}) < \mathbb{Z} \times \mathbb{Z}$ of the second series are isomorphic refinements.
7. $\{0\} < \langle 16\rangle < \langle 8\rangle < \langle 4\rangle < \langle 2\rangle < \mathbb{Z}_{48}$
$\{0\} < \langle 24\rangle < \langle 8\rangle < \langle 4\rangle < \langle 2\rangle < \mathbb{Z}_{48}$
$\{0\} < \langle 24\rangle < \langle 12\rangle < \langle 4\rangle < \langle 2\rangle < \mathbb{Z}_{48}$
$\{0\} < \langle 24\rangle < \langle 12\rangle < \langle 6\rangle < \langle 2\rangle < \mathbb{Z}_{48}$
$\{0\} < \langle 24\rangle < \langle 12\rangle < \langle 6\rangle < \langle 3\rangle < \mathbb{Z}_{48}$
9. $\{(\rho_0, 0)\} < A_3 \times \{0\} < S_3 \times (0) < S_3 \times \mathbb{Z}_2$
$\{(\rho_0, 0)\} < \{\rho_0\} \times \mathbb{Z}_2 < A_3 \times \mathbb{Z}_2 < S_3 \times \mathbb{Z}_2$
$\{(\rho_0, 0)\} < A_3 \times \{0\} < A_3 \times \mathbb{Z}_2 < S_3 \times \mathbb{Z}_2$
11. $\{\rho_0\} \times \mathbb{Z}_4$ **13.** $\{\rho_0\} \times \mathbb{Z}_4 \leq \{\rho_0\} \times \mathbb{Z}_4 \leq \{\rho_0\} \times \mathbb{Z}_4 \leq \cdots$.
17. **a.** *T* **c.** *T* **e.** *F* **g.** *F* **i.** *T*
i. The Jordan-Hölder theorem applied to the groups $\mathbb{Z}_n$ implies the Fundamental Theorem of Arithmetic.
19. Yes. $\{\rho_0\} < \{\rho_0, \rho_2\} < \{\rho_0, \rho_1, \rho_2, \rho_3\} < D_4$ is a composition (actually a principal) series and all factor groups are isomorphic to $\mathbb{Z}_2$ and are thus abelian.
21.

Chain (3)	*Chain (4)*
$\{0\} \leq \langle 12\rangle \leq \langle 12\rangle \leq \langle 12\rangle$	$\{0\} \leq \langle 12\rangle < \langle 12\rangle \leq \langle 6\rangle$
$\leq \langle 12\rangle \leq \langle 12\rangle \leq \langle 4\rangle$	$\leq \langle 6\rangle \leq \langle 6\rangle \leq \langle 3\rangle$
$\leq \langle 2\rangle \leq \mathbb{Z}_{24} \leq \mathbb{Z}_{24}$	$\leq \langle 3\rangle \leq \mathbb{Z}_{24} \leq \mathbb{Z}_{24}$

Isomorphisms

$\langle 12\rangle/\{0\} \simeq \langle 12\rangle/\{0\} \simeq \mathbb{Z}_2$, $\langle 12\rangle/\langle 12\rangle \simeq \langle 6\rangle/\langle 6\rangle \simeq \{0\}$,
$\langle 12\rangle/\langle 12\rangle \simeq \langle 3\rangle/\langle 3\rangle \simeq \{0\}$, $\langle 12\rangle/\langle 12\rangle \simeq \langle 12\rangle/\langle 12\rangle \simeq \{0\}$,
$\langle 12\rangle/\langle 12\rangle \simeq \langle 6\rangle/\langle 6\rangle \simeq \{0\}$, $\langle 4\rangle/\langle 12\rangle \simeq \mathbb{Z}_{24}/\langle 3\rangle \simeq \mathbb{Z}_3$
$\langle 2\rangle/\langle 4\rangle \simeq \langle 6\rangle/\langle 12\rangle \simeq \mathbb{Z}_2$, $\mathbb{Z}_{24}/\langle 2\rangle \simeq \langle 3\rangle/\langle 6\rangle \simeq \mathbb{Z}_2$
$\mathbb{Z}_{24}/\mathbb{Z}_{24} \simeq \mathbb{Z}_{24}/\mathbb{Z}_{24} \simeq \{0\}$

SECTION 36

1. 3 **3.** 1, 3
5. The Sylow 3-subgroups are $\langle(1, 2, 3)\rangle$, $\langle(1, 2, 4)\rangle$, $\langle(1, 3, 4)\rangle$, and $\langle(2, 3, 4)\rangle$. Also $(3, 4)\langle(1, 2, 3)\rangle(3, 4) = \langle(1, 2, 4)\rangle$, etc.

SECTION 37

1. a. The conjugate classes are $\{\rho_0\}, \{\rho_2\}, \{\rho_1, \rho_3\}, \{\mu_1, \mu_2\}, \{\delta_1, \delta_2\}$.
b. $8 = 2 + 2 + 2 + 2$

3. a. T **c.** F **e.** T **g.** T **i.** F
e. This is somewhat a matter of opinion.

9. $24 = 1 + 6 + 3 + 8 + 6$

SECTION 38

1. $\{(1, 1, 1), (1, 2, 1), (1, 1, 2)\}$

3. No. $n(2, 1) + m(4, 1)$ can never yield an odd number for first coordinate.

7. $2\mathbb{Z} < \mathbb{Z}$, rank $r = 1$

SECTION 39

1. a. $a^2b^2a^3c^3b^{-2}, b^2c^{-3}a^{-3}b^{-2}a^{-2}$ **b.** $a^{-1}b^3a^4c^6a^{-1}, ac^{-6}a^{-4}b^{-3}a$

3. a. 16 **b.** 36 **c.** 36

5. a. 16 **b.** 36 **c.** 18

11. a. *Partial answer:* $\{1\}$ is a basis for $\mathbb{Z}_4$. **c.** Yes

13. c. A blop group on S is isomorphic to the *free group* $F[S]$ on S.

SECTION 40

1. $(a : a^4 = 1); (a, b : a^4 = 1, b = a^2); (a, b, c : a = 1, b^4 = 1, c = 1)$. (Other answers are possible.)

3. *Octic group:*

	1	a	a^2	a^3	b	ab	a^2b	a^3b
1	1	a	a^2	a^3	b	ab	a^2b	a^3b
a	a	a^2	a^3	1	ab	a^2b	a^3b	b
a^2	a^2	a^3	1	a	a^2b	a^3b	b	ab
a^3	a^3	1	a	a^2	a^3b	b	ab	a^2b
b	b	a^3b	a^2b	ab	1	a^3	a^2	a
ab	ab	b	a^3b	a^2b	a	1	a^3	a^2
a^2b	a^2b	ab	b	a^3b	a^2	a	1	a^3
a^3b	a^3b	a^2b	ab	b	a^3	a^2	a	1

Quaternion group: The same as the table for the octic group except that the 16 entries in the lower right corner are

a^2	a	1	a^3
a^3	a^2	a	1
1	a^3	a^2	a
a	1	a^3	a^2

5. $\mathbb{Z}_{21}$. $(a, b : a^7 = 1, b^3 = 1, ba = a^2b)$

SECTION 41

1. a. $2P_1P_3 - 3P_1P_4 + P_1P_6 - 3P_2P_3 + 3P_2P_4 - 5P_3P_4 + 4P_3P_6 - 5P_4P_6$
b. No **c.** Yes

3. $C_i(P) = Z_i(P) = B_i(P) = H_i(P) = 0$ for $i > 0$. $B_0(P) = 0$. $Z_0(P) \simeq \mathbb{Z}$ and is generated by the 0-cycle P. $H_0(P) \simeq \mathbb{Z}$.

5. $C_i(X) = Z_i(X) = B_i(X) = H_i(X) = 0$ for $i > 0$. $B_0(X) \simeq \mathbb{Z}$ and is generated by the 0-chain $P_2 - P_1$. $Z_0(X) \simeq \mathbb{Z} \times \mathbb{Z}$ and is generated by the two 0-cycles P_1 and P_2. Since $Z_0(X)/B_0(X)$ "indentifies P_1 with P_2," $H_0(X) \simeq \mathbb{Z}$ and is generated by the coset $P_1 + B_0(X)$.

7. a. An **oriented *n*-simplex** is an ordered sequance $P_1P_2 \cdots P_{n+1}$.
b. The **boundary of** $P_1P_2 \cdots P_{n+1}$ is given by

$$\partial_n(P_1P_2 \cdots P_{n+1}) \sum_{i=1}^{n+1} (-1)^{i+1} P_1P_2 \cdots P_{i-1}P_{i+1} \cdots P_{n+1}.$$

c. Each individual *summand* of the boundary of an oriented n-simplex is a **face of the simplex.**

11. a. $\delta^{(n)}\left(\sum_i m_i\sigma_i\right) = \sum_i m_i\delta^{(n)}(\sigma_i)$

13. $H^{(n)}(X) = Z^{(n)}(X)/B^{(n)}(X)$
$H^{(0)}(S) \simeq \mathbb{Z}$ and is generated by $(P_1 + P_2 + P_3 + P_4) + \{0\}$
$H^{(1)}(S) = 0$
$H^{(2)}(S) \simeq \mathbb{Z}$ and is generated by $P_1P_2P_3 + B^{(2)}(S)$

SECTION 42

1. $H_0(X) \simeq \mathbb{Z}$. $H_1(X) \simeq \mathbb{Z} \times \mathbb{Z}$. $H_n(X) = 0$ for $n > 1$.
3. $H_0(X) \simeq \mathbb{Z} \times \mathbb{Z}$. $H_1(X) \simeq \mathbb{Z}$. $H_2(X) \simeq \mathbb{Z}$. $H_n(X) = 0$ for $n > 2$.
5. $H_0(X) \simeq \mathbb{Z}$. $H_1(X) \simeq \mathbb{Z}$. $H_2(X) \simeq \mathbb{Z}$. $H_n(X) = 0$ for $n > 2$.
7. a. T **c.** F **e.** T **g.** T **i.** F
9. $H_0(X) \simeq \mathbb{Z}$. $H_1(X) \simeq \mathbb{Z} \times \mathbb{Z} \times \mathbb{Z}$. $H_2(X) \simeq \mathbb{Z} \times \mathbb{Z}$. $H_n(X) = 0$ for $n > 2$.
11. $H_0(X) \simeq \mathbb{Z}$. $H_1(X) \simeq \mathbb{Z} \times \mathbb{Z} \times \mathbb{Z} \times \mathbb{Z}$. $H_2(X) \simeq \mathbb{Z}$. $H_n(X) = 0$ for $n > 2$.

SECTION 43

1. Both counts show that $\chi(X) = 1$.
3. It will hold for a square region, for such a region is homeomorphic to E^2. It obviously does not hold for two disjoint 2-cells, for each can be mapped continuously onto the other, and such a map has no fixed points.
5. $H_0(X) \simeq \mathbb{Z} \times \mathbb{Z}$. $H_1(X) \simeq \mathbb{Z} \times \mathbb{Z} \times \mathbb{Z}_2 \times \mathbb{Z}_2$. $H_n(X) = 0$ for $n > 1$.
7. $2 - 2n$

9. $H_0(X) \simeq \mathbb{Z}$. $H_1(X) \simeq \underbrace{\mathbb{Z} \times \mathbb{Z} \times \cdots \times \mathbb{Z}}_{(q-1)\text{ factors}} \times \mathbb{Z}_2$. $H_n(X) = 0$ for $n > 1$.

11. Let Q be a vertex of b, and let c be the 2-chain consisting of all 2-simplexes of X, all oriented the same way, so that $c \in Z_2(X)$.

a. f_{*0} is given by $f_{*0}(Q + B_0(X)) = Q + B_0(b)$.
f_{*1} is given by $f_{*1}((ma + nb) + B_1(X)) = nb + B_1(b)$.
f_{*2} is given by $f_{*2}(c + B_1(X)) = 0$.

b. f_{*0} is as in (a).
f_{*1} is given by $f_{*1}((ma + nb) + B_1(X)) = 2nb + B_1(b)$.
f_{*2} is as in (a).

13. Let Q be a vertex on b.
f_{*0} is given by $f_{*0}(Q + B_0(X)) = Q + B_0(b)$.
f_{*1} is given by $f_{*1}((ma + nb) + B_1(X)) = nb + B_1(b)$, where $m = 0, 1$.
f_{*2} is trivial, since both $H_2(X)$ and $H_2(b)$ are 0.

SECTION 44

5. For Theorem 44.4, the condition $f_{k-1}\partial_k = \partial'_k f_k$ implies that

$$f_{k-1}(B_{k-1}(A)) \subseteq B_{k-1}(A').$$

Then Exercise 14.39 shows that f_{k-1} induces a natural homomorphism of $Z_{k-1}(A)/B_{k-1}(A)$ into $Z_{k-1}(A')/B_{k-1}(A')$. This is the correct way to view Theorem 44.4.
For Theorem 44.7, if we use Exercise 14.39, the fact that $\partial_k(A'_k) \subseteq A'_{k-1}$ shows that ∂_k induces a natural homomorphism $\bar{\partial}_k : (A_k/A'_k) \to (A_{k-1}/A'_{k-1})$.

7. The exact homology sequence is

$$[H_2(a) = 0] \xrightarrow{i_{*2}} [H_2(X) \simeq \mathbb{Z}] \xrightarrow{j_{*2}} [H_2(X, a) \simeq \mathbb{Z}] \xrightarrow{\partial_{*2}} [H_1(a) \simeq \mathbb{Z}]$$
$$\xrightarrow{i_{*1}} [H_1(X) \simeq \mathbb{Z} \times \mathbb{Z}] \xrightarrow{j_{*1}} [H_1(X, a) \simeq \mathbb{Z}] \xrightarrow{\partial_{*1}} [H_0(a) \simeq \mathbb{Z}]$$
$$\xrightarrow{i_{*0}} [H_0(X) \simeq \mathbb{Z}] \xrightarrow{j_{*0}} [H_0(X, a) = 0].$$

j_{*2} maps a generator $c + B_2(X)$ of $H_2(X)$ onto the generator

$$(c + C_2(a)) + B_2(X, a)$$

of $H_2(X, a)$ and is an isomorphism. Thus (kernel j_{*2}) = (image i_{*2}) = 0.
∂_{*2} maps everything onto 0, so (kernal ∂_{*2}) = (image j_{*2}) $\simeq \mathbb{Z}$.
i_{*1} maps the generator $a + B_1(a)$ onto $(a + 0b) + B_1(X)$, so i_{*1} is an isomorphism *into*, and (kernal i_{*1}) = (image ∂_{*2}) = 0.
j_{*1} maps $(ma + nb) + B_1(X)$ onto $(nb + C_1(a)) + B_1(X, a)$, so (kernal j_{*1}) = (image i_{*1}) $\simeq \mathbb{Z}$.
∂_{*1} maps $(nb + C_1(a)) + B_1(X, a)$ onto 0, so (kernal ∂_{*1}) = (image j_{*1}) $\simeq \mathbb{Z}$.
For a vertex Q of a, i_{*0} maps $Q + B_0(a)$ onto $Q + B_0(X)$, so i_{*0} is an isomorphism, and (kernal i_{*0}) = (image ∂_{*1}) = 0.
j_{*0} maps $Q + B_0(X)$ onto $B_0(X, a)$ in $H_0(X, a)$, so (kernal j_{*0}) = (image i_{*0}) $\simeq \mathbb{Z}$.

9. The answer is formally identical with that in Exercise 44.7.

11. *Partial answer:* The exact homology sequence is

$$[H_2(Y) = 0] \xrightarrow{i_{*2}} [H_2(X) = 0] \xrightarrow{j_{*2}} [H_2(X, Y) \simeq \mathbb{Z}] \xrightarrow{\partial_{*2}} [H_1(Y) \simeq \mathbb{Z} \times \mathbb{Z}]$$
$$\xrightarrow{i_{*1}} [H_1(X) \simeq \mathbb{Z}] \xrightarrow{j_{*1}} [H_1(X, Y) \simeq \mathbb{Z}] \xrightarrow{\partial_{*1}} [H_0(Y) \simeq \mathbb{Z} \times \mathbb{Z}]$$
$$\xrightarrow{i_{*0}} [H_0(X) \simeq \mathbb{Z}] \xrightarrow{j_{*0}} [H_0(X, Y) = 0].$$

The verification of exactness is left to you. Note that the edge P_1Q_1 of Fig. 42.11 gives rise to a generator of $H_1(X, Y)$. Starting with ∂_{*2}, these maps are very interesting.

SECTION 45

1. Yes **3.** No **5.** No. **7.** Yes

9. In $\mathbb{Z}[x]$: only $2x - 7, -2x + 7$

In $\mathbb{Q}[x]$: $4x - 14, x - \frac{7}{2}, 6x - 21, -8x + 28$

In $\mathbb{Z}_{11}[x]$: $2x - 7, 10x - 2, 6x + 1, 3x - 5, 5x - 1$

11. 26, −26 **13.** 198, −198

15. It is already "primitive" because every nonzero element of $\mathbb{Q}$ is a unit. Indeed $18ax^2 - 12ax + 48a$ is primitive for all $a \in \mathbb{Q}, a \neq 0$.

17. $2ax^2 - 3ax + 6a$ is primitive for all $a \neq 0$ in $\mathbb{Z}_7$ because every such element a is a unit in $\mathbb{Z}_7$.

21. **a.** *T* **c.** *T* **e.** *T* **g.** *F* **i.** *F*

i. Either p or one of its associates must appear in every factorization *into irreducibles.*

23. $2x + 4$ is irreducible in $\mathbb{Q}[x]$ but not in $\mathbb{Z}[x]$.

31. *Partial answer:* $x^3 - y^3 = (x - y)(x^2 + xy + y^2)$

SECTION 46

1. Yes **3.** No. (1) is violated. **5.** Yes

7. 61 **9.** $x^3 + 2x - 1$ **11.** 66

13. **a.** *T* **c.** *T* **e.** *T* **g.** *T* **i.** *T*

23. *Partial answer:* The equation $ax = b$ has a solution in $\mathbb{Z}_n$ for nonzero $a, b \in \mathbb{Z}_n$ if and only if the positive gcd of a and n in $\mathbb{Z}$ divides b.

SECTION 47

1. $5 = (1 + 2i)(1 - 2i)$ **3.** $4 + 3i = (1 + 2i)(2 - i)$

5. $6 = (2)(3) = (-1 + \sqrt{-5})(-1 - \sqrt{-5})$ **7.** $7 - i$

15. **c.** **i)** order 9, characteristic 3 **ii)** order 2, characteristic 2

iii) order 5, characteristic 5

SECTION 48

1. $\sqrt{2}, -\sqrt{2}$ **3.** $3 + \sqrt{2}, 3 - \sqrt{2}$ **5.** $\sqrt{2} + i, \sqrt{2} - i, -\sqrt{2} + i, -\sqrt{2} - i$

7. $\sqrt{1 + \sqrt{2}}, -\sqrt{1 + \sqrt{2}}, \sqrt{1 - \sqrt{2}}, -\sqrt{1 - \sqrt{2}}$ **9.** $\sqrt{3}$

11. $-\sqrt{2} + 3\sqrt{5}$ **13.** $-\sqrt{2} + \sqrt{45}$

15. **a.** $\mathbb{Q}$ **b.** $\mathbb{Q}(\sqrt{6})$ **c.** $\mathbb{Q}$

17. $\mathbb{Q}(\sqrt{2}, \sqrt{3}, \sqrt{5})$ **19.** $\mathbb{Q}(\sqrt{3}, \sqrt{10})$ **21.** $\mathbb{Q}$

25. **a.** $3 - \sqrt{2}$ **b.** They are the same maps.

27. $\sigma_3(0) = 0, \sigma_3(1) = 1, \sigma_3(2) = 2, \sigma_3(\alpha) = -\alpha, \sigma_3(2\alpha) = -2\alpha,$
$\sigma_3(1 + \alpha) = 1 - \alpha, \sigma_3(1 + 2\alpha) = 1 - 2\alpha, \sigma_3(2 + \alpha) = 2 - \alpha,$
$\sigma_3(2 + 2\alpha) = 2 - 2\alpha; \mathbb{Z}_3(\alpha)_{\{\sigma_3\}} = \mathbb{Z}_3$

29. **a.** *F* **c.** *T* **e.** *F* **g.** *T* **i.** *T*

37. Yes

SECTION 49

1. The identity map of E onto E;
τ given by $\tau(\sqrt{2}) = \sqrt{2}$, $\tau(\sqrt{3}) = -\sqrt{3}$, $\tau(\sqrt{5}) = -\sqrt{5}$

3. τ_1 given by $\tau_1(\sqrt{2}) = \sqrt{2}$, $\tau_1(\sqrt{3}) = \sqrt{3}$, $\tau_1(\sqrt{5}) = -\sqrt{5}$;
τ_2 given by $\tau_2(\sqrt{2}) = \sqrt{2}$, $\tau_2(\sqrt{3}) = -\sqrt{3}$, $\tau_2(\sqrt{5}) = \sqrt{5}$;
τ_3 given by $\tau_3(\sqrt{2}) = -\sqrt{2}$, $\tau_3(\sqrt{3}) = \sqrt{3}$, $\tau_3(\sqrt{5}) = \sqrt{5}$;
τ_4 given by $\tau_4(\sqrt{2}) = -\sqrt{2}$, $\tau_4(\sqrt{3}) = -\sqrt{3}$, $\tau_4(\sqrt{5}) = -\sqrt{5}$

5. The identity map of $\mathbb{Q}(\sqrt[3]{2}, \sqrt{3})$ into itself;
τ_1 given by $\tau_1(\alpha_1) = \alpha_1$, $\tau_1(\sqrt{3}) = -\sqrt{3}$ where $\alpha_1 = \sqrt[3]{2}$;
τ_2 given by $\tau_2(\alpha_1) = \alpha_2$, $\tau_2(\sqrt{3}) = \sqrt{3}$ where $\alpha_2 = \sqrt[3]{2}(-1 + i\sqrt{3})/2$;
τ_3 given by $\tau_3(\alpha_1) = \alpha_2$, $\tau_3(\sqrt{3}) = -\sqrt{3}$;
τ_4 given by $\tau_4(\alpha_1) = \alpha_3$, $\tau_4(\sqrt{3}) = \sqrt{3}$ where $\alpha_3 = \sqrt[3]{2}(-1 - i\sqrt{3})/2$;
τ_5 given by $\tau_5(\alpha_1) = \alpha_3$, $\tau_5(\sqrt{3}) = -\sqrt{3}$;

7. **a.** $\mathbb{Q}(\pi^2)$ **b.** τ_1 given by $\tau_1(\sqrt{\pi}) = i\sqrt{\pi}$, τ_2 given by $\tau_2(\sqrt{\pi}) = -i\sqrt{\pi}$

SECTION 50

1. 2 **3.** 4 **5.** 2 **7.** 1 **9.** 2 **13.** $1 \le [E : F] \le n!$

15. Let $F = \mathbb{Q}$ and $E = \mathbb{Q}(\sqrt{2})$. Then

$$f(x) = x^4 - 5x^2 + 6 = (x^2 - 2)(x^2 - 3)$$

has a zero in E, but does not split in E.

23. **a.** 6

SECTION 51

1. $\alpha = \sqrt[6]{2} = 2/(\sqrt[3]{2}\sqrt{2}).\sqrt{2} = (\sqrt[6]{2})^3$, $\sqrt[3]{2} = (\sqrt[6]{2})^2$. (Other answers are possible.)

3. $\alpha = \sqrt{2} + \sqrt{3}$. $\sqrt{2} = (\frac{1}{2})\alpha^3 - (\frac{9}{2})\alpha$, $\sqrt{3} = (\frac{11}{2})\alpha - (\frac{1}{2})\alpha^3$. (Other answers are possible.)

7. $f(x) = x^4 - 4x^2 + 4 = (x^2 - 2)^2$. Here $f(x)$ is not an irreducible polynomial. Every irreducible factor of $f(x)$ has zeros of multiplicty 1 only.

15. **b.** The field F **c.** $F[x^P]$

SECTION 52

1. $\mathbb{Z}_3(y^3, z^9)$ **3.** $\mathbb{Z}_3(y^4, z^2)$

5. **a.** F **c.** F **e.** F **g.** T **i.** T

SECTION 53

1. 8 **3.** 8 **5.** 4 **7.** 2

9. The group has two elements, the identity automorphism ι of $\mathbb{Q}(i)$ and σ such that $\sigma(i) = -i$.

11. **a.** Let $\alpha_1 = \sqrt[3]{2}$, $\alpha_2 = \sqrt[3]{2}\dfrac{-1 + i\sqrt{3}}{2}$, and $\alpha_3 = \sqrt[3]{2}\dfrac{-1 - i\sqrt{3}}{2}$.

The maps are
ρ_0, where ρ_0 is the identity map;
ρ_1, where $\rho_1(\alpha_1) = \alpha_2$ and $\rho_1(i\sqrt{3}) = i\sqrt{3}$;

ρ_2, where $\rho_2(\alpha_1) = \alpha_3$ and $\rho_2(i\sqrt{3}) = i\sqrt{3}$;
μ_1, where $\mu_1(\alpha_1) = \alpha_1$ and $\mu_1(i\sqrt{3}) = -i\sqrt{3}$;
μ_2, where $\mu_2(\alpha_1) = \alpha_3$ and $\mu_2(i\sqrt{3}) = -i\sqrt{3}$;
μ_3, where $\mu_3(\alpha_1) = \alpha_2$ and $\mu_3(i\sqrt{3}) = -i\sqrt{3}$.

b. S_3. The notation in (a) was chosen to coincide with the notation for S_3 in Example 8.7.

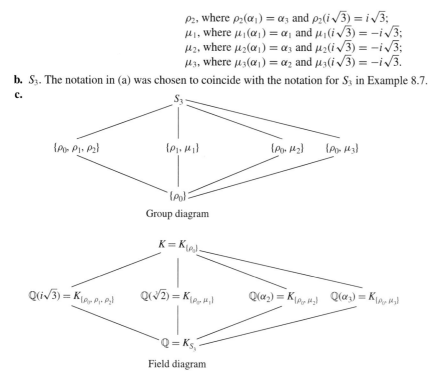

c.

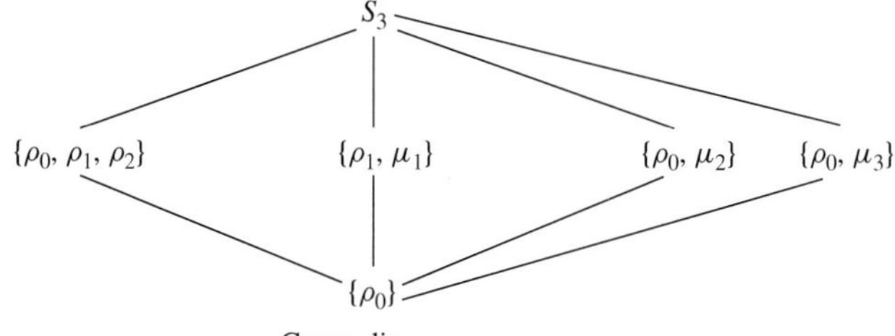

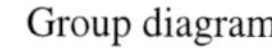

Group diagram

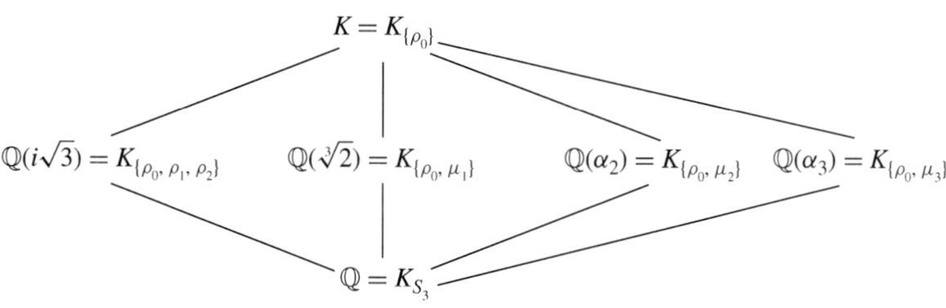

Field diagram

13. The splitting field of $(x^3 - 1) \in \mathbb{Q}[x]$ is $\mathbb{Q}(i\sqrt{3})$, and the group is cyclic of order 2 with elements: ι, where ι is the identity map of $\mathbb{Q}(i\sqrt{3})$, and σ, where $\sigma(i\sqrt{3}) = -i\sqrt{3}$.

15. **a.** *F* **c.** *T* **e.** *T* **g.** *F* **i.** *F*

25. *Partial answer:* $G(K/(E \vee L)) = G(K/E) \cap G(K/L)$

SECTION 54

3. $\mathbb{Q}(\sqrt[4]{2}, i)$: $\sqrt[4]{2} + i$, $x^8 + 4x^6 + 2x^4 + 28x^2 + 1$;
$\mathbb{Q}(\sqrt[4]{2})$: $\sqrt[4]{2}$, $x^4 - 2$;
$\mathbb{Q}(i\sqrt[4]{2})$: $i(\sqrt[4]{2})$, $x^4 - 2$;
$\mathbb{Q}(\sqrt{2}, i)$: $\sqrt{2} + i$, $x^4 - 2x^2 + 9$;
$\mathbb{Q}(\sqrt[4]{2} + i(\sqrt[4]{2}))$: $\sqrt[4]{2} + i(\sqrt[4]{2})$, $x^4 + 8$;
$\mathbb{Q}(\sqrt[4]{2} - i(\sqrt[4]{2}))$: $\sqrt[4]{2} - i(\sqrt[4]{2})$, $x^4 + 8$;
$\mathbb{Q}(\sqrt{2})$: $\sqrt{2}$, $x^2 - 2$;
$\mathbb{Q}(i)$: i, $x^2 + 1$;
$\mathbb{Q}(i\sqrt{2})$: $i\sqrt{2}$, $x^2 + 2$;
$\mathbb{Q}$: 1, $x - 1$

5. The group is cyclic of order 5, and its elements are

	ι	σ_1	σ_2	σ_3	σ_4
$\sqrt[5]{2} \rightarrow$	$\sqrt[5]{2}$	$\zeta(\sqrt[5]{2})$	$\zeta^2(\sqrt[5]{2})$	$\zeta^3(\sqrt[5]{2})$	$\zeta^4(\sqrt[5]{2})$

where $\sqrt[5]{2}$ is the real 5th root of 2.

7. The splitting field of $x^8 - 1$ over $\mathbb{Q}$ is the same as the splitting field of $x^4 + 1$ over $\mathbb{Q}$, so a complete description is contained in Example 54.7. (This is the easiest way to answer the problem.)

9. **a.** $s_1{}^2 - 2s_2$ **b.** $\frac{s_1 s_2 - 3s_3}{s_3}$

SECTION 55

3. **a.** 16 **b.** 400 **c.** 2160
5. 3^0
7. $\Phi_3(x)$ over $\mathbb{Z}_2$ is $x^2 + x + 1$.
   $\Phi_8(x)$ over $\mathbb{Z}_3$ is $x^4 + 1 = (x^2 + x + 2)(x^2 + 2x + 2)$.
9. **a.** T **c.** F **e.** T **g.** T **i.** F
11. $\Phi_1(x) = x - 1$
    $\Phi_2(x) = x + 1$
    $\Phi_3(x) = x^2 + x + 1$
    $\Phi_4(x) = x^2 + 1$
    $\Phi_5(x) = x^4 + x^3 + x^2 + x + 1$
    $\Phi_6(x) = x^2 - x + 1$

SECTION 56

1. No. Yes, K is an extension of $\mathbb{Z}_2$ by radicals.
3. **a.** T **c.** T **e.** T **g.** T **i.** F ($x^3 - 2x$ over $\mathbb{Q}$ gives a counterexample.)

APPENDIX

1. $\begin{bmatrix} 2 & 1 \\ 2 & 7 \end{bmatrix}$ **3.** $\begin{bmatrix} -3+2i & -1-4i \\ 2 & -i \\ 0 & -i \end{bmatrix}$

5. $\begin{bmatrix} 5 & 16 & -3 \\ 0 & -18 & 24 \end{bmatrix}$ **7.** $\begin{bmatrix} 1 & -i \\ 4-6i & -2-2i \end{bmatrix}$

9. $\begin{bmatrix} 8 & -8i \\ 8i & 8 \end{bmatrix}$ **11.** $\begin{bmatrix} 0 & -1 \\ 1 & 0 \end{bmatrix}$ **13.** -48

Index